Student Solutions Manual and Study Guide

for
Hornback's

Organic Chemistry

Second Edition

Joseph M. Hornback
University of Denver

Balasingam Murugaverl
University of Denver

BROOKS/COLE
CENGAGE Learning

Australia • Brazil • Japan • Korea • Mexico • Singapore • Spain • United Kingdom • United States

BROOKS/COLE
CENGAGE Learning™

For product information and technology assistance, contact us at **Cengage Learning Customer & Sales Support, 1-800-354-9706**

For permission to use material from this text or product, submit all requests online at **www.cengage.com/permissions**
Further permissions questions can be emailed to **permissionrequest@cengage.com**

ISBN-13: 978-0-534-39710-4
ISBN-10: 0-534-39710-7

Brooks/Cole
20 Davis Drive
Belmont, CA 94002
USA

Cengage Learning is a leading provider of customized learning solutions with office locations around the globe, including Singapore, the United Kingdom, Australia, Mexico, Brazil, and Japan. Locate your local office at **www.cengage.com/global**

Cengage Learning products are represented in Canada by Nelson Education, Ltd.

To learn more about Brooks/Cole, visit **www.cengage.com/brookscole**

Purchase any of our products at your local college store or at our preferred online store **www.cengagebrain.com**

Printed in the United States of America
4 5 6 7 8 18 17 16 15 14

TABLE OF CONTENTS

Chapter 1
A SIMPLE MODEL FOR CHEMICAL BONDS

1.1   Lewis structures depict the electrons in the outer shell (valence shell) as dots around the symbol of the element. The number of valence shell electrons of an atom is the same as its group number in the periodic table.

a) :Br·   b) ·Ca·   c) ·Ge·

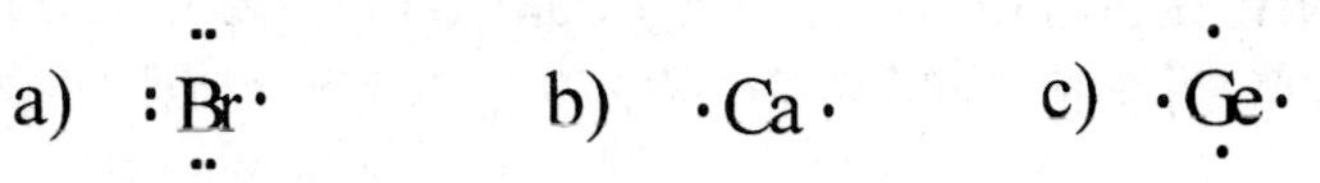

1.2   Calcium chloride and sodium sulfide are ionic compounds because they are formed between a metal and a nonmetal. Electrons are transferred from the metal atoms to the nonmetal atoms to form charged atoms (ions) that satisfy the octet rule. These ions with opposite charges form ionic compounds.

a) ·Ca· + 2 :Cl· $\longrightarrow$ Ca^{2+} + 2 :Cl:$^-$

b) 2 Na· + ·S· $\longrightarrow$ 2 Na$^+$ + :S:$^{2-}$

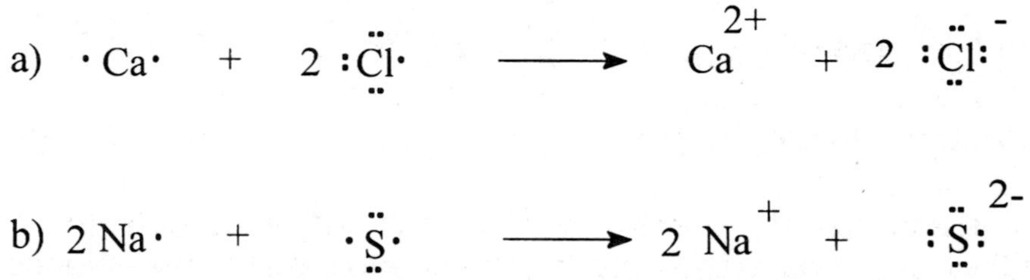

1.3   Consult Table 1.1 in the text to answer this problem. Carbon has four valence shell electrons and forms four bonds to satisfy the octet rule. Chlorine and bromine each have seven valence shell electrons and form one bond. Hydrogen also forms one bond.

a)

:Cl:
:Cl:C:Cl:
:Cl:

b)   H:Br:

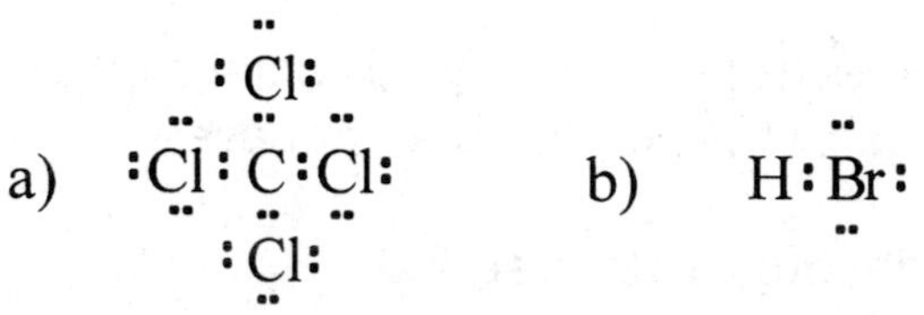

1.4   Sulfur is beneath oxygen in the periodic table, so it prefers to form two bonds.

H:S:H

1.5 The octet rule is the most important factor in determining the stability of a species.
a) octet rule satisfied, stable   b) ten electrons around N, unstable

1.6 Follow these steps for writing a Lewis structure for a molecule.

(1) Start with the neutral atoms in the molecule and bond the nonhydrogen atoms together. (If the molecule is charged, add [if negative] or subtract [if positive] one additional electron for each unit of charge.)

Consider (a), for example. In C_3H_8, each carbon has four valence electrons and each hydrogen has one. The compound is not charged so we do not change the number of electrons.

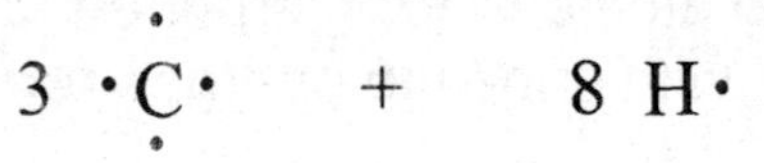

$$3 \; \cdot \overset{\textstyle \cdot}{\underset{\textstyle \cdot}{C}} \cdot \quad + \quad 8 \; H \cdot \qquad \text{(20 valence electrons total)}$$

Bond the carbons together. $\cdot \overset{\cdot \; \cdot \; \cdot}{\underset{\cdot \; \cdot \; \cdot}{C:C:C}} \cdot$

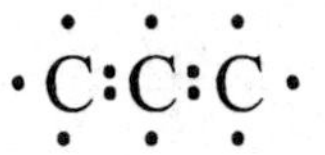

(2) Add the hydrogens.

In this example there are 8 hydrogens to add. This matches the 8 additional bonds to the carbons needed to satisfy their valences.

$$\begin{array}{c} \text{H H H} \\ \text{H:C:C:C:H} \\ \text{H H H} \end{array}$$

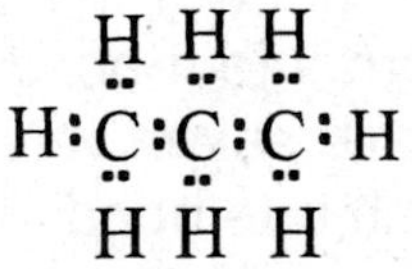

(3) If there are not enough electrons to satisfy the octet rule at all of the atoms, form additional bonds between atoms as necessary.

In this example, no additional bonds are needed.

(4) Check to see that the octet rule is satisfied at all atoms.

Each C has 8 electrons and each H has 2 electrons in this structure.

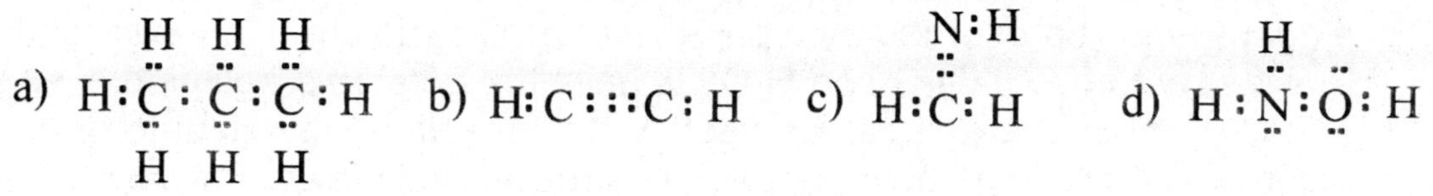

1.7 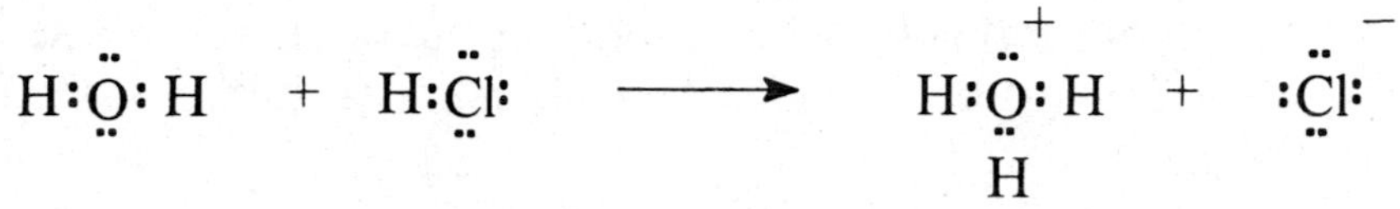

1.8

$$\text{Formal Charge} = \left[\begin{array}{l}\text{valence electrons} \\ \text{in the atom}\end{array}\right] - \left[\begin{array}{l}\text{number of} \\ \text{unshared electrons}\end{array}\right] - \frac{1}{2}\left[\begin{array}{l}\text{number of} \\ \text{shared electrons}\end{array}\right]$$

The number of <u>valence electrons</u> is the same as the group number of the atom. <u>Unshared electrons</u> are those electrons not involved in bonding. <u>Shared electrons</u> are those electrons involved in bonding. The total charge (TC) of the molecule is the sum of all the formal charges of atoms in that molecule.

For example, in (a):

FC of carbon = 4 - 0 - (8/2) = 0; FC of oxygen = 6 - 2 - (6/2) = +1.

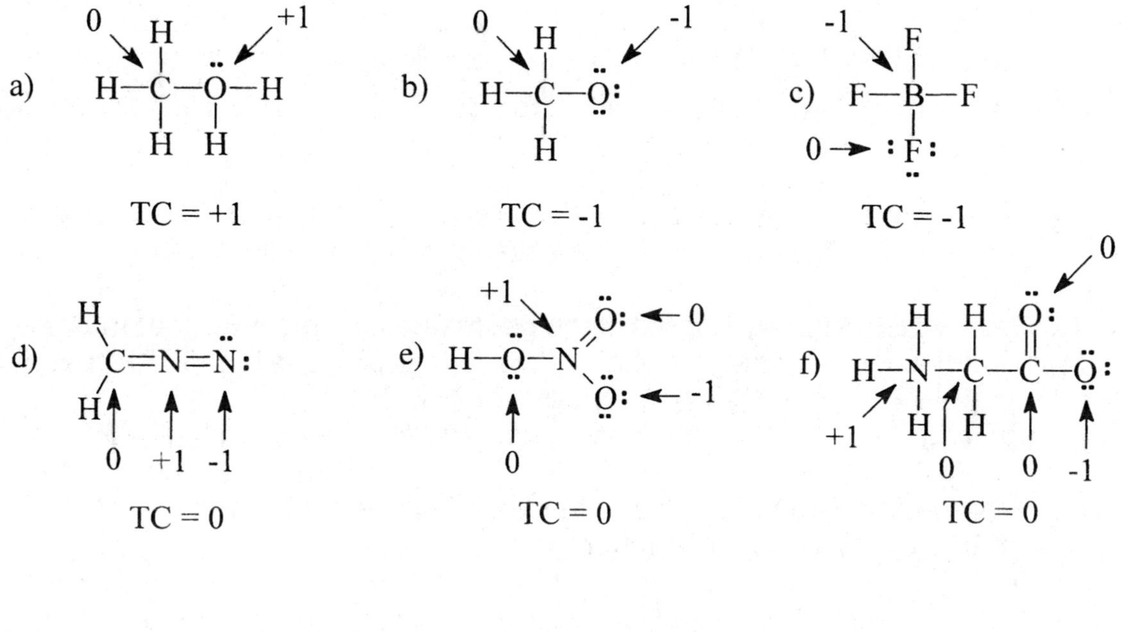

1.9 The octet rule is the most important criterion for estimating the stability of a compound represented by a particular Lewis structure. However, both of these structures, H-Cl-O and H-O-Cl, have the octet rule satisfied at Cl and O. Formal charges can be used to further refine the estimate of stability. Other things being equal, the structure with fewer formal charges is more stable. In H-Cl-O, the formal charge on chlorine is +1, on oxygen is -1, and on hydrogen is 0, whereas the formal charges on each of the atoms in H-O-Cl is zero. Therefore H-O-Cl is more stable due to less formal charges.

1.10 The actual structure is a resonance hybrid of the following two structures. Both carbon-oxygen bonds are identical, intermediate between a double bond and a single bond, and each oxygen bears a formal charge of -1/2.

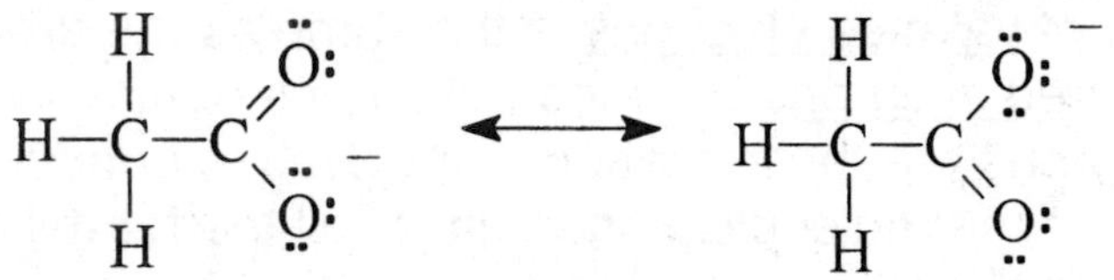

1.11 A bond dipole is due to unequal electronegativities (see Table 1.2) of the two atoms involved in the bond. A bond is polarized so that the negative end of the dipole is on the more electronegative element. An arrow pointing from the positive end of the dipole to the negative end is used to show the direction of polarization.

a) C–N b) O–N c) O–Cl d) C–Cl e) B–O f) C–Mg

The electronegativities of C and H are very similar, so the C-C bond of (g) and the C-H bond of (h) are considered to be nonpolar.

1.12 Use VSEPR theory to predict the geometries. Remember that multiple bonds count as one pair of electrons for the purpose of VSEPR theory.

a) linear b) trigonal planar c) tetrahedral

Three-dimensional models of these molecules are available at http://now.brookscole.com/hornback2.

1.13 a) tetrahedral at C, bent at O (tetrahedral)
b) trigonal planar at C, bent at N (trigonal planar)
c) trigonal planar at C, bent at O (trigonal planar)

1.14 First determine the geometry of the molecule by VSEPR theory. Then find the individual bond dipoles of the molecule. The overall dipole moment is the vector sum of the individual bond dipoles. All of the examples in this problem have only one polar bond, so the dipole moment lies along the direction of that bond.

a)   b)

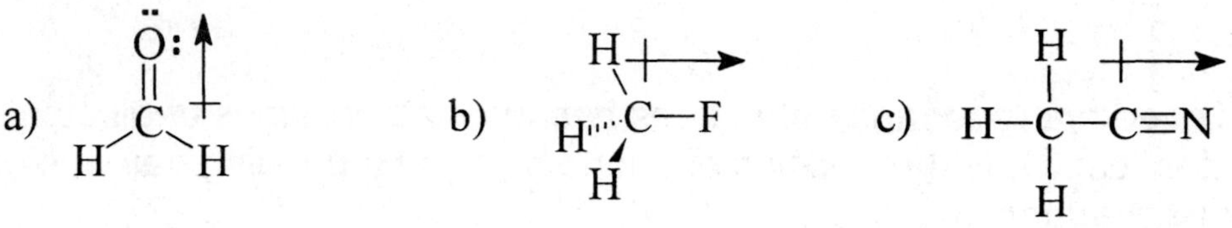

1.15 Ionic bonds are formed when metals combine with non-metals. Covalent bonds are formed between two non-metals.

a) K⁺ :C̈l:⁻   b) :C̈l:N̈:C̈l: c) Na⁺ :C⋮⋮N:⁻   d) K⁺ :Ö:H⁻
                              :C̈l:

   (ionic)   (covalent)   (both)   (both)

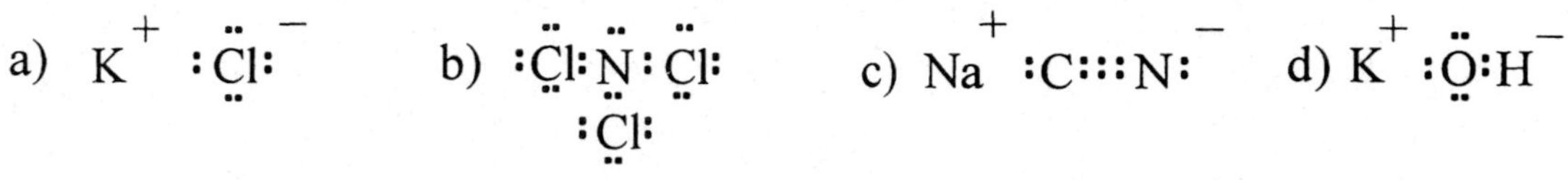

1.16 Like N, phosphorus has 5 electrons in its and valence shell and prefers to form three bonds. PH₃ is pyramidal.

H:P̈:H
   H

1.17

a) H:C̈:N̈:H
       H H
    with H above C and H below (H, H)

b) H H
    H:C̈:C̈:C̈l:
    H H

c) :N⋮⋮⋮N:

d) H
    H:C̈ :: S̈:

e) H H
    H:C̈::C̈:F̈:

f) H
    H:C̈:S̈:H
       H

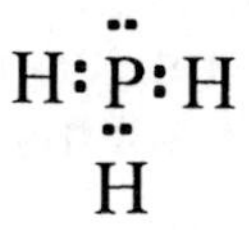

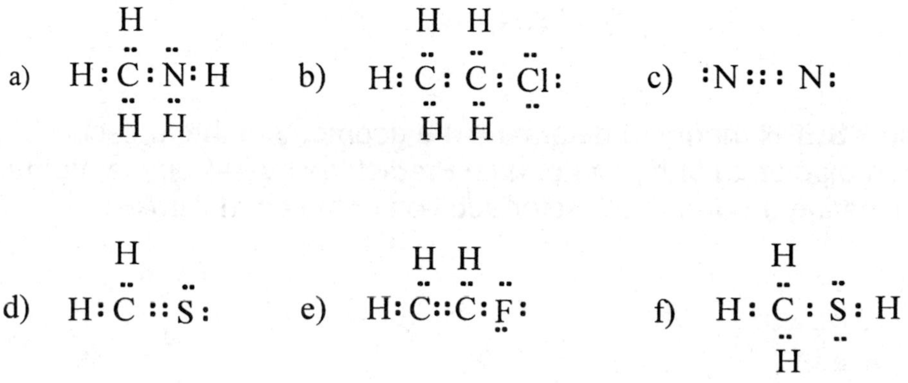

1.18

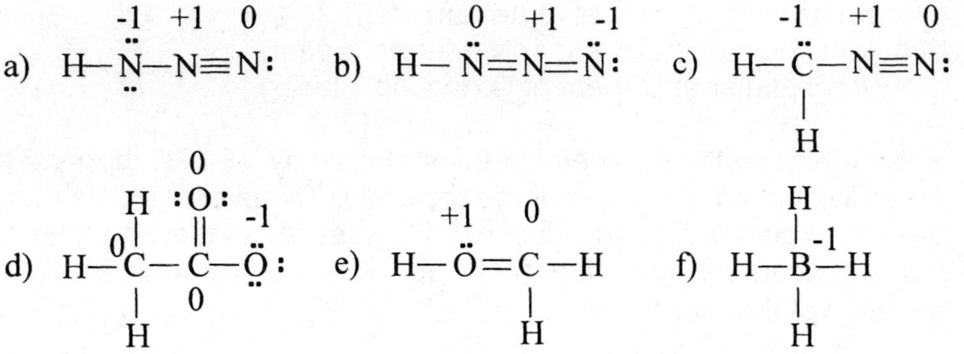

1.19 These are resonance structures because the positions of all atoms are identical. Only the position of a lone pair and a bonding pair of electrons are changed.

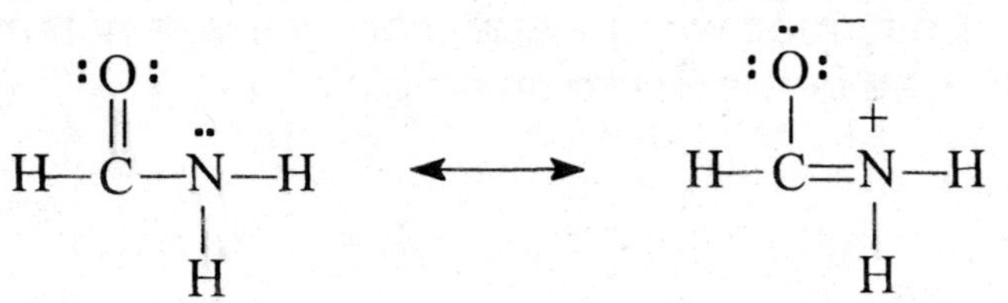

Because the resonance structure on the left has no formal charges, it is more stable than the one on the right.

1.20 Although the atoms in this structure have formal charges, the octet rule is satisfied at all the atoms. Therefore, carbon monoxide is stable.

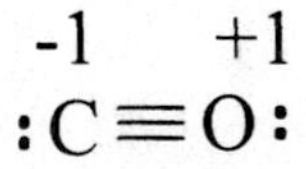

1.21 Use VSEPR theory to determine the geometry of the molecule. Assign bond dipoles to any polar bonds. Predict the overall dipole moment by estimating the result of vector addition of the bond dipoles.

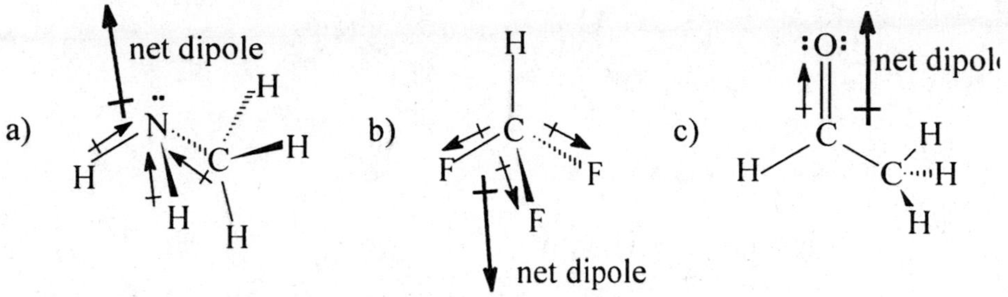

1.22

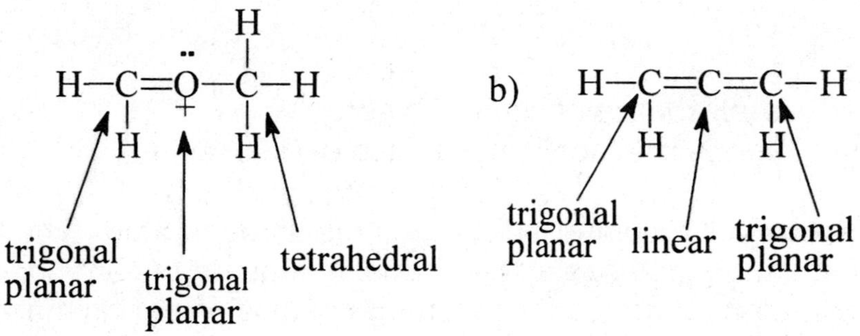

1.23 A doubly bonded oxygen needs 4 more electrons to satisfy the octet rule. A sulfur with one covalent bond needs six more electrons to satisfy the octet rule. This gives the S a formal charge of -1. The resonance structure has the double bond to the S and the negative formal charge on the O. The situation is very similar to acetate anion (see problem 1.10).

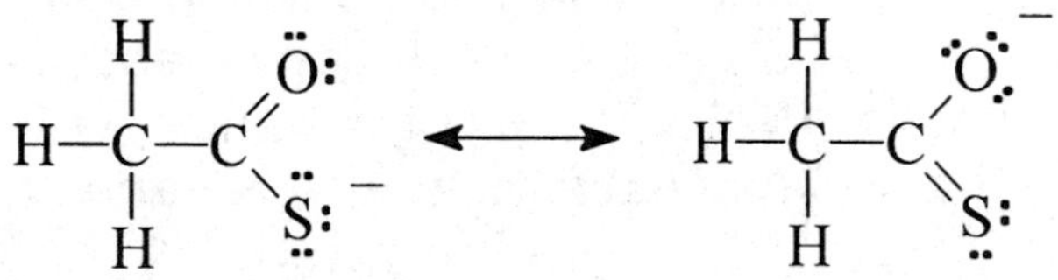

1.24

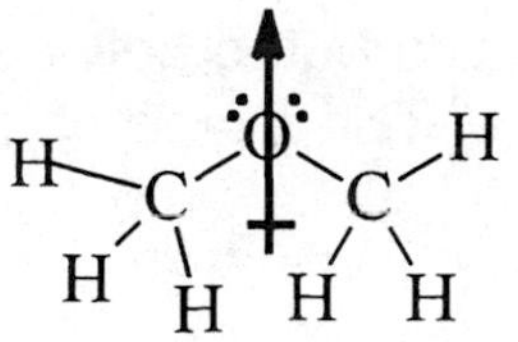

bent geometry at oxygen
(tetrahedral electron pairs)

1.25

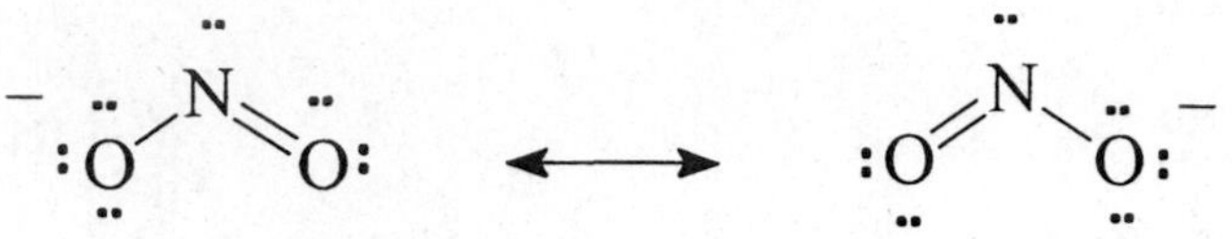

1.26 a)

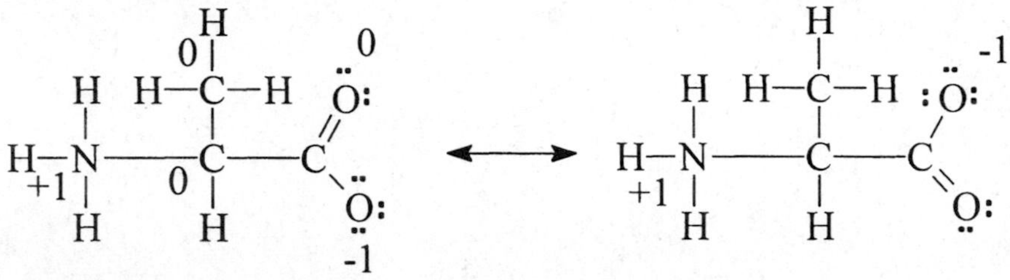

b) The overall charge of alanine is zero.
c) The oxygens are identical because of resonance.

1.27 The bonds will be polarized towards the nitrogen atom. The N atom in the ammonia molecule has a zero formal charge. In the ammonium cation the nitrogen atom carries a formal charge of plus one. This makes the N of the ammonium cation more electronegative and its hydrogens have a larger partial positive charge than the respective atoms in ammonia.

1.28 Ionic compounds normally have very high melting points. Because $CaCl_2$ is ionic, one would expect it to have a much higher melting point than 68°C. Therefore, it would be wise not to trust your lab partner and look it up yourself.

1.29

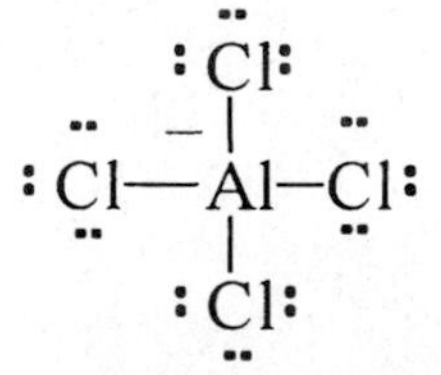

The formal charge on the aluminum atom is 3 - 0 - (8/2) = -1; that on each Cl atom is 7 - 6 - (2/2) = 0. The shape of this anion, with 4 electron pairs (bonds) around Al, is tetrahedral.

1.30 In the cyanate anion the negative charge is distributed between the nitrogen and the oxygen atoms by resonance. The shape of this anion is linear.

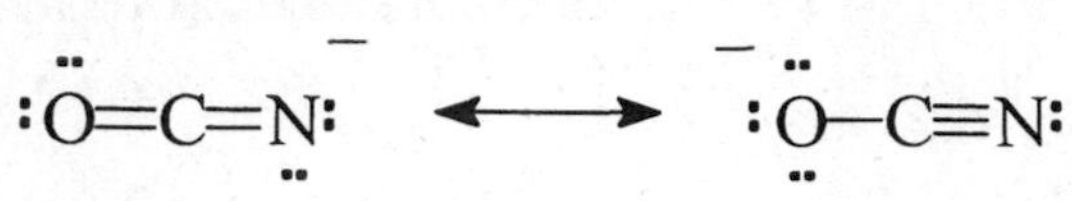

1.31

a) 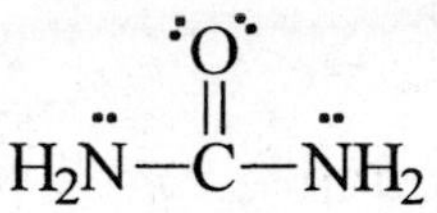

urea

b) 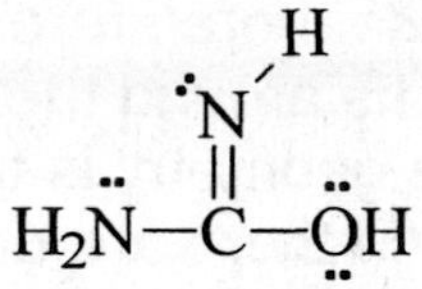

an isomer of urea

1.32 Period 2 elements can accommodate only 8 valence electrons; hence the octet rule. Atoms in periods beyond 2 can hold more than 8 electrons in their valence shell. Phosphorus is a period 3 element can accommodate more than 8 electrons in its valence shell, which allows it to form more than 3 bonds. Nitrogen (a period 2 element) has a capacity for only 8 electrons in its outer shell, limiting it to 3 bonds and an unshared pair of electrons, or 4 bonds and a positive charge.

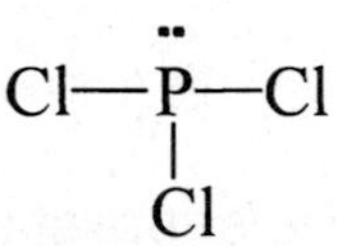

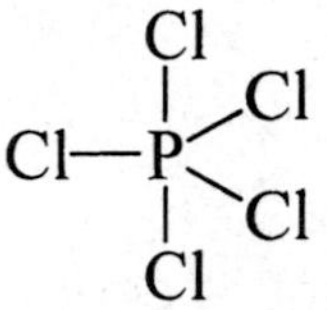

1.33 The absence of the two unpaired electrons in the unstable species NH_3 results in a formal charge of +2 on the nitrogen.

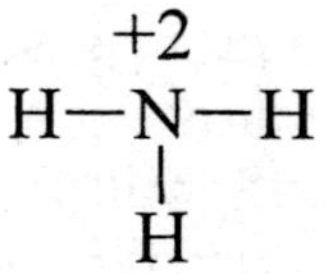

1.34

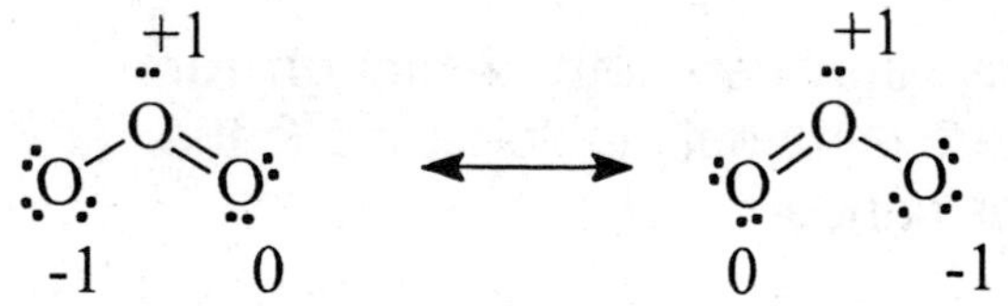

Ozone is bent at the central O. The actual structure of ozone is a hybrid of the two resonance structures shown, so the bonds are identical

1.35

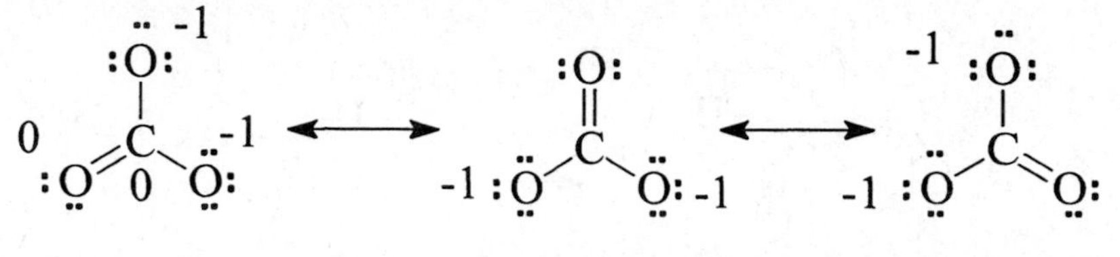

The carbonate anion has trigonal planar geometry at its carbon. All the oxygens are identical because there are three resonance structures.

1.36 a) The charge on this species is +1.
b) Because there are only three pairs of electrons around the central carbon atom, the geometry is trigonal planar.

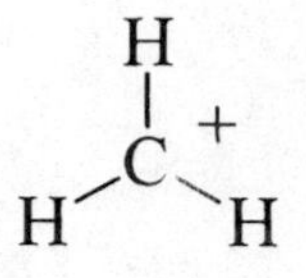

c) Because there are only 6 electrons around the carbon, the octet rule is not satisfied and the cation is not stable. It is expected to be even less stable that the carbon species from practice problem 1.4, which has 7 electrons in its valence shell.
d) The C needs another bond. An unshared pair of electrons from hydroxide ion is used to form a bond to the carbon, resulting in the formation of a stable species with the octet rule satisfied at all atoms and no formal charges.

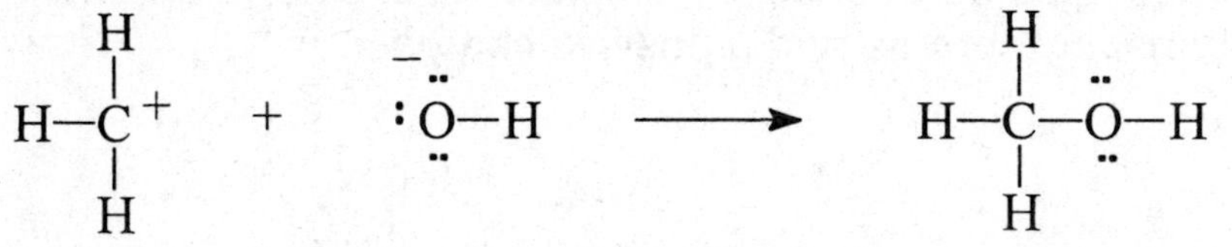

1.37 The dipole moment is equal to the charge separation times the distance of the charge separation. The charge separation is larger for FCl than for ICl because the electronegativity difference is larger for F and Cl than for Cl and I, but the FCl bond is shorter than the ICl bond because bonds become longer as one goes down a column of the periodic table.

1.38 PCl_5 has a trigonal bipyramid geometry. Although each P-Cl bond is polar, there is no net dipole because the individual bond dipoles cancel.

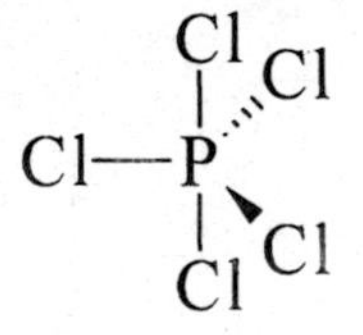

1.39 The actual structure of benzene is a hybrid of two resonance structures. Each of the C-C bonds in benzene is between a double bond and a single bond and is identical.

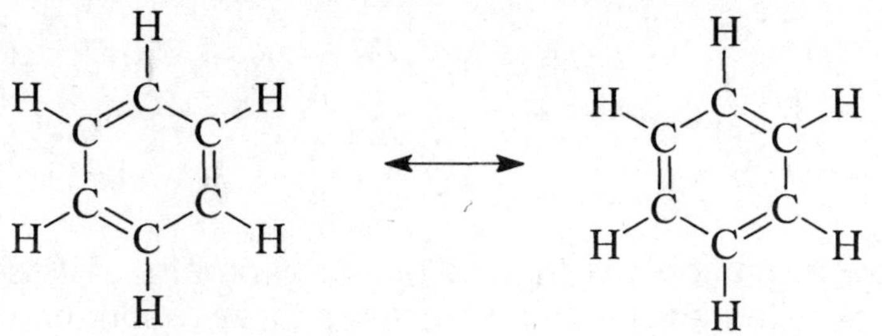

1.40 a) CBr_4 has polar bonds but its geometry is tetrahedral so the bond dipoles cancel. Therefore, the molecule has no dipole moment and is nonpolar.

b) NH_3 has polar bonds and a pyramidal geometry, so it has a dipole moment. It is a polar molecule.

c) The C-O bonds of CH_3OCH_3 are polar and the geometry is bent at the O so it has a dipole moment. It is a polar molecule.

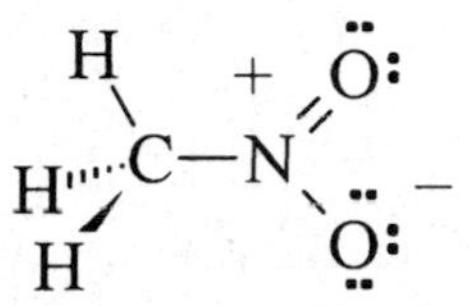

d) The C-Cl bonds of CH_2Cl_2 are polar and its geometry is tetrahedral. The bond dipoles do not cancel so it has a dipole moment and is a polar molecule.

e) The C-O bonds of CO_2 are polar but it has a linear geometry, so the bond dipoles cancel. It has no dipole moment and is a nonpolar molecule.

1.41 a) C-1 has two single bonds and one double bond, so it has trigonal planar geometry. C-2 has two double bonds, so it has linear geometry.

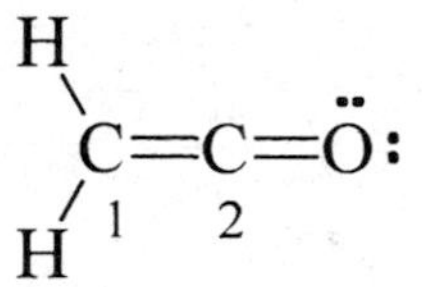

b) The C has four bonds, so it has tetrahedral geometry. N-1 has a single bond, a double bond, and an unshared pair, so it has bent (trigonal planar) geometry. N-2 has two double bonds, so it has linear geometry. There is a resonance structure for this molecule that has two unshared pairs on N-1, a single bond between N-1 and N-2, a triple bond between N-2 and N-3, and only one unshared pair on N-3. The formal charges are -1 on N-1, +1 on N-2, and 0 on N-3.

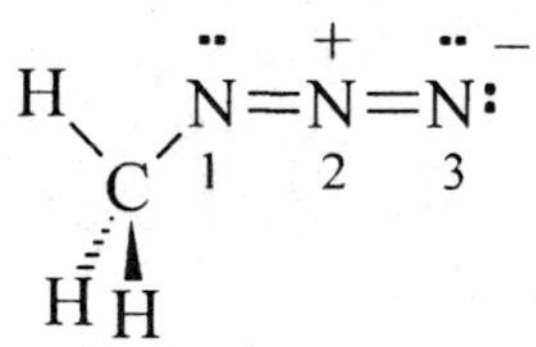

c) The sulfur has a single bond, a double bond, and an unshared pair, so it has bent (trigonal planar) geometry. Resonance makes both S-O bonds identical.

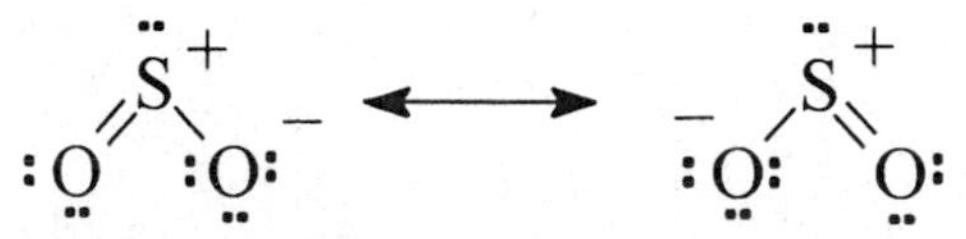

d) The C has four single bonds, so it has tetrahedral geometry. The N has two single bonds and a double bond, so it has trigonal planar geometry.

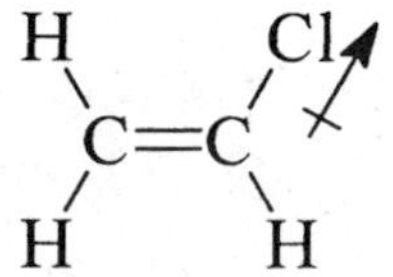

1.42 a) This molecule has a dipole moment because of the polar C-Cl bond.

b) This molecule has a dipole moment that bisects the Cl-C-Cl bond angle.

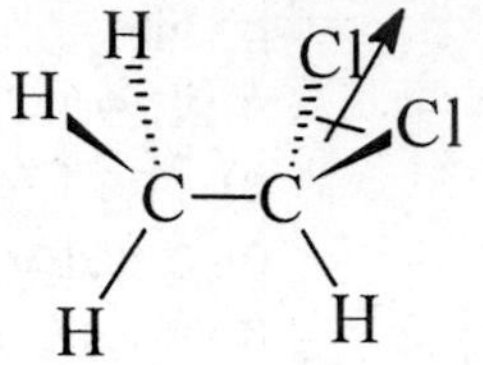

c) This linear molecule does not have a dipole moment because the bond dipoles of its polar C-O bonds cancel.

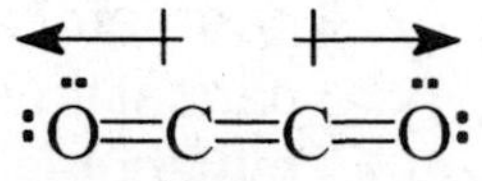

d) This planar molecule does not have a dipole moment because the bond dipoles of its polar C-Cl bonds cancel.

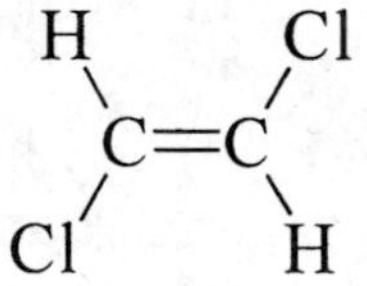

1.43

a) 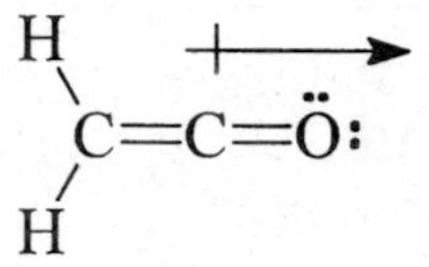

b) 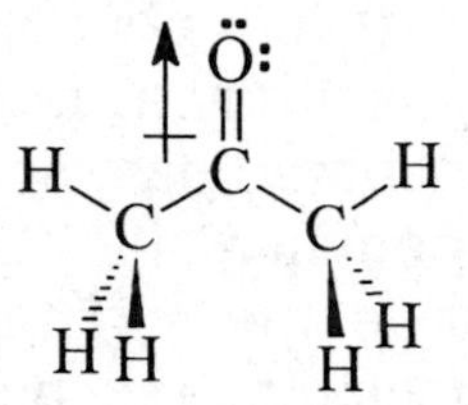

c) 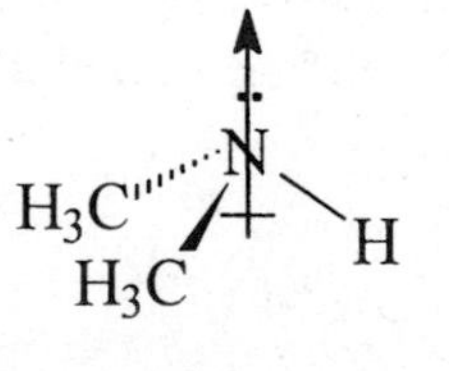

d) 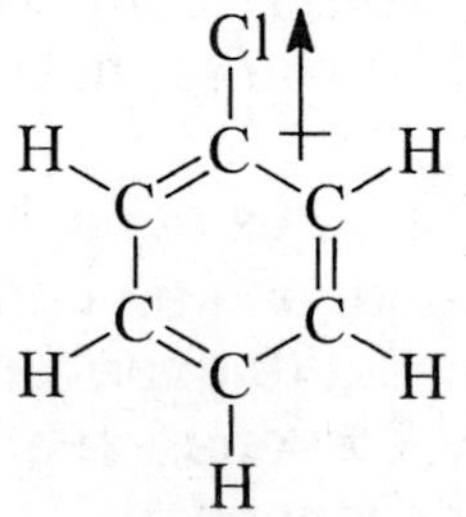

1.44 The molecule is nonpolar because the dipoles of the 4 C-Cl bonds cancel.

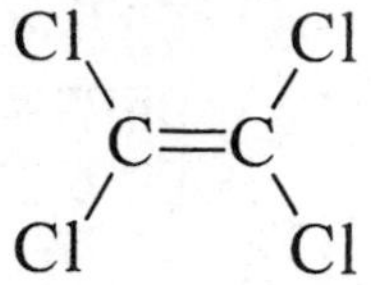

Review of Mastery Goals

After completing this chapter, you should be able to:

Write the best Lewis structure for any molecule or ion. This includes determining how many electrons are available, whether multiple bonds are necessary, and satisfying the octet rule if possible. For complex molecules, however, the connectivity must be known.
(Problems 1.2, 1.3, 1.4, 1.6, 1.15, 1.16, 1.17, 1.24, 1.29, 1.30, 1.31)

Calculate the formal charge on any atom in a Lewis structure. (In fact, you should be starting to recognize the formal charges on some atoms in some situations without doing a calculation.)
(Problems 1.8, 1.18, 1.26, 1.30)

Estimate the stability of a Lewis structure by whether it satisfies the octet rule and by the number and the distribution of the formal charges in the structure.
(Problems 1.5, 1.9, 1.19, 1.20, 1.36)

Recognize some simple cases in which resonance is necessary to describe the actual structure of a molecule. However, a better understanding of resonance will have to wait until Chapter 3.
(Problems 1.10, 1.23, 1.25, 1.34, 1.35, 1.39)

Arrange the atoms that are of most interest to organic chemistry in order of their electronegativities and assign the direction of the dipole of any bond involving these atoms.
(Problems 1.11, 1.21, 1.37)

Determine the shape of a molecule from its Lewis structure by using VSEPR theory.
(Problems 1.12, 1.13, 1.21, 1.22, 1.24, 1.29, 1.34, 1.35, 1.36)

Determine whether a compound is polar or not and assign the direction of its dipole moment.
(Problems 1.14, 1.21, 1.24, 1.38)

Chapter 2
ORGANIC COMPOUNDS --- A FIRST LOOK

2.1 In estimating the stability of a species, the first thing to consider is the octet rule. Species that satisfy the octet rule are likely to be stable. If the octet rule is not satisfied, the species is not very stable. Next look for the presence of formal charges. Their presence causes some destabilization. For example, the structure in part (a) satisfies the octet rule at all atoms. However, one of the carbon atoms has a formal charge so it is destabilized somewhat.

a) less stable; octet rule satisfied but charge on C
b) stable; octet rule satisfied at all atoms
c) very unstable; octet rule not satisfied at N
d) and e) stable; octet rule satisfied at all atoms

2.2

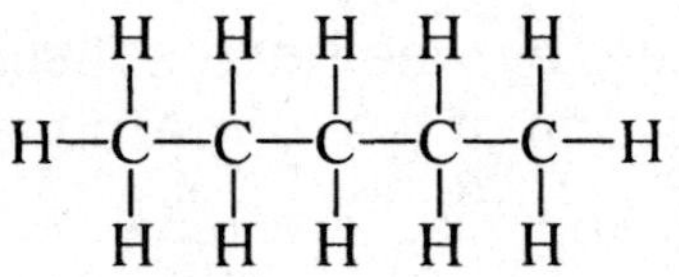
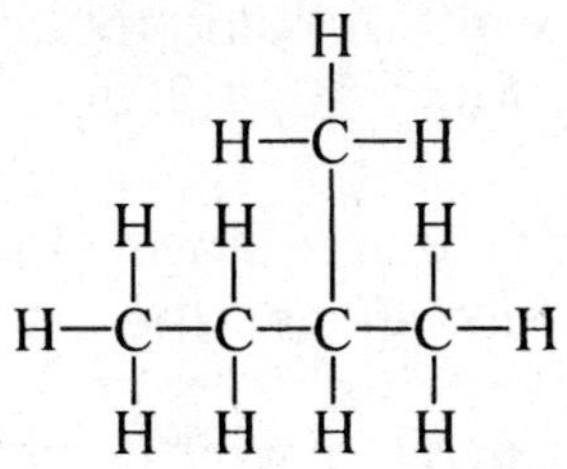
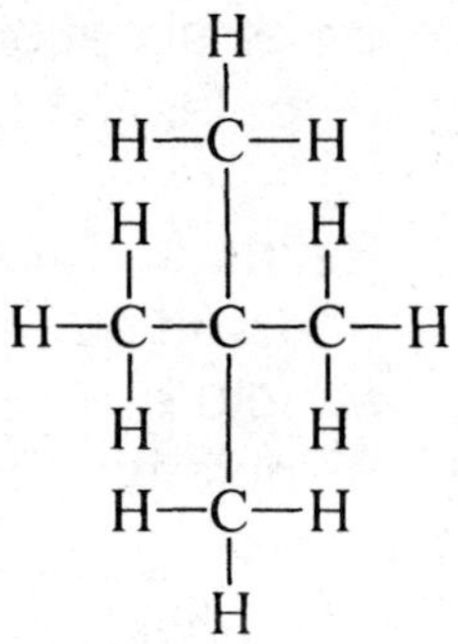

2.3

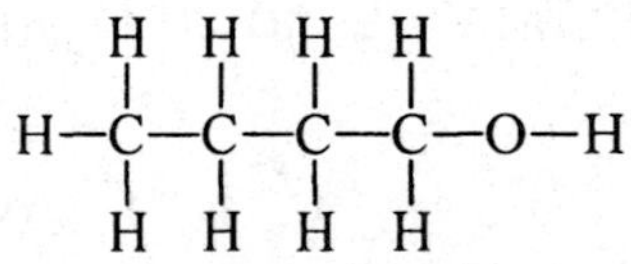
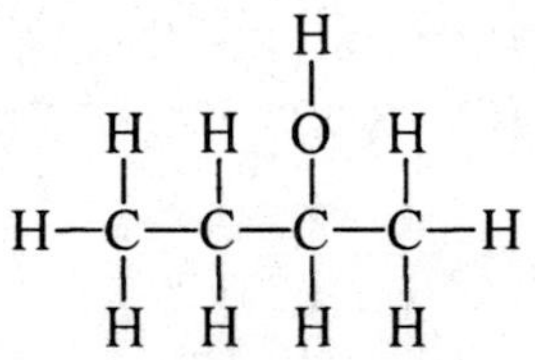
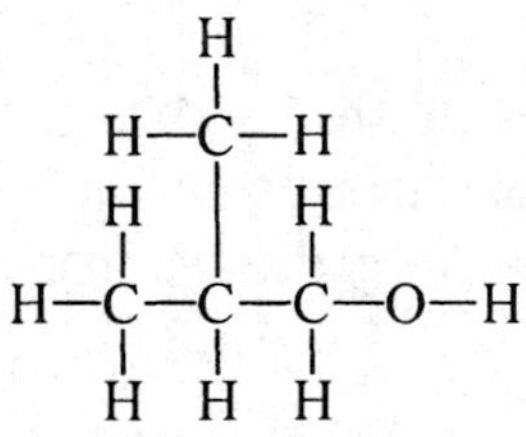
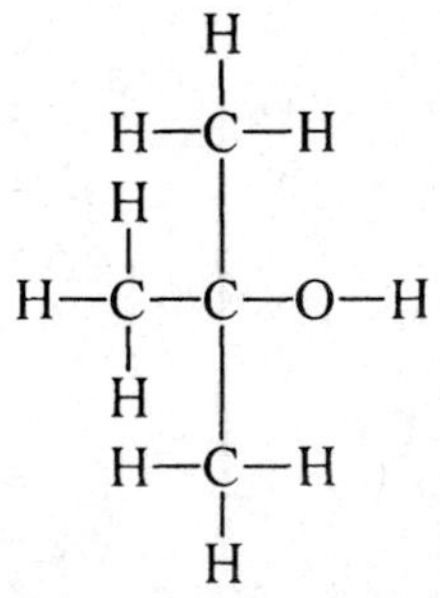
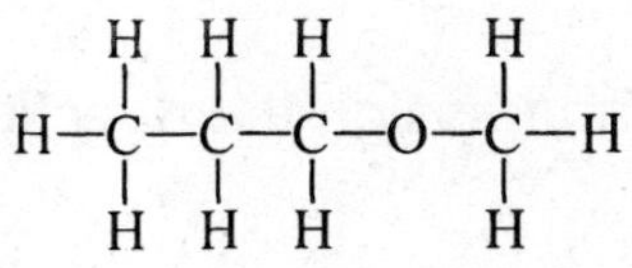
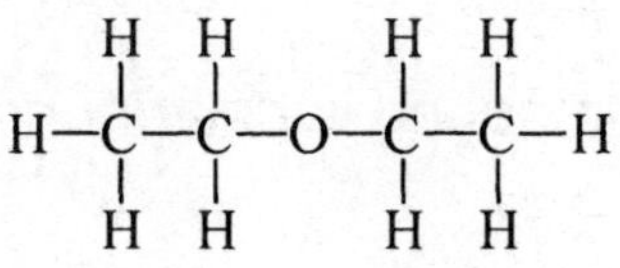
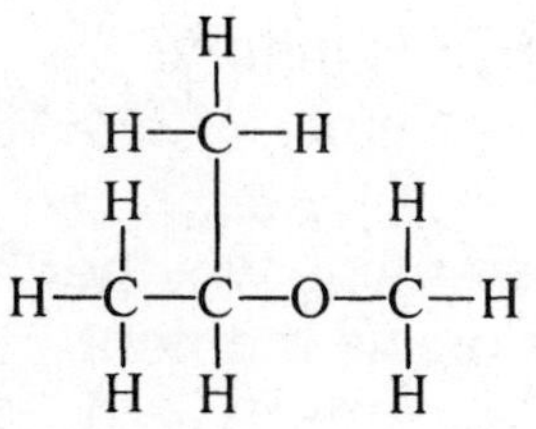

2.4 a) same b) same c) isomers d) same

2.5 For a hydrocarbon with the formula C_nH_x, the degree of unsaturation (DU) is equal to $\frac{(2n+2) - x}{2}$.

a) DU = 0 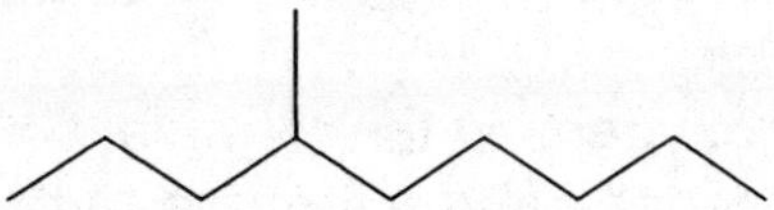 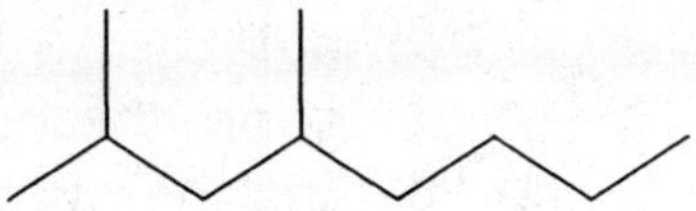

(There are many, many other possible answers for this problem.)

b) DU = 2 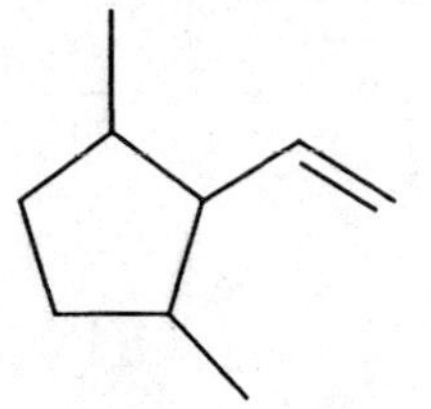 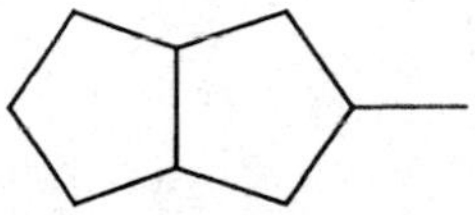

c) DU = 4  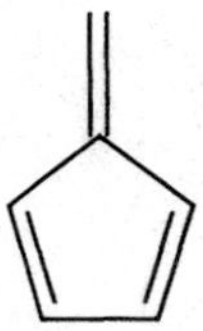

2.6

a) 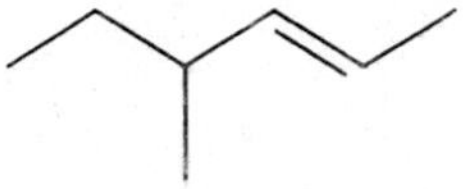 b) 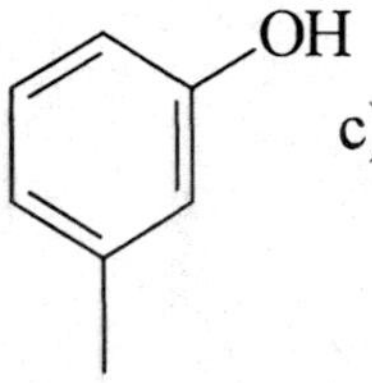

c) 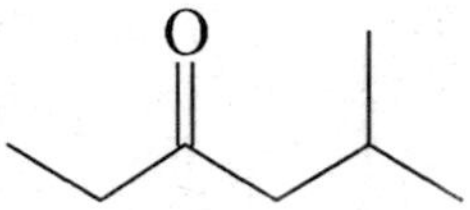

d) 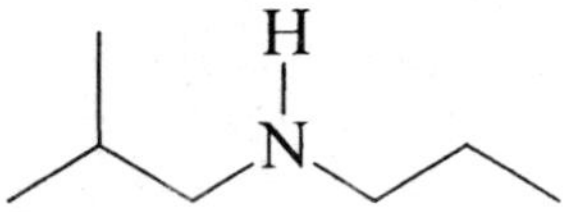

2.7

a) 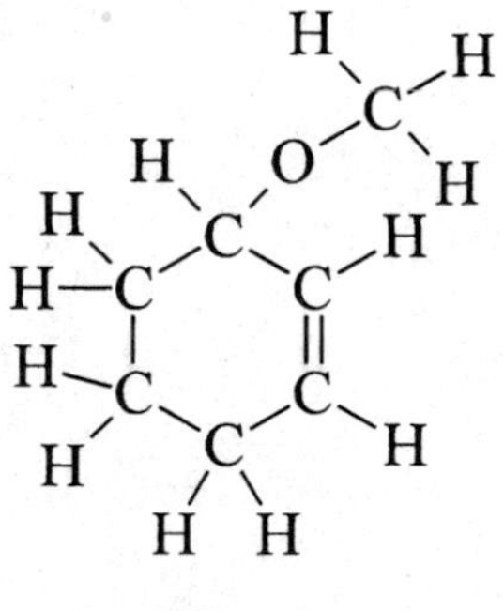

b) 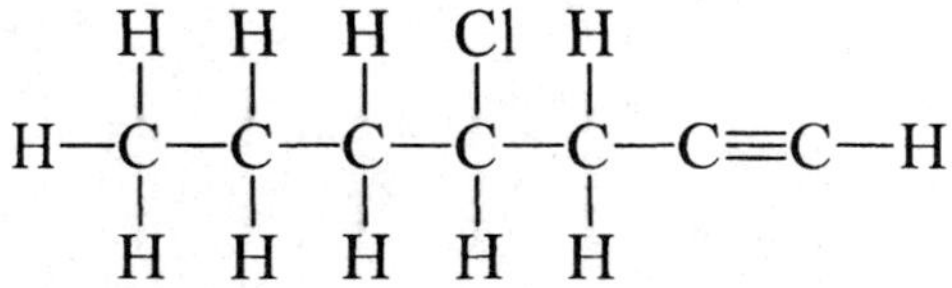

c) 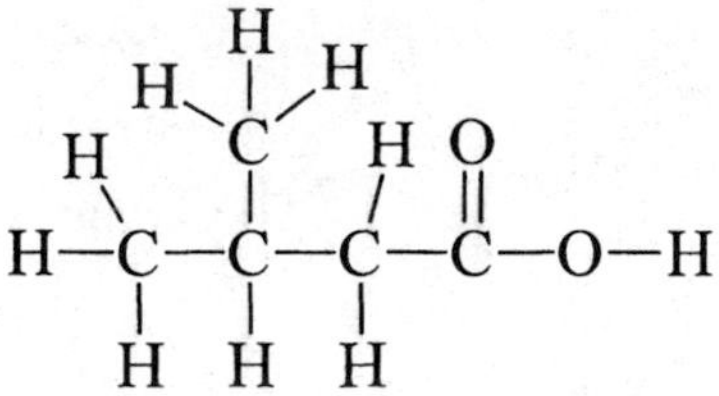

d) 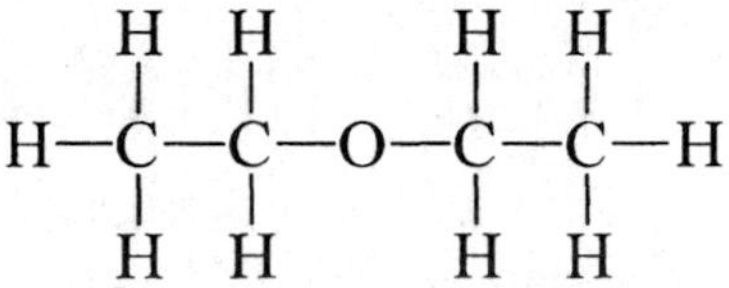

2.8 a) same b) same c) isomers d) same e) same f) isomers

2.9 Use the following procedure to calculate the DU of formulas that have atoms other than C and H:
1) Add the number of halogens in the formula to number of hydrogens in the formula;
2) Ignore the number of oxygens in the formula;
3) Add the number of nitrogens in the formula to the maximum number of hydrogens;

For example in C_6H_8ClN, the DU = $\frac{[2(6) + 2 + 1 - (8 + 1)]}{2} = 3$.

a) DU = 1 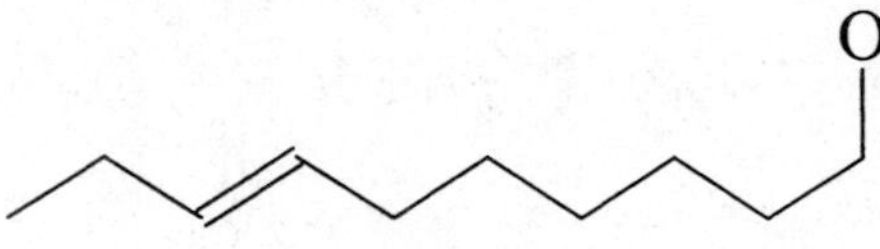 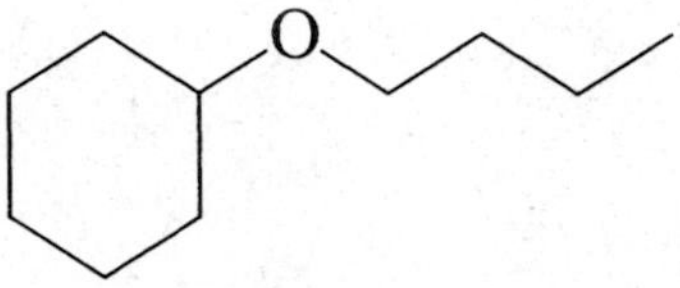

b) DU = 3 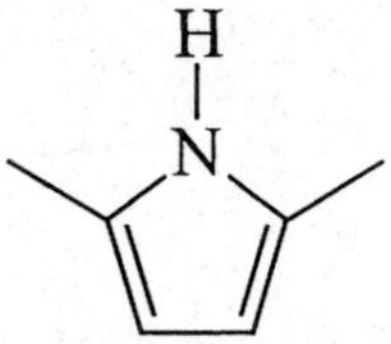 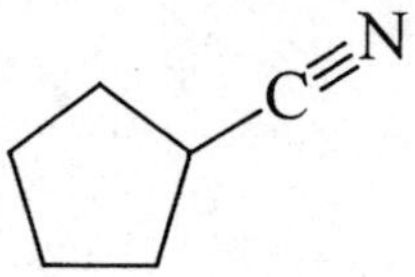

c) DU = 0 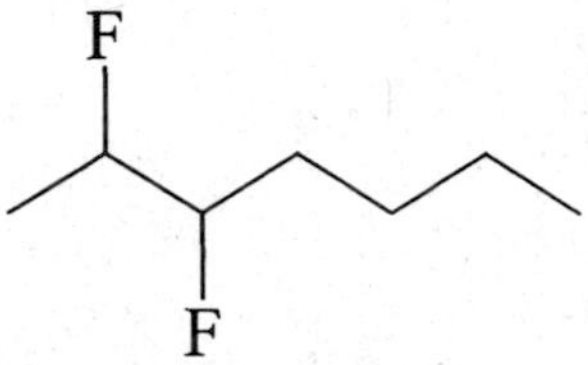 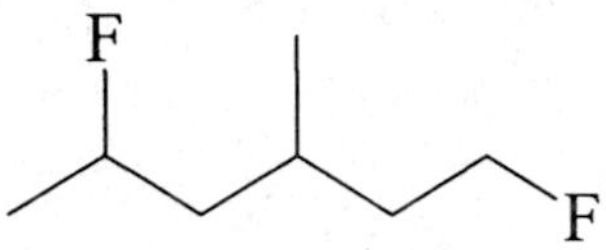

d) DU = 3 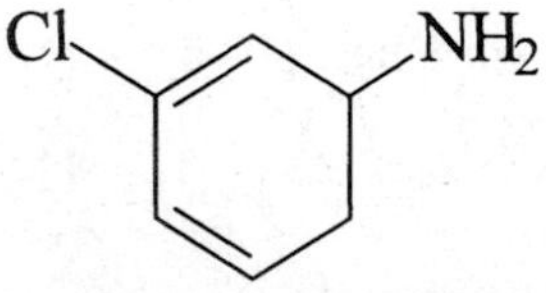 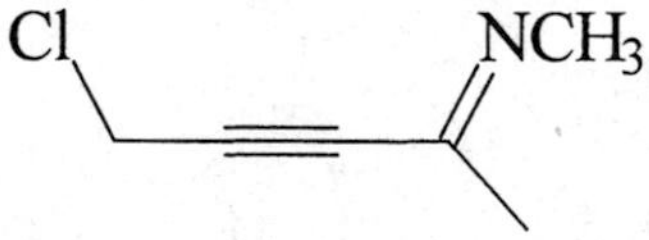

e) DU = 3 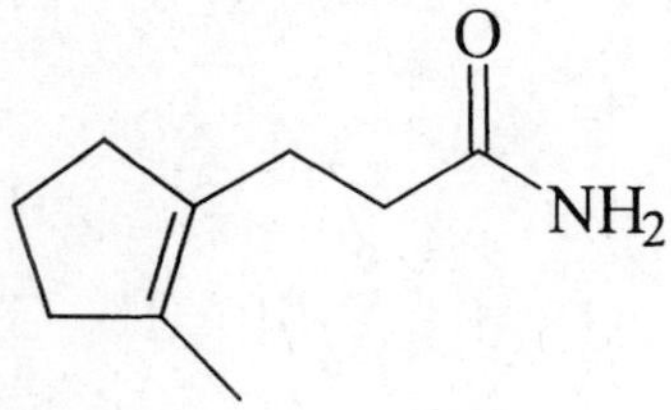 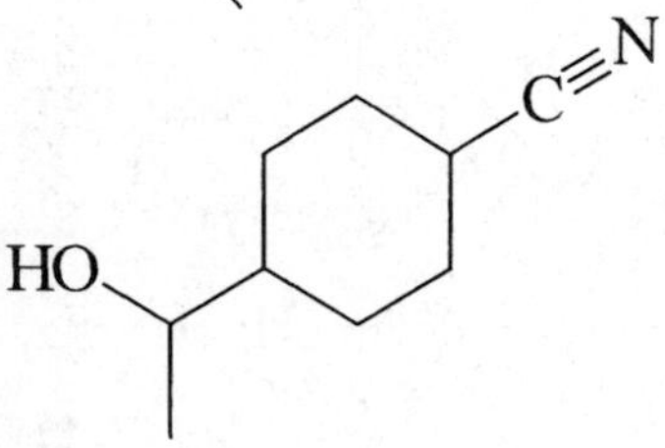

2.10 To determine the DU of a structure, add the number of rings plus the number of double bonds plus twice the number of triple bonds.

a) 5 b) 4 c) 3 d) 7 e) 6

2.11 Intermolecular forces between molecules result from the attraction of opposite charges in the molecules. The different attractive forces are:

- **Ion-ion**: the attractive force between two oppositely charged ions. This is the strongest attraction of all.
- **Ion-dipole**: the attractive force between an ion and a polar molecule. This attractive force is much weaker than an ion-ion interaction.
- **Dipole-dipole**: the attractive force between two polar molecules. This force is weaker than an ion-dipole interaction.
- **Dipole-induced dipole**: attraction between a polar molecule and a dipole that the polar molecule induces in a nonpolar molecule. This force is weaker than a dipole-dipole attraction.
- **Instantaneous dipole-induced dipole (London force)**: attraction between two nonpolar molecules that results from the interaction of an instantaneous dipole in one with an induced dipole in the other. This is the weakest force of all, but there can be a large number of them. These last three are called **van der Waals forces**.
- **Hydrogen bonding**: a special type of dipole-dipole attraction between a hydrogen bonded to an electronegative atom in one molecule and an electronegative atom in another molecule. This force is stronger than a regular dipole-dipole attraction.

a) London b) van der Waals c) ion-ion d) van der Waals and hydrogen bonding

2.12

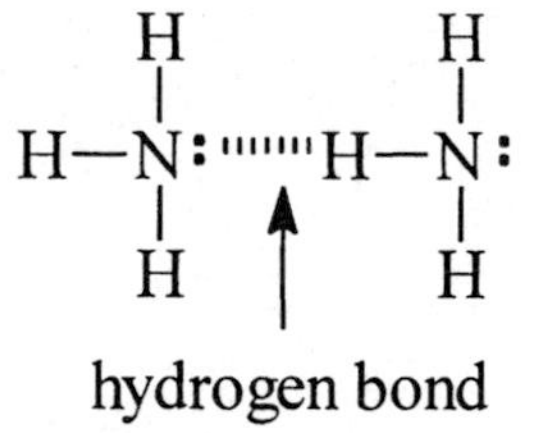

2.13 Stronger intermolecular attractive forces result in a higher melting point. KBr is an ionic compound and has strong ion-ion interactions. CH_3Br is a covalent molecule and has only weak van der Waals forces of attraction. Therefore KBr has a higher melting point.

2.14 $CH_3CH_2CH_2OH$ has the higher boiling point because it has hydrogen bonding in addition to van der Waals attractive forces. $CH_3CH_2OCH_3$ has only van der Waals attractive forces.

2.15 Both compounds have a polar, hydrophilic part and a nonpolar, hydrophobic part. Water is a polar solvent and dissolves polar compounds better than nonpolar ones. As the nonpolar part of the molecule gets larger, its solubility in water decreases. Thus, $CH_3CH_2CH_2CH_2CO_2H$ has a higher solubility because it has a smaller nonpolar part.

2.16 a) ether b) alcohol c) carboxylic acid
d) amide e) ester f) arene and aldehyde

2.17

a) 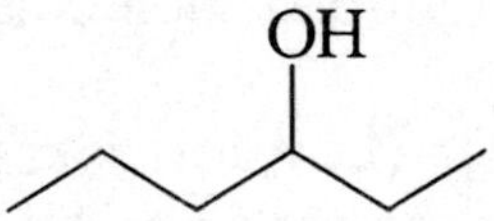

b) 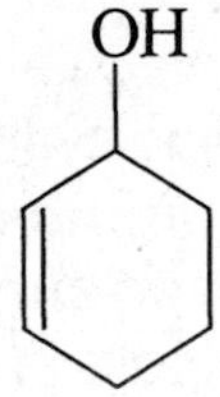

c) 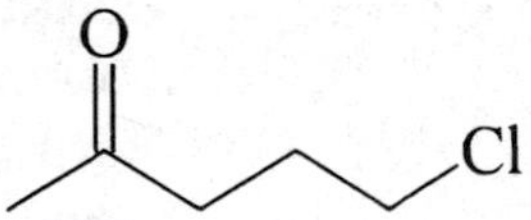

d) 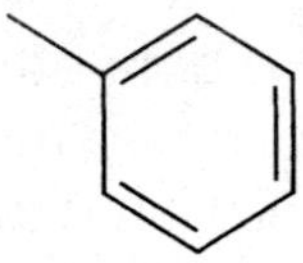

e) 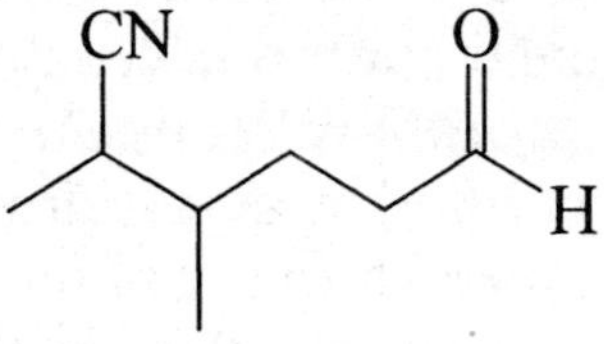

f) 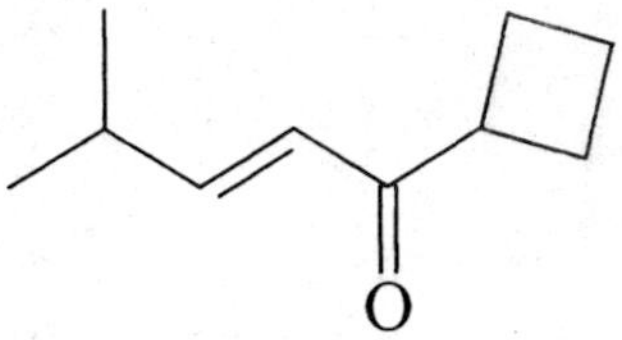

g) 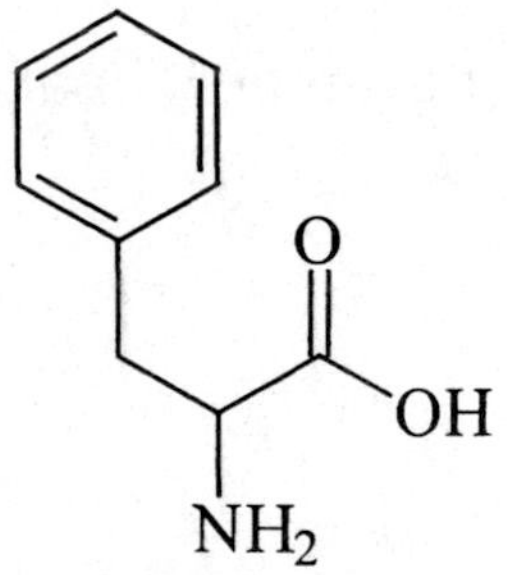

2.18

a) 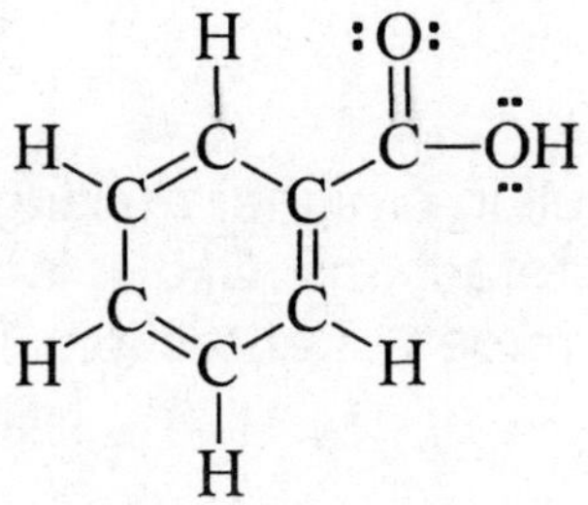

b) 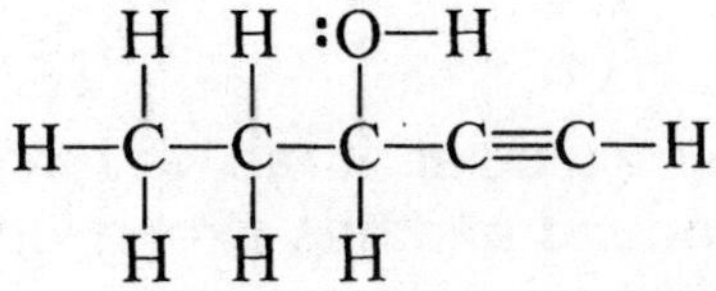

c) 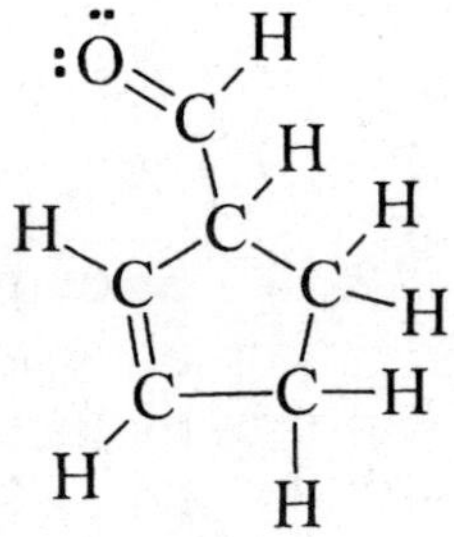

d) 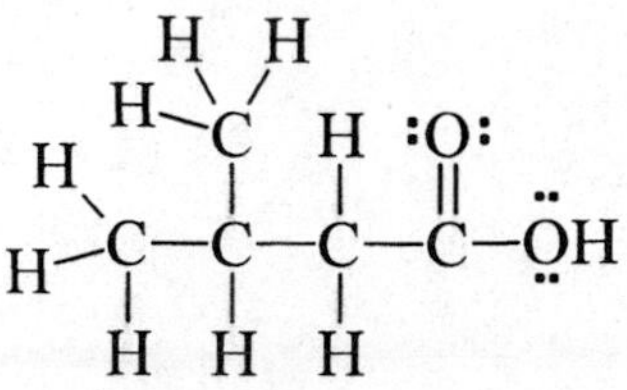 e) 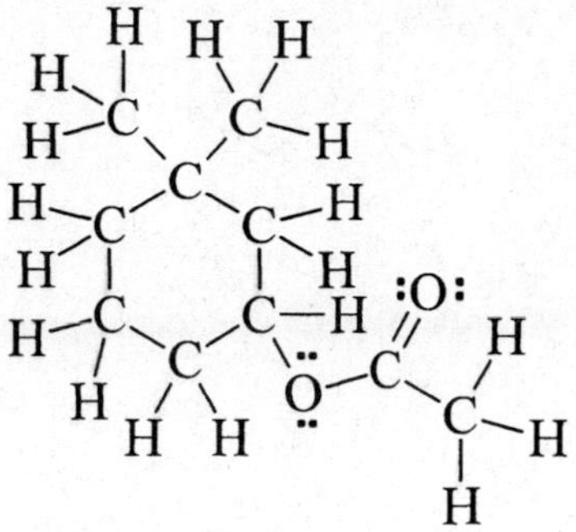 f) 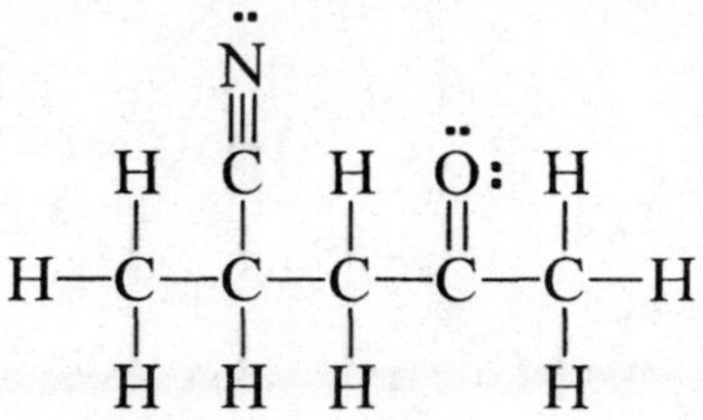

2.19 a) alcohol b) alkene and alcohol c) ketone and alkyl chloride
d) arene e) nitrile and aldehyde f) alkene and ketone
g) amine, arene, and carboxylic acid

2.20 a) arene and carboxylic acid b) alcohol and alkyne
c) alkene and aldehyde d) carboxylic acid
e) ester f) nitrile and ketone

2.21 a) same b) isomers c) same d) same e) isomers f) isomers

2.22 a) DU = 0 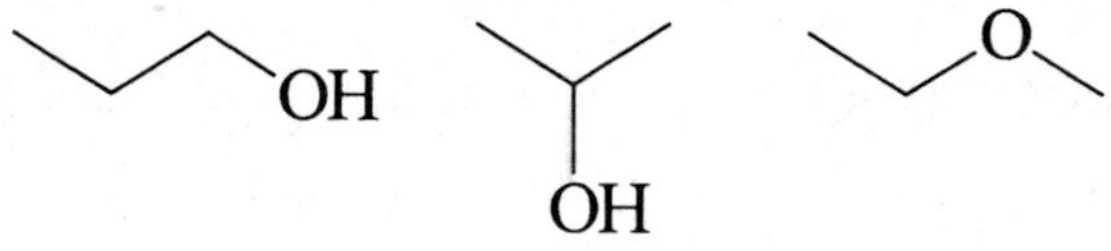

b) DU = 0 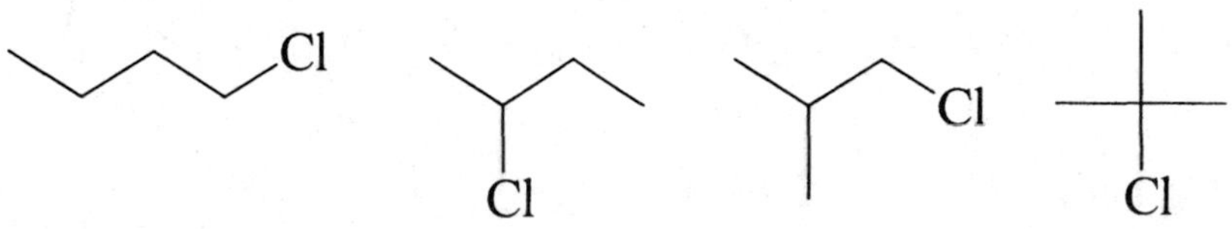

c) DU = 1 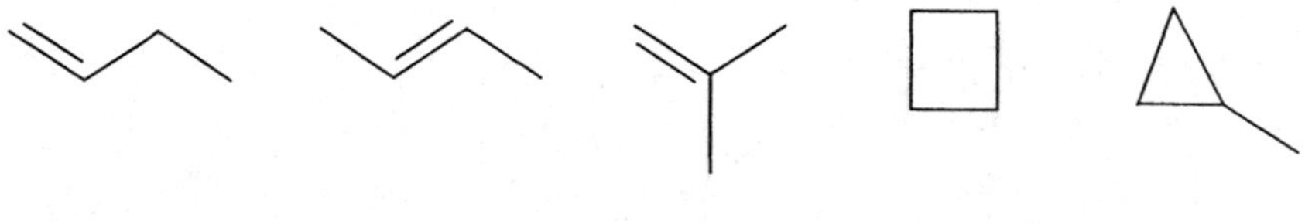

d) DU = 0

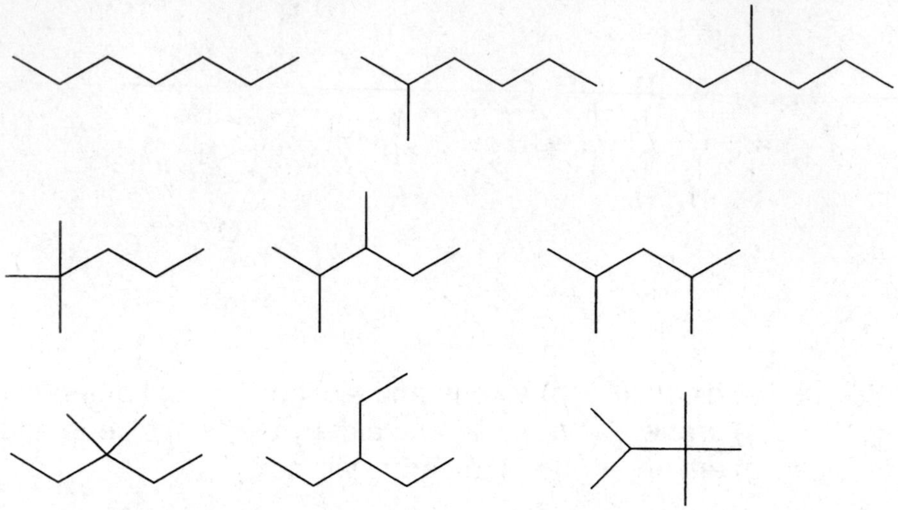

2.23 The DU for C_4H_8O = 1.

a)

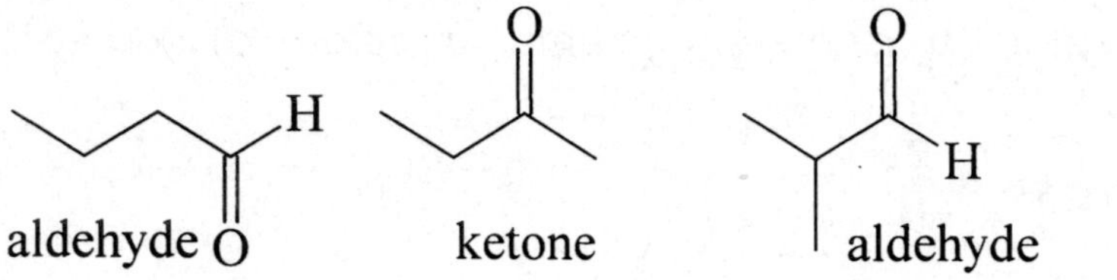

b)

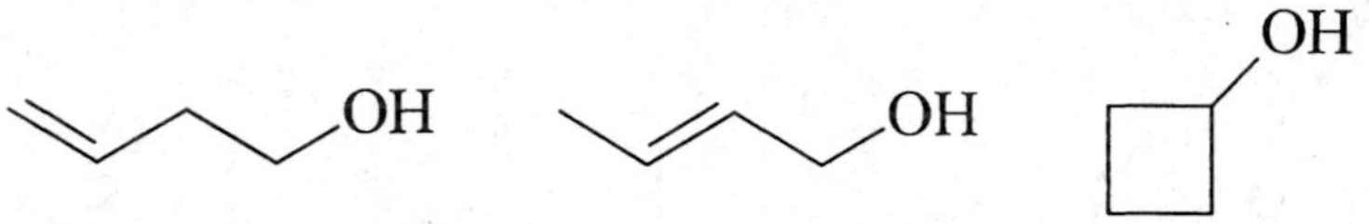

c)

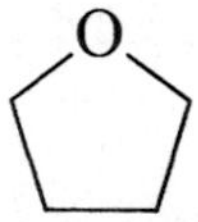

2.24 The DU for C_5H_{10} = 1.

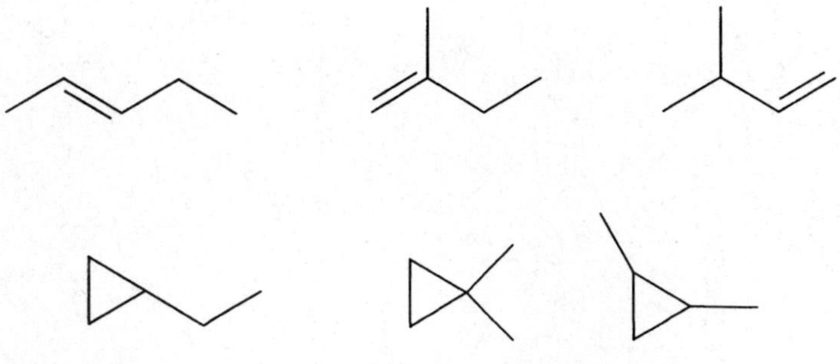

2.25

a) DU = 2 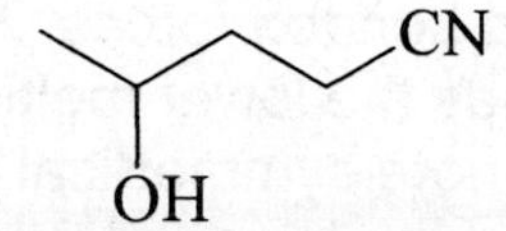

b) DU = 2 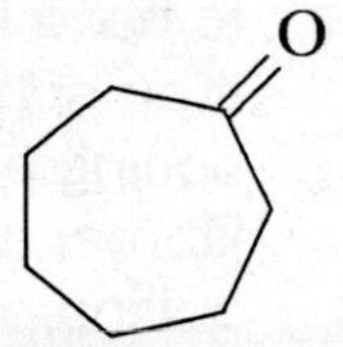

c) DU = 0 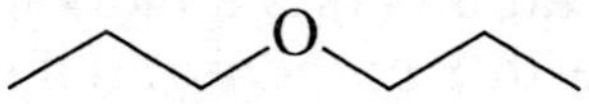

d) DU = 4 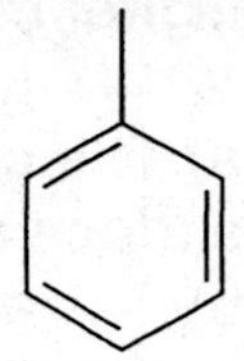

2.26 The nitrile functional group (CN) has a triple bond between the C and the N, which requires a DU of 2. The DU for $C_8H_{17}N$ = 1, so it cannot have a nitrile functional group.

2.27

a) 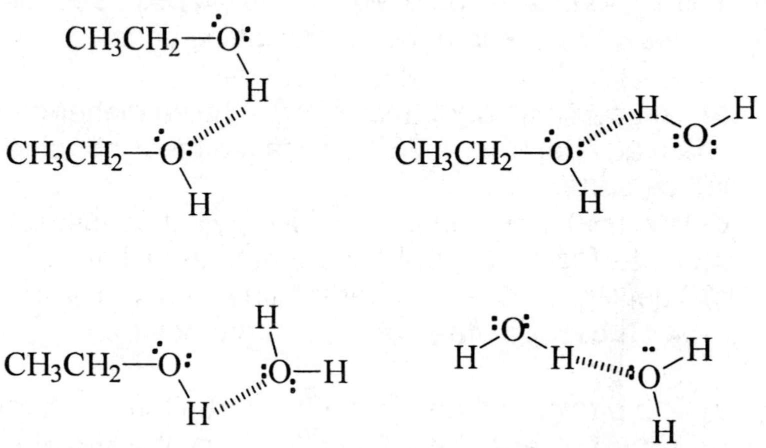

b) 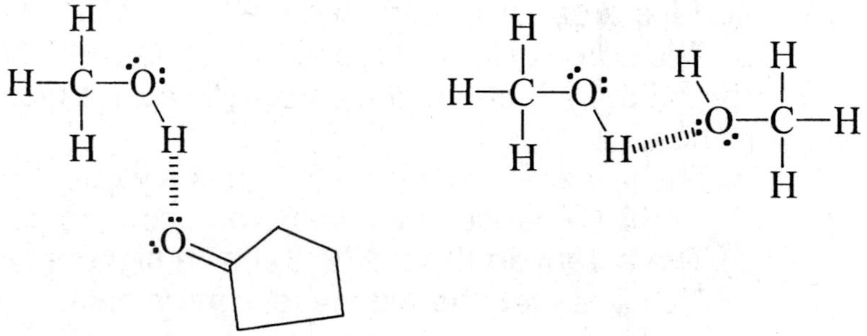

2.28 More symmetrical molecules can pack into the crystal lattice better, so they tend to have higher melting points. Rod-shaped molecules have more surface area than spherical molecules of similar molecular mass, so they have slightly higher boiling points due to increased London forces. Therefore, the rod-shaped isomer, $CH_3CH_2CH_2CH_2OH$, has the lower melting point (-90°C) and the higher boiling point (117°C) and the more symmetrical isomer, $(CH_3)_3COH$, has the higher melting point (26°C) and the lower boiling point (82°C) .

2.29 a) The compound with 10 carbons has a higher molecular weight and larger surface area than the one with 8 carbons, so it has a higher boiling point.
b) The alcohol has a higher boiling point than the ether because it can form hydrogen bonds.
c) The ether has a slightly higher boiling point because it is a polar molecule and has dipole-dipole interactions. The alkane is nonpolar and has only weaker London forces holding its molecules together.

2.30 The cycloalkane has a higher melting point because its more symmetrical shape allows it to pack into the crystal lattice better.

2.31 a) The cyclic alcohol (mp = -19°C) has a higher melting point than the straight chain alcohol (mp = -78°C) because the cyclic alcohol packs into the crystal lattice better.
b) Beeswax has a higher melting point because it is a larger compound and has more surface area for London force attraction.
c) The ketone has a higher melting point (mp = -78°C) than the alkane (mp = -154°C) because the ketone is more polar.

2.32 a) The straight chain alkane (bp = 70°C) has a higher boiling point than the branched alkane (bp = 58°C) because the straight chain alkane has more surface area for London force attraction.
b) The carboxylic acid (bp = 141°C) has a higher boiling point than the aldehyde (bp = 75°C) because the carboxylic acid is more polar and forms hydrogen bonds.
c) The larger ether (bp = 90°C) has a higher boiling point than the smaller ether (bp = 35°C) because it has more London forces.
d) The aldehyde (bp = 179°C) has a higher boiling point than the alkene (bp = 146°C) because the aldehyde is more polar.

2.33 The salt with the nonpolar hydrocarbon chains bonded to the nitrogen will be more soluble in hexane than ammonium chloride because hexane is a nonpolar solvent.

2.34 Like dissolves like. Both benzene and hexane are nonpolar compounds, and are miscible. Water is a polar compound. Therefore, benzene and water are immiscible.

2.35 The carboxylic acid is miscible with water because it is more polar and can act as a hydrogen bond donor and acceptor with water. The ester is less polar and can only act as a hydrogen bond acceptor. It is nearly insoluble in water.

2.36

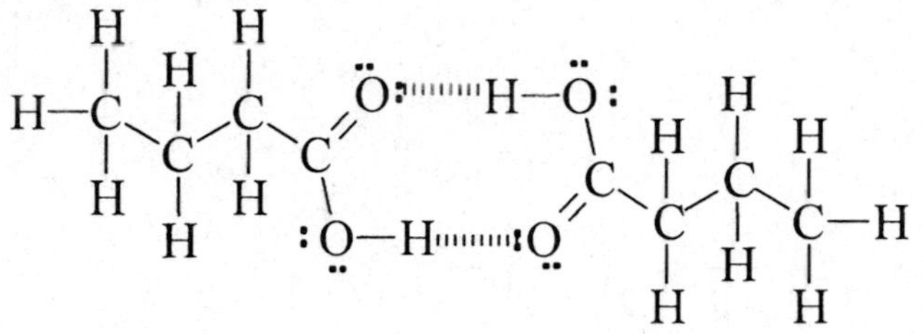

2.37 Use Table 2.1 in the text to calculate total bond energies of compounds. The aldehyde has 4 C-H bonds, 1 C-C bond, and 1 C=O bond, giving it a total bond energy of 646 kcal/mol [4(98) + 81 +173]. The alcohol has 3 C-H bonds, 1 C=C bond, 1 C-O bond, and 1 O-H bond, giving it a total bond energy of 627 kcal/mol [3(98) + 145 + 79 + 109). Therefore, the aldehyde is more stable.

2.38

$CH_2{=}CH_2$	+	H-H	$\longrightarrow$	$CH_3\text{-}CH_3$
537 kcal/mol		104 kcal/mol		669 kcal/mol

The total bond energy of the product (669 kcal/mol) is larger than that of the reactants (537 + 104 = 641 kcal/mol). This means that the product is more stable than the reactants and the reaction is energetically favorable.

2.39 Draw all the isomers (three) of C_5H_{12}. Then draw the possible $C_5H_{11}Cl$ isomers that could be produced from each. Only the straight chain isomer of C_5H_{12} can produce three isomers of $C_5H_{11}Cl$.

Isomers of C_5H_{12} Isomers of $C_5H_{11}Cl$

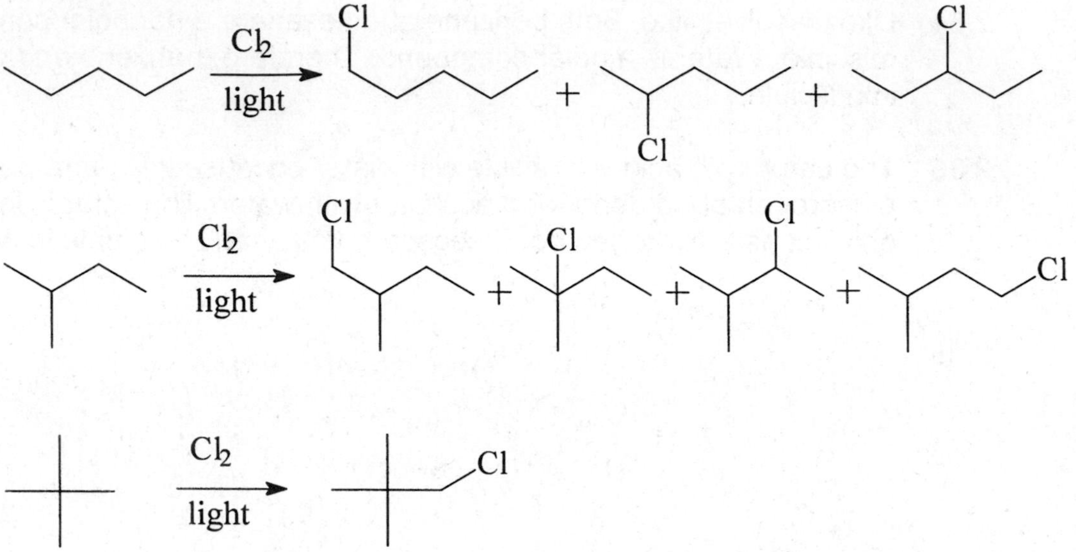

2.40 The DU of C_5H_{10} = [(2)5 +2 - 12]/2 = 1, so the unknown has one ring or one double bond. All of the carbons of the isomer that gives only one C_5H_9Cl must be identical. The five-membered cyclic alkane is the only isomer that has all of its carbons identical.

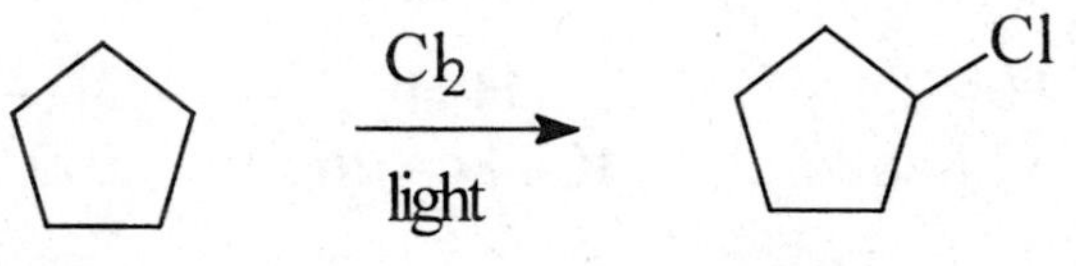

C_5H_{10} C_5H_9Cl

2.41 From Chapter 1,
μ = (*e*)(*d*). Because F is more electronegative than Cl, *e* is larger for CH_3F than for CH_3Cl. But the C-Cl bond (1.78 Å) is longer than the C-F bond (1.38 Å) so *d* is larger for CH_3Cl than for CH_3F.

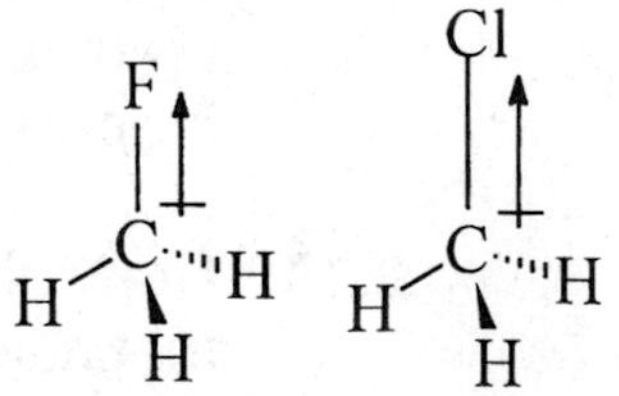

2.42 amine, alcohol, amide, carboxylic acid

2.43 Glucose contains alcohol and aldehyde functional groups. Because it contains a large number of polar functional groups that can hydrogen bond extensively with water, glucose is very soluble in water. This water solubility is why nature uses carbohydrates as energy sources. Alkanes would be insoluble in blood and could not be readily transported throughout the organism.

2.44 The zwitterion form of alanine should have a very high melting point because it has strong ion-ion interactions with other molecules.

2.45 The Du of estrone is 8 (4 pi bonds plus 4 rings). It has 18 carbons, so it must have 22 hydrogens (38 - 16).

2.46
a) alcohol, DU = 1
b) arene and amine, DU = 5
c) arene and amide, DU = 5
d) phenol, ether, amide, alkene, DU = 6
e) phenol, aldehyde, ether, DU= 5
f) ketone, alkene, alcohol, DU = 6
g) arene, carboxylic acid, DU = 5
h) phenol, amide, DU = 5

2.47 a) isomers

 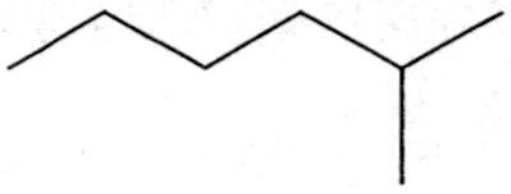

b) same

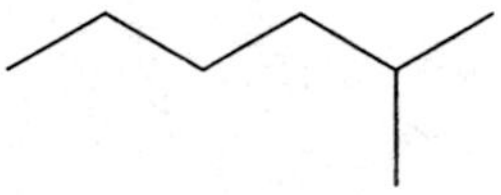 (both)

c) isomers

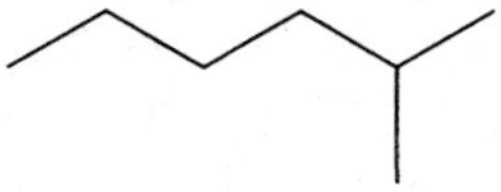 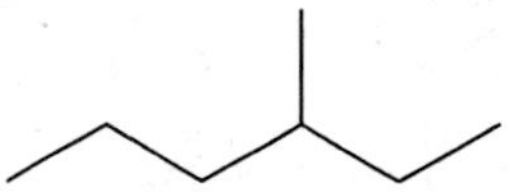

2.48 a) isomers

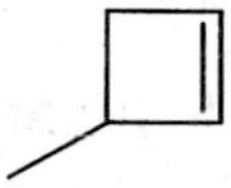 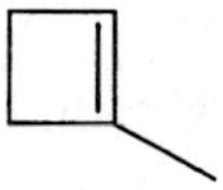

b) isomers

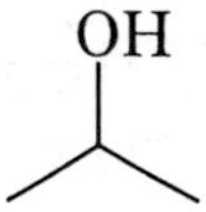

c) same

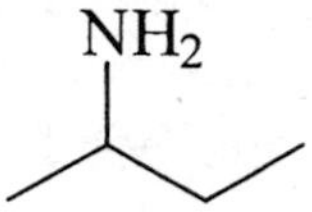

(both)

d) same

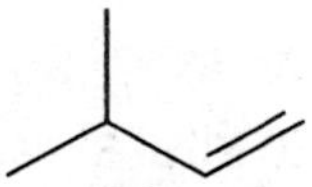 (both)

e) isomers

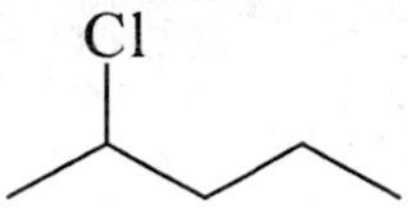

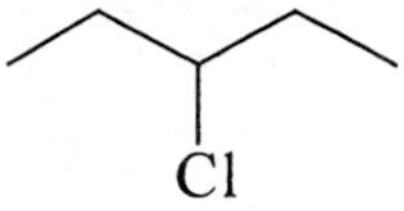

2.49 The bond dipoles of the polar C-Cl bonds of the isomer on the right cancel, so this isomer has no dipole moment.

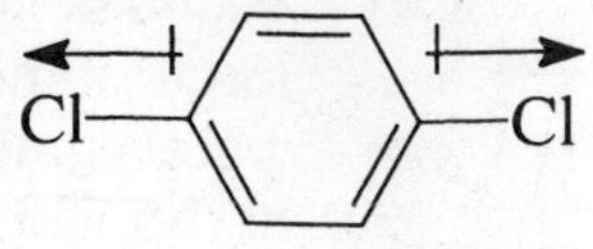

2.50 The alcohol with three carbons has a higher water solubility than the alcohol with four carbons because it has a smaller nonpolar part.

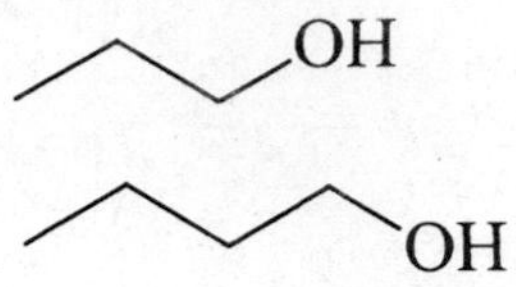

2.51 The cyclic compound, benzene, has a higher melting point because it packs better into a crystal lattice than does hexane.

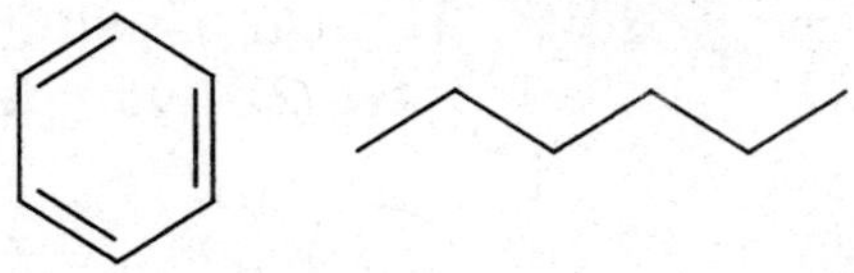

2.52 The straight-chain isomer (left structure) gives 3 $C_5H_{11}Cl$ products; the middle isomer gives 4 products; and the isomer on the right gives only one product.

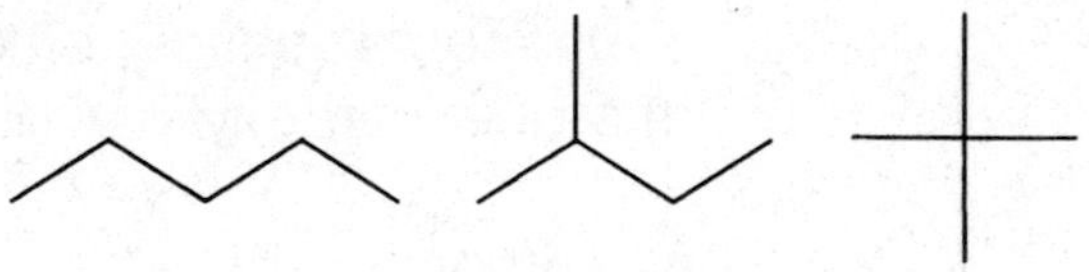

2.53

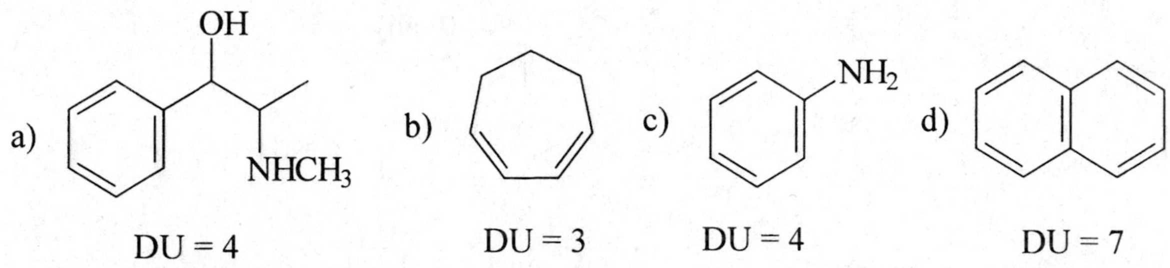

After completing this chapter, you should be able to:

Quickly recognize the common ways in which atoms are bonded in organic compounds. You should also recognize unusual bonding situations and be able to estimate the stability of molecules with such bonds. (Problem 2.1)

Know the trends in bond strengths and bond lengths for the common bonds. (Problems 2.37 and 2.38)

Recognize when compounds are structural isomers and be able to draw structural isomers for any formula. (Problems 2.2, 2.3, 2.4, 2.8, 2.21, 2.22, 2.23, 2.24, 2.39, 2.40, 2.47, and 2.48)

Calculate the degree of unsaturation for a formula and use it to help draw structures for that formula. (Problems 2.5, 2.9, 2.10, 2.25, 2.26, 2.45, 2.46, and 2.53)

Draw structures using any of the methods we have seen. You should also be able to examine a shorthand representation for a molecule and recognize all of its features. (Problems 2.6, 2.7, 2.17, and 2.18)

Examine the structure of a compound and determine the various types of intermolecular forces that are operating. On this basis, you should be able to crudely estimate the physical properties of the compound. (Problems 2.11, 2.12, 2.13, 2.14, 2.15, 2.27, 2.28, 2.29, 2.30, 2.31, 2.32, 2.33, 2.34, 2.35, 2.36, 2.43, 2.44, 2.50, and 2.51)

Recognize and name all of the important functional groups. (Problems 2.16, 2.19, 2.20, 2.42, 2.43, and 2.46)

Chapter 3
ORBITALS AND BONDING

3.1 Like the 2s orbital, the 3s orbital is spherical in shape but larger. The 3s orbital has two spherical nodes, whereas the 2s orbital has only one node.

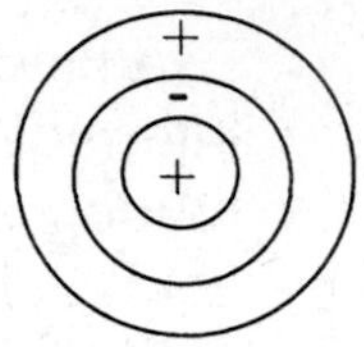

3s

3.2 An orbital energy diagram for an atom shows the relative energies of orbitals and how electrons are distributed in these orbitals. Put the electrons in the lowest energy orbital available while observing the Pauli exclusion principle and Hund's rule.

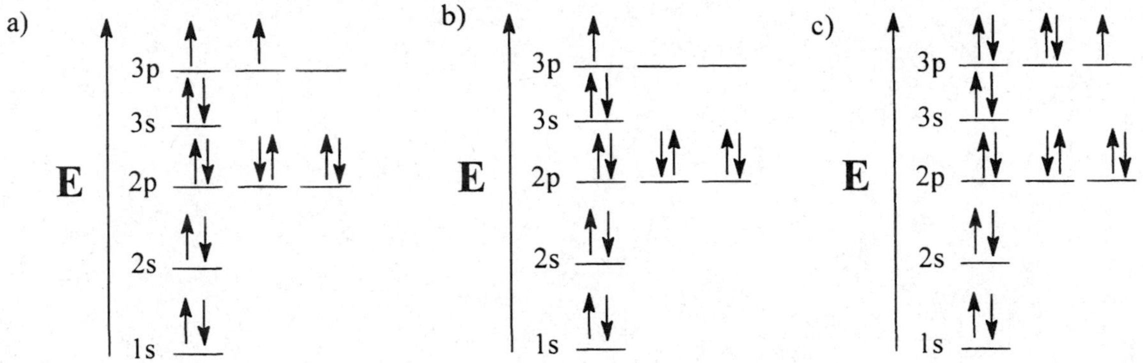

3.3 The electron arrangement that is in accord with the basic rules mentioned in problem 3.2 is the lowest-energy one and is termed the **ground state** electron configuration. An electron arrangement that is not in accord with these rules is higher energy and is an **excited state**.

a) In this electron configuration one of the three electrons in the singly filled 2p orbitals has its spin opposite to that of the other two electrons. This is in violation of Hund's rule and therefore is an excited state.
b) Higher energy orbitals contain electrons while the low-energy 1s orbital is only half filled, so this is an excited state.

3.4

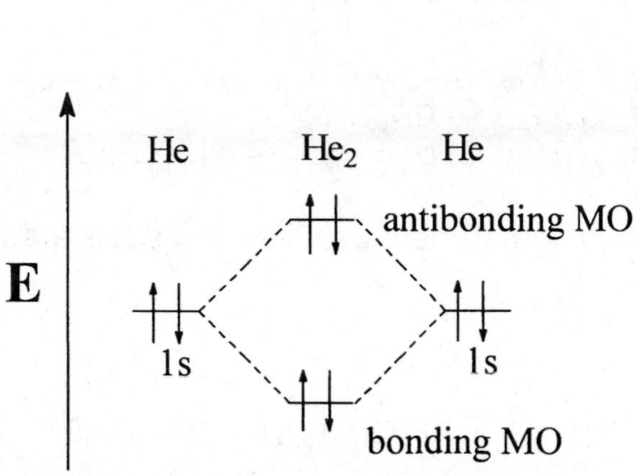

3.5 The unshared electrons on the oxygen are in nonbonding sp^3-hybridized atomic orbitals.

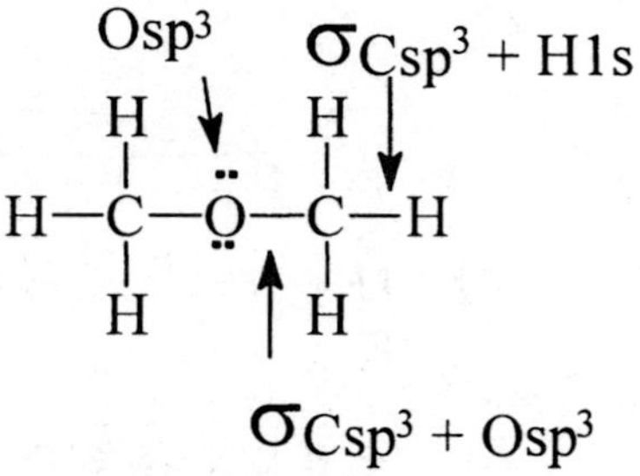

3.6 1) $\sigma_{Csp2 + H1s}$ 2) $\sigma_{Csp3 + H1s}$ 3) $\sigma_{Csp3 + Csp2}$
4) $\sigma_{Csp2 + Nsp2}$, $\pi_{C2p + N2p}$ 5) $\sigma_{Csp3 + Nsp2}$ 6) $\sigma_{Csp3 + H1s}$

3.7 a) Both the N and the C have linear geometry and are *sp* hybridized.
b) The bonds between the C and the N are one sigma bond and two pi bonds. The sigma bond results from the overlap of an *sp*-hybrid AO on C with an *sp*-hybrid AO on N. Each pi bond results a p orbital on C overlapping with a p orbital on N.
c) The unshared electrons on N occupy a nonbonding *sp*-hybridized AO.

d)

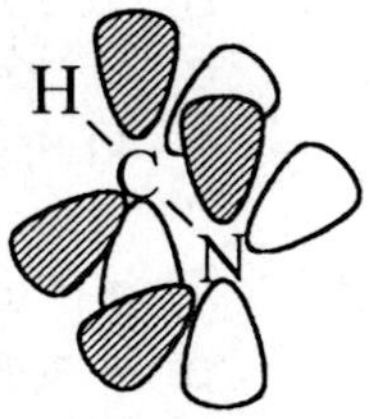

3.8 1) $\sigma_{Csp2 + H1s}$ 2) $\sigma_{Csp3 + H1s}$ 3) $\sigma_{Csp3 + Csp2}$ 4) $\sigma_{Csp2 + Csp2}$, $\pi_{C2p + C2p}$
5) $\sigma_{Csp2 + Csp}$ 6) $\sigma_{Csp + Csp}$, 2 $\pi_{C2p + C2p}$ 7) $\sigma_{Csp + H1s}$

3.9 a) 1) $\sigma_{Csp3 + H1s}$ 2) $\sigma_{Csp3 + Csp2}$ 3) $\sigma_{Csp2 + H1s}$ 4) $\sigma_{Csp2 + Osp2}$, $\pi_{C2p + O2p}$
b) 1) $\sigma_{Csp3 + Csp2}$ 2) $\sigma_{Csp2 + Csp2}$, $\pi_{C2p + C2p}$ 3) $\sigma_{Csp2 + Csp}$
4) $\sigma_{Csp + Nsp}$, 2 $\pi_{C2p + N2p}$ 5) $\sigma_{Csp2 + H1s}$

c) 1) $\sigma_{Csp2 + Osp2}$, $\pi_{C2p + O2p}$ 2) $\sigma_{Csp + H1s}$ 3) $\sigma_{Csp + Csp}$, 2 $\pi_{C2p + C2p}$
4) $\sigma_{Csp2 + Csp}$ 5) $\sigma_{Csp3 + Csp2}$ 6) $\sigma_{Csp3 + H1s}$
d) 1) $\sigma_{Csp3 + H1s}$ 2) $\sigma_{Csp3 + Csp3}$ 3) $\sigma_{Csp3 + Nsp3}$ 4) $\sigma_{Nsp3 + H1s}$

3.10 a) The π bonds are not conjugated because they are separated by CH_2 groups at each end.

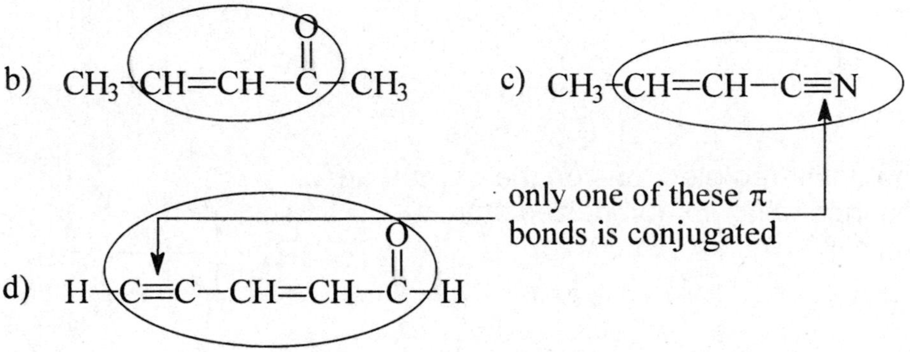

e) The π bonds are not conjugated because they are separated by a CH_2 group.

3.11 An atom with an unshared pair of electrons that is next to a π bond will assume a hybridization that puts the unshared electrons in a p orbital so that the electrons can be conjugated with the p orbitals of the π bond.

a) 1) sp^3	2) sp^2	3) sp^2	4) sp^2	5) sp^3
b) 1) sp^3	2) sp^2	3) sp^2	4) sp^3	
c) 1) sp^2	2) sp^2	3) sp^2	4) sp^3	
d) 1) sp^2	2) sp^2			

3.12 The nuclei of atoms must be in identical positions in resonance structures; only electrons in conjugated p orbital can be moved.
a) These are not resonance structures because a H has moved from oxygen to carbon.
b) These are resonance structures.
c) These are not resonance structures because a H has moved.

3.13 Each resonance structure must have the same number of electrons and the same total charge.
a) The structure on the right has two fewer electrons. It would have a formal charge of +1 on the left C.
b) The structure on the right has 5 bonds to C and 2 more electrons.
c) The center N of the structure on the right has only 6 electrons.

d) The C bonded to O in the right structure has 5 bonds (10 electrons).

3.14 The relative stability of resonance structures can be estimated by whether they satisfy the octet rule and the number and location of any formal charges.
a) The right structure is much less important because the octet rule is not satisfied at the positively charged carbon atom. It also has formal charges.
b) The first two structures are equally important. The last structure is not important because the octet rule is not satisfied at the N and it has more formal charges.

3.15 a) The right structure makes only a minor contribution to the resonance hybrid, so the actual structure resembles the left structure and the compound has only a small amount of resonance stabilization. This means that the left structure is a good representation of the overall structure and stability of the compound.
b) This compound has three equivalent resonance structures that contribute equally to the resonance hybrid. (Only two of these are shown in problem 3.14.) The ion has a large amount of resonance stabilization. All the N-O bonds are identical.

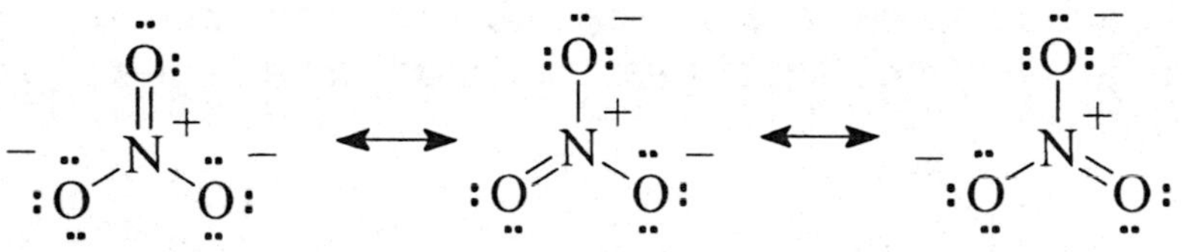

3.16 a) b)

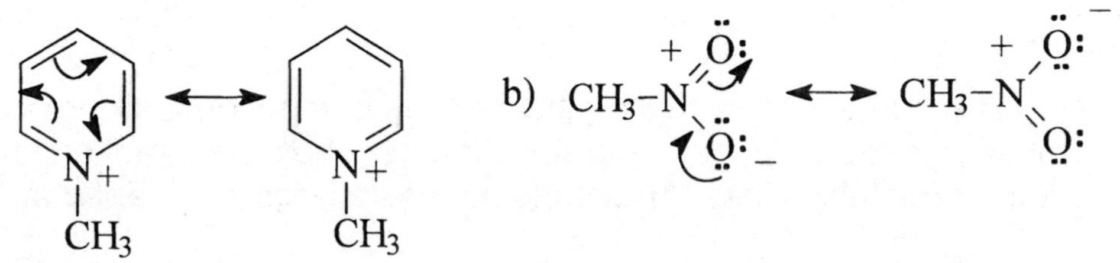

c) 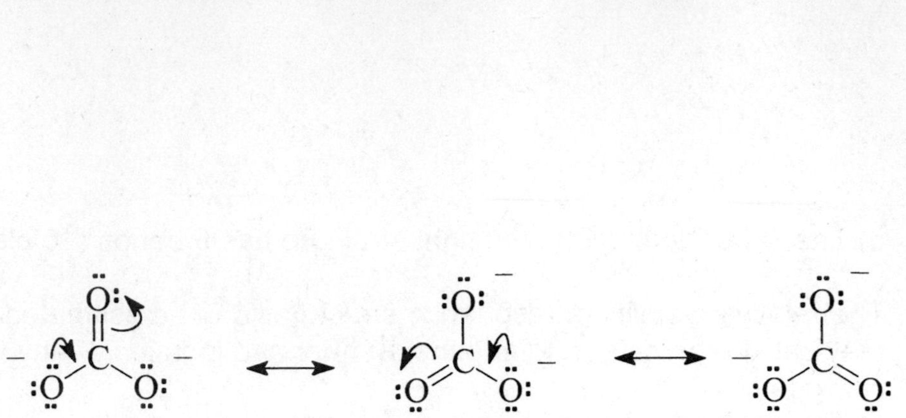

d) 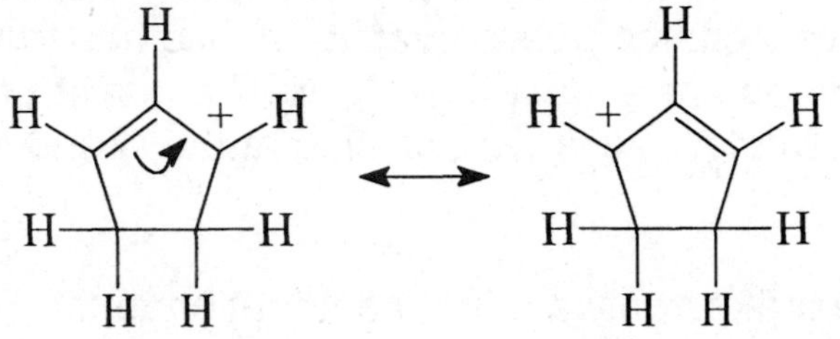

e) 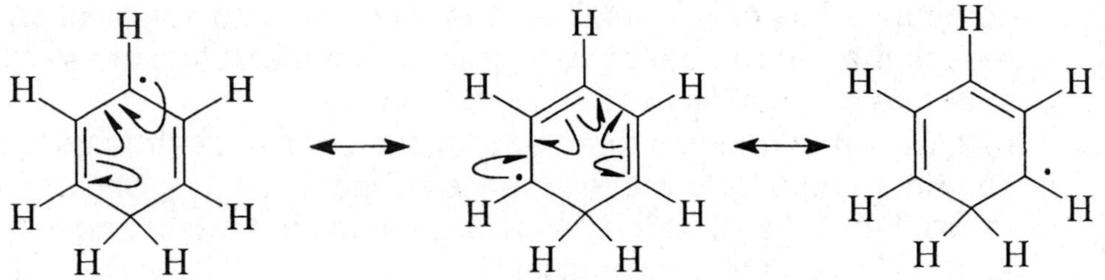

3.17 a) Each of these structures makes a similar contribution to the resonance hybrid because they have similar stabilities. Because there are three important resonance structures, this anion has a large resonance stabilization.

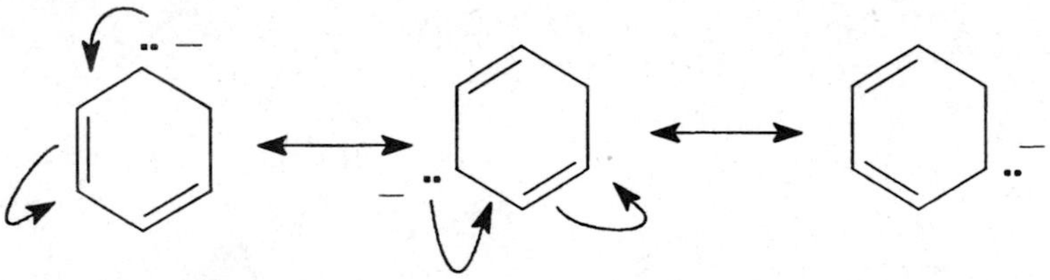

b) The first structure makes the largest contribution to the resonance hybrid because it has no formal charges. The others make similar but lesser contributions. The compound has a large resonance stabilization.

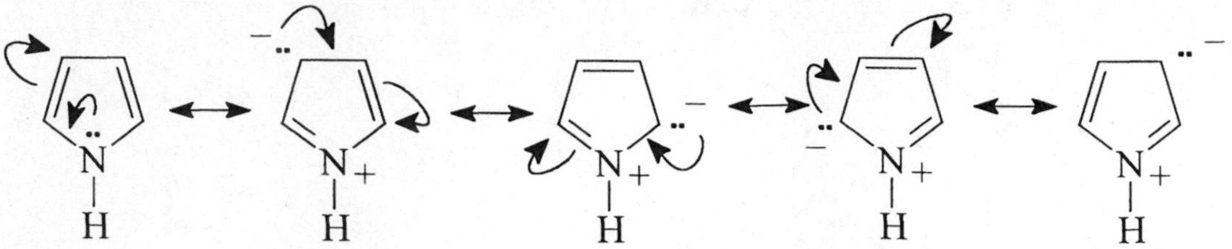

c) The structures have equal stabilities, so they contribute equally to the resonance hybrid. The compound has a large resonance stabilization.

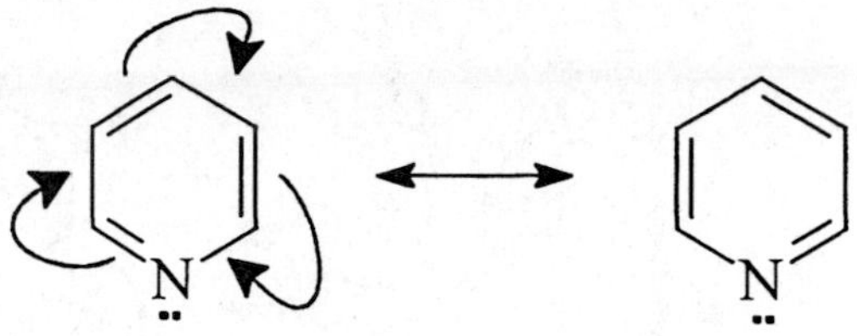

d) The situation is similar to part b. The first structure makes the largest contribution to the resonance hybrid because it has no formal charges. The others make similar but lesser contributions. The compound has a large resonance stabilization.

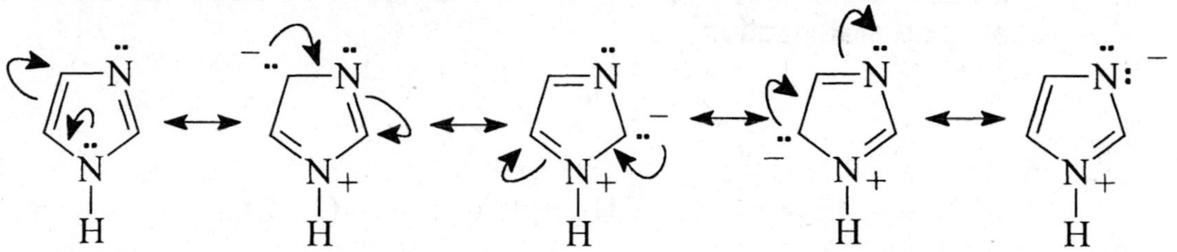

e) The first two resonance structures are equal in stability and have no formal charges so they contribute more to the resonance hybrid. The other five resonance structures are of similar stability. They contribute less to the resonance hybrid than the first two because they have formal charges. This compound has a large resonance stabilization because it has seven resonance structures.

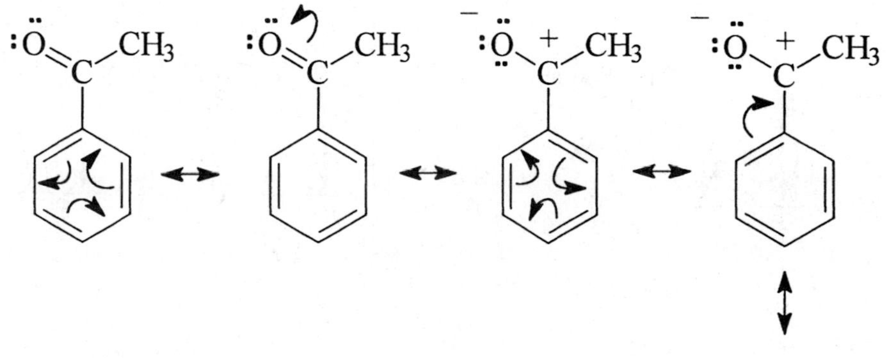

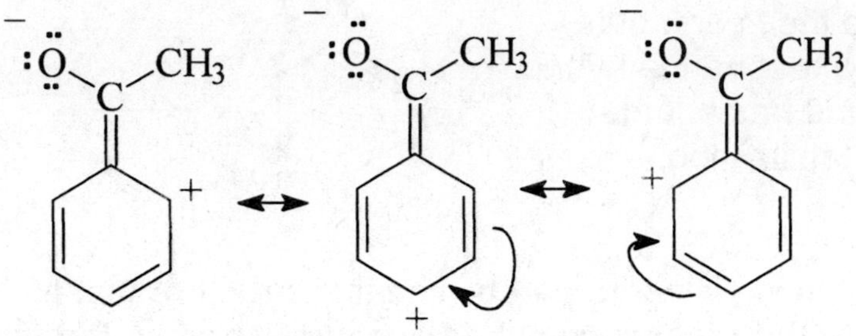

f) The structure on the left contributes more to the resonance hybrid because it does not have any formal charges. This ester has only a small resonance stabilization.

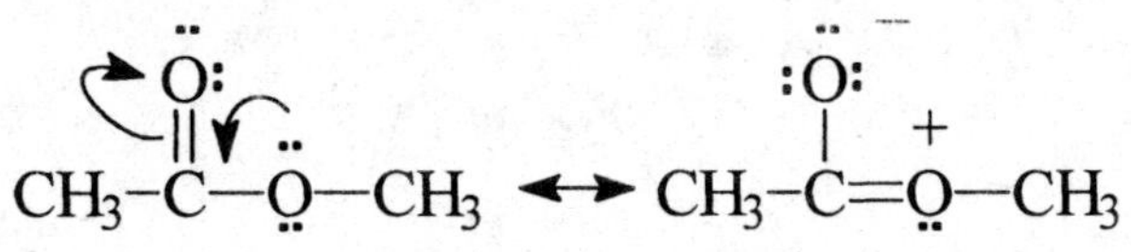

3.18

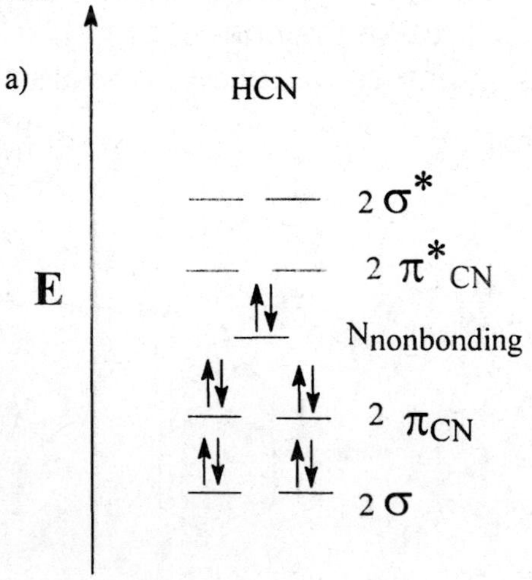

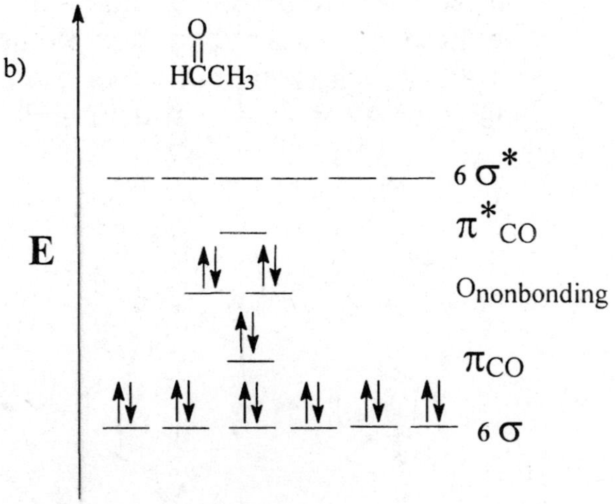

c)

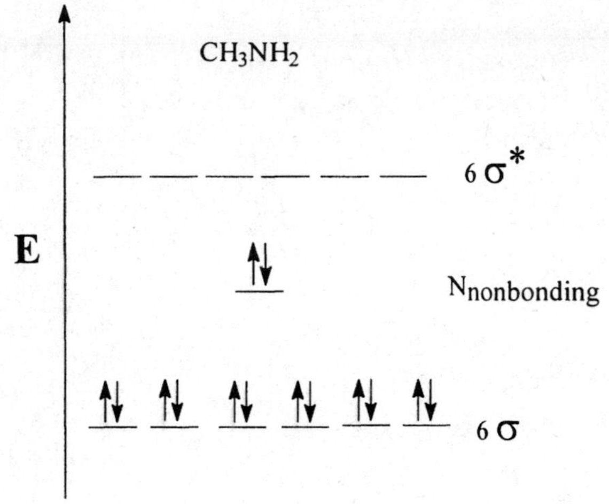

3.19

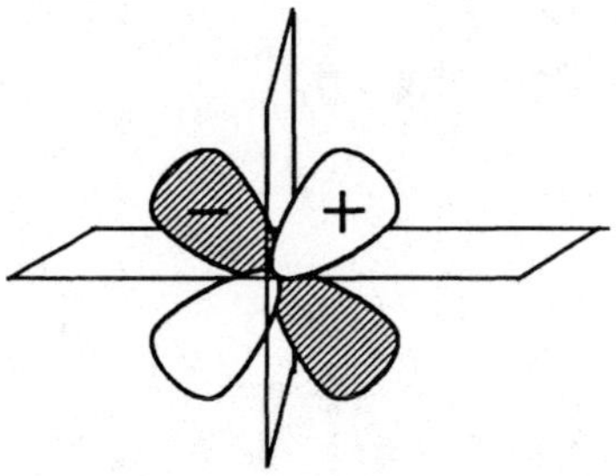

3.20

a) $CH_3{-}NH_2$
sp³ sp³

b) $CH_2{=}CHCH_2C{\equiv}N$
sp² sp² sp³ sp sp

c) 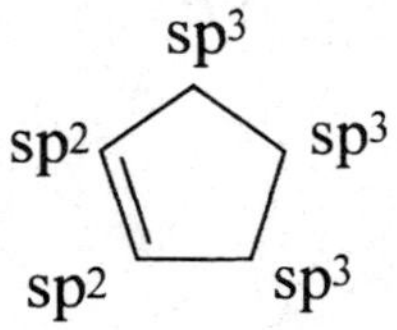

d) 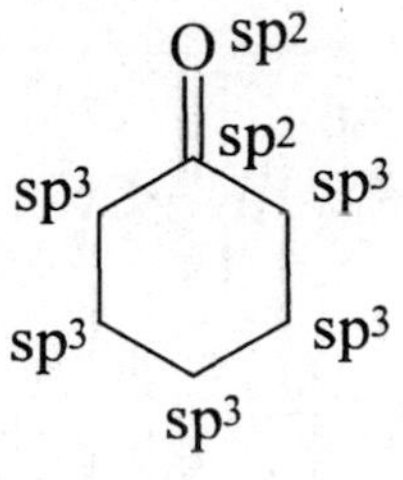

e) 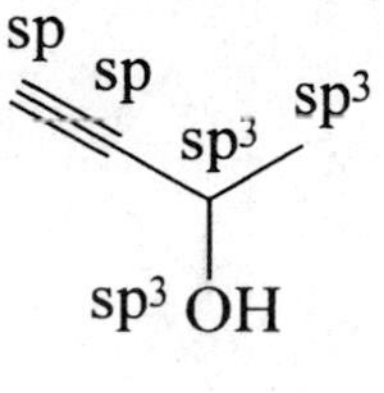

f) 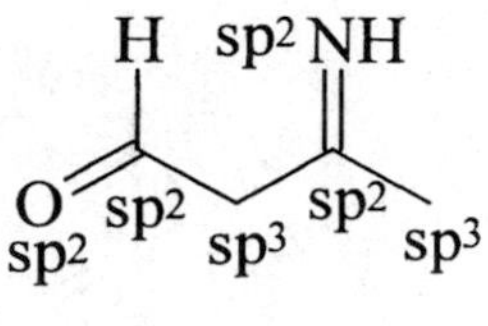

3.21

a) 1) $\sigma_{Csp2+Osp2}$ and $\pi_{C2p+O2p}$ 2) $\sigma_{Csp2+H1s}$, 3) $\sigma_{Csp2+Csp2}$ and $\pi_{C2p+Cp2}$, 4) $\sigma_{Csp2+Csp2}$, 5) $\sigma_{Csp2+Csp3}$, 6) $\sigma_{Csp3+H1s}$

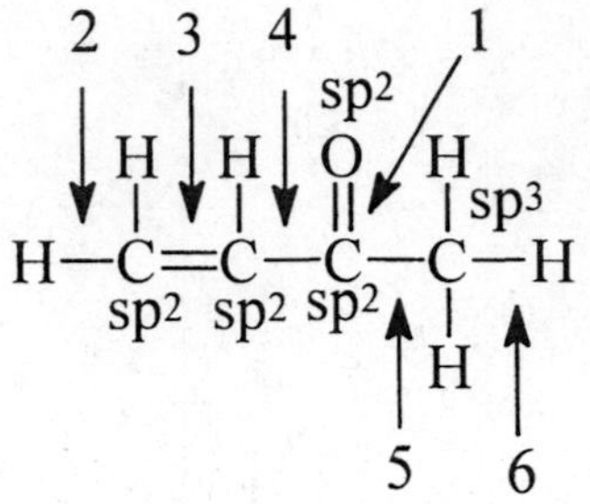

b) 1) $\sigma_{Csp3+H1s}$, 2) $\sigma_{Csp3+Csp3}$, 3) $\sigma_{Csp3+Osp3}$, 4) $\sigma_{Osp3+H1s}$

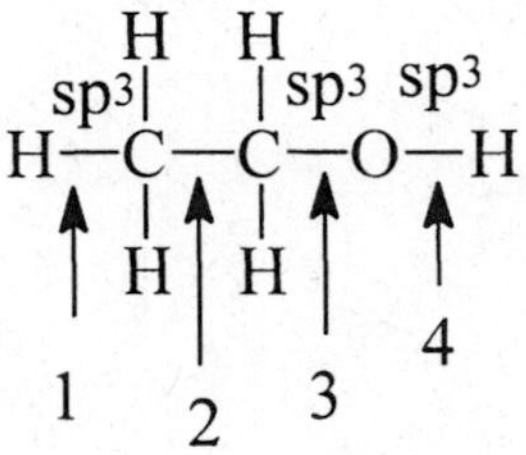

c) 1) $\sigma_{Csp3+Csp}$, 2) $\sigma_{Csp+Nsp}$ and two $\pi_{C2p+N2p}$

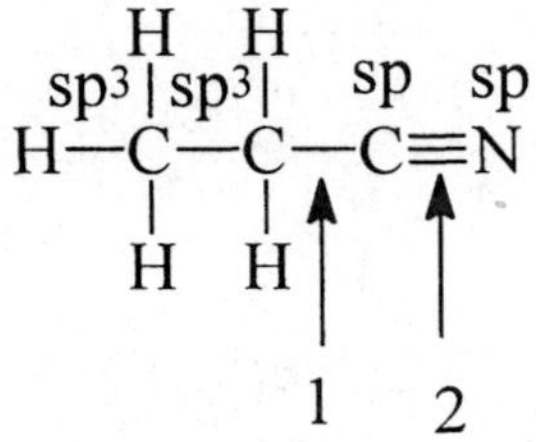

d) 1) $\sigma_{Csp2+Osp2}$ and $\pi_{C2p+O2p}$, 2) $\sigma_{Csp3+Csp}$, 3) $\sigma_{Csp+Csp}$ and two $\pi_{C2p+C2p}$, 4) $\sigma_{Csp+Csp2}$, 5) $\sigma_{Csp2+H1s}$

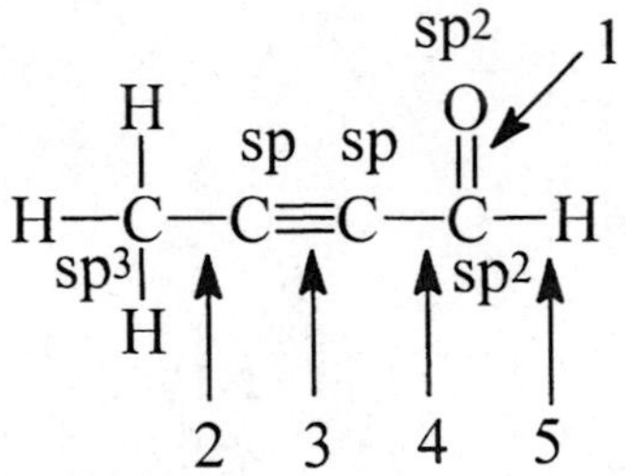

3.22

a) 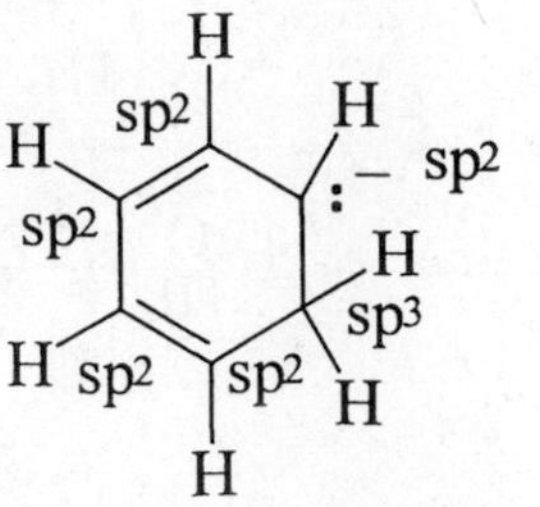

b) 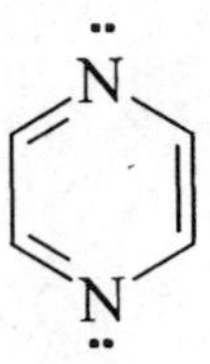

all sp2

c) 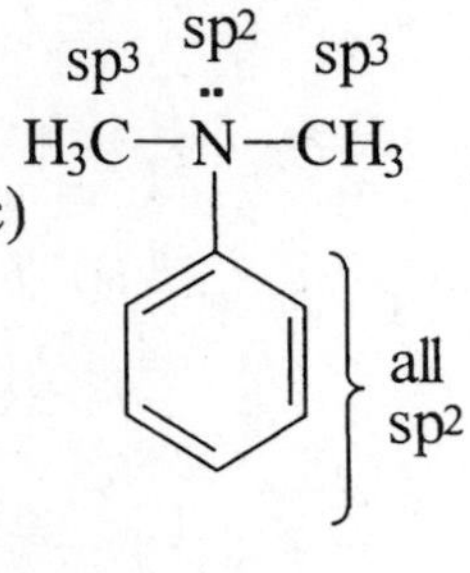

d) 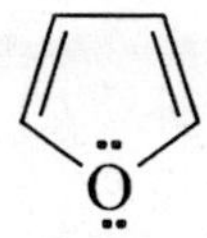

all sp2

e) 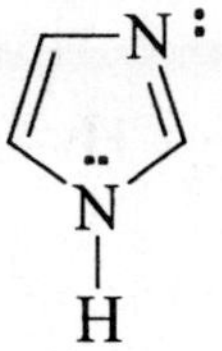

all sp2

f) 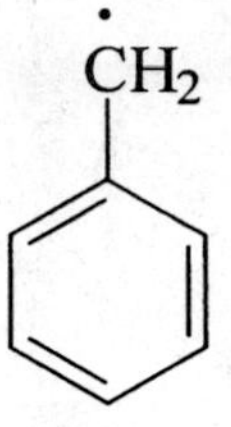

all sp2

g) 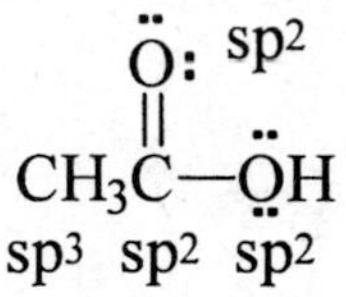

h) 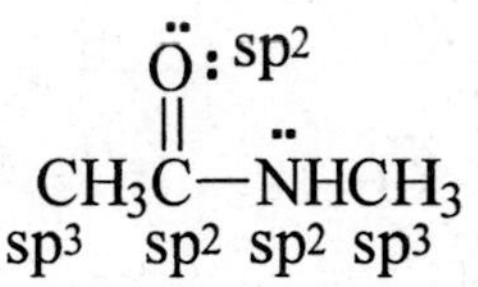

3.23

a) 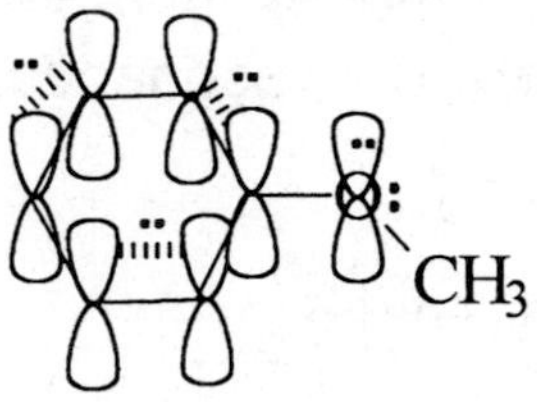

b) 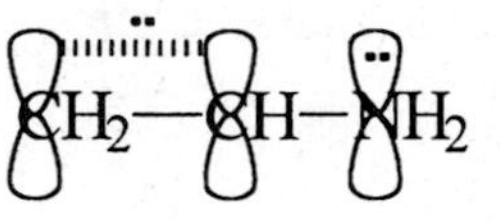

c) 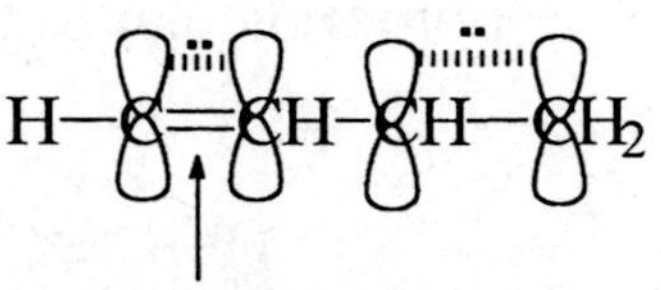

The p orbitals of the second π bond are perpendicular to these orbitals and are not part of the conjugated system.

3.24 a) The unshared electron pairs on the nitrogens are not part of the conjugated pi system.

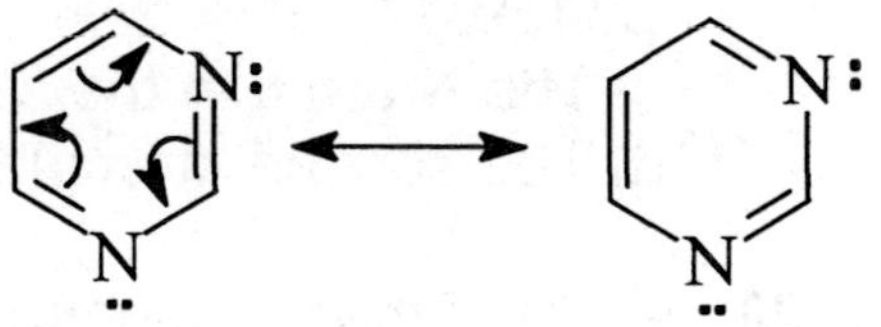

b) One unshared electron pair on the oxygen is part of the conjugated system and is involved in resonance.

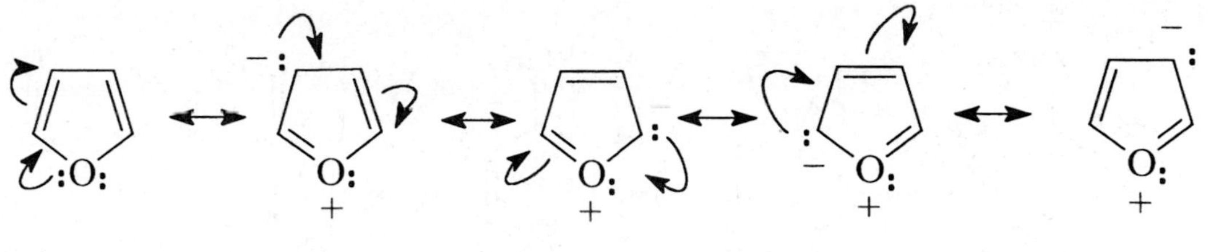

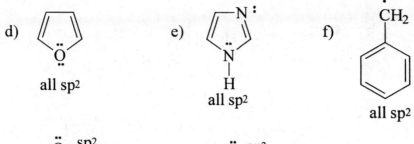

c)

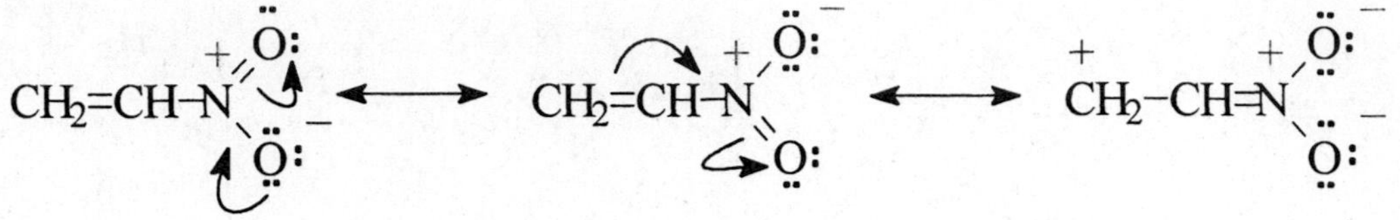

d)

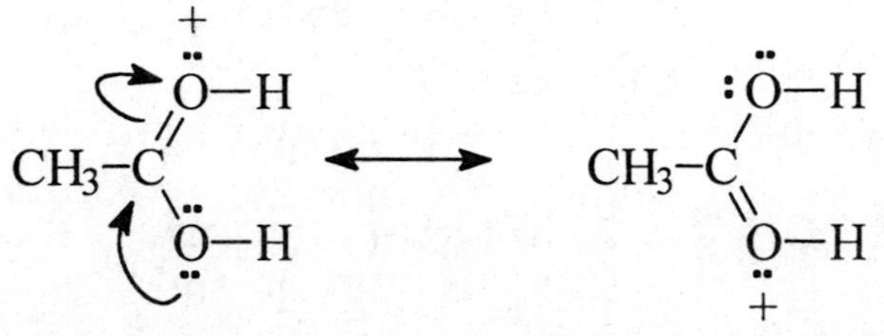

e) The unshared electron pair on the carbon is conjugated with one of the pi bonds of the triple bond. The p orbitals of the other pi bond are perpendicular to these orbitals and are not involved in resonance.

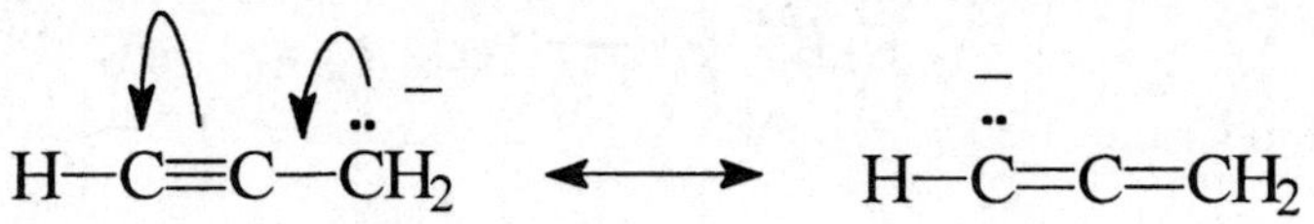

3.25 a) The N has six electrons in the structure in the middle and it has ten in the structure on the right. Both structures violate the octet rule.
b) The second structure is not a resonance structure for the first structure because a hydrogen has moved. These structures represent constitutional isomers.
c) The N atom in the second structure has only six electrons.
d) The second structure has ten electrons at one carbon.

3.26 a) All of these resonance structures are important. Although the carbocation is unstable because it does not satisfy the octet rule at one C, it is much more stable than most carbocations because it has a large amount of resonance stabilization.

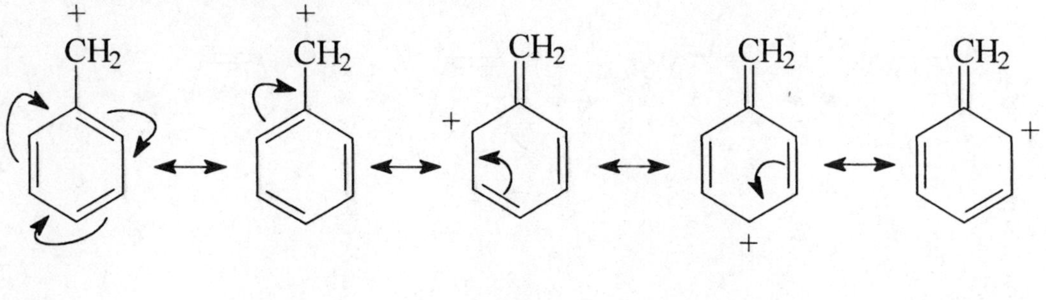

b) Resonance structure **C** contributes more to the resonance hybrid because the negative charge is on the more electronegative oxygen. Structures **A** and **B** are similar in stability. This anion has a large amount of resonance stabilization.

$$CH_2^--CH=CH-\overset{\overset{\ddot{O}:}{\|}}{C}-CH_3 \longleftrightarrow CH_2=CH-\overset{-}{CH}-\overset{\overset{\ddot{O}:}{\|}}{C}-CH_3 \longleftrightarrow CH_2=CH-CH=\overset{:\ddot{O}:^-}{\underset{|}{C}}-CH_3$$

$$\text{A} \qquad\qquad\qquad \text{B} \qquad\qquad\qquad \text{C}$$

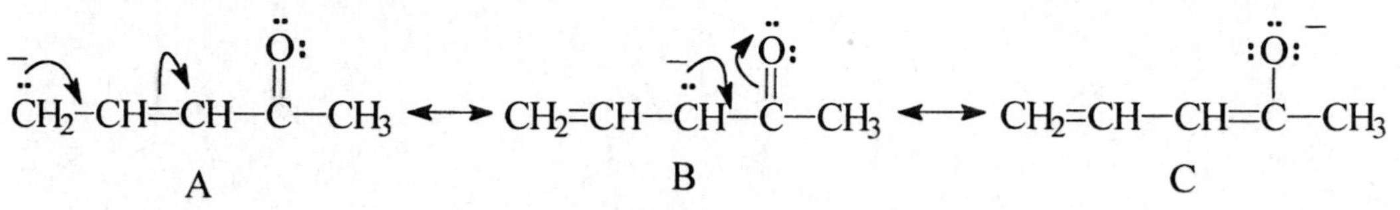

c) The second structure contributes less to the resonance hybrid because it has formal charges. The compound looks more like the first structure. However, the second structure makes a significant contribution, so the compound has some resonance stabilization.

$$CH_3-\overset{\overset{\ddot{O}:}{\|}}{C}-\ddot{N}H_2 \longleftrightarrow CH_3-\overset{:\ddot{O}:^-}{\underset{|}{C}}=\overset{+}{N}H_2$$

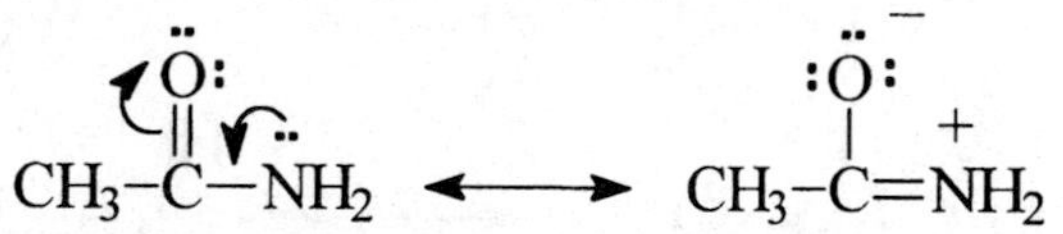

d) The second structure contributes slightly more to the resonance hybrid because the negative charge is on the more electronegative atom. This anion has a large resonance stabilization.

$$CH_3-\overset{\overset{\ddot{O}:}{\|}}{C}-\overset{-}{\ddot{N}}H \longleftrightarrow CH_3-\overset{:\ddot{O}:^-}{\underset{|}{C}}=\ddot{N}H$$

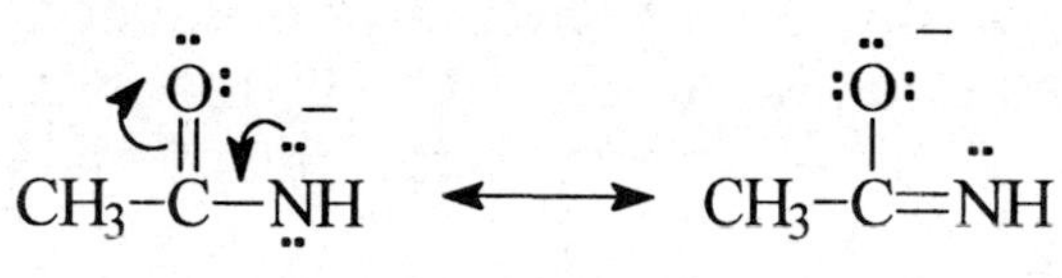

e) Resonance structure **A** is somewhat more stable than the others because it has the negative charges on the oxygens. This anion has a large amount of resonance stabilization.

$$CH_2^--CH=CH-\overset{+}{N}\overset{\ddot{O}:}{\underset{\underset{\ddot{O}:^-}{}}{}} \longleftrightarrow CH_2=CH-\overset{-}{CH}-\overset{+}{N}\overset{\ddot{O}:}{\underset{\underset{\ddot{O}:^-}{}}{}} \longleftrightarrow CH_2=CH-CH=\overset{+}{N}\overset{:\ddot{O}:^-}{\underset{\underset{\ddot{O}:^-}{\text{A}}}{}}$$

$$CH_2^--CH=CH-\overset{+}{N}\overset{\ddot{O}:^-}{\underset{\underset{\ddot{O}:}{}}{}} \qquad\qquad CH_2^--CH=CH-\overset{+}{N}\overset{\ddot{O}:^-}{\underset{\underset{\ddot{O}:}{}}{}}$$

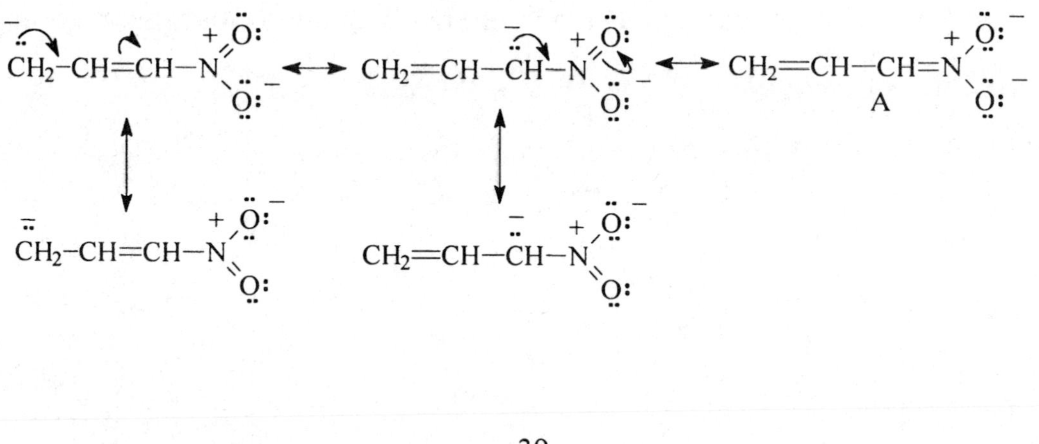

f) Resonance structures **B** and **C** are equal in stability and contribute more to the resonance hybrid than **A** because they have the negative charge on the more electronegative atom. This anion has a large resonance stabilization.

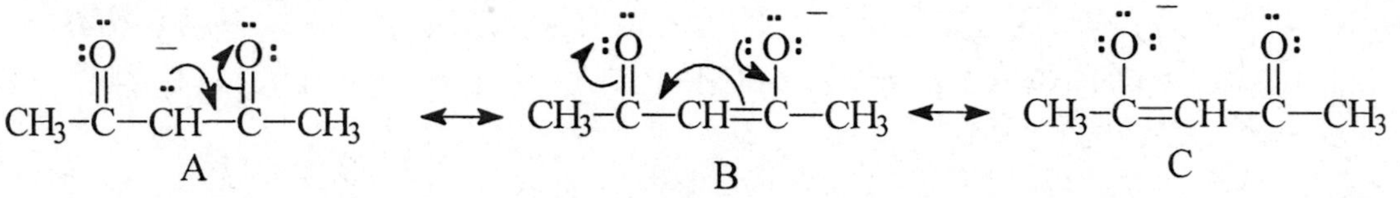

3.27 One of the lone pairs of electrons on the oxygen atom of this carbocation is conjugated with the empty p orbital of the positively charged carbon. The resonance structure on the right is more stable than the structure on the left, because the octet rule is satisfied at both C and O.

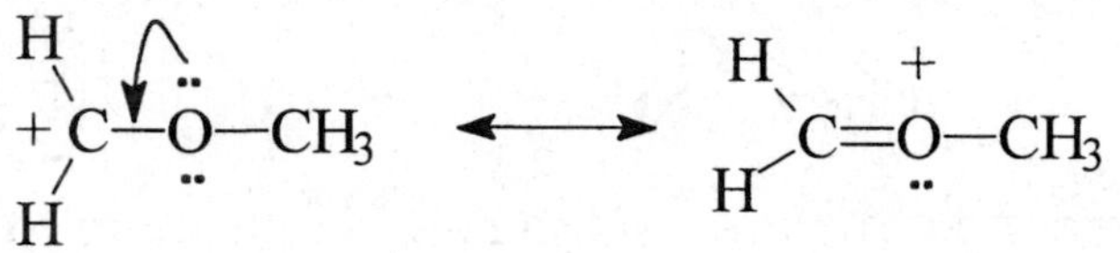

3.28 a) The anion on the left is more stable because the negative charge can be delocalized onto the oxygen atom by resonance. The other anion is not conjugated and therefore has no resonance stabilization.

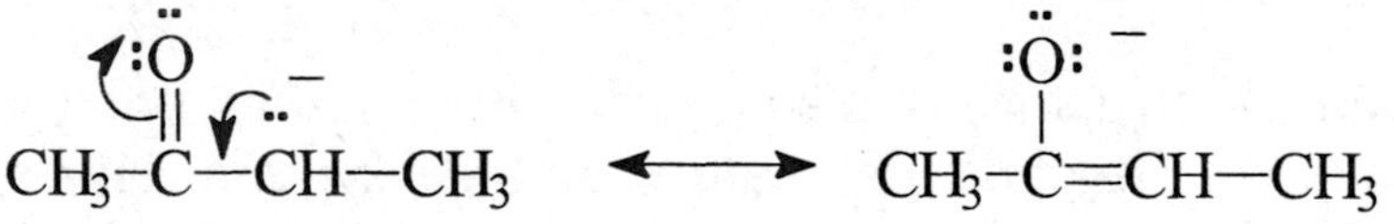

b) The anion on the right is stabilized by resonance whereas the anion on the left is not.

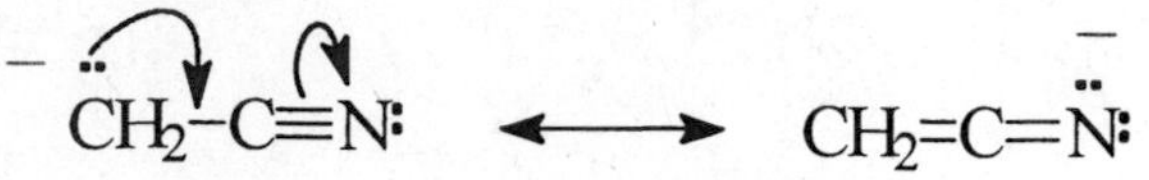

3.29 The p orbital of the positively charged carbon of the carbocation on the left is conjugated with the p orbitals on the carbons of the benzene ring. Therefore this carbocation is resonance stabilized. The carbocation on the right has no resonance stabilization because of the sp^3 hybridized carbon that separates the p orbital of the positive carbon from the p orbital of the carbon of the benzene ring.

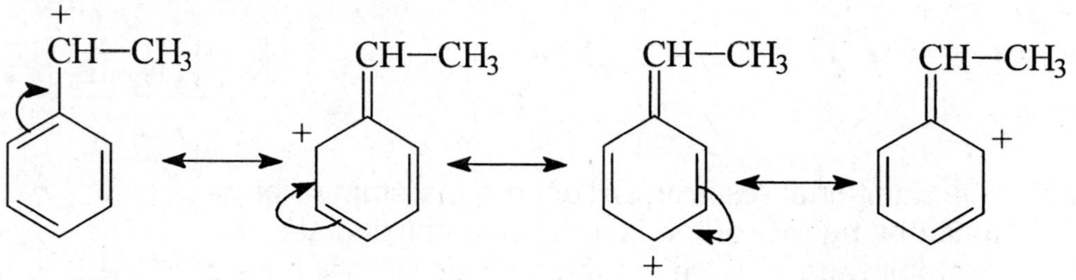

3.30

a) $H{-}C{\equiv}C{-}CH_3$

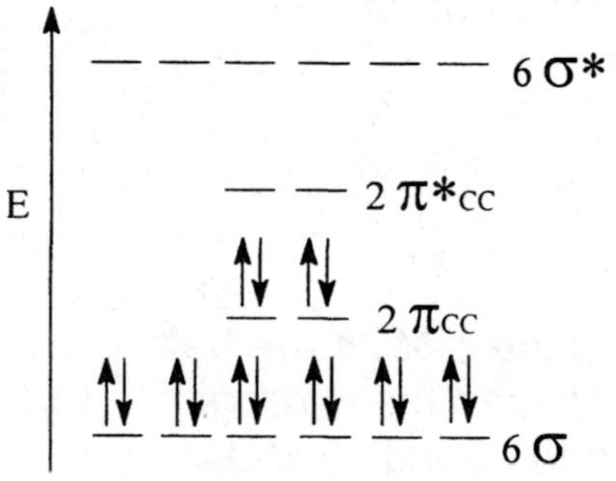

b) $CH_3{-}\ddot{O}{-}H$

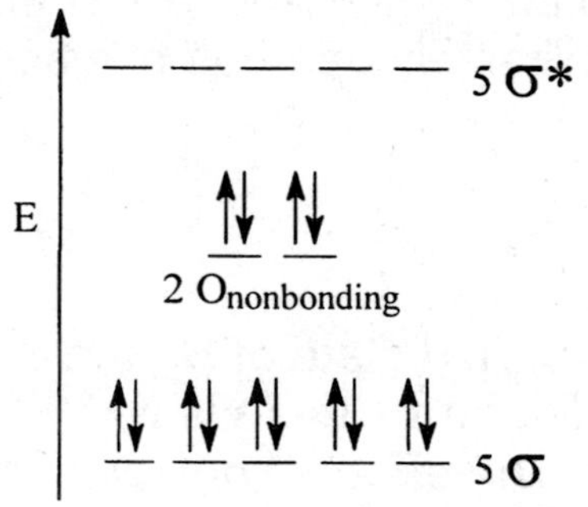

c)

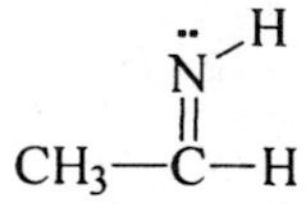

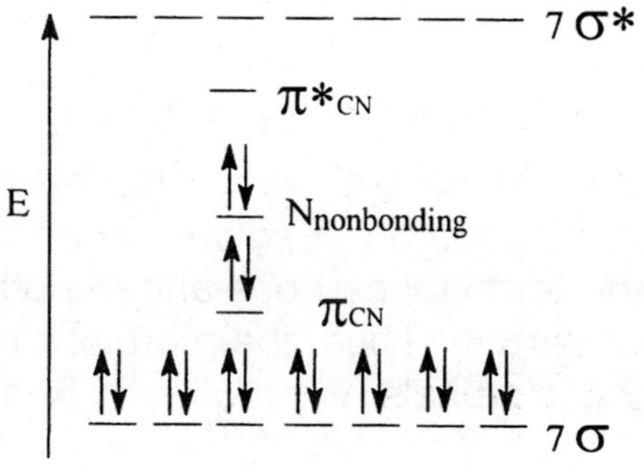

3.31 The lowest energy excited state of CH_2=CH_2 has one electron promoted from the π MO to the π* MO (don't worry about the spin for now).

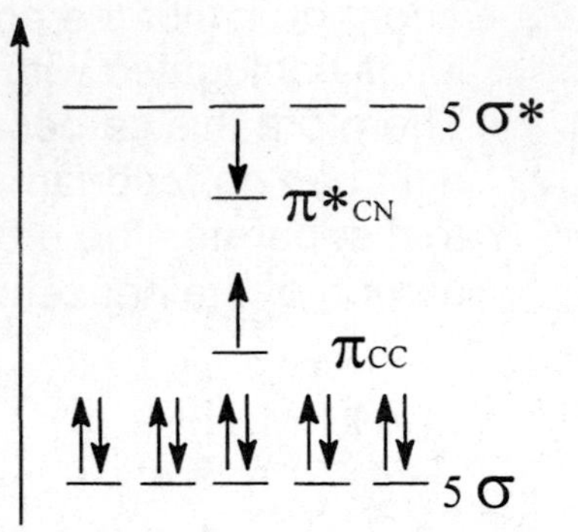

3.32 The additional electron will go into the empty sigma antibonding MO of the H_2 molecule causing some destabilization. However, the species will still have a bond between the two hydrogens. Because the bond is destabilized due to the electron in the antibonding MO, it should be weaker and longer than the bond of H_2 (approximately one-half of a bond).

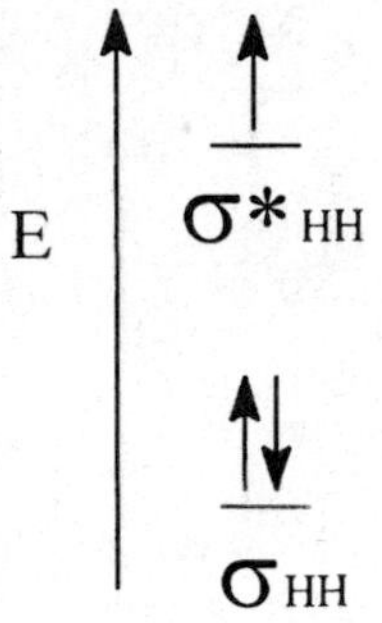

3.33 In the excited state of H_2, the σ* and the σ MOs are half-filled. Therefore there is no bond between the hydrogen atoms. The effect of one electron in the bonding MO is approximately canceled by the effect of the one electron in the antibonding MO.

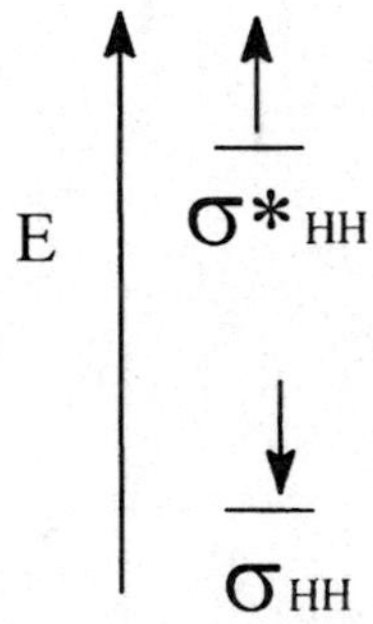

3.34 The two p orbitals on the central carbon (sp hybridized) are perpendicular to each other. One of these perpendicular p orbitals is parallel to the p orbital on one of the terminal carbons and the other is parallel to the p orbital on the other carbon. Thus, the p orbitals of one π bond are perpendicular to the p orbitals of the other π bond. The molecule is not conjugated.

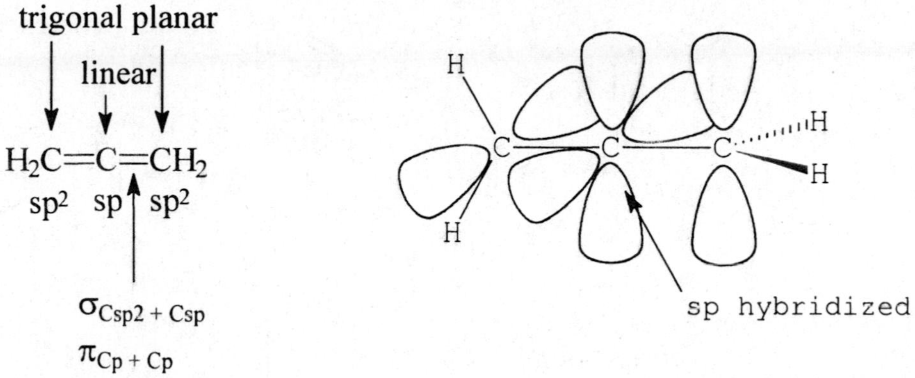

3.35

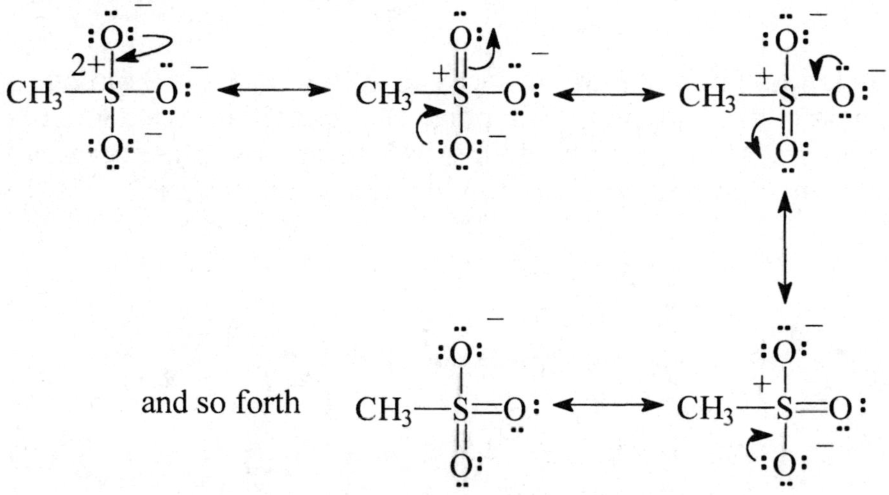

3.36 The five resonance structures are quite similar and are expected to make nearly equal contributions to the resonance hybrid. However, the number of times that a particular C-C bond appears as a double bond in these structures varies. The shortest bond is a double bond in four of the resonance structures and a single bond in only one.

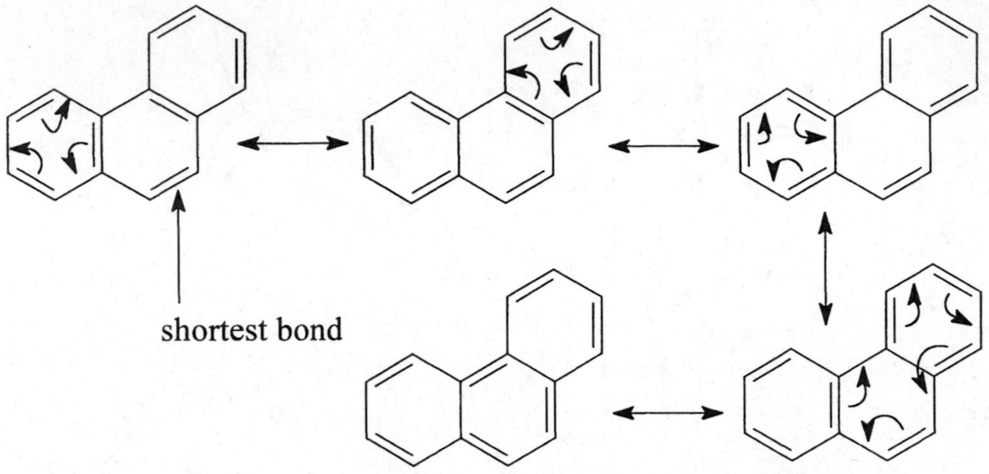

3.37 The longest bonds (1.44 Å and 1.43 Å) are single in 3 structures and double in one structure. The bond of intermediate bond length (1.40 Å) is single in 2 structures and double in 2 structures. The shortest bond (1.37 Å) is single in one structure and double in 3 structures.

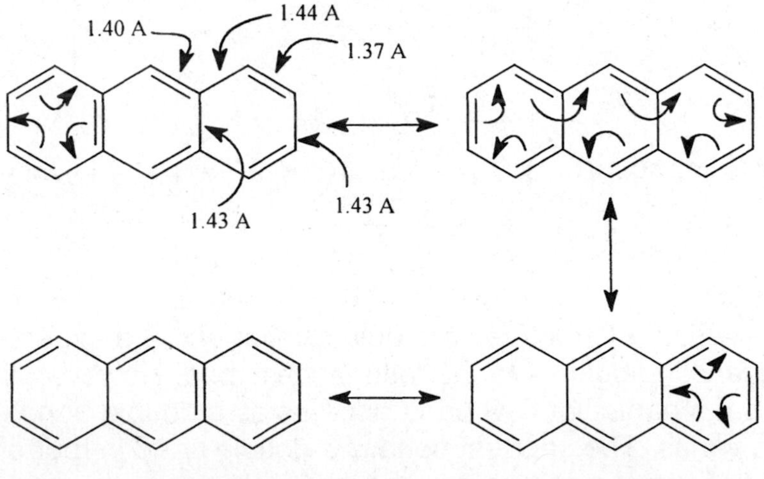

3.38 The odd electron in this radical is in a p orbital, so the species is conjugated. It has two important resonance structures. The odd electron is located on a different carbon in the two resonance structures, providing two different sites for the coupling with the chlorine radical.

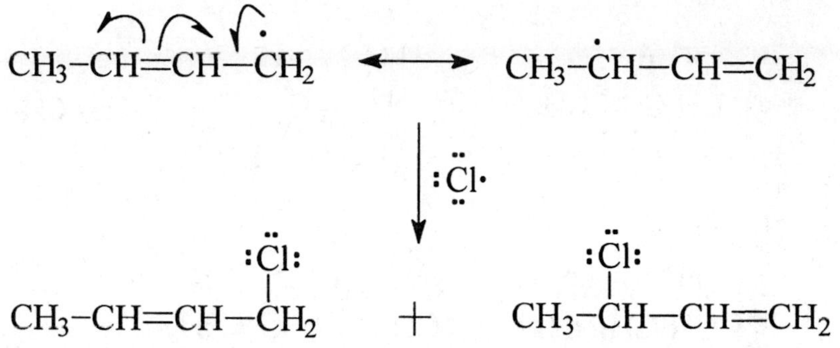

3.39 There are a total of three electrons in the two MOs of the HeH molecule, two electrons in the bonding MO and one in the antibonding MO. The molecule is predicted to be more stable than the separate atoms on the basis of this simple picture.

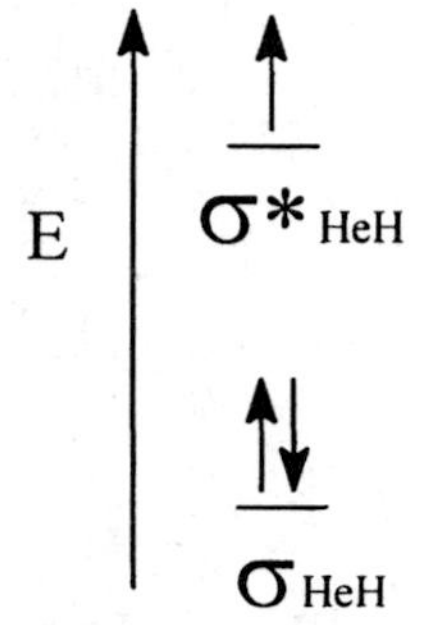

3.40 Nitrogen 1 is sp^2 hybridized and its unshared pair of electrons is in an sp^2 nonbonding AO. Nitrogen 2 is also sp^2 hybridized, but its "unshared" pair of electrons is in a p orbital and is part of the conjugated pi system. Nitrogen 3 is sp^3 hybridized and its unshared pair of electrons is in an sp^3 nonbonding AO.

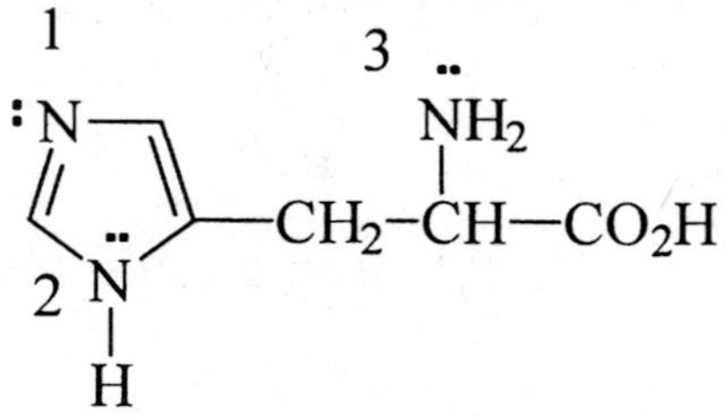

3.41 The cation has three resonance structures that contribute about equally to the resonance hybrid because they are similar in stability.

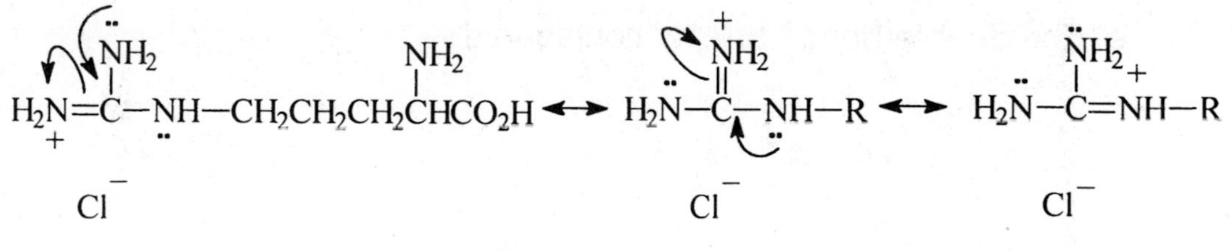

3.42

a) sp³ sp
$CH_3C\equiv CCH_3$

b) 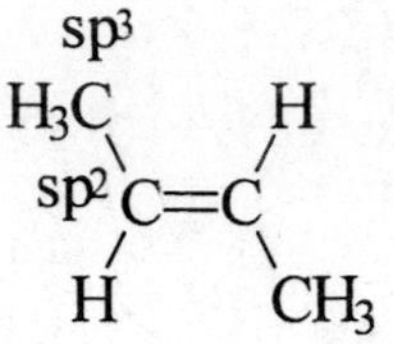

c) sp³ sp sp
$CH_3{-}C\equiv N$

d) sp³ sp³ sp³
$CH_3CH_2NH_2$

e) sp² sp² sp² (C=O) sp²
$CH_2{=}CH{-}C({=}O){-}NH_2$

f) 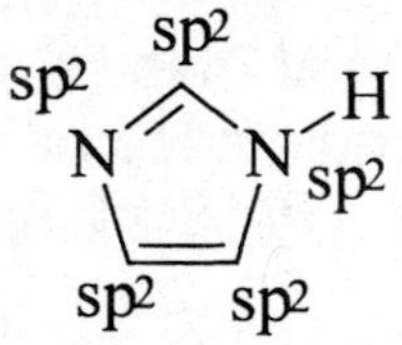

g) sp² sp² sp²
$CH_2{=}CH{-}NH_2$

3.43 a) The double bonds are conjugated.

b) All three double bonds are conjugated.

$CH_2{=}CHCH{=}CHCH{=}O$

c) The double bond is conjugated with one pi bond of the triple bond.

$H{-}C\equiv C{-}CH{=}CH_2$

d) The two double bonds and the electron pair on N are conjugated.

$CH_2{=}CH{-}CH{=}CH_2{-}\ddot{N}H_2$

e) The double bonds are not conjugated.

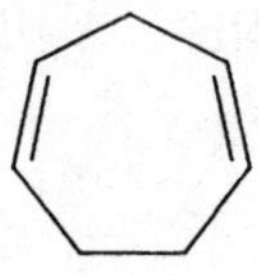

3.44

a) 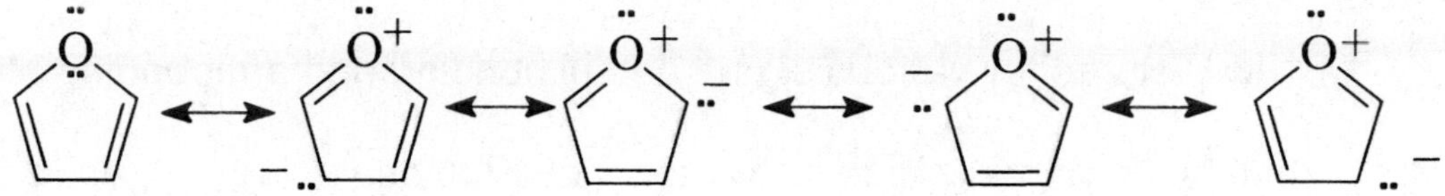

b) 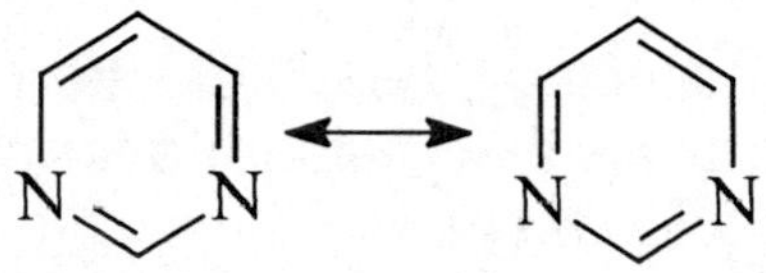

c) 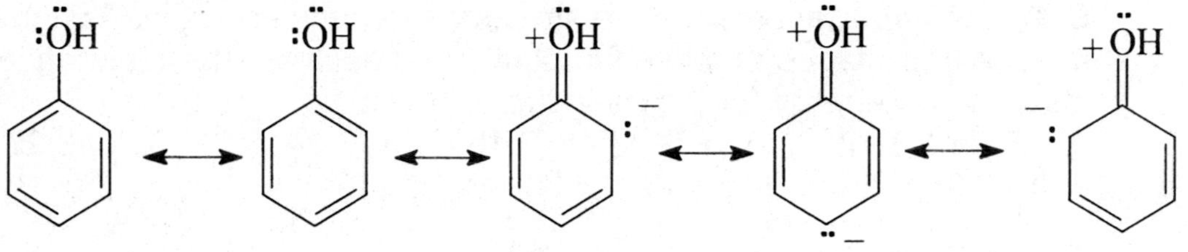

d) 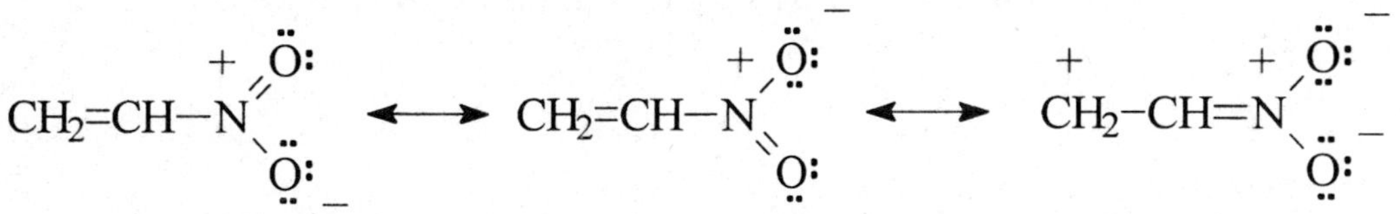

3.45 The octet rule is not satisfied at the radical carbon (the one with the odd electron).

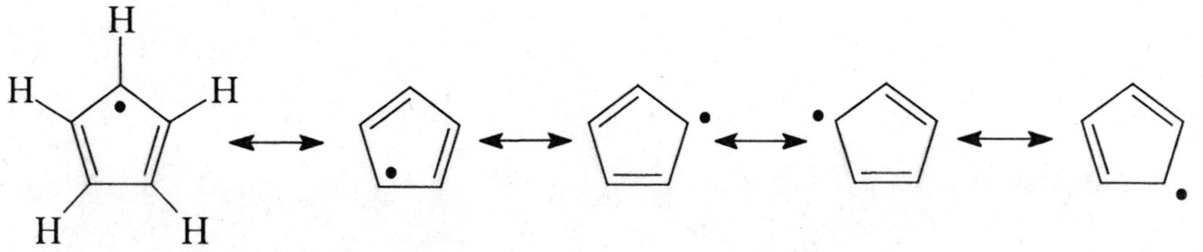

Review of Mastery Goals

After completing this chapter, you should be able to:

Assign the ground-state electron configuration for simple atoms. (Problems 3.2 and 3.3)

Identify any bond as sigma or pi. (Problem 3.34)

Draw pictures for various sigma and pi bonding and antibonding MOs.

Identify the hybridization of all atoms of a molecule.
(Problems 3.6, 3.8, 3.9, 3.11, 3.20, 3.21, 3.22, 3.40, and 3.42)

Identify the type of molecular orbital occupied by each electron pair in a molecule and designate the atomic orbitals that overlap to form that MO.
(Problems 3.5, 3.7, 3.8, 3.9, 3.21, and 3.40)

Draw the important resonance structures for any molecule. Assign the relative importance of these structures and estimate the resonance stabilization energy for the molecule.
(Problems 3.12, 3.13, 3.14, 3.15, 3.16, 3.17, 3.24, 3.25, 3.26, 3.27, 3.28, 3.29, 3.35, 3.36, 3.37, 3.38, 3.41, 3.44, and 3.45)

Show a MO energy level diagram for all the orbitals for any molecule.
(Problems 3.4, 3.18, 3.30, 3.31, 3.32, 3.33, and 3.39)

Chapter 4
THE ACID-BASE REACTION

4.1 According to the Bronsted-Lowry definition, an acid is a proton donor and a base is a proton acceptor. Compounds that have both a hydrogen and an unshared pair of electrons can potentially react as either an acid or a base, depending on the reaction conditions. For example, water, has both unshared pairs of electrons and hydrogens. Therefore it can act like an acid or a base.

a) acid b) both c) acid d) base e) both f) both g) both

4.2 A conjugate acid is formed by adding a proton (H^+) to the base, using an unshared pair of electrons on the base to form a bond to the proton. The conjugate acid has one unit more of positive charge than the base. For example, addition of a proton to the base hydroxide anion (charge = -1) produces its conjugate acid, H_2O (charge = 0).

a) $CH_3-\overset{+}{\underset{H}{\overset{\cdot\cdot}{O}}}-H$ b) $H-\overset{\cdot\cdot}{\underset{\cdot\cdot}{O}}-H$ c) $CH_3-\overset{+}{N}H_3$

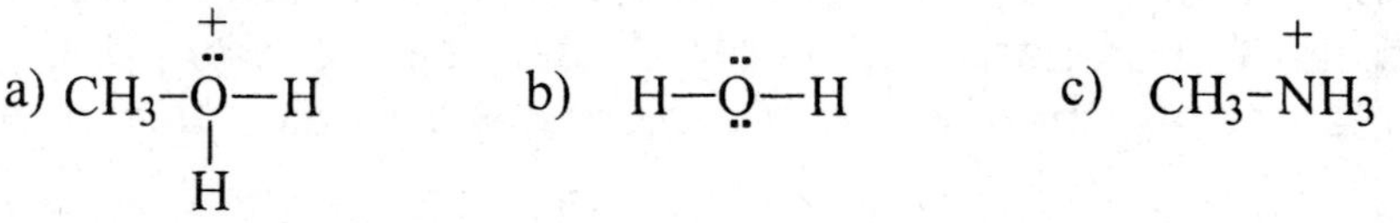

4.3 A conjugate base is formed by removing a proton from the acid, leaving the electrons of the bond as an unshared pair on the base. The conjugate base has one unit less of positive charge than the acid.

a) $H-\overset{\cdot\cdot}{\underset{\cdot\cdot}{O}}{:}^{-}$ b) $H-\overset{\cdot\cdot}{\underset{H}{O}}{:}$ c) $H-\overset{\cdot\cdot}{\underset{H}{N}}{:}^{-}$ d) $H-\overset{H}{\underset{H}{C}}-\overset{H}{\underset{H}{C}}{:}^{-}$

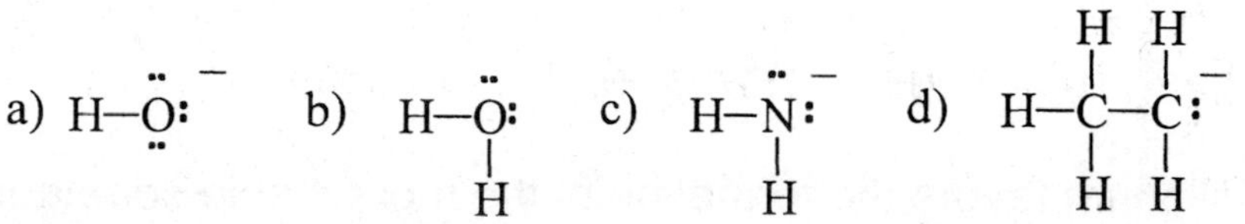

4.4 The reaction of an acid with a base is in equilibrium with the conjugate acid and the conjugate base. Remember, in an acid-base reaction the conjugate acid is the protonated form of the corresponding base and the conjugate base is the deprotonated form of the corresponding acid.

a) $:NH_2^-$ + H—$\ddot{O}$—H $\rightleftharpoons$ $:NH_3$ + $^-:\ddot{O}$—H

| | base | acid | | conjugate acid | | conjugate base |

b) $CH_3\ddot{O}:^-$ + H—$\overset{+}{\ddot{O}}$—H $\rightleftharpoons$ $CH_3\ddot{O}$—H + H—$\ddot{O}$—H

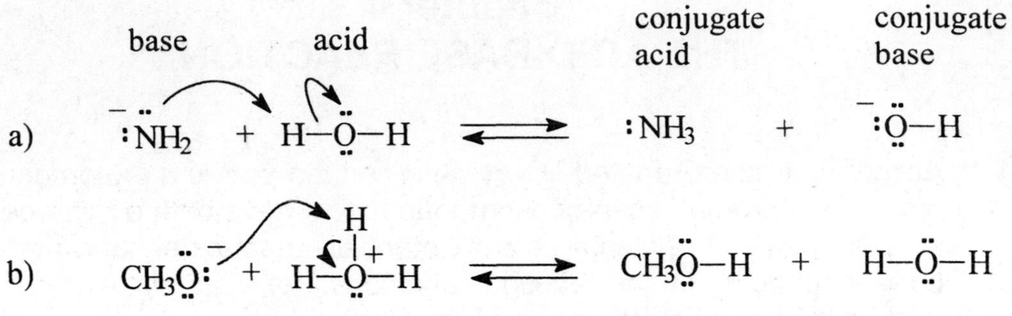

4.5 According to the Lewis acid-base definition, an acid is an electron pair acceptor and a base is an electron pair donor.

a) Lewis acid   b) Lewis base   c) Lewis acid   d) Lewis base   e) both

4.6 a) 10^4 b) 16   c) 10^{-38} d) -6

4.7 The acidity constant, K_a, is a measure of the acidic strength of a compound. The pK_a is by definition $-\log K_a$. As the strength of an acid increases, its K_a increases and its pK_a decreases. For example, C_2H_2 in (b) has a larger pK_a than water. Therefore it is a weaker acid than water.

a) stronger   b) weaker   c) stronger   d) weaker

4.8 The weaker the acid, the stronger its conjugate base. If the K_a of a compound's conjugate acid is larger than that of water (or the pK_a of the compound's conjugate acid is smaller than that of water), then the compound is a stronger base than hydroxide ion.

a) stronger   b) stronger   c) weaker   d) weaker

4.9 The equilibrium favors the formation of the more stable compounds. In the case of acid-base reactions, the weaker acid and the weaker base are favored. For example, in (a) the acid CH_3NH_2 has a lower K_a than the acid CH_3CO_2H. Thus CH_3NH_2 is a weaker acid than CH_3CO_2H. Therefore the equilibrium favors the reactants.

a) favors reactants   b) favors products

4.10 The equilibrium favors the products in each case because the weaker acid and the weaker base are on the product side of the equation.

4.11

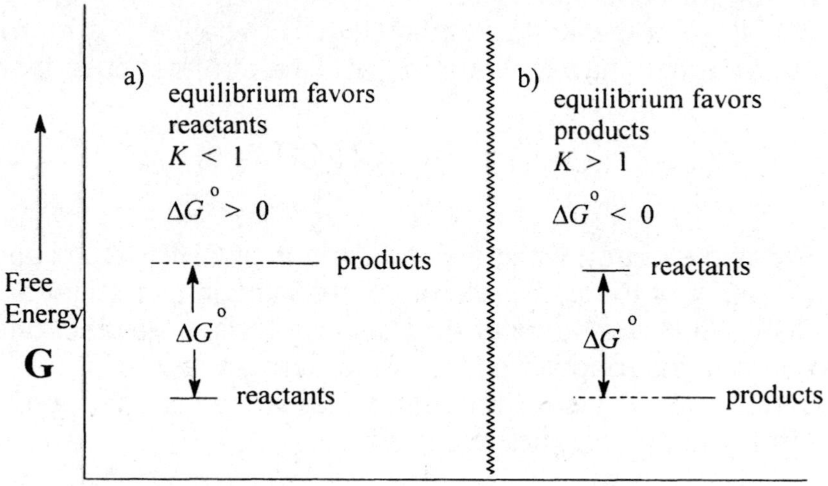

4.12

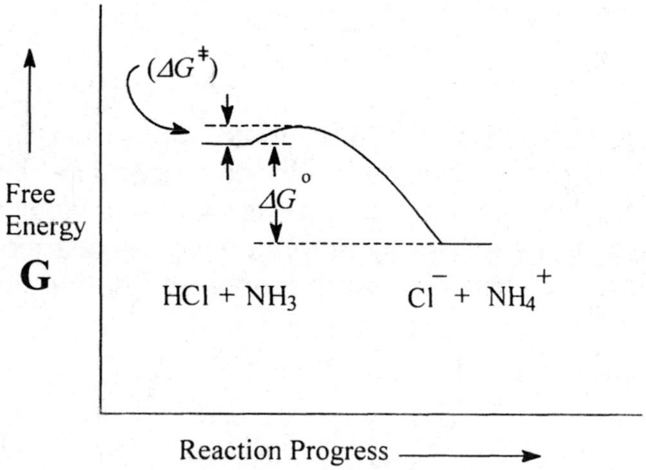

4.13 a) HCl is the stronger acid because Cl is more electronegative than S and both atoms are from the same period of the periodic table.
b) PH_4^+ is the stronger acid because P is beneath N in the periodic table.
c) H_2S is the stronger acid. Because S is beneath O in the periodic table, H_2S is a stronger acid than H_2O and H_2O is a stronger acid than CH_3CH_3 because O is more electronegative than C.

4.14 a) Remember, a stronger base is a produced from a weaker conjugate acid. H_2O, the conjugate acid of OH^-, is a weaker acid than H_2S, the conjugate acid of HS^- (S is beneath O in the periodic table). Therefore OH^- is a stronger base than HS^-.
b) CH_3NH_2 is a weaker acid than CH_3OH because O is more electronegative than N. Thus CH_3NH^- is a stronger base than CH_3O^-.

4.15 a) $H_2NCH_2CH_2O{-}H$ b) $CH_3CH_2O{-}H$ c) $CH_3S{-}H$

4.16 a) The two electron withdrawing fluorines of CHF_2CO_2H cause a larger stabilization of the conjugate base, resulting in a stronger acid.
b) CHF_2CO_2H is a stronger acid because F is more electronegative than Br and has a stronger electron withdrawing effect.
c) $CH_3OCH_2CO_2H$ is a stronger acid because the OCH_3 group is an electron withdrawing group compared to H.

4.17 The hydrogen bonded to the sp hybridized carbon is most acidic.

$$CH_3CH_2C{\equiv}C{-}H$$

4.18 In the resonance structures of the conjugate base of the meta isomer of nitrophenol, the electron pair is never on the carbon atom to which the nitro group is attached, so the electrons cannot be delocalized onto the oxygens of the nitro group. Thus there are fewer resonance structures and less resonance stabilization for the meta isomer than for the para isomer.

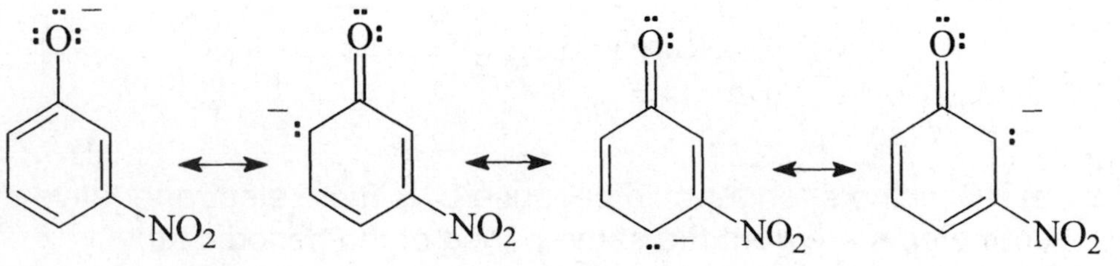

4.19 a) When an H is removed from the CH_2 group of the compound on the right, the electron pair of the conjugate base is stabilized by resonance with both the C=O and the C≡N. Therefore this compound is the stronger acid.

$$CH_3-\overset{\overset{\ddot{O}:}{\|}}{C}-\overset{-}{\underset{}{CH}}-C\equiv N: \longleftrightarrow CH_3-\overset{\overset{:\ddot{O}:^-}{|}}{C}=CH-C\equiv N:$$

$$CH_3-\overset{\overset{\ddot{O}:}{\|}}{C}-CH=C=\ddot{N}:^-$$

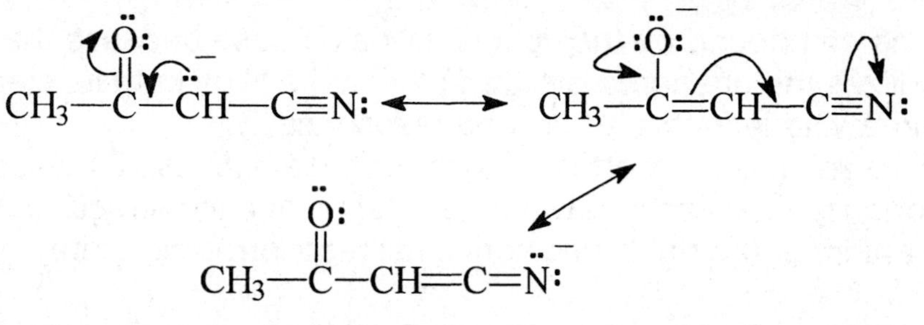

b) The most acidic H in both compounds is the one bonded to the O. The compound on the left is a stronger acid because its conjugate base is stabilized by the electron withdrawing effect of the $CH_3C=O$ group. More importantly, the $CH_3C=O$ group also stabilizes the conjugate base by resonance.

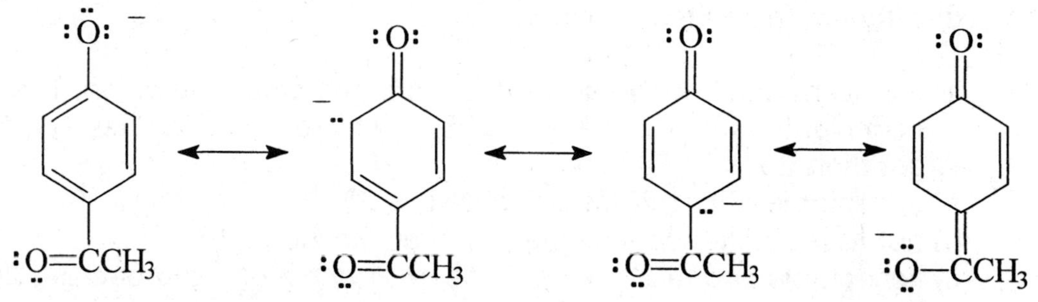

c) This is similar to problem 4.18. The para isomer (left) is more acidic because the electron-withdrawing group is more effective in stabilizing the conjugate base as a result of additional resonance stabilization (see the last resonance structure of problem 4.19 b). The other isomer (right) has no resonance structure with the electrons delocalized onto the group.

d) The H on the $CH_3C=O$ is the most acidic because the conjugate base formed by removal of this H is stabilized by resonance.

$$\overset{-}{\ddot{C}H_2}-\overset{\overset{\ddot{O}:}{\|}}{C}-\ddot{O}CH_3 \longleftrightarrow CH_2=\overset{\overset{:\ddot{O}:^-}{|}}{C}-\ddot{O}CH_3$$

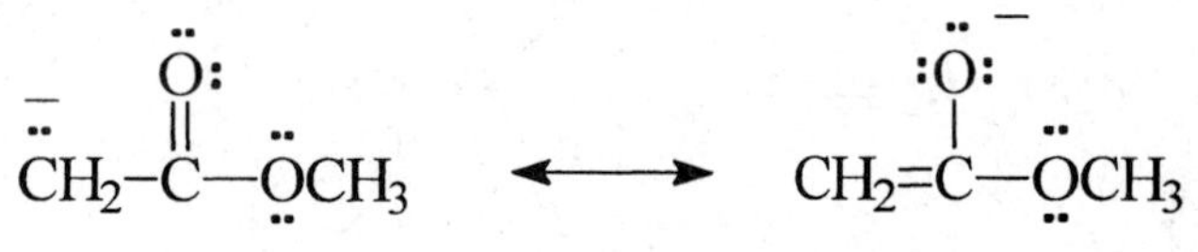

e) Phenol (left) is a stronger acid than aniline (right) because the H on the more electronegative O is more acidic than the H on the less electronegative N.

4.20 a) The compound on the right is a weaker base because the nitro group stabilizes the unshared electron pair on the N by both its electron-withdrawing inductive effect and resonance.
b) The anion on the left is the weaker base because it has a larger resonance stabilization (two additional resonance structures) than does the anion on the right (one additional resonance structure).

4.21 Remember, the equilibrium favors the formation of the weaker acid and the weaker base. The larger the pK_a , the weaker the acid.
a) Ammonia (pK_a = 38) is a weaker acid than a ketone (pK_a = 20), so the equilibrium favors the products.
b) Dimethylsulfoxide (pK_a = 38) is a weaker acid than an alcohol (pK_a = 16), so the equilibrium favors the products.
c) HCl (pK_a = -7) is a much stronger acid than phenol (pK_a = 10), so the equilibrium favors the reactants.

4.22 The solvent must not be acidic enough to protonate the anion. The pK_a of the conjugate acid of the anion is 25, so the solvent must have a pK_a larger than 25.
a) acceptable because the pK_a of NH_3 is 38
b) not acceptable because the pK_a of an alcohol is 16
c) acceptable because an ether has no H with a pK_a approaching 25

4.23 To get the conjugate acid, use the most basic pair of electrons to form a bond to a proton.

a) 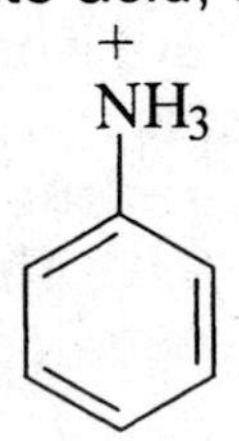

b) 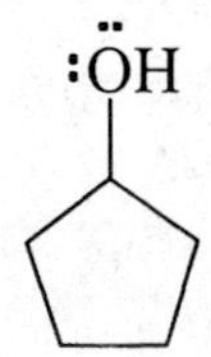

c) $CH_3CH_2CH_2\ddot{O}H_2^+$

d) $CH_3C{\equiv}CH$

e) 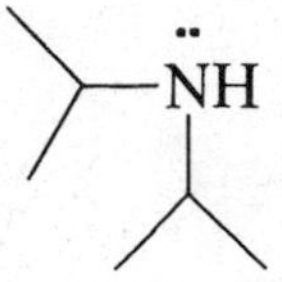

4.24 The stability of the conjugate base depends on factors such as the electronegativity and the hybridization of the atom where the electrons are located, resonance delocalization of the basic electrons, and inductive effects of nearby groups.

a) O is more electronegative than C. $CH_3CH_2\overset{..}{\underset{..}{O}}{:}^{\,-}$

b) A carboxylic acid is a stronger acid than an alcohol because of resonance stabilization of its conjugate base.

$$:\overset{-}{\underset{..}{O}}-\overset{\overset{\displaystyle \overset{..}{O}:}{\|}}{C}CH_2CH_2\overset{..}{\underset{..}{O}}H$$

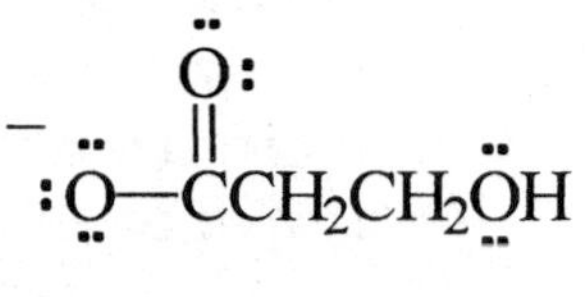

c) O is more electronegative than C or N. $H_2NCH_2CH_2-\overset{..}{\underset{..}{O}}{:}^{\,-}$

4.25 According to the Lewis acid-base definition, an acid is an electron pair acceptor. Any compound that has an unfilled valence orbital is a potential Lewis acid.
a) The boron atom in BCl_3 has an empty orbital that can accept electrons. Therefore it is a Lewis acid.
b) Methane is not a Lewis acid since there are no empty orbitals in the valence shell of any of its atoms.
c) The positively charged carbon atom of this species is sp^2 hybridized, and has an empty p orbital that can accept electrons. Therefore it is a Lewis acid.

4.26 According to the Lewis definition, a base is an electron pair donor. Any compound that has an unshared electron pair can behave as a Lewis base.
a) can   b) cannot   c) can   d) cannot   e) can

4.27 $pK_a = -\log K_a$ a) 3.76   b) 50

4.28 $K_a = 10^{-pK_a}$ (antilog of $-pK_a$)   a) 1×10^{-25} b) 4.9×10^{-10}

4.29 a) In both cases the conjugate bases are stabilized by an inductive effect and by resonance. However, the compound on the right is the stronger acid because the hydrogen is removed from the more electronegative oxygen atom.
b) The hydrogen on sulfur is more acidic than the one on oxygen, so the compound on the left is a stronger acid.

c) The compound on the right is the stronger acid because the conjugate base produced by the removal of a proton from the CH_3 group is stabilized by resonance. There is no such stabilization in butane.
d) The compound on the right is the stronger acid because the CF_3 group is a strong electron-withdrawing group. The CH_3 group is weakly electron donating and destabilizes the conjugate base.
e) The compound on the right is more acidic because the nitrogen in this molecule is sp^2 hybridized. The unshared pair of electrons of the conjugate base will more stable in a sp^2 hybridized orbital.

4.30 a) The anion on the left is a stronger base because the nitro group of the anion on the right provides additional resonance delocalization of the basic electrons.
b) The nitrogen compound is a stronger base than the phosphorus compound because acid strength increases (and base strength decreases) down a column of the periodic table.
c) The anion on the left is a stronger base because Cl is a stronger electron withdrawing group than Br as a result of its higher electronegativity.
d) The anion on the right is a stronger base because O is more electronegative than N.

4.31 The acidic hydrogen is the one bonded to the oxygen for each compound. Compound B is a stronger acid than A because of resonance stabilization of its conjugate base. Compound C is a stronger acid than B because of the inductive electron withdrawing effect of the CN group. Compound D is a stronger acid than C because of the additional resonance stabilization of the conjugate base when the CN is attached to the para position.

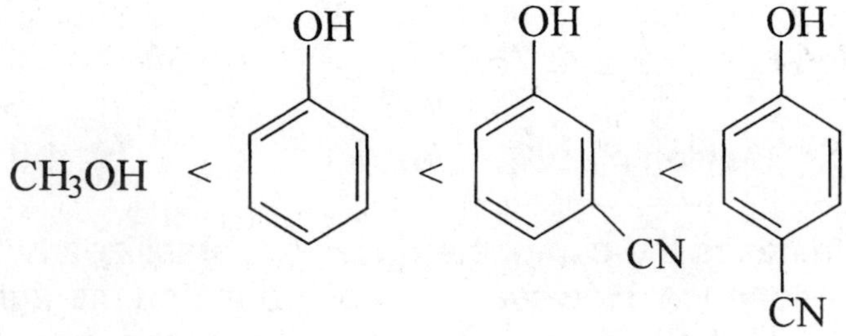

A B C D

4.32 The basic electron pair in each compound is the one on the nitrogen atom. Compound C is a weaker base than D because it has resonance stabilization of the basic electron pair. The nitro group of B provides even more resonance stabilization of the basic electrons. Compound A is the weakest base because the C=O is one of the best groups at stabilizing a pair of electrons on an adjacent atom by resonance.

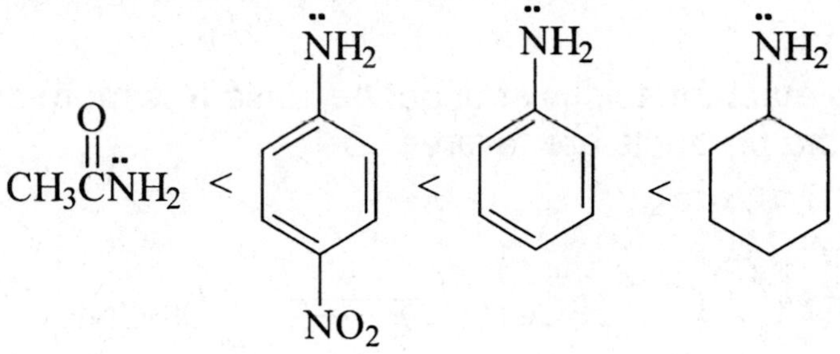

A B C D

4.33

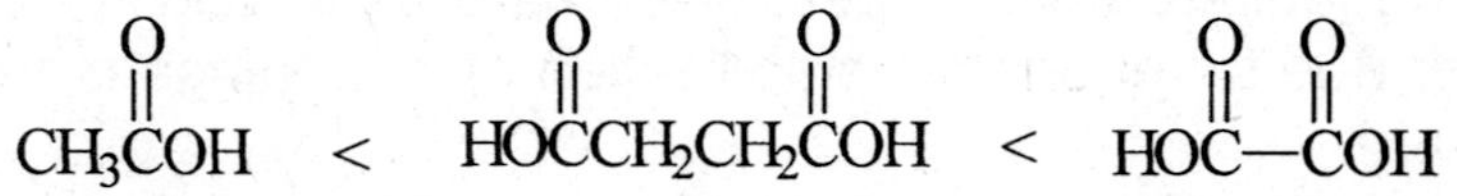

weakest acid strongest acid

4.34 a) The products are favored because butyllithium is a stronger base than LDA (see Table 4.4).
b) The equilibrium favors the products because ammonia (pK_a = 38) is a weaker acid than an alkyne (pK_a = 25).
c) The reactants are favored because an alcohol (pK_a = 16), is a weaker acid than an alkyne (pK_a = 25).

4.35 a) CH_3CH_3 is a weaker acid than CH_3CN, because the conjugate base of CH_3CN is stabilized by resonance, so the reactants are favored.

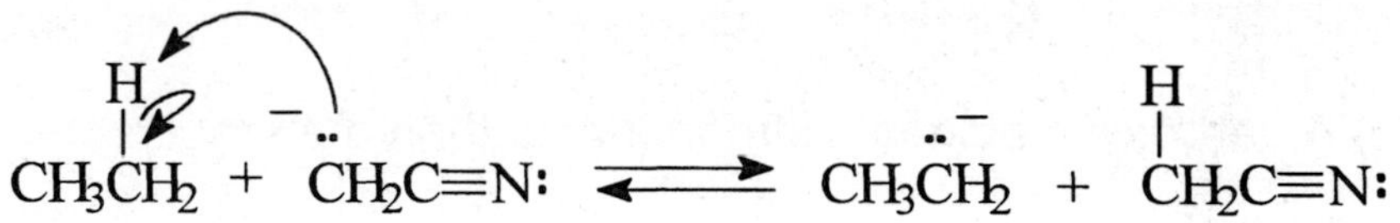

b) The electron withdrawing inductive effect of the CF_3 group stabilizes the base $CF_3CO_2^-$, so the reactants are favored.

$$CF_3\overset{\overset{\displaystyle :\ddot{O}:}{\|}}{C}\ddot{O}:^- \quad + \quad H-\overset{\overset{\displaystyle :\ddot{O}:}{\|}}{\ddot{O}CCH_3} \quad \rightleftharpoons \quad CF_3\overset{\overset{\displaystyle :\ddot{O}:}{\|}}{C}\ddot{O}H \quad + \quad CH_3\overset{\overset{\displaystyle :\ddot{O}:}{\|}}{C}\ddot{O}:^-$$

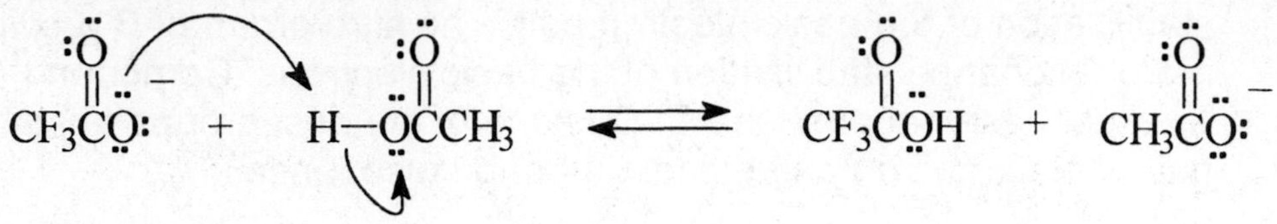

c) The reactant acid is stronger because N is more electronegative than C, so the products are favored.

$$CF_3\overset{\overset{\displaystyle :\ddot{O}:\ \ H}{\|}}{C}\ddot{N}H \quad + \quad CH_2\overset{\overset{\displaystyle :\ddot{O}:}{\|}}{C}CH_3 \quad \rightleftharpoons \quad CF_3\overset{\overset{\displaystyle :\ddot{O}:}{\|}}{C}\ddot{N}H^- \quad + \quad CH_3\overset{\overset{\displaystyle :\ddot{O}:}{\|}}{C}CH_3$$

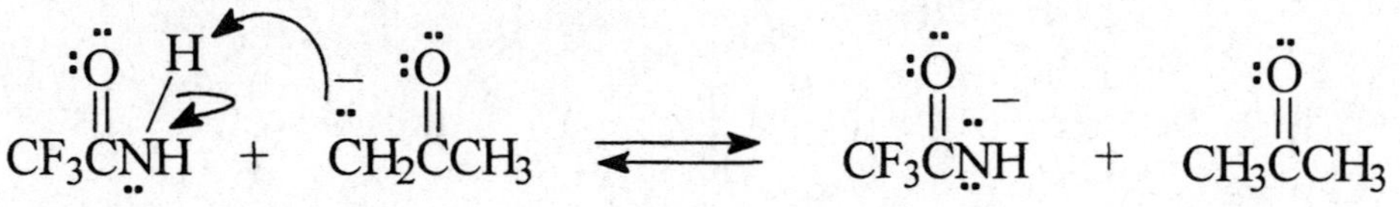

d) The products are favored because a carboxylic acid is a stronger acid than water due to resonance stabilization of its conjugate base.

$$CH_3CH_2CH_2\overset{\overset{\displaystyle \ddot{O}:}{\|}}{C}\ddot{O}-H \quad + \quad :\ddot{O}H^- \quad \rightleftharpoons \quad CH_3CH_2CH_2\overset{\overset{\displaystyle \ddot{O}:}{\|}}{C}\ddot{O}:^- \quad + \quad H\ddot{O}H$$

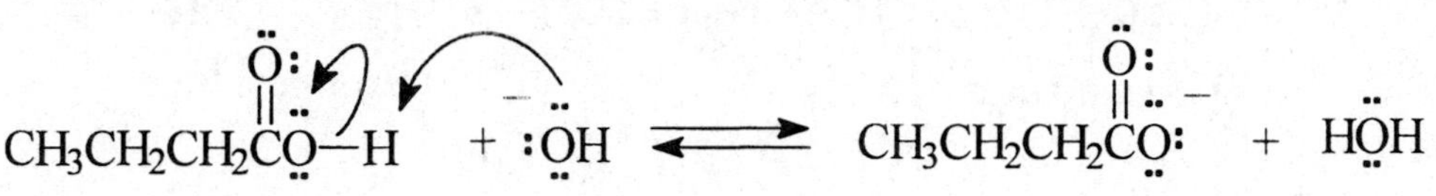

4.36 a) A sulfonic acid is a stronger acid than a carboxylic acid.

$$HO\overset{\overset{\displaystyle O}{\|}}{C}CH_2CH_2\overset{\overset{\displaystyle O}{\|}}{\underset{\underset{\displaystyle O}{\|}}{S}}O-\boxed{H}$$

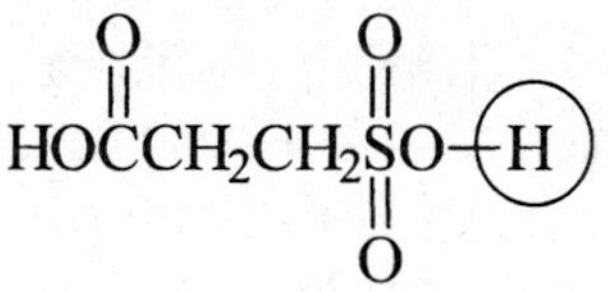

b) The circled hydrogen is more acidic because the conjugate base is stabilized by resonance.

$$CH_3CH_2\underset{\boxed{H}}{CH}C\equiv N$$

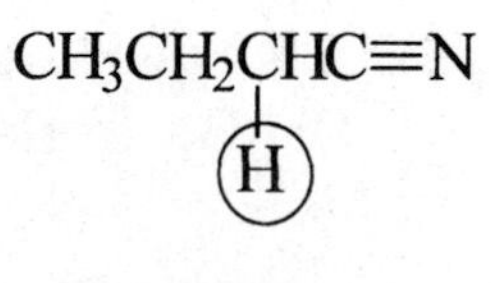

c) A carboxylic acid is a stronger acid than a phenol.

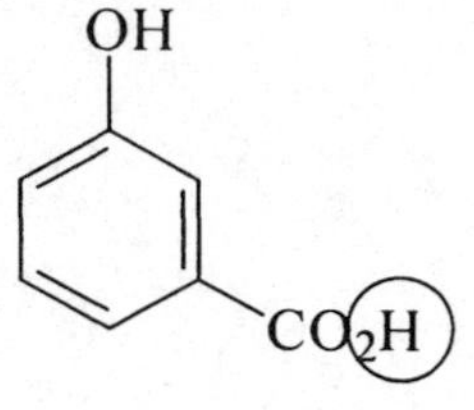

d) A phenol is a stronger acid than an alcohol.

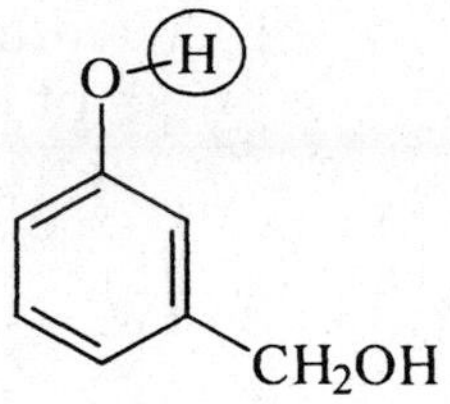

e) The circled hydrogen is more acidic because the conjugate base is stabilized by resonance involving both carbonyl groups.

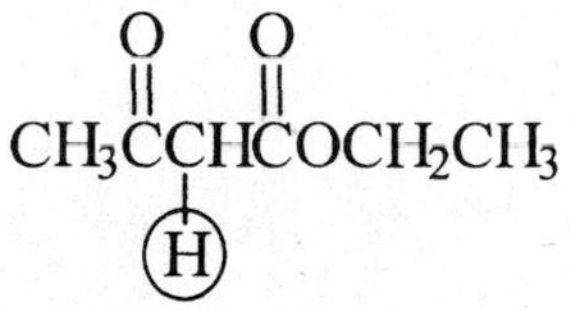

f) An ammonium cation is a stronger acid than a carboxylic acid.

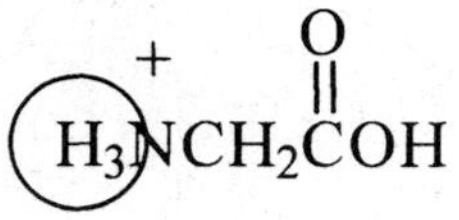

4.37 a) The equilibrium favors the products.

$$CH_3CCH_2CCH_2CH_3 + :\ddot{O}CH_2CH_3 \rightleftharpoons CH_3CCHCCH_2CH_3 + H\ddot{O}CH_2CH_3$$

$pK_a = 9$ $\qquad\qquad\qquad\qquad\qquad\qquad pK_a = 16$

b) The equilibrium favors the products.

$$CH_3CH_2NO_2 + CH_3\ddot{O}:^- \rightleftharpoons CH_3\ddot{C}HNO_2 + CH_3\ddot{O}H$$

$pK_a = 10$ $\qquad\qquad\qquad\qquad\qquad pK_a = 16$

c) The equilibrium favors the products.

$$CH_3COCH_3 + :\ddot{N}(CHCH_3)_2^- \rightleftharpoons \ddot{C}H_2COCH_3^- + H\ddot{N}(CHCH_3)_2$$

$pK_a = 25$ $\qquad\qquad\qquad\qquad\qquad\qquad pK_a = 38$

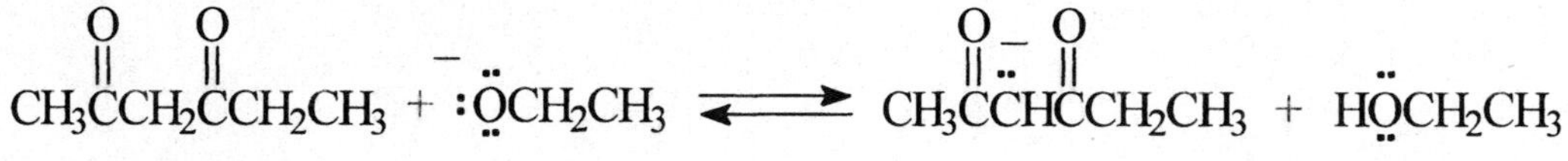

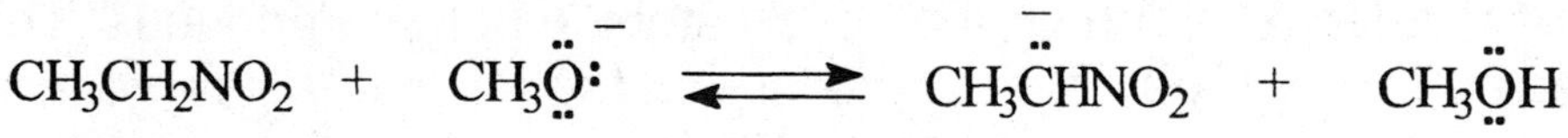

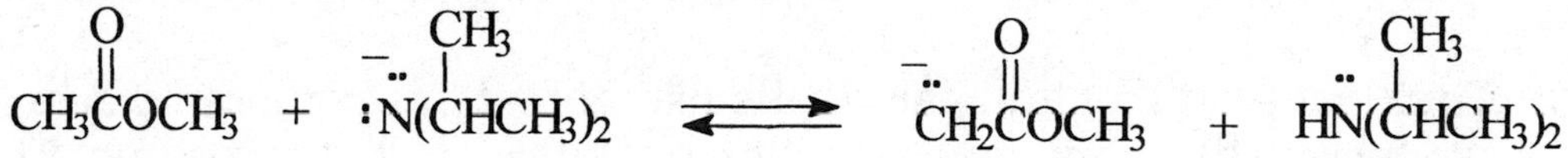

d) The equilibrium favors the reactants because the electron withdrawing effect of the F is stronger when it is closer to the basic O.

$$\underset{\text{FCH}_2\text{CH}_2\overset{\overset{\text{O}}{\|}}{\text{C}}\text{OH}}{} + \underset{\text{CH}_3\overset{\overset{\text{F}}{|}}{\text{CH}}-\overset{\overset{\text{O}}{\|}}{\text{C}}\text{O}^-}{} \rightleftharpoons \underset{\text{FCH}_2\text{CH}_2\overset{\overset{\text{O}}{\|}}{\text{C}}\text{O}^-}{} + \underset{\text{CH}_3\overset{\overset{\text{F}}{|}}{\text{CH}}-\overset{\overset{\text{O}}{\|}}{\text{C}}\text{OH}}{}$$

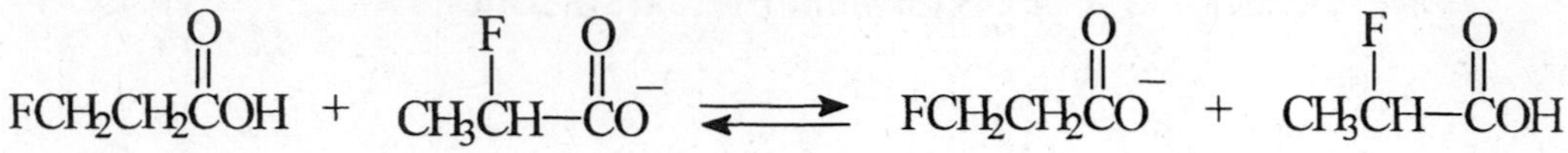

e) The equilibrium favors the products.

$$\underset{\text{CH}_3\overset{\overset{\text{OH}}{|}}{\text{CH}}-\overset{\overset{\text{O}}{\|}}{\text{C}}\text{OH}}{} + \overset{+}{\text{Na}}\ \overset{-}{\text{OH}} \rightleftharpoons \underset{\text{CH}_3\overset{\overset{\text{OH}}{|}}{\text{CH}}-\overset{\overset{\text{O}}{\|}}{\text{C}}\text{O}^-\ \overset{+}{\text{Na}}}{} + \text{HOH}$$

$$\text{p}K_a = 4 \qquad\qquad\qquad\qquad\qquad\qquad \text{p}K_a = 16$$

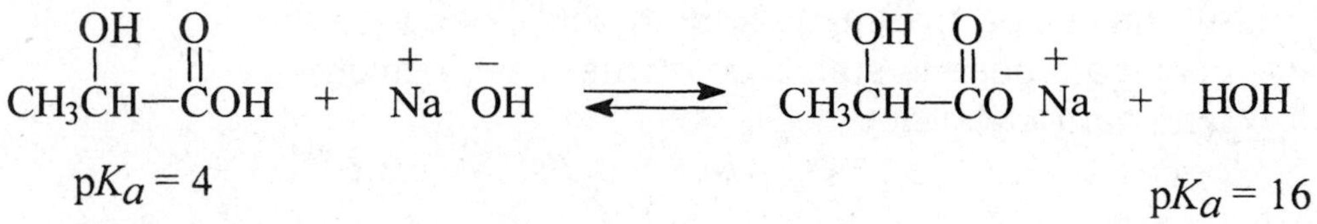

f) The equilibrium favors the products.

$$\underset{\text{CH}_3\overset{\overset{\text{CH}_3}{|}}{\underset{\underset{\text{CH}_3}{|}}{\text{C}}}-\ddot{\underset{\cdot\cdot}{\text{O}}}\text{:}^-}{} + \text{H}_2\text{O} \rightleftharpoons \underset{\text{CH}_3\overset{\overset{\text{CH}_3}{|}}{\underset{\underset{\text{CH}_3}{|}}{\text{C}}}-\ddot{\underset{\cdot\cdot}{\text{O}}}\text{H}}{} + {}^-\!\ddot{\underset{\cdot\cdot}{\text{O}}}\text{H}$$

$$\text{p}K_a = 16 \qquad\qquad\qquad\quad \text{p}K_a = 19$$

4.38 $AICl_3$ is the Lewis acid because Al has an empty p orbital, which can accept an electron pair. $CH_3CH_2OCH_2CH_3$ is the Lewis base because the O has lone pairs of electrons to donate.

4.39 The basic electrons are on a nitrogen atom in both compounds. The unshared electron pair of the structure on the left is in a sp^2-hybridized orbital, while in the structure on the right they are in a sp^3-hybridized orbital. The sp^2 orbital is lower in energy than the sp^3 orbital, so the structure on the left is a weaker base. Note that the compound on the left is stabilized by resonance, but the basic electrons are not involved in the resonance, so resonance has no effect on the basicity of the electrons.

4.40 The unshared electron pair on the nitrogen of this compound is in a p orbital which is conjugated with the p orbitals of the C=O group. These electrons are stabilized by resonance. If the nitrogen is protonated, this resonance stabilization is lost. The electron pairs on the oxygen atom are in atomic orbitals that are not involved in resonance.

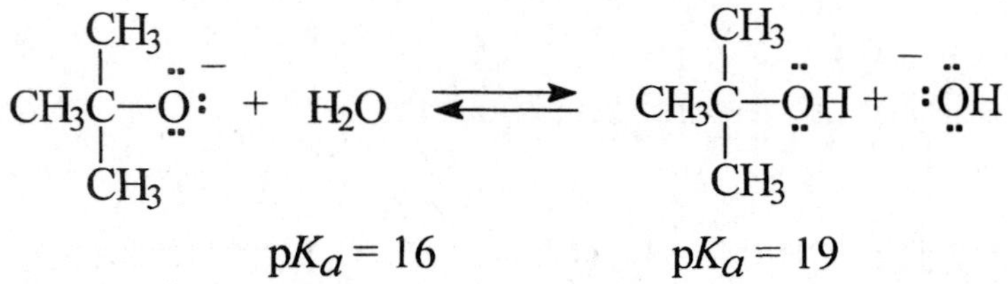

4.41 From Table 4.2, the pK_a for NH_4^+ is 9.24 and the pK_a for a carboxylic acid is about 5. Therefore a carboxylic acid is a stronger acid than ammonium ion and the equilibrium favors the product. The product is an internal salt and should have salt-like properties, such as a high melting point and a high solubility in water.

4.42 The N that is bonded to the C=O group is not very basic because its electron pair is stabilized by resonance with the C=O group. For this reason, the N of an amide is not very basic. The other N has no special stabilization for its unshared electron pair and is approximately as basic as the N of ammonia.

4.43 The electrons on the N with the H are part of the conjugated pi system and are not very basic. The electrons on the N without the H are in an AO that is perpendicular to the pi system. These electrons are not involved in resonance and are much more basic.

4.44 The NH_3^+ is an inductive electron-withdrawing group and increases the acid strength of the nearby carboxylic acid group.

4.45 The anion has two resonance structures. The conjugate acid of this anion can have the proton on the O or on the C.

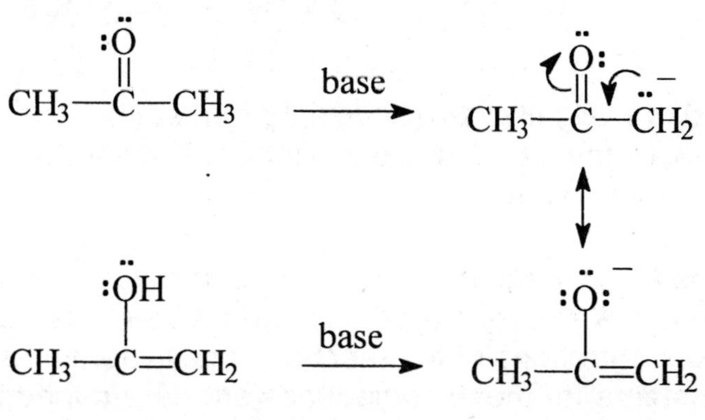

4.46 a) Isomer **b** is more stable than **a** because it is conjugated and has resonance stabilization.
b) The conjugate base of **a** has three resonance structures. Protonation at one of the carbons bearing a partial negative charge produces **a** and protonation at the other produces **b**.

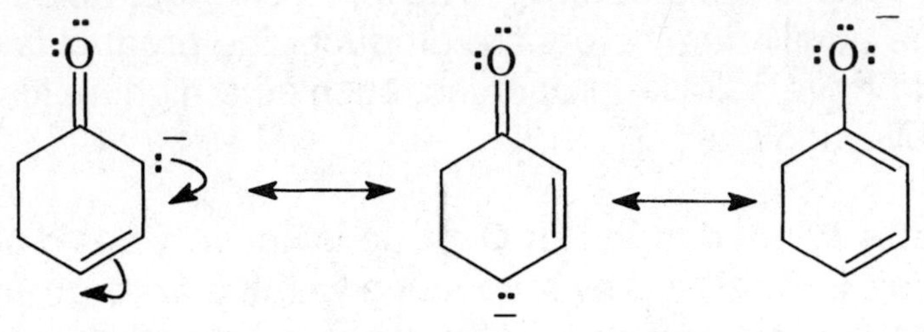

4.47 a) The CH_3O group has an electron-withdrawing inductive effect which destabilizes the acid and stabilizes the conjugate base. Therefore the inductive effect of this group should increase the acidity of compound **d**.
b) The CH_3O group has an electron-donating resonance effect which destabilizes the conjugate base, weakening the acidity of **d**.

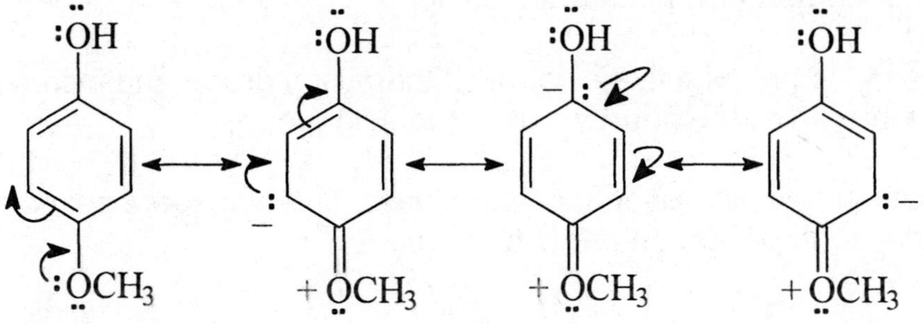

c) Compound **d** is a slightly weaker acid than **c**, so the acid-weakening resonance effect of the CH_3O group is slightly stronger than the acid-strengthening inductive effect.

4.48 The compounds at the beginning of Table 4.2 are very strong acids. Their equilibrium constants are very large and cannot be measured accurately because the concentrations of the reactants are extremely small. The equilibrium constants for these compounds are determined by some indirect method and only approximate values can be obtained. Because the pK_a values cannot be determined very precisely, they are listed without any figures right of the decimal place. A similar problem occurs with the extremely weak acids at the end of the table.

4.49 a) The circled compound is the stronger acid because the electron withdrawing fluorines are closer to the acidic hydrogen.

$CH_3CF_2CO_2H$ $CHF_2CH_2CO_2H$

b) The circled compound is the stronger acid because the conjugate base (H is removed from N) is stabilized by resonance.

$CH_3CH_2NH_2$ $CH_3\overset{O}{\overset{\|}{C}}NH_2$

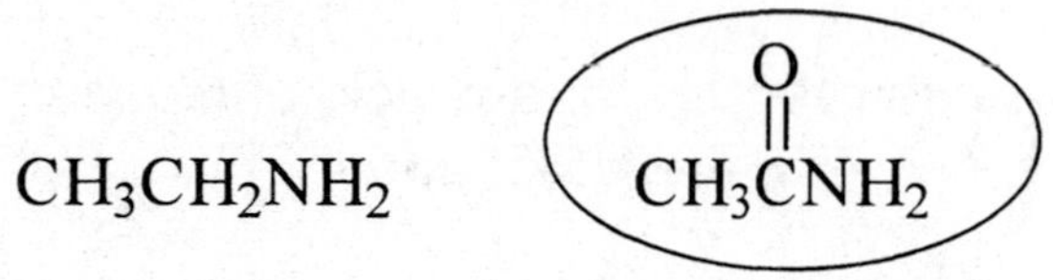

c) The circled compound is the stronger acid because the conjugate base (H is removed from O) is stabilized by resonance.

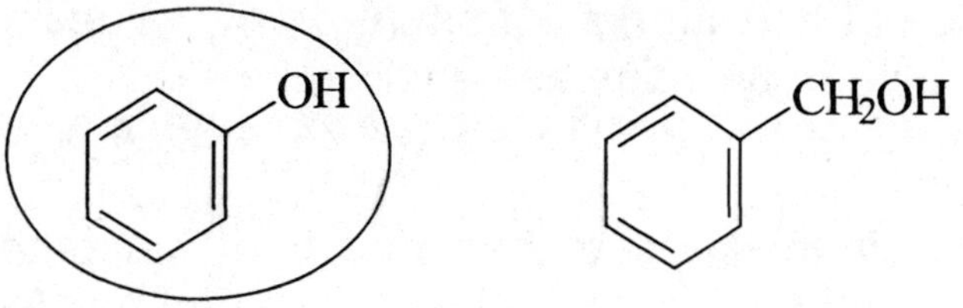

d) A carboxylic acid is a stronger acid than a phenol.

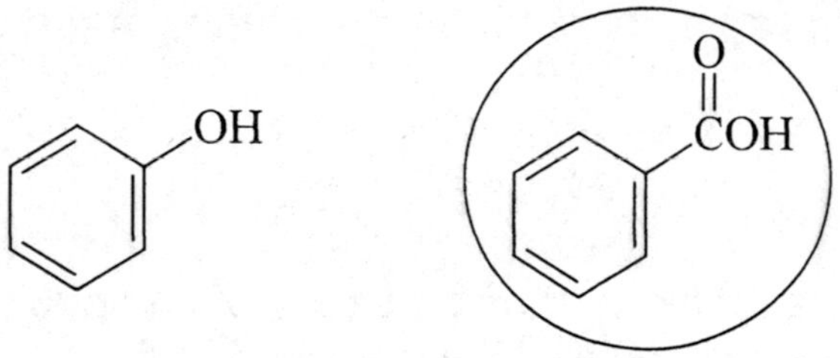

e) An H on an O is more acidic than an H on an N.

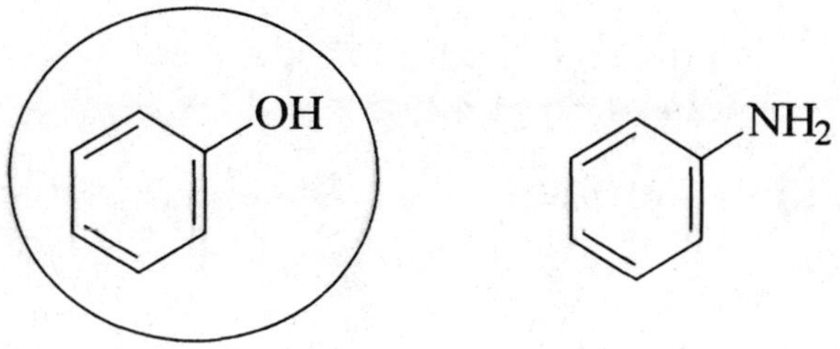

4.50 The NO_2 group stabilizes the conjugate base by resonance and increases acid strength. The compound on the right has less resonance stabilization because the p orbital on the nitrogen is not parallel to the p orbital on the ring because the oxygens of the NO_2 group bump into the CH_3 groups.

Review of Mastery Goals

After completing this chapter you should be able to:

Write an acid-base reaction for any acid and base.
(Problems 4.1, 4.2, 4.3, 4.4, 4.35, and 4.37)

Recognize Lewis acids, Lewis bases, and the reactions between them.
(Problems 4.5, 4.25, 4.26, and 4.38)

Recognize acid or base strengths from K_a or pK_a values and use these to predict the position of an acid-base equilibrium.
(Problems 4.6, 4.7, 4.8, 4.9, 4.10, 4.21, 4.34, 4.35, 4.36, 4.37, and 4.41)

Predict and explain the effect of the structure of the compound, such as the atom bonded to the hydrogen, the presence of an electron donating or withdrawing group, hydrogen bonding, the hybridization of the atom attached to the hydrogen, or resonance on the strength of an acid or base.
(Problems 4.13, 4.14, 4.16, 4.18, 4.29, 4.30, 4.35, 4.39, 4.40, 4.41, 4.42, and 4.43)

Using the same reasoning, arrange a series of compounds in order of increasing or decreasing acid or base strength.
(Problems 4.19, 4.20, 4.31, 4.32, 4.33)

Identify the most acidic proton in a compound.
(Problems 4.15, 4.17, 4.24, 4.36, 4.42, and 4.43)

Chapter 5
FUNCTIONAL GROUPS AND NOMENCLATURE I

5.1 Follow steps 1 through 5 outlined in Section 5.3 in the text for naming these alkanes.

Step 1: Find the longest, continuous carbon chain in the compound. The number of carbons in this backbone chain determines the root. If there is more than one chain of equal length, choose the one with the more branches. For example, in (e) there are two different 6 carbon chains, one with two branches and the other with three branches. Choose the one with the three branches as the root chain.

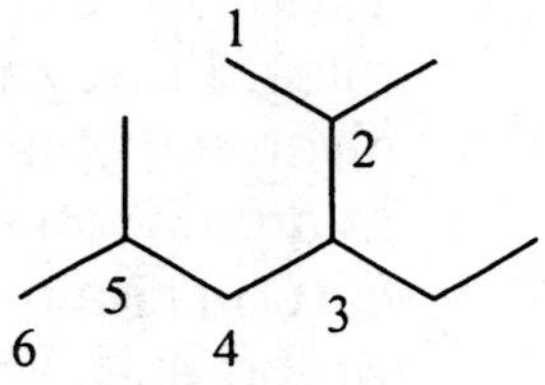

Step 2: Attach the suffix (-ane for alkanes). In the above example, the root chain has six carbons, so this compound is named as a hexane.

Step 3 : Number the root carbon chain starting from the end that will give the lower number to the first branch. If the first branch occurs at an equal distance from either end, then choose the end that will give the lower number to the second branch. In the above example, starting at either end of the root chain, the first branch occurs at position 2. In the correct numbering, shown above, the second branch occurs at position 3. If the other end were chosen as position 1, the second branch would be at position 4.

Step 4: Name the groups attached to the root.

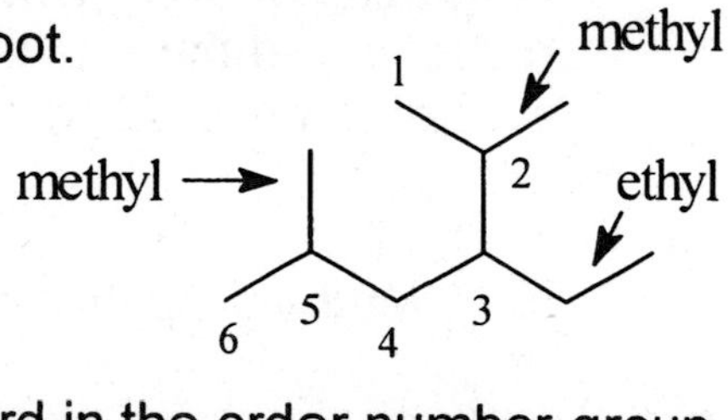

Step 5: Assemble the name as a single word in the order number-group root suffix. List the groups in alphabetical order. The name for (e) is 3-ethyl-2,5-dimethylhexane.

a) 2-Methylpentane
b) 2-Methylpentane
c) 2,4-Dimethylhexane
d) 5-Ethyl-3-methyl-5-propylnonane
e) 3-Ethyl-2,5-dimethylhexane
f) 3-Ethyl-2,6-dimethylheptane

5.2

a) 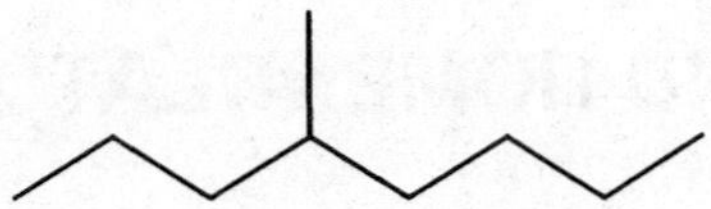

b) 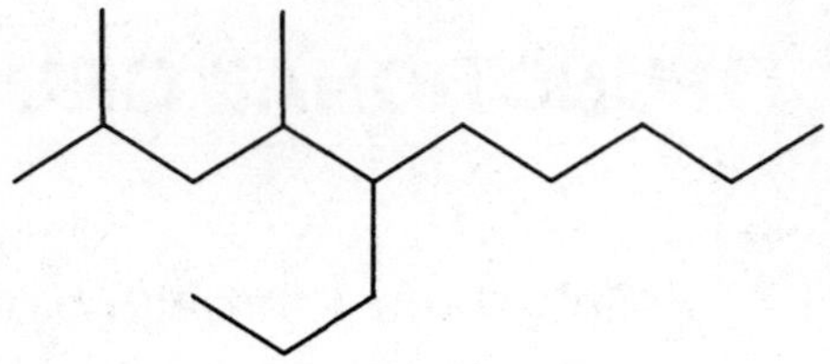

5.3 First draw the structure suggested by the name. Then name the structure according to the IUPAC rules. For example, the structure for the name given in (a) is as follows:

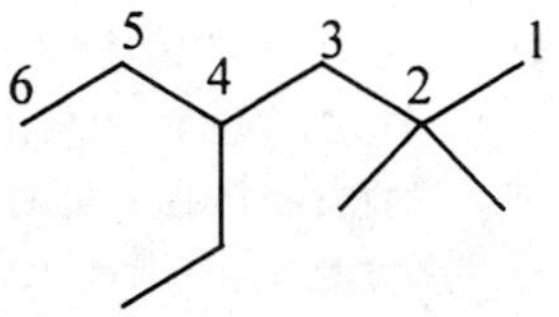

Examination of this structure shows that the root carbon chain has been numbered from the wrong end. The proper numbering is shown. The correct name for (a) is 4-ethyl-2,2-dimethylhexane.

b) The name must indicate the positions of all groups, so the number "2" must be repeated in the name. The correct name is 2,2-dimethylpentane.

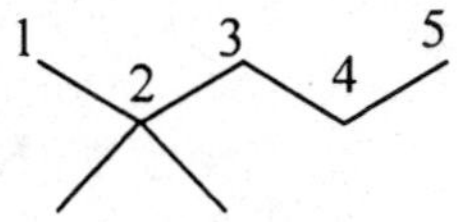

5.4 When naming complex groups, choose the longest chain beginning with the carbon that is attached to the root chain. This carbon always receives number 1. Do not forget to use the suffix -yl and to put the name in parentheses.

a) In this complex group the longest chain has 5 carbons and there is a methyl group on carbon 2, so the name is (2-methylpentyl).

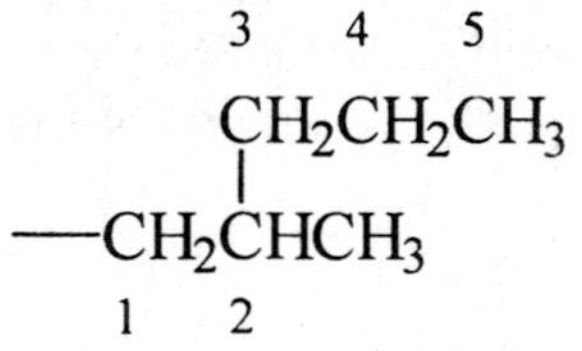

b) (1-Methylpropyl) c) (2,2-Dimethylpropyl)

5.5 a) 4-(1-Methylethyl)heptane b) 5-(1,2-Dimethylpropyl)decane

5.6

a) 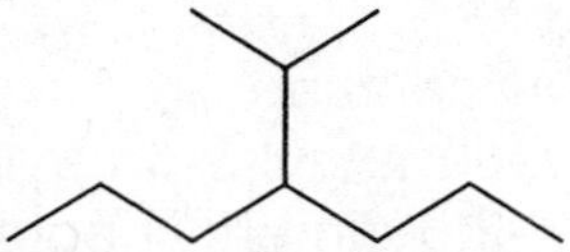

b) 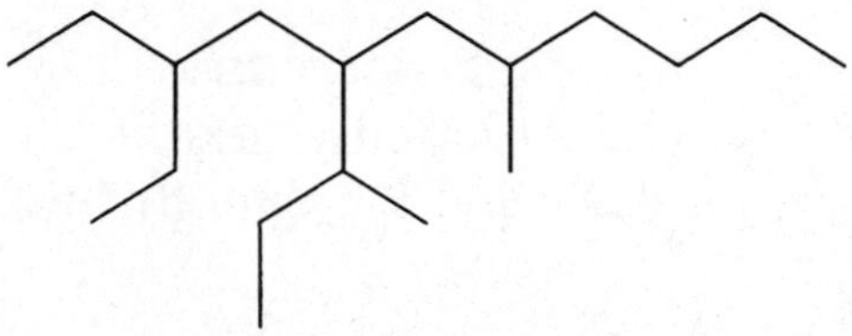

5.7 Carbon atoms in a compound can be designated according to how many other carbons are directly bonded to them. A primary carbon (p or 1°) is bonded to one other carbon, a secondary carbon (s or 2°) is bonded to two other carbons, a tertiary carbon (t or 3°) is bonded to three carbons, and a quaternary carbon (q or 4°) is bonded to four carbons.

a) 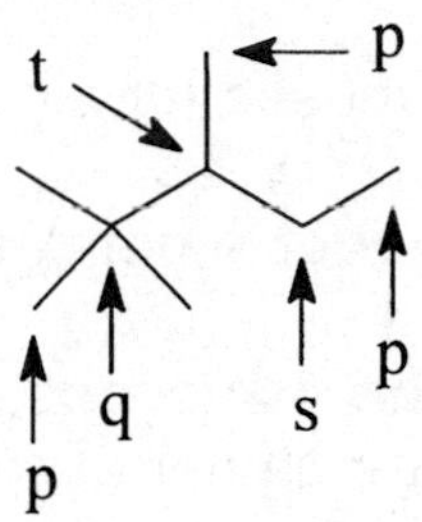

b) 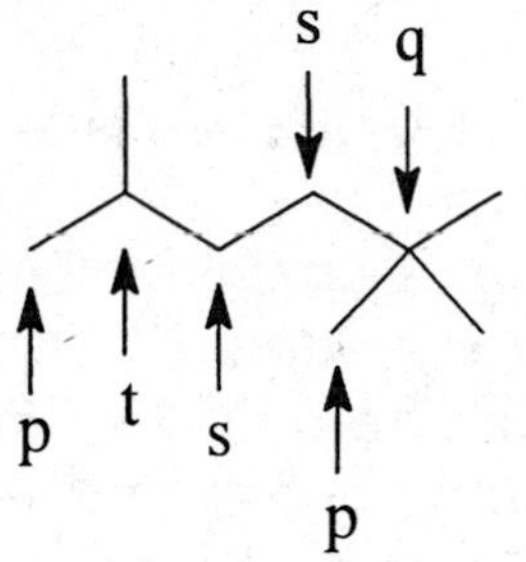

c) 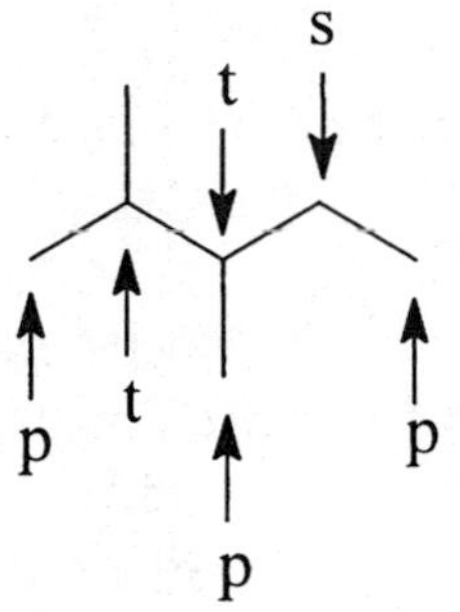

5.8 4-Isopropylheptane

5.9 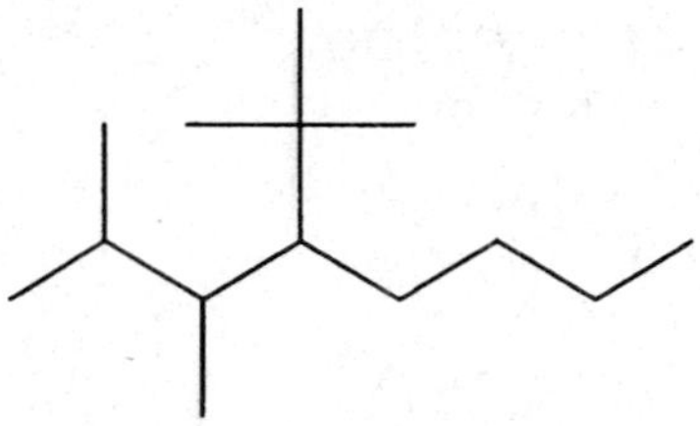

5.10 The procedure for naming cycloalkanes is very similar to that used for alkanes. The root is given by the number of carbons in the ring and the prefix cyclo- is attached to indicate the ring. Remember that the ring is named as a group if it is attached to a chain that has more carbons.
a) The ring has 5 carbons, so the root is cyclopentane. To keep the numbers as low as possible, begin at one of the methyl groups and proceed by the shortest possible path to the other methyl group. The correct name for this compound is 1,2-dimethylcyclopentane.
b) The name is (1-methylpropyl)cyclohexane. No number is needed to locate the group because all positions of the ring are identical.
c) The name is 5-cyclopentyl-2-methylheptane. There are 7 carbons in the longest alkyl chain while the ring has only 5 carbons, so the ring is named as a group.
d) The name is 1-ethyl-3,5-dimethylcyclooctane. The numbers would be the same beginning with the ethyl on one end or the methyl on the other, so begin with the ethyl group because it comes first alphabetically.

5.11

a) 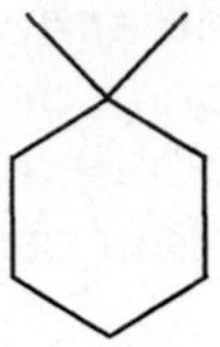

b) 

5.12 The procedure for naming alkenes is similar to that used for alkanes except for the following:
1) The longest continuous chain must include the double bonded carbons.
2) The suffix for alkenes is -ene.
3) The root is numbered from the end that gives the lowest possible number to the first carbon of the double bond. The number of the first carbon is used in designating the position of the double bond.

a) The longest carbon chain containing the double bond has 6 carbons, so the root name is hexene. The root is numbered from the end that will give the lowest possible number to the double bond. The correct name of this compound is 2,4-dimethyl-2-hexene.

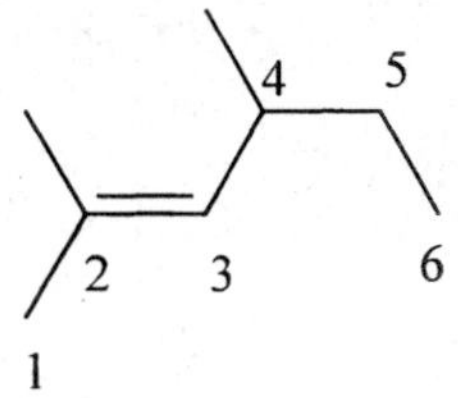

b) Here the root is a 7 carbon ring with three double bonds, so the root name is cycloheptatriene. The numbering must start with one of the carbons of a double bond and the other carbon of that double bond must be at position 2. The numbering shown gives the lowest numbers to the double bonds and then the substituent, so the name of the compound is 2-methyl-1,3,5-cycloheptatriene.

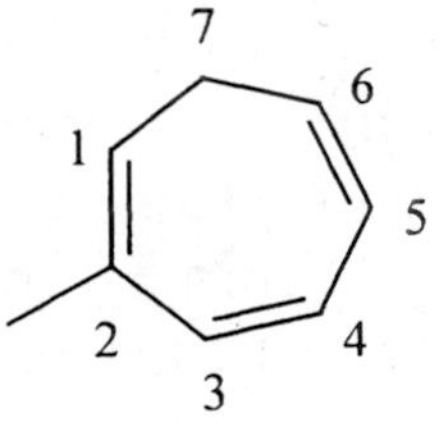

c) In a cycloalkene, the numbering must start with one of the carbons of the double bond and the other carbon of the double bond must be at position 2. In this case, the end of the double bond to use as position 1 is chosen to give the ethyl group number a lower number. The name is 3-ethyl-1,2-dimethylcyclopentene.

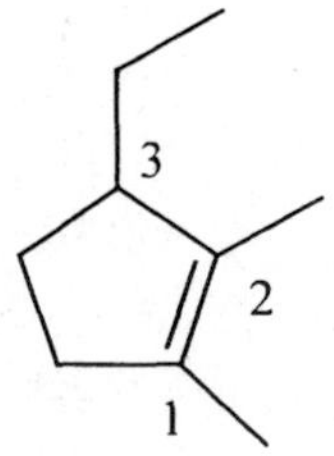

5.13

a) 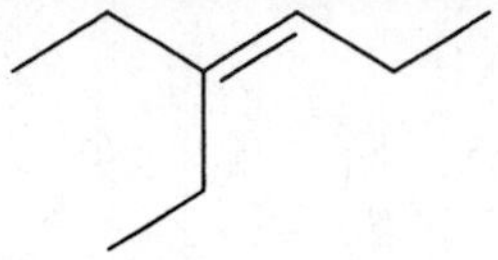 b) 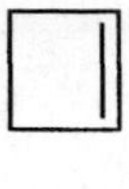 c) 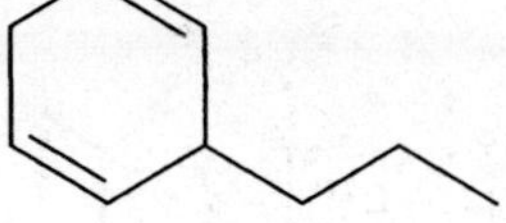

5.14 The procedure for the naming of alkynes similar to that used for alkenes, except the suffix is -yne.
a) 3-Isopropyl-1-heptyne or 3-isopropylhept-1-yne
b) 2-Methylpent-1-en-3-yne c) 3-(2-Methylpropyl)-1,4-hexadiyne

5.15

a) $HC{\equiv}CCH_2CH_2CH_3$ b) 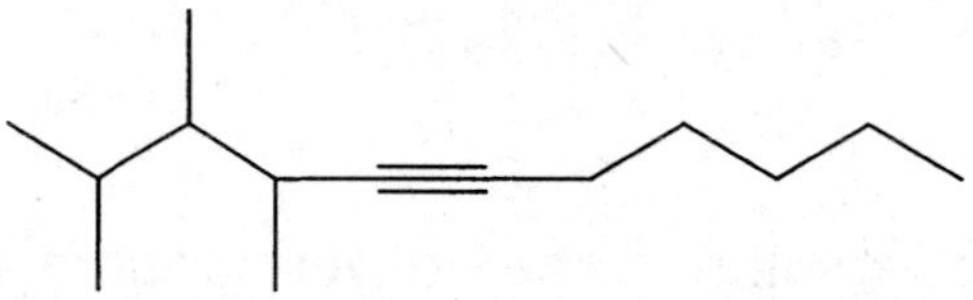

5.16 Alkyl halides are named as alkanes with the halogen as a substituent, using the group names fluoro, chloro, bromo, and iodo.
a) 5-Bromo-2,4,4-trimethylheptane
b) 1-Chloro-3-ethyl-1-methylcyclopentane

5.17

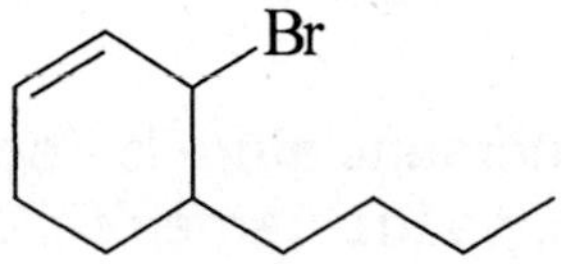

5.18 Alcohols are given the name of the hydrocarbon from which they are derived, with the suffix -ol. The longest chain must include the carbon bearing the hydroxy group and it is numbered to give the lowest possible number to this carbon.
a) 2-Butanol b) 3-Methyl-3-hexanol c) 3-Cyclopentyl-1-propanol
d) 3-Bromo-3-methylcyclohexanol (In rings, the C bearing the OH must be at position 1, so it is unnecessary to list its number.)

5.19

a) 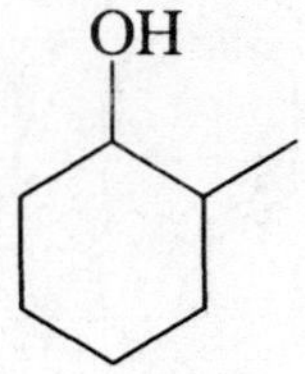

b) 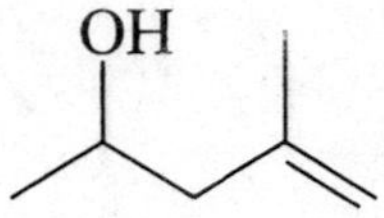

5.20 For simple ethers, common names are often used. In these, each alkyl group is named, followed by the word ether. Complex ethers are named using the IUPAC system. The smaller alkyl group and the oxygen are designated as an *alkoxy* substituent on the larger group.

a) ethyl methyl ether $\underbrace{CH_3}_{\text{methyl}}O\underbrace{CH_2CH_3}_{\text{ethyl}}$

b) 1-chloro-3-methoxycyclopentane. Of the two alkyl groups on the O, methyl is smaller, so the name uses a methoxy group on the cyclopentane root. The position bonded to the Cl is given number 1 because the numbers will be 1 and 3 in either case and the Cl appears first alphabetically.

5.21 In the common names of amines, the suffix -amine is appended to the name of the alkyl group.

a) This is a simple amine, so common nomenclature is used. Here a propyl group is attached to the nitrogen, so the name of the compound is propylamine.

b) The ethyl group attached to the nitrogen is given the prefix *N*-. The common name for this compound is *N*-ethylcyclopentylamine.

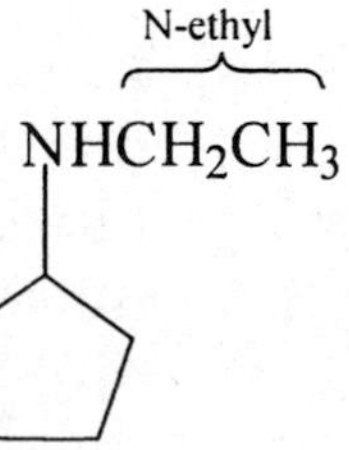

5.22 Systematic nomenclature is used for more complex amines and is similar to that used in naming alcohols except that the suffix is -amine.

a) *N*,5-Dimethyl-2-hexanamine

b) This compound has both an alcohol and an amine functional group. Only one of these can be designated with the suffix, and the other must be named as a group. The hydroxy group has higher priority than the amino group and is used to determine both the suffix and the numbering. The name of this compound is 5-amino-2-hexanol.

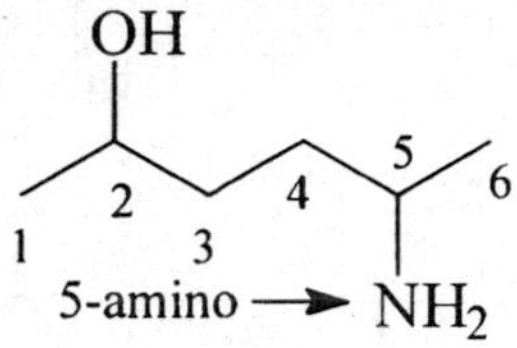

5.23 a) 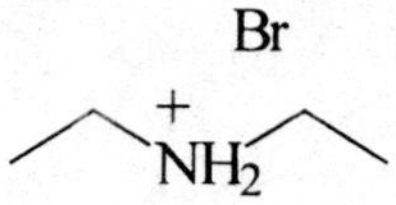

b) 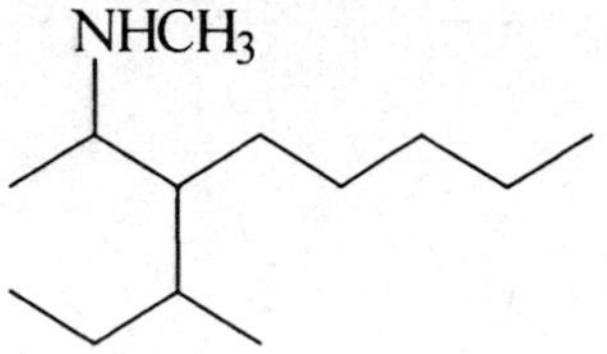

5.24 a) 3,4-Dimethylheptane
b) 4-Ethyl-2-hexene
c) 3-Chloro-1-pentyne
d) Ethylcyclopentane
e) 3-Methyl-2-pentanol
f) Ethyl propyl ether
g) Diethylamine

5.25

a) 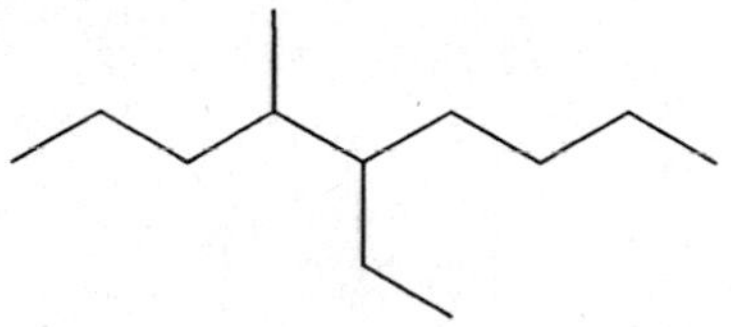

b) 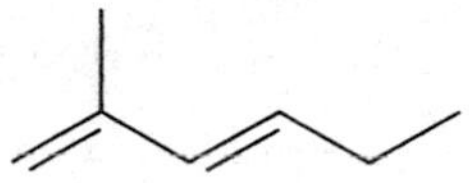

c) 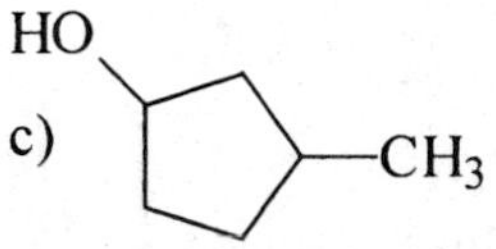

d) 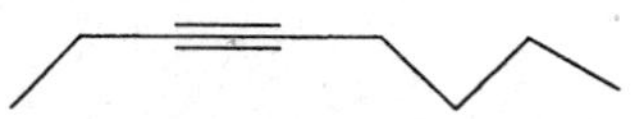

e) 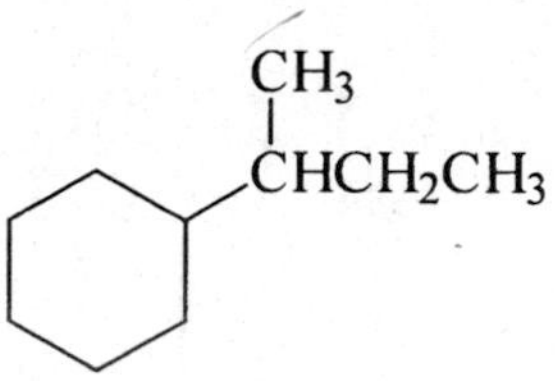

f) 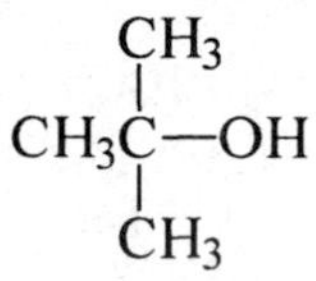

5.26

hexane

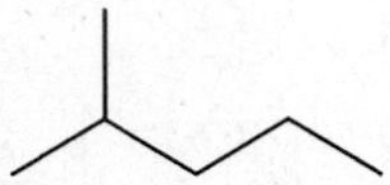

2-methylpentane

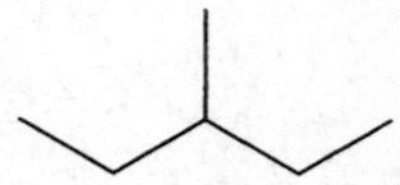

3-methylpentane

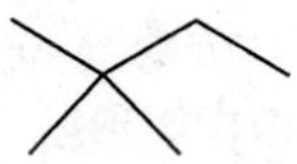

2,2-dimethylbutane

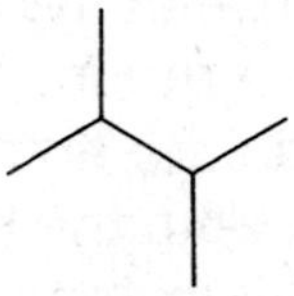

2,3-dimethylbutane

5.27

a) 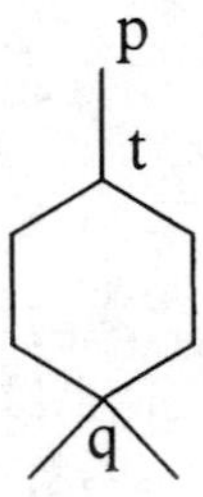

b)
$$\begin{array}{c} CH_3 \\ | \\ CH_3CH_2CH_2CHCH_2CH_3 \end{array}$$
s t s p

5.28

a) CH_3CH_2OH
a primary alcohol

b)
$$\begin{array}{c} CH_3 \\ | \\ CH_3COH \\ | \\ CH_3 \end{array}$$
a tertiary alcohol

c) 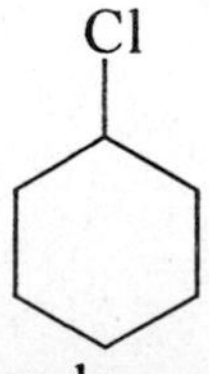

a secondary alkyl chloride

d) 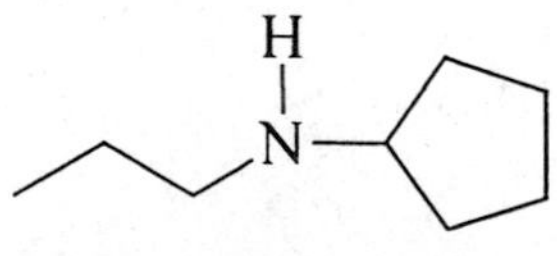

a secondary amine

5.29 a) The longest chain has 6 carbons. The correct name is 3-methyl-3-hexene.

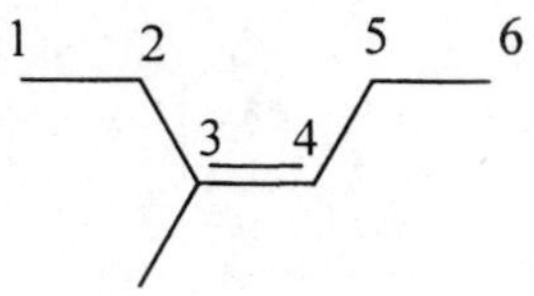

b) The smallest possible number should be given to the first branch. The correct name is 2,7,8-trimethyldecane.

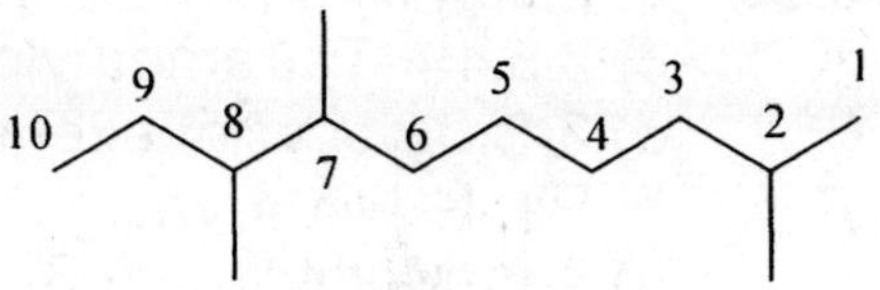

c) The ring has 6 carbons, so the root is cyclohexane. The correct name is propylcyclohexane.

d) The branches on the root should get the smallest numbers possible. The correct name is 1,2,4-trimethylcyclopentane.

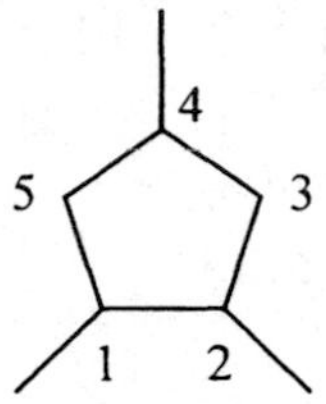

e) The double bond has higher priority than the chlorine, and its carbons must be numbered 1 and 2, so the correct name is 3-chlorocyclopentene.

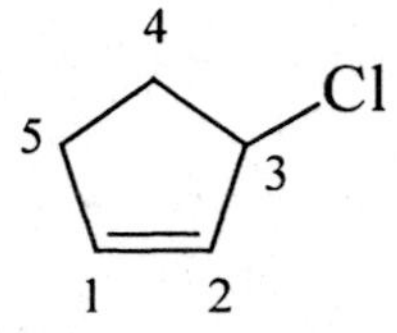

f) The alcohol has a higher priority than the triple bond and controls the numbering, so the correct name is 4-pentyn-2-ol.

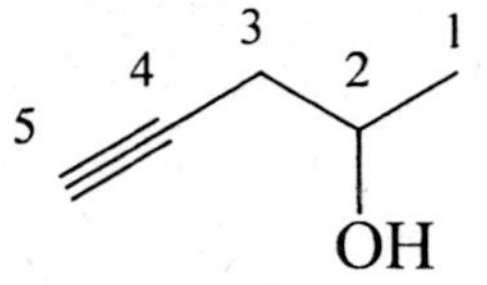

g) The carbon of the alkyl substituent attached to the ring should get number 1 and the group name is placed in parentheses. The correct name is 2-(3-methylbutyl)-1-cyclopentanol.

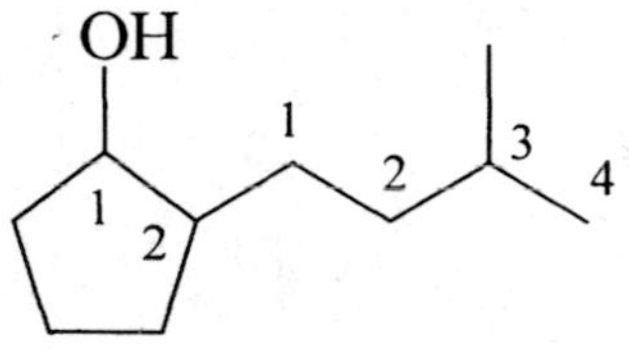

h) The amine has priority over the double bond and the methyl group attached to the nitrogen should get a prefix *N*-. The correct name is *N*-methylbut-3-en-1-amine.

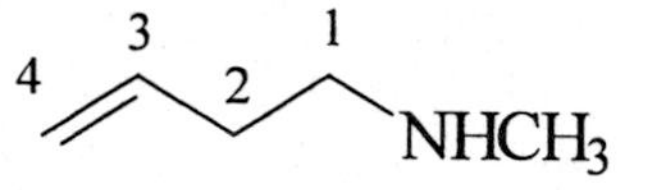

i) The longest chain has six carbons. The correct name is 2-hexanol.

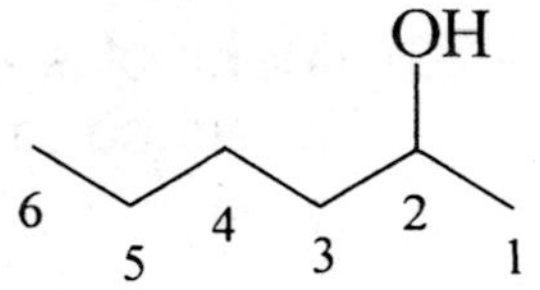

5.30 a) 2,3,6-Trimethyloctane b) 1,1,2,3-Tetramethylcyclohexane
c) 2,3,4,4-Tetramethylcyclohexene
d) 3-Bromocyclohexene e) 3-Cyclopentenol
f) Pent-1-en-4-yne g) 5-Methylhex-5-en-3-ol
h) 4-(1-Methylbutyl)-3-cyclohexenol
i) 5,5-dichloro-2-hexyne j) 3-methoxycyclohexanol
k) 3-isopropylcycloheptanamine
l) butoxycyclopentane or butyl cyclopentyl ether

5.31

a) 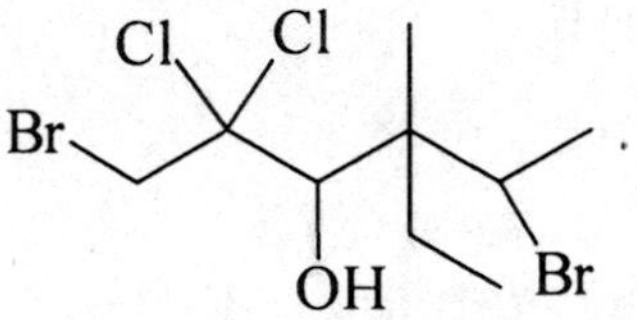

b) 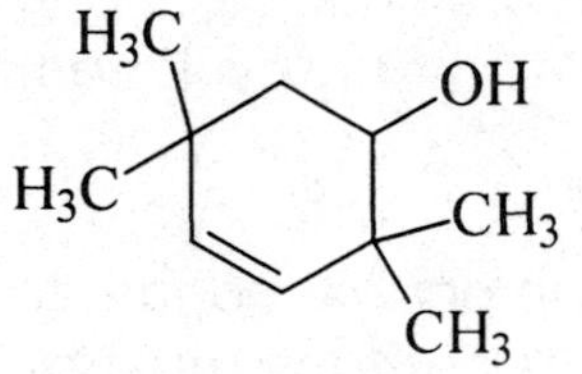

c) 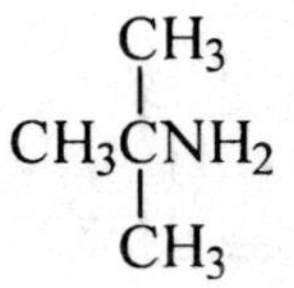

d) 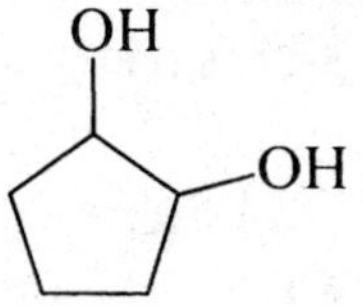

e) 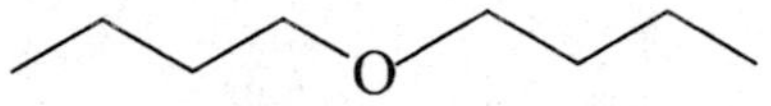

f) 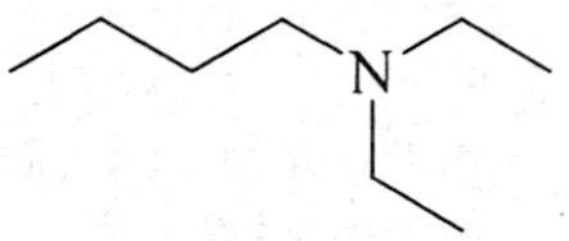

g) 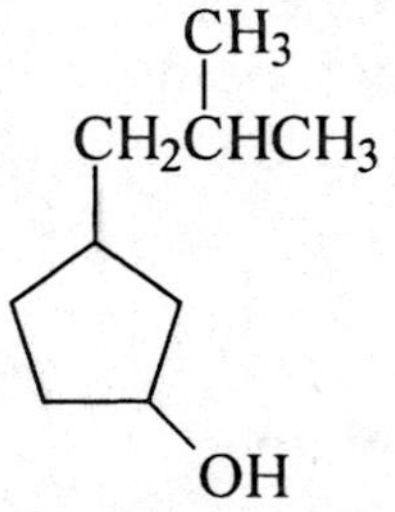

h) 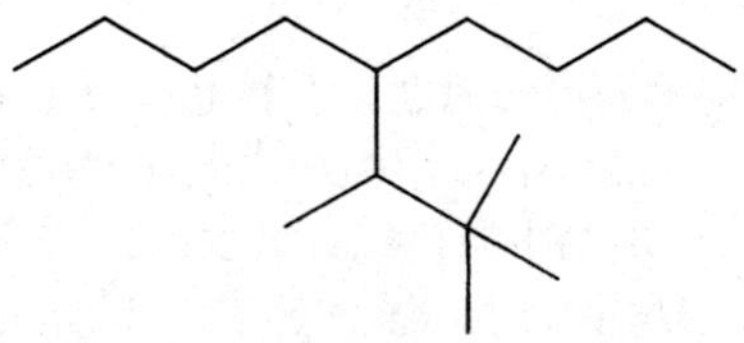

5.32 Menthol is a secondary alcohol because the OH group is attached to a secondary carbon.

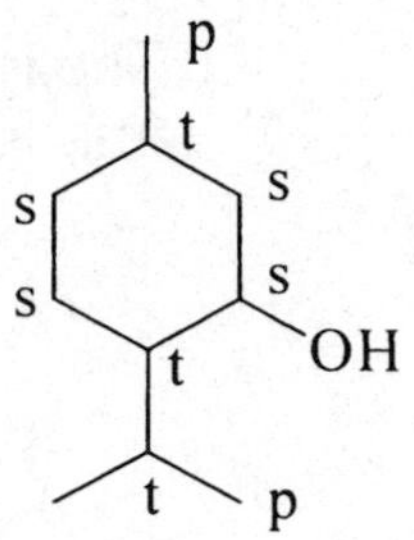

5.33 a) 2-Methyl-1,3-butadiene b) 1,2,3-Propanetriol
c) 1-Isopropyl-4-methyl-3-cyclohexenol
d) 2-Isopropyl-5-methylcyclohexanol
e) 3,7-Dimethylocta-2,6-dien-1-ol f) 1,5-Pentanediamine
g) 10,12-Hexadecadien-1-ol

5.34 1-Methyl-4-(1-methylethenyl)cyclohexene

5.35

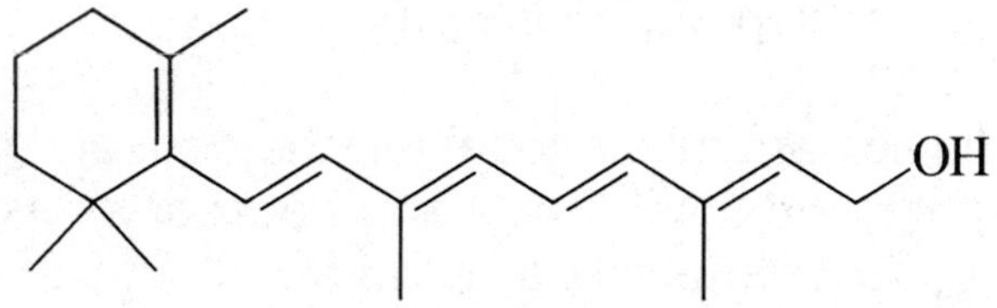

5.36 a) The melting point of a compound depends on how well the individual molecules can pack into the crystal lattice. The cyclic shape of cyclopentane allows it to pack better than the straight chain of pentane. Therefore cyclopentane has a higher melting point (-94°C) than pentane (-130°C).
b) Intermolecular attractive forces are much larger in 1-pentanol than in pentane because of the polarity and hydrogen bonding of the hydroxy group. Therefore 1-pentanol has a higher melting point.

5.37 The boiling point of a compound depends on the polarity, hydrogen bonding, surface area, and molecular weight of the molecule.
a) Nonane and octane are nonpolar hydrocarbons of similar shape. Nonane has the higher boiling point because it has a larger molecular weight than octane.
b) The polarity of an alkene is not much different from that of an alkane. Therefore the physical properties of an alkene are similar to those of the corresponding alkane. The boiling points of nonane and 3-nonene are approximately the same.
c) 1-Nonanol has the higher boiling point because it is more polar than 1-nonyne and it is capable of hydrogen bonding.
d) Propylamine is higher boiling because it can hydrogen bond. Trimethylamine does not have any hydrogens bonded to the nitrogen and is incapable of hydrogen bonding.
e) Cyclopentanol is higher boiling because it can form hydrogen bonds, whereas diethyl ether cannot.

f) The physical properties of alkynes are very similar to those of alkenes and alkanes with the same number of carbons. Therefore 1-butene and 1-butyne will have very similar boiling points.
g) 1-Pentanol is higher boiling because it can hydrogen bond.
h) Cyclopentanol and cyclopentylamine are both polar, both can form hydrogen bonds, and they have similar molecular masses. However, because oxygen is more electronegative than nitrogen, cyclopentanol forms stronger hydrogen bonds. Therefore, cyclopentanol has a higher boiling point (141°C) than cyclopentylamine (108°C).

5.38 When two immiscible liquids are mixed together, the denser liquid forms the lower layer. Chloroform is denser than water because of its more massive chlorine atoms, so it forms the lower layer.

5.39 Remember, like dissolves like. Water is very polar and can hydrogen bond, so more polar compounds and compounds capable of hydrogen bonding will be more soluble in water.
a) 1-Butanol has a higher solubility in water because it is more polar and can form more hydrogen bonds.
b) 1-Butanol has a higher solubility in water because it has a smaller nonpolar hydrophobic hydrocarbon chain than 1-hexanol.
c) Diethylamine is more soluble in water than pentane because it is polar and can hydrogen bond with water. Pentane is nonpolar and therefore hydrophobic.

5.40 Amines often have unpleasant, fishy odors. Therefore it is possible that the fishy smelling unknown compound is an amine. Amines are bases, so they react with strong acids to form water soluble salts. Testing the solubility of the unknown compound in dilute aqueous hydrochloric acid is one way of confirming the presence of the amine functional group.

5.41 Turpentine should not mix with water because it is composed of nonpolar hydrocarbons. If a paint dissolves in turpentine, the paint is also composed of nonpolar molecules.

5.42 A higher the ratio of C and H to O results in a compound with a higher energy content. A fat has a lower ratio of C and H to O than an alkane, but a higher ratio than ethanol, so its energy content should be between the values for these compounds.

5.43 a) 2-Methylhexane b) 2-Methylhexane
c) 1,1-Dichlorocyclopentane d) 5-Methyl-1-heptyne
e) 4,4-Dimethylcyclohexene f) 2-Cyclopentenol
g) *N,N*-Dimethyl-3-pentanamine h) Ethoxycyclohexane

5.44 a) Hexane has a higher boiling point than pentane because it has a larger molecular weight.
b) The primary amine on the left has a higher boiling point than the isomeric tertiary amine on the right because the primary amine can form hydrogen bonds.
c) The alcohol (right structure) has a higher boiling point than the isomeric ether (left structure) because the alcohol can form hydrogen bonds.

5.45 a) Cyclopentane has a higher melting point than pentane because its symmetrical shape allows it to pack better into the crystal lattice.
b) Propylamine has a higher melting point than butane because it is more polar and can form hydrogen bonds.

5.46 a) 1-Pentanol has a higher water solubility than hexane because it is more polar and can form hydrogen bonds.
b) 2-Butanol has a higher water solubility than 1-pentanol because it has a smaller nonpolar alkyl group.

Review of Mastery Goals

After completing this chapter, you should be able to:

Provide the systematic (IUPAC) name for an alkane.
(Problems 5.1, 5.3, 5.24, and 5.26)

Draw the structure of an alkane whose name is provided.
(Problems 5.2, 5.6, and 5.25)

Name a complex group.
(Problems 5.4, 5.5, and 5.30)

Name a cycloalkane, an alkene, an alkyne, an alkyl halide, an alcohol, an ether, or an amine.
(Problems 5.10, 5.12, 5.14, 5.16, 5.18, 5.20, 5.21, 5.22, 5.24, 5.29, 5.30, 5.33, 5.34, and 5.43)

Draw the structure of a compound containing one of these functional groups when the name is provided.
(Problems 5.11, 5.13, 5.15, 5.17, 5.19, 5.23, 5.25, 5.31, and 5.35)

Predict the approximate physical properties of a compound containing one of the above functional groups.
(Problems 5.36, 5.37, 5.38, 5.39, 5.41, 5.44, 5.45, and 5.46)

Chapter 6
STEREOCHEMISTRY I
CIS-TRANS ISOMERS AND CONFORMATIONS

The use of models is an invaluable aid in understanding the material presented in this chapter. You are strongly encouraged to use models to solve the problems in this chapter. You should also take advantage of the three-dimensional computer models that are available on the web at http://now.brookscole.com/hornback2.

6.1 Cis-trans isomers are stereoisomers that differ in the placement of groups about a double bond. In order for an alkene to exhibit cis-trans isomerism, the two groups on each end of the double bond must be different.

a) 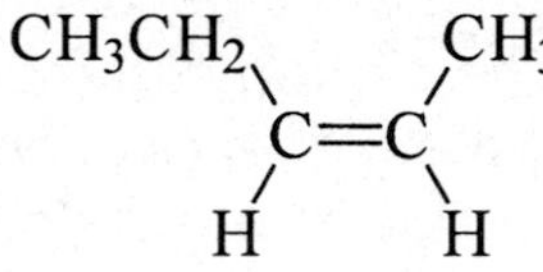 and 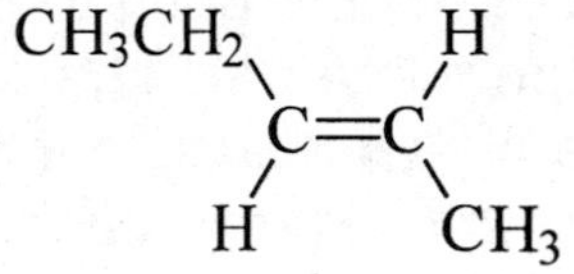

b) none 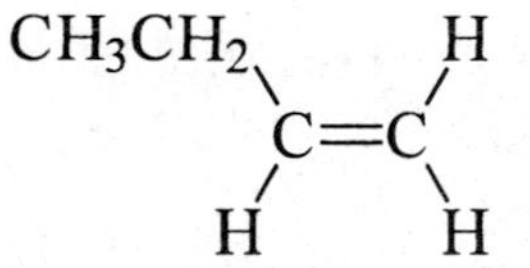 These substituents are identical.

c) none These groups are identical. 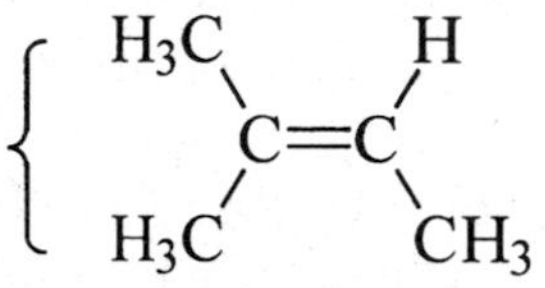

d) 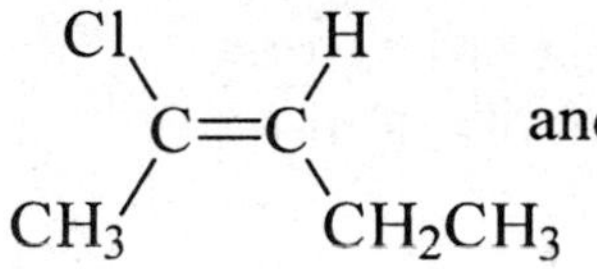 and 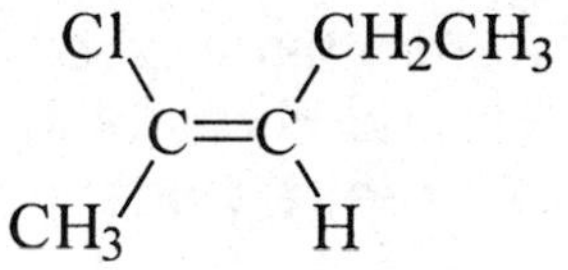

6.2 The relative stability of cis-trans isomers depends on the steric strain energy associated with each isomer. Steric strain destabilizes a molecule by forcing it to deviate from its optimum bonding geometry. The isomer with the larger groups on opposite sides of the double bond (trans) is more stable.

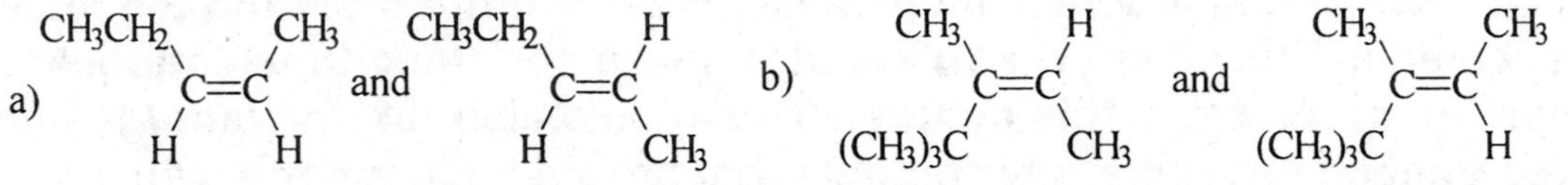

In both cases the isomer on the right is more stable because the larger groups are trans.

6.3 The Cahn-Ingold-Prelog sequence rules are used to assign priorities to groups. These rules use the atomic numbers of the atoms attached to the carbons of the double bond to determine the priority of groups.
Rule 1: Of the two atoms attached to one carbon of the double bond, the one with the higher atomic number has the higher priority.
Rule 2: If the two atoms attached to the carbon are the same, compare the atoms attached to them, in order of decreasing priority. The decision is made at the first point of difference.
Rule 3: Multiple-bonded atoms are treated as though they are equivalent to the same number of single bonds to the same atoms.

a) The carbon of the ethyl group is bonded to C, H, and H. The carbon of the methyl group is bonded to 3 H's. So the ethyl group has higher priority than a methyl group. —CH_2CH_3

b) The carbon bonded to O, O, and O has higher priority than carbon bonded to O, O, and N. —$\overset{\overset{O}{\|}}{C}OH$

c) These two groups differ only in the bonding arrangement at the third carbon. The third carbon bonded to C, H, and H has higher priority than that bonded to three hydrogens. —$CH_2CH_2CH_2CH_3$

d) The carbon bonded to three nitrogens has higher priority than the carbon bonded to C, H, and H. —$C\equiv N$

e) The carbon on the aromatic ring is bonded to three carbons, whereas the carbon of the isopropyl group is bonded to two carbons and a hydrogen.

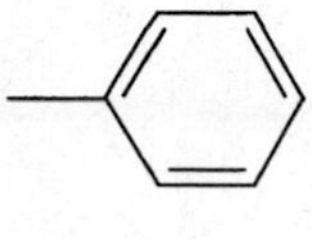

6.4 First choose the higher priority group attached to each of the carbons of the double bond. If the high priority groups are on the same side of the double bond, the configuration is designated as *Z*. If the high priority groups are on opposite sides of the double bond, the configuration is designated as *E*.

a) The atoms attached to the left carbon of the double bond are both carbons. The upper C is bonded to three H's. The lower C is attached to two H's and a C, so it has higher priority. The atoms attached to the right carbon of the double bond are also both carbons. However, the upper group has higher priority because it is bonded to a chlorine. The higher priority groups are on opposite sides of the double bond, so this is the *E*-isomer.

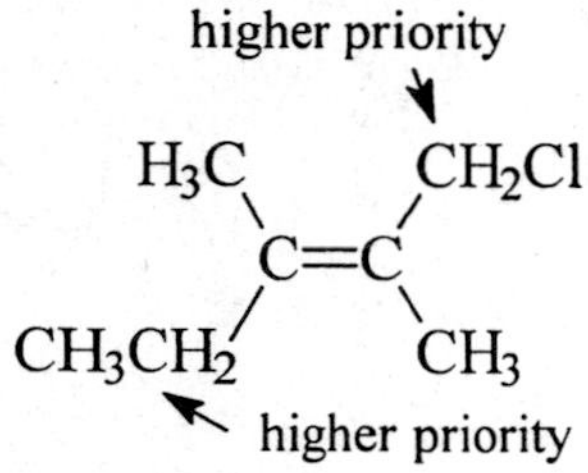

b) Remember, double and triple bonds are treated as though they are two or three single bonds to the same atom. The higher priority groups (F and C≡CH) are on opposite sides of the double bond, so the compound is the *E*-isomer.

c) On the left C of the double bond, the lower group has higher priority because its C is bonded to three Cs whereas the upper C is bonded to two C's and an H. On the right side of the double bond, the lower group has the higher priority because the C is bonded to an O, rather than an N. This is the *Z*-stereoisomer.

6.5 For propane, all the eclipsed conformations are of the same energy while all the staggered conformations are of the same energy. The eclipsed conformation has a higher energy due to torsional strain and steric strain (due to eclipsing CH_3 and H). These strains are minimized in the staggered conformation, so it is lower in energy. The plot is identical to that for ethane with the exception that the energy difference is 3.3 kcal/mol (13.8 kJ/mol) due to the extra steric strain from the eclipsed methyl and hydrogen.

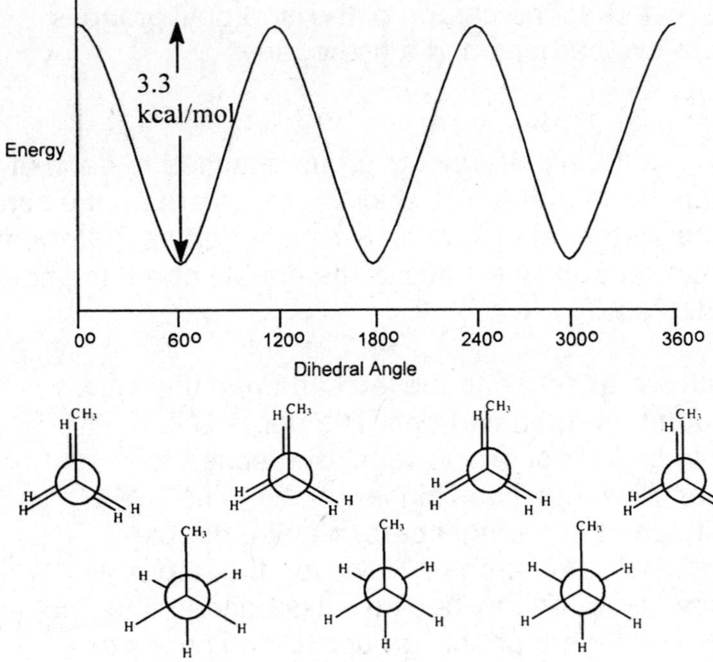

6.6

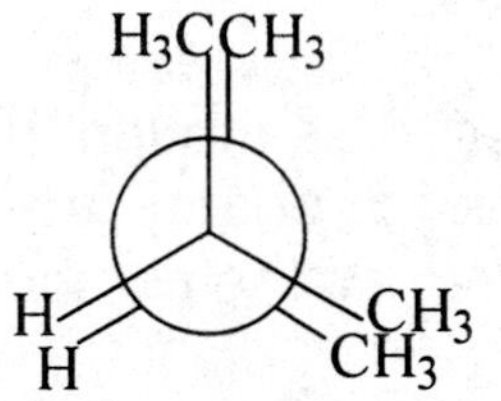

6.7

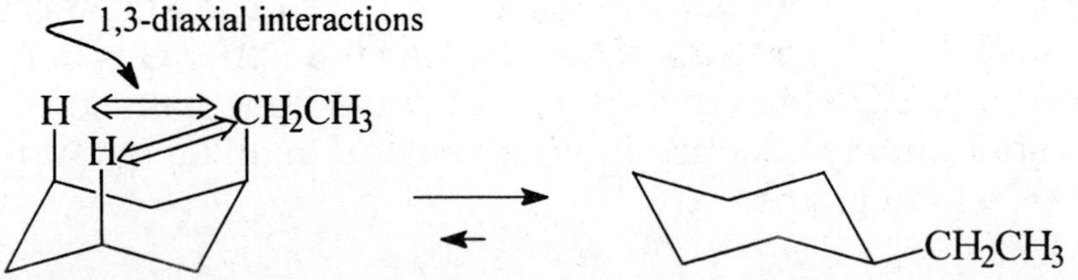

The conformation on the left has steric crowding between the axial ethyl group and the two axial hydrogens. The steric strain resulting from these 1,3-diaxial interactions destabilizes this conformer. In general, substituents larger than hydrogen prefer to be equatorial on a cyclohexane ring to avoid 1,3-diaxial interactions. The conformer on the right is produced by the ring flip, and does not have any 1,3-diaxial interactions because the ethyl group is equatorial. Therefore, the conformation on the right is more stable.

6.8 Substituents larger than H prefer to be equatorial on a cyclohexane ring. Larger substituents have a smaller amount of the conformation with the substituent axial present at equilibrium than smaller substituents. Therefore, compare the axial strain energies (see Table 6.2) of the substituents. The one with the smaller axial destabilization energy will have a larger amount of the conformation with the substituent in the axial position present at equilibrium.

a) The axial destabilization energy of the CN group is smaller than that of the CH_3 group. Therefore, cyanocyclohexane will have more of the conformation with the cyano group axial present at equilibrium compared to methylcyclohexane.

b) Ethylcyclohexane will have more of the conformation with the ethyl group axial present at equilibrium because the phenyl group has a larger axial destabilization energy than the ethyl group.

c) Chlorocyclohexane will have more of the conformation with the axial Cl present at equilibrium because the ethyl group has a larger axial destabilization energy than Cl.

6.9 The C-Br bond is longer than the C-Cl bond so the larger Br is farther from the axial H's. The two effects (larger atom, longer bond) approximately cancel in this case.

6.10 Stereoisomers are compounds that have the same chemical formula and connectivity but a different arrangement of the bonds in space. In cycloalkanes, they are similar to cis-trans isomers in that they cannot interconvert without breaking a bond. It is easiest to show the stereoisomers of a cyclic compound by drawing the ring flat and not worrying about conformations.

a)
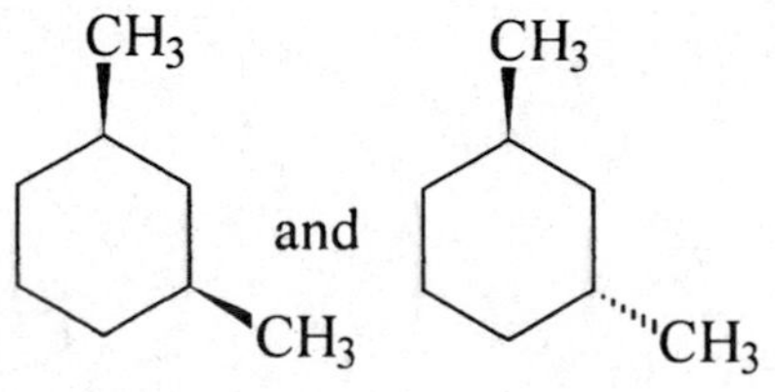

b)
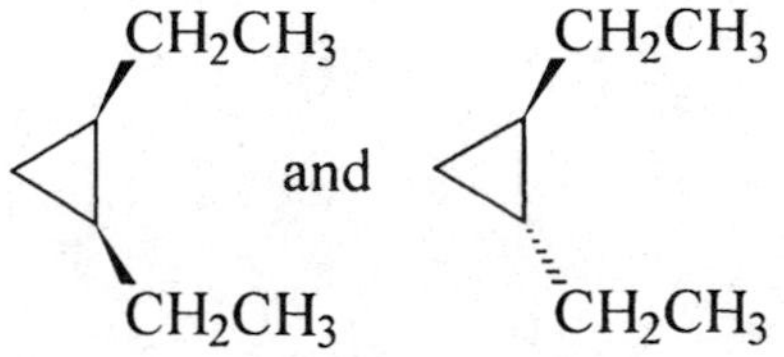

c)
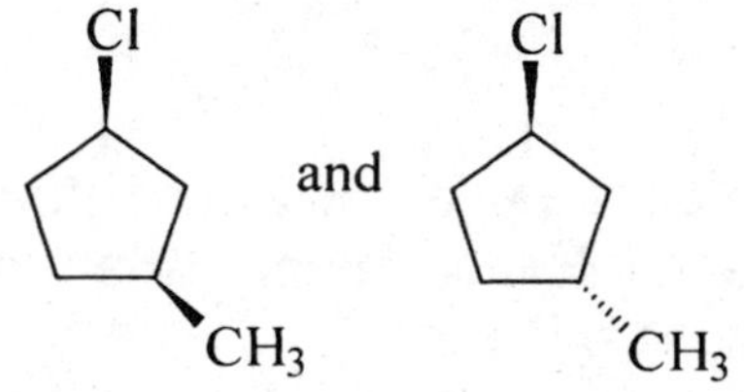

6.11

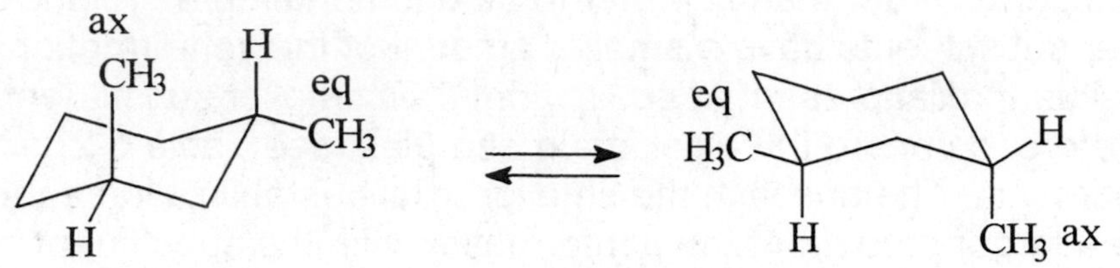

6.12 This flat representation of a cyclohexane ring shows the relation of axial and equatorial bonds. The bonds on one side of the ring (all above or all below) are cis and they alternate between axial and equatorial as one proceeds around the ring. Similarly, bonds on opposite sides of the ring are trans.

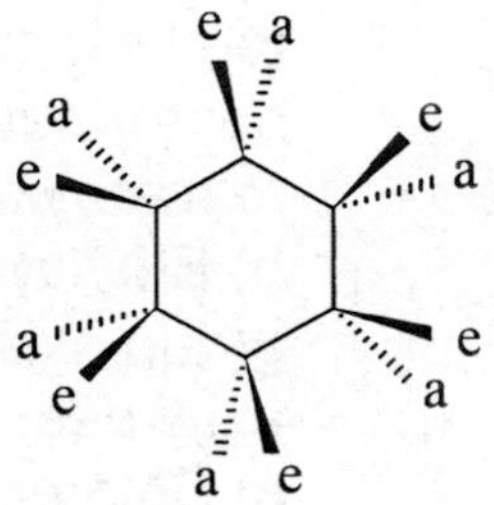

a) In *cis*-1,4-dimethylcyclohexane one methyl group is axial and one is equatorial. The ring flip of one conformation converts the axial methyl to equatorial and the equatorial methyl to axial.

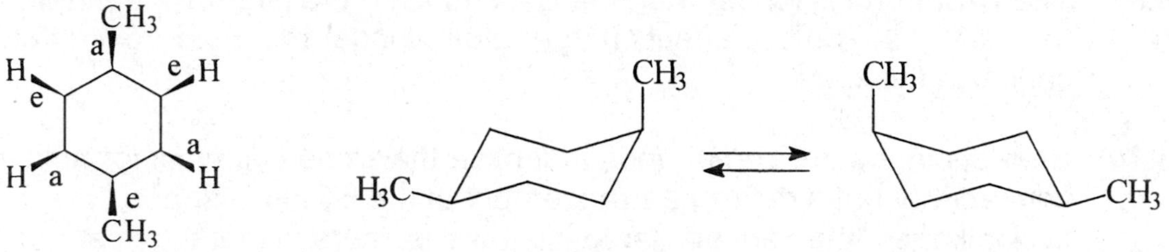

b) In one conformation of the trans isomer both methyl groups are axial and in the other conformation, produced by the ring flip, both methyl groups are equatorial.

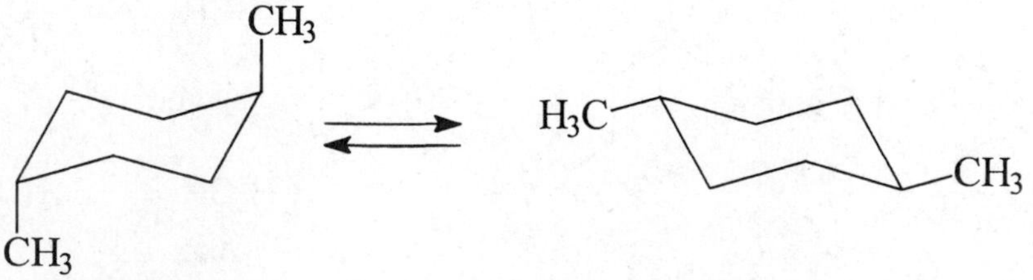

c) The *trans*-isomer is more stable because it has the conformation with both methyl groups equatorial. Steric interactions are at a minimum when bulky substituents are in equatorial positions.

d) The more stable conformation of the more stable isomer is the conformation of the *trans*-isomer shown in (b) with both methyl groups equatorial.

6.13 a) The conformation with both groups equatorial is the more stable conformation of the *cis*-stereoisomer. The conformation with the isopropyl group equatorial and the hydroxy group axial is the more stable conformation of the *trans*-stereoisomer because the isopropyl group has a larger axial strain energy than the hydroxy group.

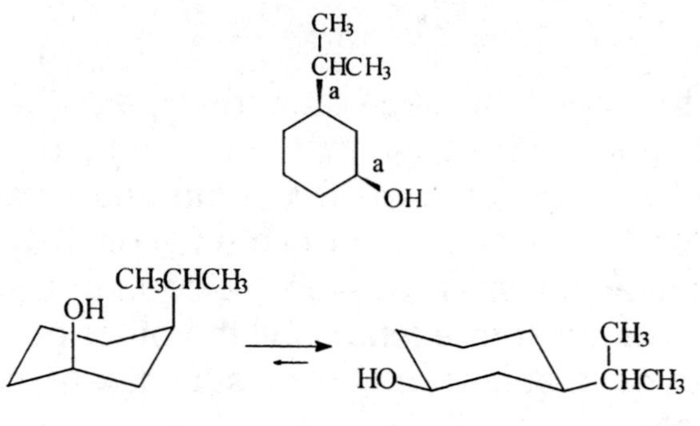

cis-3-isopropylcyclohexanol

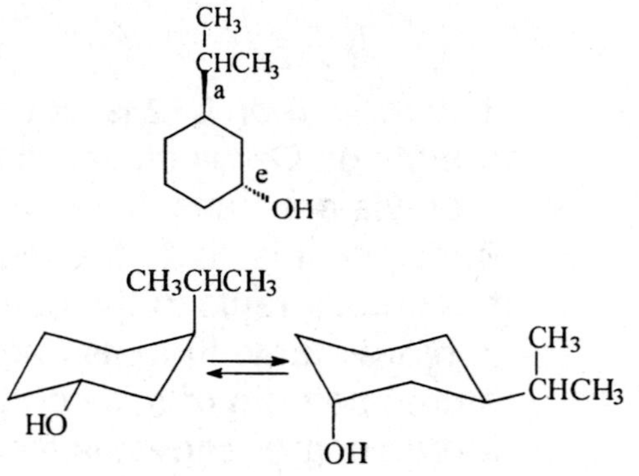

trans-3-isopropylcyclohexanol

b) The *cis*-isomer exists almost entirely in the conformation with both substituents equatorial and has no steric strain. The lower energy conformation of the *trans*-isomer is the one with the isopropyl group equatorial because the axial destabilization energy for the isopropyl group is larger than that for the OH group. The steric strain for this conformation is the axial strain energy for the hydroxy group [0.9 kcal/mol (3.8 kJ/mol) from Table 6.2]. Therefore, the *cis*-stereoisomer is more stable than the *trans*-stereoisomer by 0.9 kcal/mol (3.8 kJ/mol).

6.14 In the representation of a cyclohexane chair conformation, all bonds projecting closer to the top of the page at each carbon are on the top side of the ring and are cis to each other. Similarly, the bonds projecting closer to the bottom of the page at each carbon are on the bottom side of the ring and are cis to each other but trans to the bonds on the top of the ring.

a)

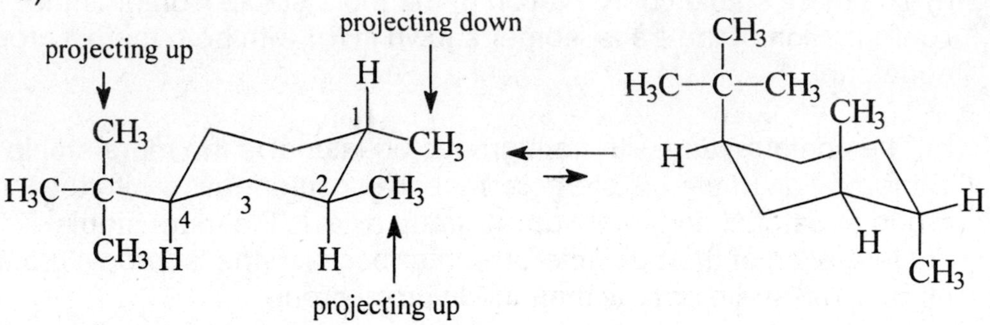

The methyl on C-2 is closer to the top of the page than the H, and the methyl on C-1 is closer to the bottom of the page than the H, so the methyls are trans. The *t*-Bu on C-4 is closer to the top of the page than the H, so it is cis to the methyl on C-2 and trans to the methyl on C-1. All the substituents in the conformation shown on the left are equatorial, so it is more stable than its ring-flip. The conformations of all the other possible stereoisomers of this compound have at least one axial substituent, so the stereoisomer shown is most stable.

b)

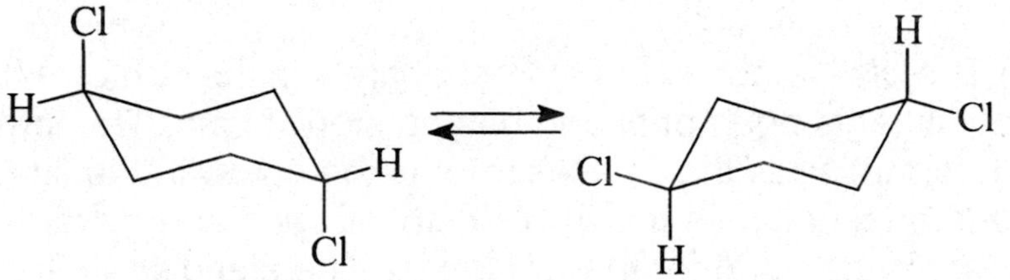

The chlorines are on opposite sides of the ring, so they are trans. The chlorines are both axial in the conformation on the left. The conformation produced by the ring flip (right) is more stable because both chlorines are equatorial. In the other stereoisomer of this compound, the chlorines are cis. One of the chlorines will be axial in each conformation of the *cis*-stereoisomer. Therefore, the stereoisomer shown is more stable.

c)

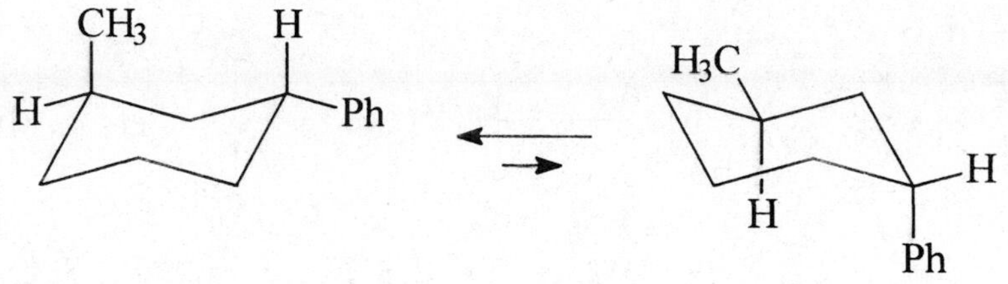

The phenyl and methyl groups are trans to each other. The conformation on the left is more stable because the axial destabilization energy of the methyl group is smaller than that for the phenyl group. In the *cis*-isomer of this compound, the more stable conformation will have both the phenyl and the methyl groups equatorial. Therefore the *cis*-stereoisomer is more stable.

6.15

a)

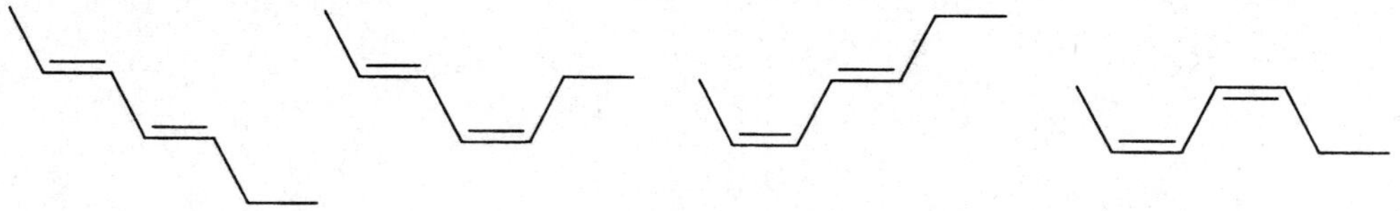

b)

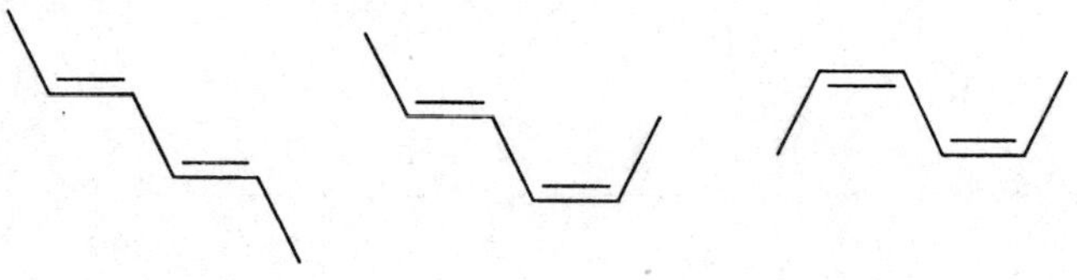

c)

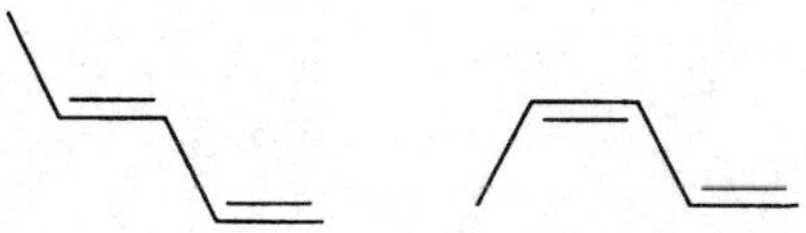

6.16

a) b) $—C≡N$ c) $—\overset{\displaystyle CH_3}{\underset{\displaystyle CH_3}{\overset{|}{\underset{|}{C}}}}—CH_3$ d) $—C≡CH$

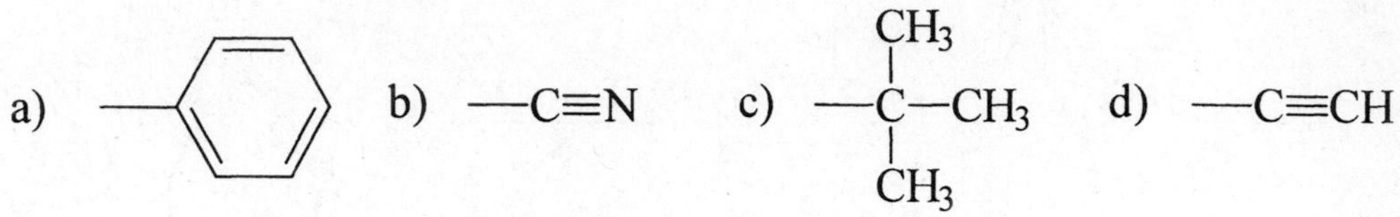

6.17 a) *Z*   b) *Z*   c) *Z*   d) *E*

6.18 The energy versus dihedral angle plot for 2-methylpropane is similar to that of propane shown in problem 6.5, with the exception that the energy difference between the high and the low energy conformers of 2-methylpropane is larger than that for propane because it has two methyl-hydrogen eclipsing interactions.

6.19

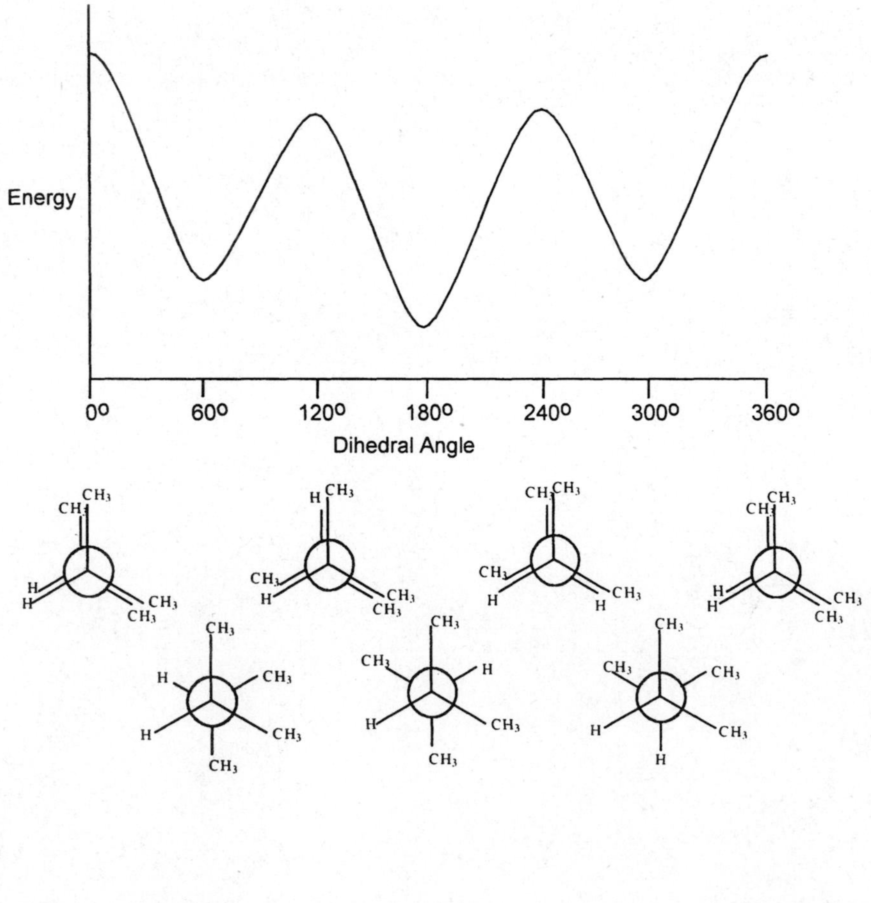

6.20 a) The three sp^3 carbons of cyclopropane form an equilateral triangle with an internuclear angle of 60°. This angle is much smaller than the tetrahedral angle of 109.5° required for the atomic orbitals of the sp^3 hybridized carbons, causing a large amount of angle strain in the molecule. In addition to the angle strain, the rigid and flat cyclopropane ring also has a significant amount of torsional strain due to the eclipsed conformation about each C-C bond.

b) Planar cyclobutane has angle strain and torsional strain due to its eclipsed conformations about each C-C bond. As the ring distorts from planarity, torsional strain decreases while angle strain increases. At first, the increase in angle strain is less than the decrease in torsional strain, resulting in a net decrease in ring strain. Eventually, angle strain increases faster than torsional strain decreases. As a result, the lowest energy conformer of cyclobutane is slightly nonplanar.

c) The internuclear angle of planar cyclopentane ring is 108°, so it has little or no angle strain. However, planar cyclopentane does have a considerable amount of torsional strain due to eclipsed conformations at all C-C bonds. By distorting from planarity the cyclopentane ring can relieve most of its torsional strain without increasing its angle strain significantly. Therefore the low energy conformation of cyclopentane is nonplanar and shaped somewhat like an envelope.

d) Planar cyclohexane has angle strain because its internuclear angle is 120°, which is larger than the tetrahedral angle of 109.5°. It also has a substantial amount of torsional strain because all its bonds are eclipsed. When the ring distorts from planarity its angle strain is decreased. There are two nonplanar conformations of cyclohexane, called the chair and the boat, that are free of angle strain. The chair conformer is the lowest energy conformation because it has neither torsional nor angle strain. The boat conformer has torsional and steric strain.

e) Cyclodecane is nonplanar. It has some angle strain and some torsional strain. In addition, cyclodecane has transannular strain that results from steric interactions of C-H bonds that point towards the center of the ring.

6.21 Both the -C≡N and -C≡CH groups have linear geometry. In the axial position both are aligned parallel to the ring axis and their 1,3-diaxial interactions with other groups are not very large. The $-CH_3$ group has tetrahedral geometry. One hydrogen of the methyl group must be pointed over the ring to interact with the other axial substituents. Therefore, the axial strain energy is larger for the methyl group.

6.22 The chair conformation with two of the three methyl groups in equatorial positions is more stable.

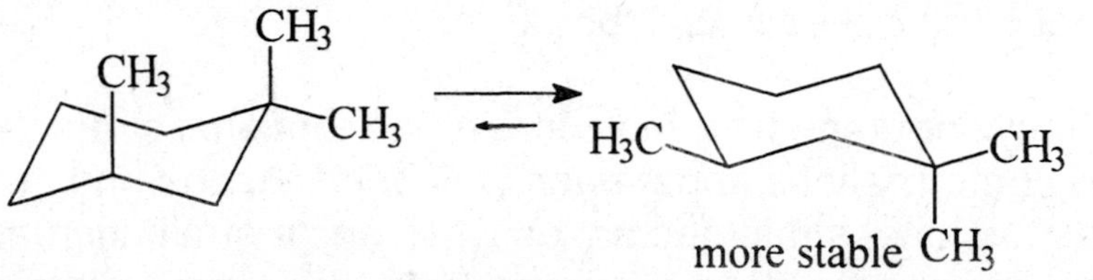

The axial strain energies listed in Table 6.2 of this chapter are for 1,3-diaxial interactions between a group and two axial hydrogens on the same side of the ring. The 1,3-diaxial interaction between two bulky groups, such as two methyl groups in this case, will be considerably larger. Therefore, it is not possible to determine the exact energy difference between the conformations shown above using the axial strain values listed in this chapter.

6.23

cis-stereoisomer *trans*-stereoisomer

The *trans*-stereoisomer is more stable than the *cis* -stereoisomer because it has the conformation with both substituents equatorial.

6.24 The axial destabilization energy of the phenyl group (2.9 kcal/mol, 12.1 kJ/mol) is larger than that of the methyl group (1.7 kcal/mol, 7.1 kJ/mol). Therefore the chair conformer with the phenyl group equatorial is more stable than the other conformer by 1.2 kcal/mol (5.0 kJ/mol).

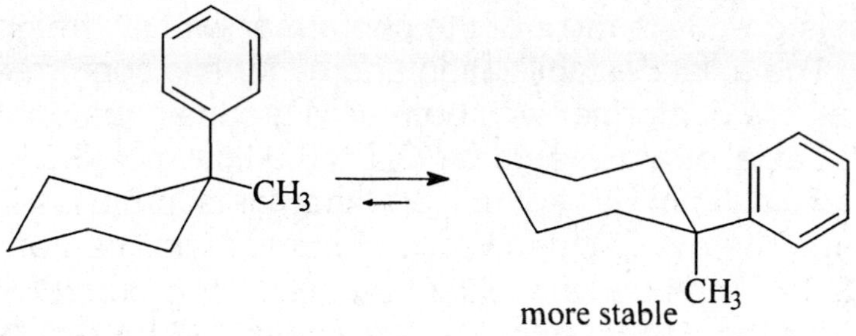

6.25 The axial destabilization energy of the bulky *t*-butyl group is so large that it is almost always equatorial. In this compound, when the *t*-butyl group is equatorial, the methyl group is axial.

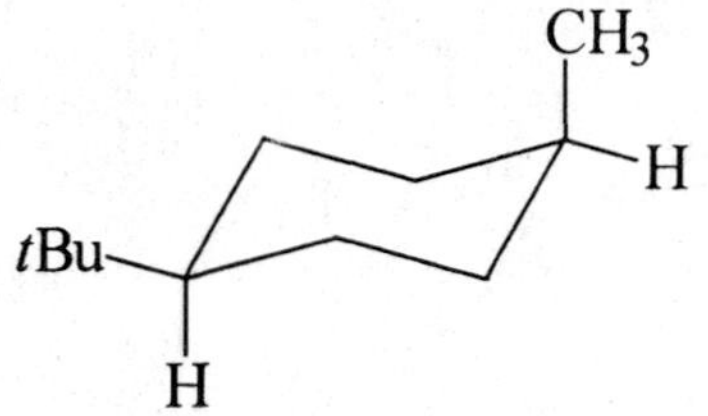

6.26

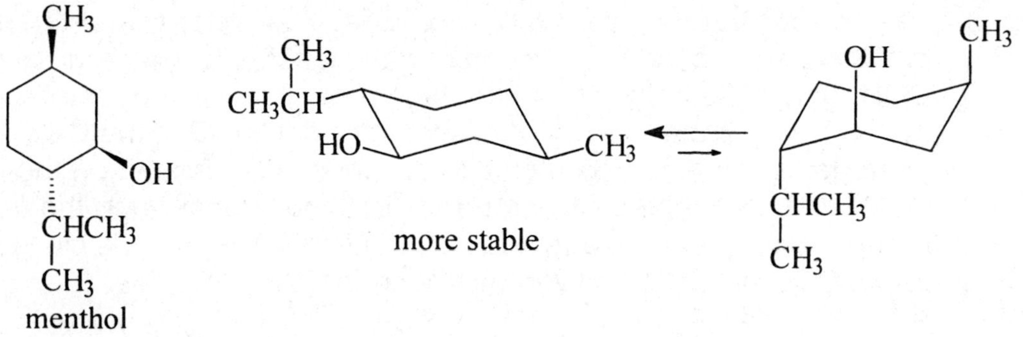

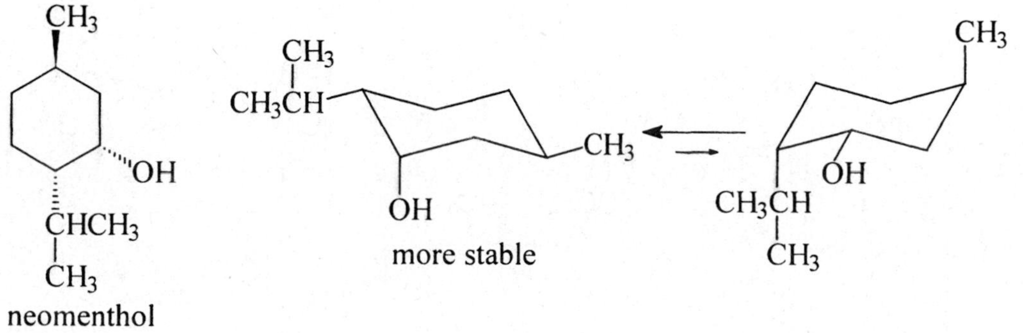

For menthol, the groups are either all axial or all equatorial. Obviously the equilibrium is greatly in favor of the conformer with all the groups equatorial. The axial destabilization energy for this conformer is zero. For neomenthol, the conformer with both alkyl groups equatorial is more stable. (The axial strain energy for OH is 0.9 kcal/mol [3.8 kJ/mol], that for methyl is 1.7 kcal/mol [7.1 kJ/mol, and that for isopropyl is 2.2 kcal/mol [9.2 kJ/mol]). The axial destabilization energy of this conformation is 0.9 kcal/mol (3.8 kJ/mol) because it has an axial OH group. Overall, menthol is more stable than neomenthol by 0.9 kcal/mol (3.8 kJ/mol).

6.27

a) 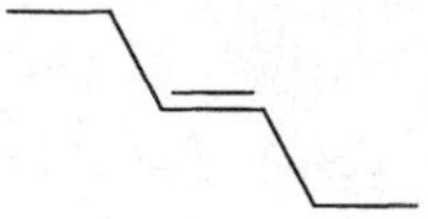

b) 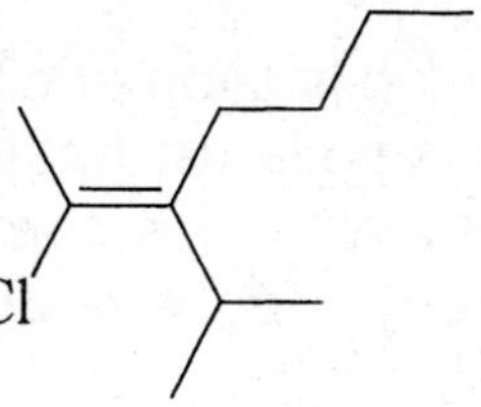

6.28 The *tert*-butyl group has a very large axial strain energy and, therefore, it is always equatorial. In the isomer on the left, the CO_2H group and the OCH_3 group are both axial when the *tert*-butyl group is equatorial. As a result, the H of the CO_2H group is too far from the O of the OCH_3 to hydrogen bond to it. In contrast, in the case of the isomer on the right, the CO_2H group and the OCH_3 group are both equatorial when the *tert*-butyl group is equatorial. As a result, the H of the CO_2H group is close enough to the O of the OCH_3 to hydrogen bond to it.

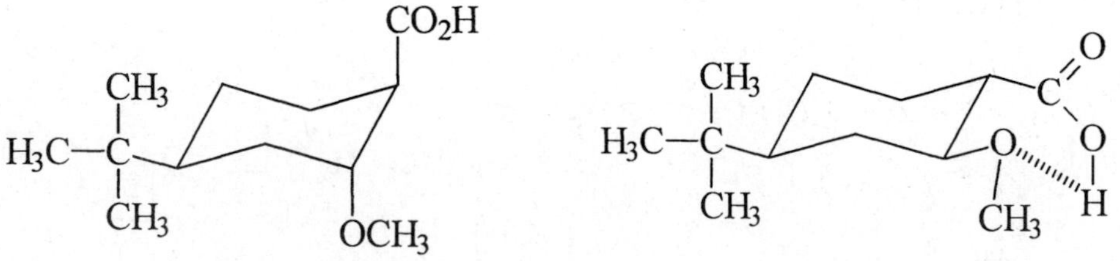

6.29

a)

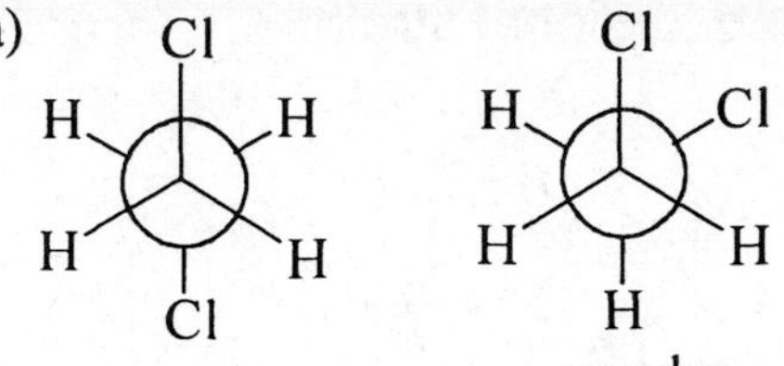

b)

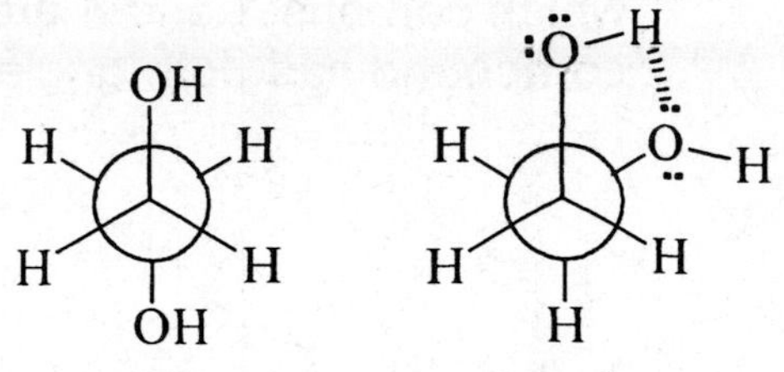

Besides steric and torsional strain, the stabilities of the conformations are influenced by dipole-dipole interactions (a and b) and hydrogen bonding (b).

6.30 Because its two methyl groups are eclipsed, *cis*-1,2-dimethylcyclopropane has more steric strain than *trans*-1,2-dimethylcyclopropane. Therefore *trans*-1,2-dimethylcyclopropane is more stable.

6.31 Linoleic acid has two double bonds, so there are four cis-trans isomers. α-Linolenic acid has three double bonds, so there are eight stereoisomers.

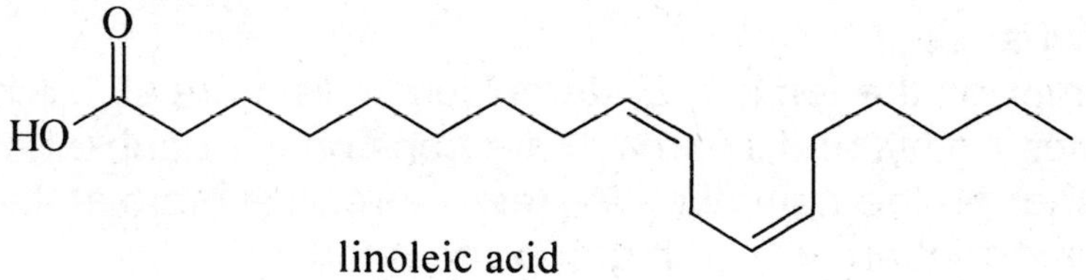

linoleic acid

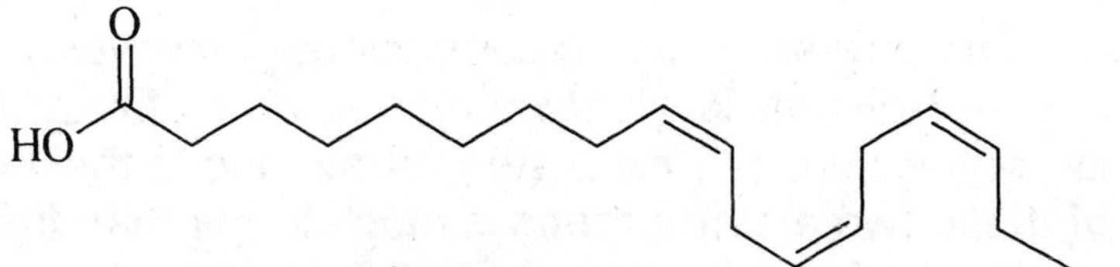

α-linolenic acid

6.32 Angle and torsional strain in its four-membered ring causes penicillin G to be relative unstable. One of the bonds in the four-membered ring will break relatively easily.

6.33 This compound has a conformationally locked boat cyclohexane ring in which carbons 1 and 4 are connected by a CH_2 group (C-7). This compound has both angle strain and torsional strain.

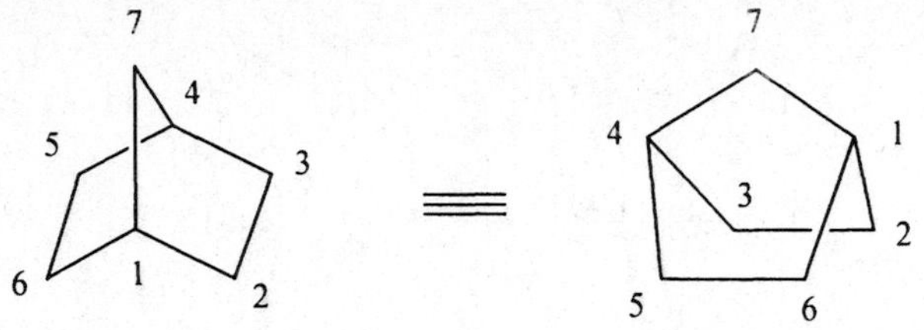

6.34 a) *Z* b) *Z*

6.35 a) This is the most stable conformation of pentane because it is anti about all of the C-C bonds.
b) This is a less stable conformation of pentane because it is eclipsed about one of the bonds.
c) This is a less stable conformation of 2-methylbutane because the methyl group on one carbon is gauche to both of the methyl groups on the other carbon.

6.36 a) The stereoisomer on the right (*trans*-1,2-dimethylcyclopropane) is more stable because the *cis*-isomer has more steric strain due to its eclipsed methyl groups.
b) The isomer on the left (the *E*-stereoisomer) is more stable because the larger groups (ethyl and isopropyl) are trans on the double bond.
c) The isomer on the right (the *cis*-stereoisomer) is more stable because it has the conformation with both groups equatorial.

6.37 Both methyl groups are axial in both compounds. However, in the isomer on the right (*cis*-1,3-dimethylcyclohexane), the two methyl groups are on the same side of the ring. The extra steric strain due to the 1,3-diaxial interaction of these two methyl groups is much larger than the 1,3-diaxial interactions between the methyl groups and the hydrogens in the other isomer, so the conformation on the right has more strain energy.

6.38 a) The Cl and the Br are cis and both are axial. The ring-flip conformation is more stable because both groups are equatorial. This compound is more stable than the *trans*-stereoisomer because it has the conformation with both groups equatorial.
b) The groups are cis. The methyl group is equatorial and the phenyl group is axial. The ring-flip conformation is more stable because the phenyl group has a larger axial strain energy than the methyl group. The *trans*-stereoisomer is more stable because it has the conformation with both groups equatorial.
c) The groups are cis. The methyl group is equatorial and the OH group is axial. This conformation is more stable than the ring-flip conformation because the methyl group has a larger axial strain energy than the hydroxy group. The *trans*-stereoisomer is more stable because it has the conformation with both groups equatorial.
d) The groups are trans. The ethyl group is equatorial and the methyl group is axial. This conformation is more stable because the ethyl group has a larger axial strain energy than the methyl group. The cis-stereoisomer is more stable because it has the conformation with both groups equatorial.

Review of Mastery Goals

After completing this chapter, you should be able to:

Recognize compounds that exist as cis-trans isomers and estimate the relative stabilities of these isomers.
(Problems 6.1, 6.2, 6.15, and 6.31)

Use the *Z* and *E* descriptors to designate the configurations of cis-trans isomers.
(Problems 6.3, 6.4, 6.16, 6.17, 6.27, and 6.34)

Determine the conformations about a C-C single bond and estimate their relative energies.
(Problems 6.5, 6.6, 6.18, 6.19, 6.29, and 6.35)

Determine the types and relative amounts of strain present in cyclic molecules.
(Problems 6.20, 6.30, 6.32, and 6.33)

Draw the two chair conformations of cyclohexane derivatives and determine which is more stable.
(Problems 6.7, 6.9, 6.11, 6.21, 6.22, and 6.28)

Use analysis of conformations to determine the relative stabilities of stereoisomeric cyclohexane derivatives.
(Problems 6.10, 6.12, 6.13, 6.14, 6.23, 6.24, 6.25, 6.26, and 6.38)

Chapter 7
STEREOCHEMISTRY II
CHIRAL MOLECULES

7.1 Try to see if the object is identical to its mirror image or not.

a) Achiral b) Chiral c) Chiral d) Achiral e) Chiral f) Chiral

7.2 Carbon atoms in a compound that are attached to four different groups are chirality centers. Carbons that are doubly bonded or triply bonded are not chirality centers. The presence of chirality centers in a molecule does not necessarily mean that the molecule is chiral. However, any compound with a single chirality center is chiral.

a) This compound has no chirality center, so it is achiral.
b) This compound has no chirality center, so it is achiral.

c) This molecule has one chirality center, so it is chiral.

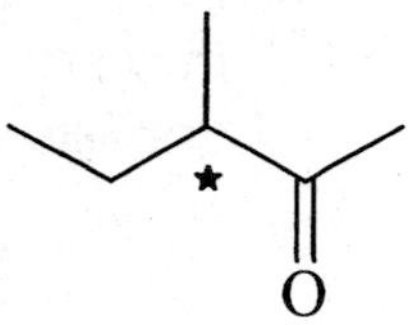

d) This molecule has one chirality center, so it is chiral.

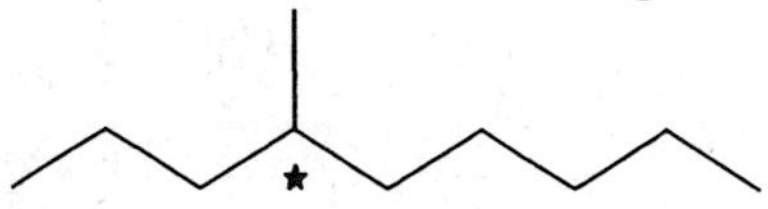

e) The carbon bonded to the chlorine is bonded to a hydrogen and to two other carbons that are part of the ring. Proceeding around the ring in one direction, the second carbon encountered is attached to two methyl groups, while in the other direction the second carbon is bonded to two hydrogens. Therefore these groups are not identical. This molecule is chiral.

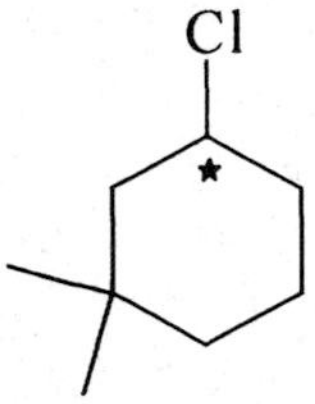

f) This molecule has no chirality center because identical substituents are encountered as one proceeds in either direction around the ring.

7.3 a) Has a plane of symmetry
b) Has a plane of symmetry
c) No plane of symmetry

d) Has a plane of symmetry passing through Cl and bisecting the bond on the opposite side of the ring connecting the two quaternary carbons
e) No plane of symmetry
f) Has a plane of symmetry passing through C=O perpendicular to plane of H_3C-C-CH_3
g) Has a plane of symmetry passing through H-C-CH_3 and bisecting the Cl-C-Cl angle
h) No plane of symmetry

7.4 First assign priorities from 1 through 4 to the four groups bonded to the chirality center using the Cahn-Ingold-Prelog sequence rules outlined in Section 6.2 in the text. The group with the highest priority gets number 1 and the lowest priority group gets number 4. View the molecule at the chirality center with group number 4 pointed directly away from you. If the priority numbers of the remaining groups cycle in a clockwise direction (1 → 2 → 3 → 1), the chirality center has the *R* configuration. If the cycle is in a counterclockwise direction then the chirality center has the *S* configuration.

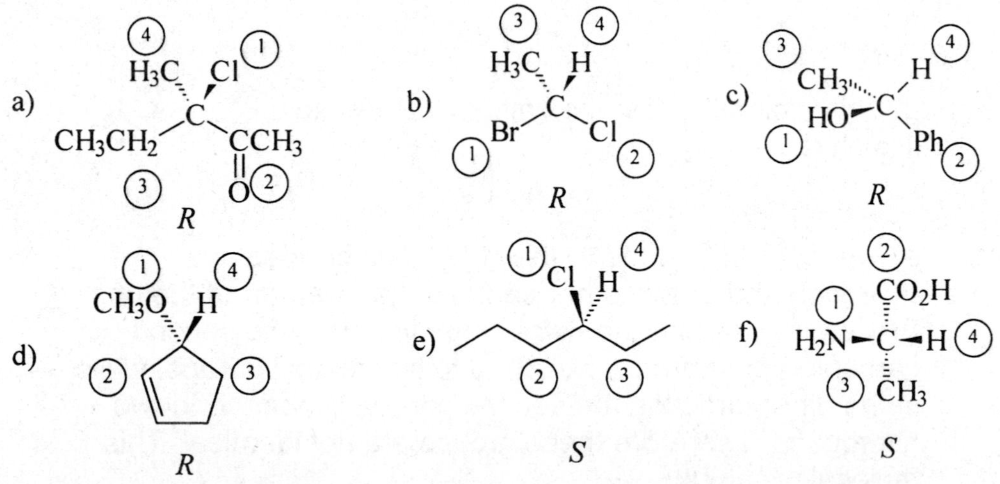

7.5 Draw the structure without stereochemistry and identify the chirality center. Assign priorities to the groups attached to the chirality center. Draw a tetrahedral carbon and place the lowest priority group on the bond pointed away from you. Place group number 1 on any one of the other three bonds. If the configuration is *R* at the chiral center, place groups 2 and 3 in a clockwise direction from group 1. If the configuration is *S*, place groups 2 and 3 in a counterclockwise direction from group 1.

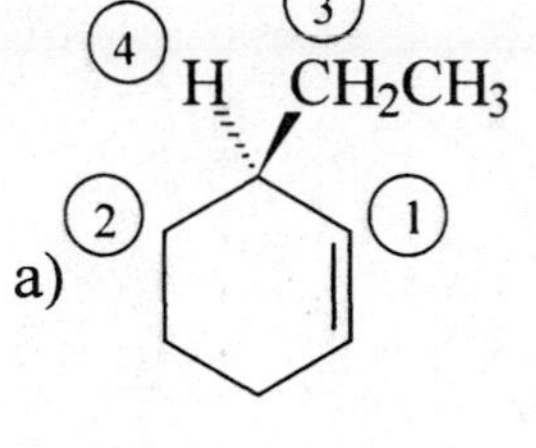

R (clockwise)

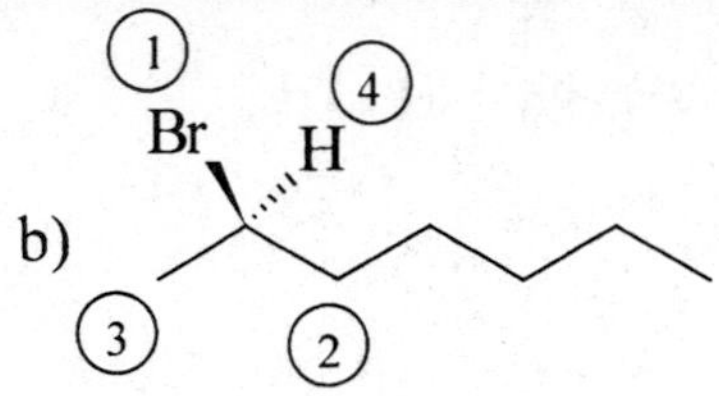

R (clockwise)

7.6 a) False. Enantiomeric molecules exhibit different properties only in a chiral environment.
b) False. Enantiomers have identical physical properties.
c) True. Water is not chiral.
d) Cannot be determined. The direction of rotation of plane polarized light by a chiral molecule has no relationship to the assigned configuration of the molecule and can only be determined by experiment.
e) True. The assignment of *d* to a chiral molecule denotes that the compound rotates plane polarized light in the clockwise or + direction.

7.7 Enantiomers are nonsuperimposable mirror images of a molecule. Diastereomers are non-mirror image stereoisomers of a molecule.

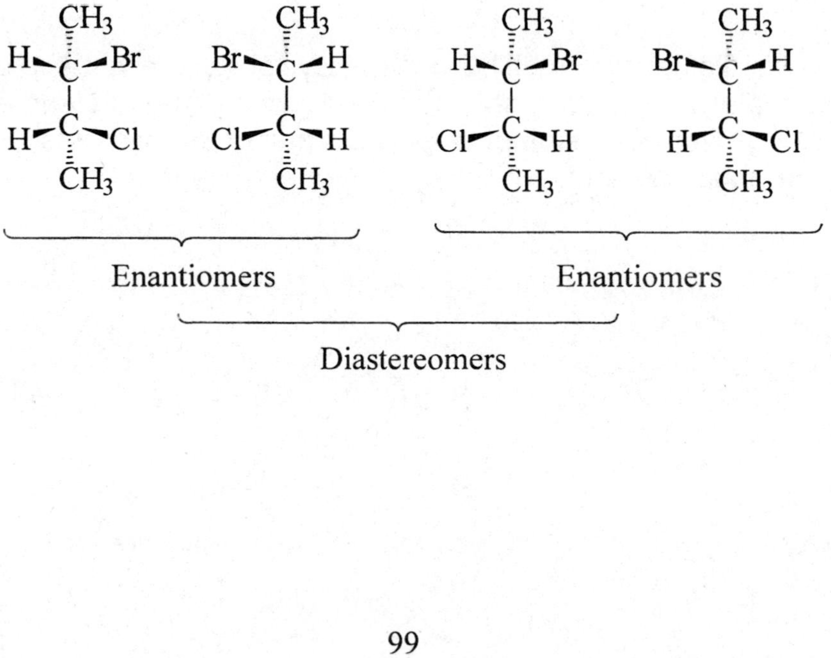

7.8

a) 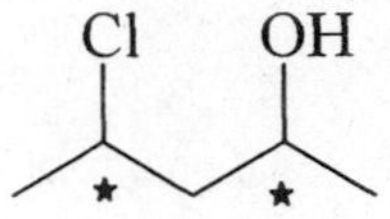 4

b) 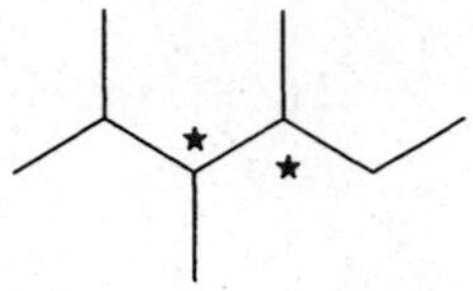 4

c) 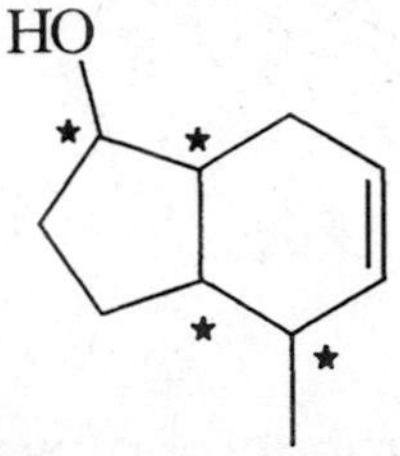 16

d) None

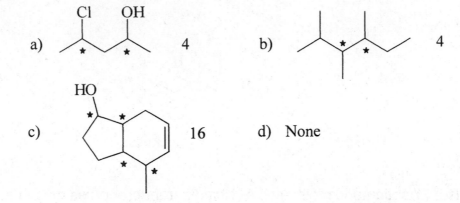

7.9 The meso stereoisomer does not rotate plane polarized light because it has a plane of symmetry. The other two stereoisomers do rotate plane polarized light.

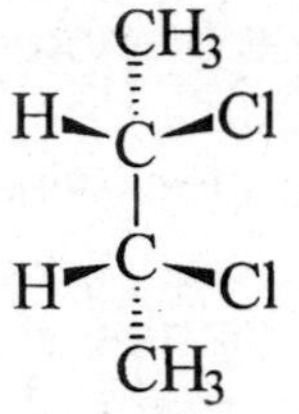

Meso

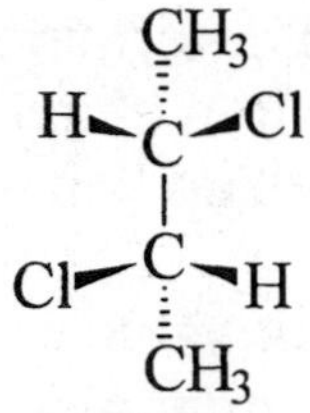

Rotates

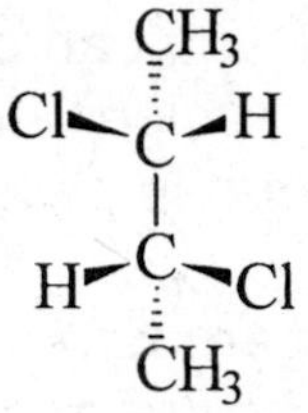

Rotates

7.10 The *cis*-diastereomer of 1,2-dimethylcyclopropane is *meso* and does not rotate plane polarized light because it has a plane of symmetry bisecting the ring. The *trans*-isomer exists as a pair of enantiomers that do rotate plane polarized light.

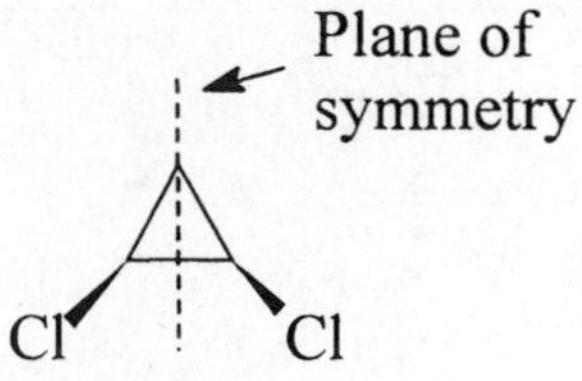

Meso

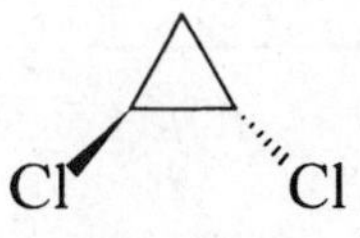

Rotates

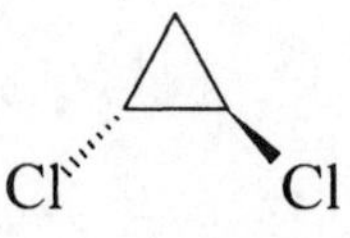

Rotates

7.11 a) Yes b) No, meso c) Yes d) No, no chirality center

7.12 To construct a Fisher projection, the molecule is first arranged with the horizontal bonds to its chirality center projecting out of the page and the vertical bonds projecting into the page. In the Fisher projection a tetrahedral carbon is represented by a cross. The horizontal line of the cross represent bonds projecting above the page and the vertical line represent the bonds projecting into the page.

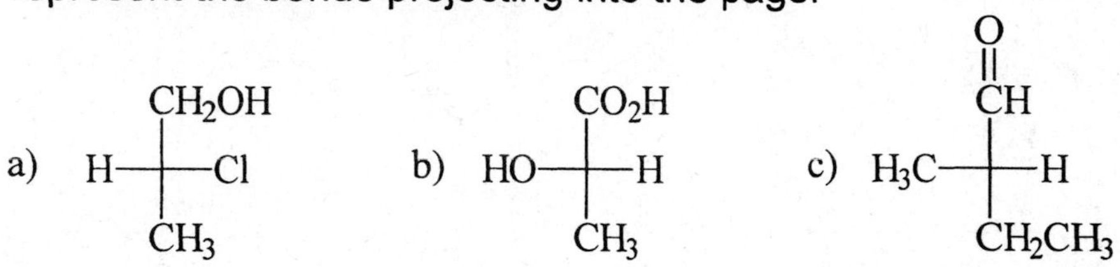

7.13

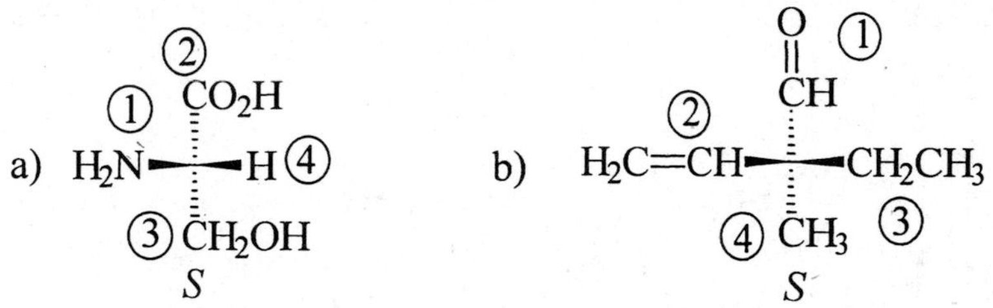

7.14 a) This compound is chiral because it is an allene with two different groups on each end.
b) This compound is not chiral. The N is bonded to two ethyl groups.
c) This allene is not chiral because two groups on one end, the H's, are the same.
d) The silicon is a chirality center because it is bonded to four different groups, so this compound is chiral.
e) This biphenyl is not chiral because one ring is symmetrically substituted with two methyl groups. The plane of the lower ring, which bisects the plane of the upper ring, is a plane of symmetry.
f) This biphenyl is chiral. The CO_2H group on the upper ring destroys the symmetry plane that is present in example (e).

7.15 The compound does not have a symmetry plane, so it is chiral. However, because it has only hydrogens in the ortho positions, the two enantiomers interconvert rapidly by rotation about the bond connecting the rings, and the compound cannot be resolved.

7.16 a) *R* b) *R* c) *R* d) *R* e) *S*
f) *S* g) *R* h) *S* i) *S*

7.17

a) 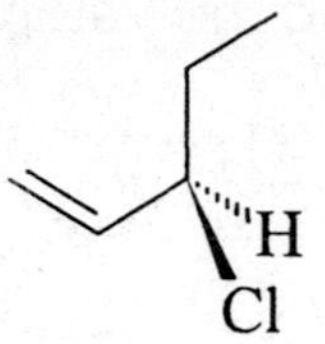

b) 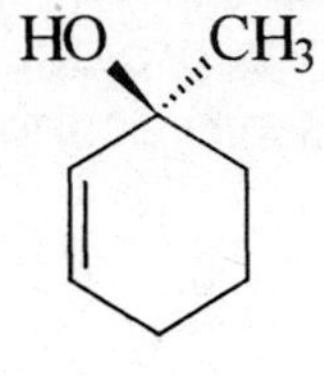

7.18 a) 8 b) 3 (one is meso) c) 8 d) none (not chiral)

7.19

a)

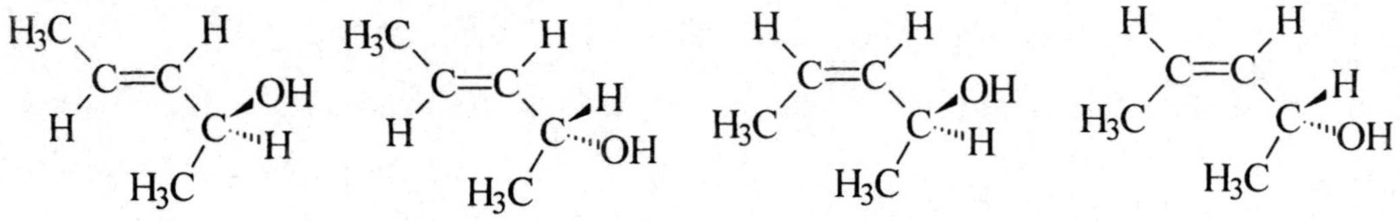

b) 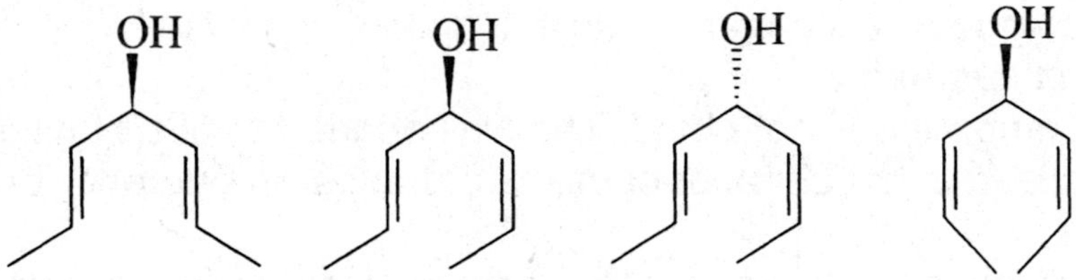

7.20 a) This compound has a plane of symmetry passing through the C between the C's bonded to the OH groups and the C on the opposite side of the ring, so it is a meso compound and it will not rotate plane polarized light.
b) This is a chiral compound, so it will rotate plane polarized light.
c) This compound does not have a chirality center, so it will not rotate plane polarized light.
d) This is a chiral compound, so it will rotate plane polarized light.
e) This has two planes of symmetry (horizontal and vertical), so it will not rotate plane polarized light.

f) This has one plane of symmetry (horizontal), so it will not rotate plane polarized light.
g) This has no plane of symmetry, so it will rotate plane polarized light.
h) This has a plane of symmetry passing horizontally through the center of the molecule. It is a meso compound and it will not rotate plane polarized light
i) This does not have a plane of symmetry, so it will rotate plane polarized light.

7.21 Enantiomers exhibit different properties only when they are in a chiral environment.
a) True b) True c) True (Water is achiral.)
d) True (They rotate plane polarized light the same magnitude but in opposite directions.)
e) False f) True g) True (Methanol is achiral.)
h) True i) False [(*S*)-2-Butanol is chiral.]
j) Cannot be determined (There is no relationship between the direction of rotation of plane polarized light and the absolute configuration.)
k) False (The reagents are achiral, so a 50:50 mixture of enantiomers must be produced.)

7.22 a) Identical b) Identical c) Identical
d) Enantiomers e) Diastereomers f) Diastereomers
g) Enantiomers h) Diastereomers i) Identical

7.23 a) Diastereomers b) Diastereomers c) Enantiomers
d) Enantiomers e) Identical f) Enantiomers
g) Enantiomers

7.24 a) +66.5 b) +0.7°
c) The observed rotation is directly proportional to the concentration. For example, if the concentration of the solution is halved, the observed rotation will be one half that of the rotation observed for the concentrated solution. The results can be distinguished for each case: one half of +160° is + 80°; one half of -200° is -100°; and one half of +520° is + 260°.

7.25 The resolution of this amine using one enantiomer of a chiral carboxylic acid is similar to the scheme described in Section 7.7 of the text.

The mixture of the *R* and *S* enantiomers of the amine is reacted with the *S* enantiomer of 2-chloropropanoic acid to produce a mixture of diastereomers of the salt.

Since diastereomers have different physical properties, the mixture can be separated by conventional separation techniques.

Each of the pure diastereomers of the salt is then treated with a strong base to regenerate the pure enantiomers of the amine.

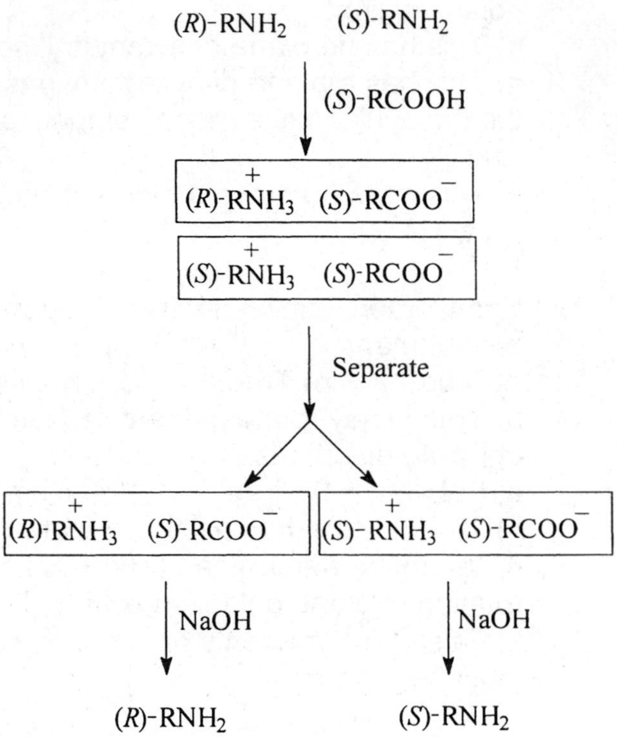

7.26 a) The DU of **X** = 1. Therefore **X** has one ring or one pi bond.
b) Unknown **X** must have a double bond because it reacts with H_2 to form a saturated compound, **Y**, with DU = 0.
c) Compound **X** has a chirality center because it rotates plane polarized light. Compound **Y** is achiral, so hydrogenation of the double bond destroys the chirality center by converting the double bond group into an alkyl group that is the same as another alkyl group on the chirality center.

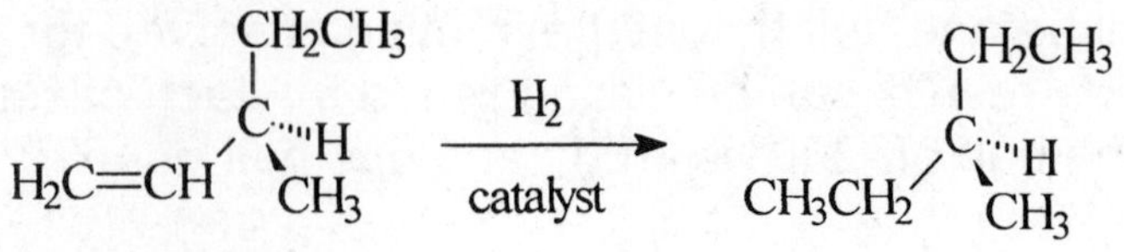

X (or enantiomer) **Y**

7.27 There are four stereoisomers of this compound.

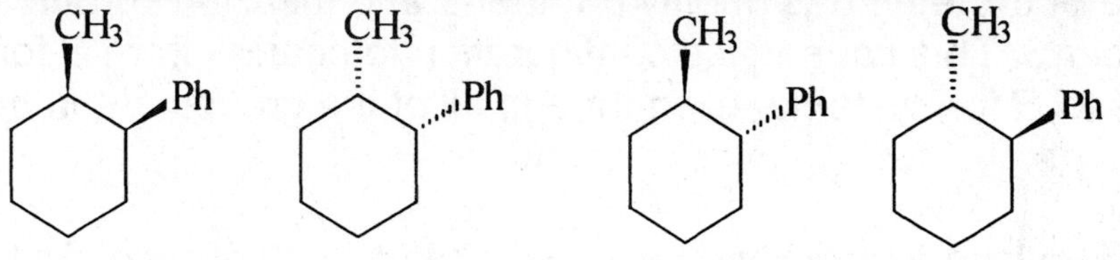

Two enantiomers of the *cis*-diastereomer

Two enantiomers of the *trans*-diastereomer

Both enantiomers of the *cis*-diastereomer are of equal stability as are both enantiomers of the *trans*-diastereomer. The *trans*-diastereomer is more stable than the *cis*-diastereomer because it has the conformation with both substituents equatorial. The methyl group is axial in the more stable conformer of the less stable diastereomer (cis) because the axial strain energy of the phenyl group is larger than that of the methyl group.

7.28

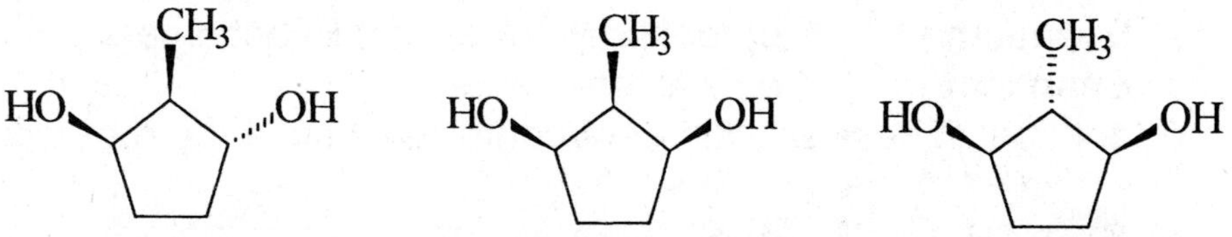

Chiral | Not chiral | Not chiral

7.29 a) This tripeptide has four chirality centers and 16 stereoisomers.
b) Sucrose has nine chirality centers and 512 stereoisomers.
c) Pancratistatin has six chirality centers and 64 stereoisomers.
d) Testosterone has six chirality centers and 64 stereoisomers.
e) This prostaglandin has four chirality centers and 16 stereoisomers (without any isomers involving the double bonds).
f) Vitamin E has three chirality centers and 8 stereoisomers.
g) Vitamin D_2 has six chirality centers and 64 stereoisomers (without any isomers involving the double bonds).
h) Vitamin C has two chirality centers and 4 stereoisomers.
I) Apoptolidin has 25 chirality centers and 33,554,432 stereoisomers (without any isomers involving the double bonds).

7.30 In all of the amino acids, except cysteine, the CO_2H group has priority number 2. Because S has a higher atomic number than O, the CH_2SH group of cysteine has priority number 2 and the CO_2H group has priority number 3. This causes the configuration to change from *S* for most amino acids to *R* for cysteine, even though all of the groups are in the same positions.

7.31 a) This allene is not chiral because it has two identical groups on one end.
b) This allene is chiral because the groups bonded to each end are different.
c) This biphenyl is not chiral because the two methyl groups on one ring are identical. It has a plane of symmetry that bisects the ring substituted with the methyl groups.
d) This biphenyl is chiral because it does not have a plane of symmetry.
e) The N is bonded to three identical groups, so this ammonium salt is not chiral.
f) The S is pyramidal and bonded to three different groups, so this compound is chiral.

7.32 a) Enantiomers b) Identical c) Enantiomers
d) Enantiomers e) Diastereomers
f) Identical (These are different conformations of the meso diastereomer.)

7.33 a) *S* b) *R* c) *R* d) *S*

7.34 a) This molecule is not chiral because it has a plane of symmetry. It is meso.
b) This molecule is chiral because it does not have a plane of symmetry.
c) This molecule is chiral because it does not have a plane of symmetry after rotation so the methyl groups are aligned.
d) This molecule is not chiral because it has a plane of symmetry passing through the CH_2 between the two OH groups. It is meso.
e) This molecule is chiral because it does not have a plane of symmetry.

7.35 The model and the Fischer projection represent enantiomers.

7.36 Estradiol has five chirality centers and 32 stereoisomers.

Review of Mastery Goals

After completing this chapter you should be able to:

Identify chiral compounds, locate chirality centers, and determine how many stereoisomers exist for a particular compound.
(Problems 7.2, 7.7, 7.8, 7.18, 7.19, 7.22, 7.27, 7.29, 7.32, 7.34, and 7.36)

Locate any symmetry planes that are present in a molecule.
(Problems 7.3 and 7.20)

Designate the configuration of chirality centers as *R* or *S*.
(Problems 7.4, 7.5, 7.13, 7.16, 7.17, 7.30, and 7.33)

Recognize the circumstances under which enantiomers have different properties.
(Problems 7.6 and 7.21)

Understand when a compound, mixture or solution is optically active and what information this provides about the sample.
(Problems 7.20, 7.21, and 7.24)

Recognize *meso*-stereoisomers.
(Problems 7.9, 7.10, 7.11, 7.20, 7.28, 7.32, and 7.34)

Be able to use Fischer projections properly.
(Problems 7.12, 7.23, and 7.35)

Understand the principles behind the process of separating enantiomers.
(Problem 7.25)

Identify other chiral molecules, such as biphenyls and allenes.
(Problems 7.14, 7.15, and 7.31)

Chapter 8
NUCLEOPHILIC SUBSTITUTION REACTIONS
REACTIONS OF ALKYL HALIDES, ALCOHOLS, AND RELATED COMPOUNDS

8.1 S_N2 reactions always occur with inversion of configuration at the reaction center. The nucleophile approaches the carbon from the side opposite the leaving group.

a) The nucleophile (OH^-) replaces the leaving group (Cl) with inversion of configuration. One way to show the product with configuration inverted from that of the reactant is to replace the Cl with OH and interchange the position of the OH with one of the other groups. (Interchanging two groups at a chirality center inverts the configuration at that center.)

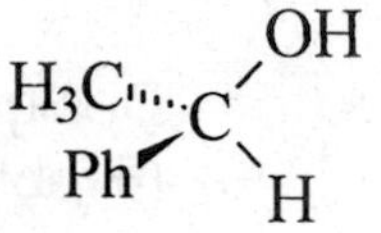

b) The nucleophile is the oxygen of CH_3O^-. The position of the nucleophile and the H are opposite the position of the Cl and the H in the reactant because the reaction has occurred with inversion.

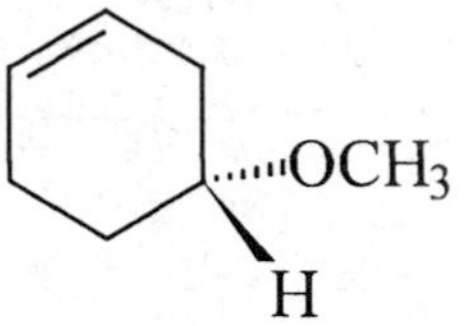

c) In this case the product has the same relative stereochemistry as the reactant because none of the bonds to the chirality centers are broken in the reaction. The S_N2 reaction occurs with inversion of configuration at the carbon bonded to the leaving group. Configuration at other chirality centers is unchanged.

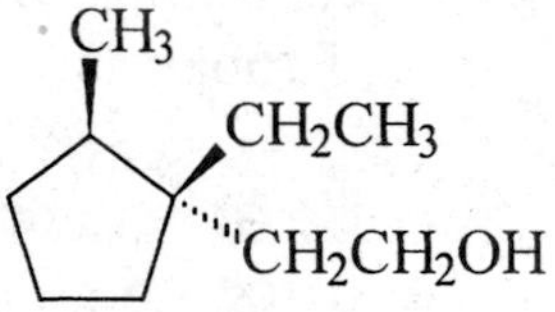

8.2 The rate of the S_N2 reaction is controlled by steric factors at the electrophilic carbon. The approach of the nucleophile is greatly slowed by bulky groups near the electrophilic carbon.

a) The compound on the right reacts faster because it has its leaving group on a secondary carbon and has less steric hindrance.

b) The compound on the left has two bulky methyl groups on the carbon adjacent to the electrophilic carbon. Therefore the compound on the right reacts faster because it has less steric hindrance.

c) A phenyl substituent on the reaction center provides added resonance stabilization to the transition state. Therefore the compound on the left reacts faster.

8.3 In general the rate of an S_N2 reaction depends on steric effects and follows the trend: methyl > primary > secondary >> tertiary.

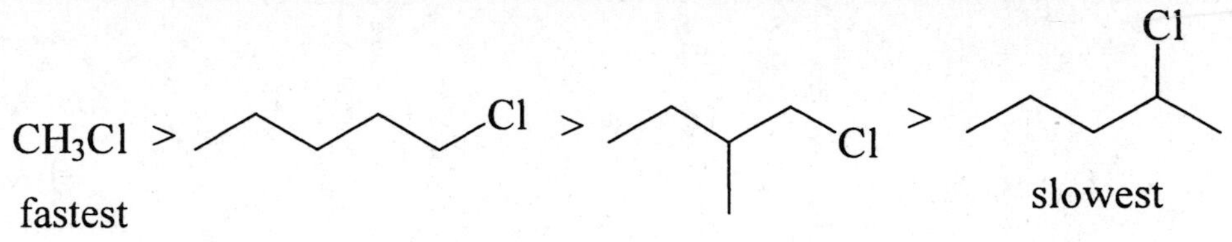

8.4

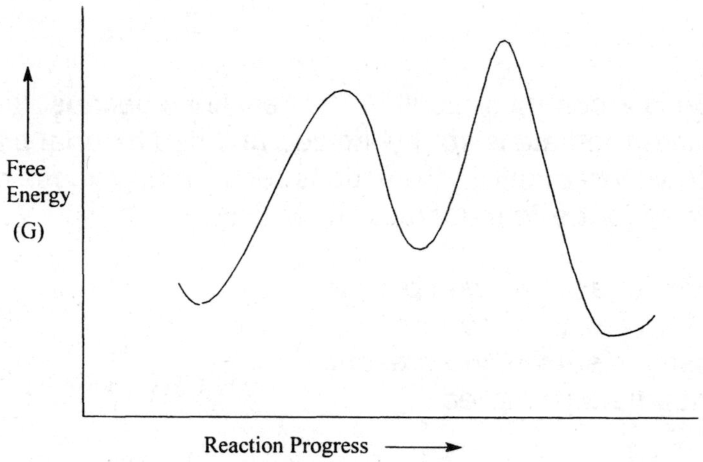

8.5 The transition state for the S_N2 reaction shown in Figure 8.1 is much higher in energy than either the reactant or the product. Because its energy is not significantly closer to the reactant or product, its structure is approximately half way between the reactant and the product, so the bond is approximately half broken.

8.6 For the S_N1 reaction, formation of the carbocation is the rate limiting step. The transition state resembles the carbocation intermediate. Any effect that helps to stabilize the carbocation will also stabilize the transition state and will increase the rate.
a) The compound on the left reacts faster because it will produce a more stable tertiary carbocation.
b) The compound on the left has a faster rate because the carbocation intermediate will be resonance stabilized.
c) The compound on the left has a faster rate because the carbocation intermediate will be resonance stabilized.
d) The compound on the right has a faster rate because the methoxy group provides extra resonance stabilization of the carbocation.

8.7

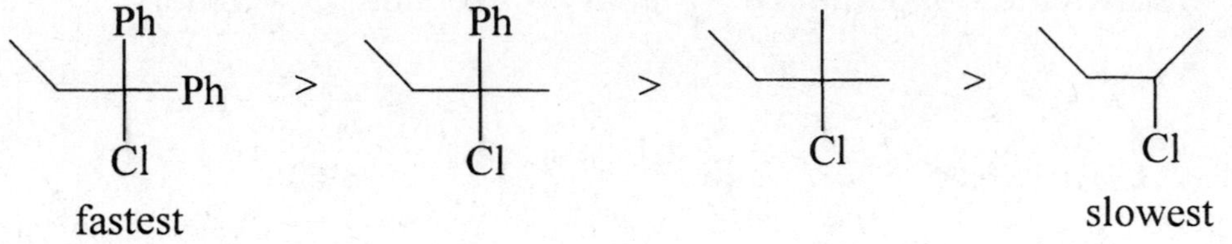

8.8 Racemization is a common result in S_N1 reactions because the carbocation intermediate is *sp*^{2} hybridized and has trigonal planar geometry. However, many S_N1 reactions result in an excess of the product of inversion due to the formation of an ion-pair.

a) The product is racemic, with perhaps some excess inversion. When no stereochemistry is shown, the presence of both enantiomers is implied.

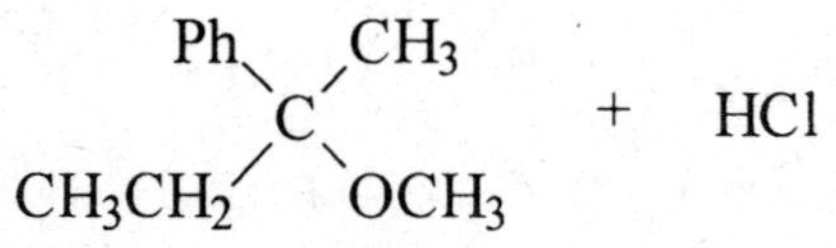

b) The reaction does not occur at the chirality center, so its stereochemistry is unchanged.

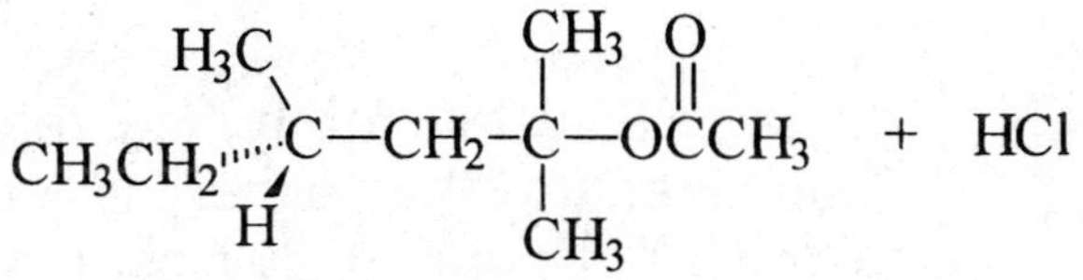

8.9 a) Primary substrates with a strong nucleophile (hydroxide ion) react by the S_N2 mechanism. The right reaction is faster because mesylate ion is a better leaving group than chloride ion.
b) Tertiary substrates react by the S_N1 mechanism. The right reaction is faster because iodide ion is a better leaving group than bromide ion.

8.10 a)

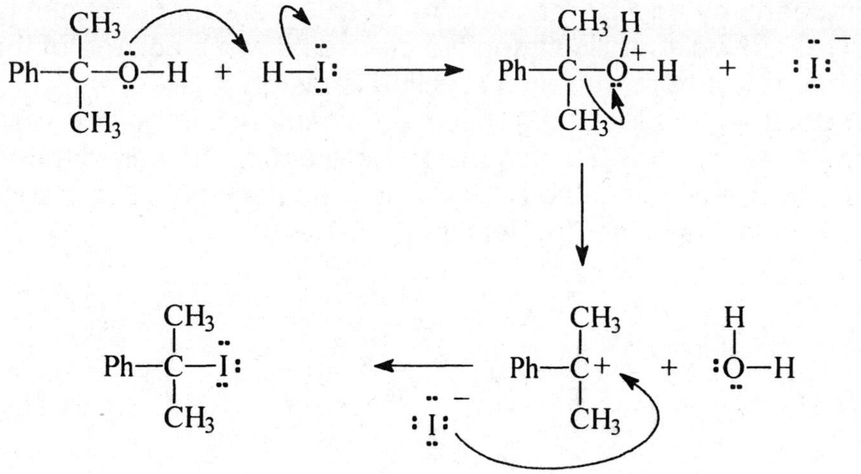

b)

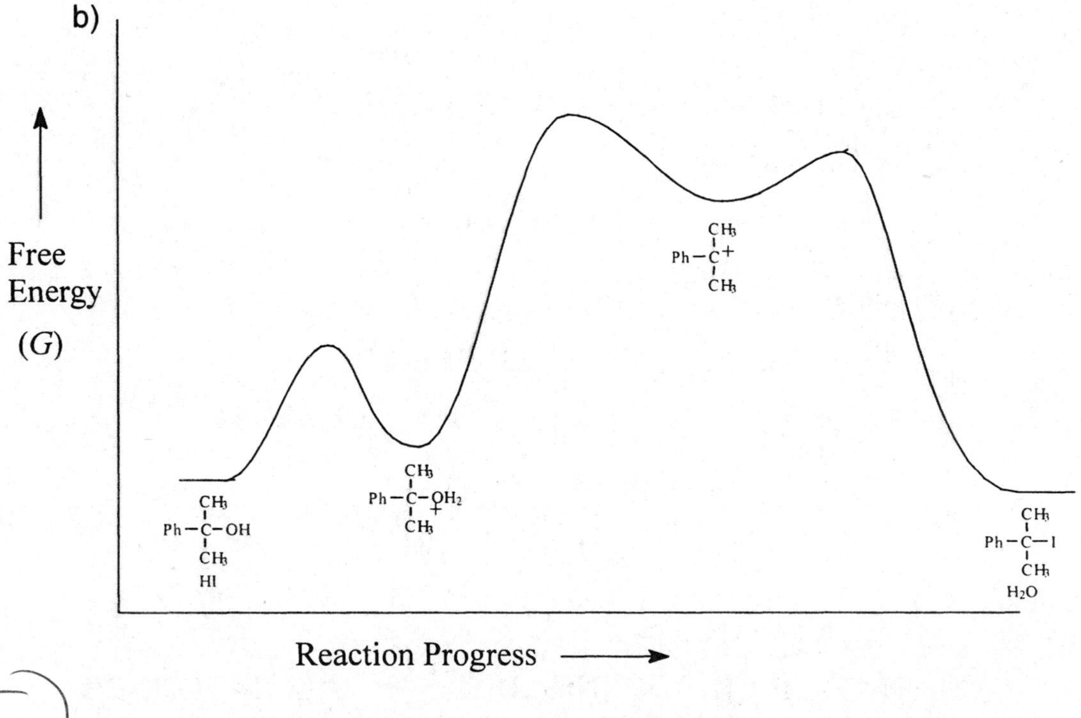

8.11 a) Because the leaving group is on a tertiary carbon, the reaction proceeds by an S_N1 mechanism. The rate determining step is the formation of the carbocation intermediate, so the reactivity of the nucleophile does not affect the rate of an S_N1 reaction. Therefore both reactions proceed at the same rate.

b) Because the leaving group is on a primary carbon, the reaction proceeds by an S_N2 mechanism. CH_3S^- is a stronger nucleophile than CH_3O^-. (Nucleophilic strength increases down a column of the periodic table.) Therefore, the right reaction is faster.

c) Because the leaving group is on a primary carbon, the reaction proceeds by an S_N2 mechanism. Here again, the only difference between the two reactions is the nucleophile. The nucleophile in the left reaction is a stronger base, so that reaction is faster.

8.12

a) $CH_3CH_2CHCH_3$ (:Cl:) $\quad+\quad$:S̈H $^-\quad\xrightarrow{S_N2}\quad CH_3CH_2CHCH_3$ (:S̈H) $\quad+\quad$:C̈l: $^-$

b)

$$CH_3CH_2\underset{:\ddot{B}r:}{\overset{CH_3}{C}}CH_3 \quad\xrightarrow{S_N1}\quad CH_3CH_2\overset{CH_3}{\underset{+}{C}}CH_3 \quad+\quad :\ddot{B}r:^-$$

$CH_3CH_2\ddot{O}H$

$$CH_3CH_2\overset{CH_3}{C}CH_3 \quad:\ddot{O}-CH_2CH_3$$

$$CH_3CH_2\overset{CH_3}{C}CH_3 \quad H-\overset{+}{\underset{}{O}}-CH_2CH_3$$

$+\ \ H-\ddot{B}r:$

:B̈r: $^-$

c) $CH_3CH_2CH_2CH_2CH_2$ (:ÖTs) $\quad+\quad$:N̈H$_2$CH$_3$ $\quad\xrightarrow{S_N2}\quad CH_3CH_2CH_2CH_2CH_2$ $\quad$ TsÖ: $^-$ $\quad\overset{+}{N}H_2CH_3$

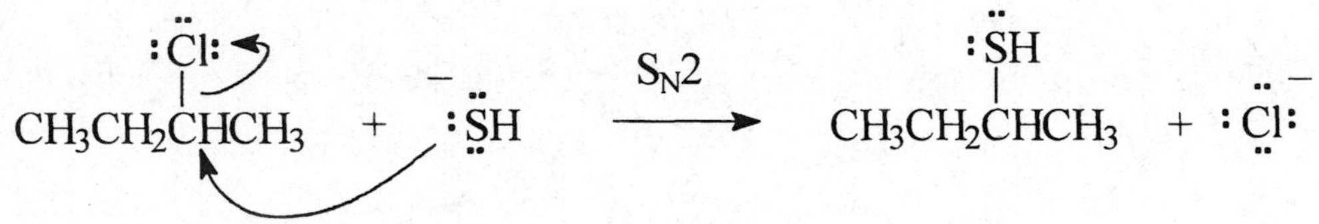

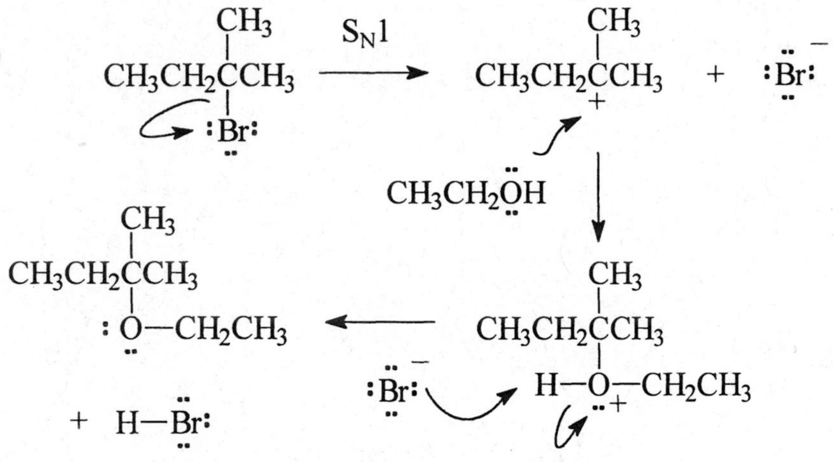

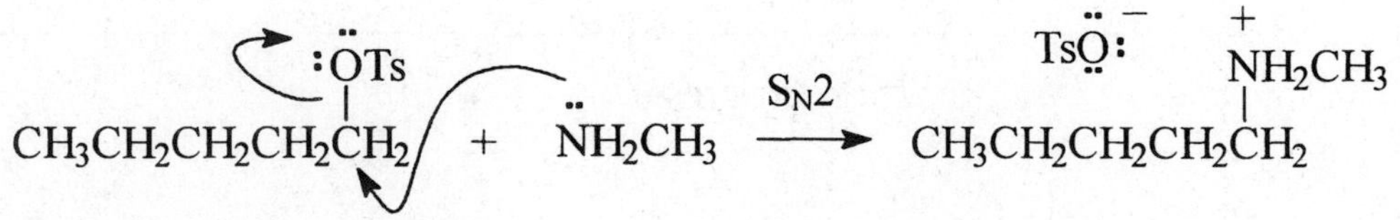

d) 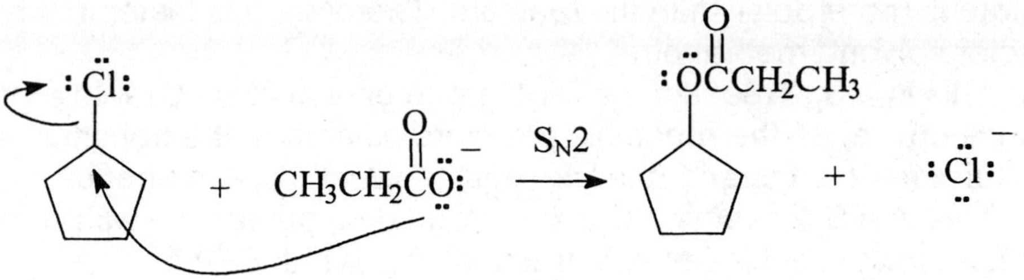

8.13 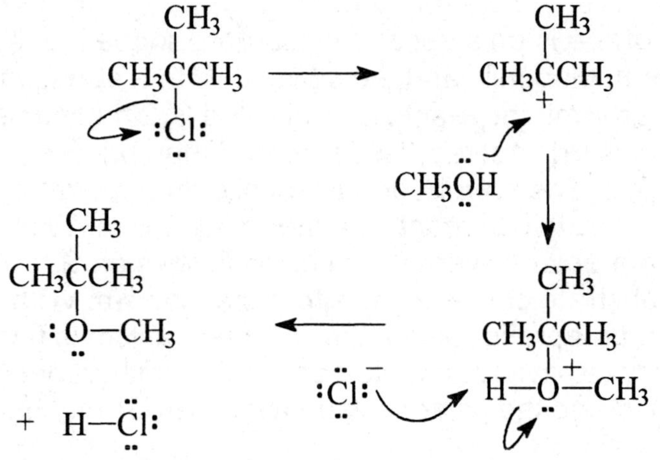

8.14 The conditions of Figure 8.8 have a more polar solvent (water v. acetic acid) and a weaker nucleophile (water v. acetate anion) than the conditions of Figure 8.9 so the carbocation has a longer lifetime under the conditions of Figure 8.8.

8.15 Solvent polarity dramatically affects the reaction rate. To predict the effect of solvents on reaction rates, the polarity of the reactants is compared with that of the transition state. If the transition state is more polar than the reactants, the transition state will be more stabilized than the reactants in a polar solvent. This will decrease $\Delta G^{\ddagger}$, resulting in a faster reaction. On the other hand, if the reactants are more polar than the transition state, increasing solvent polarity will stabilize reactants more, resulting in an increase in $\Delta G^{\ddagger}$ and a decrease in the reaction rate. In general, rates of S_N1 reactions are faster in polar solvents while those of S_N2 reactions involving negative nucleophiles are faster in aprotic solvents.

a) This is a S_N1 reaction (leaving group on a tertiary C), so the transition state is more polar than the reactant. Therefore it is faster in the more polar solvent, methanol.
b) This is a S_N2 reaction (leaving group on a primary C) with a negative nucleophile, so the reactants are more polar than the transition state. Therefore it is faster in the less polar solvent, pure methanol.
c) This is a S_N2 reaction (leaving group on a primary C) with a negative nucleophile, so it is faster in the aprotic solvent, DMSO.

8.16 a) The substrate is a tertiary alkyl halide, so the reaction follows the S_N1 mechanism.
b) The leaving group is on a secondary carbon, so the mechanism depends on the nucleophile and the solvent. With a strong nucleophile and an aprotic solvent, the reaction follows the S_N2 mechanism.
c) The substrate is unhindered (less hindered than primary), so the mechanism is S_N2. The solvent or nucleophile do not matter.
d) The allylic substrate can react by either mechanism. With a weak nucleophile and a polar solvent the reaction follows an S_N1 mechanism.
e) The allylic substrate can react by either mechanism. With a strong nucleophile and an aprotic solvent the S_N2 mechanism is favored.
f) The secondary benzylic substrate can react by either mechanism. The weak nucleophile and the polar solvent favor the S_N1 mechanism.

8.17 The first step of both of these reactions is an acid-base reaction. The resulting conjugate bases undergo intramolecular S_N2 reactions to form cyclic products.

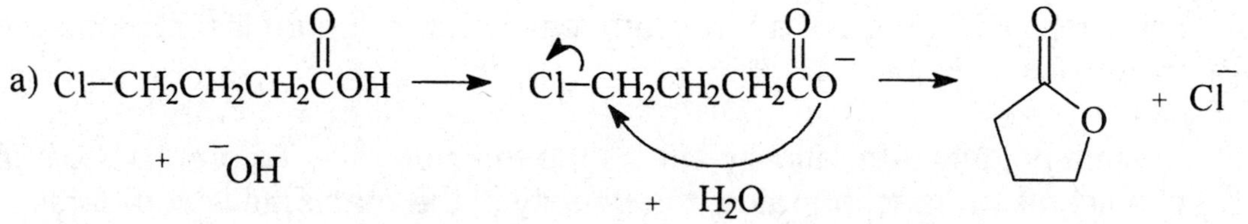

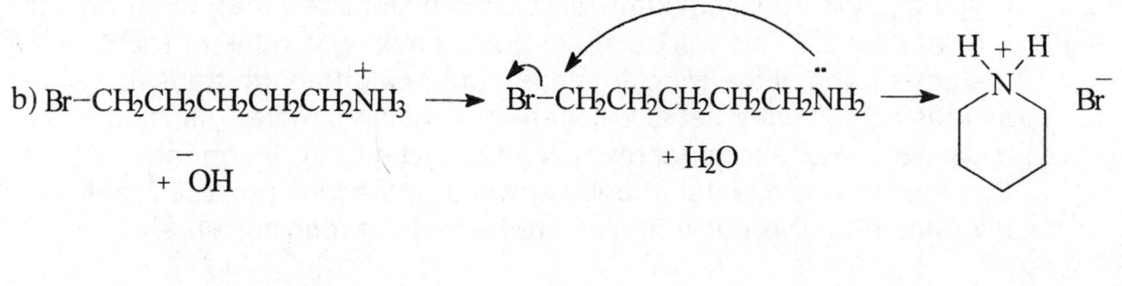

8.18 Elimination reactions compete with substitution reactions. The competition occurs because the nucleophile is also a base. When it reacts as a base, it removes a proton from the carbon adjacent to the leaving group, resulting in the formation of the elimination product.

a) 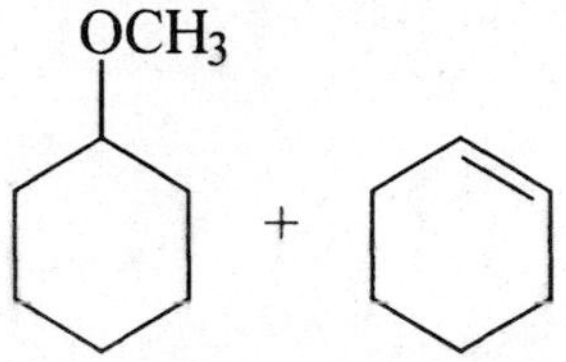

b) 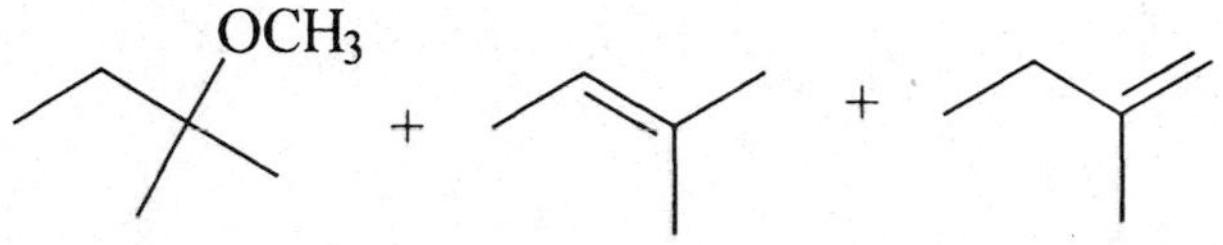

c) 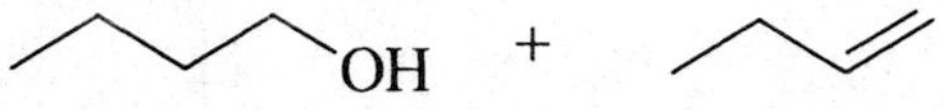

8.19 A carbocation may rearrange to form a more stable carbocation. Such rearrangement occurs by migration of a hydrogen or an alkyl group, with its bonding pair of electrons, from an adjacent carbon to the positively charged carbon.

a) 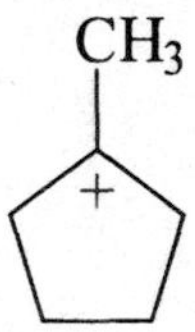

b) $CH_3\underset{+}{\overset{CH_3}{C}}CH_2CH_2CH_3$

c) 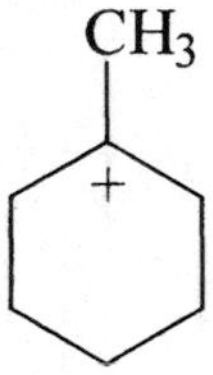

8.20 a) The carbocation formed in this reaction is stabilized by resonance. The nucleophile can bond to either of the carbons that have the positive charges in the two resonance structures.

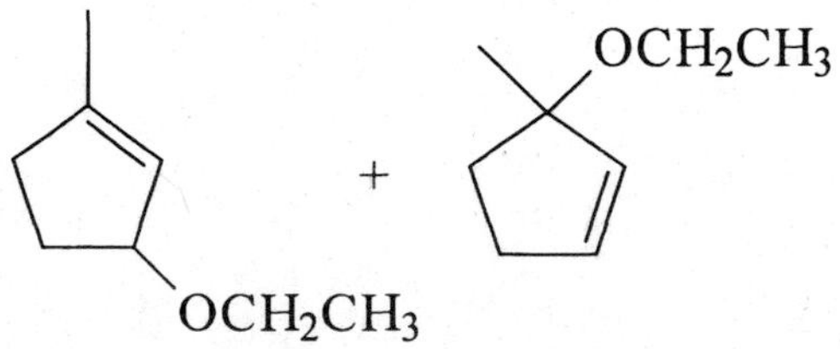

b) Migration of a methyl group converts the initially formed secondary carbocation to a much more stable tertiary carbocation. Products from both of these carbocations may be formed.

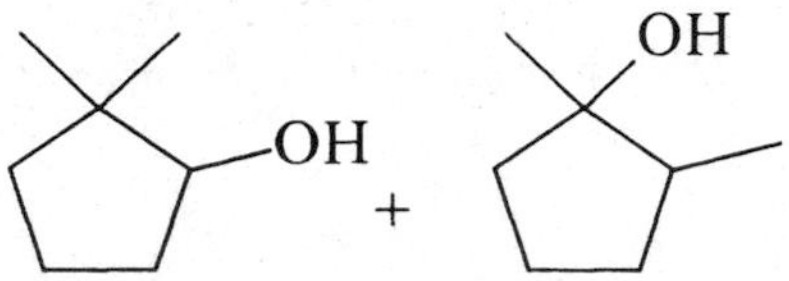

8.21

a)

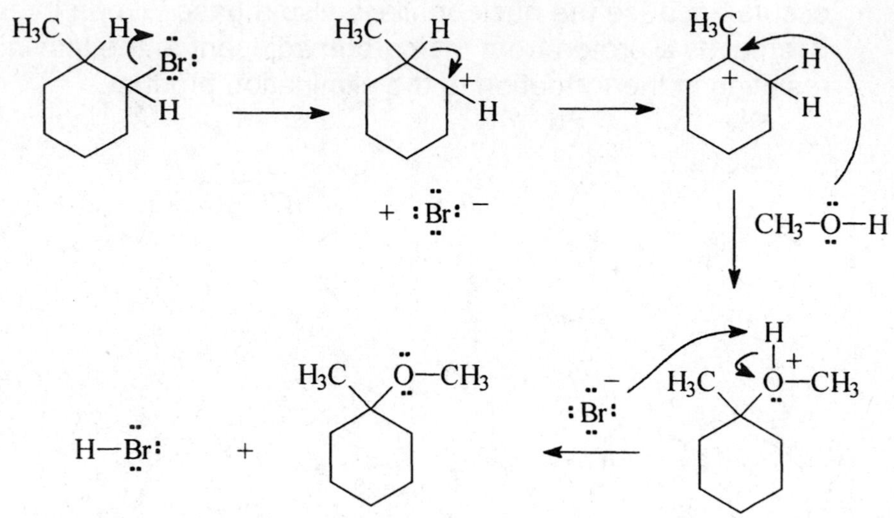
H₃C
H
Br
H
H₃C
H
+
H
+ Br −
H₃C
H
+
H
CH₃–O–H
H
+
H₃C
O–CH₃
Br −
H₃C
O–CH₃
H–Br
+

b)

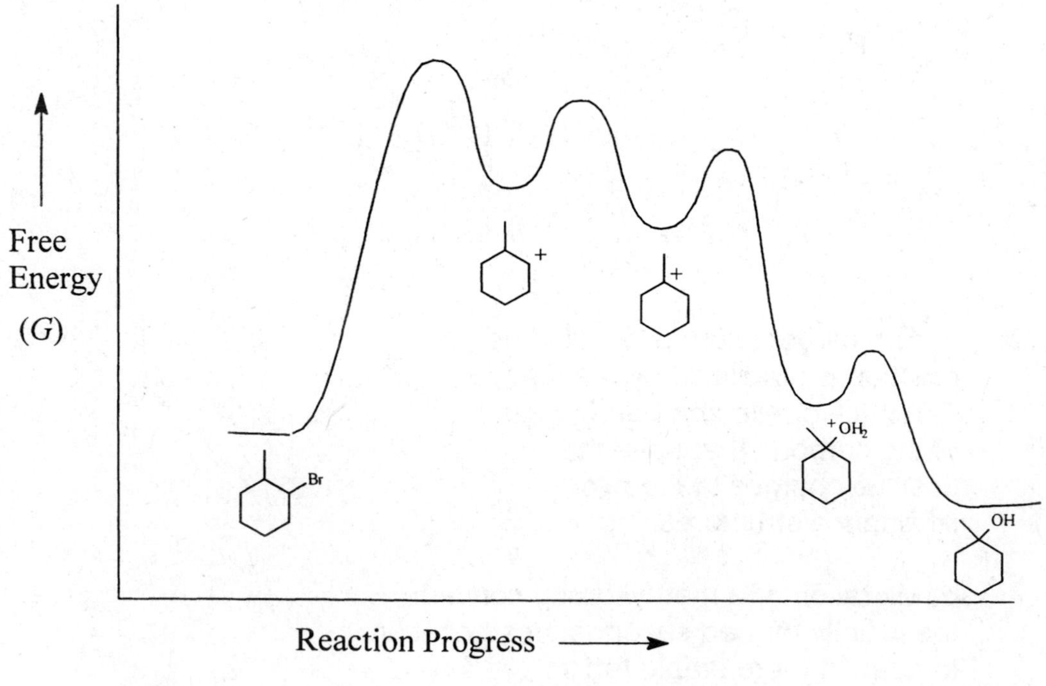
Free
Energy
(G)
Br
+
+
+OH₂
OH
Reaction Progress

8.22 a) The right compound has a faster rate because iodide ion is a better leaving group than bromide ion.
b) The left compound has a faster rate because the transition state is resonance stabilized.
c) The left compound has a faster rate because the leaving group is on a less hindered primary carbon.
d) The right compound has a faster rate because the leaving group is on a primary carbon.
e) The left compound has a faster rate because the transition state is stabilized by resonance.

8.23 a) The right compound reacts faster because it will produce a more stable tertiary carbocation.
b) The right compound reacts faster because iodide ion is a better leaving group than chloride ion.
c) The left compound reacts faster because the carbocation intermediate has greater resonance stabilization.
d) The left compound reacts faster because the carbocation intermediate is resonance stabilized.
e) The carbonyl group adjacent to the electrophilic carbon will destabilize the carbocation intermediate by its electron withdrawing inductive effect, so the right compound reacts faster.

8.24

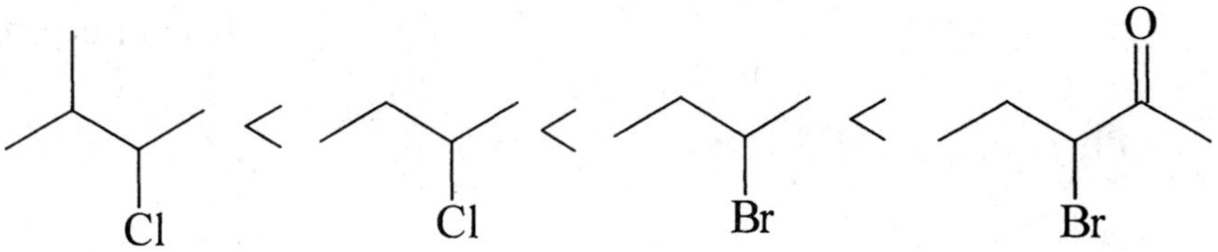

8.25

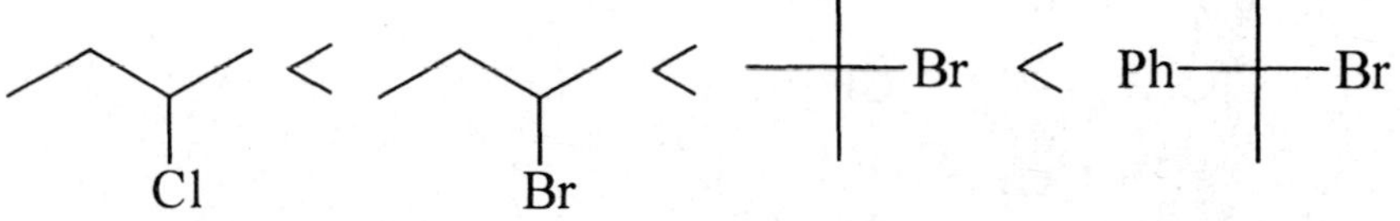

8.26 a) The substrate is a secondary alkyl halide so the mechanism depends on the nucleophile and the solvent. With a strong nucleophile and an aprotic solvent, the reaction follows the S_N2 mechanism.

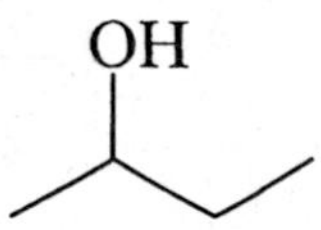

b) The substrate is a secondary alkyl halide so the mechanism depends on the nucleophile and the solvent. A weak nucleophile and a protic solvent favor the S_N1 mechanism.

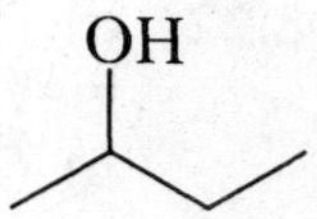

c) The substrate is a tertiary alkyl halide, so the reaction follows an S_N1 mechanism even though the nucleophile is very strong.

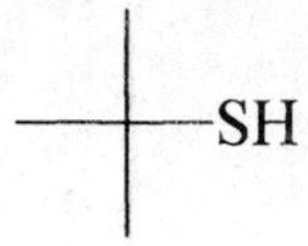

d) The leaving group is on a primary carbon, so S_N2 is the preferred mechanism regardless of the solvent or the nucleophile.

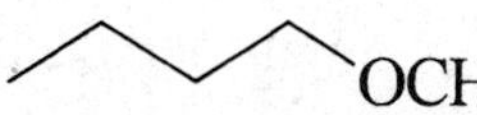

e) Like a secondary halide, this allylic halide can react by either mechanism. With a weak nucleophile and a polar solvent the reaction follows an S_N1 mechanism.

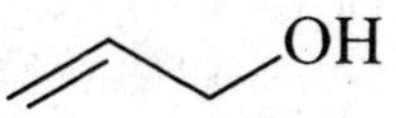

f) The reaction follows an S_N2 mechanism because of the strong nucleophile in an aprotic solvent.

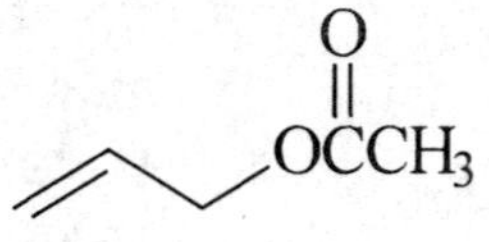

8.27

a) 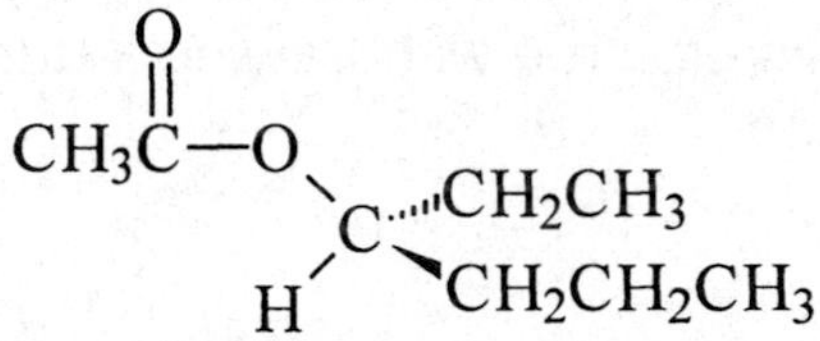

b) 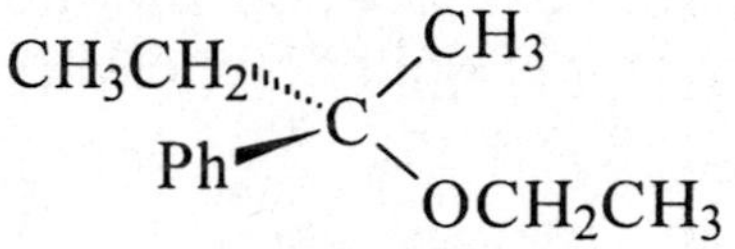

plus enantiomer

c) 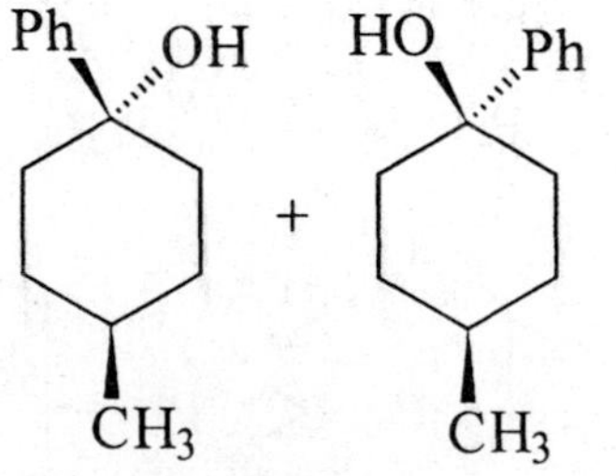

d) 

e) 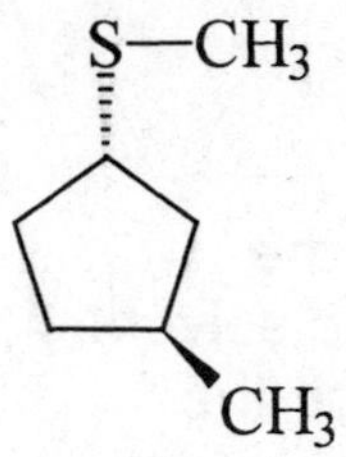

f) 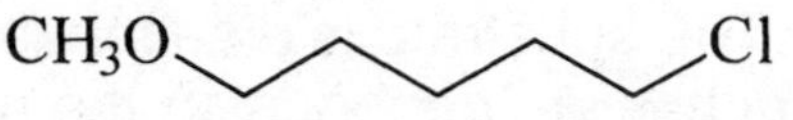

g)

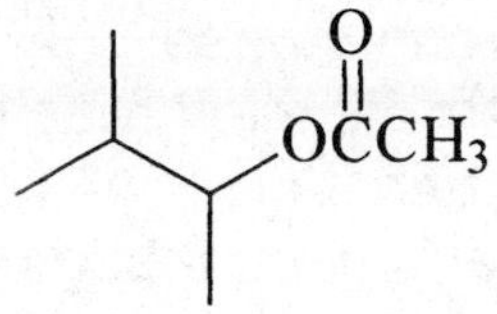

h) $CH_3\overset{\displaystyle CH_3}{\underset{\displaystyle CH_3}{C}}\!\!-\!\!\overset{\displaystyle OCH_3}{CH}CH_2CH_3$ + $CH_3\overset{\displaystyle CH_3}{C}\!\!-\!\!\overset{\displaystyle CH_3}{\underset{\displaystyle OCH_3}{CH}}CH_2CH_3$

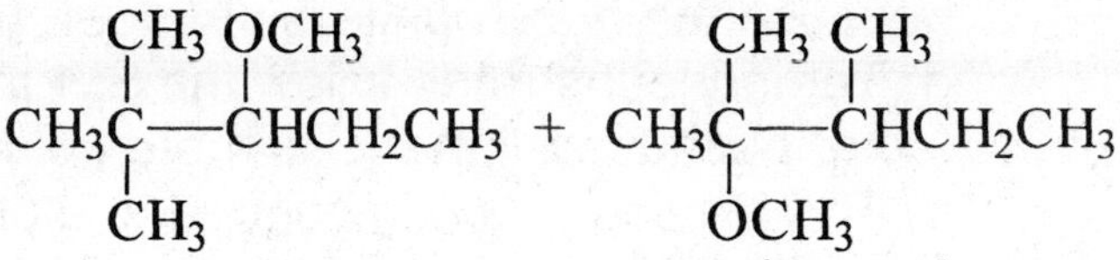

major product

i)

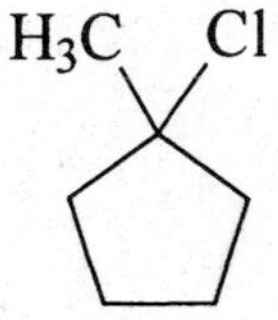

j)

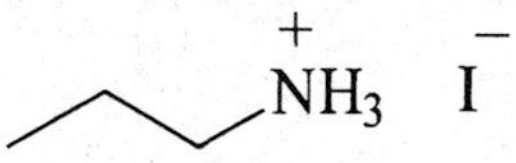

k)

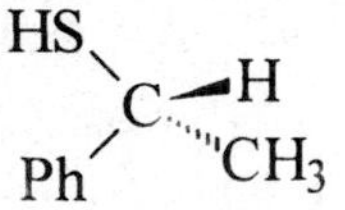

l) first step   second step

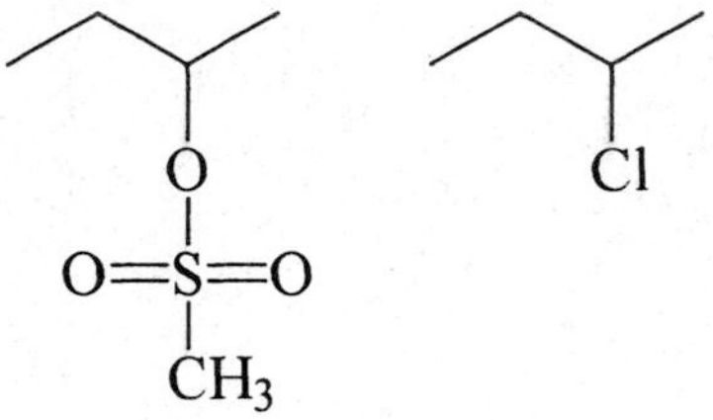

m) $CH_3CH_2CH_2CH_2I$

8.28 a) The substrates are tertiary, so they follow the S_N1 mechanism. The right reaction is faster because bromide ion is a better leaving group than chloride ion.

b) The substrates are primary, so they follow the S_N2 mechanism. The left reaction is faster because the nucleophile is more reactive in an aprotic solvent.

c) The leaving group is on a secondary carbon and the benzene ring can provide resonance stabilization of the carbocation. With a weak nucleophile and a polar solvent, the S_N1 mechanism is favored. The right reaction is faster because the methyl group provides added stability to the resonance stabilized cation.

d) Primary substrates follow the S_N2 mechanism. The right reaction is faster because it has less steric hindrance at the electrophilic carbon.

e) Secondary substrates with a weak nucleophile and a polar solvent favor the S_N1 mechanism. The left reaction is faster because the carbocation is resonance stabilized.

f) Primary substrates follow the S_N2 mechanism. The left reaction is faster because the nucleophile is stronger. (It is a stronger base.)

g) Tertiary substrates follow the S_N1 mechanism. The left reaction will be faster because methanol is more polar than ethanol.

h) The leaving group is bonded to a primary carbon, so these reactions follow the S_N2 mechanism. The left reaction is faster because intramolecular reactions are favored by entropy over intermolecular reactions.

i) Primary substrates follow the S_N2 mechanism. The left reaction is faster because the nucleophile is less hindered.

8.29 a) S_N1

$$CH_3-\underset{\underset{CH_3}{|}}{\overset{\overset{CH_3}{|}}{C}}-Br \longrightarrow CH_3-\underset{\underset{CH_3}{|}}{\overset{\overset{CH_3}{|}}{C}}+ \longrightarrow CH_3-\underset{\underset{CH_3}{|}}{\overset{\overset{CH_3}{|}}{C}}\overset{+}{-}\ddot{O}CH_2CH_3$$

$$H\ddot{O}CH_2CH_3$$

$$:\ddot{O}CH_2CH_3$$
$$H$$

$$CH_3-\underset{\underset{CH_3}{|}}{\overset{\overset{CH_3}{|}}{C}}-\ddot{O}CH_2CH_3 + \overset{+}{H}\ddot{O}CH_2CH_3$$
$$H$$

b) S_N2

$$CH_3CH_2CH_2CH_2-Cl \longrightarrow CH_3CH_2CH_2CH_2 + Cl^-$$
$$:\ddot{O}CH_3 \qquad\qquad :\ddot{O}CH_3$$

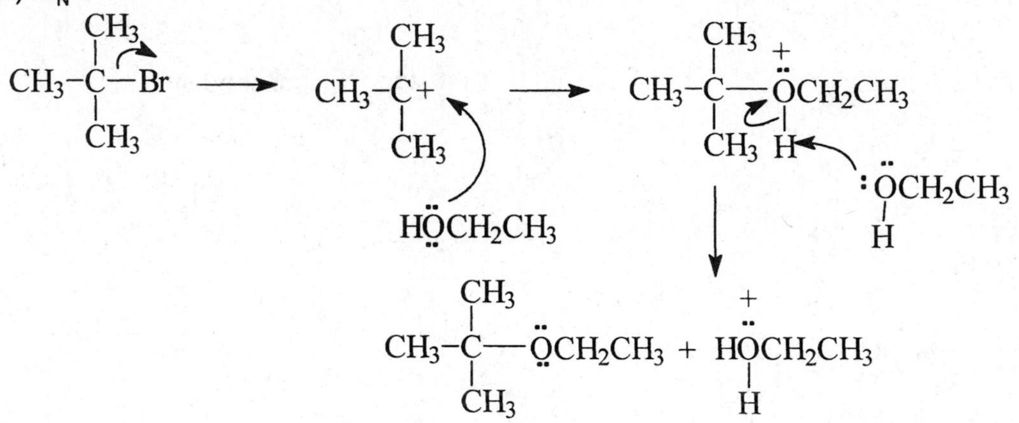

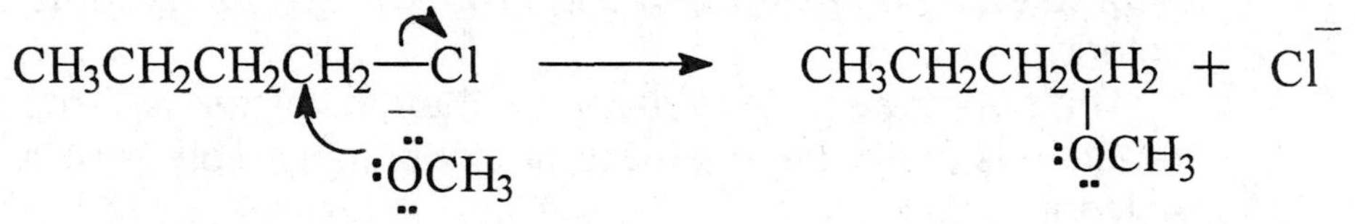

c) S_N1

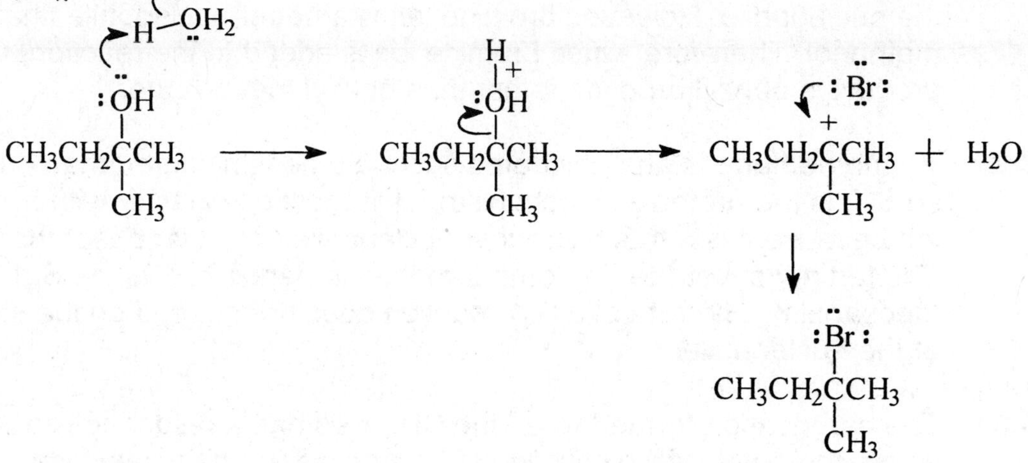

8.30 This is an S_N1 reaction. The nucleophile can bond to any of the carbons that have a positive charge in the resonance structures.

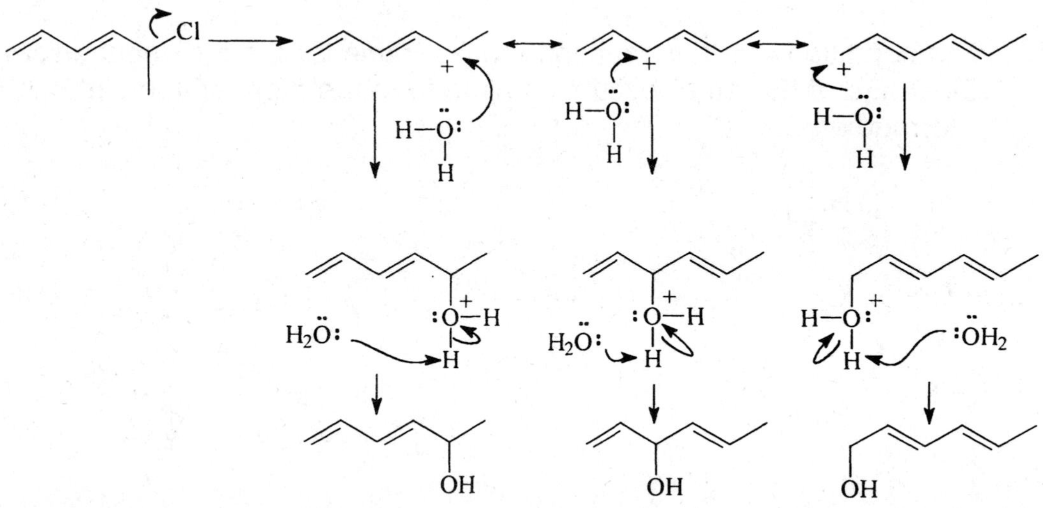

8.31 The trifluorosulfonate anion is a better leaving group because it is a weaker base due to the inductive stabilization effect of the electron-withdrawing CF_3 group.

8.32 The reaction follows the S_N1 mechanism, so the rate does not depend on the nucleophile. However, bromide ion is a better nucleophile than methanol. Therefore, when bromide ion is added to the reaction, the product is benzyl bromide rather than benzyl methyl ether.

8.33 Bromomethane is an unhindered substrate (less hindered than primary), so S_N2 is the preferred mechanism. This reaction is faster with hydroxide ion because it is a much stronger nucleophile (stronger base) than water. The tertiary substrate, 2-bromo-2-methylpropane, follows the S_N1 mechanism. The rate of a S_N1 reaction does not depend on the strength of the nucleophile.

8.34 The carbocation formed when the Cl leaves has a resonance structure where the octet rule is satisfied at all atoms and has a very large resonance stabilization.

$$CH_3{-}\overset{+}{\underset{\cdot\cdot}{O}}{-}\overset{+}{C}H_2 \quad \longleftrightarrow \quad CH_3{-}\overset{+}{\underset{\cdot\cdot}{O}}{=}CH_2$$

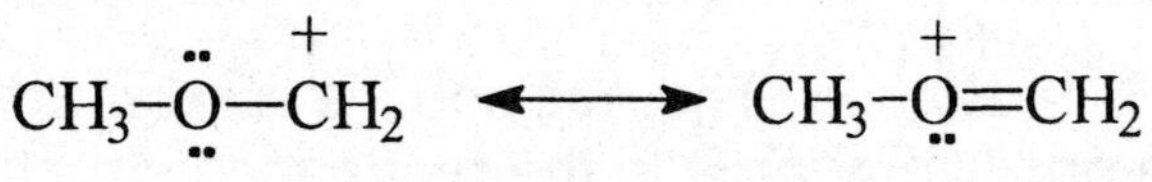

8.35 The conditions for this reaction are favorable for the S_N1 mechanism. Cleavage of the ether occurs so as to form the more stable carbocation intermediate.

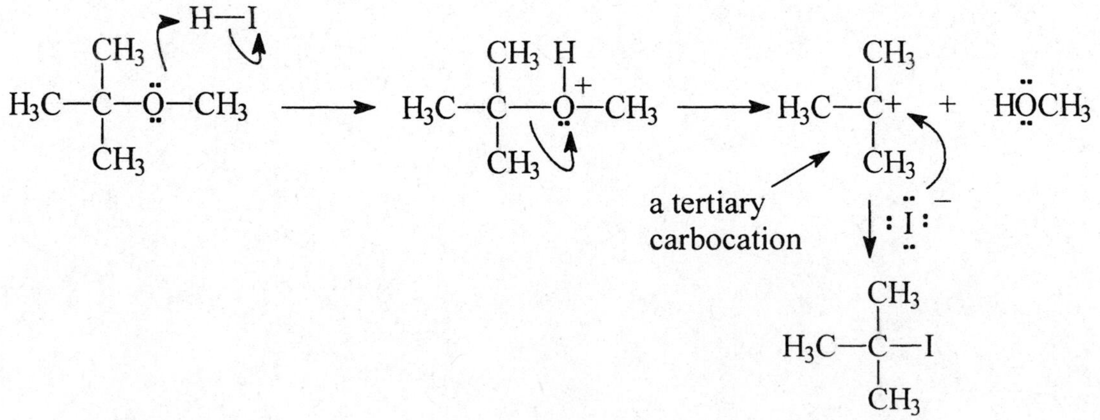

8.36 a) The solvent, ethanol, is polar, and there are no strong nucleophiles present, so the S_N1 mechanism is favored.
b) i) The left compound will give a precipitate more rapidly because bromide ion is a better leaving group than chloride ion.
ii) The left compound will give a precipitate more rapidly because the carbocation is resonance stabilized.
iii) The right compound will give a precipitate more rapidly because a tertiary carbocation is more stable than a secondary carbocation.
iv) The right compound will give a precipitate more rapidly because its carbocation is resonance stabilized. Bromobenzene does not undergo nucleophilic substitution reactions because the Br is bonded to an sp^2 hybridized carbon.

8.37 a) A strong nucleophile (iodide anion) and an aprotic solvent (acetone) favor the S_N2 mechanism.
b) i) The right compound will give a precipitate more rapidly because the chlorine is attached to a less hindered secondary carbon.
ii) The right compound will give a precipitate more rapidly because bromide ion is a better leaving group than chloride ion.
iii) The right compound will give a precipitate more rapidly because the chlorine is bonded to a less hindered primary carbon.
iv) The left compound will give a precipitate more rapidly because the compound on the right has more steric hindrance due to the two methyl groups on the carbon adjacent to the electrophilic carbon.

8.38 a) An S_N1 mechanism is favored because the solvent is polar and there is not a strong nucleophile present.
b) i) The right compound reacts faster because a secondary carbocation is more stable than a primary carbocation.
ii) The right compound reacts faster because a tertiary carbocation is more stable than a secondary carbocation.
iii) The left compound reacts faster because the carbocation produced from this compound is stabilized by resonance.

8.39 The carbocation, which is formed at the carbon bonded to the hydroxy group, is destabilized by the inductive electron-withdrawing inductive effect of the chlorine on the adjacent carbon.

8.40 a) 1-Propanol as both nucleophile and as solvent (S_N1)
b) Acetate ion as nucleophile in an aprotic solvent such as DMF (S_N2)
c) Acetate ion as nucleophile in a protic solvent such as acetic acid (S_N1)
d) Methylamine as nucleophile in a polar solvent such as water (S_N2)
e) Hydrobromic acid as catalyst and nucleophile with water as the solvent (S_N1)

8.41 This reaction occurs because the negative oxygen of the leaving group is stabilized by resonance with the benzene ring and the nitro group. As a result, the negative oxygen is much less basic than usual.

8.42 The substrate in both cases is a primary alkyl halide, so the S_N2 mechanism is favored. The reaction on the right is faster because the nucleophile is less sterically hindered.

8.43

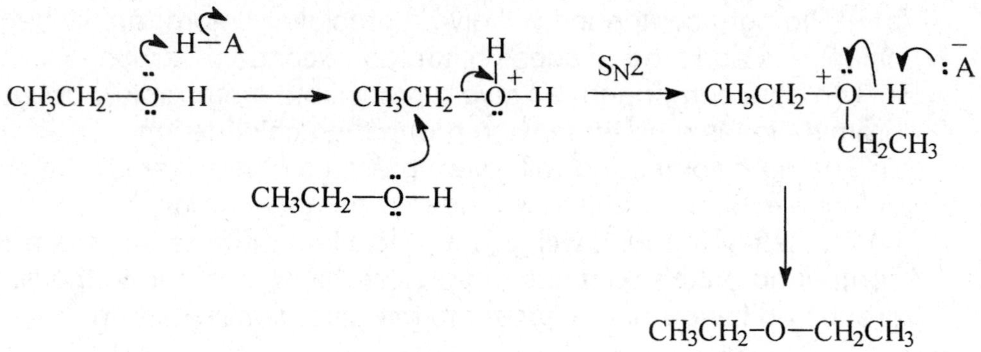

8.44

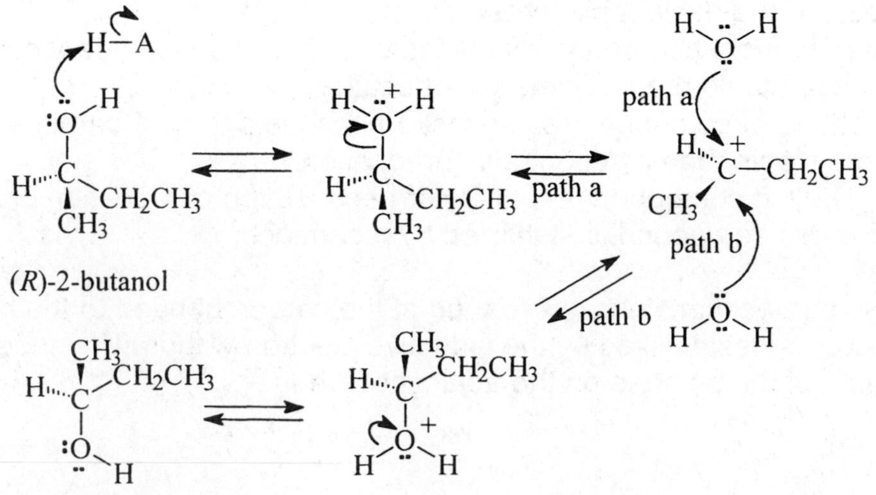

8.45 This is an intramolecular S_N2 reaction. As in any S_N2 reaction, the nucleophile must approach the carbon from the side opposite the leaving group, so the reactant must first undergo a conformational change so that the OH and the Br are anti.

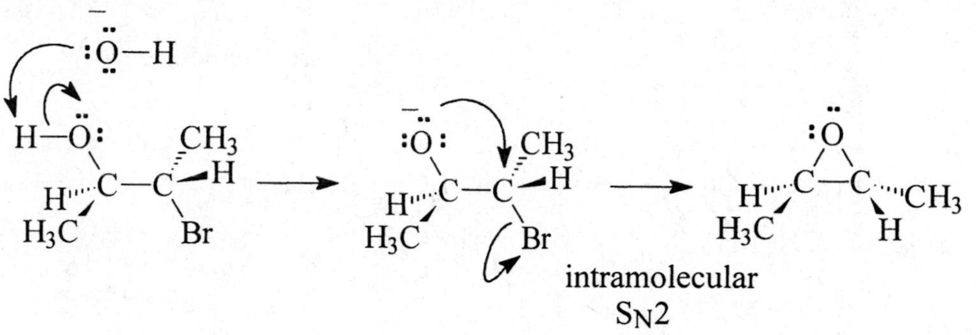

8.46 substitution product elimination products

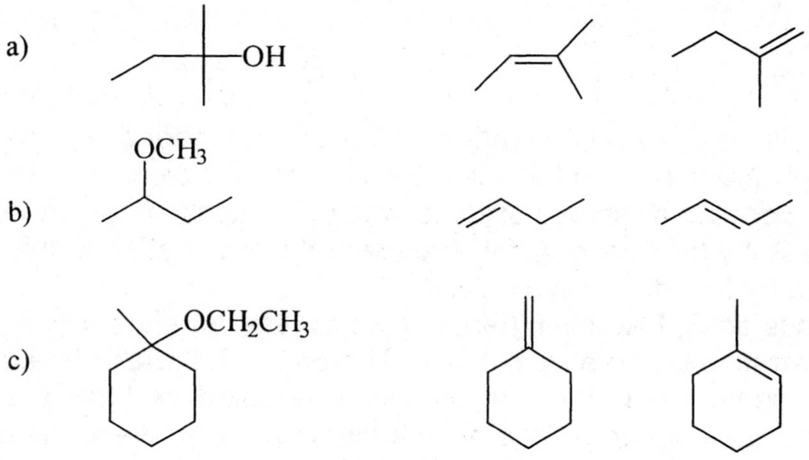

8.47 The substitution products are as follows:

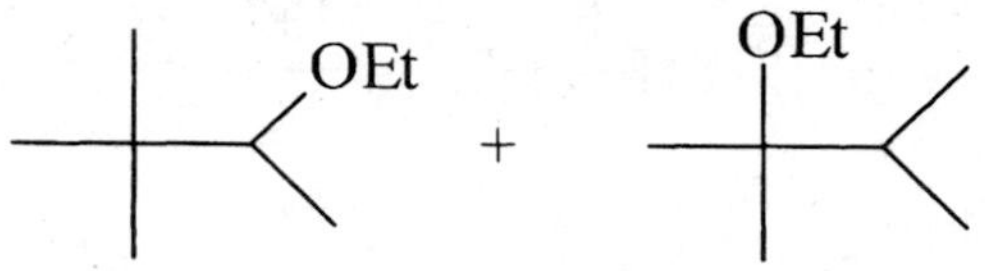

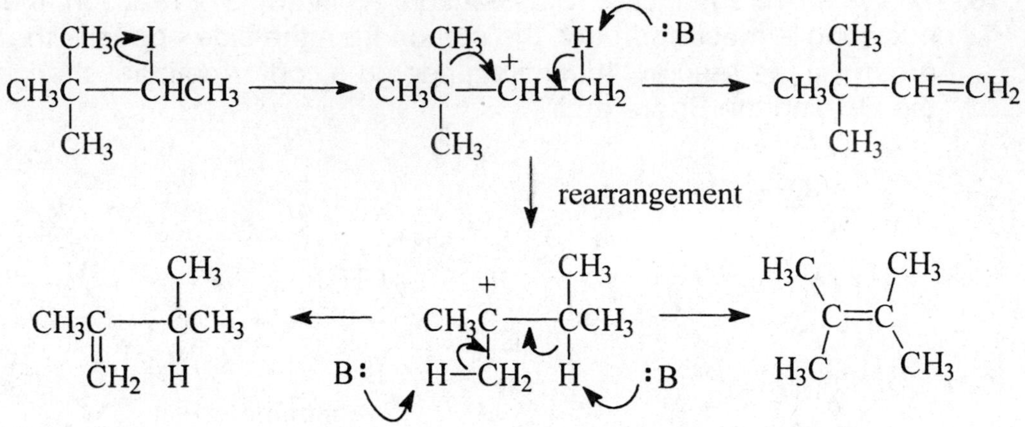

8.48

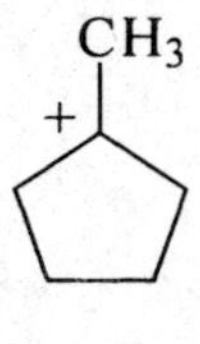

8.49 a) This is an S_N2 reaction (primary), so its rate depends on the concentrations of both the alkyl chloride and hydroxide ion. Therefore, the rate will double if the hydroxide ion concentration is doubled, and will decrease by a factor of 4 if the concentrations of both the alkyl chloride and the hydroxide ion are halved.

b) This is an S_N1 reaction (tertiary), so its rate depends only on the concentration of the alkyl chloride. Therefore, the rate will be unaffected if the concentration of the hydroxide ion is doubled, and the rate will be halved if the concentrations of both the alkyl chloride and the hydroxide ion are halved.

8.50

a) 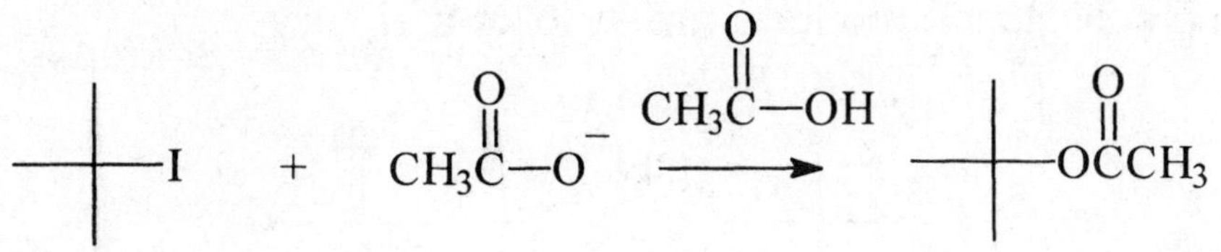

b) 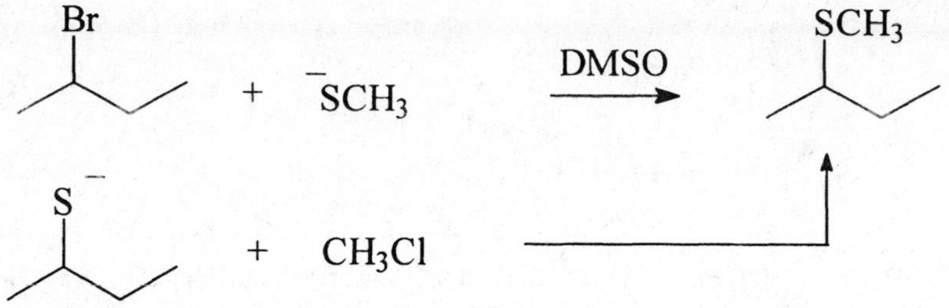

c) 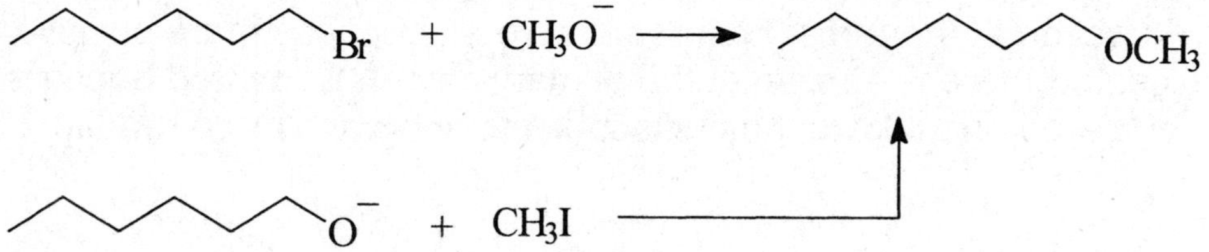

8.51

a) 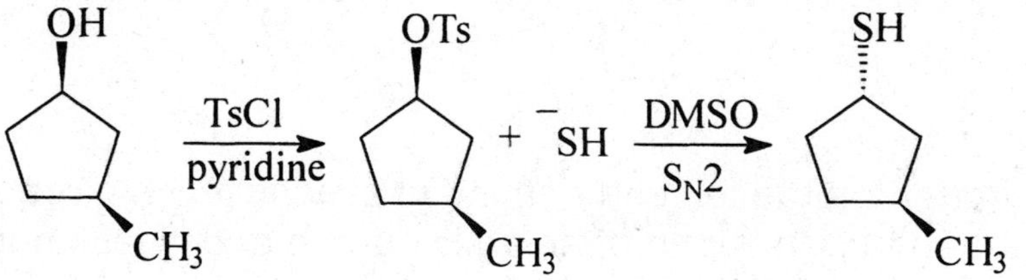

b) 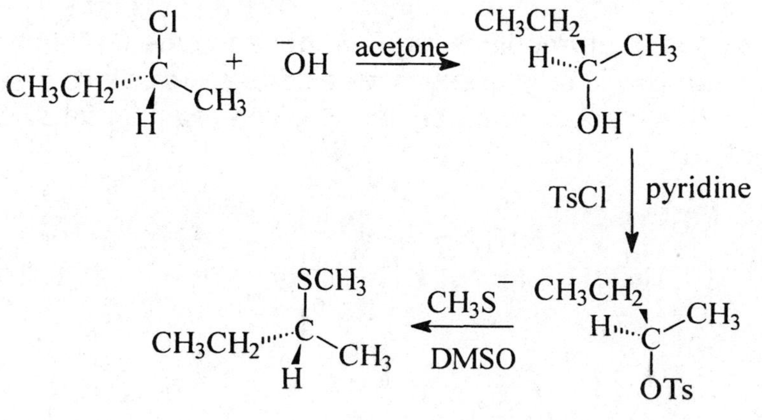

8.52 The carbon of the cyanide anion is the stronger nucleophile because it is more basic.

8.53

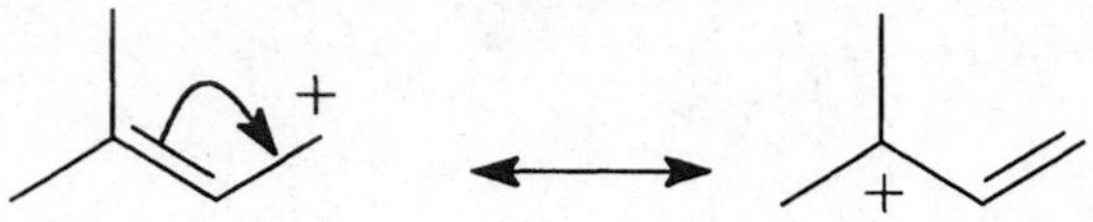

The carbocation formed in this reaction has two resonance structures. The resonance structure on the right contributes more to the resonance hybrid because it is more stable (a tertiary carbocation). The nucleophile can bond to either carbon where some positive charge is located. Reaction at the tertiary carbon is favored because more positive charge is located there. Reaction at the primary carbon is favored because it is less hindered and allows the nucleophile to approach more readily.

8.54

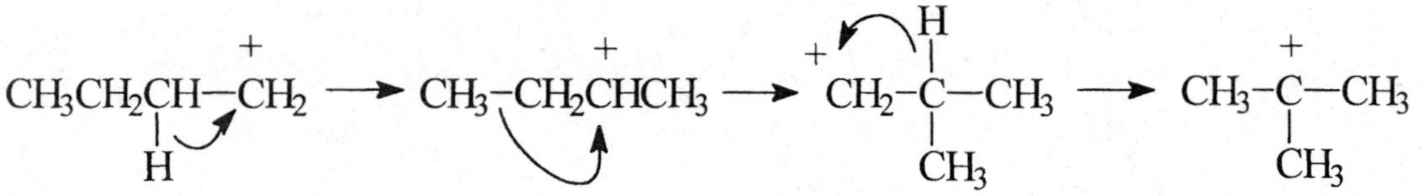

8.55 The nitrogen on the left end of the dipeptide is much more nuceophilic. The other nitrogen is part of an amide group and is less nucleophilic because its electron pair is stabilized (delocalized) by resonance with the carbon-oxygen double bond.

8.56 a) An excess of ammonia is required because the first ammonia reacts with the carboxylic acid group in an acid-base reaction. A second ammonia then reacts as a nucleophile. The reaction follows an S_N2 mechanism.

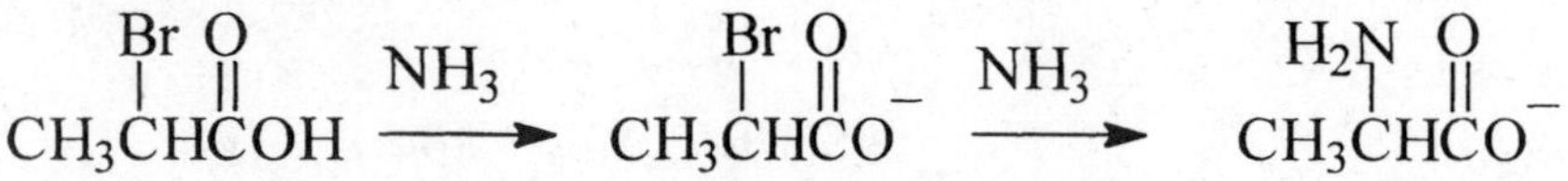

b)

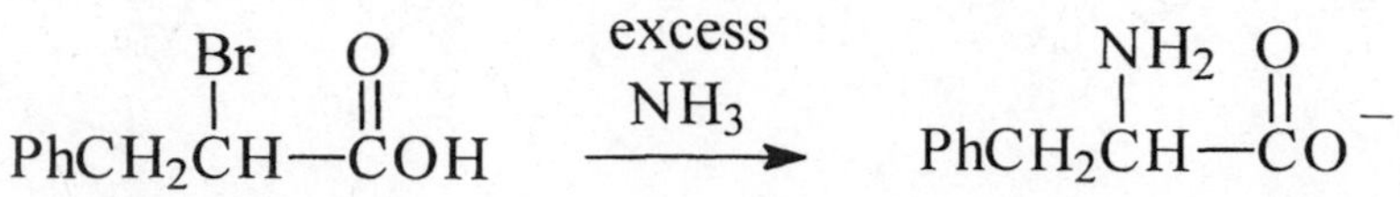

8.57 These are both S_N2 reactions because the leaving chlorine is on a secondary carbon, acetate anion is a good nucleophile, and DMSO is an aprotic solvent. S_N2 reactions occur with inversion of configuration.
a) Product 1 has inverted configuration.
b) The Cl is cis to the methyl group in the reactant, so the acetate group must be trans to the methyl in the product, as it is in Product 2.

8.58 This is an S_N1 reaction because the leaving group is on a tertiary carbon, the nucleophile (methanol) is weak, and the solvent (methanol) is polar. S_N1 reactions occur with racemization, so both products are formed in this reaction.

8.59 These are both S_N2 reactions because the leaving chlorine is on a secondary carbon, acetate anion is a good nucleophile, and DMSO is an aprotic solvent. The alkyl chloride on the left has less steric hindrance at the electrophilic carbon because the two methyl groups are farther away, so it reacts faster. The alkyl chloride on the right has the two methyl groups bonded to the carbon directly adjacent to the electrophilic carbon.

8.60 The nitrogen is the nucleophilic atom in each of these compounds. The methyl groups cause steric hindrance as the nucleophile approaches the electrophilic carbon and slow the reaction. The compound without any methyl groups is the least hindered and reacts fastest, followed by the compound with one methyl group. The compound with two methyl groups on the ring reacts the slowest.

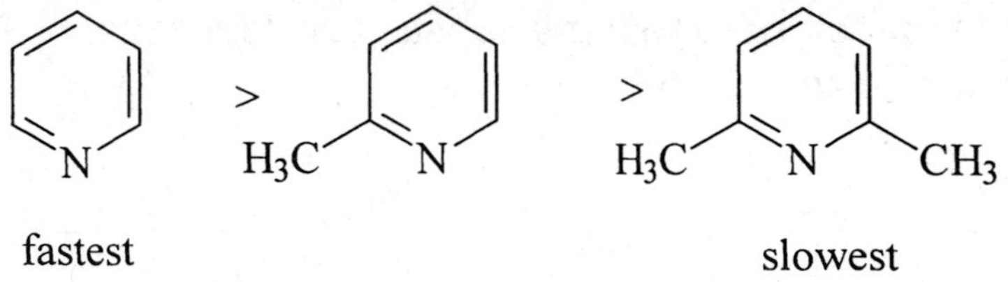

Review of Mastery Goals

After completing this chapter, you should be able to:

Write mechanisms for the S_N1 and S_N2 reactions.
(Problems 8.10, 8.12, 8.13, 8.21, 8.29, 8.35, 8.43, 8.44, and 8.45)

Recognize the various nucleophiles and leaving groups and understand the factors that control their reactivities.
(Problems 8.9, 8.11, 8.31, 8.32, 8.33, 8.41, 8.42, and 8.52)

Understand the factors that control the rates of these reactions, such as steric effects, carbocation stabilities, the nucleophile, the leaving group, and solvent effects.
(Problems 8.2, 8.3, 8.6, 8.7, 8.15, 8.22, 8.23, 8.24, 8.25, 8.34, 8.39, 8.55, 8.59, and 8.60)

Be able to use these factors to predict whether a particular reaction will proceed by an S_N1 or an S_N2 mechanism and to predict what effect a change in reaction conditions will have on the reaction rate.
(Problems 8.9, 8.11, 8.16, 8.28, 8.36, 8.37, and 8.38)

Show the products of any substitution reaction.
(Problems 8.12, 8.17, 8.18, 8.26, 8.27, 8.30, 8.40, 8.50, 8.51, and 8.56)

Show the stereochemistry of the products.
(Problems 8.1, 8.8, 8.27, 8.56, 8.57, and 8.58)

Recognize when a carbocation rearrangement is likely to occur and show the products expected from the rearrangement.
(Problems 8.19, 8.20, and 8.47.)

Show the structures of the products that result from the elimination reactions that compete with the substitution reactions.
(Problems 8.18 and 8.46)

Chapter 9
ELIMINATION REACTIONS
REACTIONS OF ALKYL HALIDES, ALCOHOLS, AND RELATED COMPOUNDS

9.1 An elimination reaction occurs when a proton and the leaving group are lost from adjacent carbons, resulting in the formation of a double bond.

a) $CH_3CH{=}CHCH_2CH_3$
(cis and trans)

b) 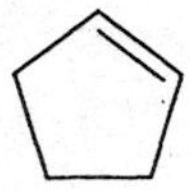

c) 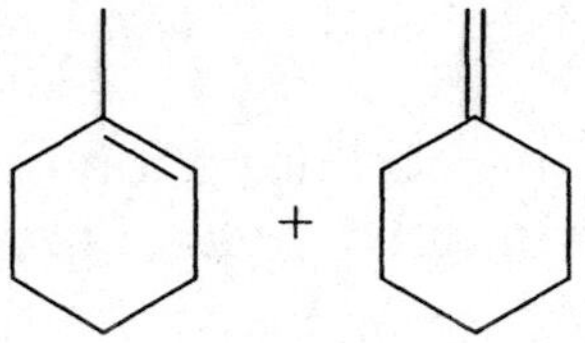

9.2 (1*R*,2*S*)-1-Bromo-1,2-diphenylpropane is the enantiomer of the compound shown at the bottom of Figure 9.2. Enantiomers produce the same stereoisomer of the alkene, so this compound also produces (*E*)-1,2-diphenyl-1-propene.

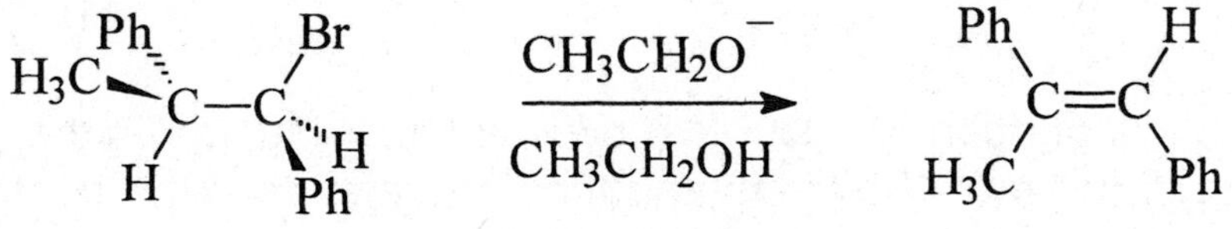

9.3 In E2 reactions, an anti-periplanar geometry of the leaving group and the hydrogen is required. Therefore, when trying to determine the elimination product, first draw the conformation with the hydrogen and the leaving group anti.
a) The hydrogen and the bromine are in the anti-coplanar geometry in the conformation shown. Anti elimination gives the (*Z*)-isomer of the product.

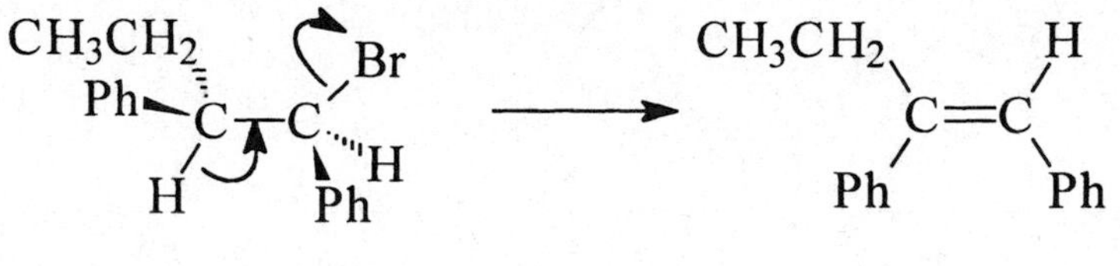

b) In this case a rotation of 60° about the C-C bond is needed to bring the hydrogen and the bromine to the required anti-coplanar geometry. Anti elimination then gives the (*E*)-isomer of the product. Note that the elimination of either Br gives the same product.

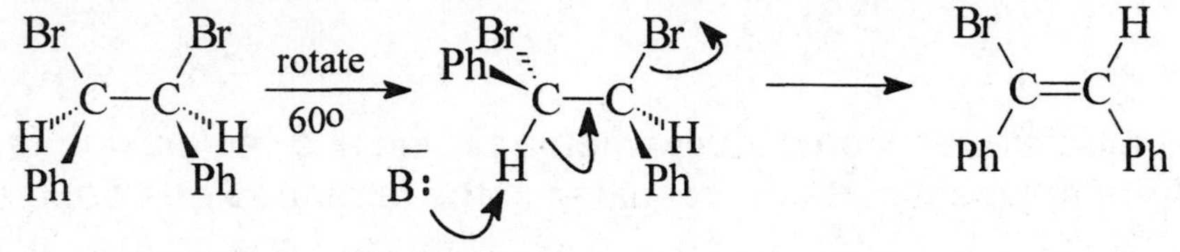

9.4 The conformation that produces the *trans*-alkene is more stable because the bulky phenyl groups are anti, so more of the *trans*-alkene is formed.

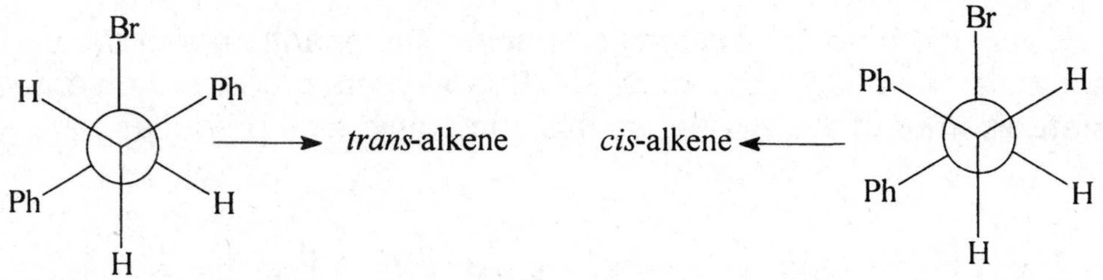

9.5 For anti elimination to occur in a cyclohexane ring, the hydrogen and the leaving group must be **trans** and the ring must be in the conformation where both the hydrogen and the leaving group are **axial**.

a) In this compound only one hydrogen (circled) can be in the required ***trans*-diaxial** arrangement with the chlorine. This results in the production of a single alkene.

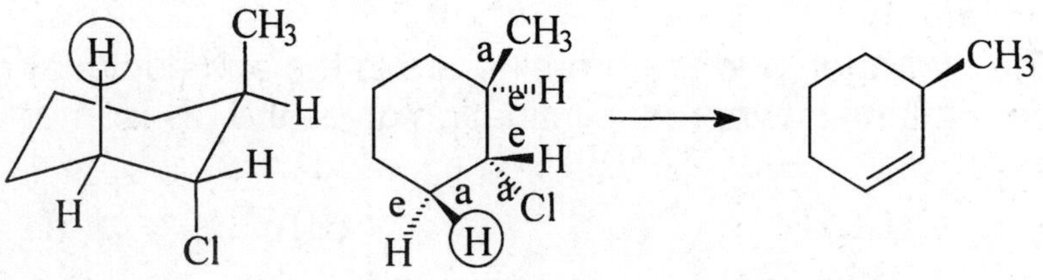

b) In this compound there are two hydrogens (circled) that can have a *trans*-diaxial geometry with the bromine. This results in the formation of two isomeric alkenes.

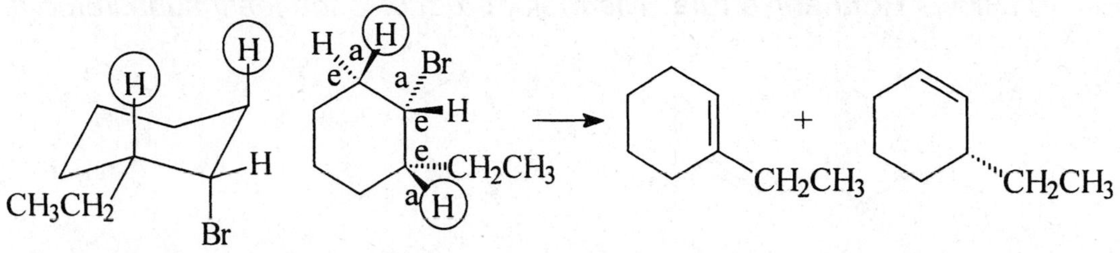

9.6 The bulky *t*-Bu group must be equatorial and locks the ring in a single conformation. For elimination to occur the leaving group, OTs, and a hydrogen on an adjacent carbon must be in a *trans*-diaxial arrangement. The isomer on the left has the OTs group axial when the *t*-butyl group is equatorial, so it can readily undergo E2 elimination. The isomer on the right has the OTs group equatorial when the *t*-butyl group is equatorial, so it cannot readily react.

9.7 For an elimination reaction that can produce structural isomers which have the double bond in different positions, the major product can be predicted using Zaitsev's rule: *The major product is the more highly substituted alkene.*

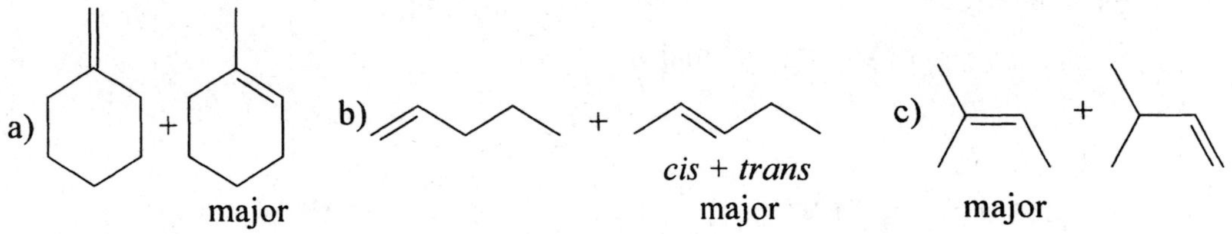

9.8 (*E*)-2-Butene is the major product because it is more stable and the conformation leading to it is more stable.

9.9 Most E2 reactions follow Zaitsev's rule. However, the reaction known as the Hofmann elimination is an exception. This reaction is observed with compounds that have a quaternary nitrogen atom as the leaving group and follows Hofmann's rule: *The major product is the least substituted alkene.*

a) 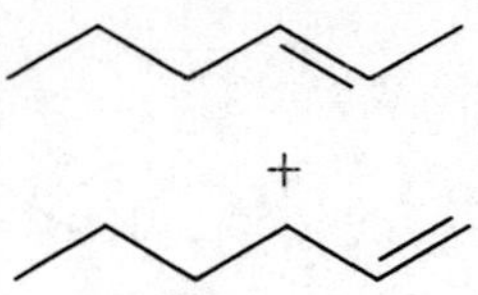

major

b) 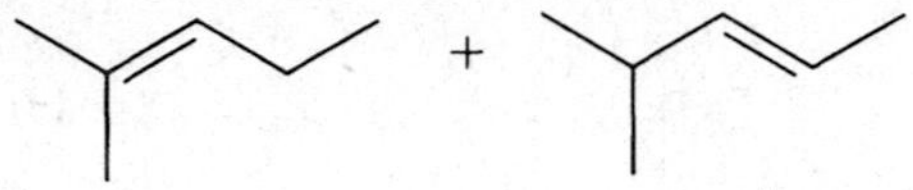

major

c) 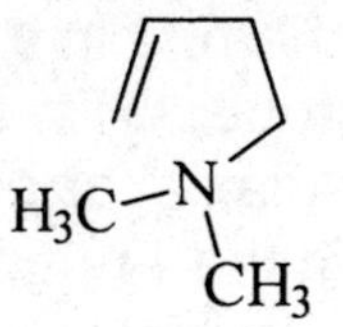

d) 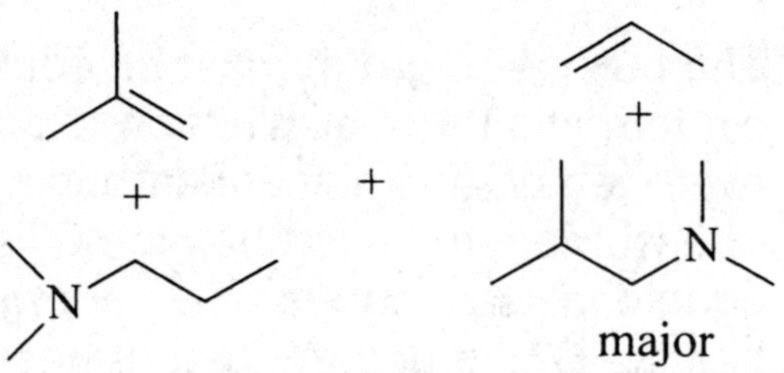

9.10 a) 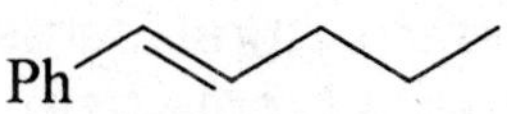

b) 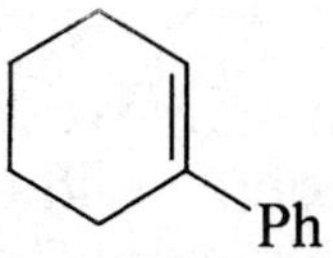

c) 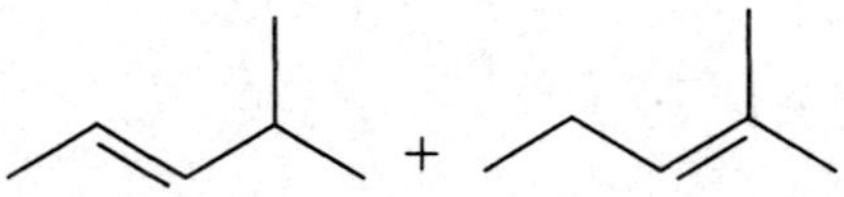

major

d) 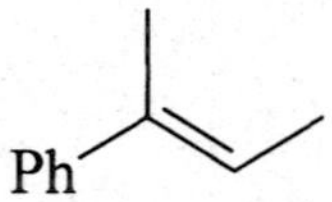

9.11

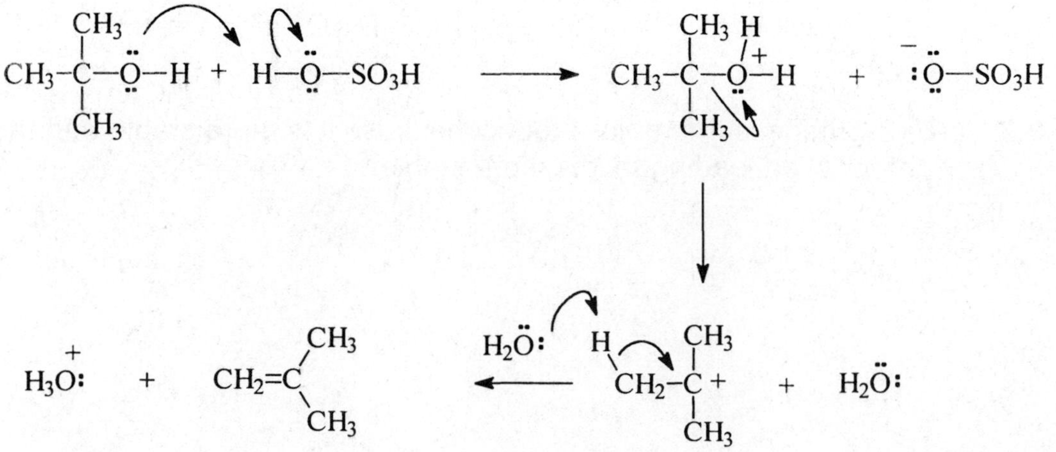

9.12 A secondary carbocation is formed first in this reaction. This carbocation then rearranges to form a more stable tertiary carbocation. Substitution and elimination products from both of these carbocations may also be formed.

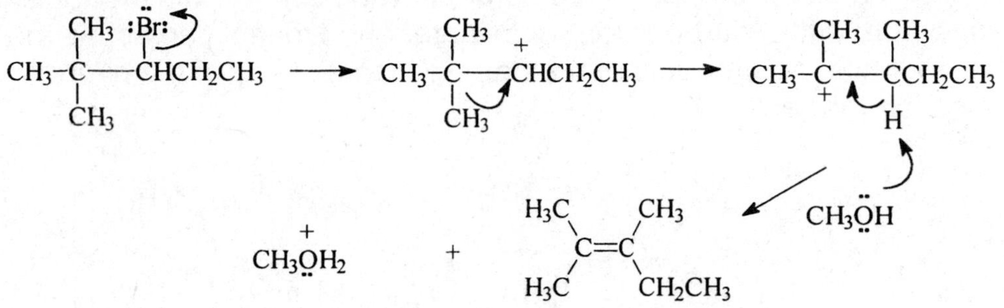

other products

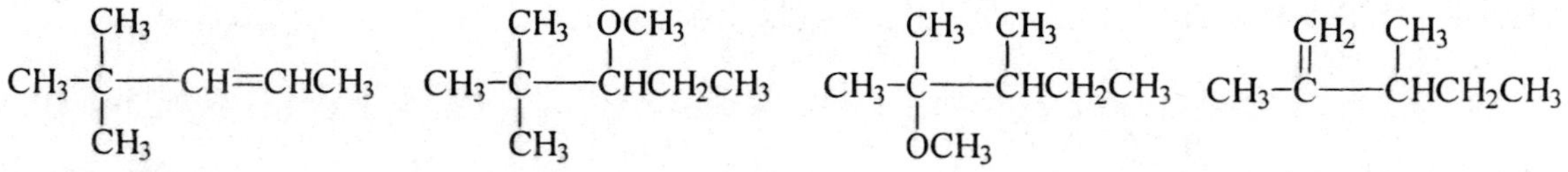

9.13 All of these reactions are run under conditions that favor the S_N1/E1 pathway. Usually the S_N1 product is major. The E1 products follow Zaitsev's rule.

a)

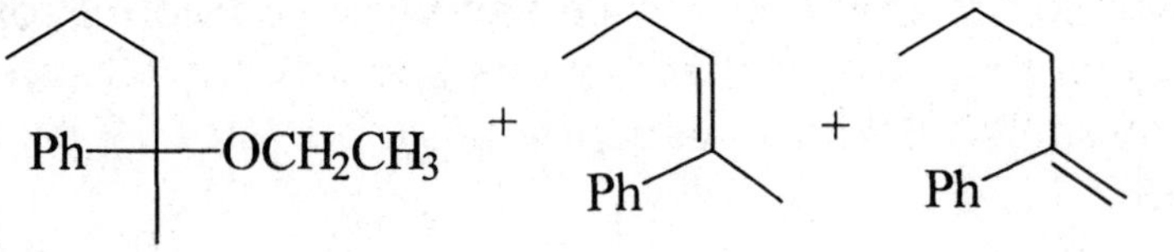

major minor more minor

b)

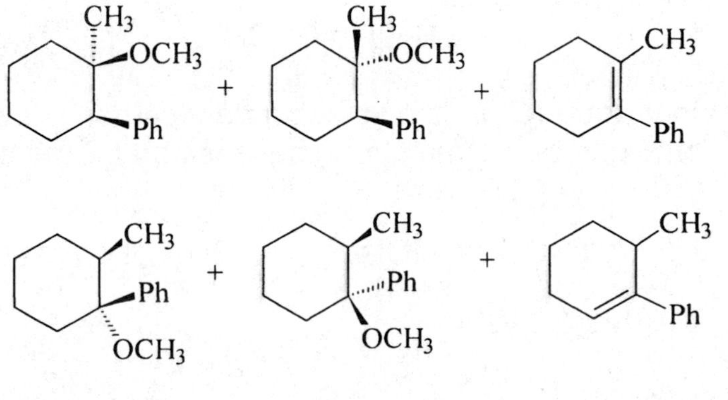

c) This compound produces the same carbocation as the one formed in part (d), so the products will be the same.

d) The secondary carbocation that is formed initially can rearrange to a tertiary carbocation. Elimination products and substitution products can form from either carbocation. Both water and ethanol can act as the nucleophile, so products with ethoxy groups in place of the hydroxy groups are also formed.

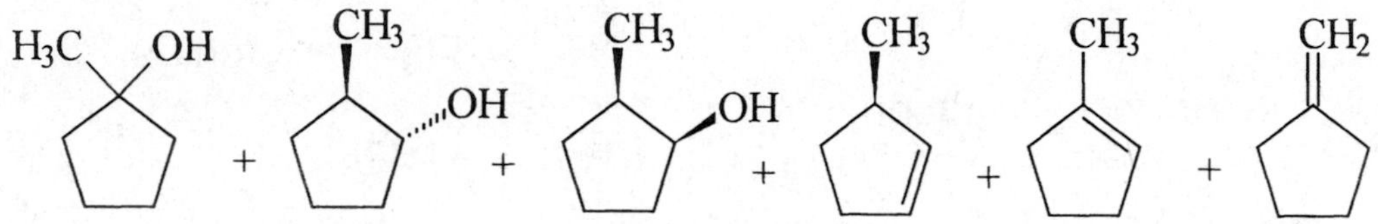

9.14 These reactions follow the E1cb mechanism.

a) 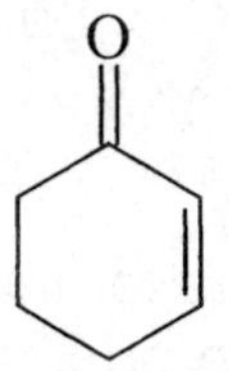

b) $PhCH{=}C(Br)C(=O)CH_3$

9.15 a) The leaving group (OTs) is on a secondary carbon. Propanoate anion is a good nucleophile that is also a weak base. So an S_N2 mechanism is favored, even in a protic solvent. Minor amounts of E2 products may also be formed.

$$CH_3CH_2CH(OC(=O)CH_2CH_3)CH_3 \quad + \quad CH_3CH{=}CHCH_3 \quad + \quad CH_3CH_2CH{=}CH_2$$

b) The leaving group (OTs) is on a secondary carbon. However, in contrast to part (a), ethoxide anion is a strong base, so an E2 mechanism is favored.

$$\underset{\text{major}}{CH_3CH{=}CHCH_3} \quad + \quad CH_3CH_2CH{=}CH_2$$

c) The leaving group (Cl) is on a tertiary carbon. There is no strong base present, so the reaction follows the S_N1/E1 mechanisms.

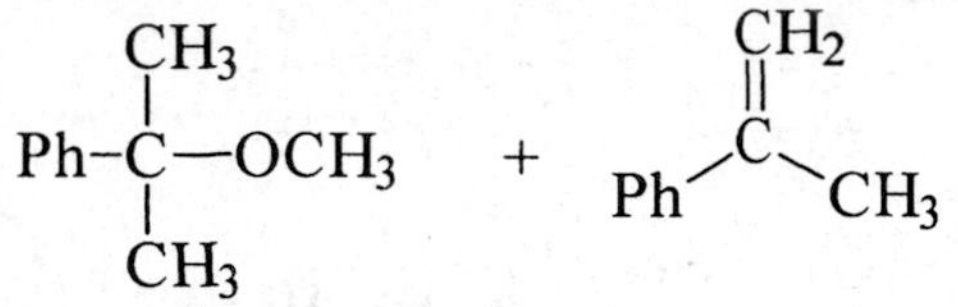

major

9.16 a) The leaving group is on a primary carbon and the base is not sterically hindered, so the S_N2 mechanism is favored.

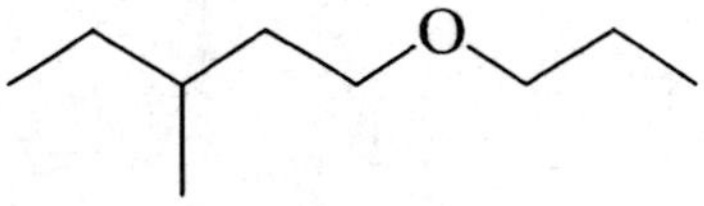

b) The leaving group is on a primary carbon. The hindered base (*t*-BuO$^-$) favors an E2 elimination reaction.

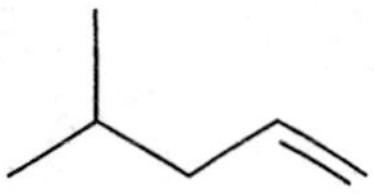

c) The leaving group is on a secondary carbon. Hydroxide ion is a strong base, so the E2 mechanism is favored.

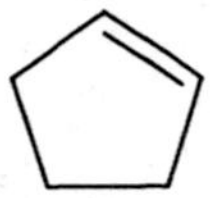

d) The leaving group is on a secondary carbon. Acetate anion is a good nucleophile and a weak base, so an S_N2 mechanism is favored, especially in an aprotic solvent (DMF).

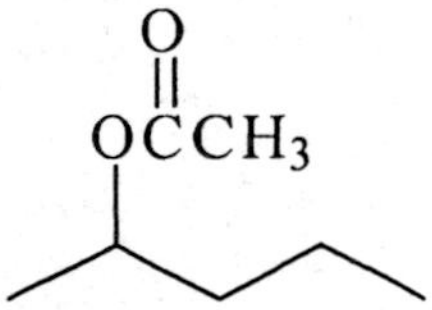

e) The leaving group is on a tertiary carbon. Ethoxide anion is a strong base, so the E2 mechanism is favored.

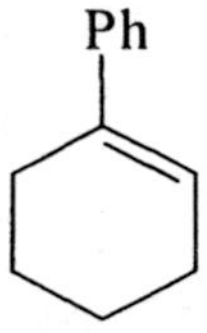

f) The leaving group is on a secondary carbon. In the absence of a good nucleophile or a strong base, the S_N1/E1 mechanisms are favored, especially in a polar solvent. Both H_2O and EtOH can act as the nucleophile.

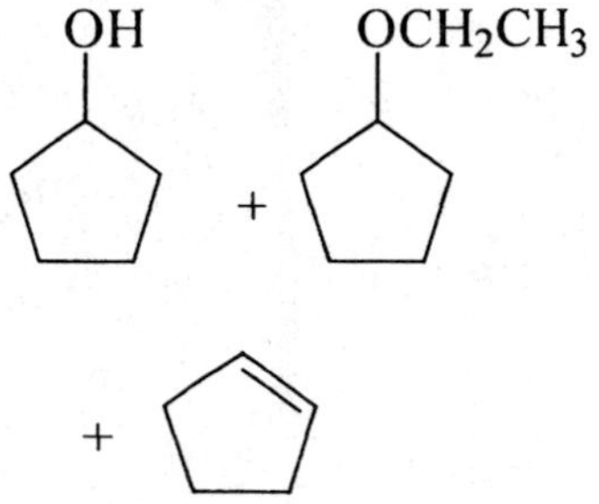

9.17

a) 

major

b) $CH_2{=}CH_2$ + $CH_2{=}CHCH_2CH_3$

major

c) 

d) 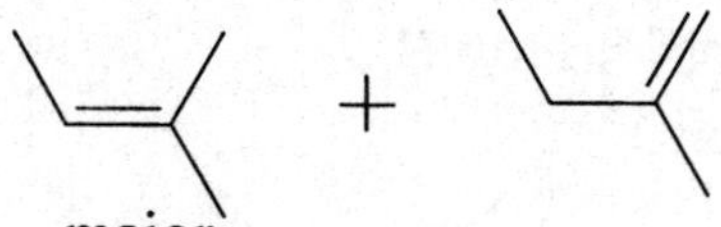

major

e) $Ph_2C{=}CH_2$

f) 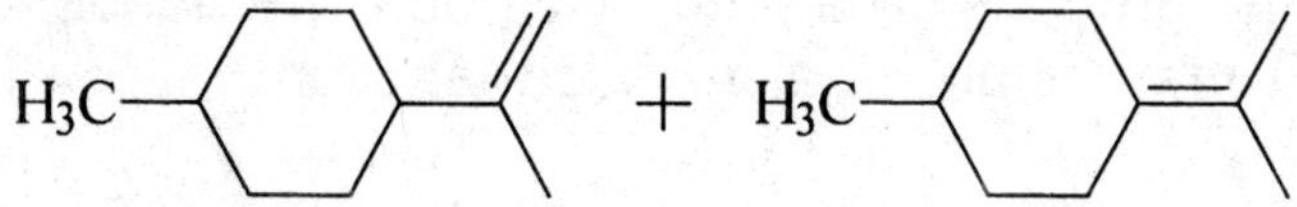

major

g) 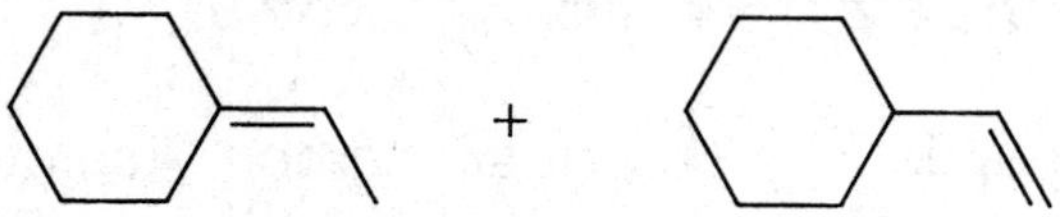

major

h) 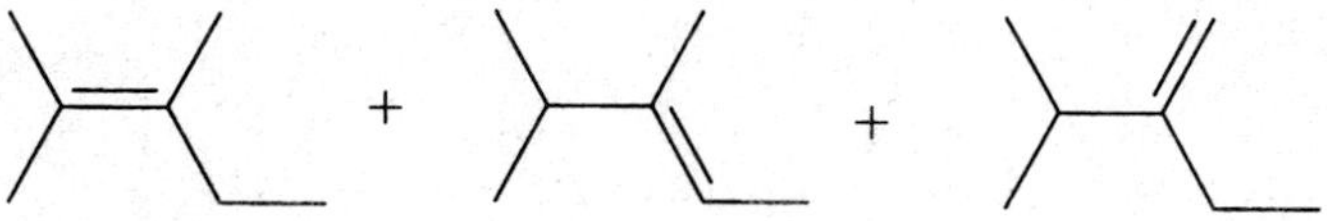

major

i) 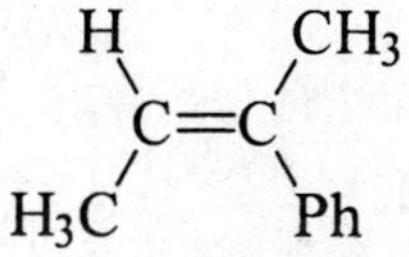

major

j) 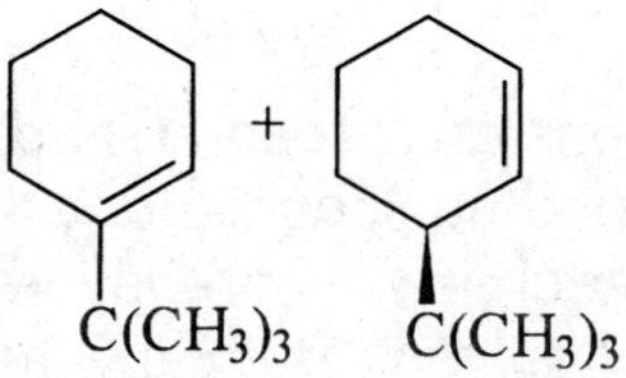

major

k) 

l) 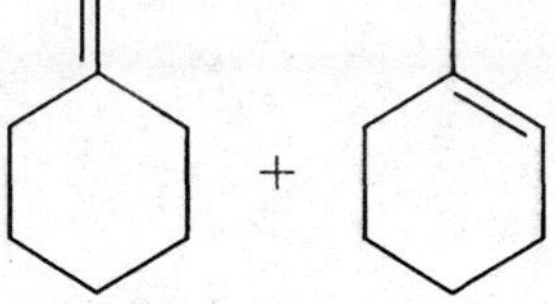

major

m) 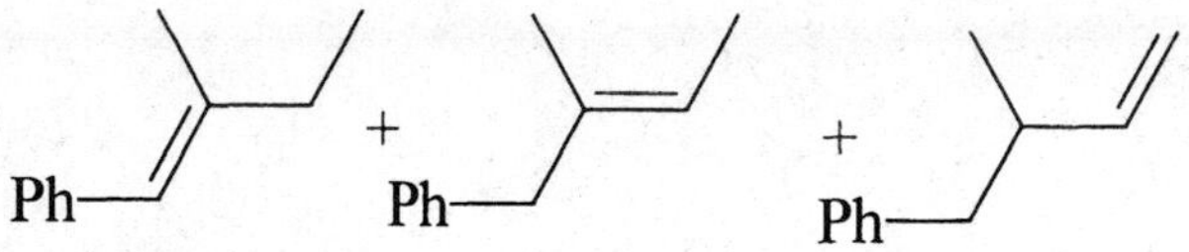

major

9.18 a) A secondary substrate with a moderate nucleophile and an aprotic solvent follows the S_N2 mechanism.

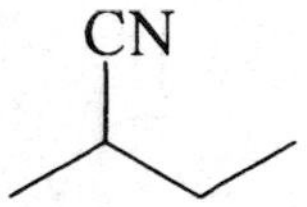

b) A secondary substrate with a strong base reacts predominantly by the E2 mechanism. The reaction follows Zaitsev's rule.

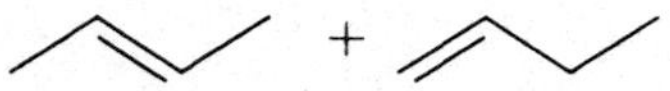

major

c) A secondary substrate with a weak nucleophile and a polar solvent (CH_3OH is both the nucleophile and the solvent) follows the S_N1 mechanism. A small amount of the E1 product may also be formed.

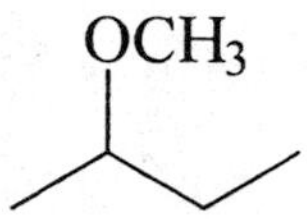

d) A primary substrate with a good nucleophile follows the S_N2 mechanism.

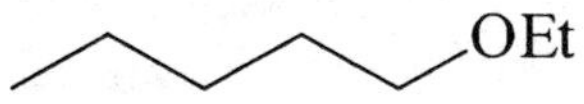

e) A primary substrate with a hindered strong base follows the E2 mechanism.

f) A tertiary substrate with a strong base follows the E2 mechanism.

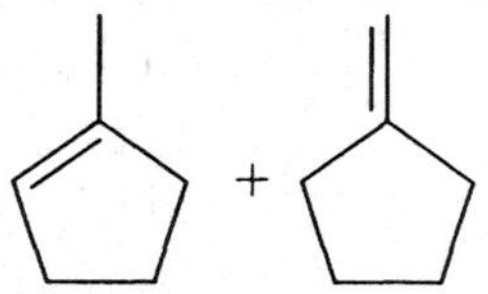

major

g) A tertiary substrate in the absence of a strong base follows theS_N1/E1 mechanisms.

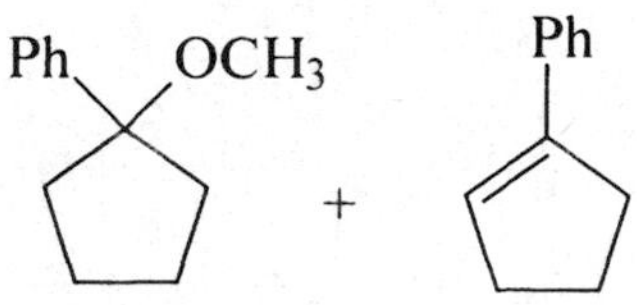

major

9.19

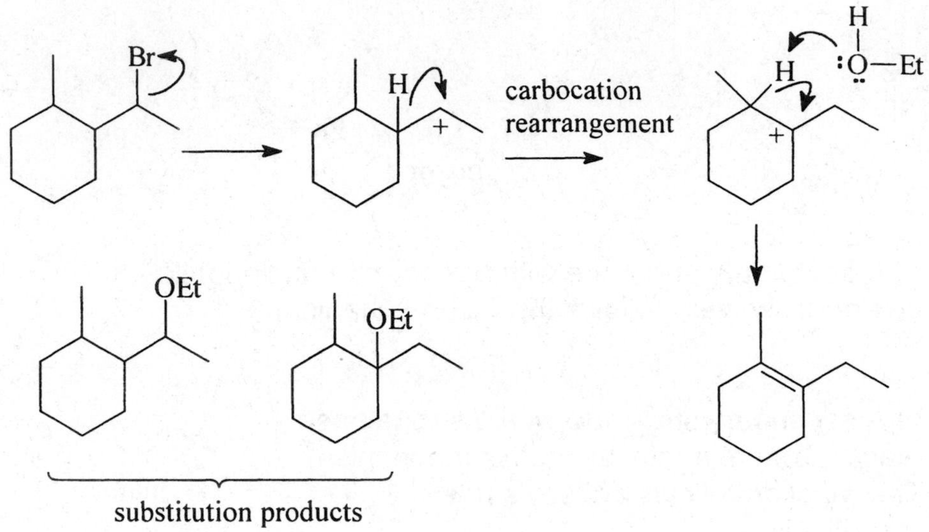

9.20 A tertiary substrate with a strong base favors E2 elimination. Normally the elimination products follow Zaitsev's rule with a less hindered base like ethoxide ion, producing more of the more highly substituted alkene. When a sterically hindered, strong base (*t*-BuO^-) is used, the proportion of the less highly substituted alkene will increase.

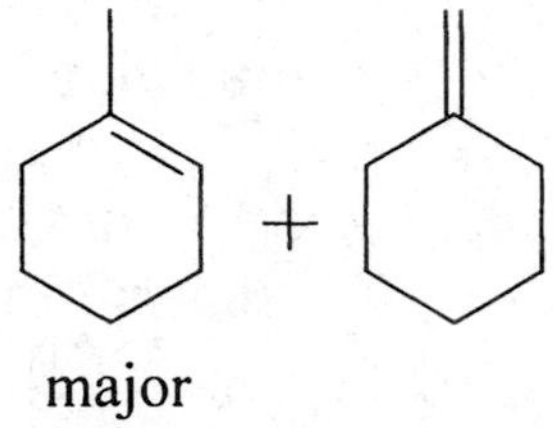

9.21 In the stereoisomer of 1,2,3,4,5,6-hexachlorocyclohexane shown, the Cl atoms are either all axial or all equatorial. Anti elimination cannot occur because there is no H trans to any Cl. When the Cl's are axial, there are no H's in a *trans*-diaxial arrangement. All of the other stereoisomers of this compound will have at least one hydrogen anti to a chlorine in one of the conformations.

9.22

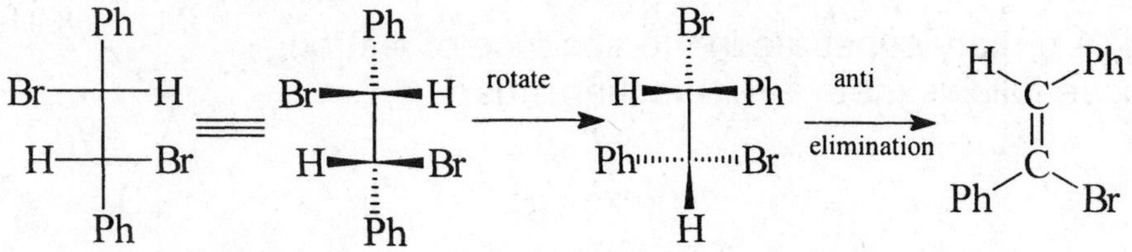

9.23 The most stable chair conformation of the *trans*-isomer has both the bromine and the *tert*-butyl group equatorial. However, elimination cannot occur from this conformation because bromine is not axial. Therefore, elimination occurs very slowly for the *trans*-isomer.

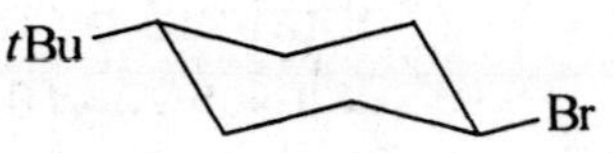

trans-isomer

For the *cis*-isomer, on the other hand, the most stable chair conformation is also the conformation needed for elimination, with the *tert*-butyl group equatorial and the bromine axial. The *cis*-isomer will react much faster than the *trans*-isomer because the conformation required for elimination is the more stable conformation.

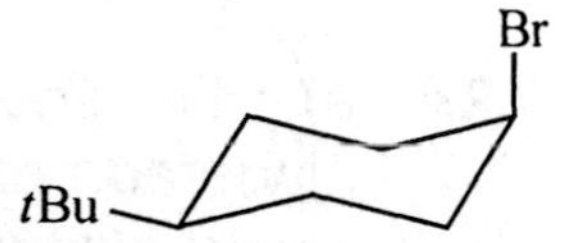

cis-isomer

9.24 The *cis*-isomer has a faster rate of E2 elimination because its more stable chair conformation has the methyl equatorial and chlorine axial. Elimination can occur from this conformer. The more stable conformation of the *trans*-isomer has both the chlorine and the methyl group equatorial. Elimination cannot occur from this conformation because chlorine is not axial. So the *trans*-isomer must ring flip to its less stable conformation before it can react.

9.25 The upper route, using the conjugate base of cyclopentanol and methyl iodide, will yield only the substitution product because methyl iodide cannot give an elimination product. In the case of the lower route, with bromocyclopentane (a secondary substrate) and methoxide anion (a strong base), E2 elimination will predominate. Therefore, the upper route will give the higher yield of cyclopentyl methyl ether.

9.26 a) An E2 elimination product can be obtained from a primary alkyl halide using a strong, hindered base. Use potassium *tert*-butoxide in *tert*-butanol.
b) An S_N1 substitution product can be obtained from a tertiary alkyl halide using a weak nucleophile in a polar solvent. Use methanol as both the solvent and the nucleophile.
c) An E2 elimination product can be obtained from a tertiary substrate by reaction with a strong base such as sodium methoxide in methanol. The conjugated product is favored.
d) An E2 elimination product can be obtained from a secondary substrate by reaction with a strong base such as sodium ethoxide in ethanol.

e) A substitution product from a unhindered primary substrate can be obtained by using most nucleophiles. This S_N2 product can be obtained using sodium ethoxide in ethanol.
f) The use of the strongly basic nucleophile, methoxide ion, with a secondary substrate results primarily in E2 elimination. To obtain the substitution product, use methanol, a weaker base, as both nucleophile and solvent. The mechanism will follow the S_N1 pathway.

9.27 a) 1-Hexene can be prepared by reaction of 1-bromohexane with a hindered base. Note that 2-bromohexane cannot be used because it would give a mixture of 1-hexene and 2-hexene, with the latter as the major product.

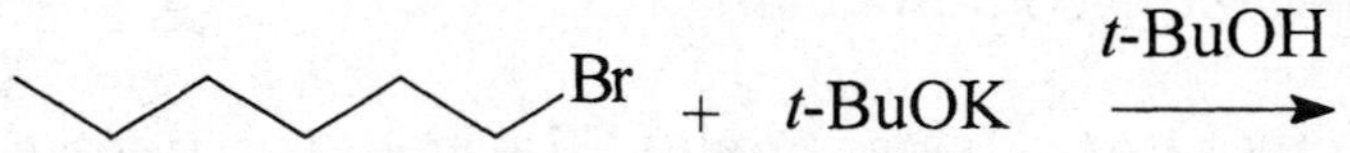

b) 3-Heptene can be prepared by reacting 4-bromoheptane with sodium methoxide in methanol. The alkyl halide is symmetrical so only a single elimination product is formed.

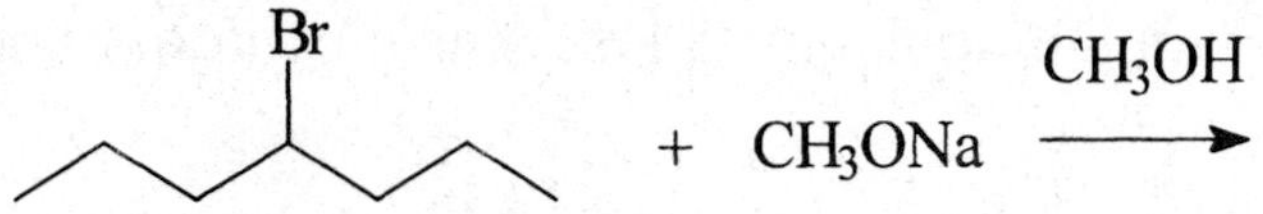

c) Cyclohexene can be prepared from chlorocyclohexane by E2 elimination using sodium methoxide in methanol.

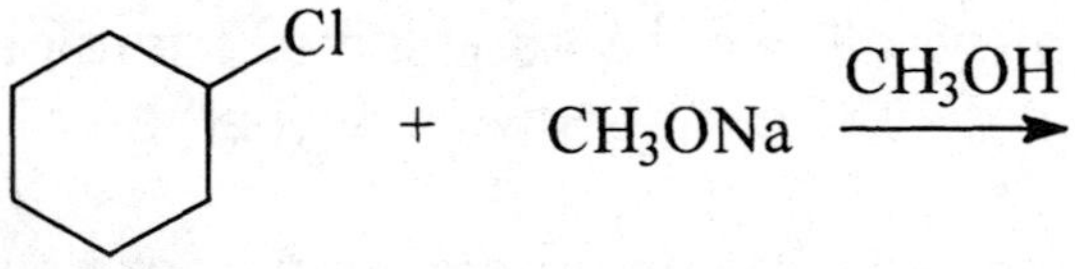

d) This alkene can be prepared from 4-bromo-1,2-dimethylcyclopentane by treating it with a strong base such as sodium ethoxide in ethanol. Because the bromide is symmetrical, only the desired alkene is produced.

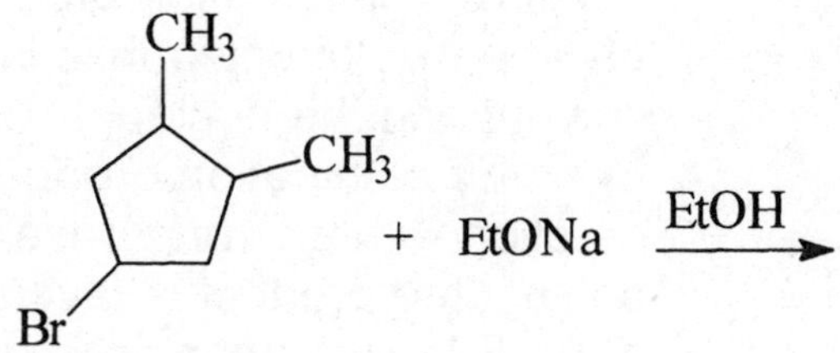

9.28 Because the carbocation is stabilized by resonance, there is not a strong nucleophile present, and the solvent is polar, this reaction follows the S_N1/E1 mechanisms. The resonance structures have the positive charge at two different locations. Therefore two elimination and substitution products are possible.

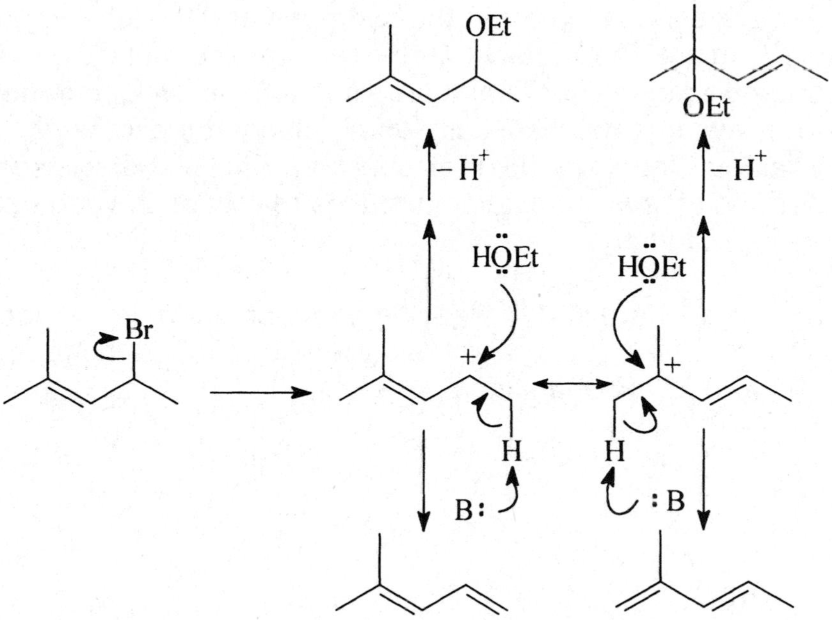

9.29 The bond dissociation energy of a C-D bond is larger than that of a C-H bond by about 1.2 kcal/mol (5.0 kJ/mol). If a C-H bond is broken during the rate-determining step in a reaction, then replacing the hydrogen with deuterium results in a significant decrease in the rate of the reaction. In E2 elimination involving cyclohexane rings, breaking of an C-H bond trans to the leaving group is required. In the chair conformation of the left compound the deuteriums are trans to the bromine, whereas in the right compound the hydrogens are trans to the bromine. Therefore, the right compound will have a faster rate of E2 elimination.

9.30 In this E2 elimination reaction, breaking of a C-H or a C-D bond is required in the rate determining step. The major product still has the deuterium because the C-H bond is easier to break than the C-D bond.

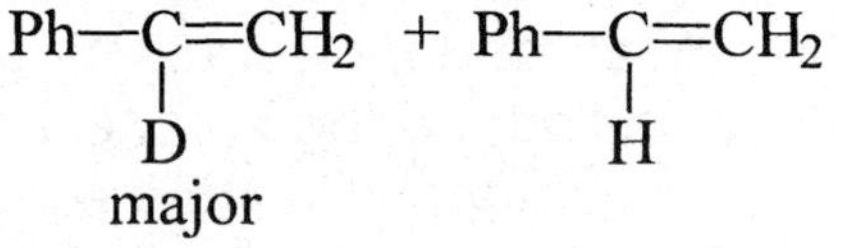

9.31 A cyclobutane ring is relatively rigid and is nearly eclipsed about its C-C bonds. It is easier for a syn elimination from an eclipsed conformation to occur that it is for the molecule to attain a staggered conformation so that an anti elimination can occur. As a result, the compound shown prefers to lose the D, which is cis to the leaving group.

9.32 For syn elimination occur the hydrogen and the leaving group must be in a syn coplanar arrangement. This requires the molecule to be in a higher-energy eclipsed conformation. Although the cyclopentane ring is nonplanar to relieve its torsional strain, a fully staggered geometry is not easily attained. Therefore the molecule can achieve the eclipsed geometry needed for syn elimination about as readily as the geometry needed for anti elimination.

9.33 a) The reactant has the leaving group on a tertiary carbon, there is no strong base present, and the solvent is polar, so the reaction follows the S_N1 and E1 mechanisms.

b)

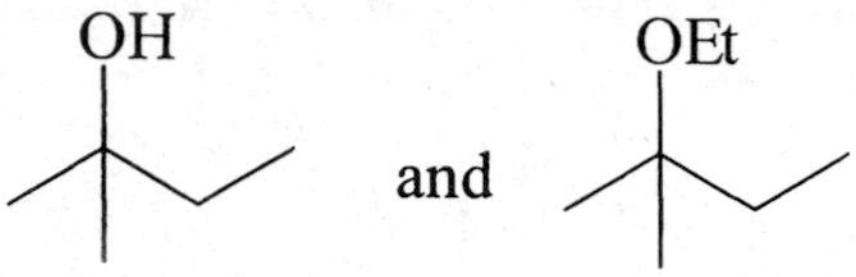

OH   OEt

         and

c)

         and

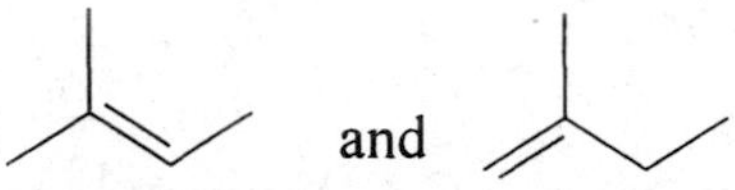

major

d) Iodide ion is a better leaving group than bromide ion, so the rate of reaction will be faster with 2-iodo-2-methylbutane. However, the amount of substitution and elimination products will not change because the leaving group is not involved in the second step of the mechanism, where the partitioning between substitution and elimination occurs.

9.34 Because of the relatively acidic hydrogen on the carbon adjacent to the carbonyl group, this compound reacts by the E1cb mechanism to produce the alkene with the double bond conjugated with the carbonyl group.

$$\underset{\quad}{CH_3\overset{\overset{\displaystyle O}{\|}}{C}CH=C}\overset{\displaystyle CH_2CH_3}{\underset{\displaystyle CH_2CH_3}{}}$$

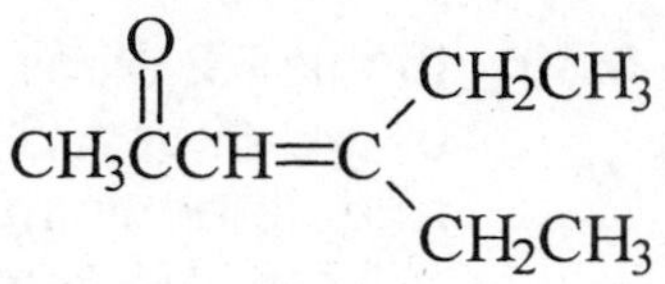

9.35 The ineffectiveness of DDT on some insects is due to the presence of an enzyme in these resistant insects that catalyzes the elimination of HCl to form DDE. This elimination reaction could be prevented by replacing the hydrogen with an alkyl group. Perhaps this would make the resulting compound an effective insecticide.

9.36 E2 elimination requires the leaving group and the hydrogen to be anti.
a) To get the H and the Br anti requires a 180° rotation about the C-C bond in this compound. Anti elimination produces the alkene with the phenyl groups trans, product 1.
b) This is the diastereomer of the compound in part (a). It is already shown in the conformation with the Br and the H anti. Elimination produces the alkene with the phenyl groups cis, product 2.

9.37 Anti elimination (E2) in a cyclohexane ring requires the leaving group and the hydrogen both to be axial. In these two diastereomers of 1-*t*-butyl-4-chlorocyclohexane, the bulky *t*-butyl group strongly prefers to be equatorial. The *cis*-diastereomer, on the left, has the chlorine axial when the *t*-butyl group is equatorial and readily undergoes anti elimination. The *trans*-diastereomer, on the right, has its chlorine equatorial and cannot readily undergo anti elimination. Therefore the cis-diastereomer, on the left, reacts faster.

9.38 Both of these problems involve E2 elimination in a cyclohexane ring. The bulky *t*-butyl group is always equatorial. The leaving chlorine must be axial for anti elimination to be possible.
a) This stereoisomer is shown in the lower energy conformation with the *t*-butyl group equatorial. Elimination can occur from this conformation because the chlorine is axial. The hydrogen that is removed in the E2 reaction must also be axial. Because the methyl group occupies the axial position on one of the carbons next to the Cl (the Cl and the CH_3 are trans), elimination can only occur forming the double bond to the other carbon, so product 1 is formed.
b) This is a diastereomer of the compound in part (a). It has the Cl and the CH_3 cis. When the *t*-butyl group is equatorial, the chlorine is axial, and the methyl group is equatorial. Anti elimination can occur on either side of the Cl, so both product 1 and product 2 will be formed. Product 2 will be the major product because the reaction follows Zaitsev's rule.

9.39 The reactant in this problem is the same as in Problem 9.38a. However, the reaction conditions are different. In Problem 9.38a, the reaction was conducted with the strong base, ethoxide anion, so the mechanism was E2 and anti elimination was preferred. In this problem, there is no strong base present, so the mechanism is S_N1/E1. First the Cl leaves to form a carbocation. Because the carbocation is planar, stereochemistry is lost and anti elimination is not required. So both products are formed. Product 2 should be the major elimination product because E1 reactions follow Zaitsev's rule.

Review of Mastery Goals

After completing this chapter, you should be able to:

Provide a detailed description, including stereochemistry, for the E2 mechanism and summarize the conditions that favor its occurrence. (Problems 9.2, 9.3, 9.4, 9.5, 9.6, 9.20, 9.21, 9.23, and 9.24)

Provide similar information about the E1 mechanism. (Problems 9.11, 9.12, 9.19, 9.28, and 9.33)

Understand Zaitsev's rule and Hofmann's rule and when each applies. (Problems 9.7, 9.8, 9.9, and 9.17)

Show the expected products, including stereochemistry and regiochemistry, of any elimination reaction. (Problems 9.1, 9.5, 9.10, 9.13, 9.14, 9.17, 9.22, 9.27, 9.34, 9.36, and 9.37)

Predict whether the major pathway that will be followed under a particular set of reaction conditions will be S_N1, S_N2, E1, or E2. (Problems 9.15, 9.16, 9.18, 9.25, and 9.26)

Chapter 10
SYNTHETIC USES OF SUBSTITUTION
AND ELIMINATION REACTIONS
INTERCONVERTING FUNCTIONAL GROUPS

10.1

a) $CH_3CH_2CH_2\overset{OH}{\underset{|}{CH_2}}$

b)

c) $CH_2=CH\overset{OH}{\underset{|}{CH_2}}$

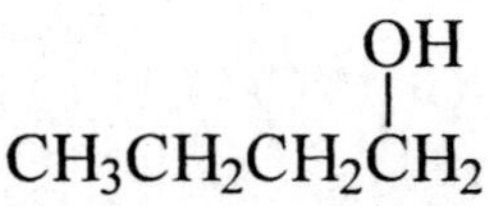

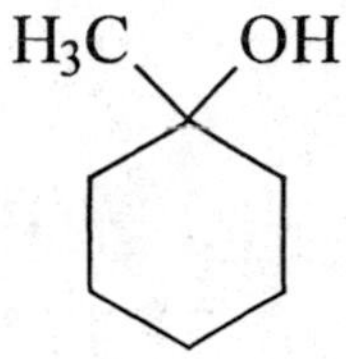

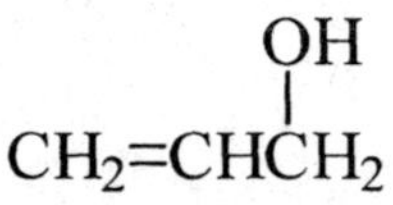

10.2 The reaction conditions (no good nucleophile, polar solvent) favor an S_N1 reaction, so only the chlorine bonded to the tertiary carbon is replaced.

10.3 The stereochemistry of the C-O bond is retained because it is not broken in the reaction.

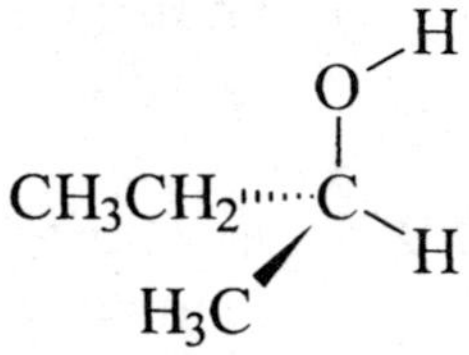

10.4

a)

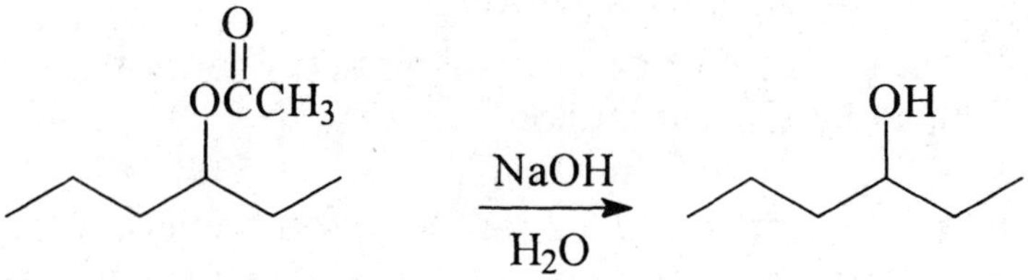

b)

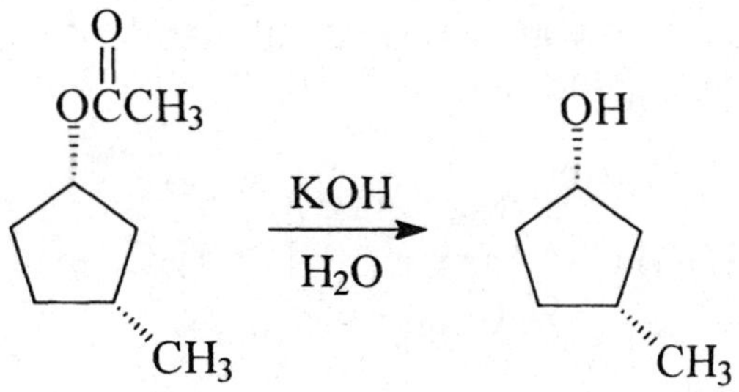

10.5

a) $CH_3CH_2\overset{\displaystyle OCH_2CH_3}{\underset{\displaystyle |}{CH_2}}$

b)

c)

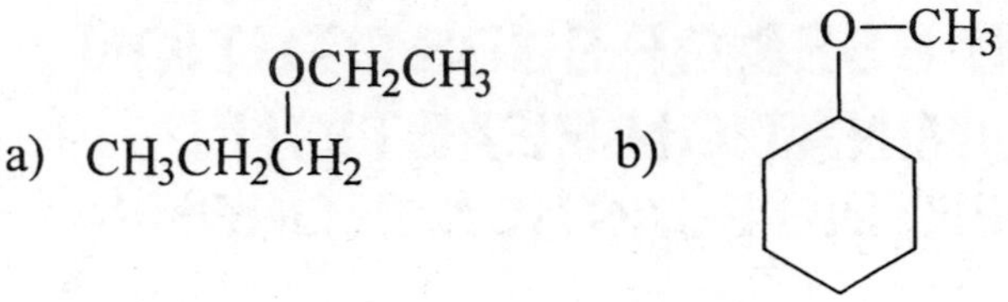

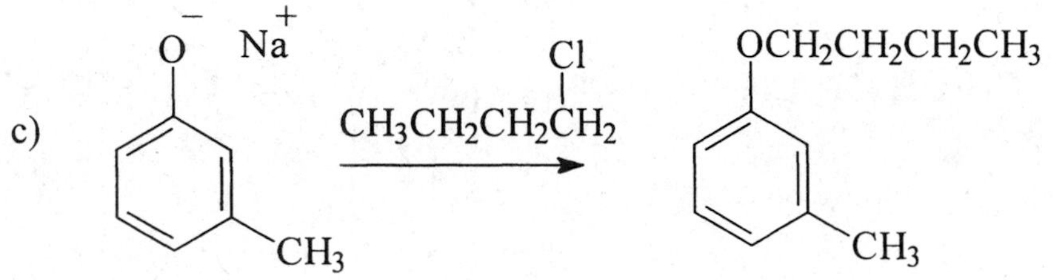

10.6 The N is a stronger base than the O, so this factor favors the N reacting as the nucleophile. However, the two methyl groups on the N decrease its reactivity. The steric effect is controlling the reactivity in this case.

10.7 a) The right route is better because methyl iodide cannot undergo an elimination reaction, so its only reaction with *t*-BuO⁻ ion is substitution. The left route gives primarily E2 elimination (leaving group on a tertiary carbon with a strong base).
b) The right route is better because elimination cannot occur. The left route gives primarily E2 elimination.

10.8 a) To minimize the competing E2 reaction, the conjugate base of the primary alcohol should be reacted with methyl iodide. (However, the reaction of methoxide anion with 1-bromobutane, a primary alkyl halide, would give an acceptable yield also.)

$$CH_3CH_2CH_2CH_2OH \xrightarrow[\text{2) } CH_3I]{\text{1) Na}} CH_3CH_2CH_2CH_2OCH_3$$

b) Because direct nucleophilic substitution cannot be effected on a benzene ring, the conjugate base of phenol must be reacted with the primary alkyl halide.

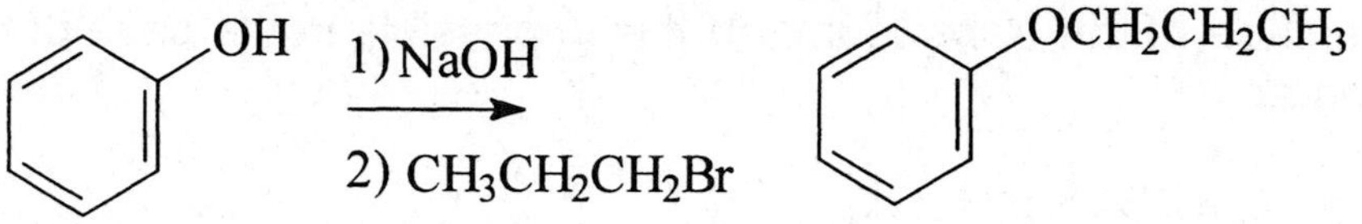

c) To avoid predominant E2 elimination, the conjugate base of the secondary alcohol must be reacted with the primary alkyl halide.

10.9

a)

b) $CH_3\overset{\displaystyle CH_3}{\underset{\displaystyle CH_3}{C}}-O-CH_2CH_3$

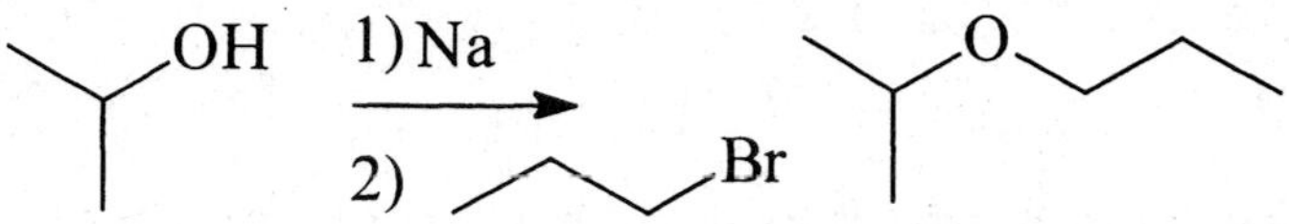

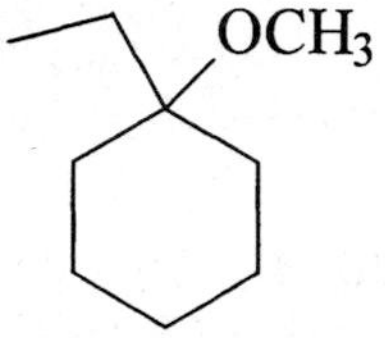

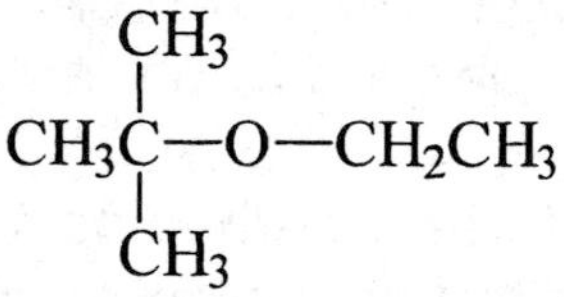

c) $Ph_2CHOCH_2CH_2Cl$

10.10

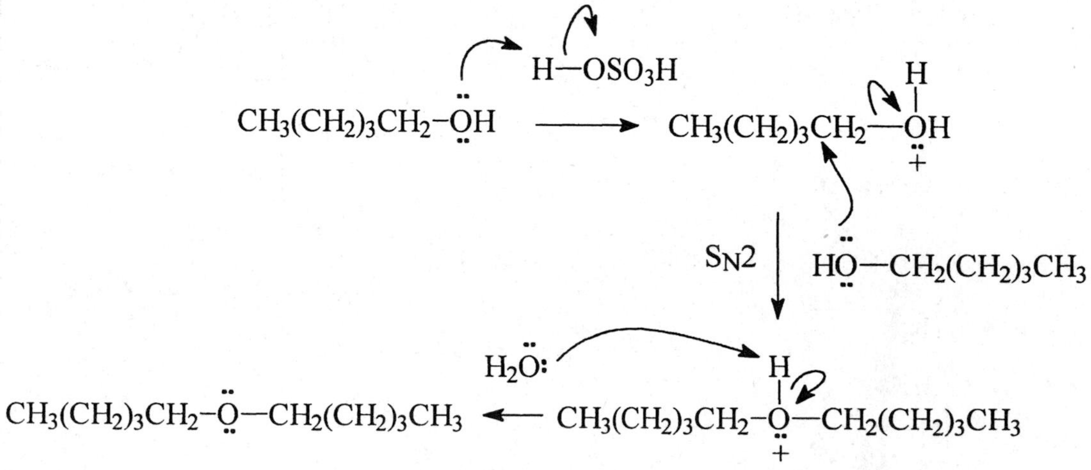

10.11 First hydroxide ion removes the proton from the hydroxy group. Then the negative oxygen acts as a nucleophile and displaces the Cl. Because this step is an intramolecular S_N2 reaction, the oxygen nucleophile must approach from the side opposite the leaving Cl. This can occur readily in the case of the *trans*-isomer but is impossible in the case of the *cis*-isomer.

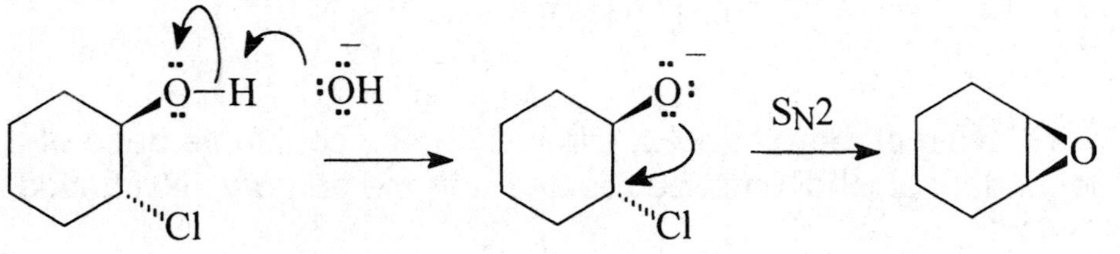

10.12 First OH^- removes the hydrogen of the hydroxy group on the substrate. The conjugate base of the substrate then undergoes an intramolecular S_N2 reaction, displacing Br with inversion. This must occur from the conformation where the negative oxygen and the Br are anti.

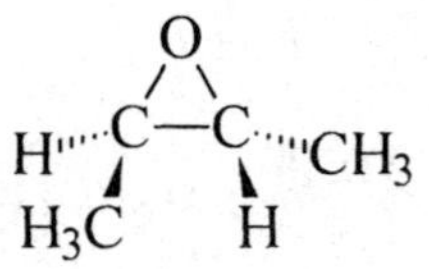

10.13 The hydroxy group on the tertiary carbon acts as the leaving group because the formation of a tertiary carbocation in an S_N1 reaction is greatly favored over the formation of a secondary carbocation.

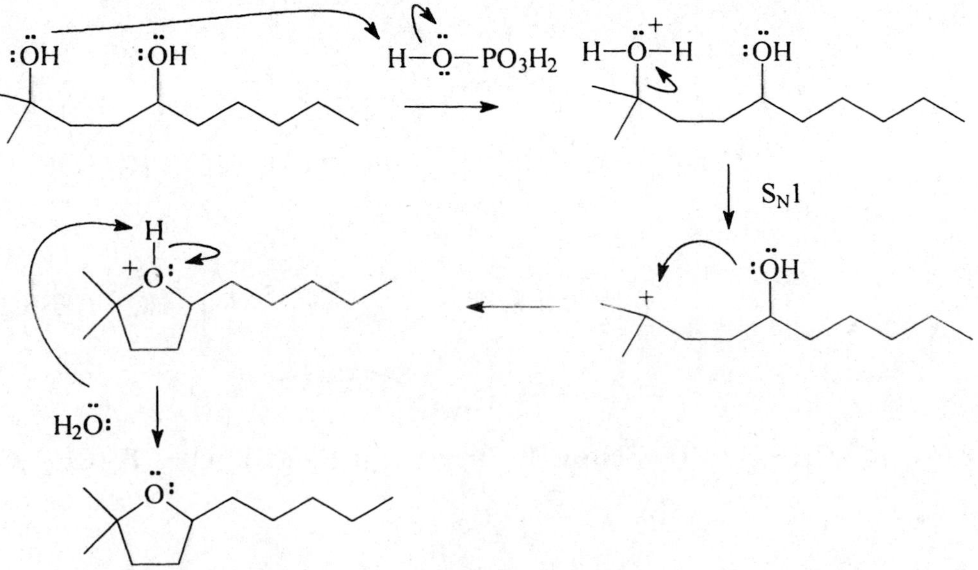

10.14

a) 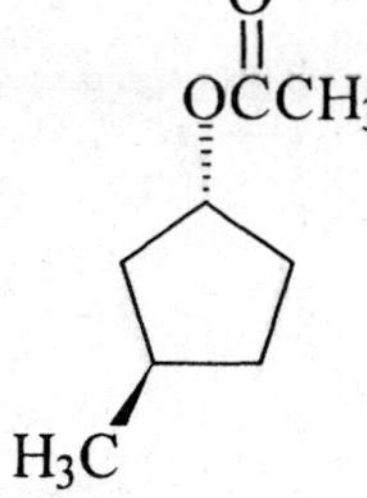

b) 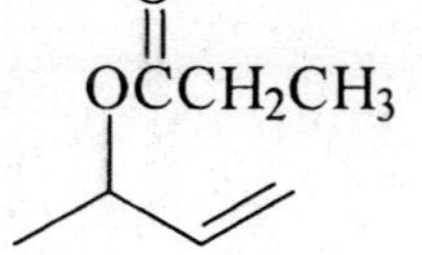

c) 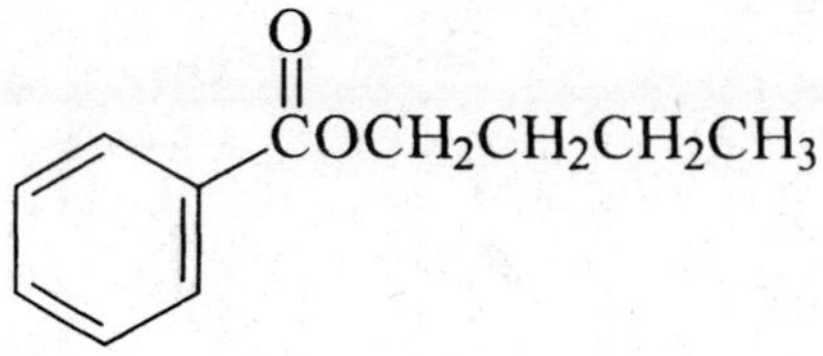

10.15

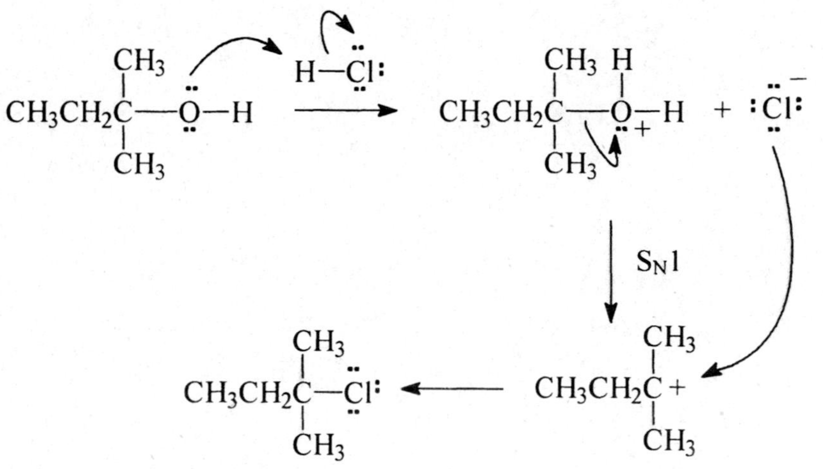

10.16

a) 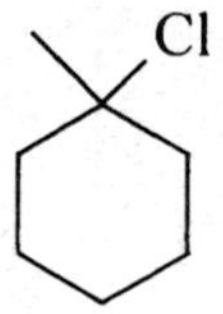

b) 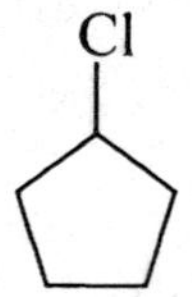

c) 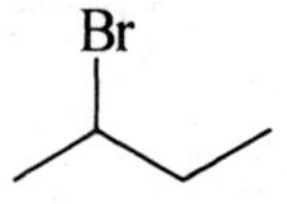

d) $CH_3CH_2CH_2I$

e) 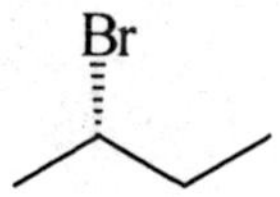

f) 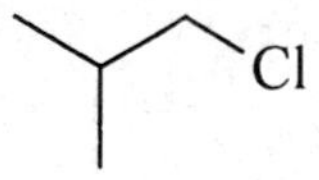

g) 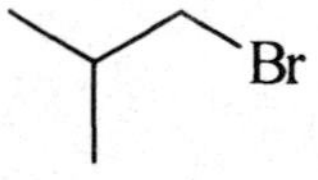

h) 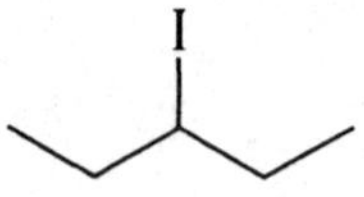

i) 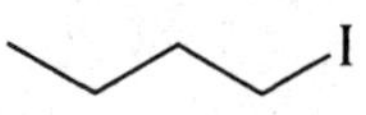

j) 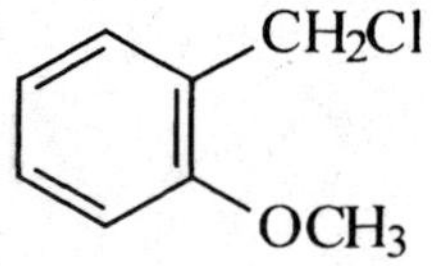

10.17

a)
$$CH_3-\underset{\underset{CH_3CH_2}{|}}{\overset{\overset{CH_3}{|}}{C}}-OH \xrightarrow{\text{HBr}}$$

b)
$$CH_3-\underset{\underset{Ph}{|}}{\overset{\overset{CH_3}{|}}{C}}-OH \xrightarrow{\text{HCl}}$$

c)
$$\xrightarrow[\text{2) NaI}]{\text{1) TsCl, pyridine}}$$

d)
$$\xrightarrow{\text{PBr}_3}$$

e)
$$\xrightarrow{\text{HBr}}$$

f)
$$\xrightarrow{\text{SOCl}_2}$$

10.18

a) $(CH_3CH_2)_2\overset{+}{\underset{+}{N}}CH_2CH_3 \quad \overset{H}{} \; Br^-$

b) $CH_3CH_2-\overset{\overset{CH_3}{|}}{\underset{\underset{CH_3}{|}}{\overset{+}{N}}}-CH_3 \quad I^-$

10.19

a)
$$\xrightarrow[\text{H}_2\text{O}]{\text{NaOH}} CH_3(CH_2)_4CH_2NH_2$$

b)
$$\xrightarrow[\text{H}_2\text{O}]{\text{NaOH}}$$

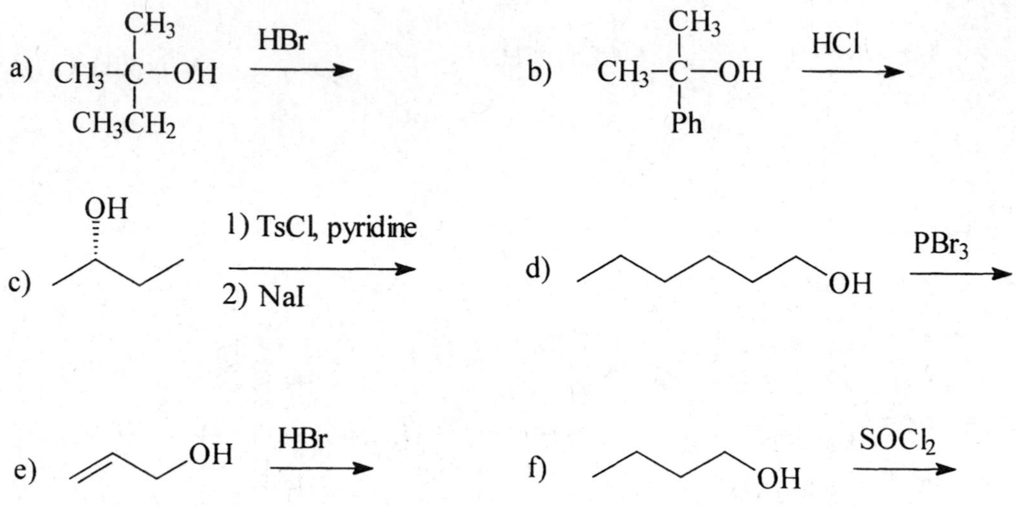

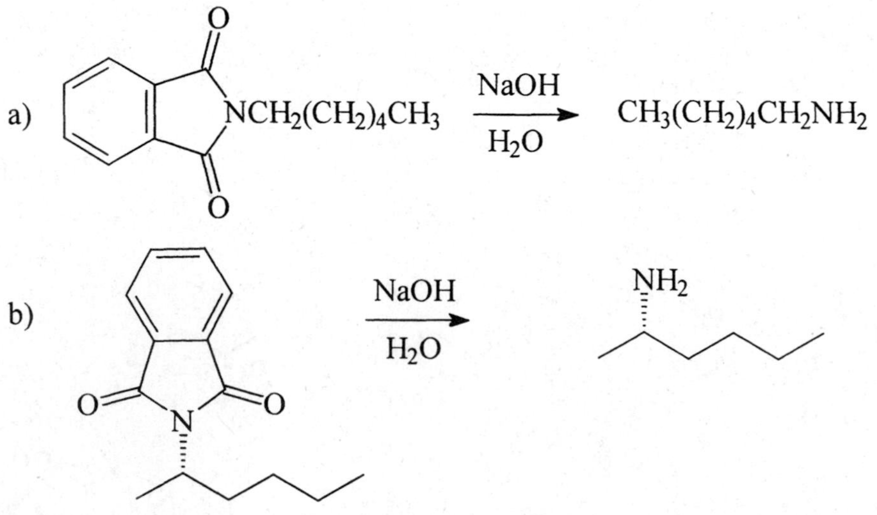

10.20

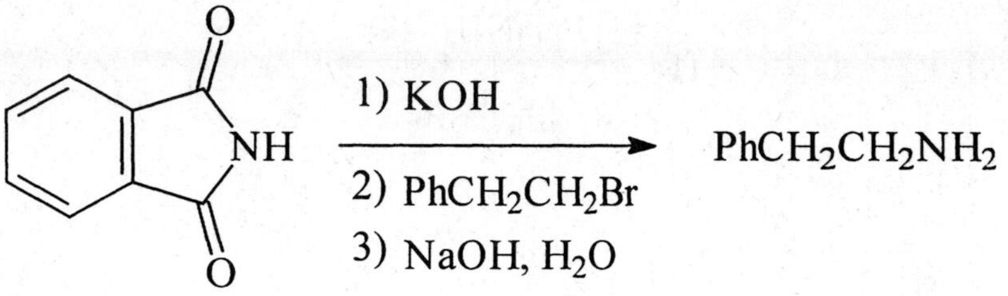

1) KOH
2) PhCH₂CH₂Br
3) NaOH, H₂O

$PhCH_2CH_2NH_2$

10.21

a) [structure of heptane]

b) [structure of 1,3-dimethylbenzene with CH₃ groups]

10.22

a) [structure of phenyl-CH₂CN]

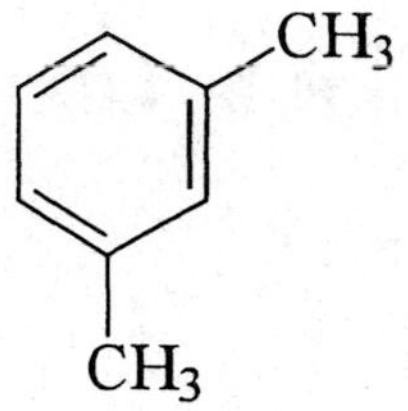

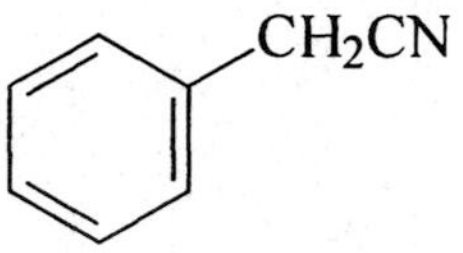

b) $CH_3C{\equiv}C-CH_2CH_2CH_3$

c) [structure with CN]

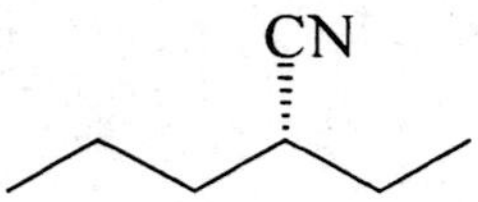

d) $HC{\equiv}C-CH_2CH_3$ $\xrightarrow{\text{1) NaNH}_2\text{ 2) CH}_3\text{I}}$ $CH_3C{\equiv}C-CH_2CH_3$

e) $Cl{-}CH_2CH_2CH_2{-}CN$

f) [cyclopentene structure] + $HC{\equiv}CH$

10.23

a) [cis-1-chloro-3-methylcyclohexane structure] $\xrightarrow[\text{DMSO}]{\text{NaCN}}$

b) $Br-CH_2CH_2CHCH_3$ (with CH₃ branch) $\xrightarrow[\text{NH}_3\ (l)]{H-C{\equiv}C{:}^- Na^+}$

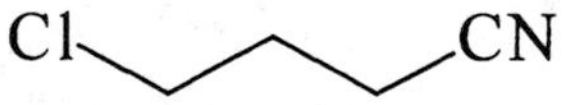

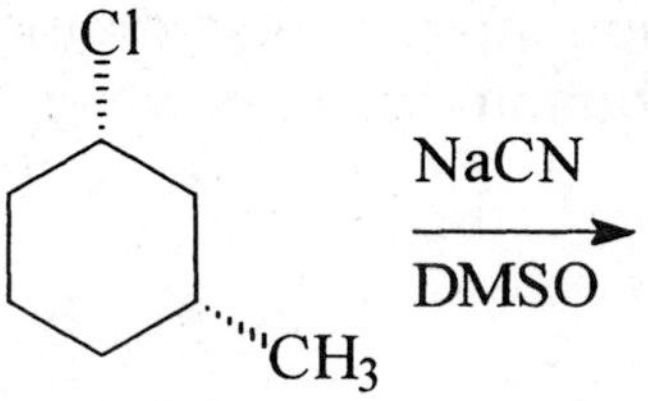

c) $CH_3-C\equiv C-H \xrightarrow[\text{2) } PhCH_2Br]{\text{1) } NaNH_2}$

10.24

a) 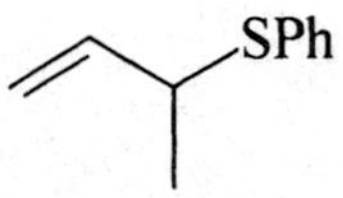 b) $Ph_3\overset{+}{P}-CH_2CH_2CH_3 \; Br^-$ c) $CH_3CH_2CH_2CH_2SCH_3$ d) 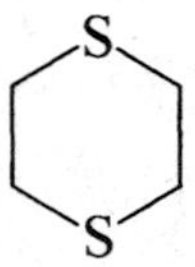

10.25

a) 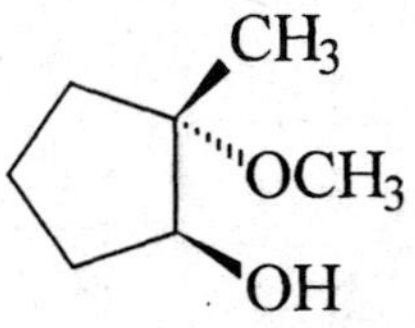

b) 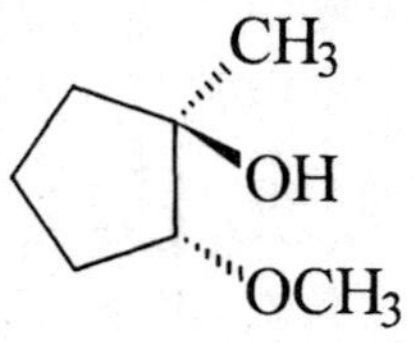

c) 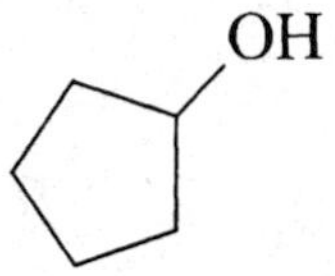

10.26

a) $CH_3CH_2CH{=}CH_2$ b) 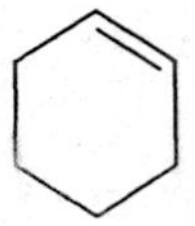 c) 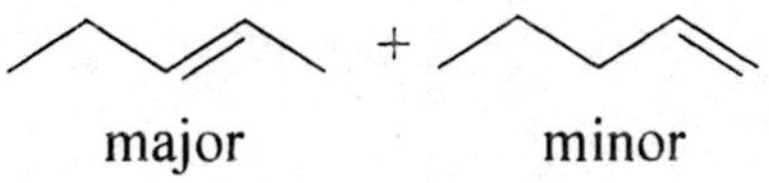

d) 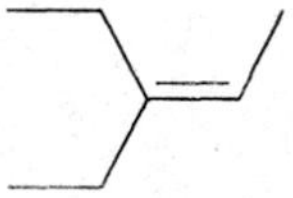

10.27 a) This is not a good way to prepare the desired alkene because the alkene shown would be a minor product according to Zaitsev's rule.
b) This is a good method to prepare the desired alkene. The alkene shown would be the only significant product because it is conjugated.

10.28 The top reaction is better because 3-bromopentane is symmetrical and will produce only the desired alkene. The bottom reaction would produce 1-pentene also.

10.29

a)

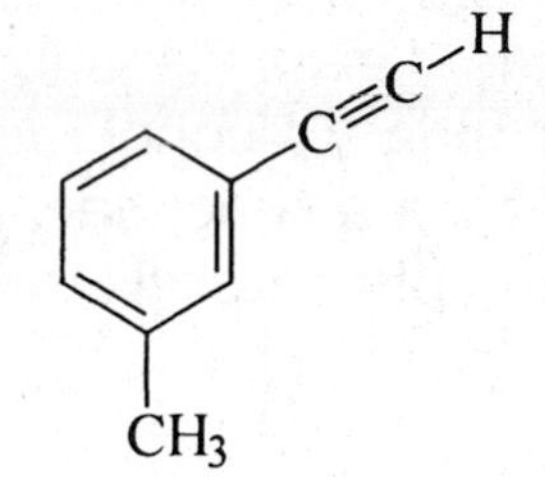

b)

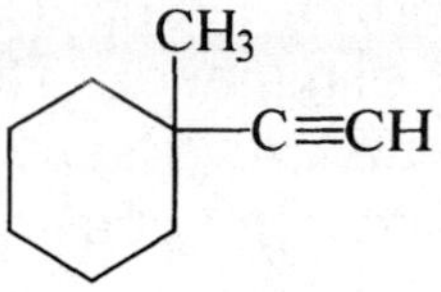

10.30

a) b) c) + +

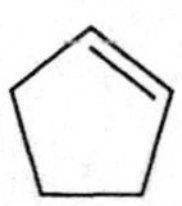

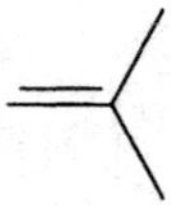

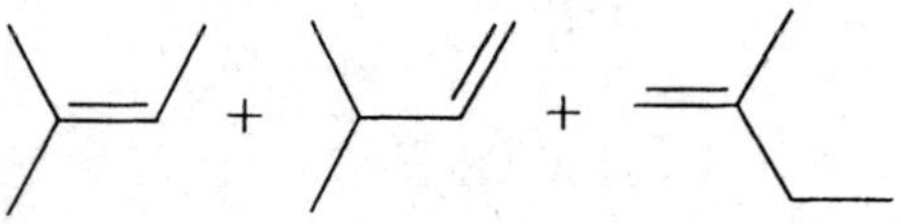

major   minor   minor

10.31

a)

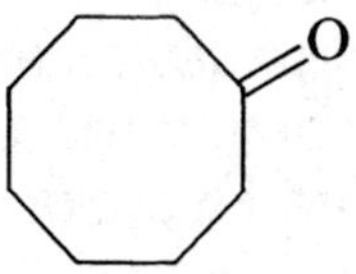

b)

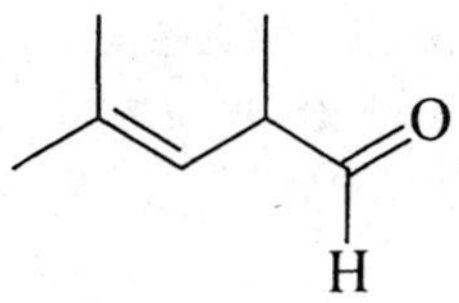

c)

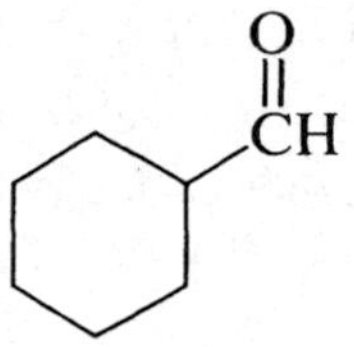

d)

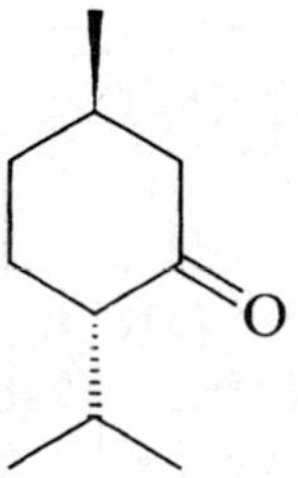

e)

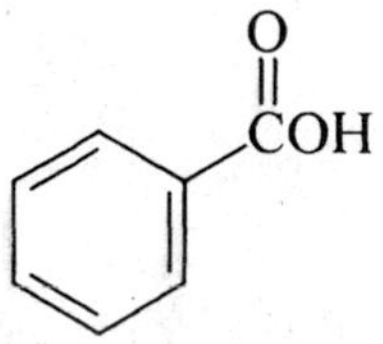

10.32 a) The three methyl groups can be added all at once to butylamine by treating it with excess iodomethane. The primary amine can be prepared by a Gabriel synthesis.

$$CH_3CH_2CH_2CH_2Br \xrightarrow[\text{2) KOH, H}_2\text{O}]{\text{1)}} CH_3CH_2CH_2CH_2NH_2 \xrightarrow[\text{CH}_3\text{I}]{\text{excess}} CH_3CH_2CH_2CH_2N(CH_3)_3 \overset{+}{} \quad I^-$$

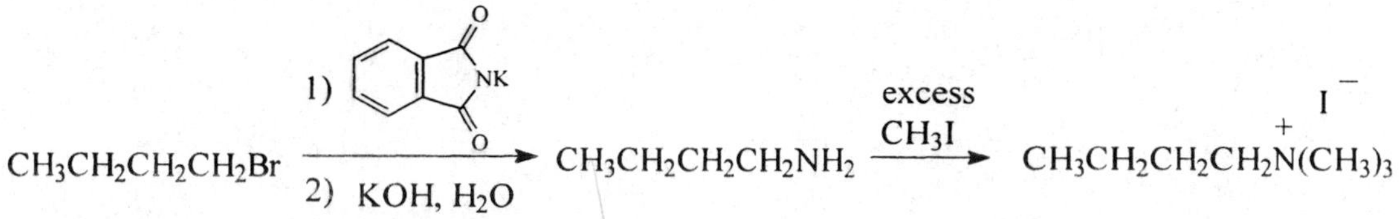

b) The product is an ether, so it can be prepared by a Williamson ether synthesis. We need to react the secondary alkoxide anion with ethyl iodide to avoid E2 elimination. The secondary alcohol required for the Williamson synthesis can be prepared from a secondary halide using acetate anion as the synthetic equivalent of hydroxide ion, again avoiding E2 elimination.

$$
\underset{BrCHCH_2CH_3}{\overset{CH_3}{|}} \quad \xrightarrow[\text{2) KOH, H}_2\text{O}]{\overset{\text{1) CH}_3\text{CO}_2^{-}}{\text{DMSO}}} \quad \underset{HOCHCH_2CH_3}{\overset{CH_3}{|}} \quad \xrightarrow[\text{2) CH}_3\text{CH}_2\text{I}]{\text{1) Na}} \quad CH_3CH_2-O-\underset{CHCH_2CH_3}{\overset{CH_3}{|}}
$$

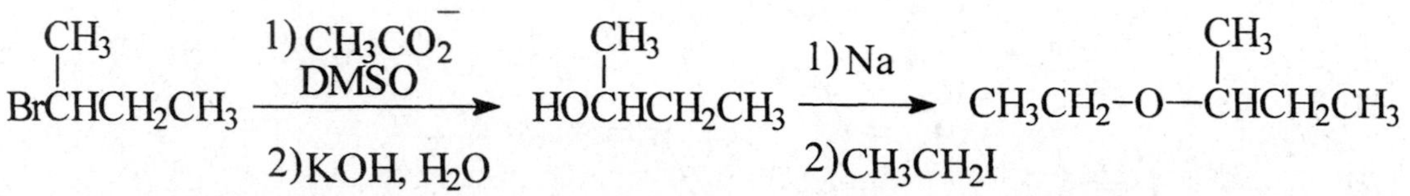

c) The Cl must be replaced by a Br with retention of configuration. We do not know any substitution reaction that proceeds with retention, but the same result can be obtained indirectly by two reactions with inversion. So convert the Cl to an OH with inversion (S_N2 using acetate anion as the equivalent of hydroxide ion to avoid E2). Then convert the OH to a leaving group (tosylate ester) and replace it with a Br with inversion (S_N2).

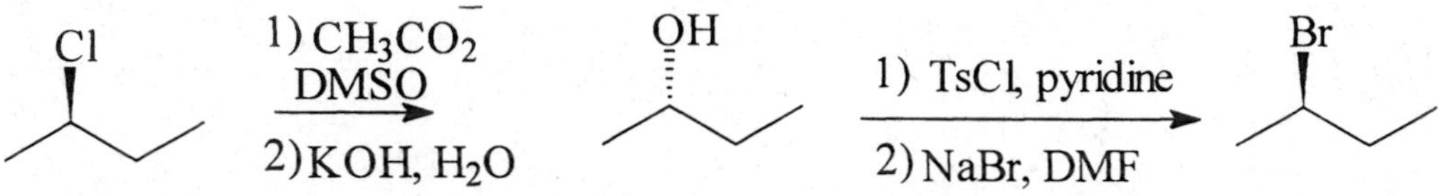

d) The OH of the target comes from the O of the epoxide. The nucleophile must be added to the less substituted carbon of the epoxide, so use basic conditions (CH_3O^-). Remember that the opening will occur with inversion of configuration.

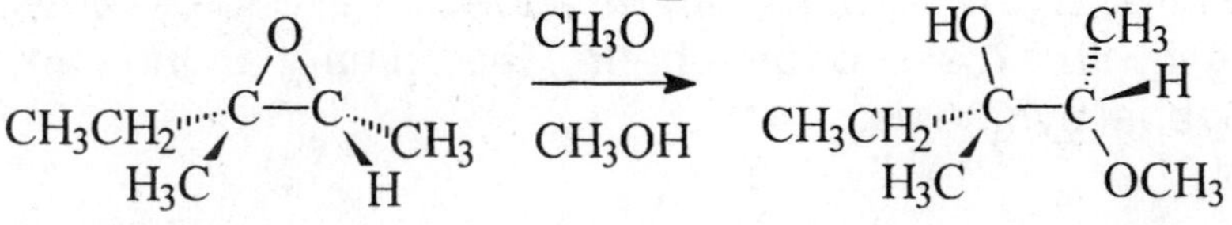

e) We need to add two alkyl groups to ethyne. Use an acetylide anion nucleophile. It does not matter which alkyl group is added first.

$$H{-}C{\equiv}C{-}H \xrightarrow[2)\,PhCH_2Cl]{1)\,NaNH_2} PhCH_2{-}C{\equiv}C{-}H \xrightarrow[2)\,CH_3I]{1)\,NaNH_2} PhCH_2{-}C{\equiv}C{-}CH_3$$

10.33

a) 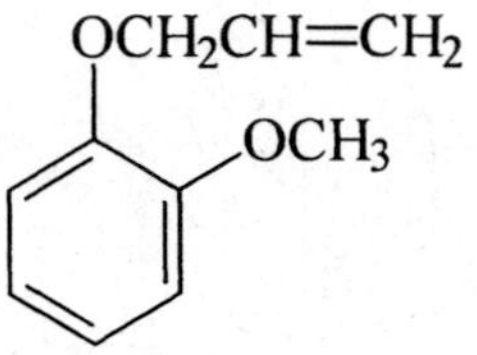

b) 

c) 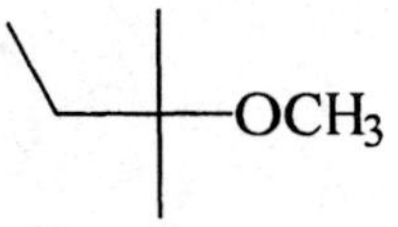

d) 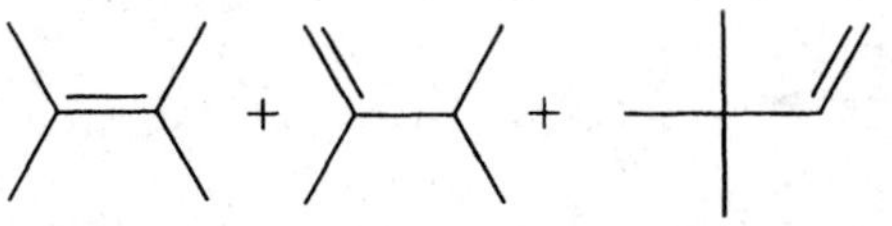

major minor minor

e) $CH_3CH(Cl)CH_3$

f) $PhCH_2NH_2$

g) 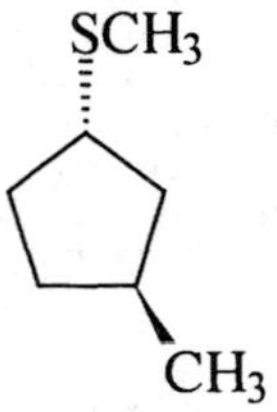

h) 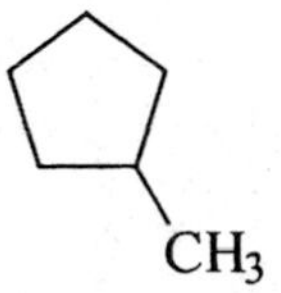

i) $CH_3CH_2CH_2CH_2CN$

j) $CH_3CH_2C{\equiv}CCH_2CH_2CH_2Ph$

k) $CH_3CH(OH){-}CH_2SH$

l) 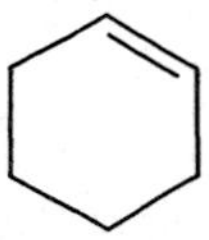

m) $Ph{-}C{\equiv}C{-}Ph$

n) 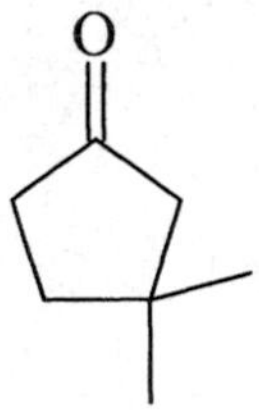

o) $CH_3CH_2CH(CH_3)CH_2CH(=O)$

10.34

a) 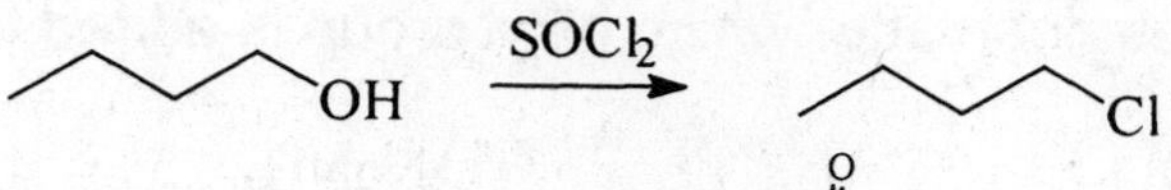

b) 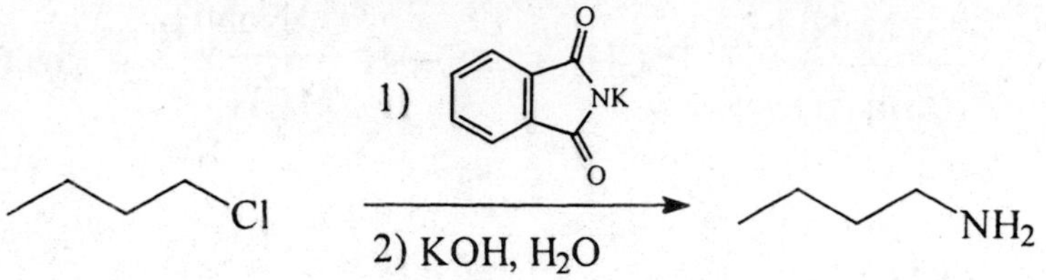

c) 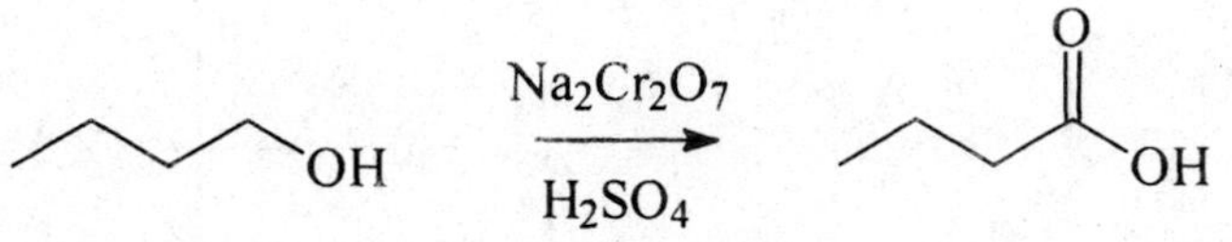

d) 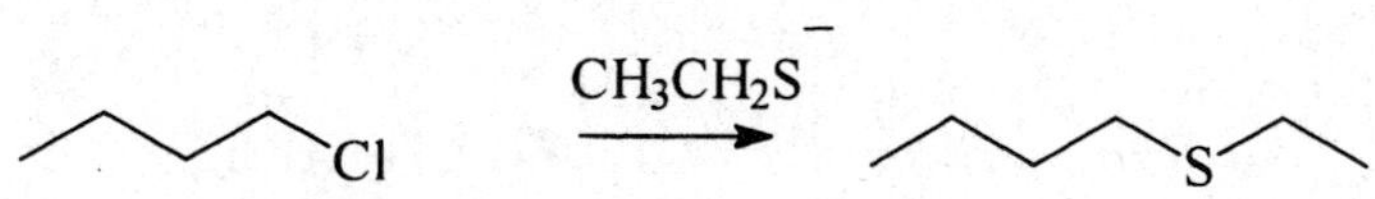

e) 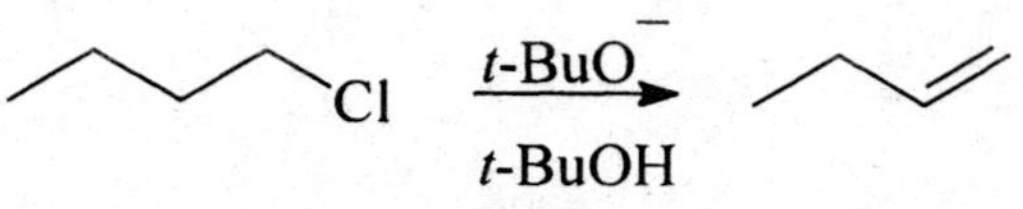

f) 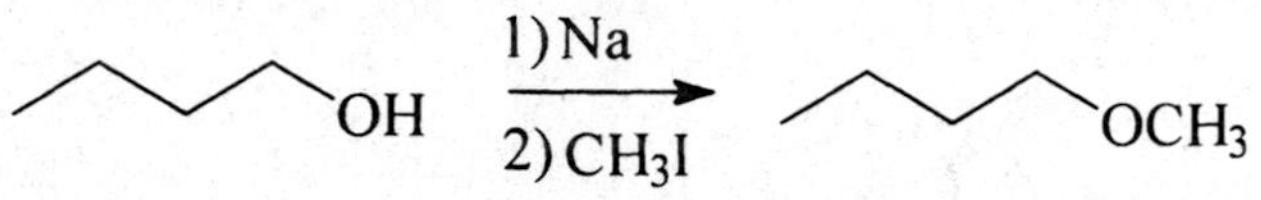

g) 

h) 

i) 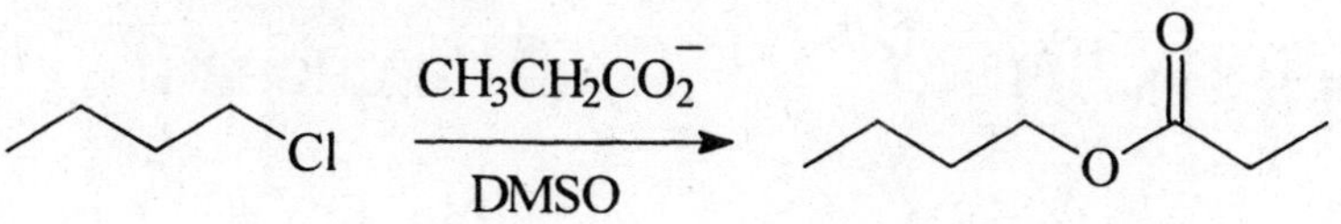

j) 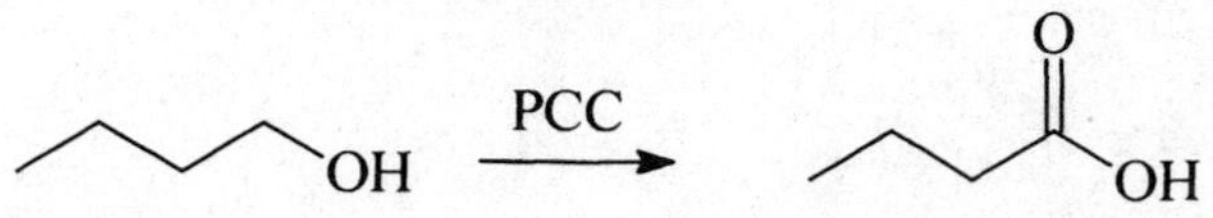

10.35

a) 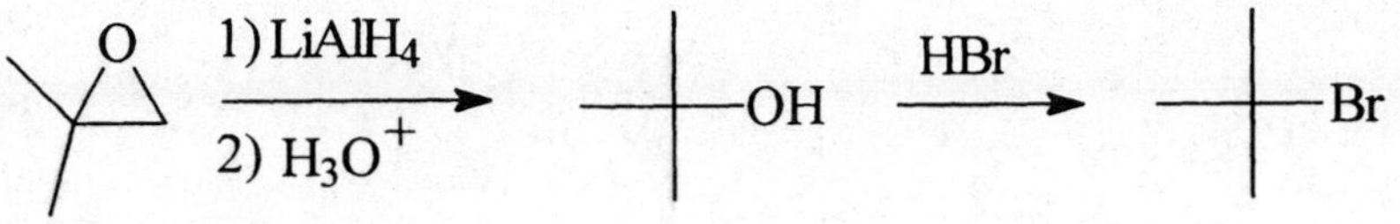

b) 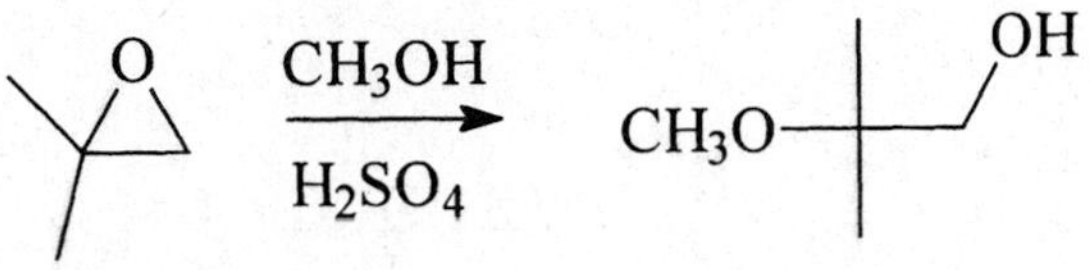

c) 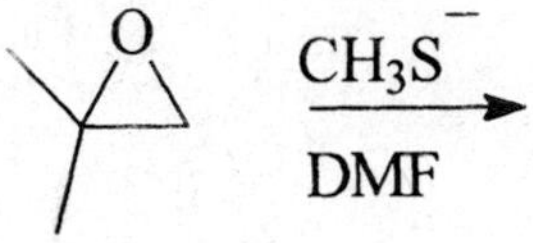

10.36

a) 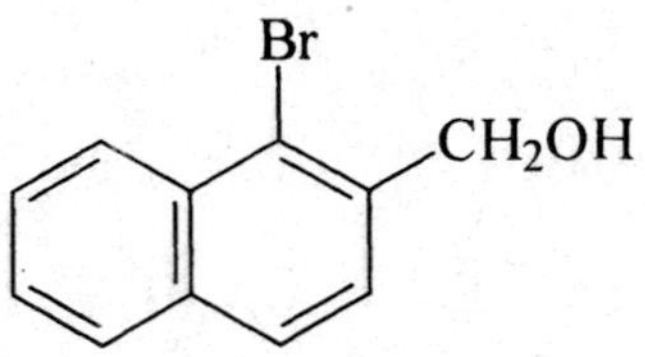

b) 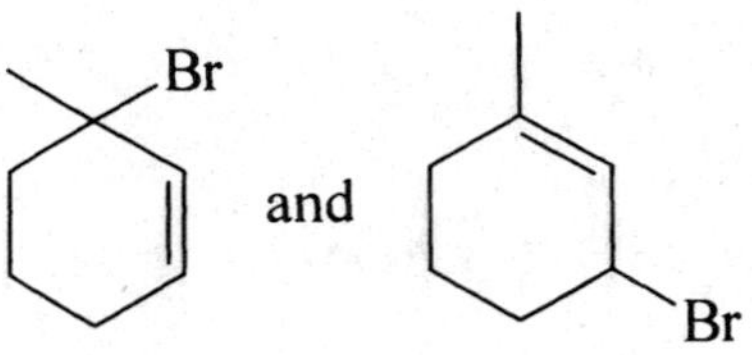

c) 

d) 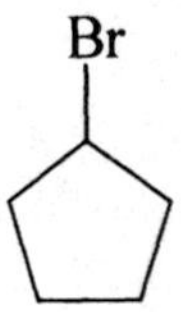

e) 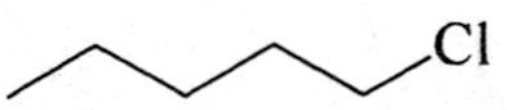

f) 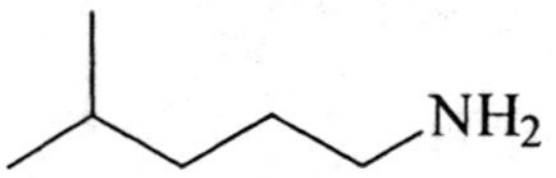

g) $CH_3CH_2\text{—}\overset{+}{P}Ph_3 \quad Br^-$

h) 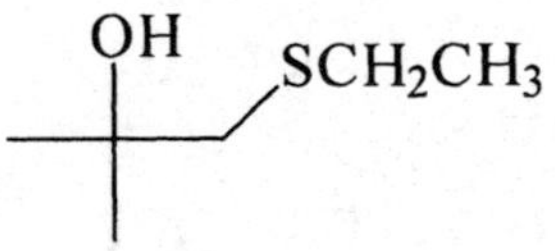

i) $PhCH{=}CH_2$

j) 

k) 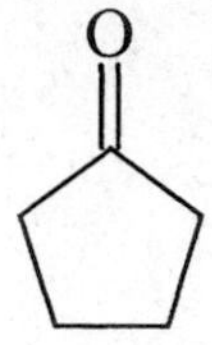

l) $PhCH_2OC(=O)CH_2CH_3$

m) 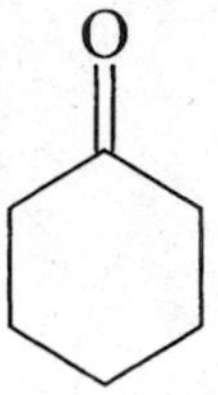

n) 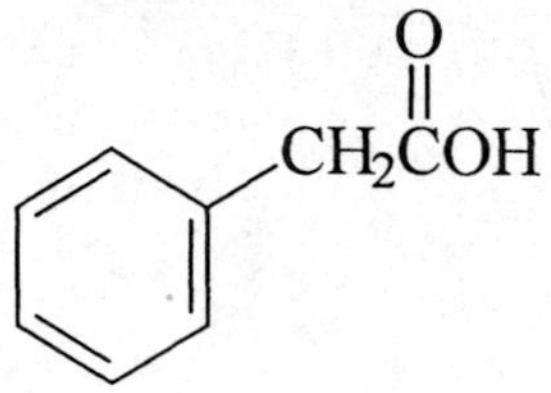

o) 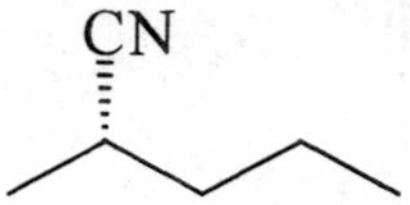

p) $CH_3CH_2CH_2C{\equiv}CCH_2CH_2CH_3$

q) 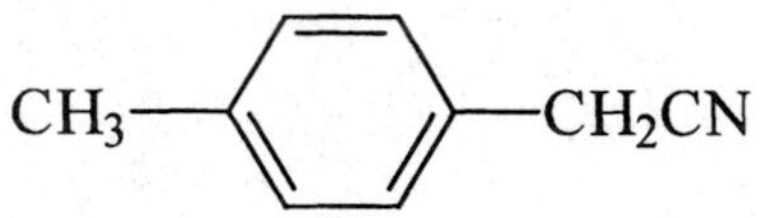

r) $PhCH_2CH_2C(=O){-}H$

10.37

a) PBr_3

b) 1) 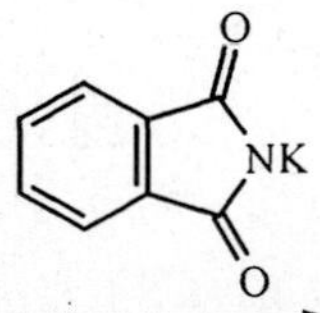
→
2) KOH, H_2O

c) KCN, DMF

d) $\xrightarrow[\text{DMF}]{CH_3CH_2CO_2^-}$

e) KOH, H_2O

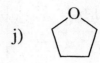

f) H_2O

g) PPh_3

h) $\xrightarrow[CH_3OH]{CH_3O^-}$

i) $\xrightarrow[EtOH,\ \Delta]{KOH}$

j) $\xrightarrow[CH_3OH]{CH_3O^-}$

k) $CH_3CH_2CH_2OH$

l) $SOCl_2$

m) $\xrightarrow[heat]{H_2SO_4}$

n) $\xrightarrow[2)\ CH_3I]{1)\ Na}$

o) $\xrightarrow[acetone]{CrO_3,\ H_2SO_4,\ H_2O}$

p) $\xrightarrow{CH_3C{\equiv}C^-\ Na^+}$

q) PhS^-

r) $\xrightarrow[CH_3CH_2OH]{CH_3CH_2O^-}$

s) PCC

t) Ag_2O, H_2O, THF

10.38

a)  $\xrightarrow[2)\ H_2O]{1)\ LiAlH_4}$ heptane

b) $PhCH_2CH_2CH_2Br$ $\xrightarrow[2)\ KOH,\ H_2O]{1)}$ 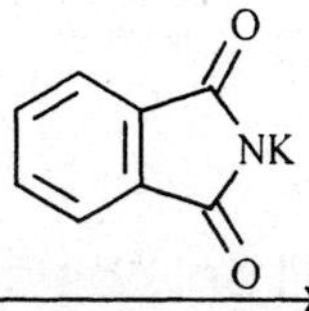 $PhCH_2CH_2CH_2NH_2$

c) $PhCH_2Cl$ $\xrightarrow{HC{\equiv}C^-\ Na^+}$ $PhCH_2{-}C{\equiv}CH$

d) 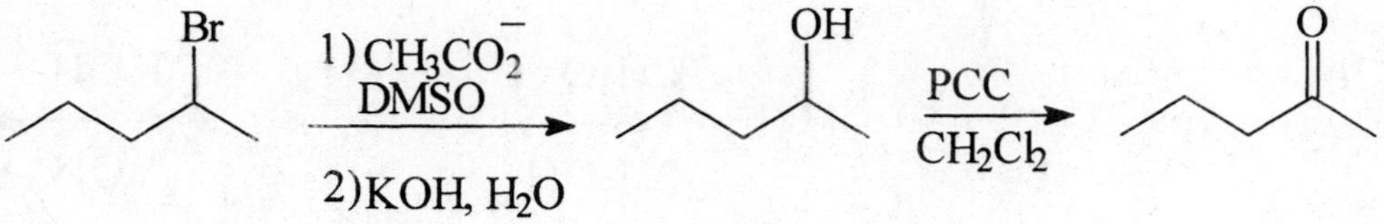

e) 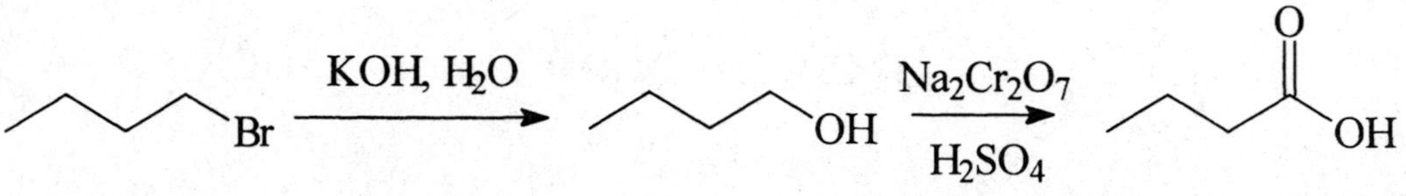

f) 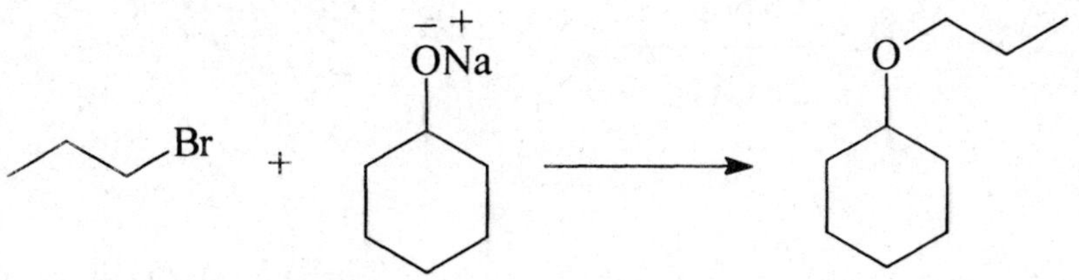

g) 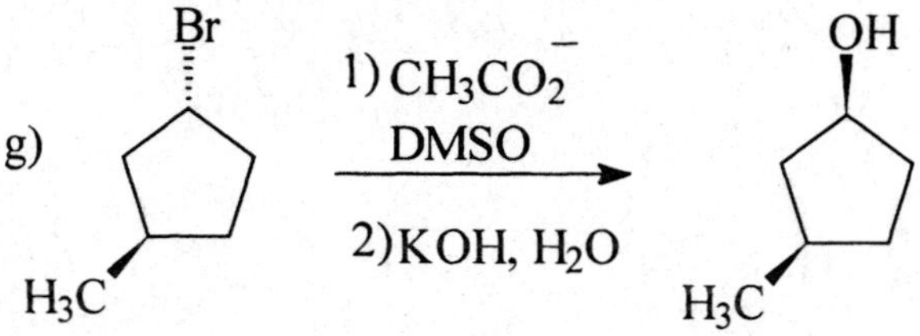

10.39

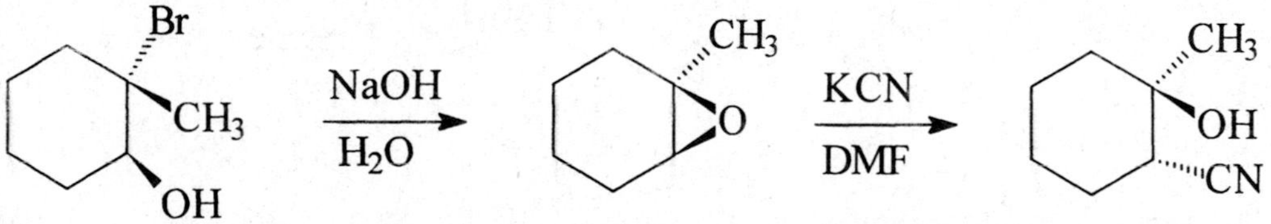

10.40 a) The product shown will only be a very minor product because the reaction of a tertiary alkyl chloride with a strong base favors E2 elimination.

b) The product shown will not form because direct nucleophilic substitution on a benzene ring does not occur.

c) This product will not form because the nucleophile will attack the less hindered carbon of the epoxide ring under basic conditions.

d) The reaction of primary alcohols with HCl requires the presence of the Lewis acid $ZnCl_2$.

10.41 a) The acetylide anion produced in the first step is a strong base, so the second step favors E2 elimination.

b) Multiple alkylation competes, giving a low yield of the desired primary amine.

c) The ether is formed in low yield because the secondary alkyl chloride and the strong base favor E2 elimination.

d) Nucleophilic substitution on a benzene ring does not occur.

e) This reaction follows the S_N1 mechanism, so the product is a mixture of stereoisomers.

f) This is an E1 elimination reaction. The initially formed secondary carbocation can rearrange to form a more stable tertiary carbocation. Therefore, this reaction produces a mixture of products in which the product shown is only a minor component.

10.42 a)

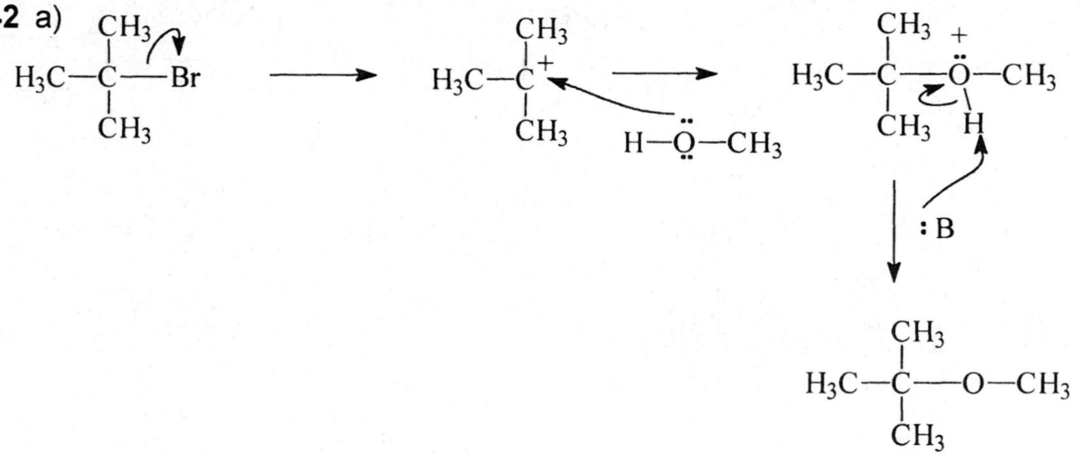

b)

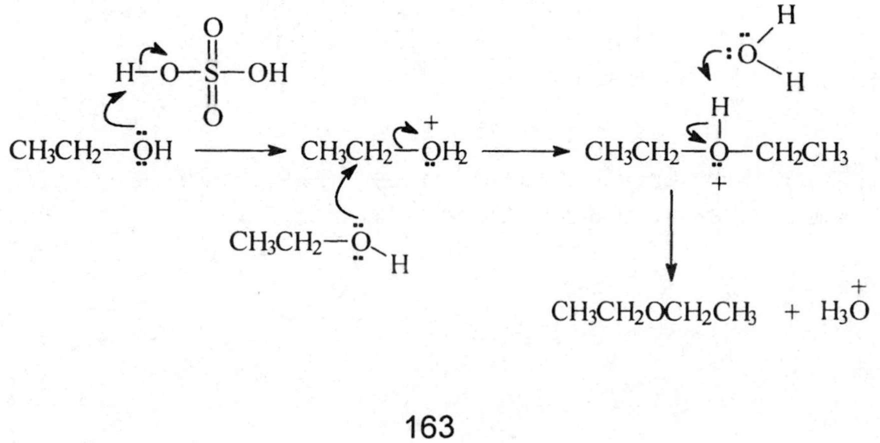

c)

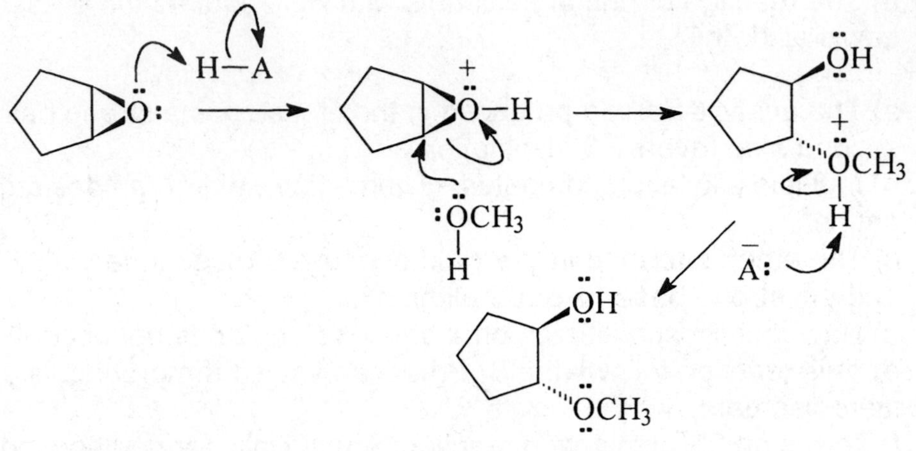

d)

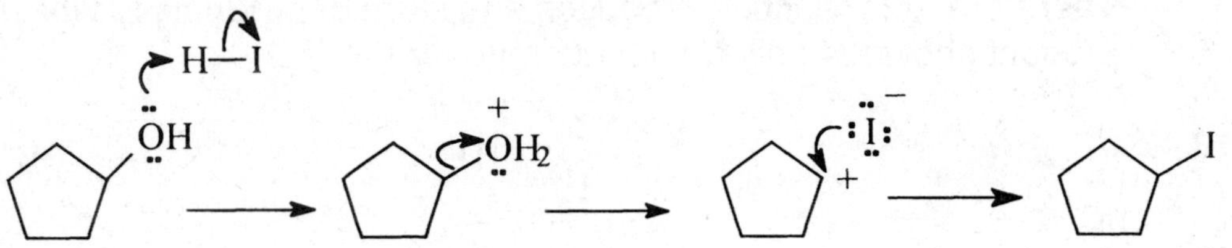

e)

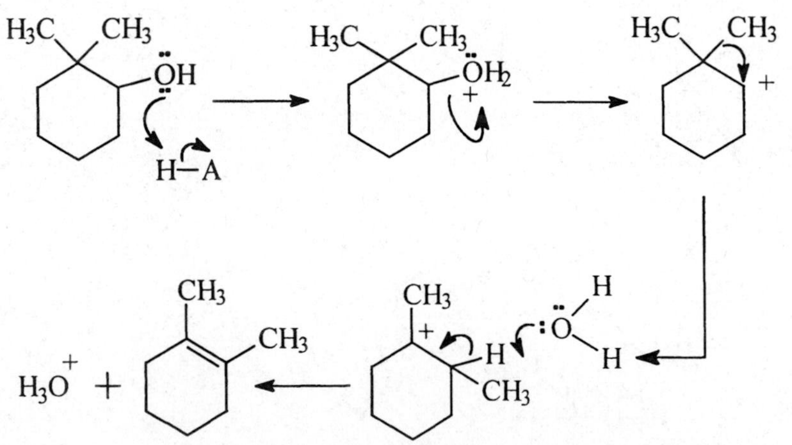

10.43 Attack of methanol at carbon a gives one enantiomer whereas attack at carbon b gives the other.

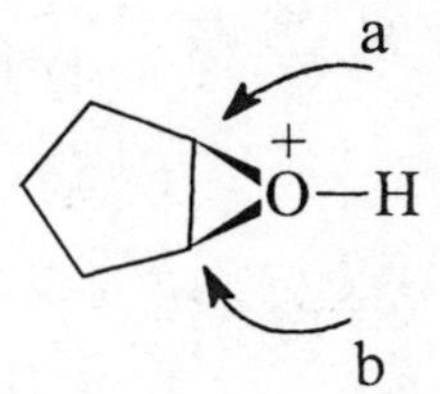

10.44

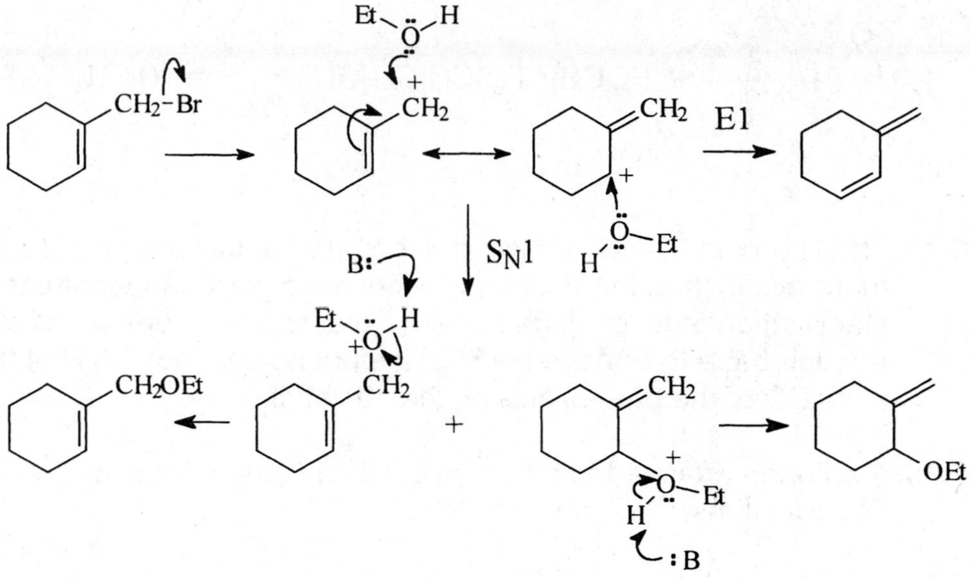

10.45 a) Hydrogens are removed, so this is an oxidation.
b) Hydrogens are added, so this is a reduction.
c) Hydrogens are removed, so this is an oxidation.
d) Hydrogens are added and an oxygen is removed, so this is a reduction.

10.46

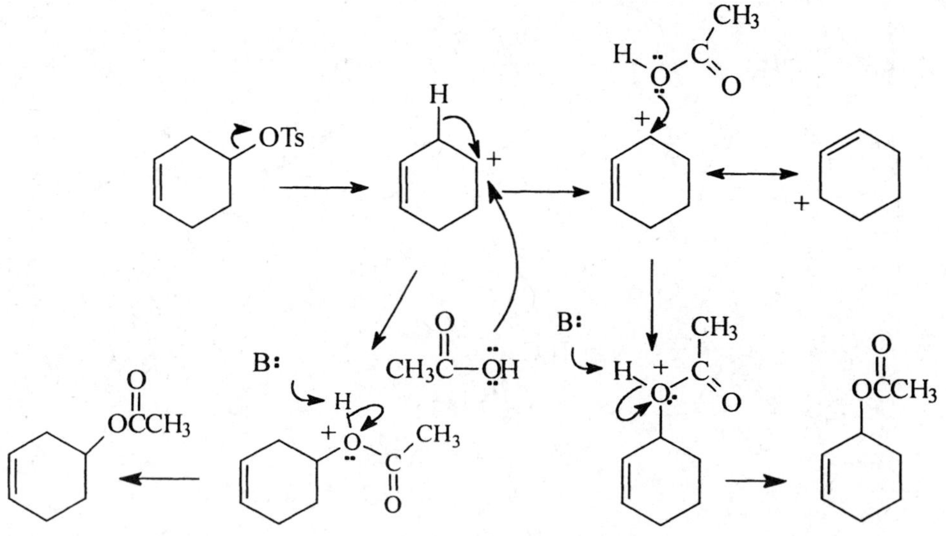

10.47

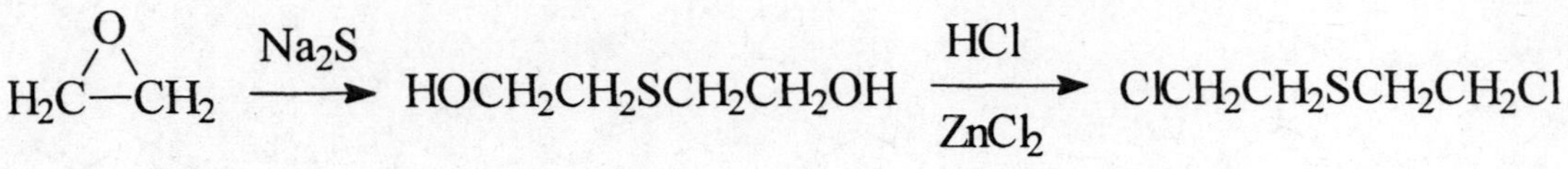

10.48 The H on the oxygen attached directly to the benzene ring (a phenol) is more acidic than the H of the alcohol because it has resonance stabilization of its conjugate base. Potassium carbonate is a strong enough base to remove the H of the phenol, but not the H of the alcohol, so the O of the phenol acts as the nucleophile.

10.49 A chlorine attached to an sp^2-hybridized carbon does not leave in S_N1 or S_N2 reactions.

10.50

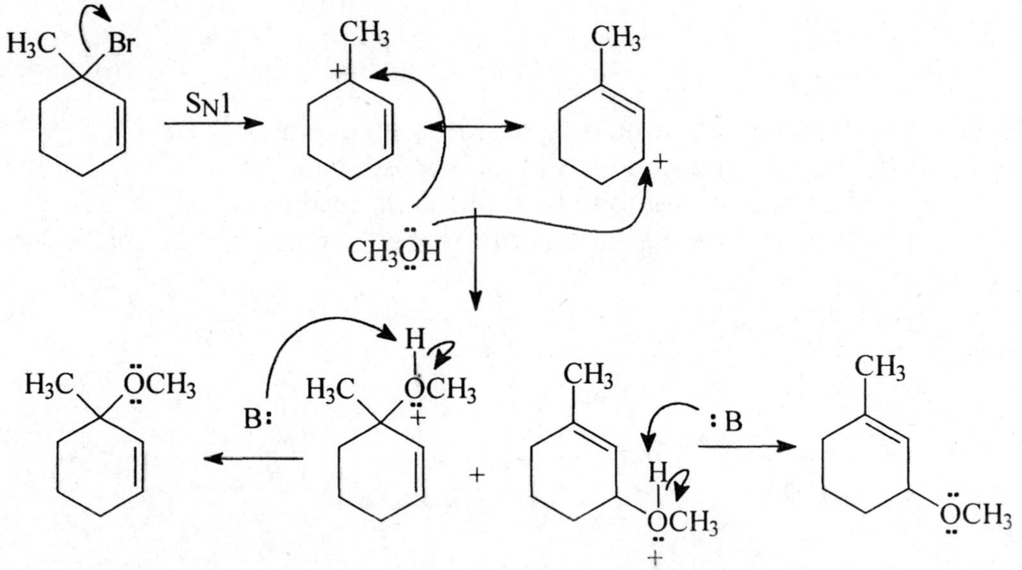

10.51

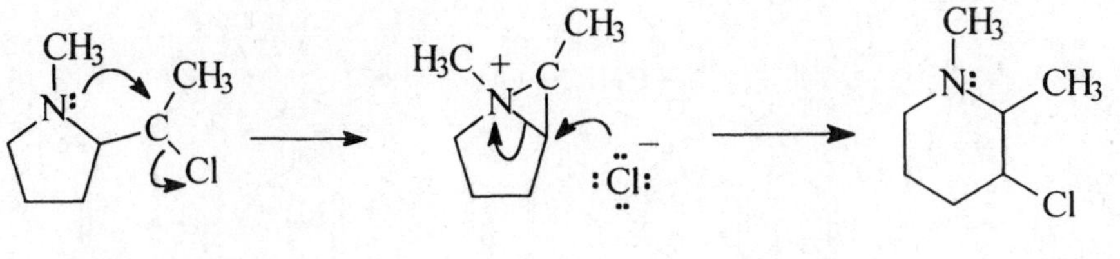

10.52

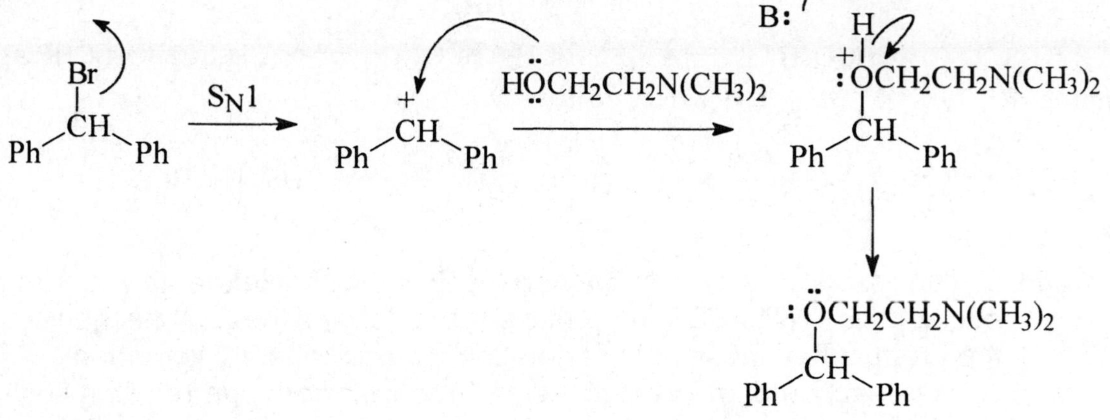

10.53 a)

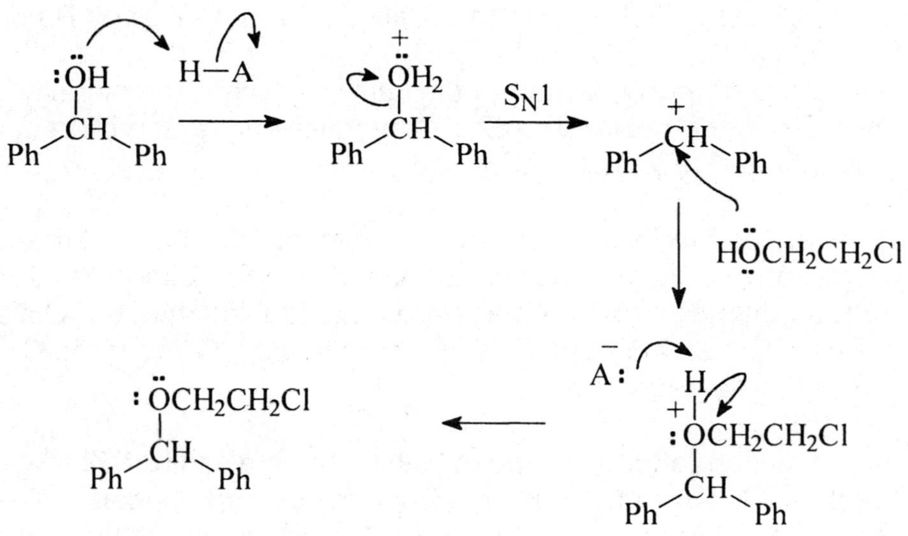

b) The second step is an S_N2 substitution of an amine nucleophile on a primary alkyl halide.

10.54

a) $PhCH_2\overset{\overset{\displaystyle Br}{|}}{C}HCO_2H \xrightarrow[\text{NH}_3]{\text{excess}} PhCH_2\overset{\overset{\displaystyle NH_2}{|}}{C}HCO_2H$

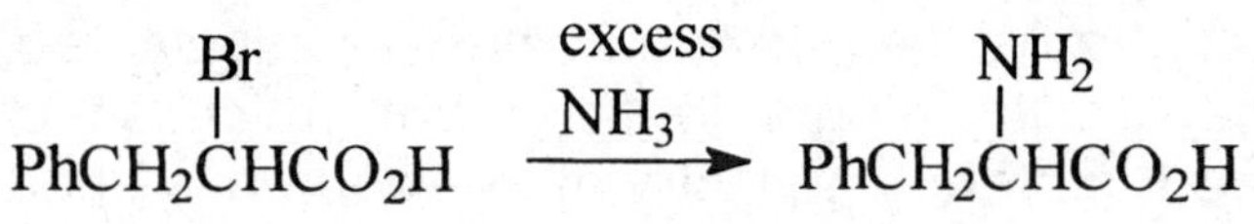

b) $HSCH_2CH_2\underset{}{C}H(NH_2)CO_2H \xrightarrow{NaOH} {}^{-}SCH_2CH_2CH(NH_2)CO_2^{-} \xrightarrow{CH_3I} CH_3SCH_2CH_2CH(NH_2)CO_2H$

c) $HOCH_2CH(NH_2)CO_2H \xrightarrow{PCl_3} ClCH_2CH(NH_2)CO_2H \xrightarrow{{}^{-}SH} HSCH_2CH(NH_2)CO_2H$

10.55 a) The reaction of a secondary alkyl chloride with acetate anion (a good nucleophile) in DMSO (an aprotic solvent) follows the S_N2 mechanism, so the product has inverted configuration. Thus, Product 2 is formed.
b) The reaction of the acetate ester, Product 2 from part (a), with sodium hydroxide in water, does not cleave the bond from the oxygen to the carbon which is the chirality center. Thus, the stereochemistry at the carbon does not change in this reaction. Product 2 gives Product 4.

10.56 The bond from the oxygen to the carbon which is the chirality center is not involved in the reaction, so the stereochemistry at the chirality center is not changed. Product 1 is formed.

10.57 The reaction to form an epoxide is an intramolecular S_N2 reaction. The nucleophile (oxygen) must approach the carbon bonded to the leaving group (chlorine) from the backside. This requires that the OH and the Cl be trans. The chlorohydrin on the right has the OH and the Cl in the proper trans geometry.

10.58 The reaction to form an epoxide is an intramolecular S_N2 reaction. The nucleophile (oxygen) must approach the carbon bonded to the leaving group (chlorine) from the backside. This requires that the OH and the Cl be in the conformation where they are anti. The reactant must be rotated 180° about its C-C bond to get the OH and the Cl anti. The methyl groups end up cis when the reaction occurs from this conformation, so Product 1 is formed.

10.59 a) Under basic conditions, the reaction follows the S_N2 mechanism, so methoxide ion attacks the epoxide at the less substituted carbon. The bond from the epoxide oxygen to the carbon which is a chirality center is not broken, so the stereochemistry at that carbon is unchanged. Product 4 is formed.

b) Under acidic conditions, the mechanism changes to borderline S_N2. The reaction occurs with S_N1 regiochemistry and S_N2 stereochemistry. Methanol bonds to the more highly substituted carbon with inversion of configuration, so Product 1 is formed.

Review of Mastery Goals

After completing this chapter, you should be able to:

Show the major product(s) of any of the reactions discussed in this chapter.
(Problems 10.1, 10.4, 10.5, 10.9, 10.14, 10.16, 10.18, 10.19, 10.21, 10.22, 10.24, 10.25, 10.26, 10.29, 10.30, 10.31, 10.33, 10.36, and 10.40)

Show the stereochemistry of the product(s).
(Problems 10.3, 10.4, 10.12, 10.14, 10.16, 10.19, 10.22, 10.25, 10.33, 10.43, 10.55, 10.56, 10.57, 10.58, and 10.59)

Write the mechanisms of these reactions.
(Problems 10.2, 10.10, 10.11, 10.13, 10.15, 10.42, 10.44, 10.46, 10.50, 10.52, and 10.53)

Synthesize compounds using these reactions.
(Problems 10.7, 10.8, 10.17, 10.20, 10.23, 10.27, 10.28, 10.32, 10.34, 10.35, 10.37, 10.38, 10.39, 10.41, 10.47, and 10.54)

Chapter 11
ADDITIONS TO CARBON-CARBON DOUBLE AND TRIPLE BONDS
REACTIONS OF ALKENES AND ALKYNES

11.1 Like an S_N1 reaction, the first step in the electrophilic addition reaction is the rate determining step. According to the Hammond postulate, the transition state for this step resembles the carbocation intermediate. Structural features that stabilize the carbocation intermediate also stabilize the transition state and accelerate the electrophilic addition reaction.

a) $CH_2{=}CH_2$ < $CH_3CH_2CH{=}CH_2$ < $CH_3CH_2C(CH_3){=}CH_2$

slowest ... fastest

b) 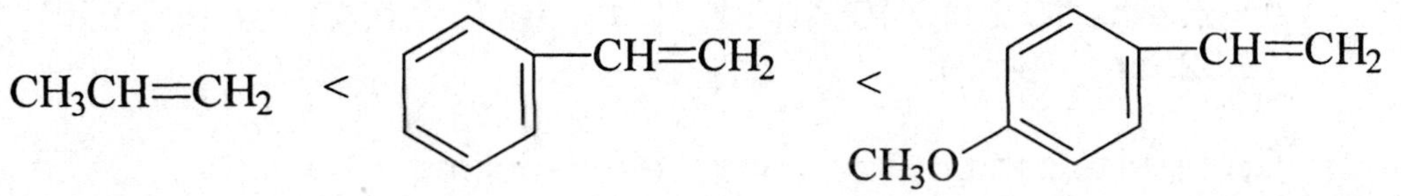

slowest ... fastest

11.2 The two secondary carbocations shown below, of approximately equal stabilities, are formed in the reaction of HBr with 2-hexene, resulting in the formation of two products.

$CH_3\overset{+}{C}HCH_2CH_2CH_2CH_3$ and $CH_3CH_2\overset{+}{C}HCH_2CH_2CH_3$

11.3

a) 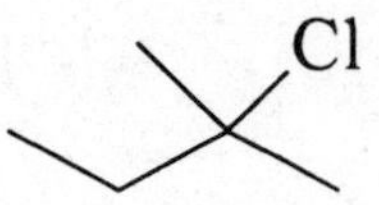

b) 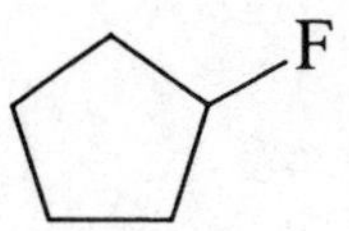

c) 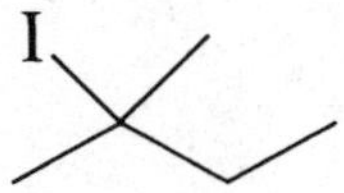

d)

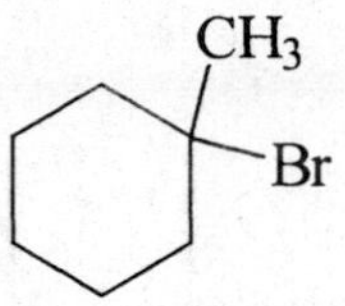

e)

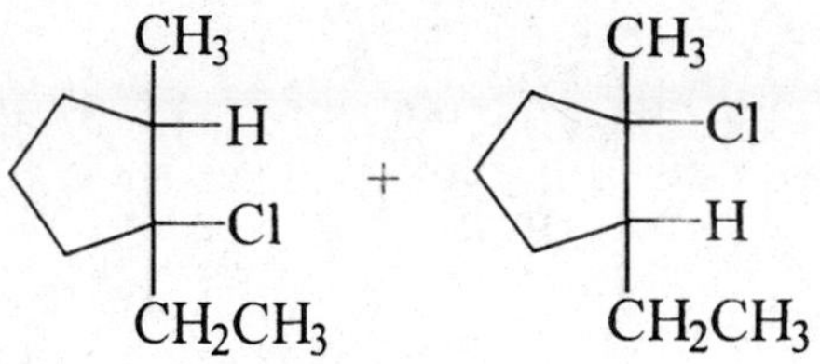

11.4

a) $CH_3\overset{\displaystyle CH_3}{\underset{\displaystyle Br}{CHCHCH_3}}$ + $CH_3\overset{\displaystyle CH_3}{\underset{\displaystyle Br}{CCH_2CH_3}}$ b) $\overset{\displaystyle Ph}{\underset{\displaystyle Br}{C}}=CH_2$ c)

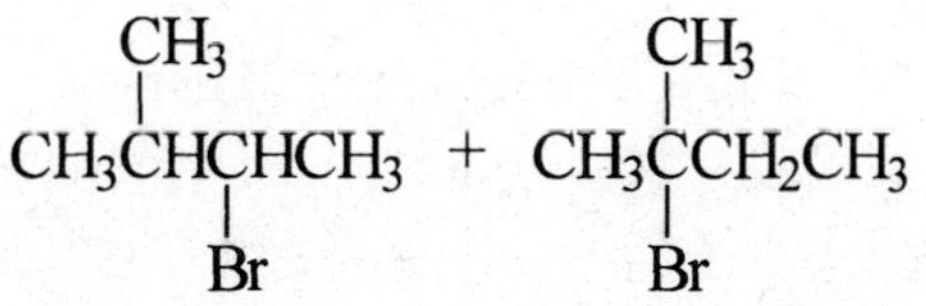

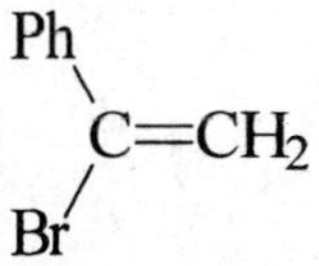

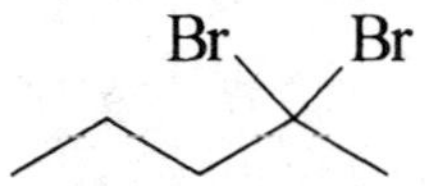

11.5 These are acid catalyzed hydrolysis reactions, where the electrophile is H⁺ ion and the nucleophile is H_2O. Remember, the first step of electrophilic addition is the formation of a carbocation intermediate, so carbocation rearrangement will occur if favorable.

a)

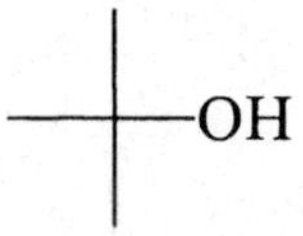

b)

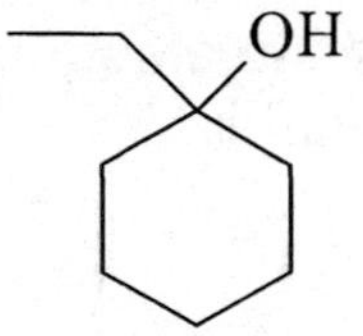

c)

+

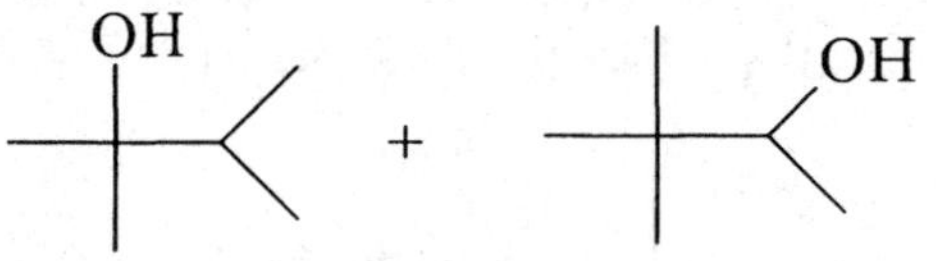

11.6

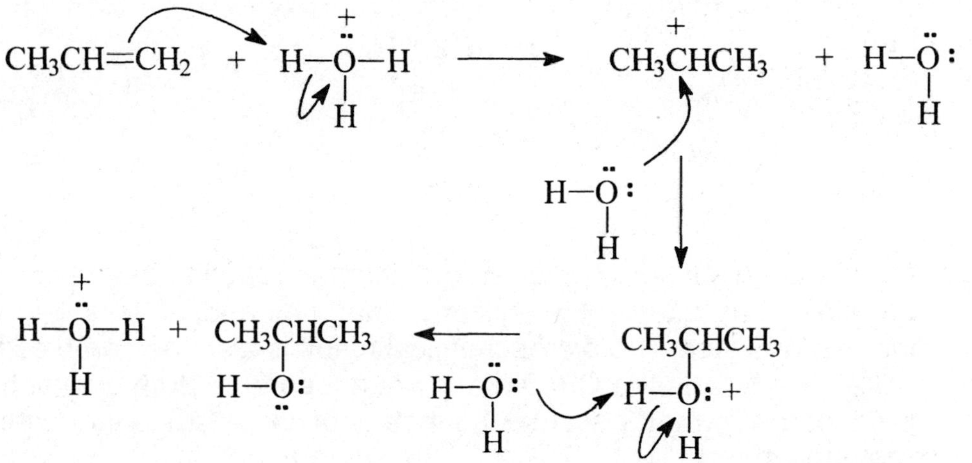

Phenylethene reacts faster than propene because the intermediate carbocation formed from phenylethene is stabilized by resonance.

11.7

a) 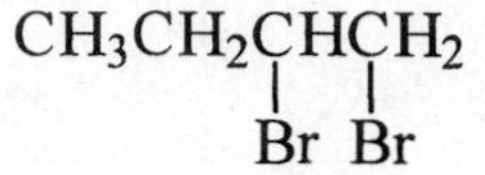

b) 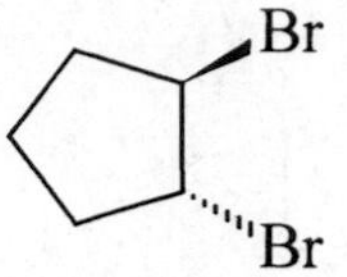

+ enantiomer

c) 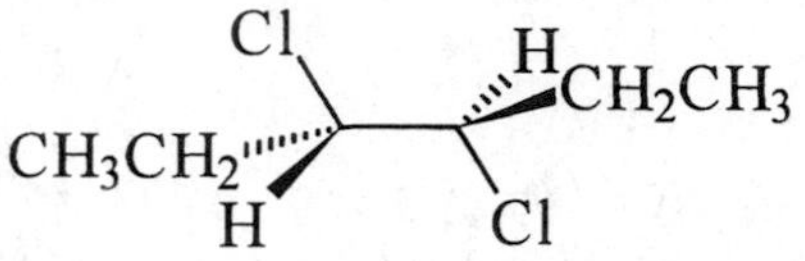

d) $CH_3CH_2CBr_2CHBr_2$

11.8

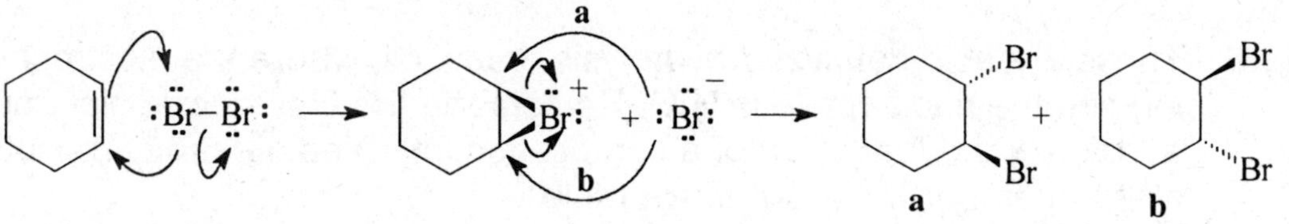

11.9 The right one reacts faster because the electron donating methyl group makes the double bond more nucleophilic.

11.10

b) 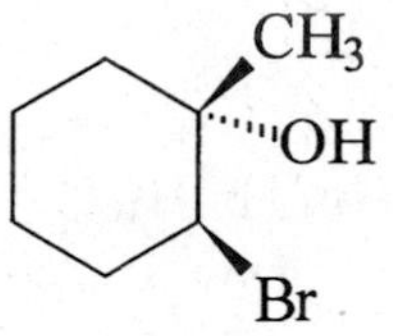

+ enantiomer

c) 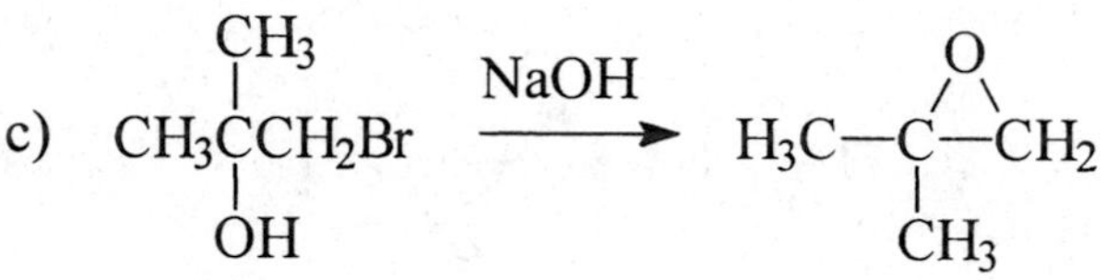

11.11 The mechanism is very similar to that of the reaction of an alkene with bromine in water, with the exception that methanol acts as the nucleophile instead of water. The stereochemical result is also the same as that in water; the Br and the CH_3O group are added anti. Both enantiomers of the product are formed because the methanol can attack either carbon of the bromonium ion.

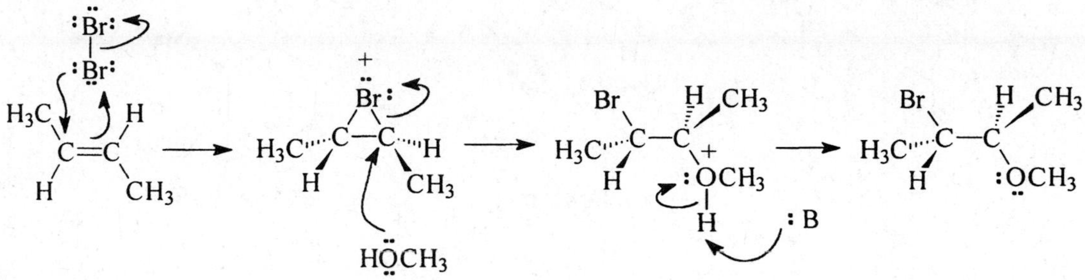

11.12

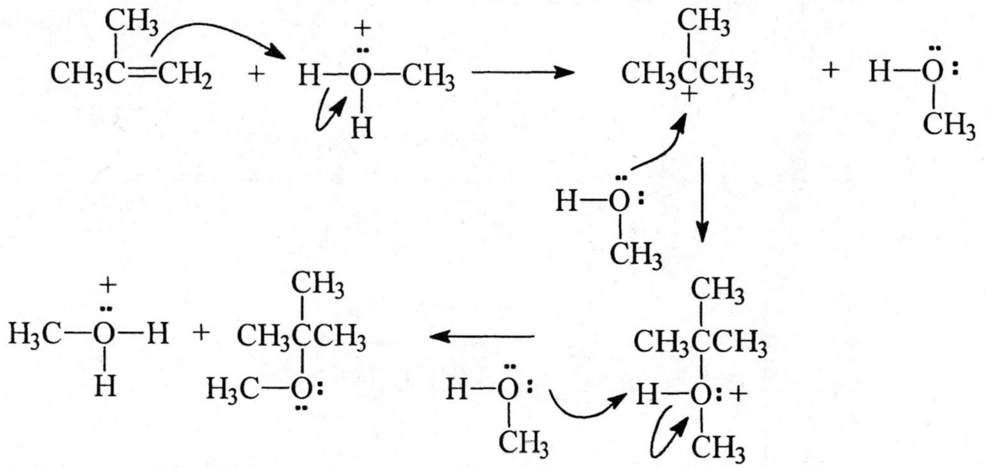

11.13

a) 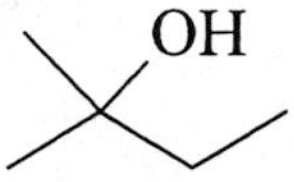

b) 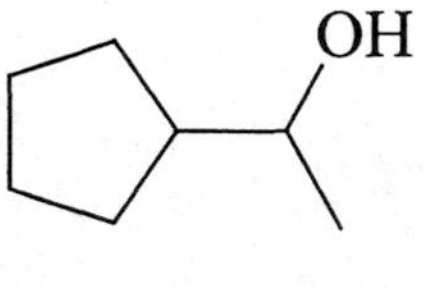

c) 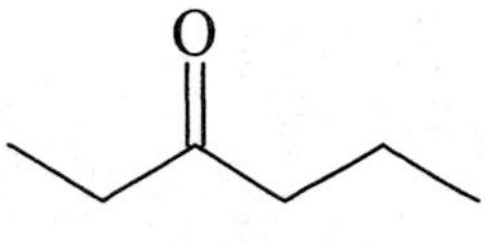

d) 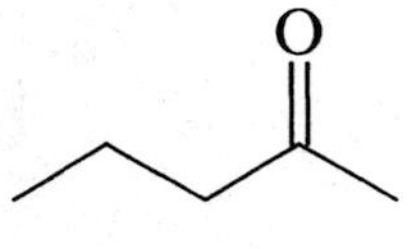

11.14 The left synthesis, starting with 3-hexyne, is better because the alkyne is symmetrical and only 3-hexanone, the desired ketone, is produced. The right reaction, starting with 2-hexyne, also produces 2-hexanone.

11.15

a)

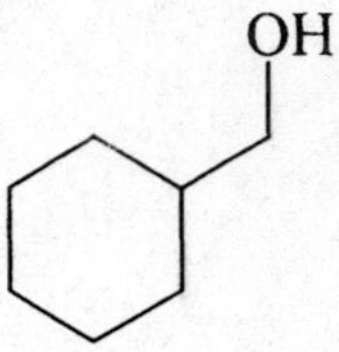

b)

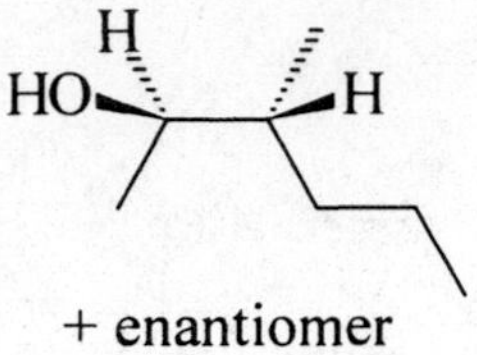

+ enantiomer

c)

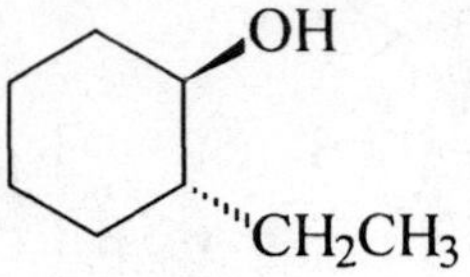

+ enantiomer

11.16 The right alcohol is the major product because the boron prefers to bond to the less hindered carbon of the alkene.

Major

11.17

a)

$$1)\,Hg(O_2CCH_3)_2,\,H_2O$$
$$2)\,NaBH_4,\,NaOH$$

b)

$$1)\,BH_3,\,THF$$
$$2)\,H_2O_2,\,NaOH$$

c)

$$H_2O$$
$$H_2SO_4$$

11.18

a)

b) Ph

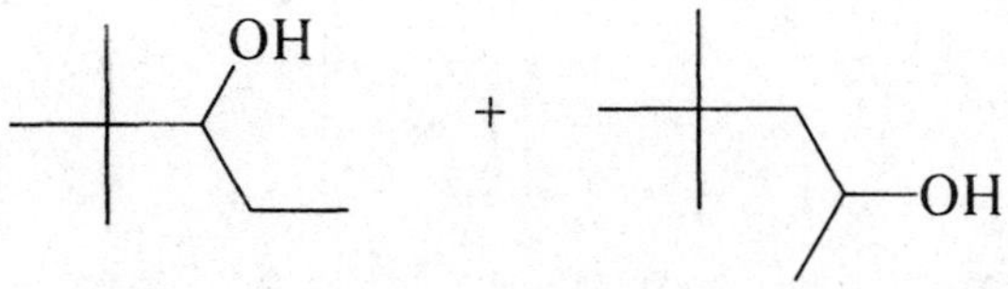

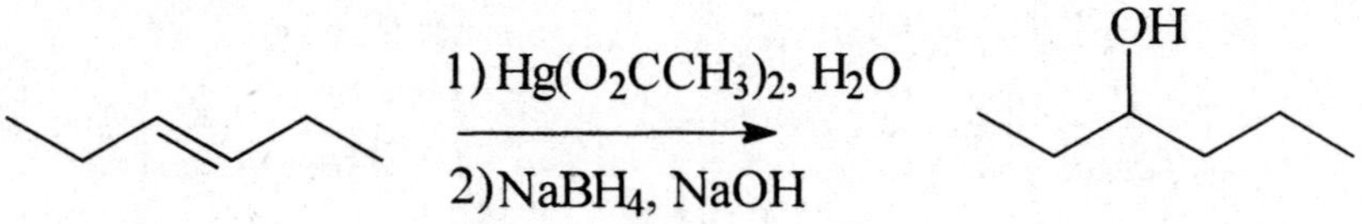

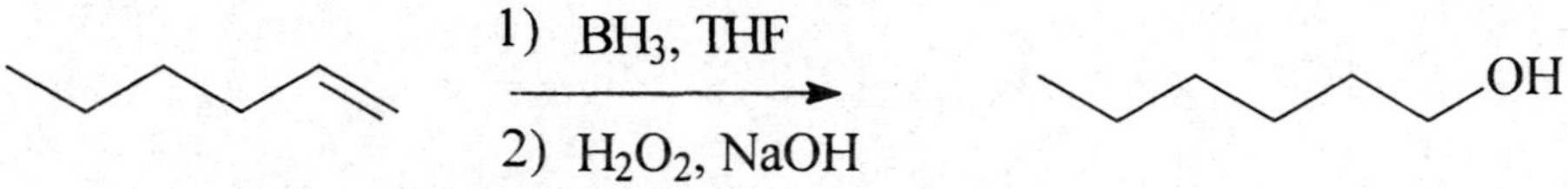

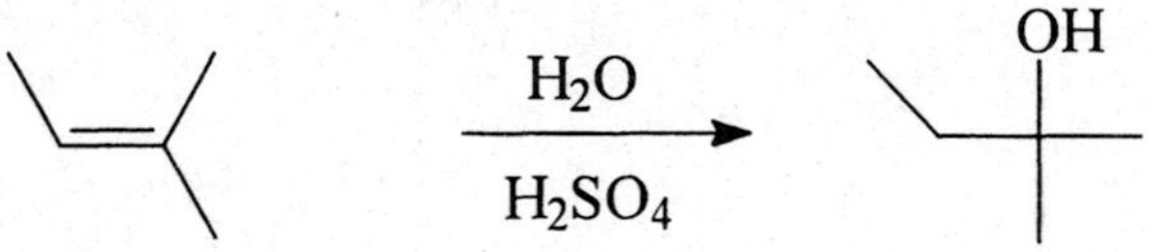

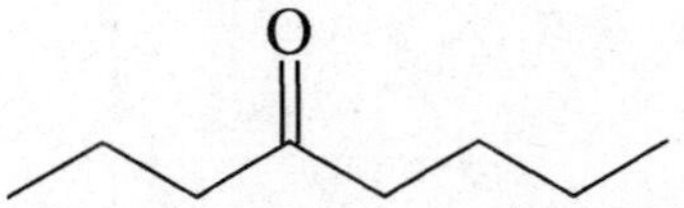

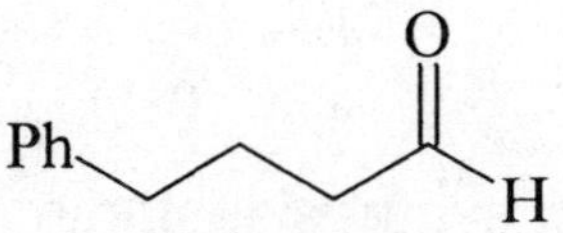

11.19

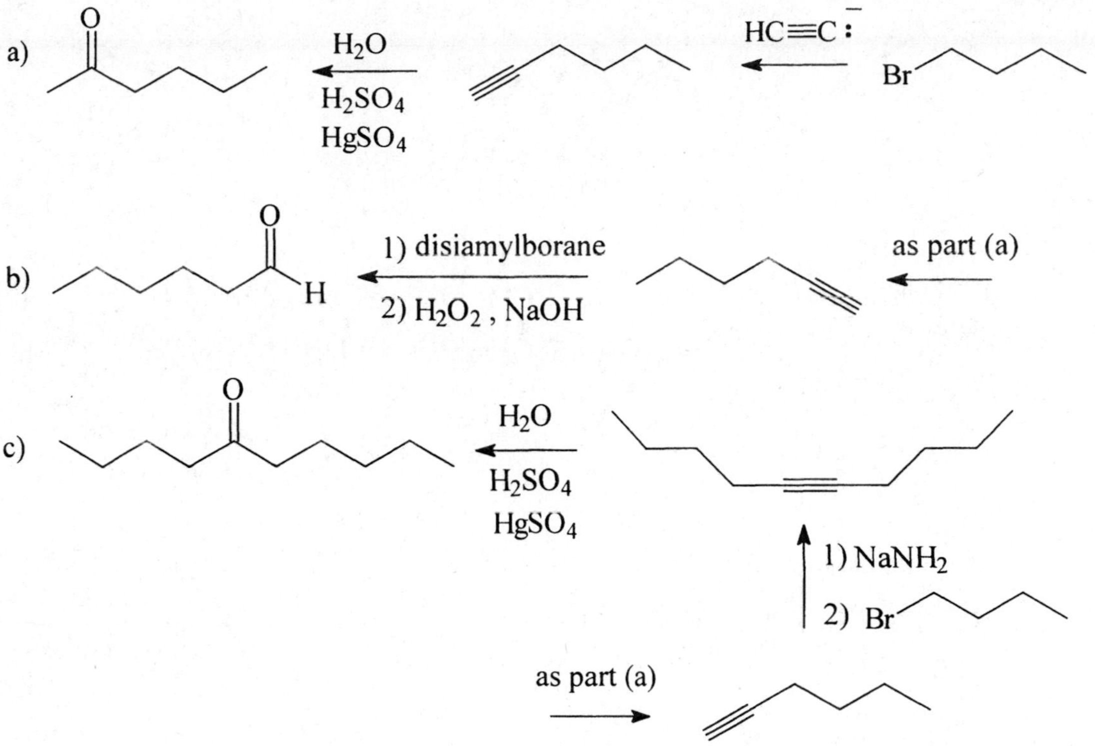

11.20 This reaction will produce the (*S*)-enantiomer of the alcohol.

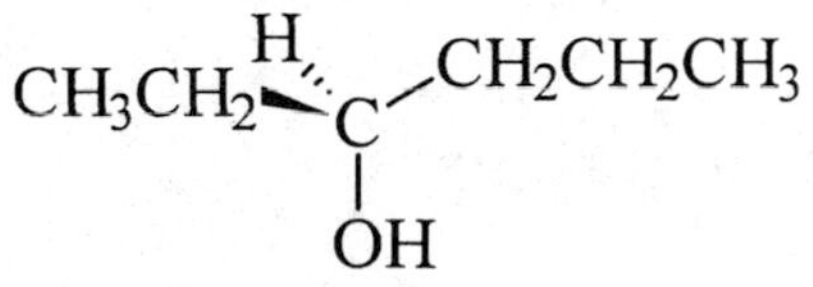

11.21 Use (*S,S*)-*trans*-2,5-dimethylborolane, the enantiomer of the borolane shown in the text, to produce (*R*)-2-butanol, the enantiomer of the alcohol shown in the text.

11.22

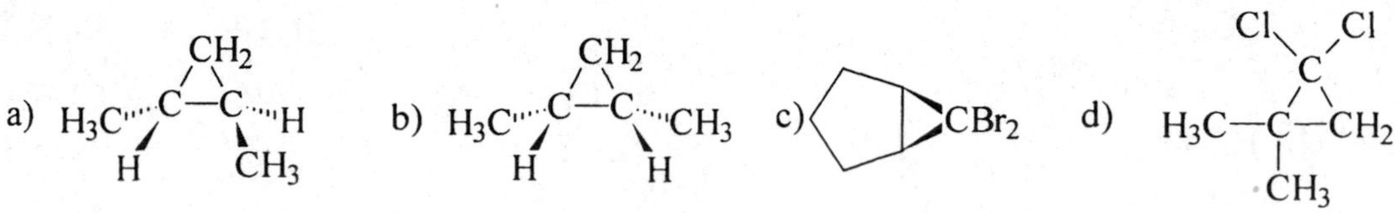

11.23

a)

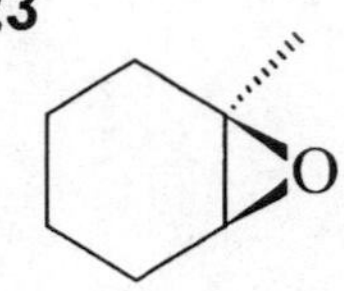

b) H₃C''''C—C''''CH₃ (epoxide with H's) → HO–C(CH₃)(H₃C)(H)–C(CH₃)(OH)(H)

$H_3C\cdots C - C\cdots CH_3$ → product

11.24

a) (cyclohexane-1,2-diol, cis)

b) $CH_3CH_2CH-CH_2$ with OH OH

c) $CH_3CH_2\cdots C - C\cdots CH_3$ with HO, OH, H, H   + enantiomer

11.25

a) $\xrightarrow[t\text{-BuOOH}]{[OsO_4]}$

b) $\xrightarrow{PhCO_3H}$ (epoxide) $\xrightarrow[H_2O]{NaOH}$

11.26

a) $CH_3CH_2\overset{O}{\overset{\|}{C}}H$ + $H\overset{O}{\overset{\|}{C}}H$

b) 2 $CH_3\overset{O}{\overset{\|}{C}}CH_3$

c) (structure with CH₃ ketone and CHO)

d) (cyclohexanone) + $H\overset{O}{\overset{\|}{C}}H$

e) 2 (butanedial with H, CHO)

11.27

a) $\underset{CH_3CH_2}{\overset{H_3C}{>}}C=CHCH_3$

b) (cyclopentene with propyl)

d) $CH_3\overset{CH_2}{\overset{\|}{C}}CH_2CH_2\overset{CHCH_3}{\overset{\|}{C}}H$ or $CH_3\overset{CHCH_3}{\overset{\|}{C}}CH_2CH_2\overset{CH_2}{\overset{\|}{C}}H$

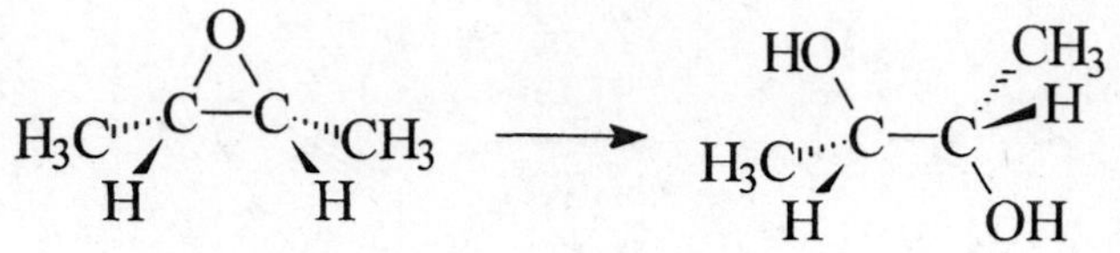

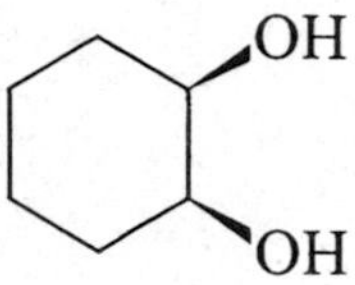

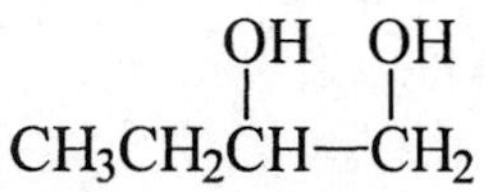

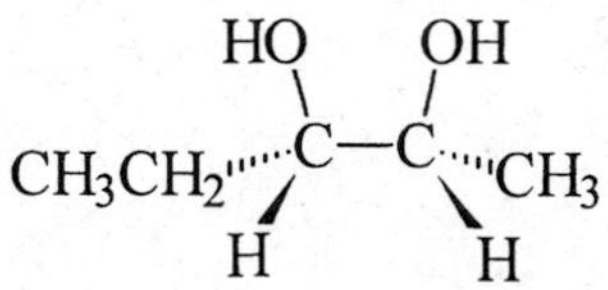

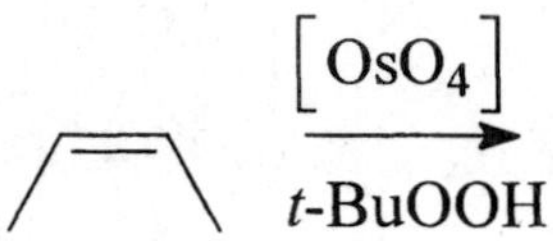

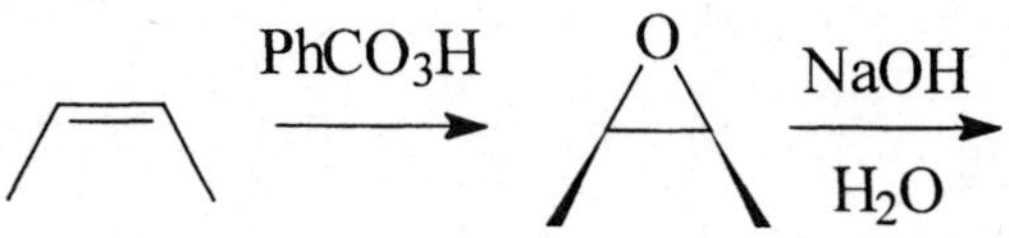

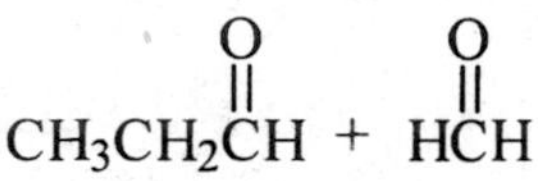

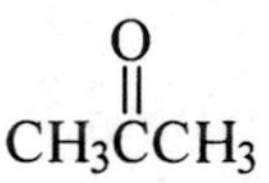

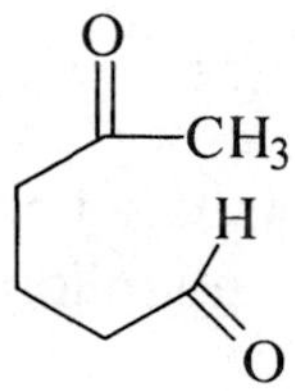

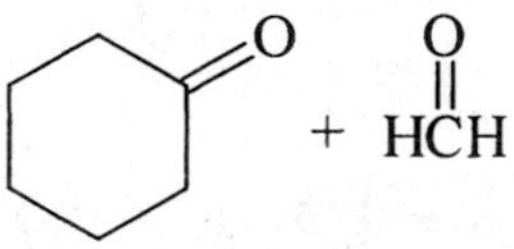

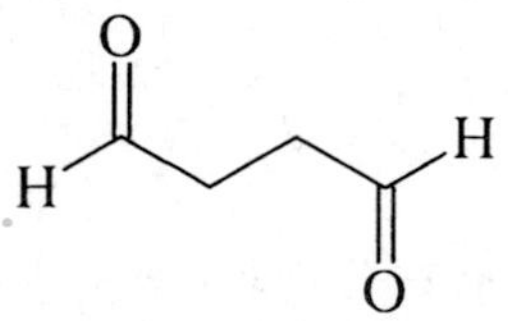

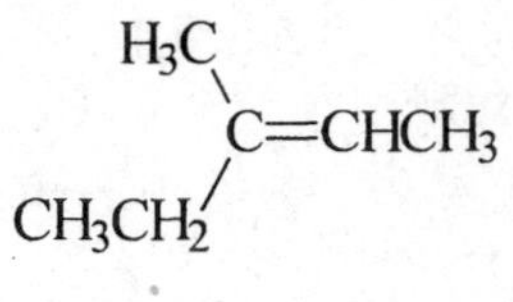

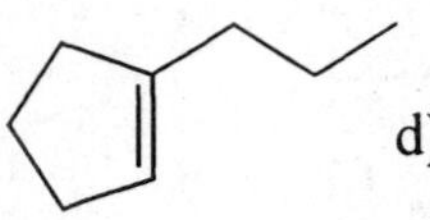

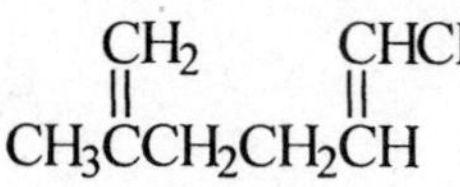

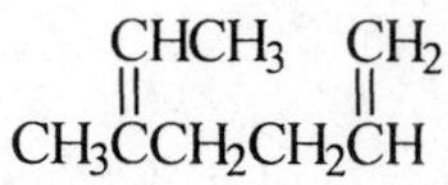

11.28

a) 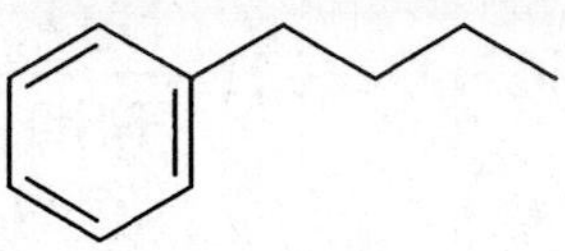

b) 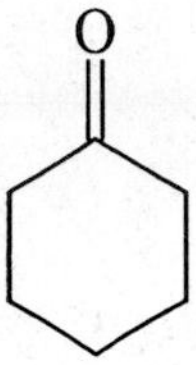

c) $CH_3CH_2CH_2CH_2CH_3$

d) 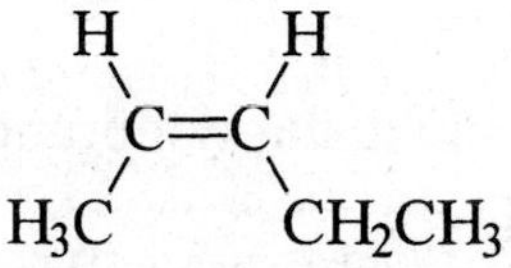

e) 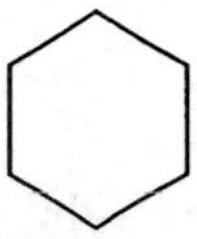

11.29 The H⁺ adds so as to give the most stable carbocation in the first step.

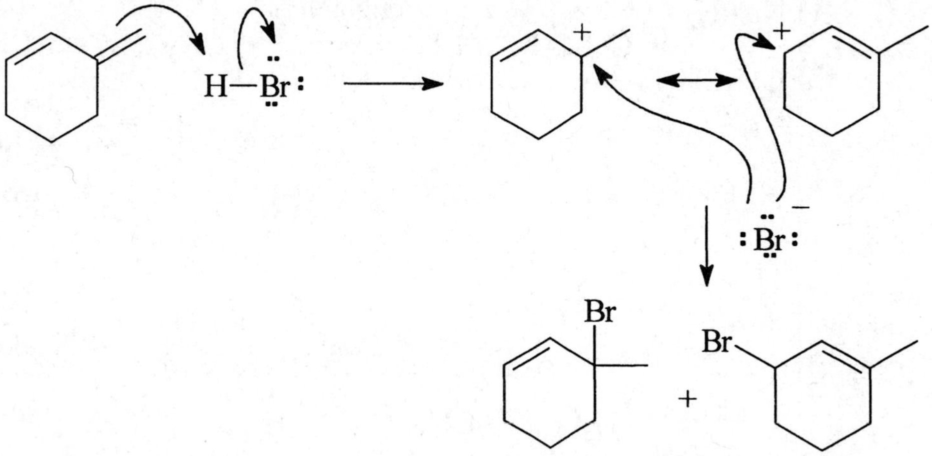

11.30

a) 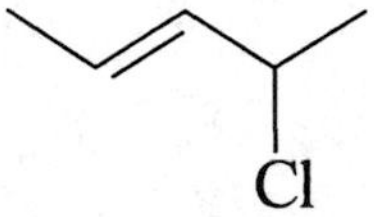

b) 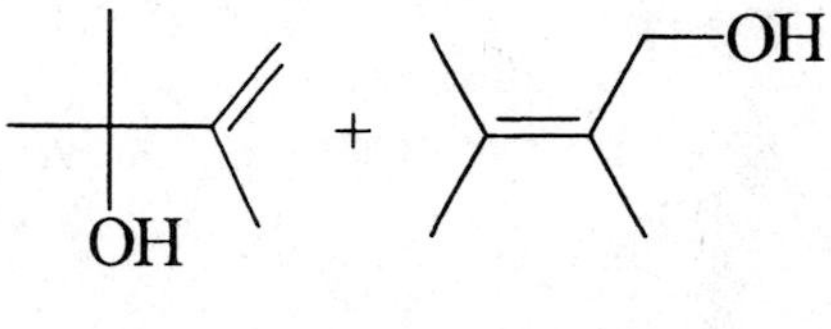

11.31

a)

$$CH_3CH_2CH_2CH{=}CH_2 \xrightarrow[\text{2) } H_2O_2\text{, NaOH}]{\text{1) } BH_3\text{, THF}} CH_3CH_2CH_2CH_2\underset{}{\overset{OH}{|}}{CH_2} \xrightarrow[\text{2) } CH_3I]{\text{1) Na}}$$

b)

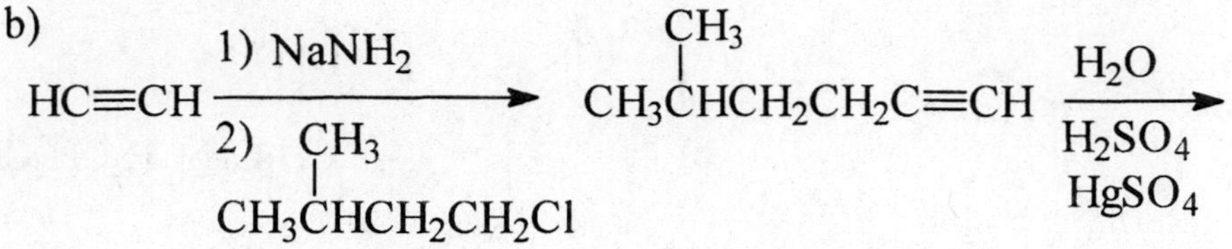
1) NaNH2
HC≡CH
2) CH3
CH3CHCH2CH2Cl
CH3
CH3CHCH2CH2C≡CH
H2O
H2SO4
HgSO4

c)

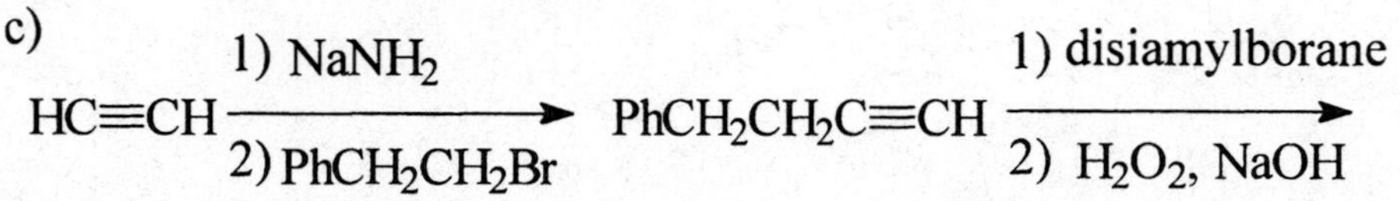
1) NaNH2
HC≡CH
2) PhCH2CH2Br
PhCH2CH2C≡CH
1) disiamylborane
2) H2O2, NaOH

d)

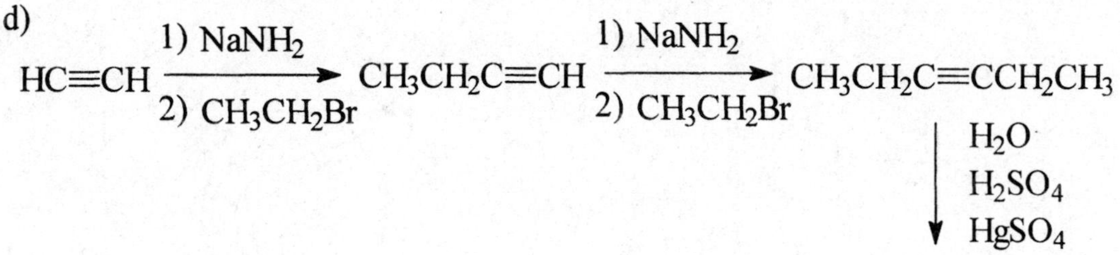
1) NaNH2
HC≡CH
2) CH3CH2Br
CH3CH2C≡CH
1) NaNH2
2) CH3CH2Br
CH3CH2C≡CCH2CH3
H2O
H2SO4
HgSO4

e)

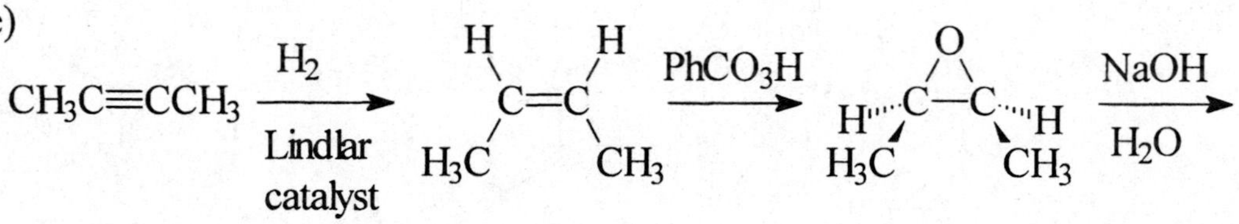
CH3C≡CCH3
H2
Lindlar
catalyst
PhCO3H
NaOH
H2O

f)

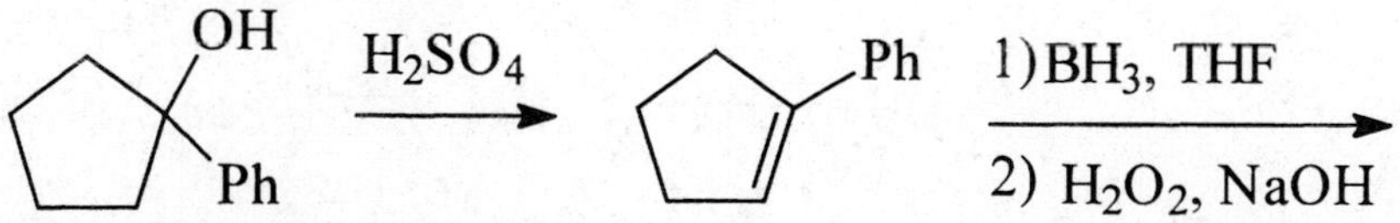
OH
Ph
H2SO4
Ph
1)BH3, THF
2) H2O2, NaOH

g)

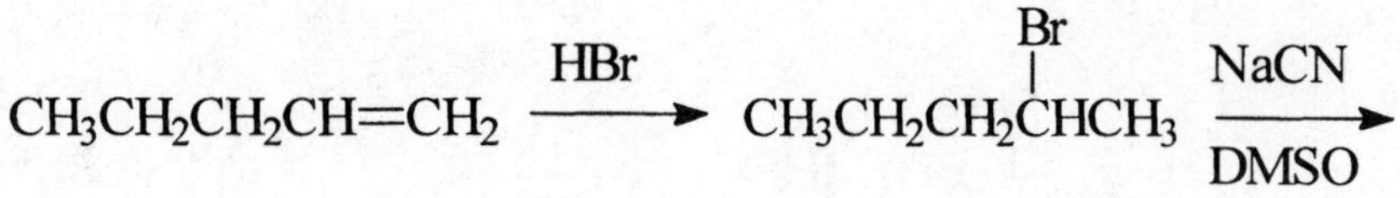
CH3CH2CH2CH=CH2
HBr
Br
CH3CH2CH2CHCH3
NaCN
DMSO

h) 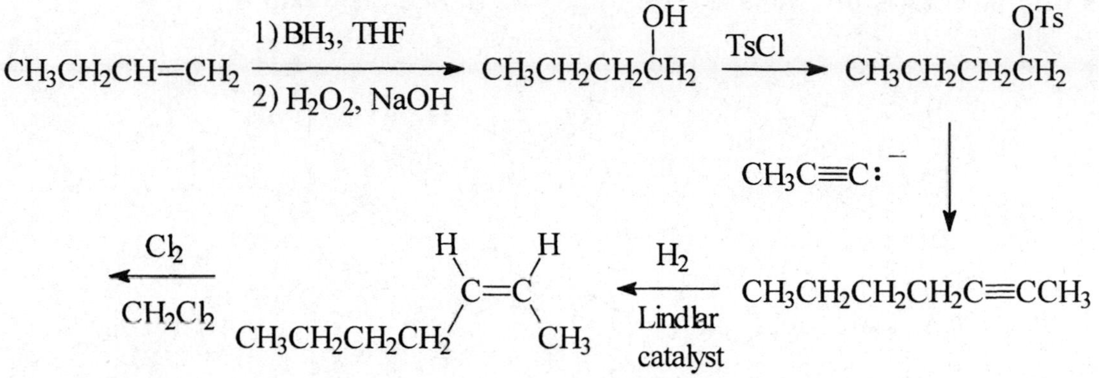

11.32 (All chiral products are produced as a racemic mixture.)

a) 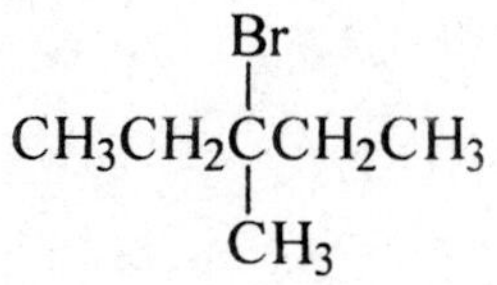 b) 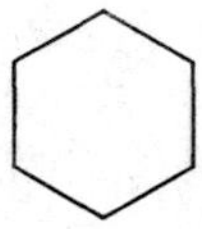 c) 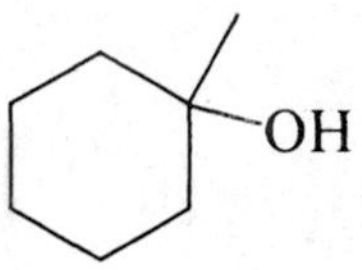 d) 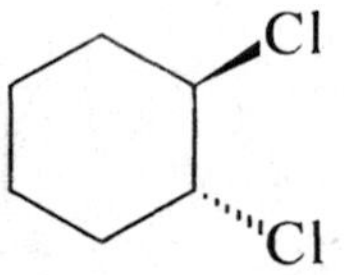

e) 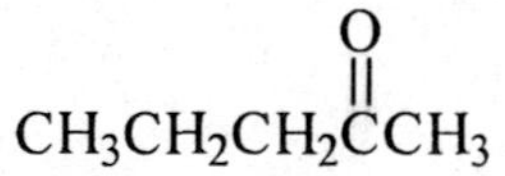 f) 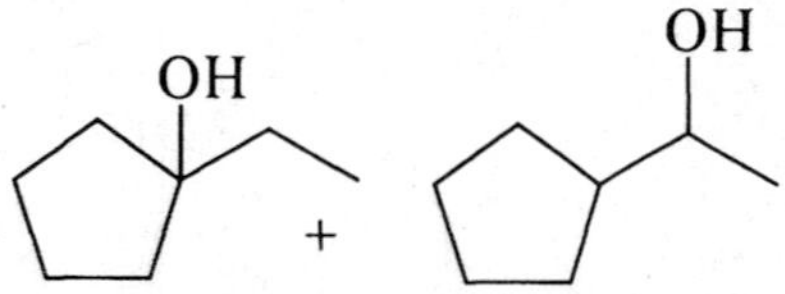 g) 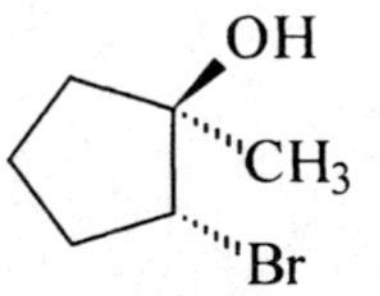

h) 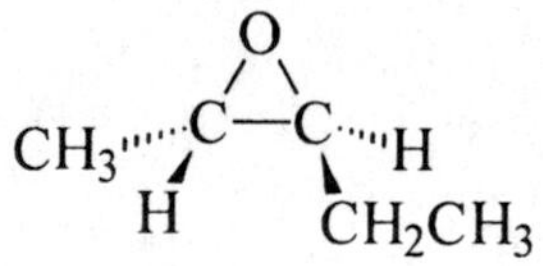 i) 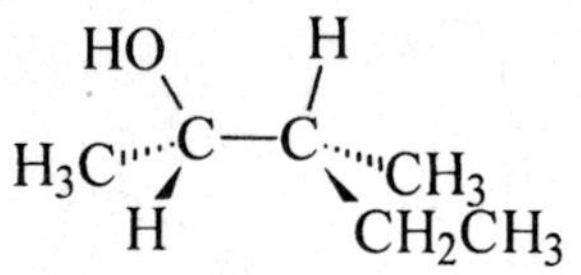 j) 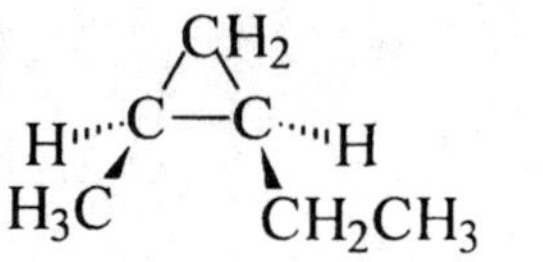

k) 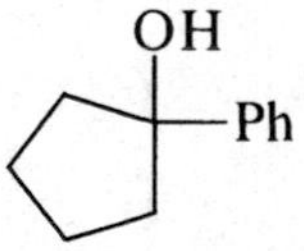 l) 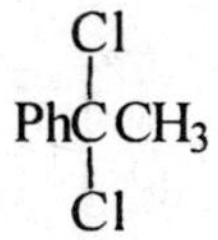 m) 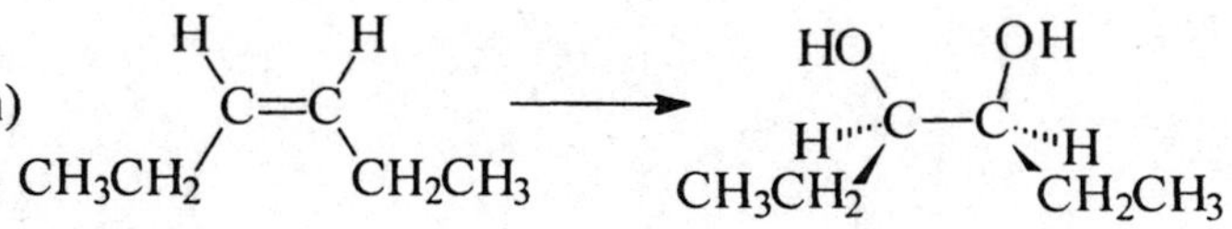

n) 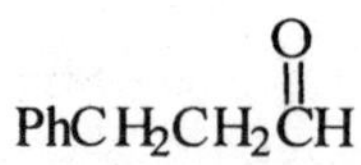

11.33 (All chiral products are produced as a racemic mixture.)

a) 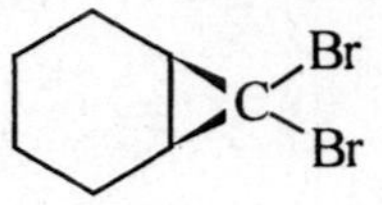

b) 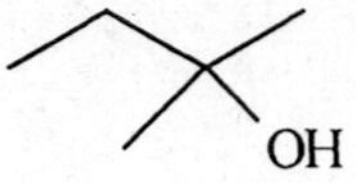

c) 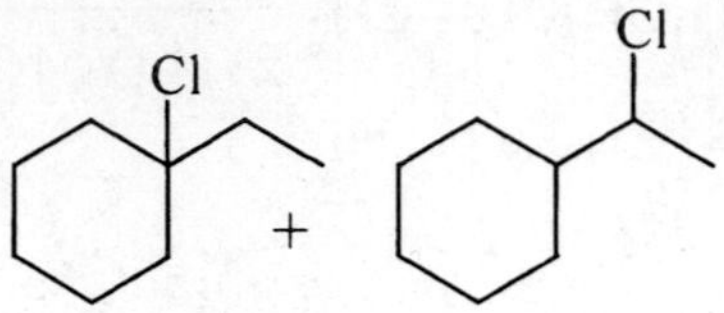

d) $PhCH_2-C(=O)-Ph$

e) 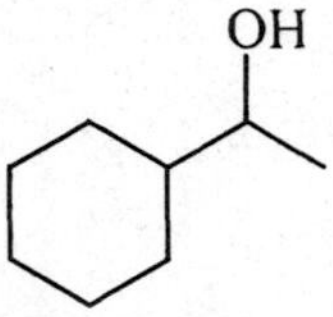

f) 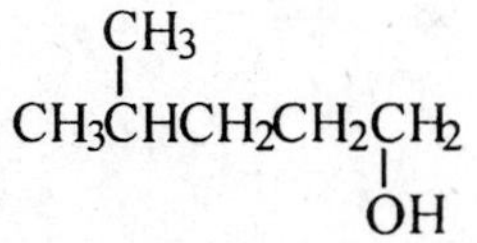

g) $CH_3CH_2CH_2CH_2C(=O)-H$

h) 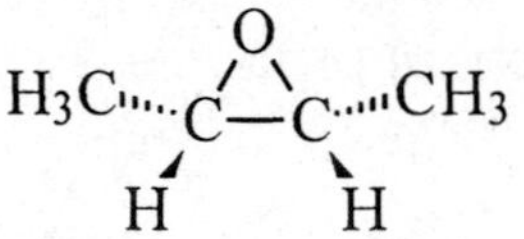

i) 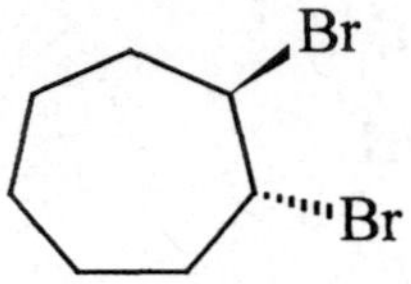

j) 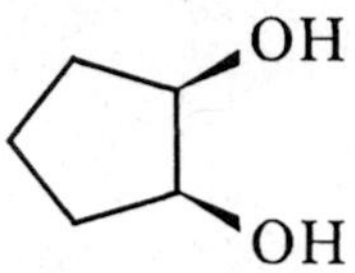

k) 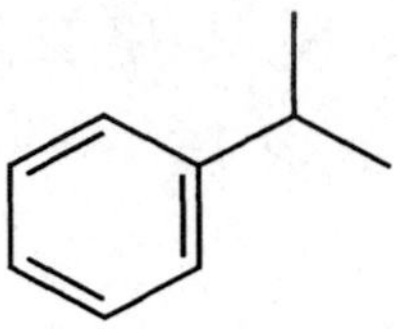

l) 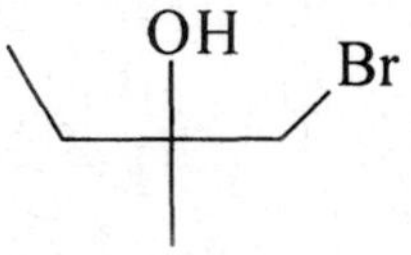

m) 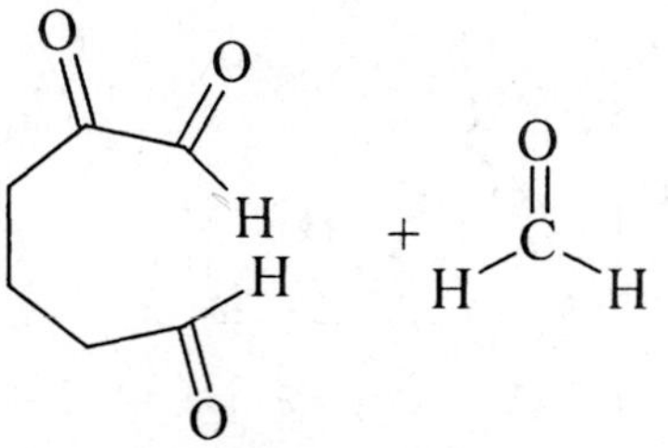

n) 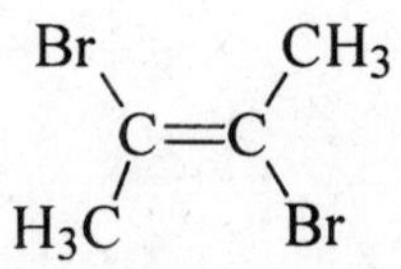

o) 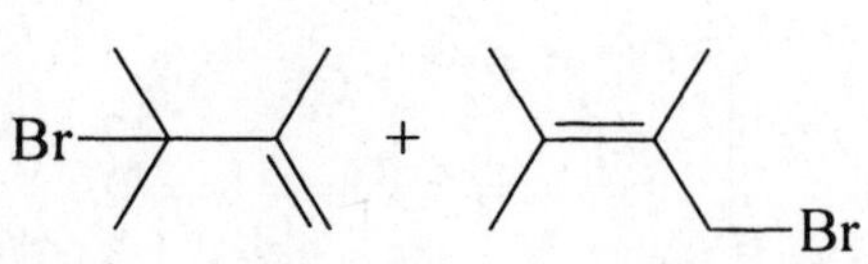

11.34 (All chiral products are produced as a racemic mixture.

a) 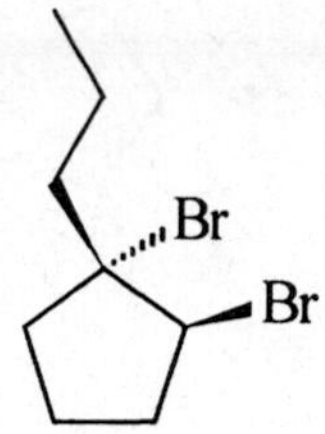

b) 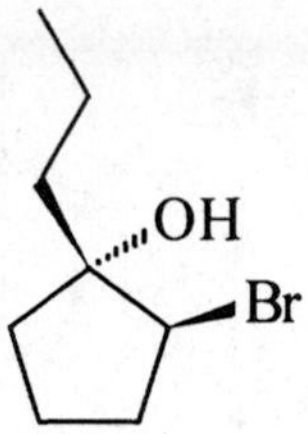

c) 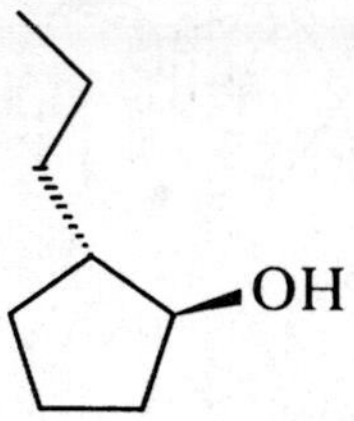

d) 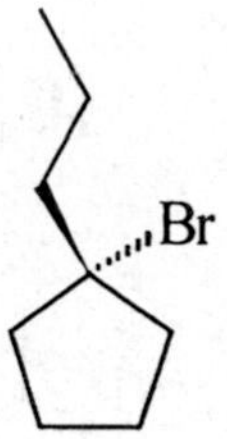

e) 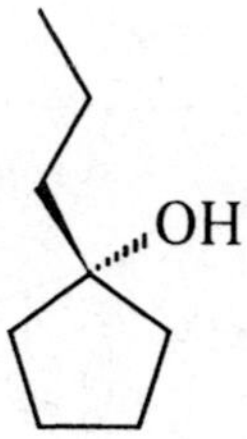

f) 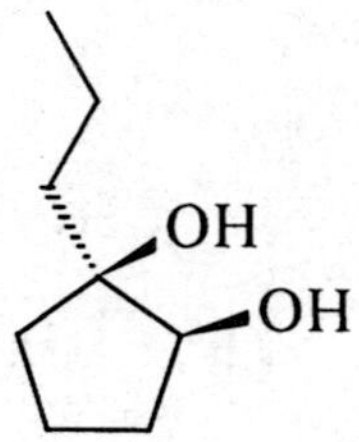

g) 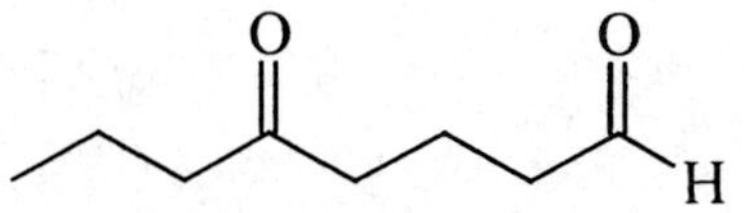

11.35

a) 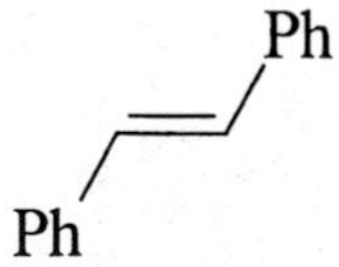

b) 

c) 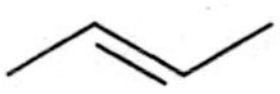 or 

d) 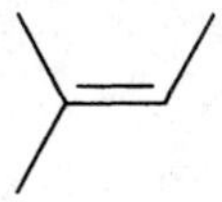

e) 

f) 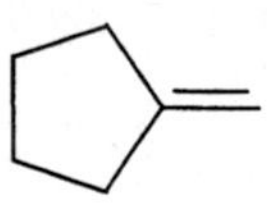 or 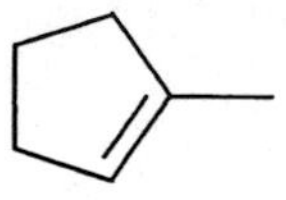 or 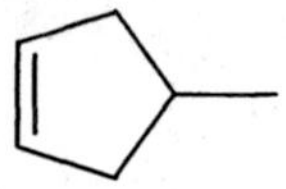

11.36

a) 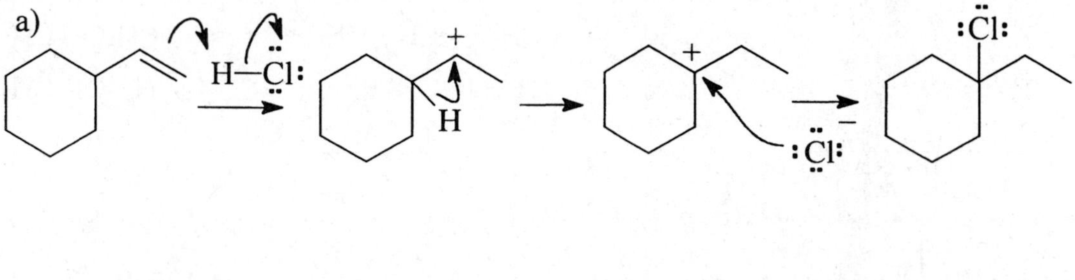

b) 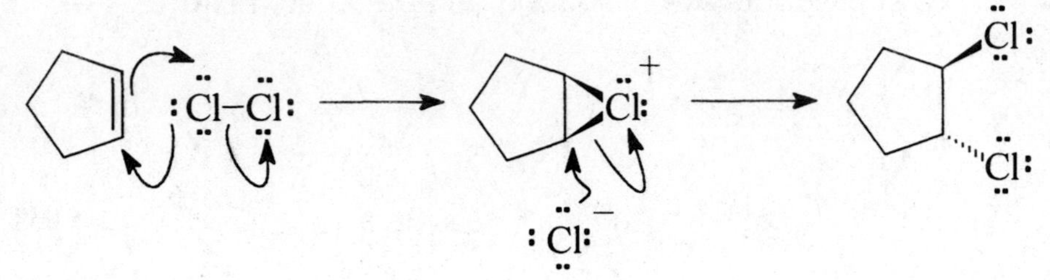
Cl–Cl
Cl
Cl
Cl
Cl
Cl

c) 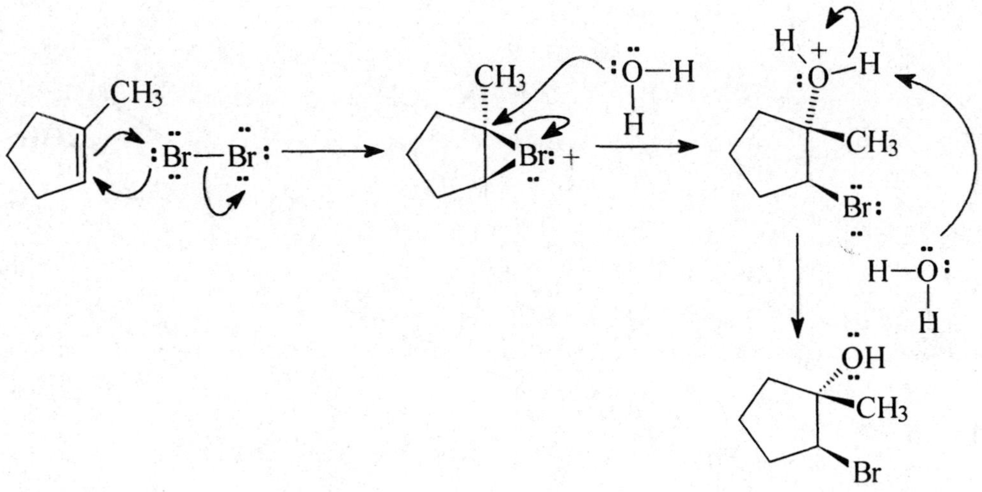
CH3
Br—Br
CH3
O—H
H
Br: +
H
O
H
CH3
Br
H—O
H
OH
CH3
Br

d) 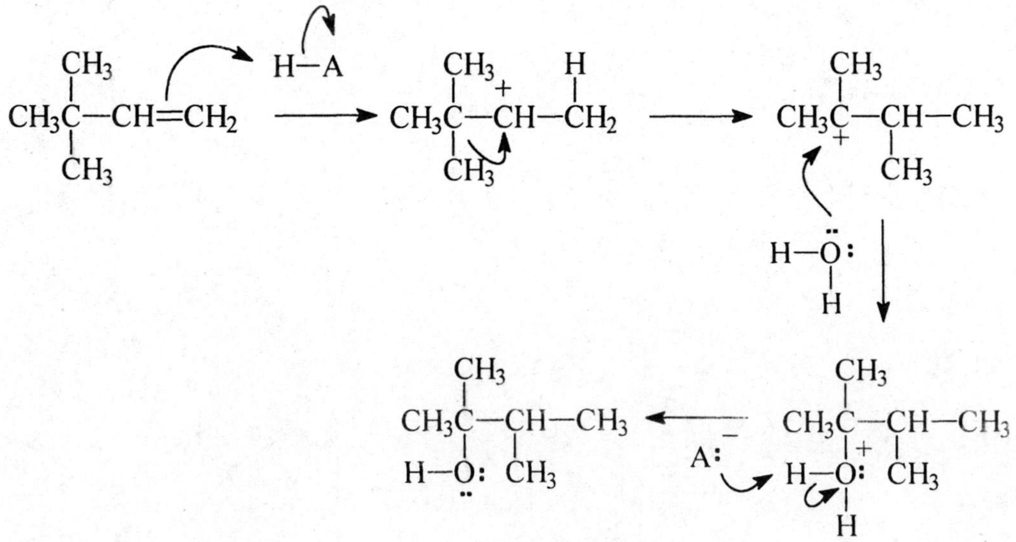
CH3
CH3C—CH=CH2
CH3
H—A
CH3
CH3C—CH—CH2
CH3
H
CH3
CH3C—CH—CH3
CH3
H—O
H
CH3
CH3C—CH—CH3
H—O CH3
H
A
CH3
CH3C—CH—CH3
H—O CH3

e)

11.37

a)

1) BH$_3$, THF
2) H$_2$O$_2$, NaOH

b)

1) Hg(O$_2$CCH$_3$)$_2$, H$_2$O
2) NaBH$_4$

c)

OsO$_4$
t-BuOOH

d)

Cl$_2$
H$_2$O

e)

HCl

f)

1) BH$_3$, THF
2) H$_2$O$_2$, NaOH

OH   PCC

g)

H$_2$
Pt

h)

$$CH_3\overset{O}{\overset{\|}{C}}OOH$$

11.38

a)   CH$_3$C≡CH $\xrightarrow[\text{2) CH}_3\text{Br}]{\text{1) NaNH}_2}$ CH$_3$C≡CCH$_3$ $\xrightarrow[\text{H}_2\text{SO}_4]{\text{HgSO}_4, \text{H}_2\text{O}}$

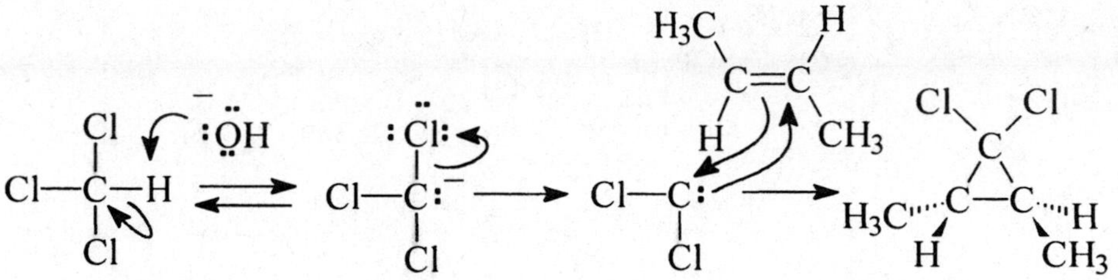

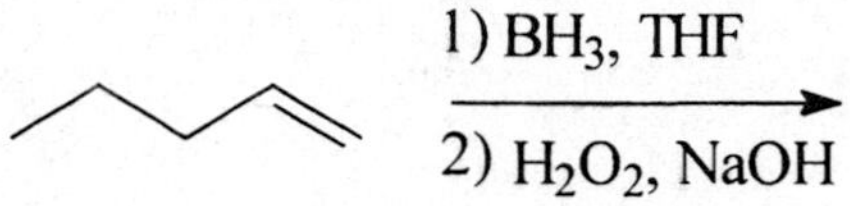

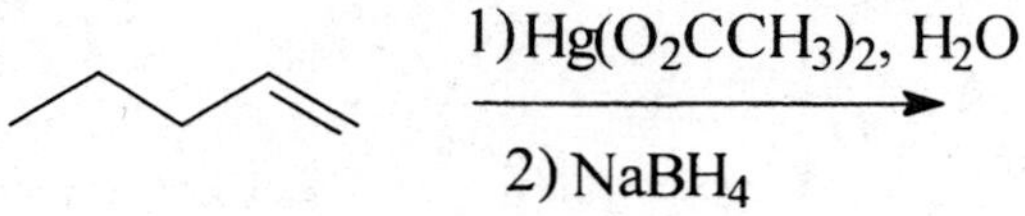

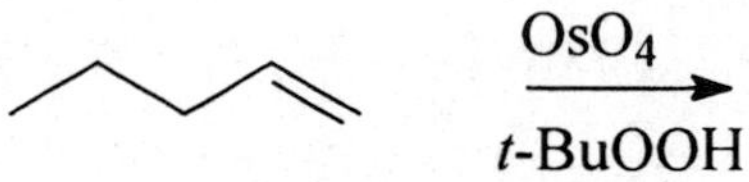

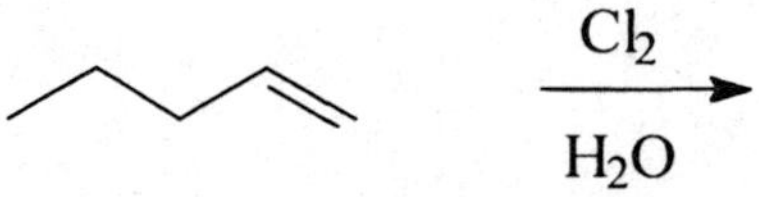

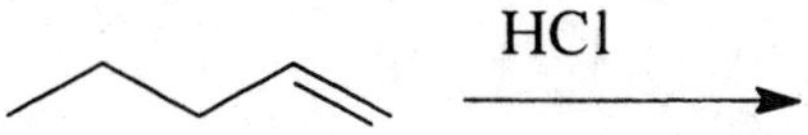

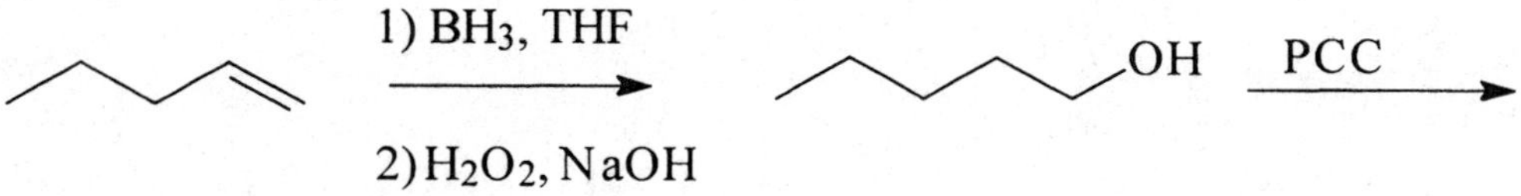

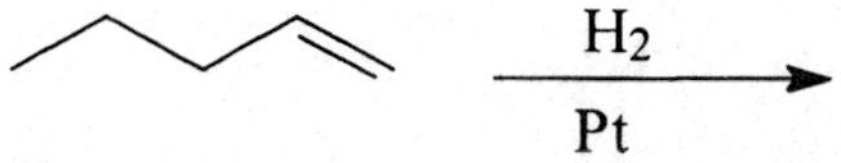

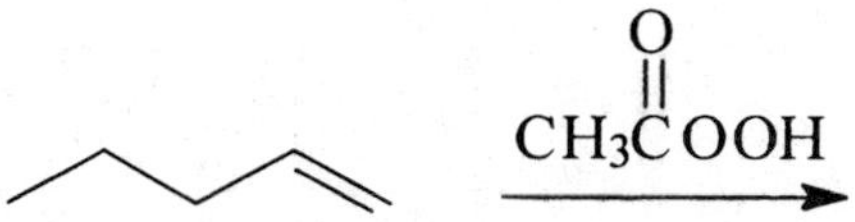

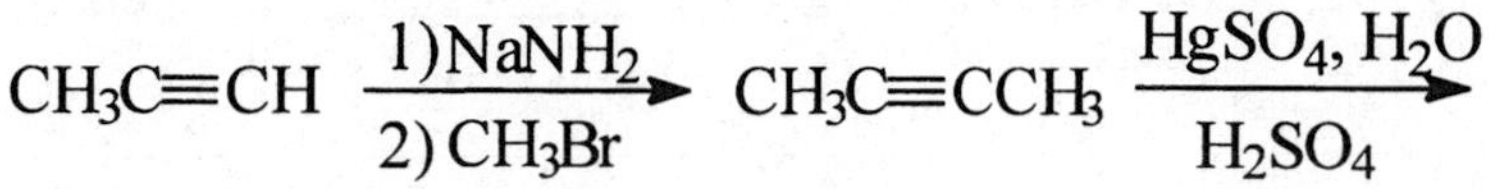

b) $CH_3C\equiv CH$ $\xrightarrow[\text{2) } CH_3CH_2CH_2Br]{\text{1) } NaNH_2}$ $CH_3C\equiv CCH_2CH_2CH_3$

$\xrightarrow[\text{Lindlar catalyst}]{H_2}$

c) $\xrightarrow{\text{as (b)}}$

$\xrightarrow[\text{$t$-BuOOH}]{OsO_4}$

d)

$\xrightarrow[\text{heat}]{H_3PO_4}$

$\xrightarrow[\text{H}_2\text{O}]{Br_2}$

e) $HC\equiv CH$ $\xrightarrow[\text{2) } CH_3CH_2CH_2Br]{\text{1) } NaNH_2}$ $HC\equiv CCH_2CH_2CH_3$ $\xrightarrow[\substack{\text{Lindlar}\\\text{catalyst}}]{H_2}$

$\xrightarrow[\text{2) } H_2O_2, NaOH]{\text{1) } BH_3, THF}$

f) $CH_3C\equiv CH$ $\xrightarrow[\text{2) } CH_3Br]{\text{1) } NaNH_2}$ $CH_3C\equiv CCH_3$ $\xrightarrow[\substack{\text{Lindlar}\\\text{catalyst}}]{H_2}$

$\xrightarrow[\text{Zn(Cu)}]{CH_2I_2}$

g)

$\xrightarrow[\text{heat}]{H_3PO_4}$

$\xrightarrow{HCl}$

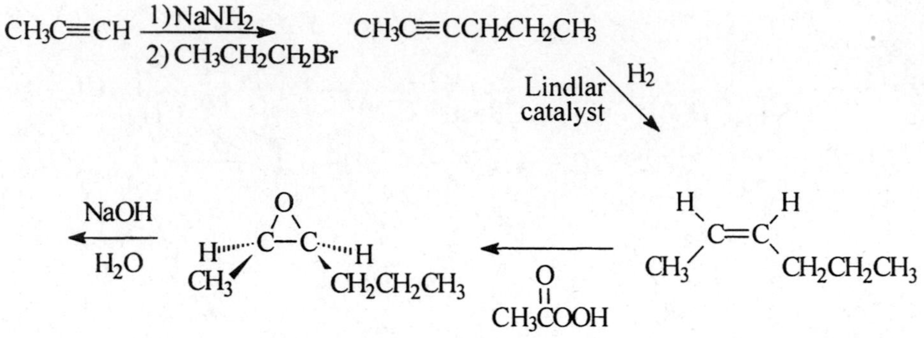

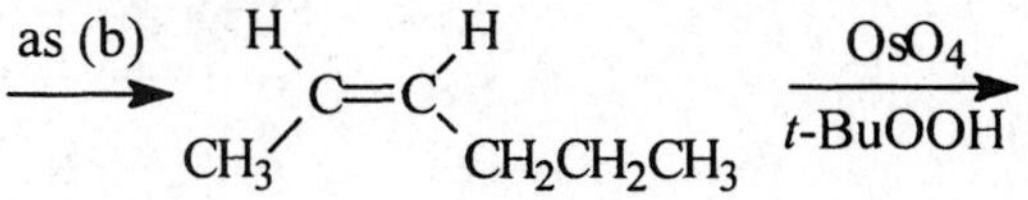

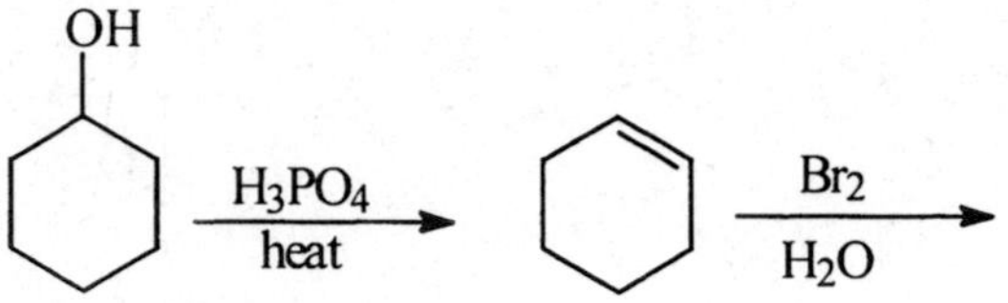

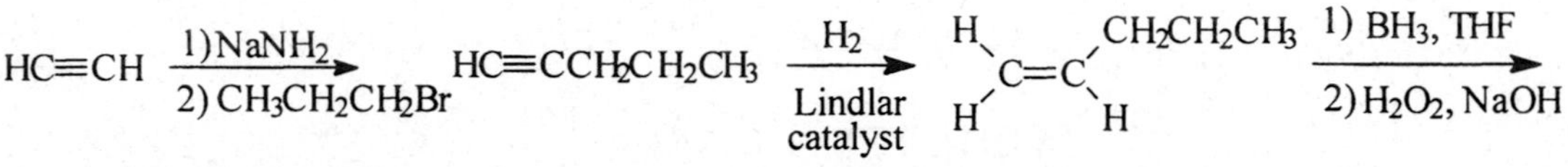

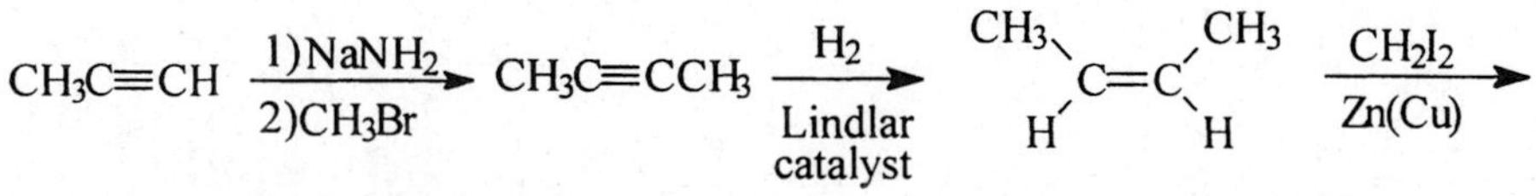

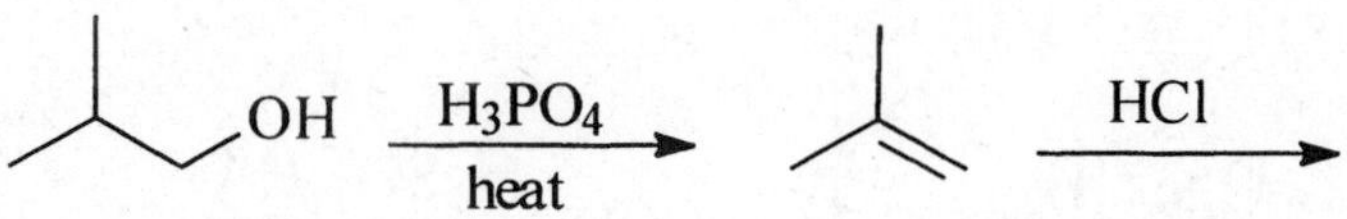

h) 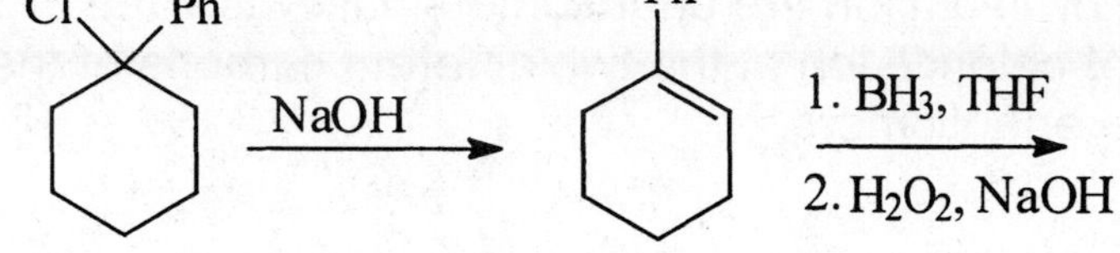

11.39

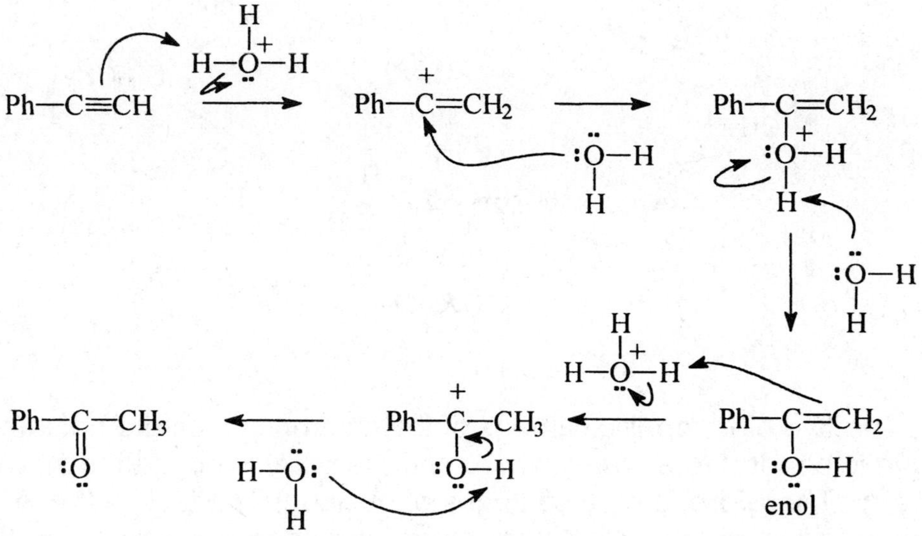

11.40 a) The left compound is has a faster rate because a tertiary carbocation intermediate is more stable than a secondary carbocation intermediate.
b) The right compound has a faster rate because the electron-withdrawing nitro group destabilizes the carbocation intermediate.
c) The right compound has a faster rate because the carbocation intermediate is resonance stabilized.

11.41 a) The two chloronium ions that are formed in this reaction are enantiomers (non-superimposable mirror images).

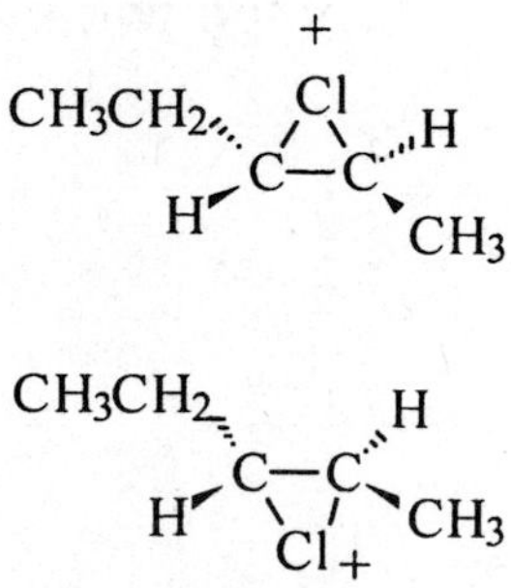

b) The products formed by attack of chloride ion at the two different carbons of one chloronium ion are enantiomers. Likewise, the products formed by attack of chloride ion at the two different carbons of the other chloronium ion are enantiomers.

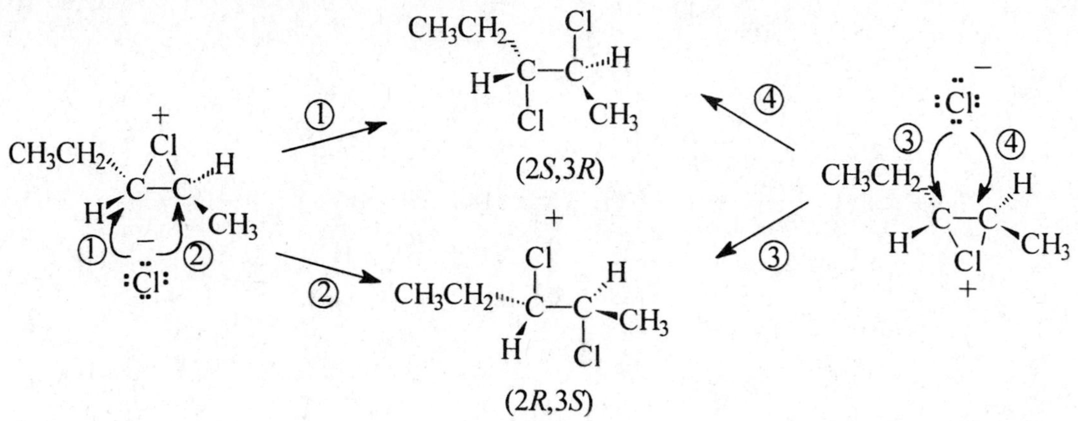

c) The two electrophilic carbons of the chloronium ion are not identical. One is bonded to a methyl group and the other is bonded to an ethyl group. Therefore, the percentages of nucleophile attack at these two carbon sites of one chloronium ion are not necessarily identical; that is, the amount of product formed by path 1 does not necessarily equal the amount of product formed by path 2.

d) The nucleophile will attack the same carbons of the two enantiomeric chloronium ions in exactly the same amounts and, thus, will produce a racemic mixture. For example, in the above scheme, paths 1 and 3 occur to an identical extent, giving equal amounts of the two enantiomers. Similarly paths 2 and 4 occur to an identical extent, giving equal amounts of the two enantiomers.

11.42

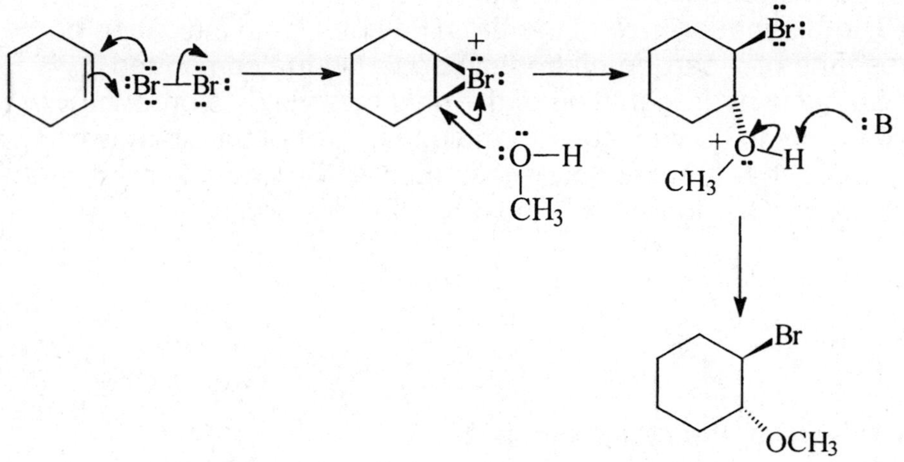

11.43 In the oxymercuration reaction, the mercury electrophile adds to the double bond to form the cyclic mercurinium ion and then water reacts as a nucleophile with the mercurinium ion. If the reaction is run in methanol as solvent, the initially formed mercurinium ion will react with methanol as the nucleophile in the same manner to give a methyl ether after workup with sodium borohydride.

$$CH_3CH_2CH_2\overset{\overset{\displaystyle OCH_3}{|}}{C}H-CH_3$$

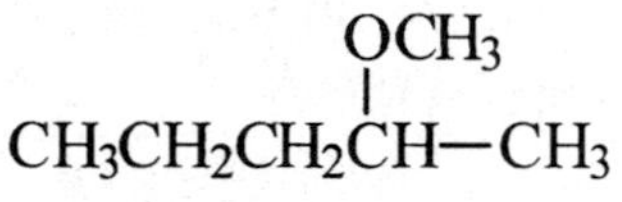

11.44

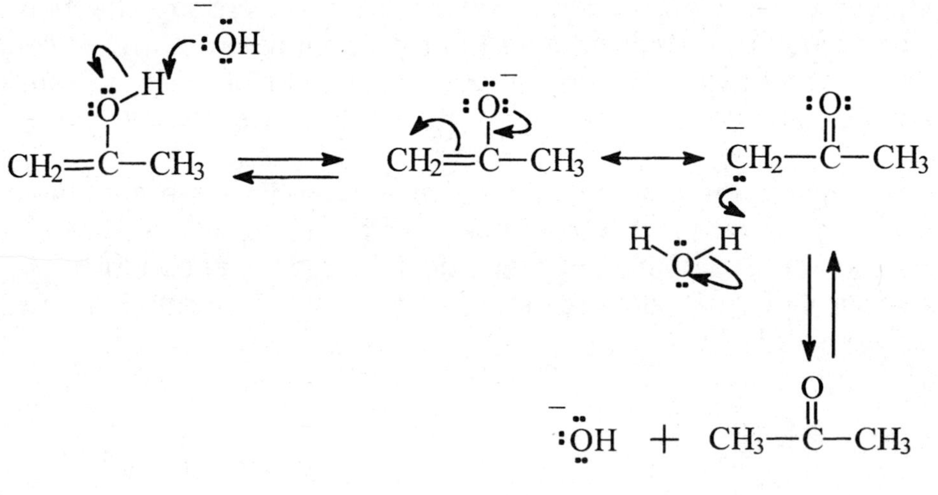

11.45 a) The DU for this compound is 2.
b) The disappearance of the bromine color is an indication of the presence of a carbon-carbon pi bond(s) in the compound.
c) Catalytic hydrogenation of the unknown yields a product with two more hydrogens in its molecular formula than that of the unknown. This indicates that there is only one carbon-carbon double bond in the unknown. The second DU must be due to a ring.
d)

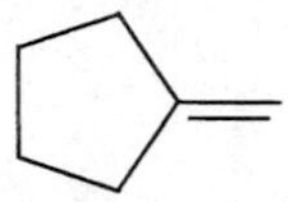

11.46 a) The DU of the compound is 2.
b) The product of the catalytic hydrogenation has DU = 0, so the unknown compound must have two pi bonds.
c) The result of hydrogenation in the presence of Lindlar catalyst shows that the unknown has a carbon-carbon triple bond.
d)

$$\begin{array}{l} \quad\;\; CH_3 \\ \quad\;\;\; | \\ CH_3CH{-}C{\equiv}C{-}CH_2CH_3 \end{array}$$

11.47 Addition reactions involving boranes are regioselective and are greatly influenced by sterics. Since the carbons of the double bond are both monosubstituted, BH_3 adds to both carbons, but gives slightly more of the product in which it has added to the less hindered carbon attached to the methyl group. This steric effect is more pronounced with disiamylborane because it has two bulky alkyl groups attached to the boron. Therefore, disiamylborane adds almost entirely to the less hindered carbon.

11.48 In this reaction, the primary carbocation intermediate is more stable than the secondary carbocation because the strong inductive electron withdrawing effect of the CF_3 group destabilizes the secondary carbocation more. Therefore, the reaction occurs with anti-Markovnikov regiochemistry.

11.49 The carbocation intermediate formed from this vinyl ethyl ether has a very large resonance stabilization because its resonance structure has the octet rule satisfied at all atoms, so hydration reaction of this alkene is much faster than that of ethene.

$$CH_3CH_2-\overset{..}{\underset{..}{O}}-\overset{+}{\underset{}{C}}H-CH_3$$

$$\updownarrow$$

$$CH_3CH_2-\overset{+}{\underset{..}{O}}=CH-CH_3$$

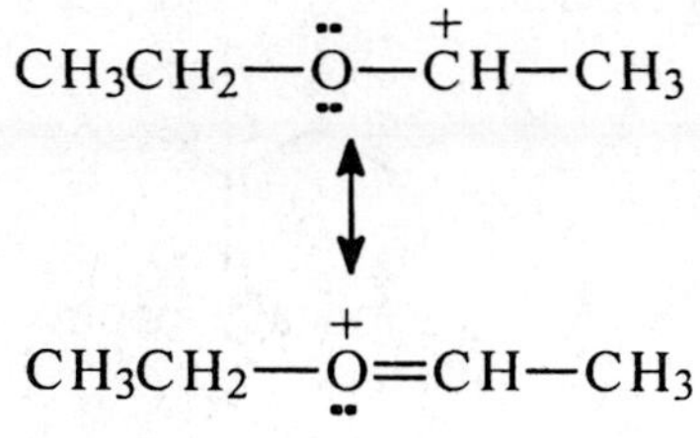

11.50 Of the two vinyl cation intermediates that could be formed from the addition of HCl to propyne, the secondary vinyl cation is more stable.

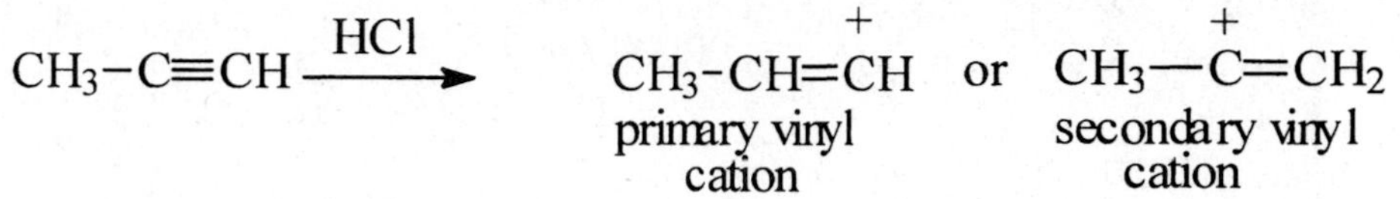

$$CH_3-C\equiv CH \xrightarrow{\text{HCl}}$$

$$CH_3\text{-}\overset{+}{CH}=CH \quad \text{or} \quad CH_3-\overset{+}{C}=CH_2$$

primary vinyl cation   secondary vinyl cation

11.51

$$CH_2=CH-CH_2CH_2CH_2-\overset{..}{\underset{..}{O}}H \longrightarrow CH_2-CH-CH_2CH_2CH_2-\overset{..}{\underset{..}{O}}H$$

:Br:
|
:Br:

$$Br\overset{+}{}$$

:B

$$BrCH_2 \diagup O \diagdown \quad \longleftarrow \quad BrCH_2 \diagup \overset{+}{O} \diagdown \overset{H}{}$$

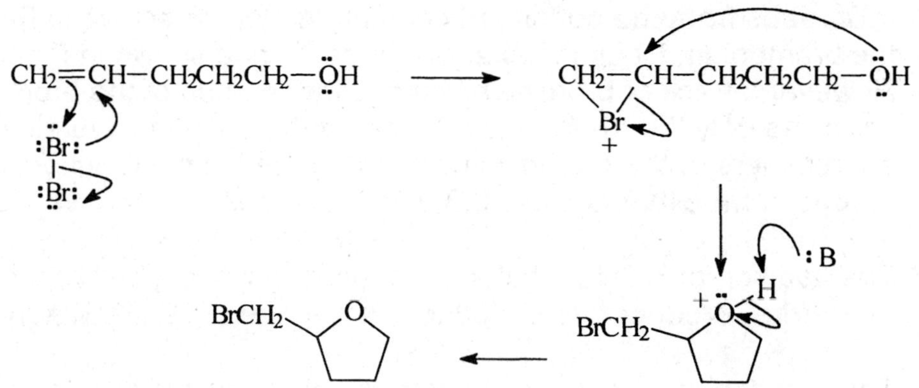

11.52 a) The DU of limonene is 3. The DU of the hydrogenation product is 1, so limonene has two pi bonds and one ring.
b) Possible structures for limonene based on the ozonolysis products are shown below. Limonene actually has the structure shown on the left.

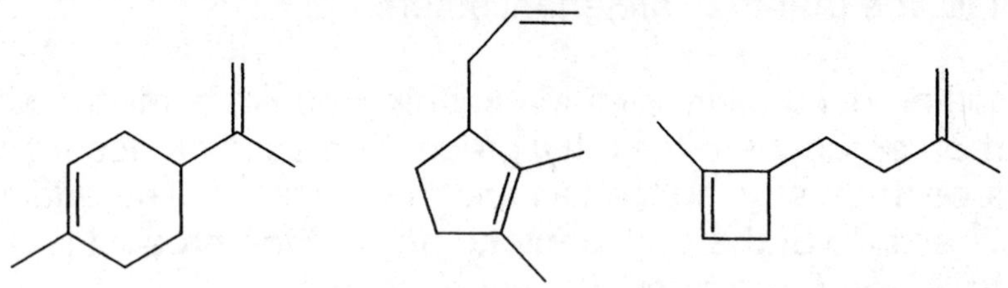

11.53

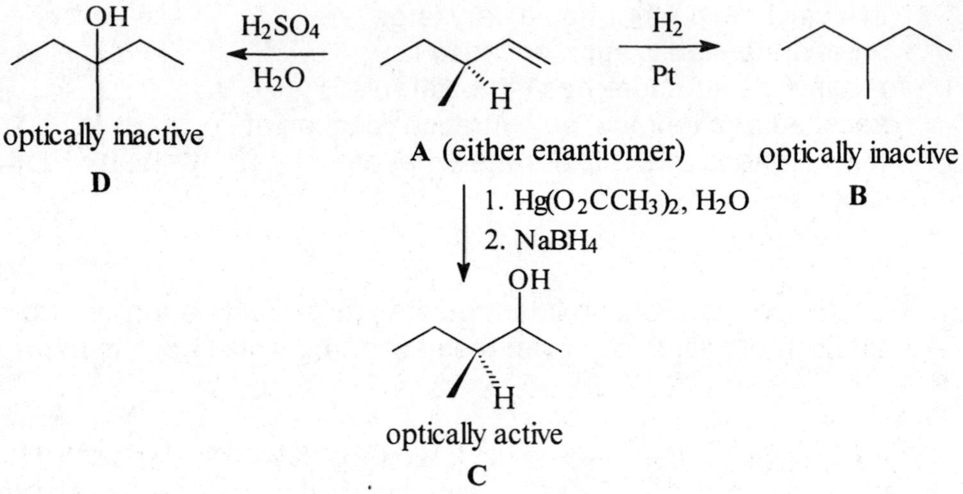

11.54 If Br_2 adds from the bottom, rather than the top as shown in Figure 11.3, the bromonium ion is the enantiomer of the one shown in Figure 11.3. However, attack of bromide anion at either carbon of the bromonium ion still gives only the *meso*-diastereomer of the product. Overall, then, both enantiomers of the bromonium ion are formed in the reaction, but the only product from either is *meso*-2,3-dibromobutane.

11.55 The addition of HCl to 1-butene produces racemic 2-chlorobutane. Therefore, Product 1 and Product 2 are formed in equal amounts.

11.56 The reaction of Cl_2 with *cis*-2-pentene occurs by anti addition. This results in the formation of Product 1 and Product 2 in a racemic (50:50) mixture.

11.57 The addition occurs with anti stereochemistry, so the OH and the Cl are trans in the product. The product is the *trans*-stereoisomer and is racemic. Product 1 and Product 2 are in different conformations so it is more difficult to see that they are enantiomers.

11.58 The addition occurs with anti-Markovnikov regiochemistry, so the OH is on the carbon adjacent to the ethyl group. The addition occurs with syn-stereochemistry, so the OH and the H are cis. The addition can occur from either side of the cyclopentene ring, so the product is racemic. Thus, products 4 and 5 are formed in equal amounts.

11.59 This is a syn addition. Racemic *trans*-1,2-dimethylcyclopropane is formed, so Products 1 and 2 are formed in equal amounts.

11.60 This reaction results in the syn addition of two OH groups to the alkene. The product is the *meso*-stereoisomer, Product 3.

11.61 The OH is on the less substituted carbon (anti-Markovnikov addition). The OH and the H are added syn, so the OH is trans to the methyl group. And the addition has occurred on the less hindered side of the ring, away from the two methyl groups on the four membered ring.

11.62 The hydrogens are added to the alkene from the less hindered side of the ring, that is, from the side opposite the methyl group. As a result, the H added to the ring is trans to the methyl group and the methyl groups are cis in the product.

Review of Mastery Goals

After completing this chapter, you should be able to:

Show the products, including regiochemistry and stereochemistry, resulting from the addition to alkenes of all of the reagents discussed in this chapter.
(Problems 11.2, 11.3, 11.4, 11.5, 11.7, 11.10, 11.13, 11.15, 11.16, 11.20, 11.22, 11.23, 11.24, 11.26, 11.27, 11.28, 11.33, 11.34, 11.35, 11.43, 11.55, 11.56, 11.57, 11.58, 11.59, 11.60, 11.61, and 11.62)

Show the products, including regiochemistry and stereochemistry, resulting from the addition to alkynes of selected reagents as discussed in this chapter. These reagents include HX, X_2, $Hg^{2+}/H^+/H_2O$, disiamylborane, and H_2.
(Problems 11.4, 11.7, 11.13, 11.18, 11.28, and 11.33)

Show the mechanisms for any of these reactions.
(Problems 11.6, 11.8, 11.11, 11.12, 11.29, 11.36, 11.39, 11.41, 11.42, and 11.44)

Show rearranged products when they are likely to occur.
(Problems 11.4, 11.5, and 11.33)

Predict how the rate of addition varies with the structure of the alkene.
(Problems 11.1, 11.9, and 11.40)

Predict the products from additions to conjugated dienes.
(Problems 11.30 and 11.33)

Use these reactions in synthesis.
(Problems 11.14, 11.17, 11.19, 11.21, 11.25, 11.31, 11.37, and 11.38)

Chapter 12
FUNCTIONAL GROUPS AND NOMENCLATURE II

12.1 a) (1-Methylpropyl)benzene b) *o*-Chlorotoluene
c) 1-Bromo-4-methoxybenzene d) 4-Butyl-3-chlorotoluene
e) 3-Ethyl-4-phenylcyclohexene f) 3-Phenyl-1-butanol

12.2

a) 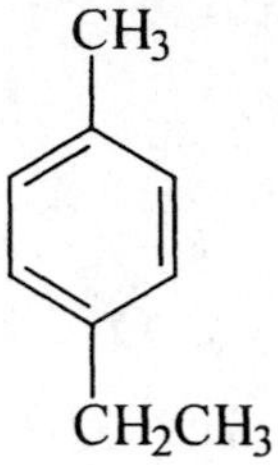

b) 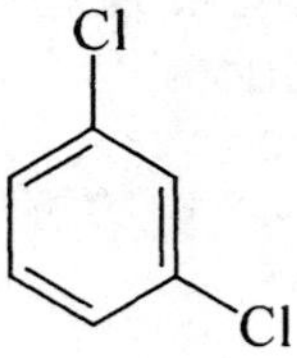

c) 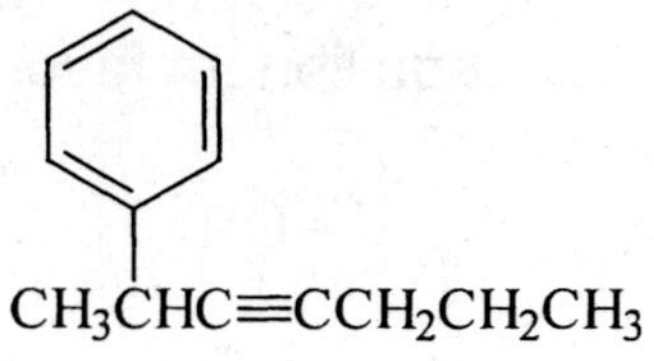

d) 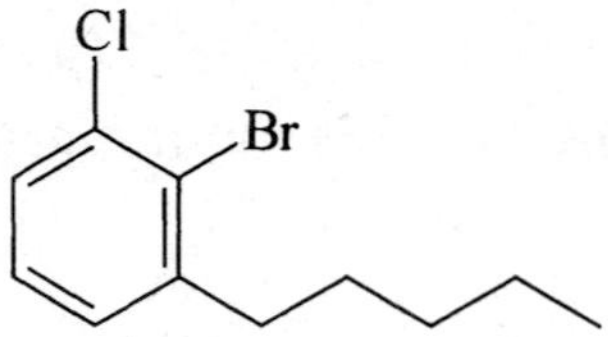

e) 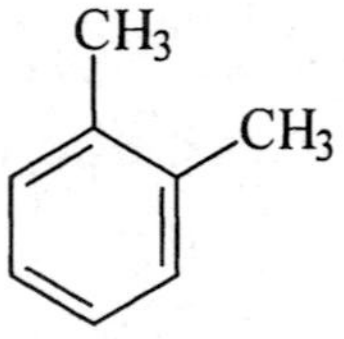

f) 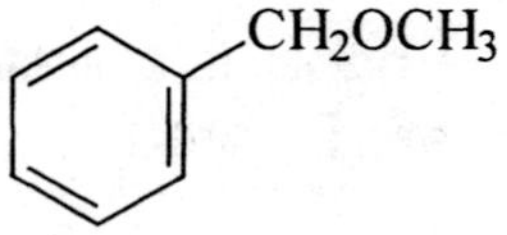

12.3 a) *m*-Propylphenol b) 3,5-Dibromophenol c) *p*-Methoxyphenol

12.4

a) 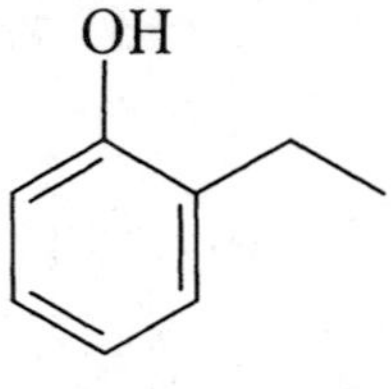

b) 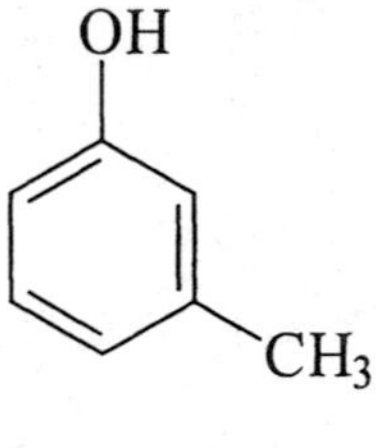

c) 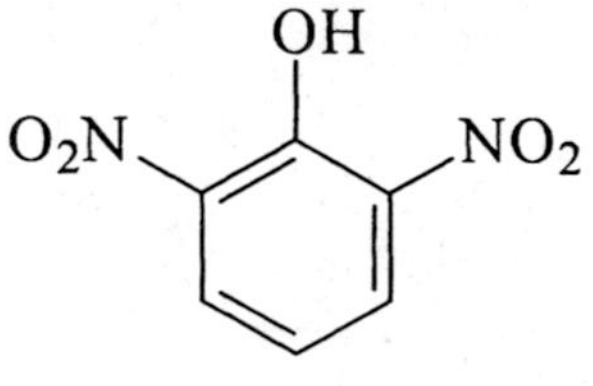

12.5

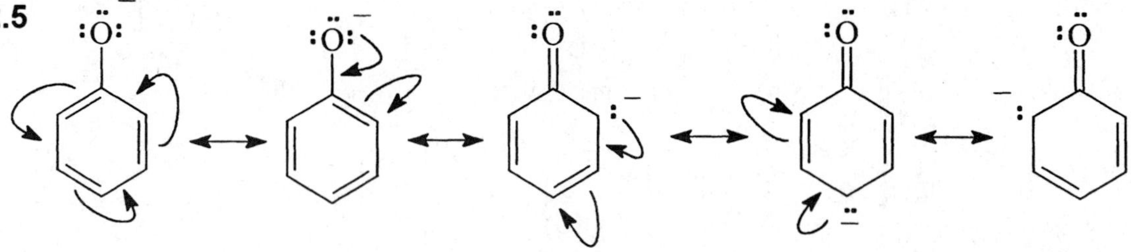

12.6 In acid-base reactions, the equilibrium favors the formation of the weaker acid and weaker base. As the strength of an acid increases, the pK_a decreases.

a) 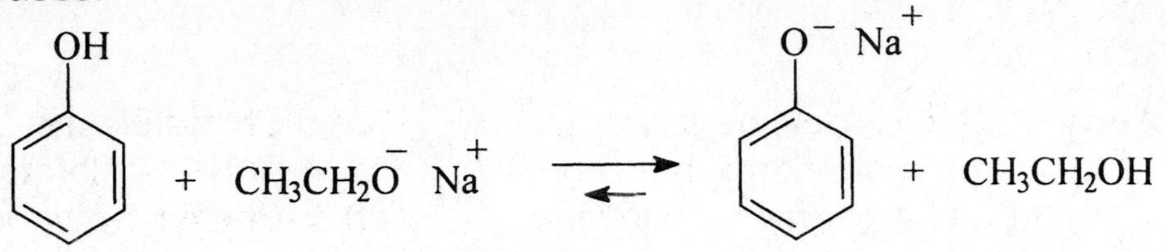

Phenol (pK_a = 10) is a stronger acid than ethanol (pK_a = 16), so the equilibrium favors the products.

b) 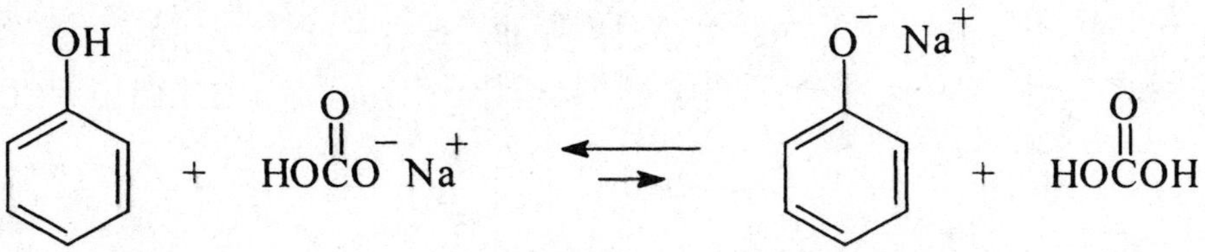

Phenol is a weaker acid than carbonic acid (pK_a = 6.35), so the reactants are favored.

12.7 a) Hexanal
b) 5-Methyl-3-hexanone
c) (*Z*)-4-Methyl-2-hexenal
d) *p*-Chlorobenzaldehyde
e) 5-(2-Methylpropyl)-3-cycloheptenone
f) 2,4-Pentanedione
g) 3-Phenylcyclopentanone
h) 4-Methylpent-3-en-2-one

12.8

a) 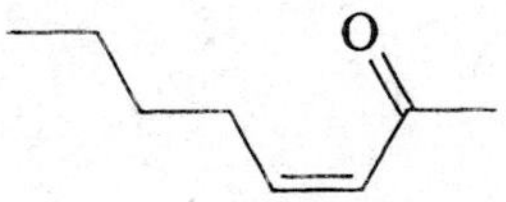

b) 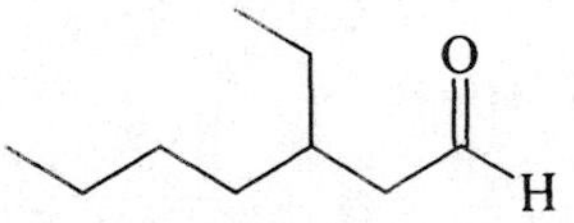

c) $CH_2{=}CHCH{=}CH\overset{O}{\overset{\|}{C}}H$

d) 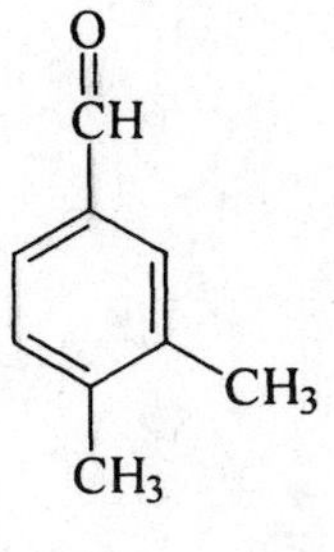

e) $Ph\overset{O}{\overset{\|}{C}}CH_2CH_3$

f) 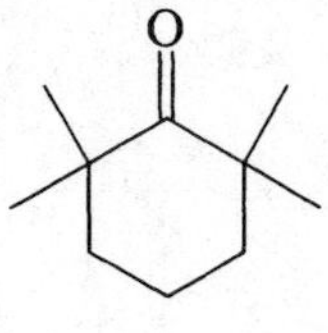

12.9

a) The circled hydrogens are more acidic because the conjugate base is stabilized by resonance.

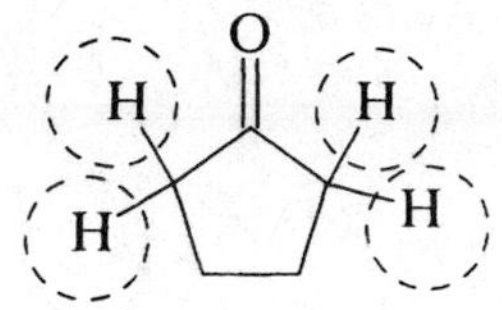

b) The circled hydrogens are more acidic because the conjugate base is stabilized by resonance with the carbonyl group and the benzene ring.

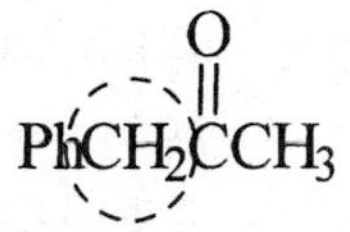

c) The circled hydrogens are more acidic because the conjugate base is stabilized by resonance with both of the adjacent carbonyl groups.

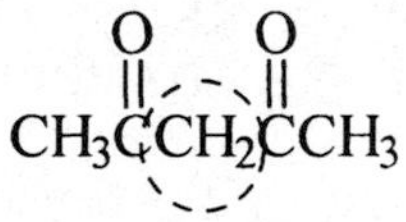

12.10 a) The equilibrium does not lie completely to the right because hydroxide ion is not a strong enough base. The conjugate acid of the base should have a larger pK_a than the substrate for complete removal of the proton. The pK_a of acetone is about 20 whereas the pK_a of water (the conjugate acid of hydroxide ion) is only 15.7, so this equilibrium favors the reactants.

b) In order to completely remove a proton from acetone, the conjugate acid of the base used must be a weaker acid than acetone; that is, it must have a pK_a greater than 20.

LDA, $NaNH_2$, $CH_3\overset{\overset{O}{\|}}{S}\ddot{C}H_2^-$, NaH

12.11 a) 3-Methylpentanoic acid
b) *p*-Bromobenzoic acid
c) (*Z*)-4,4-Dichloro-2-pentenoic acid
d) 2-Cyclohexenecarboxylic acid
e) 3-Methyl-4-nitrobenzoic acid
f) 3-Cyclopropylbutanoic acid

12.12

a) 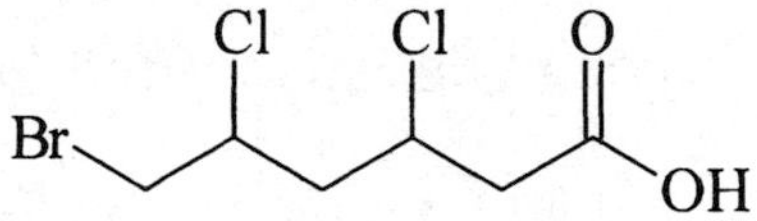
b) 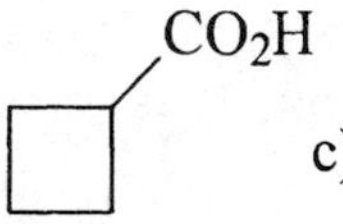
c) 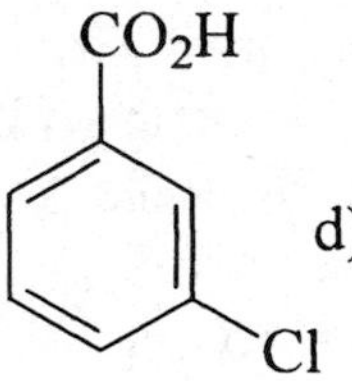
d) 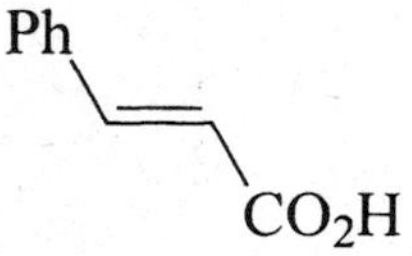

12.13

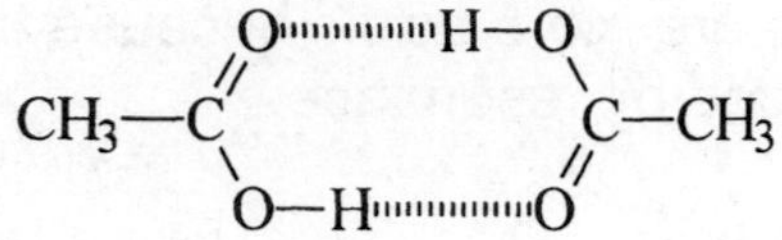

12.14 a) Benzoyl chloride
b) Propanoic anhydride
c) Methyl propanoate
d) Propyl 3-ethylbenzoate
e) Cyclopentyl 4-methylpentanoate
f) *N*-Methyl-3-pentynamide
g) Isopropyl 3-cyanocyclopentanecarboxylate
h) Pentanenitrile
i) Potassium 2-methylpentanoate

12.15

a) $CH_3CH_2\overset{\overset{O}{\|}}{C}Cl$ b) $CH_3\overset{\overset{O}{\|}}{C}N(CH_3)_2$ c) $CH_3(CH_2)_3\overset{\overset{O}{\|}}{C}O\overset{\overset{O}{\|}}{C}(CH_2)_3CH_3$ d)

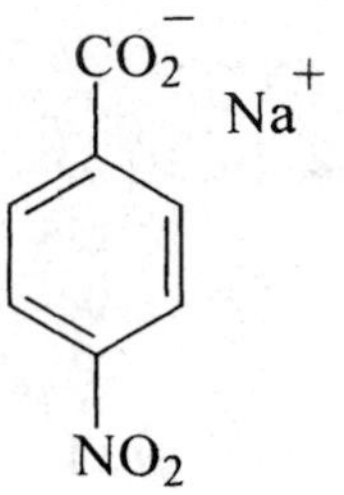

e) $CH_3(CH_2)_4\overset{\overset{O}{\|}}{C}NH_2$ f) $CH_3\overset{\overset{O}{\|}}{C}O\overset{\overset{CH_3}{|}}{C}HCH_3$ g) $Ph\overset{\overset{O}{\|}}{C}OCH_2Ph$ h)

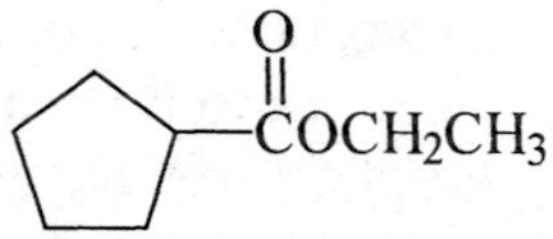

i)

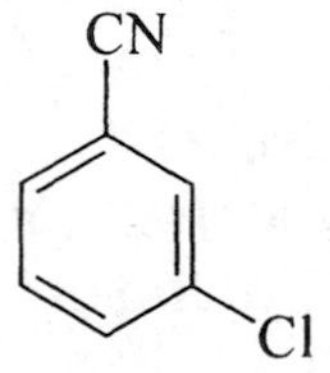

j) $CH_3CH_2CH_2CH_2\overset{\overset{CH_3}{|}}{C}HCH_2C{\equiv}N$

12.16 a) The primary amide (on the left) has a higher melting point because it can form hydrogen bonds.
b) The carboxylic acid has a higher boiling point because it can form hydrogen bonds.
c) The carboxylic acid has a higher boiling point because it can form hydrogen bonds.
d) The amide has a higher melting point because it is more polar and forms stronger hydrogen bonds.

12.17 a) Cyclohexanethiol   b) Dipropyl sulfide
c) 5-Methylhex-3-yne-1-thiol   d) Benzenesulfonic acid
e) Propyl *p*-toluenesulfonate

12.18

a) [structure: secondary butyl with SH] b) [benzene ring with SH] c) $CH_3\overset{O}{\underset{O}{S}}OCHCH_3$ with CH_3 d) [benzene ring with SO_3H top, Br bottom] e) $Ph-S-CCl_3$

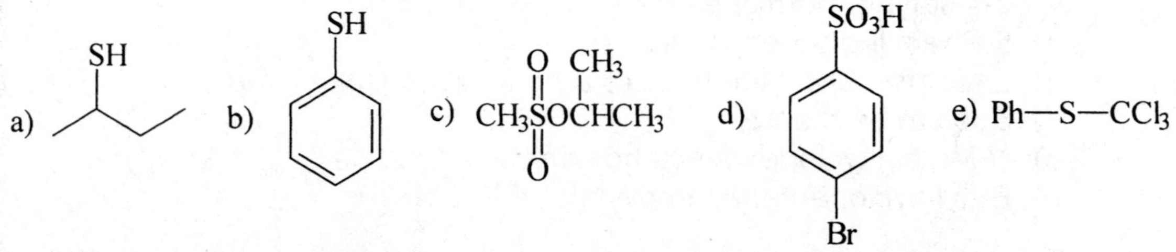

12.19 a) 4-Hydroxy-2-cyclohexenone   b) 2-Amino-4-methylpentanoic acid
c) 1,4-Butanedioic acid   d) Methyl 4-methyl-5-oxopentanoate
e) 3-Cyanobenzaldehyde
f) 4-Ethyl-3-hydroxy-*N*-methylhexanamide
g) 3-Oxobutanenitrile   h) Isopropyl 3-oxo-4-phenylhexanoate

12.20

a) $HO\overset{O}{C}(CH_2)_4\overset{O}{C}OH$ b) $CH_3CH_2\overset{OH}{\underset{CH_3CH_2}{C}}-\overset{O}{C}OCH_2CH_3$ c) [cyclohexane with $CH_2\overset{NH_2}{C}HCH_2OH$]

d) $CH_3CH_2CH=CHCH_2CH_2\overset{OH}{C}H-\overset{O}{C}-O-\overset{CH_3}{\underset{CH_3}{C}}-CH_3$ e) $CH_3\overset{CH_3}{C}H-\overset{O}{C}-\overset{O}{C}-N(CH_3)_2$

f) $PhCH_2\overset{NH_2}{C}H-\overset{O}{C}OH$

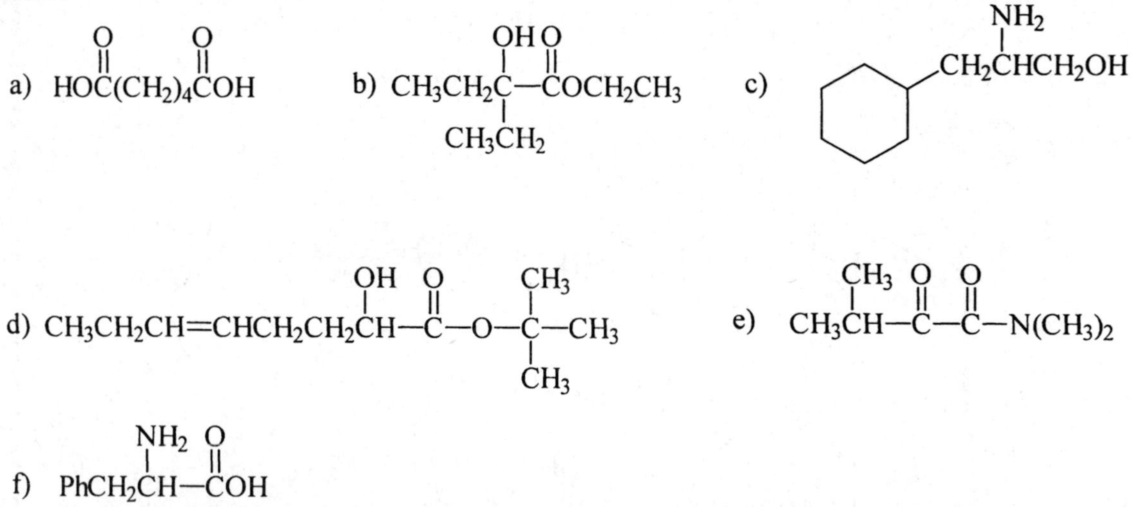

12.21 a) 1-Bromo-4-chlorobenzene or *p*-bromochlorobenzene
b) 4,5-Dimethyl-2-hexynal
c) (*Z*)-3-(1-Methylpropyl)-3-hexenoic acid
d) 4-Ethyl-6,6-dimethyloctanamide
e) 3,4-Dimethylbenzonitrile
f) Methyl acetate or methyl ethanoate
g) 1-Cyclobutyl-2-ethylbenzene or *o*-cyclobutylethylbenzene

h) 2-Ethyl-3-methyl-2-cyclopentenone
i) 2-Methylbenzoic acid or *o*-methylbenzoic acid
j) 4-Methylpentanoyl chloride
k) 3-Bromo-3-methylpentanenitrile
l) Isopropyl 3-methyl-2-butenoate
m) 2-*t*-Butyl-4-methylphenol
n) 5-Phenylhex-5-en-3-one
o) 2-Methoxybenzaldehyde or *o*-methoxybenzaldehyde
p) Sodium benzoate
q) *N*-Methylcyclohexanecarboxamide
r) Ethyl cyclopentanecarboxylate

12.22

a) 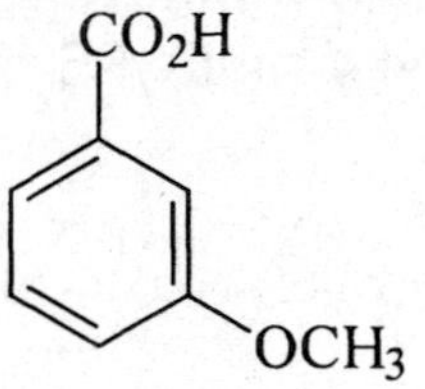

b) 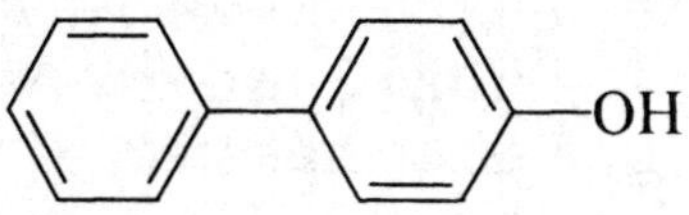

c) 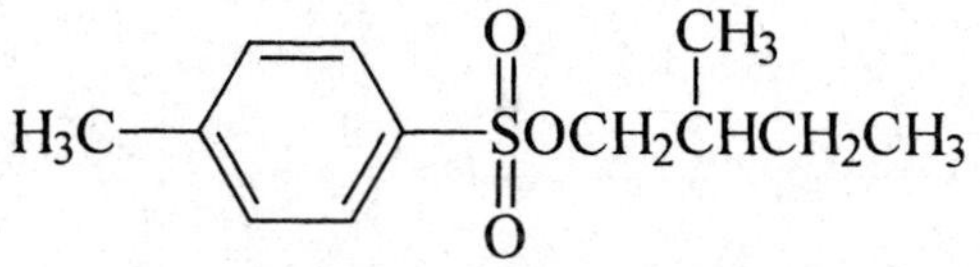

d) 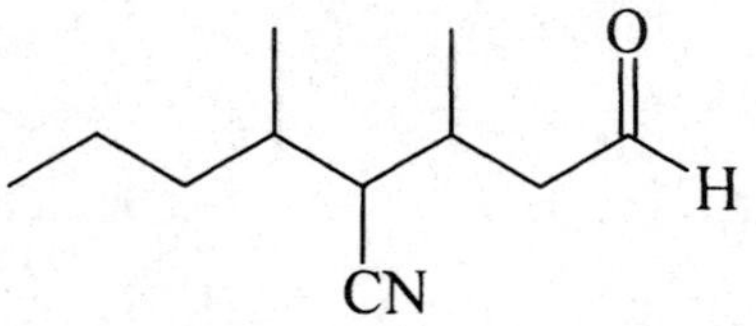

e) $CH_3\overset{\overset{O}{\|}}{C}ONa$

f) 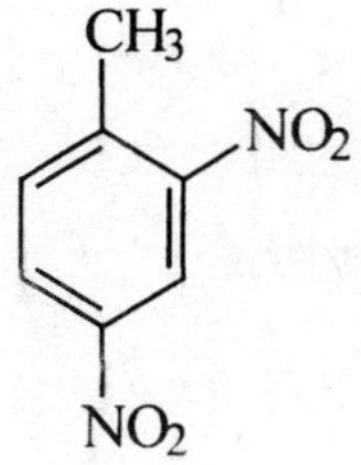

g) 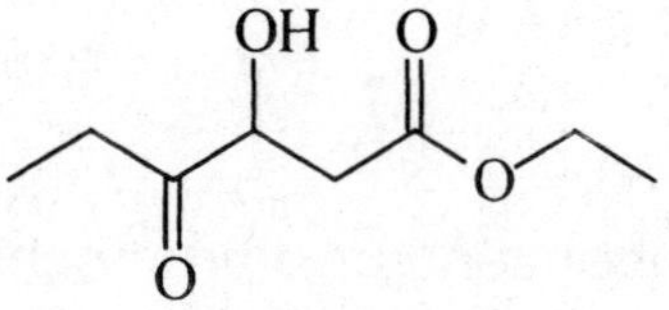

h) $CH_3\overset{\overset{O}{\|}}{C}Cl$

i) 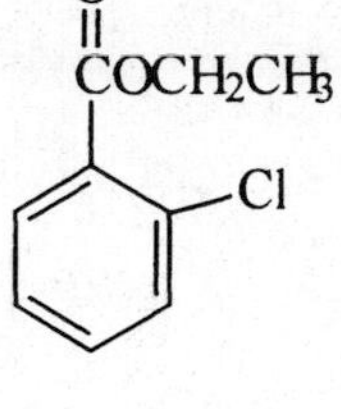

j) 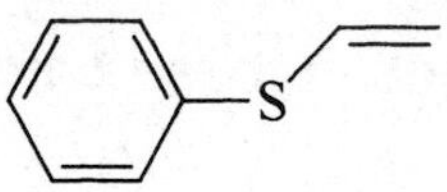

k) 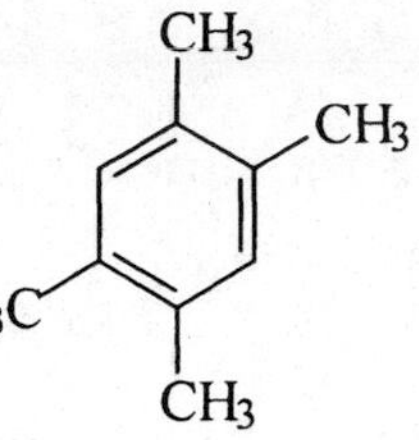

l) 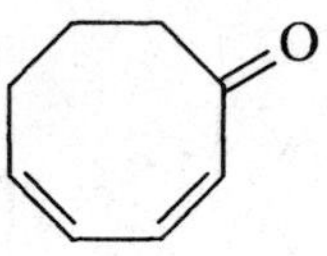

m) CH_3CN

n) 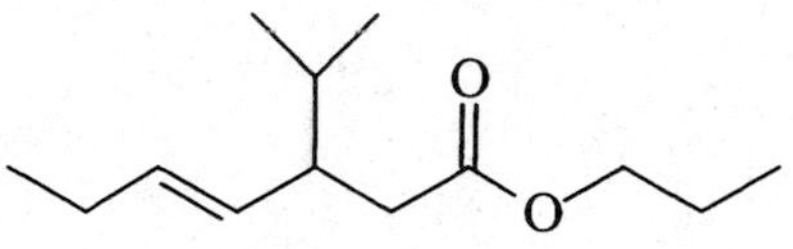

o) 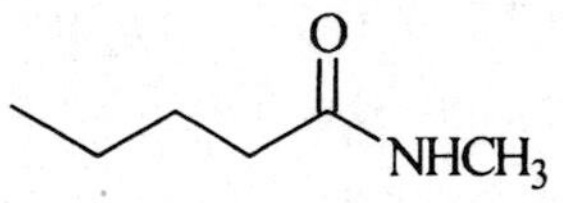

p) 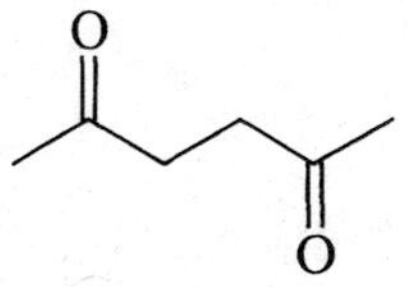

q) 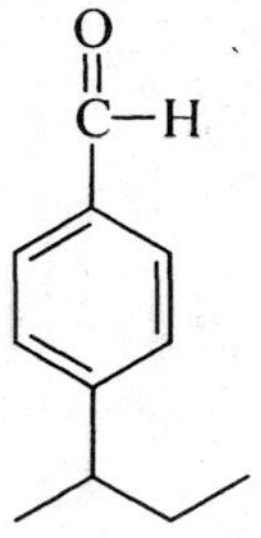

r) 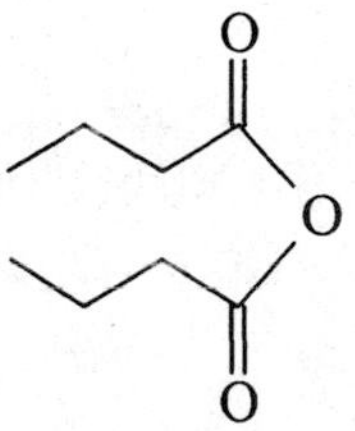

s) 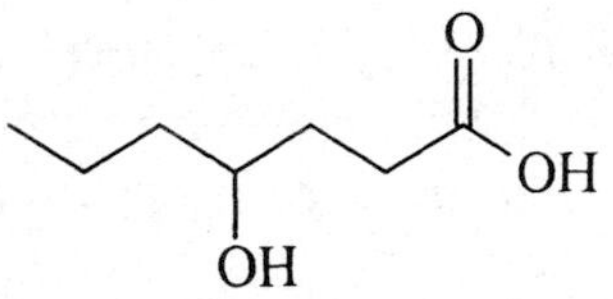

t) 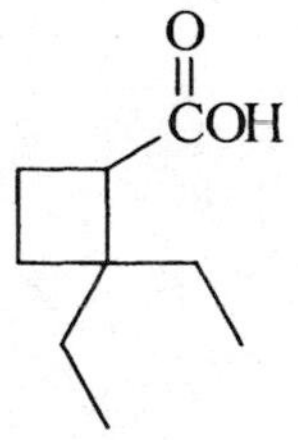

12.23 a) (*E*)-3-Oxo-4-hexenoic acid
b) 4-Hydroxypentanal
c) 3-Hydroxy-4-methoxybenzoic acid
d) (*E*)-4-Oxo-2-butenamide

12.24 a) 3-Hexanone has a higher melting point than heptane because it is more polar.
b) Cyclopentanol has a higher boiling point than cyclopentanone because it can form hydrogen bonds.
c) 2-Phenylethanoic acid has a higher boiling point than methyl benzoate because it is more polar and can form hydrogen bonds.

12.25 a) Phosphine and arene b) Thiol and alkene c) Sulfide and arene
d) Sulfoxide and arene e) Sulfone f) Sulfonic acid

12.26 The most acidic hydrogens in these compounds are circled.

a) 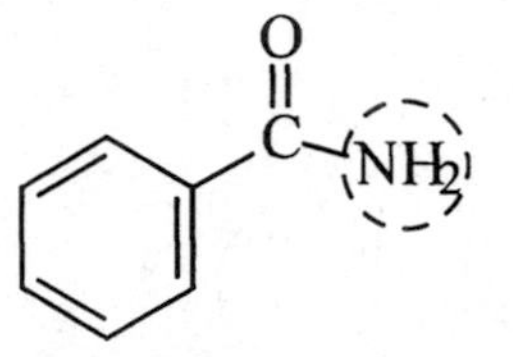

b) 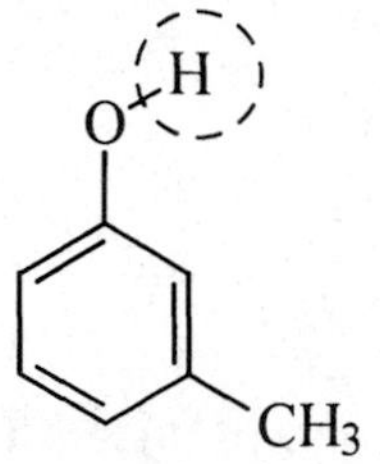

c) 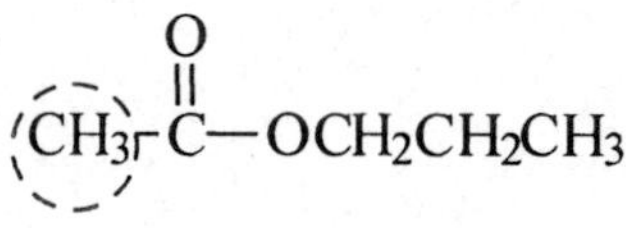

d) 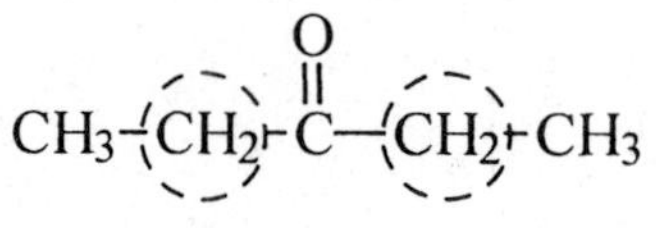

e) 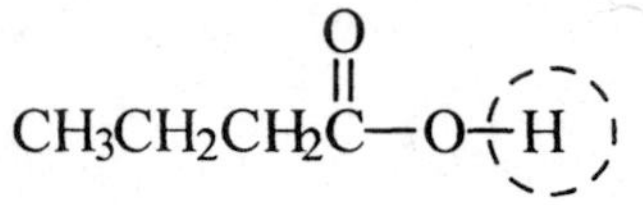

12.27 In the name ibuprofen, "ibu" , "pro", and "fen" are derived from the isobutyl, propanoic acid, and phenyl moieties in the structure. The systematic name for this compound is 2-(4-isobutylphenyl)propanoic acid.

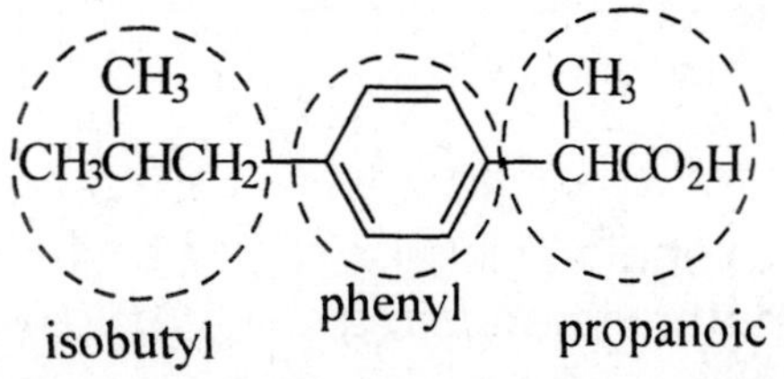

12.28 Picric acid such a strong acid because the conjugate base is highly stabilized by the strong inductive and resonance effects of the three nitro groups.

12.29 Bicarbonate ion (the conjugate base of carbonic acid) is a strong enough base to deprotonate benzoic acid but not *p*-methylphenol. A typical scheme for the separation of these compounds from a mixture using bicarbonate as the base is shown on the next page.

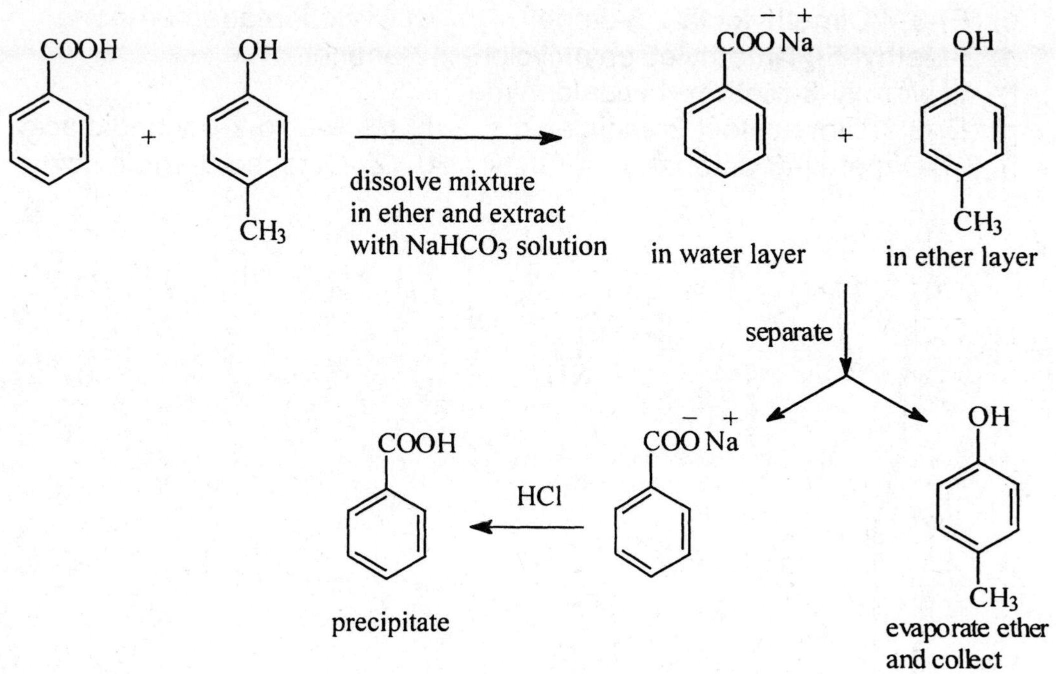

12.30

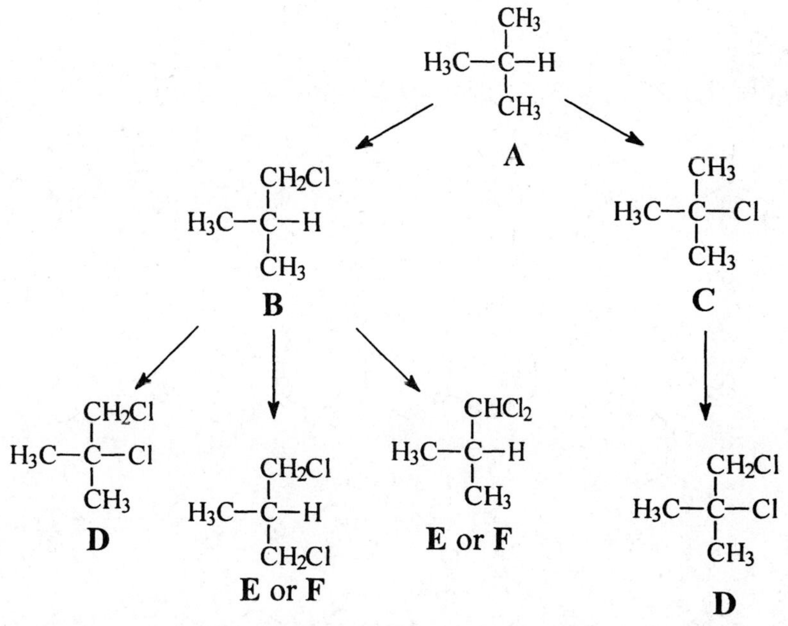

Based on these experiments it cannot be determined whether **E** is 1,3-dichloro-2-methylpropane and **F** is 1,1-dichloro-2-methylpropane or vice versa.

12.31 a) Methyl 2-hydroxybenzoate
b) (*E*)-3-Phenyl-2-propenal
c) (*E*)-3,7-Dimethylocta-2,6-dienal
d) Cycloheptadec-9-enone
e) 2-Methyl-5-(1-methylethenyl)cyclohex-2-enone
f) 4-Hydroxy-3-methoxybenzaldehyde
g) (*E*)-3,7-Dimethyloct-2-enoic acid
h) (*E*)-9-Oxo-2-decenoic acid
i) 3,7-Dimethyl-6-octenal
j) (*Z*)-Octadec-9-enoic acid

12.32

a) 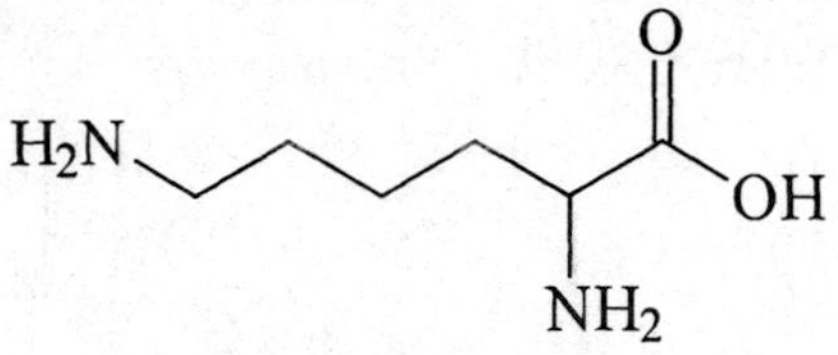

b) 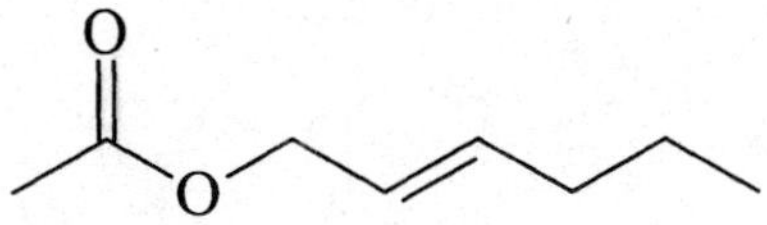

c) 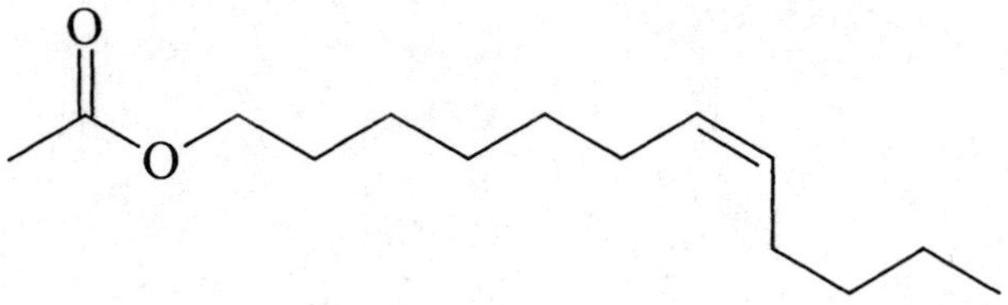

d) 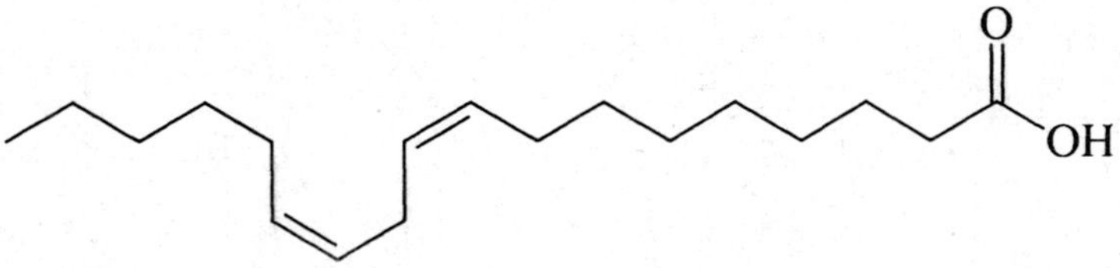

e) 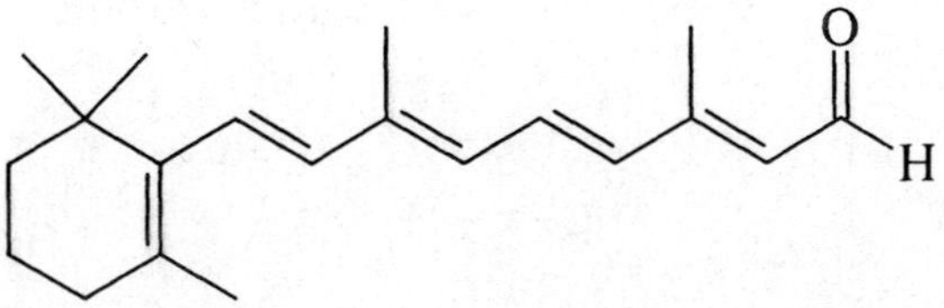

f) 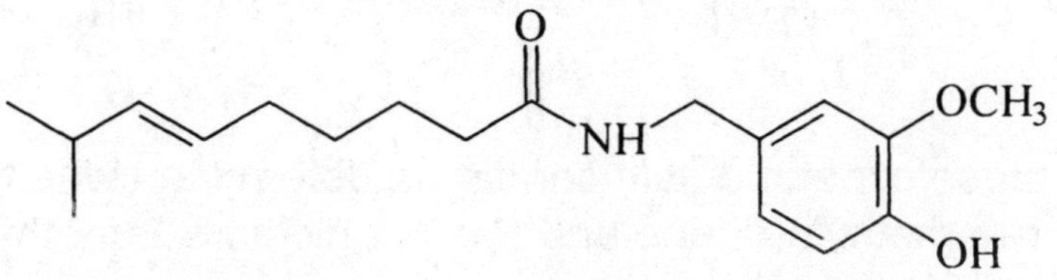

12.33 a) Isopropyl benzoate
b) 3-Pentynoic acid
c) 3-Ethylcyclopentanone
d) Cyclohept-1-enecarbaldehyde

12.34 a) Benzene has a higher melting point than hexane because its more symmetrical shape allows it to pack better into the crystal lattice.
b) Benzamide has a higher melting point than benzoic acid because the amide is more polar and forms stronger hydrogen bonds.

12.35 a) 2-Pentanol has a higher boiling point than 2-pentanone because it can form hydrogen bonds.
b) Benzamide has a higher boiling point than benzoic acid because the amide is more polar and forms stronger hydrogen bonds.

12.36 a) Benzoic acid (pK_a ≈ 5) is soluble in both aqueous sodium hydroxide (pK_a of water = 15.7) and aqueous sodium bicarbonate (pK_a of carbonic acid = 6.35).
b) Benzaldehyde has no acidic hydrogen, so it is not soluble in either of these reagents.
c) 4-Methylphenol (pK_a ≈ 10) is soluble in aqueous sodium hydroxide, but not in aqueous sodium bicarbonate.

Review of Mastery Goals

After completing this chapter, you should be able to:

Name an aromatic compound, a phenol, an aldehyde, a ketone, a carboxylic acid, an acid chloride, an anhydride, an ester, an amide, a nitrile, and a carboxylic acid salt.
(Problems 12.1, 12.3, 12.7, 12.11, 12.14, 12.21, 12.31, and 12.33)

Draw the structure of a compound containing one of these functional groups when the name is provided.
(Problems 12.2, 12.4, 12.8, 12.12, 12.15, 12.22, and 12.32)

Recognize the common functional groups that contain sulfur or phosphorus.
(Problems 12.17, 12.18, and 12.25)

Name a compound containing more than one functional group or draw the structure of such a compound when the name is provided.
(Problems 12.19, 12.20, and 12.23)

Understand how the physical properties of these compounds depend on the functional group that is present.
(Problems 12.6, 12.9, 12.10, 12.13, 12.16, 12.24, 12.34, 12.35, and 12.36)

Chapter 13
INFRARED SPECTROSCOPY

13.1 a) 3.33×10^{-3} cm b) 9.68×10^{14} s^{-1} c) 0.858 kcal/mol (3.59 kJ/mol) d) 92.3 kcal/mol (387 kJ/mol)

13.2 This frequency of light falls in the infrared region of the electromagnetic spectrum.

13.3 The energy of light with this frequency is 3.81×10^{-5} kcal/mol (1.60×10^{-4} kJ/mol).

13.4 a) The C-H bond absorbs at a higher wavenumber because hydrogen is a lighter atom than deuterium.
b) The C≡C bond absorbs at a higher wavenumber because a triple bond is stronger than a double bond.
c) The C-Cl bond absorbs at a higher wavenumber because the chlorine is a lighter atom than iodine and the C-Cl bond is stronger than the C-I bond.

13.5 a) OH, 3000 cm^{-1}, very broad; =CH, 3100-3000 cm^{-1}; -CH, 3000-2850 cm^{-1}
b) =CH, 3100-3000 cm^{-1}; -CH, 3000-2850 cm^{-1}
c) NH_2, two bands, 3400-3250 cm^{-1}; =CH, 3100-3000 cm^{-1}
d) OH, 3550-3200 cm^{-1}, broad; =CH, 3100-3000 cm^{-1}; -CH, 3000-2850 cm^{-1}
e) NH, one band, 3400-3250 cm^{-1}; -CH, 3000-2850 cm^{-1}
f) -CH, 3000-2850 cm^{-1}; CHO, 2830-2700 cm^{-1}, two bands

13.6 The band in the triple bond region at 2150-2100 cm^{-1} is much stronger for 1-hexyne than it is for the more symmetrical 3-hexyne. In addition, the ≡CH band near 3300 cm^{-1} in the spectrum of 1-hexyne confirms the presence of the triple bond.

13.7 The 3000-2900 cm^{-1} range is good region in the IR spectrum to monitor hydrocarbon emissions because most hydrocarbons have -CH bonds that absorb in this region.

13.8 a) The carbonyl group of this ketone is part of a five membered ring, so the band will be shifted about 30 cm^{-1} to higher wavenumbers from the

base position of 1715 cm^{-1}. The predicted position is 1745 cm^{-1}.
b) This is a conjugated aldehyde, so the band will be shifted to lower wavenumbers by 20-40 cm^{-1} from the base position of 1730 cm^{-1}. The predicted position is 1710-1690 cm^{-1}.
c) This is a conjugated ketone, so the band will be shifted to lower wavenumbers by 20-40 cm^{-1} from the base position of 1715 cm^{-1}. The predicted position is 1695-1675 cm^{-1}.
d) This is a conjugated ester, so the band will be shifted to lower wavenumbers by 20-40 cm^{-1} from the base position of 1740 cm^{-1}. The predicted position is 1720-1700 cm^{-1}.
e) This is a conjugated carboxylic acid, so the band will be shifted to lower wavenumbers by 20-40 cm^{-1} from the base position of 1710 cm^{-1}. The predicted position is 1690-1670 cm^{-1}.

13.9 a) The compound on the left, a ketone, has its carbonyl absorption near 1715 cm^{-1} whereas the compound on the right, an aldehyde, has its carbonyl peak near 1730 cm^{-1} and has two bands in the region of 2830-2700 cm^{-1}.
b) The ester on the left (non-conjugated) has its carbonyl absorption near 1740 cm^{-1} whereas the ester on the right (conjugated) has its carbonyl absorption near 1720-1700 cm^{-1}.
c) The compound on the right is a carboxylic acid and has a very broad band near 3000 cm^{-1} whereas the compound on the left is a conjugated ester and will not show this feature.

13.10 a) =CH, 3100-3000; -CH, 3000-2850; C=O, 1695-1675; C=C, 1660-1640.
b) $-NH_2$, 3400-3250 (two bands); =CH, 3100-3000; -CH, 3000-2850; aromatic ring, 1600-1450 (four bands), 900-675.
c) =CH, 3100-3000, -CH, 3000-2850; C=C, 1660-1640; C-O, 1300-1000.
d) OH, 3550-3200 (broad); -CH, 3000-2850; C-O, 1300-1000.
e) ≡CH, 3300; -C-H, 3000-2850; C≡C, 2150-2100; NO_2, 1550, 1380.
f) =CH, 3100-3000; -CH, 3000-2850; aldehyde CH, 2830-2700 (two bands); C=O, 1710-1690; aromatic ring, 1600-1450 (four bands), 900-675.

13.11 a) 1-Butyne has absorptions at 3300 cm^{-1} (≡CH) and 2150-2100 cm^{-1} (C≡C) that are not present in the spectrum of 1-butene.
b) Benzyl alcohol has absorptions at 3100-3000 (=CH), and 1600-1450 and 900-675 cm^{-1} (aromatic ring) that are not present in the spectrum of *t*-butanol.

c) Benzaldehyde has two bands for the aldehyde H at 2830-2700 cm^{-1} that are not present in the spectrum of the ketone, acetophenone.
d) The primary amine has two bands in the 3400-3250 cm^{-1} region whereas the secondary amine has only one.

13.12 a) The strong absorption at 1715 cm^{-1} is typical for a carbonyl group of a ketone. The absorption in the region of 3000-2850 cm^{-1} is indicative of H's bonded to *sp*^{3}-hybridized C's. There is no evidence for the presence of a C=C bond.
b) The strong and broad OH absorption centered at 3000 cm^{-1} along with the carbonyl absorption at 1716 cm^{-1} are typical for a carboxylic acid. There is no evidence for the presence of a C=C bond.
c) The two bands in the region of 2830-2700 cm^{-1} and the carbonyl absorption at 1696 cm^{-1} provide evidence for the presence of an aldehyde. The observed shift of the carbonyl band to a lower wavenumber indicates that it is a conjugated aldehyde. The absorptions in the region of 3100-3000 cm^{-1} are due to *sp*^{2}-hybridized C-H bonds, and those in the region of 3000-2850 cm^{-1} are due to *sp*^{3}-hybridized C-H bonds. The appearance of four bands in the 1600-1450 cm^{-1} region and the strong band near 750 cm^{-1} indicate the probable presence of an aromatic ring.

13.13 a) NH_2, two bands, 3400-3250 cm^{-1}; ≡CH, 3300 cm^{-1}; -CH, 3000-2850 cm^{-1}; C≡C, 2150-2100 cm^{-1}.
b) OH, 3000 cm^{-1}, very broad; -CH, 3000-2850 cm^{-1}; -C≡N, 2260-2200 cm^{-1}, medium; C=O, 1710 cm^{-1}.
c) -CH, 3000-2850 cm^{-1}; C=O, 1740 cm^{-1}; C-O, 1300-1000 cm^{-1}.
d) =CH, 3100-3000 cm^{-1}; -CH, 3000-2850 cm^{-1}; C=O, 1695-1675 cm^{-1}; aromatic ring, 1600-1450 cm^{-1} (four bands), and 900-675 cm^{-1}.

13.14 a) The broad absorption centered at 3300 cm^{-1} indicates the presence of a hydroxy group. The absorption at 2900 cm^{-1} suggests the presence of hydrogens on *sp*^{3}-hybridized C's. The compound is a saturated alcohol.
b) The two bands in the 3400-3250 cm^{-1} region indicate the presence of an NH_2 group. The absorption at 2900 cm^{-1} shows the presence of hydrogens on *sp*^{3}-hybridized C's. The compound is a saturated primary amine.
c) The absorptions in the region of 3100-3000 cm^{-1} are due *sp*^{2}-hybridized C-H bonds, and those at 3000-2850 cm^{-1} are due to *sp*^{3}-hybridized C-H bonds. The appearance of four bands in the 1600-1450 cm^{-1} region and the strong band near 750 cm^{-1} indicate an aromatic ring. The strong

absorption at 1683 cm^{-1} is due to a carbonyl group. The absence of bands for other carbonyl containing functional groups indicates that the compound is a ketone or an ester. It is often difficult to determine with certainty whether a compound is a ketone or an ester based solely on its IR spectrum. In this case there is a strong band near 1250 cm^{-1}, but it is not as broad as the C-O band of an ester normally appears. (See the spectrum in part (d) of this problem for an example of a spectrum of an ester.) Note that the carbonyl group of a conjugated ketone should appear at 1695-1675 cm^{-1} whereas that of a conjugated ester should appear at 1720-1700 cm^{-1}. This compound is actually a conjugated ketone.
d) The absorptions in the region of 3100-3000 cm^{-1} are due *sp*^{2}-hybridized C-H bonds, and those in the region of 3000-2850 cm^{-1} are due to *sp*^{3}-hybridized C-H bonds. The appearance of four bands in the 1600-1450 cm^{-1} region and the strong band near 700 cm^{-1} suggests the presence of an aromatic ring. The strong absorption at 1715 cm^{-1} is due to a carbonyl group. The strong C-O band near 1200 cm^{-1} indicates that the compound is an ester. The observed shift of the carbonyl band to a lower wavenumber is consistent with a conjugated ester.

13.15 a) The terminal alkyne (at the right) will have a sharp band at 3300 cm^{-1} due to the ≡C-H group. Both will have peaks in the 2150-2100 cm^{-1} region due to the C≡C group, but the band from the terminal alkyne will be more intense because it is less symmetrical.
b) The conjugated aldehyde (at the left) will have the absorption for its C=O group at lower wavenumbers (1710-1690 cm^{-1}) than the non-conjugated aldehyde (1730 cm^{-1}).
c) The amine will have two bands in the 3400-3250 cm^{-1} region. These bands will be of weaker intensity and less broad than the OH band of the alcohol, which will appear in the 3550-3200 cm^{-1} region.
d) The ester will show a band for its C=O group near 1740 cm^{-1}, whereas that for the ketone will appear near 1715 cm^{-1}. In addition, the ester will have a intense band for the C-O group in the 1300-1000 cm^{-1} region.
e) The alcohol will show a broad absorption for its OH group in the region of 3550-3200 cm^{-1}. The absorption for the OH group of the carboxylic acid will be broader than that for the alcohol and will be centered near 3000 cm^{-1}. In addition, the carboxylic acid will have an absorption for its C=O group near 1710 cm^{-1}.
f) The ether will have an intense band in the 1300-1000 cm^{-1} region due to the C-O bond.

13.16 The bands 3000-2850 cm^{-1} region show the presence of H's bonded to sp^3-hybridized C's. The band at 1746 cm^{-1} and the intense band near 1200 cm^{-1} suggest the presence of an ester. The unknown is a saturated ester.

13.17 The broad, intense band near 3350 cm^{-1} is characteristic of an alcohol. The bands in the 3100-3000 cm^{-1} region show the presence of H's bonded to sp^2-hybridized C's and the bands in the 3000-2850 cm^{-1} region show the presence of H's bonded to sp^3-hybridized C's. The presence of four bands in the 1600-1450 cm^{-1} region and bands in the 900-675 cm^{-1} region indicate the possible presence of an aromatic ring. Therefore, the compound is definitely an unsaturated alcohol, possibly containing an aromatic ring.

13.18 The bands in the 3100-3000 cm^{-1} region show the presence of H's bonded to sp^2-hybridized C's and the bands in the 3000-2850 cm^{-1} region show the presence of H's bonded to sp^3-hybridized C's. The two bands in the 2830-2700 cm^{-1} region and the C=O band at 1724 cm^{-1} suggest that the compound is an aldehyde. The compound has a number of bands in the 1600-1450 and 900-675 regions, indicating the possible presence of an aromatic ring. The position of the band for the aldehyde C=O indicates that it is not conjugated.

13.19 The broad peak centered near 3000 cm^{-1} (OH) and the peak at 1706 (C=O) are characteristic of a carboxylic acid. The peaks from 3000-2850 cm^{-1}, superimposed on the OH peak, indicate the presence of H's on sp^3-hybridized C's. There is no indication of the presence of a C=C so the unknown is a saturated carboxylic acid.

13.20 The intense peak at 1721 cm^{-1} shows the presence of a C=O group. The absence of any peaks for other carbonyl containing functional groups, along with the position of the C=O peak, indicate that the compound is a ketone. The peaks from 3000-2850 cm^{-1} indicate the presence of H's on sp^3-hybridized C's. There is no indication of the presence of a C=C, so the unknown is a saturated ketone.

13.21 The DU of this compound is 0, so it has no rings or pi bonds. The broad, intense peak near 3350 cm^{-1} is characteristic of an alcohol. Thus, the unknown is a 5-carbon saturated alcohol. One possibility is 2-methyl-1-butanol.

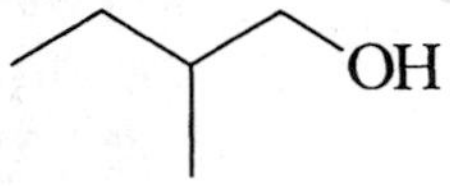

13.22 The DU of this compound is 1. The peak at 1741 cm^{-1} is due to the presence of a C=O, which accounts for the DU. The position of the C=O peak, along with the intense peak near 1250 cm^{-1}, suggests that the unknown is a saturated ester. Butyl acetate is one possibility.

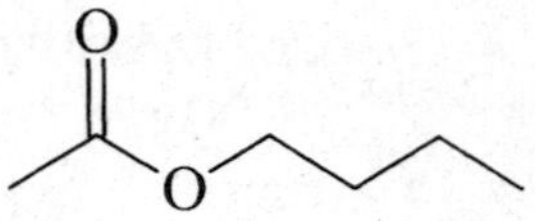

13.23 The DU of this compound is 3. The peaks in the 3000-2850 cm^{-1} region show the presence of H's bonded to sp^3-hybridized C's. There is no sign of H's bonded to sp^2-hybridized C's. The peak near 2250 cm^{-1} suggests the presence of a C≡N group, which accounts for two degrees of unsaturation. The peak at 1740 cm^{-1} is due to a C=O group and accounts for the other degree of unsaturation. The position of the C=O band and the intense band near 1200 cm^{-1} suggests that the unknown is an ester. One possibility is ethyl cyanoacetate.

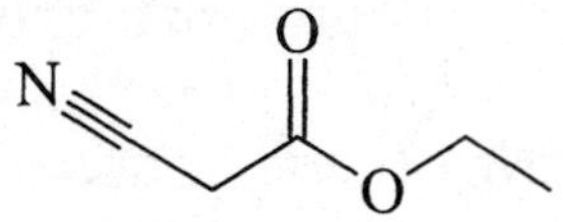

13.24 The DU of this compound is 2. The peaks in the 3000-2850 cm^{-1} region show the presence of H's bonded to sp^3-hybridized C's. There is no sign of H's bonded to sp^2-hybridized C's. The peak at 1718 cm^{-1} indicates the presence of a C=O group. The absence of peaks expected for other carbonyl containing functional groups and the position of the C=O peak indicate that the unknown is a ketone. Because there is no evidence for the presence of a C=C, the second degree of unsaturation must be due to a ring. Cycloheptanone is one possibility for the unknown.

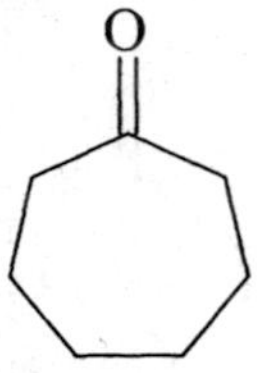

13.25 Th DU of this compound is 5. The peaks in the 3000-2850 cm^{-1} region show the presence of H's bonded to sp^3-hybridized C's. The small peaks in the region of 3100-3000 cm^{-1} suggest the presence of H's bonded to sp^2-hybridized C's. The two peaks in the 2830-2700 cm^{-1} region, along with the C=O peak at 1706 cm^{-1} indicate the presence of an aldehyde. The position for the C=O of the aldehyde suggests that it is conjugated. The bands in the 1600-1450 and 900-675 cm^{-1} regions indicate the possible presence of an aromatic ring. Note that an aromatic ring has a DU = 4, so the presence of an aromatic ring and a C=O would account of the DU of 5. One possible structure is 4-ethylbenzaldehyde.

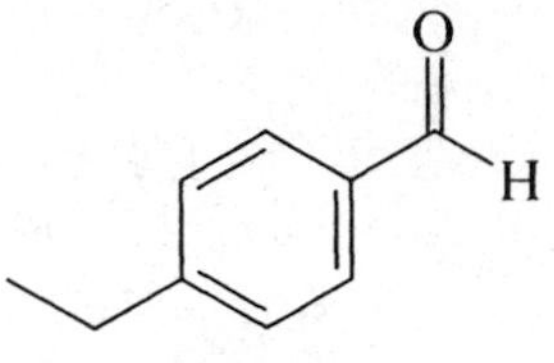

13.26 Morphine has a broad, intense band in the 3550-3200 cm^{-1} region, resulting from the presence of its OH groups. Heroin has bands near 1740 cm^{-1} due to its ester groups.

13.27 a) The compound is *p*-methylbenzoic acid. It has a very broad, intense absorption for the OH group centered near 3000 cm^{-1}. Superimposed on this are bands in the 3100-3000 cm^{-1} region due to H's on *sp*^{2}-hybridized C's and bands in the 3000-2850 cm^{-1} region due to H's on *sp*^{3}-hybridized C's. There is also a intense C=O band in the 1690-1670 cm^{-1} region. Bands for the aromatic ring appear in the 1600-1450 and 900-675 cm^{-1} regions.
b) The compound is 4-pentyn-2-ol. It has a broad, intense peak in the 3550-3200 cm^{-1} region due to the OH group. It also has a sharp peak near 3300 cm^{-1} due to the ≡C-H group. It also has bands in the 3000-2850 cm^{-1} region due to H's on *sp*^{3}-hybridized C's. The band for the C ≡C group appears in the 2150-2100 cm^{-1} region.

13.28 The compound at the left is an ester and is expected to have a carbonyl peak near 1740 cm^{-1}. The compound at the right is a ketone and is expected to have a carbonyl peak near 1715 cm^{-1}.

13.29 The intense peaks near 1520 and 1350 cm^{-1} result from the presence of a nitro group (NO_2). Nitrobenzene, the compound at the right, is expected to exhibit these peaks.

Review of Mastery Goals

After completing this chapter you should be able to:

Predict the important absorption bands in the IR spectrum of a compound. (Problems 13.5, 13.8, 13.9, 13.10, 13.11, 13.13, 13.15, 13.26, 13.28, and 13.29)

Determine the functional group that is present in a compound by examination of its infrared spectrum.
(Problems 13.12, 13.14, 13.16, 13.17, 13.18, 13.19, 13.20, 13.21, 13.22, 13.23, 13.24, and 13.25)

Chapter 14
NUCLEAR MAGNETIC RESONANCE SPECTROSCOPY

14.1 The position of absorption or chemical shift is given by:

$$\delta = \frac{\text{observed position of peak (Hz)}}{\text{operating frequency of instrument (Hz)}} \times 10^6$$

a) 7.4 δ
b) 1480 Hz on a 200 MHz and 2960 Hz on a 400 MHz instrument
c) 7.4 δ

14.2 The number of absorption signals in the NMR spectrum is equal to the number of different types of hydrogens (chemically nonequivalent) in a molecule.
a) The CH_3 groups are chemically nonequivalent. The hydrogens of the CH_2 group are enantiotopic, so they have the same chemical shift. Thus there are three absorptions.
b) The methyl groups are different. There are two different types of H's on the aromatic ring, the two H's ortho to the carbonyl group and the two H's meta to the carbonyl group. Therefore there are four absorptions.
c) The two CH_3 groups are different. At first glance, the H's of the CH_2 group appear to be the same. Careful analysis shows that they are diastereotopic, and, thus, different. The situation is similar to the CH_2 group of 2-bromobutane described in the text. (In general, H's of CH_2 groups will be diastereotopic if there is a chirality center elsewhere in the molecule.) In addition, there are the CH and the OH groups. Therefore there is a total of six absorptions.
d) The two CH_3 groups are different, as are the two vinylic hydrogens. Therefore there are four absorptions.
e) The two OH groups are identical and the two vinylic hydrogens are identical, so there are only two absorptions.

14.3 Approximate chemical shifts for different types of hydrogens can be obtained using Table 14.1. Remember that most values in this table are for CH_3 groups. The signal for a CH_2 group appears about 0.3 δ downfield and that for a CH group appears about 0.7 δ downfield from these values. The presence of an additional functional group causes an additional downfield shift. This shift is relatively large if the additional functional group is attached to the same carbon as the hydrogen under consideration. The shift is much smaller if the functional group is attached to the carbon adjacent to the carbon attached to the hydrogen and is negligible if the functional group is further away.

a) $CH_3\overset{O}{\overset{\|}{C}}CH_2CH_3$ b) $CH_3\overset{OH}{\overset{|}{C}}HCH_3$ ← 2 - 5   c) $CH_3\overset{O}{\overset{\|}{C}}OCH_2CH_3$ d) $CH_3\overset{Cl\ Cl}{\overset{|\ \ |}{C}}HCH_2$

2.2   >2.2   >0.9   >0.9   >3.3   2.0   >3.7   >0.9   >0.9   >3.0   >3.0

most downfield ⤶

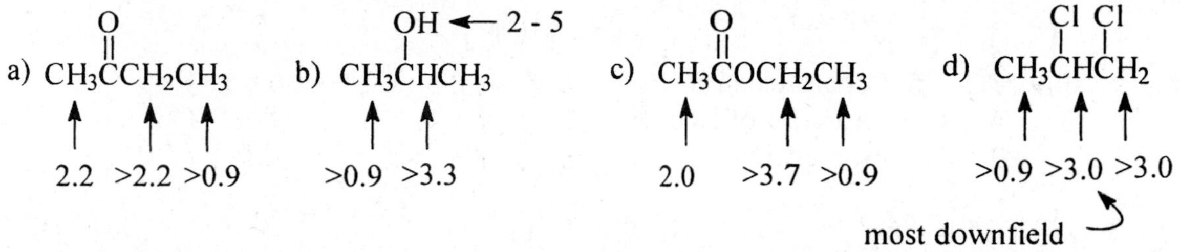

14.4 The signal from H_m will appear as a triplet (1:2:1) because the two inner lines overlap. (This is the same result as when H_m is coupled to two identical hydrogens.)

14.5

position for H_m peak in absence of H_a and H_x

coupling of H_x and H_m

J_{mx}

J_{am}

J_{mx}

coupling of H_m and H_a

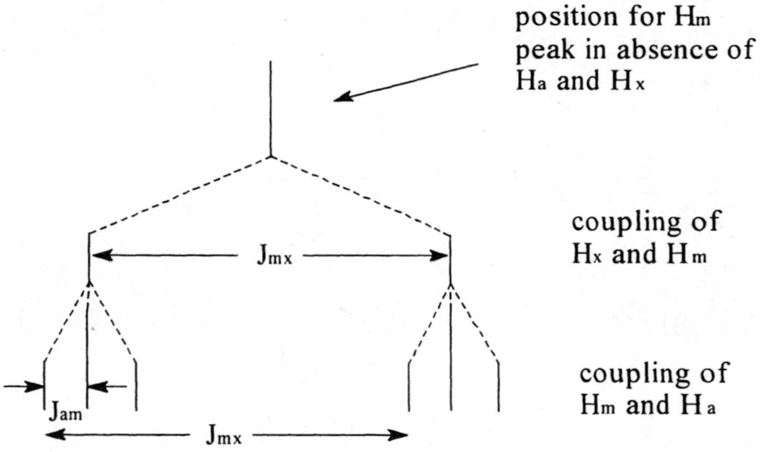

14.6 In general, the absorption peak for a hydrogen(s) that is coupled to **n** equivalent hydrogens is split into **n + 1** peaks.
a) a = x = 3   b) a = 7, x = 2   c) a = 3, x = 2
d) a = 5, x = 2   e) a = 3, m = 6, x = 3

14.7

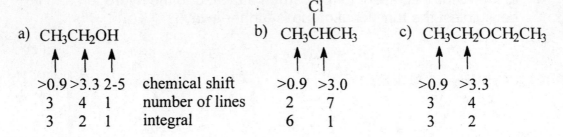

a) CH_3CH_2OH

	CH_3	CH_2	OH
chemical shift	>0.9	>3.3	2-5
number of lines	3	4	1
integral	3	2	1

b) $CH_3CH(Cl)CH_3$

	CH_3	CH
chemical shift	>0.9	>3.0
number of lines	2	7
integral	6	1

c) $CH_3CH_2OCH_2CH_3$

	CH_3	CH_2
chemical shift	>0.9	>3.3
number of lines	3	4
integral	3	2

d) 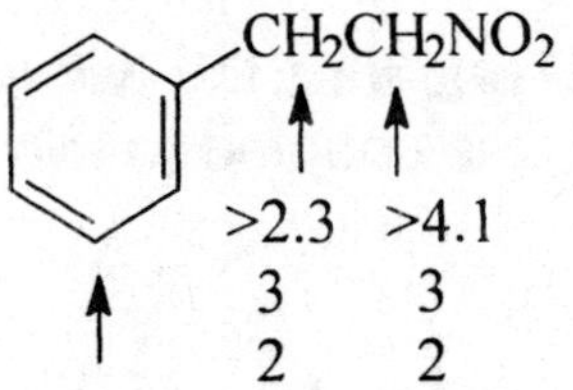

	ring H	CH_2	CH_2NO_2
chemical shift	7-8	>2.3	>4.1
number of lines	1-m	3	3
integral	5	2	2

e) $CH_3\overset{O}{\overset{\|}{C}}OCH_2CH_3$

	CH_3C	OCH_2	CH_3
chemical shift	2.0	>3.7	>0.9
number of lines	1	4	3
integral	3	2	3

f) $CH_3CH_2CH_2Cl$

	CH_3	CH_2	CH_2Cl
chemical shift	0.9	>0.9	>3.0
number of lines	3	6-m	3
integral	3	2	2

g) CH_3CHCl_2

	CH_3	$CHCl_2$
chemical shift	>0.9	>>3.0
number of lines	2	4
integral	3	1

14.8

a) $CH_3\overset{O}{\overset{\|}{C}}CH_3$

b) 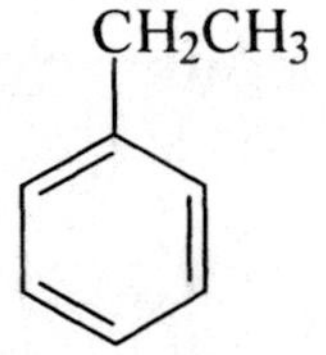

c) $CH_3\overset{O}{\overset{\|}{C}}H$

d) 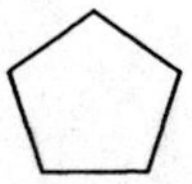

e) $CH_3\overset{O}{\overset{\|}{C}}O\overset{CH_3}{\overset{|}{C}}HCH_3$

f) $CH_3CH_2\overset{CH_3}{\overset{|}{C}}HCl$

14.9 The number of absorptions in the ^{13}C-NMR spectrum equals the number of different types of carbons in a molecule. Identifying any symmetry planes in a molecule can aid in determining if any carbons are identical.
a) 1 b) 3 c) 6 d) 3

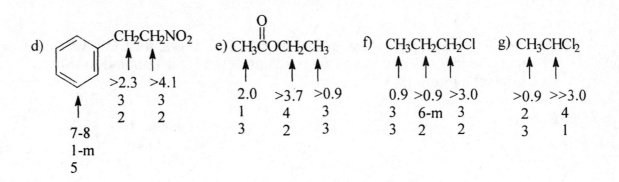

14.10 The chemical shifts for carbons are approximately 20 times larger than the chemical shifts for the hydrogens attached to that carbon. The signals are shifted downfield by nearby electron withdrawing groups, just as was the case for hydrogens.

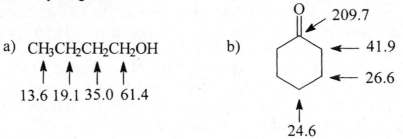

a) $CH_3CH_2CH_2CH_2OH$

13.6 19.1 35.0 61.4

b) 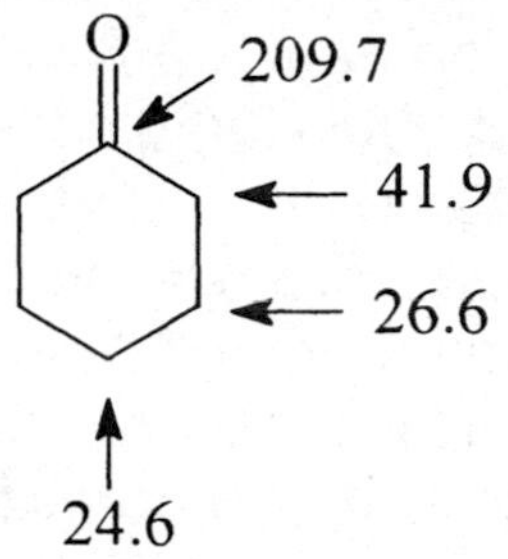

14.11 a) The DU for this compound is 5. There are 5 different types of carbons in this molecule. The signal around 190 δ is due to a carbonyl carbon bonded to one H, so it is an aldehyde. The four signals in the 140 -120 δ region are due aromatic carbons. To have only four different aromatic carbons, the two substituents must be para.

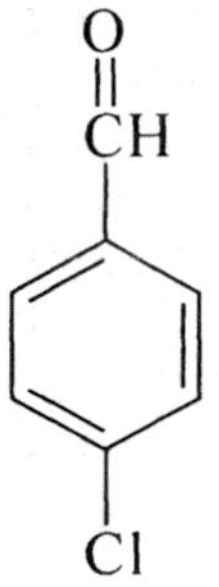

b) This compound has a DU of 1, and contains five different carbons. The signal at ~147 δ is an alkene carbon that is not bonded to any hydrogens and the one at ~110 δ is an alkene type carbon that is bonded to two hydrogens. The alkene accounts for the DU of 1. The only way to attach one CH_2 group and two CH_3 groups is to add an ethyl and a methyl group to the double bond.

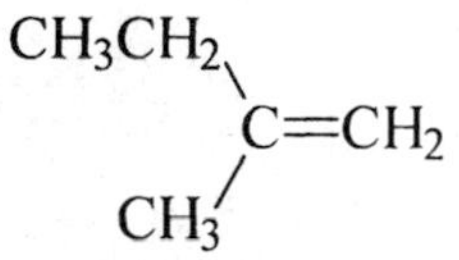

c) The compound has a DU of 2. It must have symmetry because the spectrum shows the presence of only three different types of carbons. There are probably two of each type because this sums to the correct formula, C_6H_{10}. The signal at ~128 δ is due to an alkene carbon (actually two of them) that is bonded to one hydrogen. This accounts for one C=C and for one DU, so the other DU must be due to a ring. The other C's are two sets of two identical CH_2 groups. Putting these carbons together gives the structure, cyclohexene.

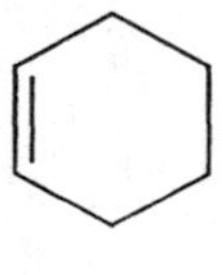

14.12 a) This compound has a DU of 5. The two bands in the region of 2830-2700 cm^{-1} and the carbonyl absorption at 1706 cm^{-1} indicate the presence of an aldehyde. The observed shift of the carbonyl band to a lower wavenumber suggests that it is a conjugated aldehyde. The absorptions in the region of 3100-3000 cm^{-1} are due sp^2-hybridized C-H bonds, and those at 3000-2850 cm^{-1} are due to sp^3-hybridized C-H bonds. The appearance of four bands in the 1600-1450 cm^{-1} region and the strong band near 800 cm^{-1} suggest the presence of an aromatic ring. (Check the NMR spectrum for confirmation. Note that an aromatic ring has DU = 4.) The ^{1}H-NMR spectrum shows presence of five different types of hydrogens in this compound. The singlet at 9.8 δ is typical of an aldehydic hydrogen. The four H's in the 8-7 δ region confirm the presence of an aromatic ring with two substituents. The pattern for these hydrogens, two rather symmetrical distorted doublets, is typical for para substitution. The hydrogens on adjacent positions couple to give a pair of doublets and the other pair of adjacent H's do the same. The 2 H's that appear as quartet at 2.7 δ must be coupled to the 3 H's appear as a triplet at 2.3 δ, indicating the presence of an ethyl group. The compound is 4-ethylbenzaldehyde.

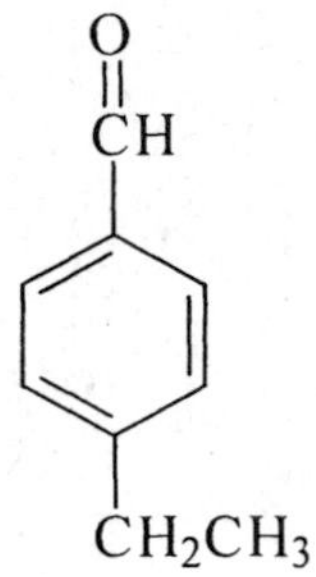

b) This compound has a DU of 1. The carbonyl absorption at 1741 cm^{-1} along with the strong absorptions in the 1300-1000 cm^{-1} region suggest the presence of an ester group. The bands in the region of 3000-2850 cm^{-1} are due to sp^3-hybridized C-H bonds. The ^{1}H-NMR spectrum shows the presence of four types of hydrogens and the integrals provide the actual number of hydrogens. The signal at 4.2 δ must be due to a CH$_2$ attached to the oxygen of the ester. This signal appears as four lines, so the CH$_2$ group must be next to a CH$_3$ group, the triplet at 1.3 δ (a -CH$_2$CH$_3$ group). The CH$_2$ protons of the two triplets (somewhat distorted) at 3.7 δ and 2.9 δ must be coupled (a -CH$_2$CH$_2$- group). The IR, NMR, and DU data fit ethyl 3-bromopropanoate.

$$\underset{\text{Br}CH_2CH_2\overset{\displaystyle O}{\overset{\displaystyle \|}{C}}OCH_2CH_3}{}$$

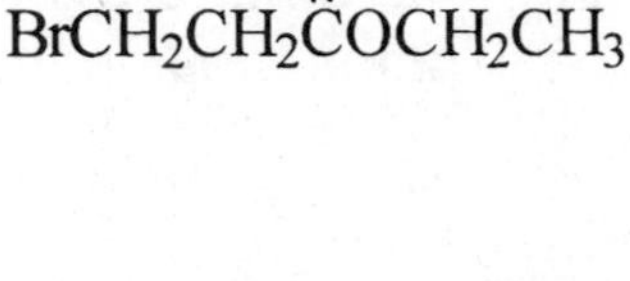

c) This compound has a DU of 4. The strong and broad absorption in the 3500-3200 cm^{-1} region indicates that the compound is an alcohol. The absorptions in the regions of 3100-3000 cm^{-1} are due to sp^2-hybridized C-H bonds, and those at 3000-2850 cm^{-1} are due to sp^3-hybridized C-H bonds. The appearance of four bands in the 1600-1450 cm^{-1} region and the strong bands in the 900-675 region suggest the presence of an aromatic ring. The NMR spectrum shows the presence of four different types of hydrogens and the integrals match the actual number of hydrogens. The signal at 7.3 δ, due to five hydrogens, indicates the compound has a monosubstituted aromatic ring. The singlet at 2.7 δ is probably due to the hydroxy hydrogen, which is not coupled due to rapid chemical exchange. The H that appears as a quartet at 4.8 δ must be coupled to the 3 H's appearing as a doublet at 1.4 δ (a $CHCH_3$ group). The unknown is 1-phenyl-1-ethanol.

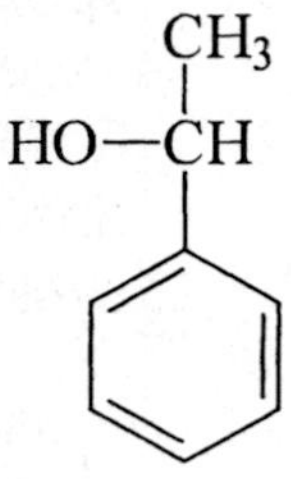

14.13 The IR spectrum of this compound shows an OH band in the 3500-3200 cm^{-1} region and a strong absorption in the C-O region. The absorptions in the region of 3000-2850 cm^{-1} are due to sp^3-hybridized C-H bonds. The IR data along with its DU of 0 indicate that this is a saturated alcohol. There are only four signals (all CH_2) in the ^{13}C-NMR spectrum indicating that the compound has symmetry. Three of the CH_2 groups must be duplicated. The CH_2 group at 63 δ must be attached to an oxygen. The absence of any CH_3 or CH groups indicates that the OH groups must be on the ends of the chain of carbons.

$HOCH_2CH_2CH_2CH_2CH_2CH_2CH_2OH$

14.14 a) The DU is 0. The IR spectrum shows the presence of an OH group, and H's on sp^3-hybridized C's, so the compound is a saturated alcohol. The coupling in the ^{1}H-NMR is complex and difficult to analyze. The singlet at 2.6 δ is probably due to the hydrogen of the OH, which is not coupled due to rapid chemical exchange. The ^{13}C-NMR spectrum is more useful in this case. It shows the presence of five different carbons. The CH_2 at 68 δ must be bonded to the O because it is furthest downfield. Because its chemical shift is further downfield, the CH is probably bonded to this CH_2. This allows the fragment $CHCH_2OH$ to be written. Attaching the remaining fragments, a CH_2 and two different CH_3 groups, results in 2-methyl-1-butanol.

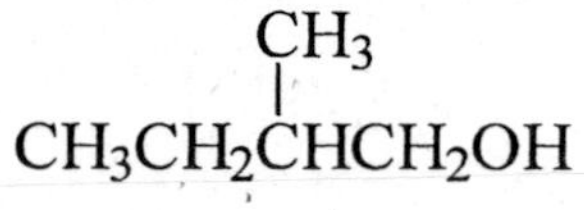

b) The DU is 2. The IR spectrum shows the presence of a C=O group and sp^3 C-H bonds. Because there is no indication of a C=C, the other DU is probably due to a ring. The ^{1}H-NMR spectrum is not very helpful. The presence of only four signals in the ^{13}C-NMR spectrum indicates that the compound has symmetry. The signal at ~214 δ is due to a carbonyl carbon that is attached to no hydrogens, so the compound is a ketone. Since there is only one carbonyl group in the compound, there are probably two of each of the three CH_2 groups shown in the spectrum. This gives the proper formula of $C_7H_{12}O$. To assemble these fragments into a compound, start with the C=O group. Bond a chain of three CH_2 groups to each side and then connect the last C's to form a ring. The unknown is cycloheptanone.

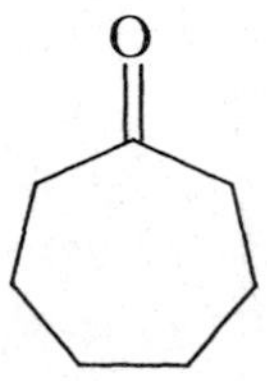

c) This compound has a DU of 5. The two bands in the region of 2830-2700 cm^{-1} and the carbonyl absorption near 1730 cm^{-1} suggest that it is an aldehyde (not conjugated). The presence of sp^2-hybridized C-H bonds (3100-3000 cm^{-1}) and the four bands in the 1600-1450 cm^{-1} region, along with the high DU, suggest the presence of an aromatic ring. In the ^{1}H-NMR spectrum, the signal for 5 H's in the aromatic region (7.2 δ) confirms the presence of a monosubstituted aromatic ring. The hydrogen of the aldehyde, at 9.7 δ, is a doublet, so the aldehyde group must be attached to a CH group. The H that appears as a quartet at 3.6 δ must be coupled to the three H's that appear as a doublet at 1.5 δ. This H must also be coupled to the aldehyde H, but the coupling is too small to be seen without expanding the peak. The ^{13}C-NMR shows the aldehyde carbon at 200 δ. The three types of aromatic CH's around 130-125 δ are part of the monosubstituted benzene ring. The C of the aromatic ring that is bonded to the substituent appears at 147 δ. We know the CH (at ~53 δ) is bonded to the aldehyde carbonyl and the methyl group (at ~ 14 δ) from the ^{1}H-NMR spectrum. Its final bond must be to the aromatic ring.

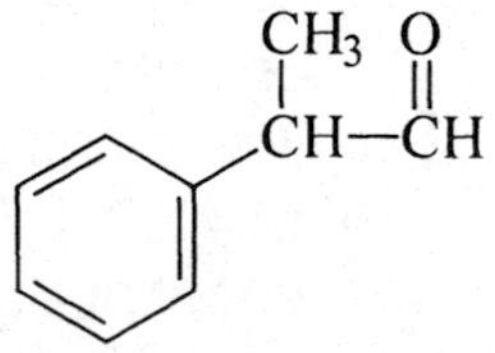

14.15

a)

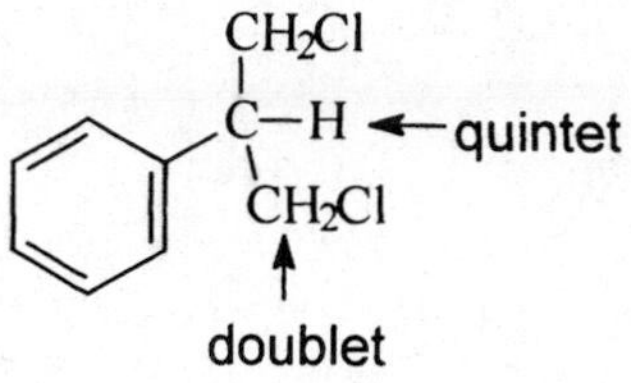

b)

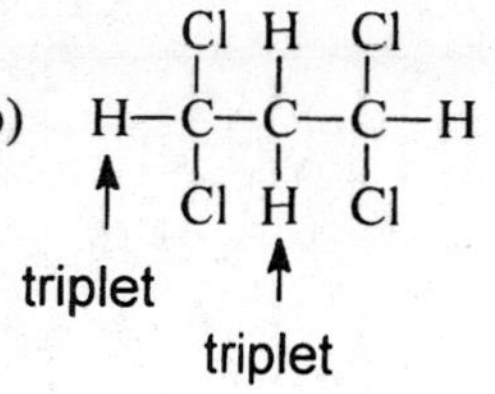

c)

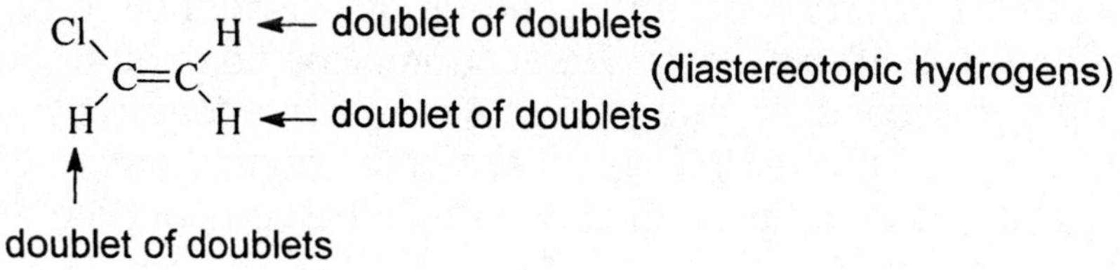

14.16

a)

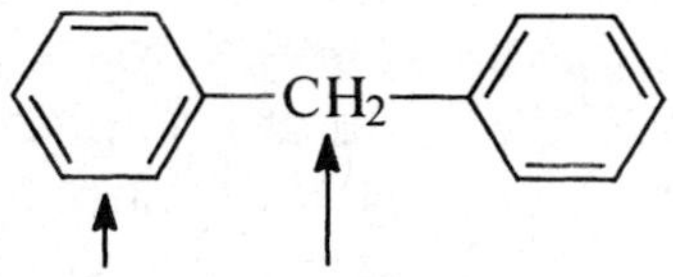

chemical shift	7-8	>>2.3
multiplicity	m	1
integral	5	1

b)

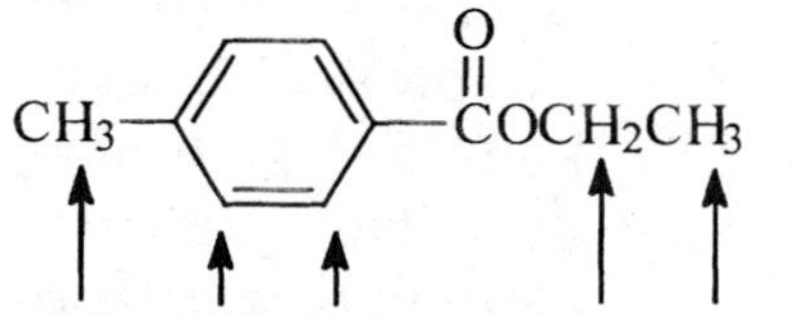

chemical shift	2.3	7-8	7-8	>3.7	>0.9
multiplicity	1	2	2	4	3
integral	3	2	2	2	3

c)

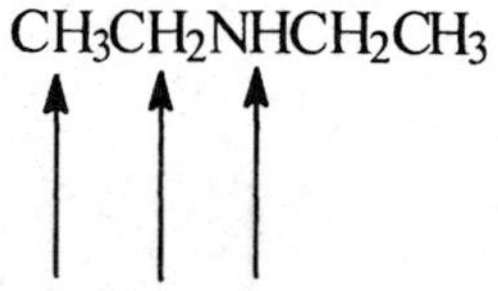

chemical shift	>0.9	>2.2	1-3
multiplicity	3	4	1
integral	6	4	1

d)

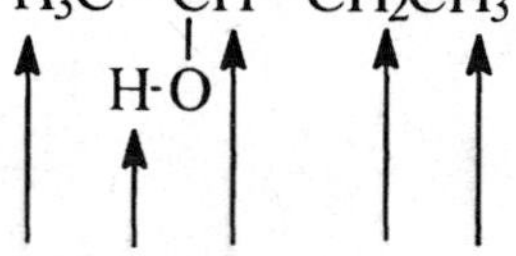

>0.9	2-5	>3.3	>0.9	0.9
2	1	6-m	5-m	3
3	1	1	2	3

14.17

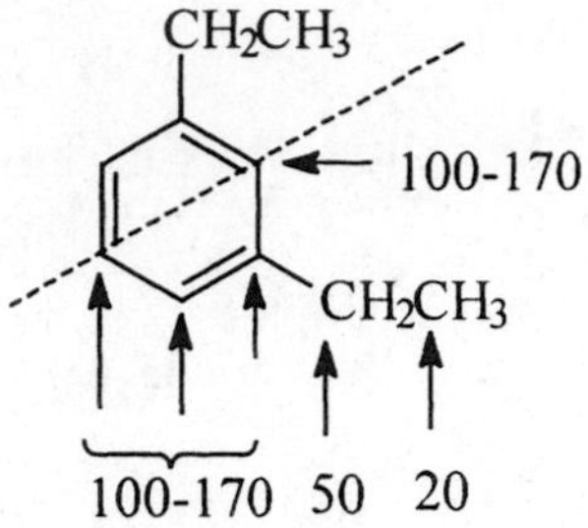

14.18 a) The absorption for the carbonyl group of the compound on the left appears near 1715 cm^{-1} in its IR spectrum. The compound on the right is a conjugated ketone, so its carbonyl absorption appears between 1695 and 1675 cm^{-1}. The ^{1}H-NMR spectrum of the compound on the left has two singlets in the upfield region, whereas the compound on the right has a triplet and a quartet.

b) The para-isomer has two peaks in its ^{13}C-NMR spectrum whereas the ortho-isomer has 3 peaks.

c) The aldehyde has two absorptions in the 2850-2700 cm^{-1} region of its IR spectrum and its carbonyl absorption appears near 1730 cm^{-1}. The ketone has no absorptions in the 2850-2700 cm^{-1} region and its carbonyl absorption appears near 1715 cm^{-1}. The ^{1}H-NMR spectrum of the aldehyde will show an peak around 10 δ which is absent in spectrum of the ketone.

d) The symmetrical ketone on the left has only two signals, a triplet and a quartet, in its ^{1}H-NMR spectrum. The less symmetrical compound on the right has four signals, with one being a singlet (methyl). The left compound has 3 peaks in its ^{13}C-NMR spectrum, whereas the right one has 5.

14.19

a) CH_3CHCH_3
    |
    CH_2Cl

b) $Br—CH_2CH_2CH_2-Br$

c) CH_3CHCH_2OH
    |
    CH_3

d) $CH_3CH—\overset{\overset{\displaystyle O}{\|}}{C}—CH_3$
    |
    CH_3

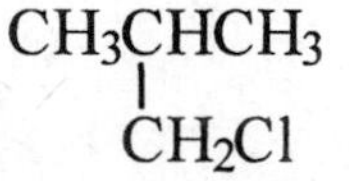

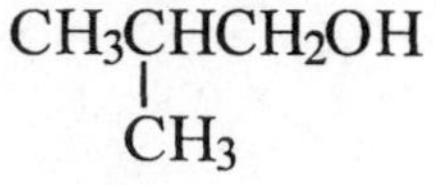

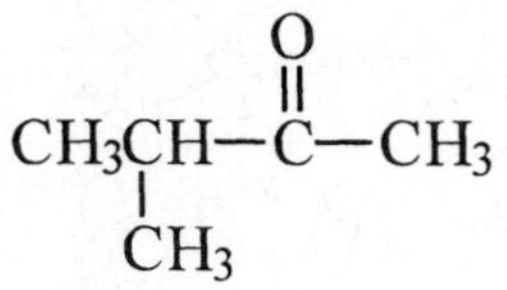

14.20 The DU equals 5. The peak at 190 δ results from a carbonyl carbon, which accounts for one DU. Since this carbon is bonded to one H, the unknown must be an aldehyde. The four signals between 140-120 δ are in the region for a benzene ring, which accounts for the other four degrees of unsaturation. There must be some symmetry because only four C's appear in this region. To account for six C's and five H's (C_7H_6O - CHO of aldehyde group), two of the CH signals must be due to two CH's each. The compound is benzaldehyde.

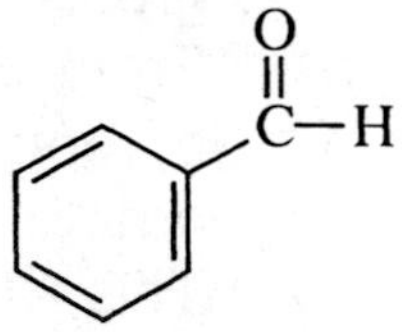

14.21 The DU is 4. The IR spectrum shows the presence of =CH (3100-3000 cm^{-1}), -CH (3000-2850 cm^{-1}), and possibly an aromatic ring (four bands in the 1600-1450 cm^{-1} region). If an aromatic ring is present, it accounts for the DU of 4. The multiplet for 5 H's at 7.2 δ in the ^{1}H-NMR spectrum confirms the presence of a monosubstituted aromatic ring. The two triplets at 3.6 δ and 3.2 δ, each representing two H's coupled to two H's, show the presence of a CH_2CH_2 group. Assembly of these fragments, including the Br shown in the formula, shows that the unknown is 1-bromo-2-phenylethane.

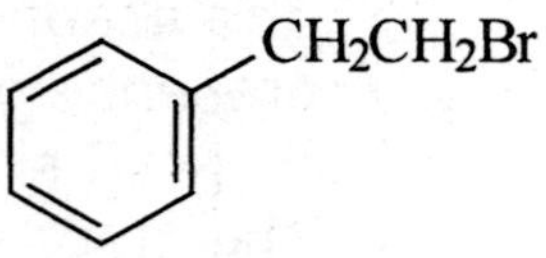

14.22 The DU for this compound is 3. The IR spectrum indicates the presence of -CH (3000-2850 cm^{-1}), C≡N (2250 cm^{-1}), and C=O and C-O of an ester (1740 and 1200 cm^{-1}). The two H's appearing as a quartet at 4.3 δ in the ^{1}H-NMR spectrum are split by 3 H's. These must be the three H's responsible for the triplet at 1.3 δ. Therefore the unknown contains an ethyl group. Because the signal for the CH_2 of the ethyl group appears so far downfield (> 3.7 δ), it must be bonded to the oxygen of the ester group. The singlet at 3.5 δ shows the presence of a CH_2 group with no nearby H's. Assembly of these fragments gives ethyl 2-cyanoethanoate. Note that the unsplit CH_2 group at 3.45 δ is downfield because it is bonded to two electron withdrawing groups.

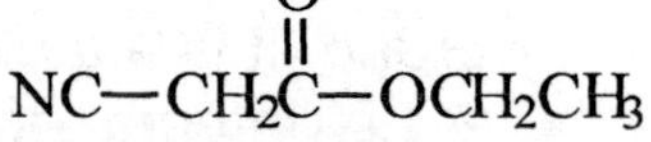

14.23 The DU for this compound is 1. The IR spectrum indicates the presence of -CH (3000-2850 cm^{-1}), and an ester group (C=O at 1733 cm^{-1} and C-O at 1200 cm^{-1}). The actual number of hydrogens under each of the signals in the ^{1}H-NMR spectrum must be double that of the integral value in order to sum to the 10 H's in the formula. The signal at 4.2 δ is due to a CH_2 group that is bonded to the oxygen of the ester group. This CH_2 group is split into a quartet, so it must be coupled to a CH_3 group. This CH_3 group must be split into a triplet by the CH_2 group. Where is the signal for this CH_3 group? Careful examination of the signal at 1.2 δ reveals that it is actually two triplets. One of these triplets must be the signal for that CH_3 group. The quartet at 2.3 δ is due to hydrogens of another CH_2 group. From its chemical shift, this is probably the CH_2 which is bonded to the carbonyl carbon of the ester group. Because this CH_2 appears as a quartet, it is also attached to a CH_3 group. This CH_3 group is responsible for the other triplet near 1.2 δ. The compound is an ester with two ethyl groups, one bonded to the oxygen of the ester and one to the carbonyl carbon of the ester. It is ethyl propanoate.

$$CH_3CH_2-\overset{\overset{\displaystyle O}{\|}}{C}-OCH_2CH_3$$

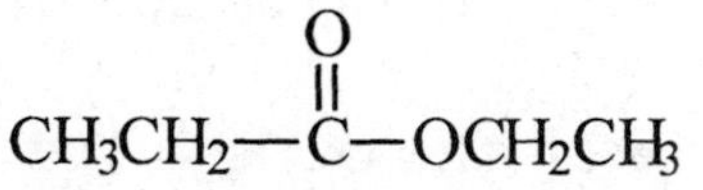

14.24 The DU is 2. The very broad OH absorption and the carbonyl peak at 1706 cm^{-1} in the IR spectrum show that the unknown is a carboxylic acid. The absence of alkene and aromatic bands (no C=C) suggest the presence of a ring (C=O and ring give DU = 2). The singlet at 11.7 δ in the ^1H-NMR spectrum is consistent with the hydrogen of a carboxylic acid. The other signals in this spectrum are too complex to be much help. The presence of only four different types of carbons in the ^{13}C-NMR spectrum indicates some symmetry in the structure. The C at 182 δ is due to the carbonyl carbon of the carboxylic acid. To sum to the total of six C's and ten H's, there must be two of both kinds of CH_2 groups. Assembly of these fragments gives cyclopentanecarboxylic acid.

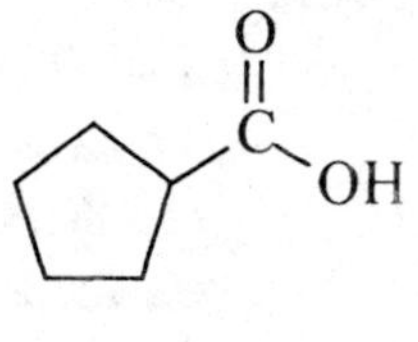

14.25 The DU is one. The carbonyl band at 1741 cm^{-1} and the C-O absorption near 1250 cm^{-1} indicate that the unknown is an ester. This accounts for the DU. The most downfield signal in the ^1H-NMR spectrum, at 4.2 δ, must be from a CH_2 (integral =2) attached to the oxygen of the ester group. It is a triplet, so this CH_2 is bonded to another CH_2. The singlet at 2.1 δ (integral =3) results from a CH_3 group with no nearby H's. Its chemical shift is consistent with a methyl group that is attached to the carbonyl group of the ester. The signal at 1.6 δ appears to contain five lines. This indicates a CH_2 group (integral =2) bonded to two CH_2 groups (at 4.2 and 1.4 δ) with similar coupling constants. The signal at 1.4 δ appears to contain six lines. This indicates a CH_2 group (integral =2) bonded to a CH_2 group (at 1.6 δ) and a CH_3 group (at 0.9 δ) with similar coupling constants. Thus, the unknown is an ester with a butyl group on the oxygen and a methyl group on the carbonyl carbon. The most downfield signal in the ^{13}C-NMR spectrum is the carbonyl carbon. The CH$_2$ group at 63 δ is the one bonded to the O of the ester. Proceeding upfield, we see the next CH$_2$ of the butyl group, the CH$_3$ bonded to the carbonyl carbon, and the final CH$_2$ and CH$_3$ of the butyl group. The compound is butyl ethanoate.

$$CH_3\overset{\displaystyle O}{\overset{\|}{C}}{-}OCH_2CH_2CH_2CH_3$$

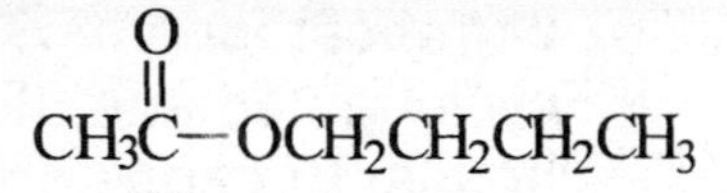

14.26 The DU = 0. The presence of a broad OH band at 3300 cm^{-1} in the IR spectrum shows that the compound is a saturated alcohol. The singlet at 2.6 δ in the ^{1}H-NMR spectrum is due to H of the hydroxy group. It is not coupled because of fast chemical exchange. The triplet (integral = 2) at 3.6 δ is due to hydrogens of a CH$_2$ group attached to the hydroxy group, which is also bonded to another CH$_2$ group. So far, the structure of the unknown can be assigned as RCH$_2$CH$_2$OH, but the rest of the ^{1}H-NMR is difficult to interpret. The ^{13}C-NMR spectrum shows the presence of seven CH$_2$ groups and one CH$_3$ group. These fragments can only be assembled into a straight chain primary alcohol. The unknown is 1-octanol.

$$CH_3CH_2CH_2CH_2CH_2CH_2CH_2CH_2{-}OH$$

14.27 The DU = 2. The IR spectrum shows the presence of a carbonyl group and sp^3-hybridized C-H bonds. The position of the carbonyl band (1721 cm^{-1}) is characteristic of a ketone. The absence of absorptions due to =C-H and C=C bands suggests that the compound has a ring. The ^{1}H-NMR spectrum is complex and not very useful. However, the doublet near 1 δ (integral = 3) indicates the presence of a methyl group attached to a carbon with one hydrogen, CHCH$_3$. The ^{13}C-NMR is much more useful in this case. It shows the presence of only five different carbons, indicating some symmetry in the structure. Since there is only one carbonyl carbon (near 210 δ) and one CHCH$_3$ (from the ^{1}H-NMR spectrum), there must be two of each of the CH$_2$ groups to sum to the formula C$_7$H$_{12}$O. To assemble these fragments, bond a CH$_2$ group to each side of the carbonyl carbon. Bond another CH$_2$ group to each of these. Next bond these CH$_2$ groups to the CH group, which is also bonded to the CH$_3$ group. The compound is 4-methylcyclohexanone.

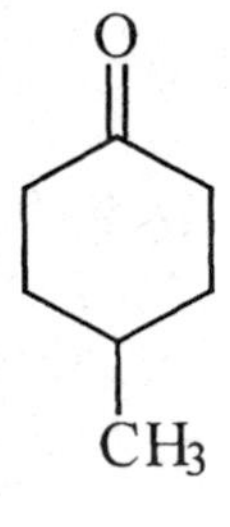

14.28 **A** (DU = 3) gives **B** (DU = 1) upon catalytic hydrogenation, so **A** must have two pi bonds and one ring and **B** has only the ring remaining. Because all of the carbons of **B** are identical, it must be cycloheptane. So **A** must be a cycloheptadiene. It only remains to establish the positions of the double bonds. Because one of the ozonolysis fragments contains only two C's, the double bonds must be adjacent. **A** is 1,3-cycloheptadiene.

$C_2H_2O_2$ $C_5H_8O_2$

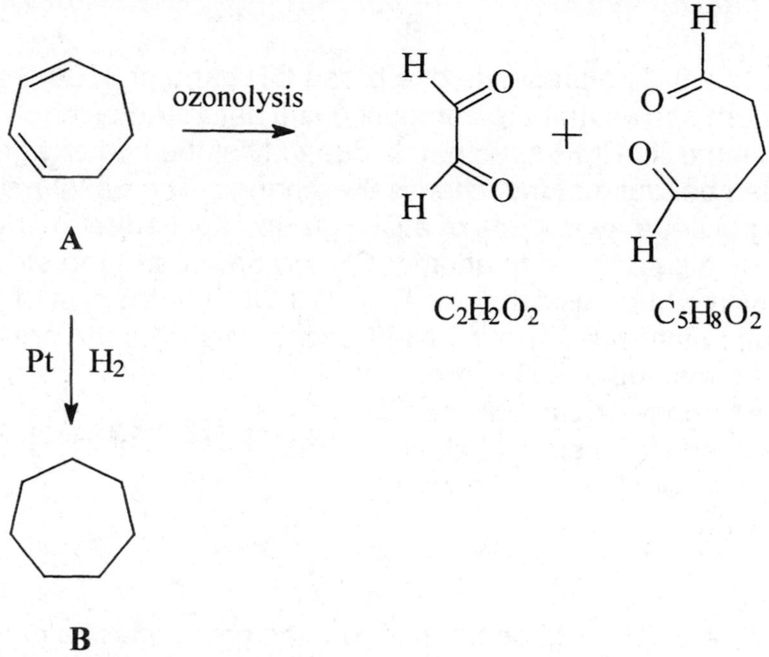

14.29 Because the product has a C with no H's and a CH_2, it must be the alcohol formed by rearrangement of the initial carbocation.

$$CH_3-\underset{\underset{OH}{|}}{\overset{\overset{CH_3}{|}}{C}}-CH_2CH_3$$

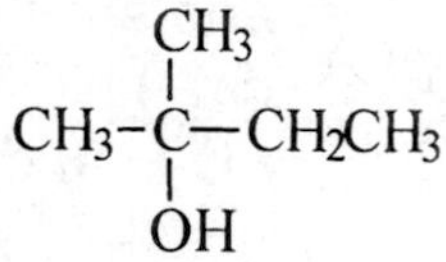

14.30 There will be three absorptions in the ¹H-NMR spectrum of this compound. The hydrogens on the carbons bonded to the chlorines are enantiotopic and will produce one signal. The hydrogens of CH_2 group are diastereotopic, and will give two signals.

14.31 The triplet (2 H's) near 3.6 δ must be split by two H's, as is also the case for the triplet (2 H's) near 3.0 δ, so they must result from a CH_2CH_2 group. The singlet at 2.2 δ (3 H's) results from a CH_3 group with no nearby H's. The product is 4-chloro-2-butanone. This is the product of anti-Markovnikov addition of HCl to the reactant. The inductive electron withdrawing effect of the carbonyl group must destabilize the secondary carbocation so much that it is less stable than the primary carbocation formed by addition with the opposite regiochemistry.

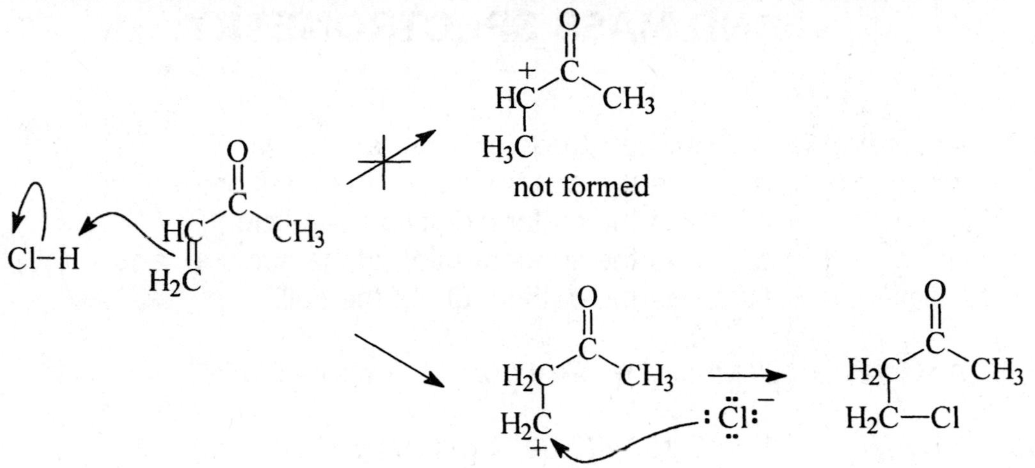

14.32 a) 1 b) 4 c) 2 d) 6 (CH_2 is diastereotopic) e) 4 f) 6

14.33 a) 2 b)3 c) 4 d) 7 e) 6 f) 8 g) 3

Review of Mastery Goals

After completing this chapter, you should be able to:

Predict the approximate chemical shifts, multiplicity, and integrals of peaks in the ^{1}H-NMR spectrum of a compound.
(Problems 14.2, 14.3, 14.4, 14.5, 14.6, 14.7, 14.15, 14.16, 14.30, and 14.32)

Predict the number and approximate chemical shifts of peaks in the ^{13}C-NMR spectrum of a compound.
(Problems 14.9, 14.10, 14.17, and 14.33)

Determine the hydrocarbon skeleton of a compound by examination of its ^{1}H and/or ^{13}C-NMR spectrum.
(Problems 14.8, 14.11, 14.19, 14.20, and 14.28)

Use a combination of IR and NMR spectra to determine the structure of an unknown compound.
(Problems 14.12, 14.13, 14.14, 14.18, 14.21, 14.22, 14.23, 14.24, 14.25, 14.26, 14.27, 14.29, and 14.31)

Chapter 15
ULTRAVIOLET-VISIBLE SPECTROSCOPY AND MASS SPECTROMETRY

15.1 According to the Lambert-Beer law, **$A = \varepsilon c l$**,

where
- **A** is the absorbance of the solution,
- **ε** is the molar extinction coefficient,
- **c** is the concentration of the solution, and
- **l** is the path length of the cell.

$A = (1.80 \times 10^5\ M^{-1}cm^{-1})(1.94 \times 10^{-6}\ M)(1\ cm) = 0.349.$

15.2 $\varepsilon = A/cl$ $\quad \varepsilon = (0.153)/[(0.0014\ g)/(182\ g\ M^{-1})][1\ cm]$

$\varepsilon = 1.99 \times 10^4\ M^{-1}cm^{-1}$

15.3 $c = A/\varepsilon l$ $\quad c = (0.643)/(27{,}000\ M^{-1}cm^{-1})(1\ cm) = 2.38 \times 10^{-5}\ M$

15.4 The small molar absorptivity and the wavelength indicate that the transition is $n \rightarrow \pi^*$.

15.5 The absorption at 213 nm is due to a $\pi \rightarrow \pi^*$ transition (higher energy, larger ε) and that at 320 nm is due to an $n \rightarrow \pi^*$ transition (lower energy, smaller ε).

15.6 a) The ketone on the left is conjugated and should absorb at longer wavelength than the ketone on the right, which is not conjugated.
b) The ketone on the right is conjugated and should absorb at longer wavelength than the aldehyde on the left, which is not conjugated.
c) The triene on the left has three conjugated double bonds and should absorb at longer wavelength than the triene on the right, which has only two conjugated double bonds.
d) Cyclohexanone has an absorption for a $n \rightarrow \pi^*$ transition in the accessible UV region, whereas the alcohol shows no such absorption.

15.7 The exact mass of butane is 58.0783 and that of acetone is 58.0419. Since the molecular masses of these compounds differ by several hundredths of a mass unit, they can be easily distinguished by a high resolution mass spectrometer.

15.8 The predicted intensity of the M+1 peak for butane is 4.4% and that for acetone is 3.3%. The intensities of these two peaks are similar and probably could not be used with confidence to distinguish between these compounds.

15.9 a) M : M+2 : M+4 = 1 : 0.67 : 0.11 (9:6:1)
b) M : M+2 : M+4 = 1 : 1.33 : 0.33 (3:4:1)

15.10 a) The similar intensities of the $M^{+\cdot}$ (m/z 164) and the M+2 (m/z 166) peaks indicate that there is one Br is present in this compound.
b) The appearance of the molecular ion at odd m/z (73) indicates the presence of an odd number of nitrogens in the compound.

15.11

a) $CH_3\overset{\cdot+}{N}HCH_3$ b) 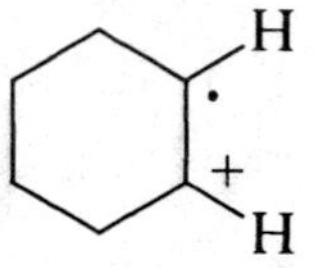

15.12

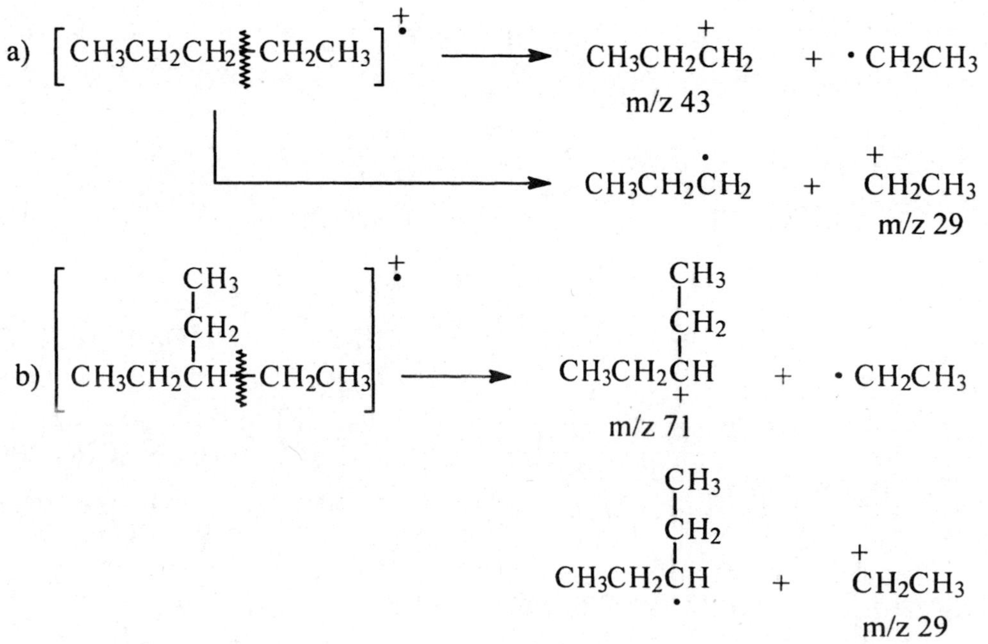

15.13

a)

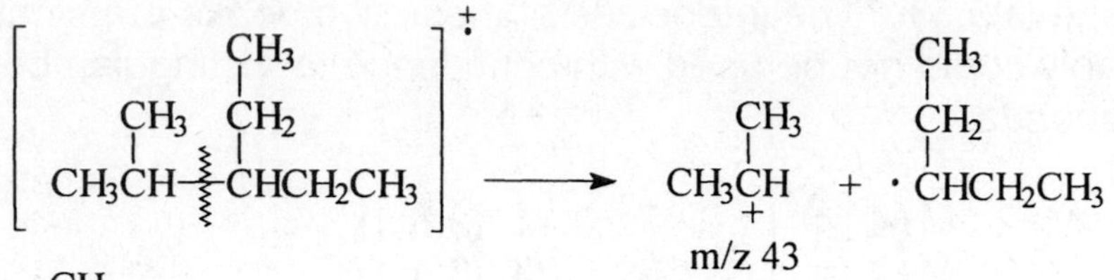

$$\begin{bmatrix} CH_3 \\ | \\ CH_3 \quad CH_2 \\ | \quad | \\ CH_3CH \overset{\displaystyle \{}{} CHCH_2CH_3 \end{bmatrix}^{\ddagger} \longrightarrow \begin{array}{c} CH_3 \\ | \\ CH_3CH \\ + \end{array} + \begin{array}{c} CH_3 \\ | \\ CH_2 \\ | \\ \cdot CHCH_2CH_3 \end{array}$$

m/z 43

b)

$$\begin{array}{c} CH_3 \\ | \\ CH_2 \\ | \\ +CHCH_2CH_3 \end{array}$$

m/z 71

15.14

a)

$$\begin{bmatrix} CH_2 \overset{\displaystyle \{}{} Cl \\ \text{[benzene ring]} \end{bmatrix}^{\ddagger} \longrightarrow \begin{array}{c} \overset{+}{CH_2} \\ \text{[benzene ring]} \end{array} + \cdot \ddot{\underset{\displaystyle ..}{C}l} :$$

m/z 91

b)

$$CH_3CH_2CH_2 \overset{\displaystyle \overset{.+}{:}OH}{\underset{\displaystyle |}{\{}} CH_2 \longrightarrow CH_3CH_2\overset{.}{C}H_2 + \begin{array}{c} \overset{+}{:}\overset{..}{O}-H \\ || \\ CH_2 \end{array}$$

m/z 31

c)

$$CH_3CH_2 \overset{\displaystyle \{}{} \overset{\displaystyle \overset{:\overset{+}{O}\cdot}{||}}{C} \overset{\displaystyle \{}{} CH_2CH_2CH_2CH_3 \longrightarrow CH_3CH_2C \overset{+}{\equiv} \overset{.}{O} : + \cdot CH_2CH_2CH_2CH_3$$

m/z 57

$$\longrightarrow CH_3\overset{.}{C}H_2 + :O \equiv \overset{+}{C}CH_2CH_2CH_2CH_3$$

m/z 85

$$\longrightarrow \begin{array}{c} \overset{.}{:}\overset{+}{O}-H \\ | \\ CH_3CH_2C = CH_2 \end{array} + CH_2 = CHCH_3$$

m/z 72 (McLafferty rearrangement)

15.15 $A/\varepsilon l = c$ $(0.5)/(25,000 \text{ M}^{-1}\text{cm}^{-1})(1 \text{ cm}) = 2 \times 10^{-5} \text{ M}$

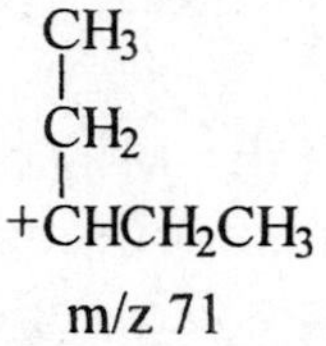

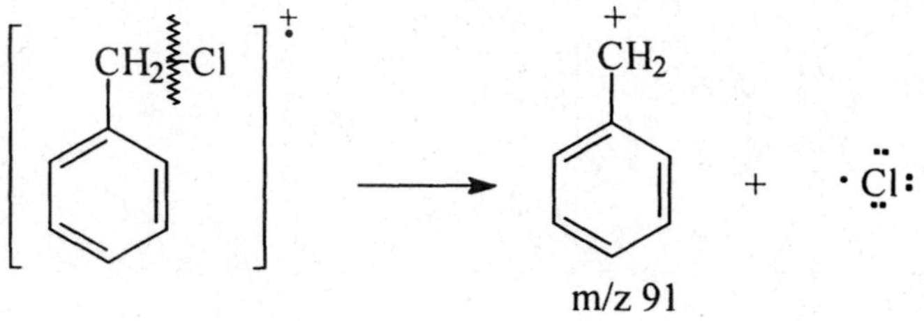

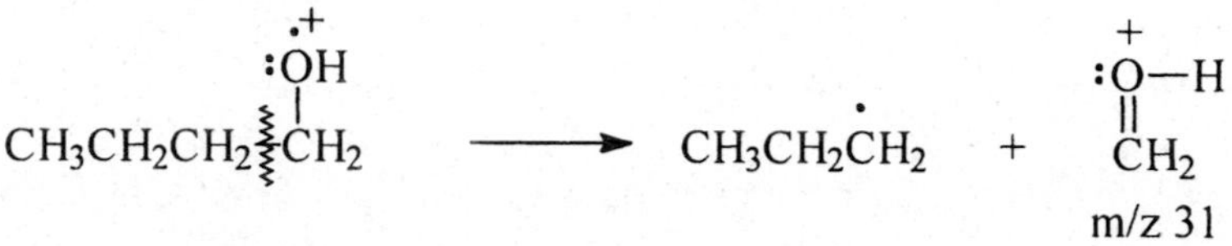

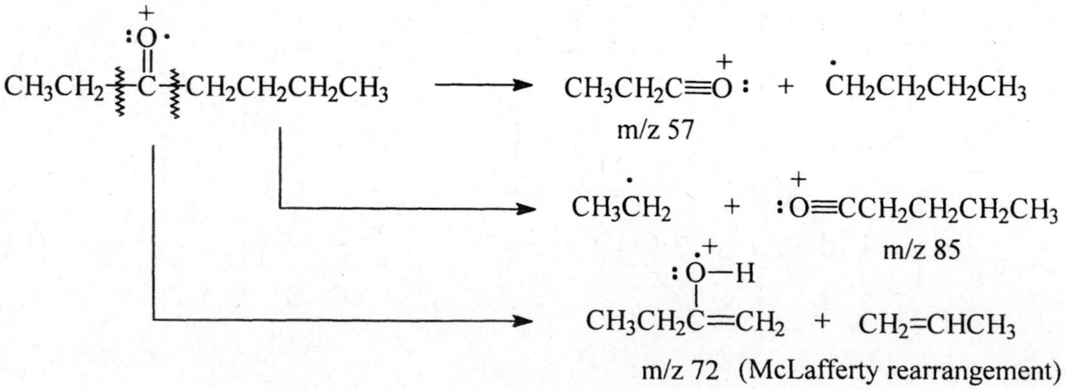

15.16 a) $\pi \to \pi^*$ at 252 nm and $n \to \pi^*$ at 325 nm

b) $\pi \to \pi^*$ c) $n \to \pi^*$ d) $\pi \to \pi^*$

15.17 a) 2-Butanone will show an absorption maximum for its $n \to \pi^*$ transition in this region.

b) The $\pi \to \pi^*$ transition of 1,3-pentadiene does not occur in the accessible UV region because it is not conjugated.

c) 1,3-Cyclohexadiene has conjugated double bonds, so its $\pi \to \pi^*$ transition will occur in this region.

d) Propylbenzene has an aromatic ring, so a $\pi \to \pi^*$ transition will occur in this region.

e) Naphthalene has two fused aromatic rings, so a $\pi \to \pi^*$ transition will occur in this region.

f) Diethyl ether will not show an absorption in this region.

g) Cyclopentanol will not show an absorption in this region.

h) 2-Cyclohexenone will show absorption maxima for $\pi \to \pi^*$ and $n \to \pi^*$ transitions in this region.

15.18 a)

$$\left[CH_3CH_2CH_2CH_2 {\Large\sim} CH_2CH_3 \right]^{+\bullet} \longrightarrow CH_3CH_2CH_2CH_2^+ + {}^\bullet CH_2CH_3$$
$$m/z\ 86 \qquad\qquad\qquad\qquad m/z\ 57$$

$$\longrightarrow CH_3CH_2CH_2CH_2^\bullet + {}^+CH_2CH_3$$
$$m/z\ 29$$

$$\longrightarrow CH_3CH_2CH_2^+ + {}^\bullet CH_2CH_2CH_3$$
$$m/z\ 43$$

b)

$$\left[CH_3CH_2CH_2 {\Large\sim} \underset{\underset{}{CH}}{\overset{\overset{CH_3}{|}}{}} {\Large\sim} CH_2CH_3 \right]^{+\bullet} \longrightarrow CH_3CH_2CH_2\underset{\underset{+}{CH}}{\overset{\overset{CH_3}{|}}{}} + {}^\bullet CH_2CH_3$$
$$m/z\ 100 \qquad\qquad\qquad\qquad\qquad m/z\ 71$$

$$\longrightarrow CH_3CH_2CH_2\underset{\underset{\bullet}{CH}}{\overset{\overset{CH_3}{|}}{}} + {}^+CH_2CH_3$$
$$m/z\ 29$$

$$CH_3CH_2CH_2^+ \qquad CH_3CH_2\underset{\underset{+}{CH}}{\overset{\overset{CH_3}{|}}{}}$$
$$m/z\ 43 \qquad + \qquad m/z\ 57$$

$$CH_3CH_2\underset{\underset{\bullet}{CH}}{\overset{\overset{CH_3}{|}}{}} \qquad CH_3CH_2CH_2^\bullet$$

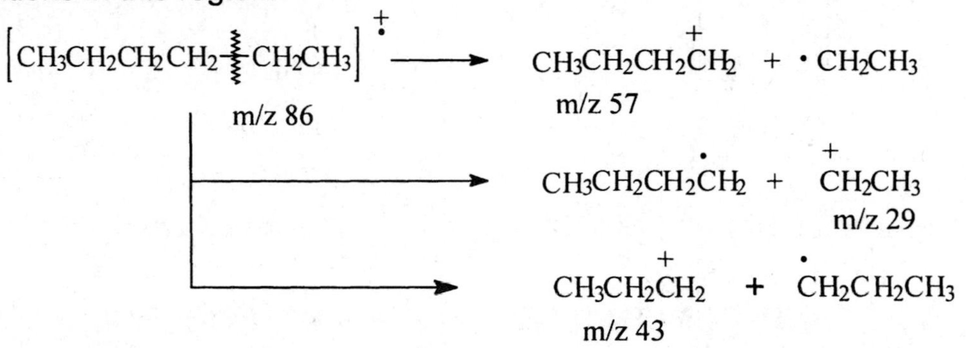

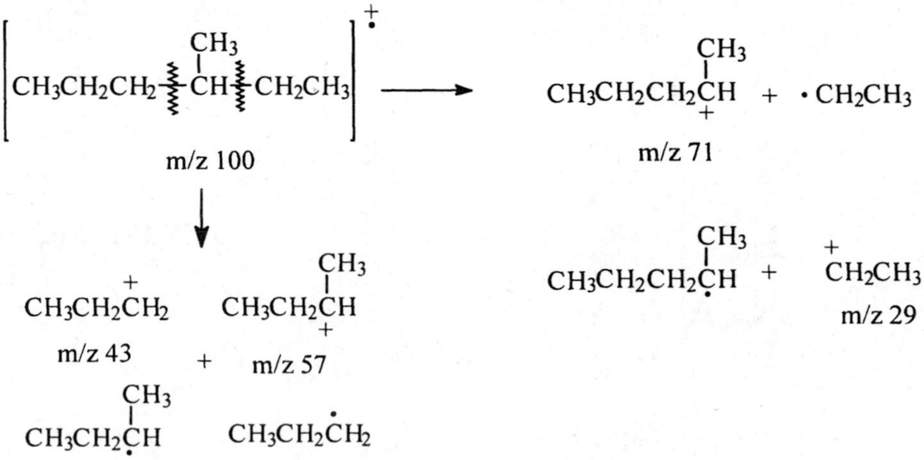

c) 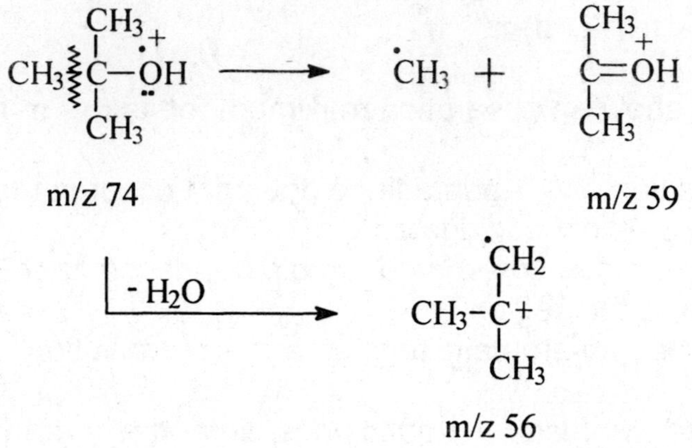
m/z 74
m/z 59
- H₂O
m/z 56

d) 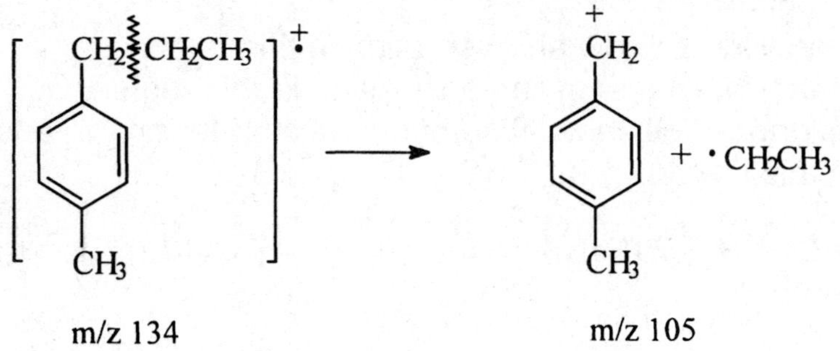
m/z 134
m/z 105

e) 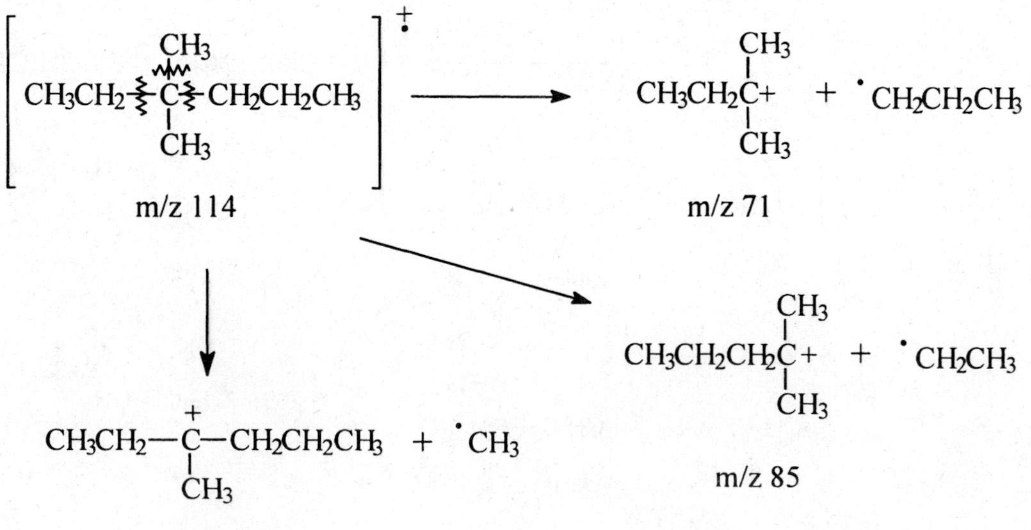
m/z 114
m/z 71
m/z 85
m/z 99

f)

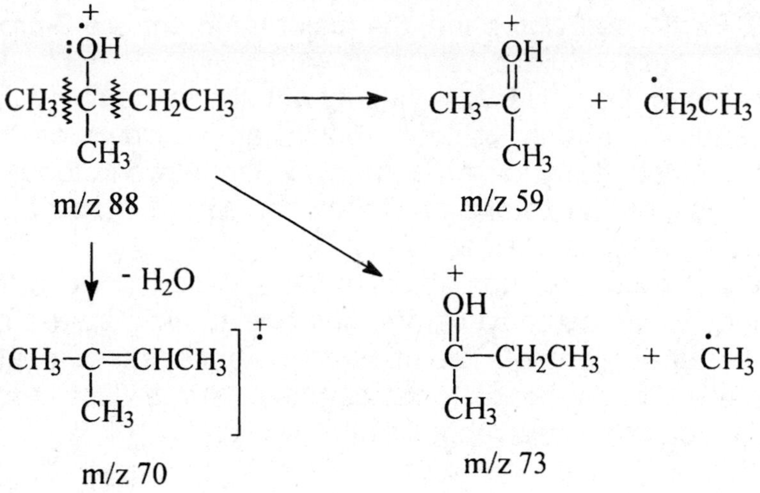

$$CH_3\overset{+}{\underset{CH_3}{\overset{:\overset{\bullet}{O}H}{\underset{|}{C}}}}CH_2CH_3 \longrightarrow CH_3\overset{\overset{+}{O}H}{\underset{CH_3}{C}} + {}^{\bullet}CH_2CH_3$$

m/z 88   m/z 59

$\downarrow$ - H_2O

$$\left[CH_3-\underset{CH_3}{\overset{|}{C}}=CHCH_3 \right]^{\overset{+}{\cdot}}$$

m/z 70

$$\overset{\overset{+}{O}H}{\underset{CH_3}{C}}-CH_2CH_3 + {}^{\bullet}CH_3$$

m/z 73

g)

$$\left[CH_3\overset{O}{\overset{||}{C}}CH_2CH_2CH_2CH_3 \right]^{\overset{+}{\cdot}} \longrightarrow CH_3C\overset{+}{\equiv}\overset{\bullet\bullet}{O}: + CH_3CH_2CH_2\overset{\bullet}{C}H_2$$

m/z 100   m/z 43

$\downarrow$ McLafferty Rearrangement

$$CH_3CH_2CH_2CH_2C\overset{+}{\equiv}\overset{\bullet\bullet}{O}: + {}^{\bullet}CH_3$$

m/z 85

$$CH_3-\underset{}{\overset{\overset{+}{\cdot}\overset{\bullet\bullet}{O}H}{C}}=CH_2 + CH_2=CHCH_3$$

m/z 58

h) $$CH_3CH_2CH_2\overset{\overset{+}{\cdot}\overset{\bullet\bullet}{:}OH}{\underset{|}{CH}}CH_3 \longrightarrow CH_3CH_2CH_2\overset{\overset{+}{O}H}{\overset{||}{CH}} + H\overset{\overset{+}{O}H}{\overset{||}{C}}CH_3 + \text{m/z 70}$$

m/z 88   m/z 73   m/z 45   (loss of H_2O)

i) $$\left[CH_3CH=CHCH_2\overset{}{C}H_2CH_3 \right]^{\overset{+}{\cdot}} \longrightarrow CH_3CH=CH\overset{+}{C}H_2 + {}^{\bullet}CH_2CH_3$$

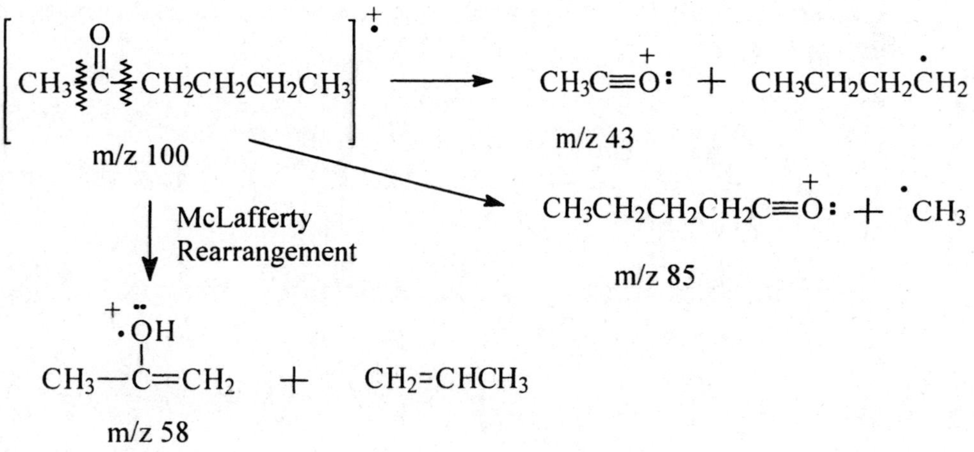

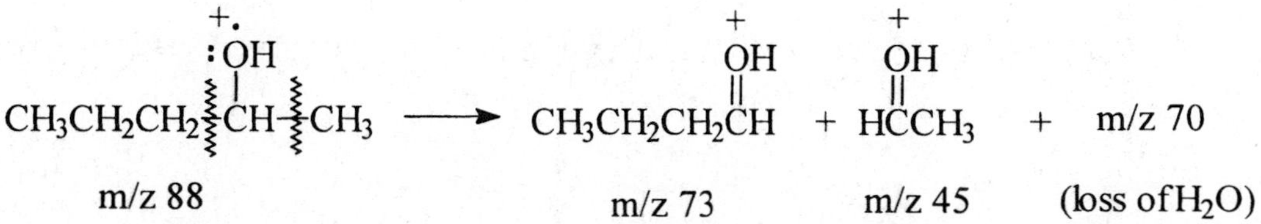

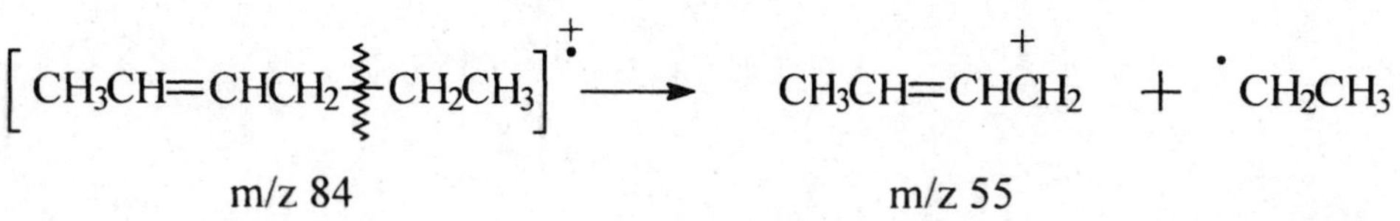

m/z 84   m/z 55

15.19 a) The major fragment in the mass spectrum of butane is an ethyl cation at m/z 29. For 2-methylpropane, the major fragment is a 2-propyl cation at m/z 43.
b) Both ketones produce fragment ions with the same m/z value due to the cleavage of the bonds to the carbonyl carbon. However, the ions produced by McLafferty rearrangements in these two ketones have different masses. The ketone on the left has a peak at m/z 58 while the one on the right has a peak at m/z 72.
c) The ketone on the left has peaks at m/z 43 and 71 due to the cleavage of the bonds to the carbonyl carbon and one at m/z 58 due to the McLafferty rearrangement. The one on the right has a peak at m/z 57 due to the cleavage of the bonds to the carbonyl carbon. This ketone cannot undergo a McLafferty rearrangement.

15.20 Neopentane does not show a molecular ion peak in its mass spectrum because fragmentation to produce a stable tertiary carbocation occurs very readily.

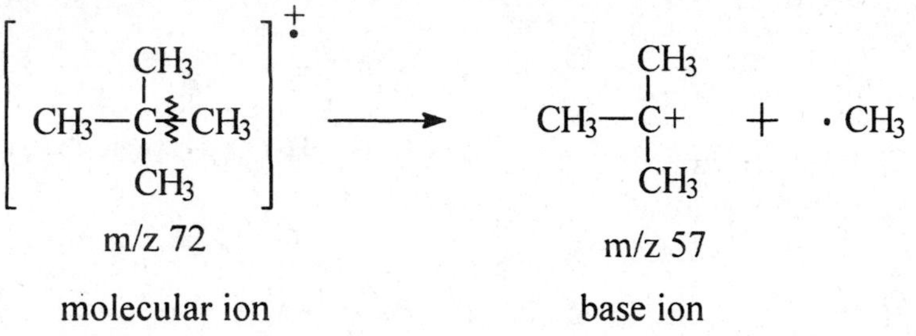

15.21

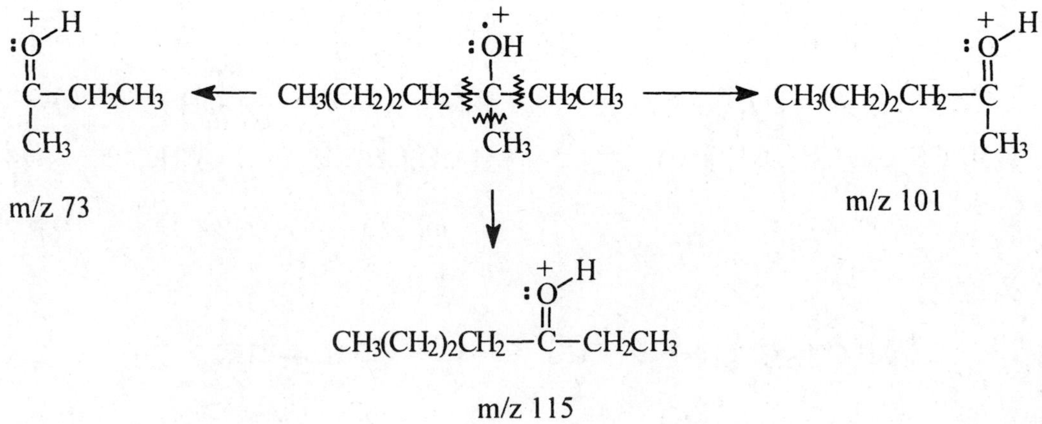

15.22 The peaks at m/z 186, 188 and 190 in a 1:2:1 ratio suggest the presence of two bromine atoms. Subtracting two ^{79}Br from 186 leaves a mass of 28, corresponding to C_2H_4. Therefore the formula is $C_2H_4Br_2$. The unknown is one of the two isomers with this formula, 1,1-dibromoethane or 1,2-dibromoethane. Determining which of these two isomers is correct is more difficult. Both are expected to lose a bromine atom to give C_2H_4Br at m/z 107 and 109, the base ions in the spectrum. However, only 1,2-dibromoethane can fragment to give CH_2Br with m/z 93 and 95. The presence of these peaks in the spectrum indicate that the unknown is 1,2-dibromoethane.

$$\left[\begin{array}{c} \overset{Br}{\underset{CH_2}{|}} \quad \overset{Br}{\underset{CH_2}{|}} \end{array} \right]^{+} \longrightarrow \quad \overset{+}{C}H_2Br \ + \ {}^{\bullet}CH_2Br$$

m/z 186, 188, 190   m/z 93, 95

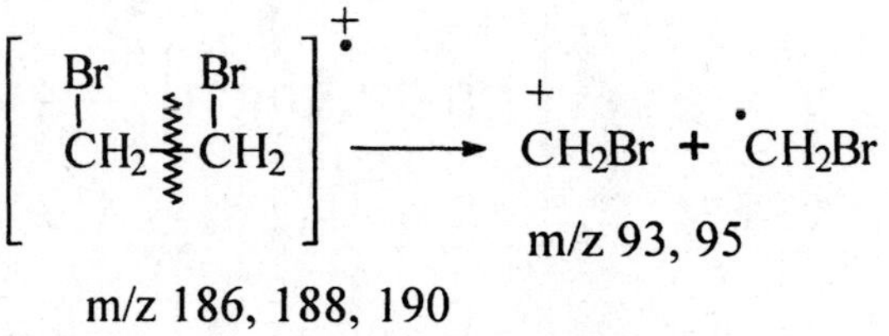

15.23 Both the ketones produce fragment ions at m/z 43 and 85 due to the cleavage of the bonds to the carbonyl carbon. However, the fragment ions resulting from the McLafferty rearrangement are different for each of these ketones. In the case of 3-methyl-2-pentanone, McLafferty rearrangement produces an ion with m/z 72, while in the case of 4-methyl-2-pentanone the ion has m/z 58. Therefore, the first spectrum is that of 4-methyl-2-pentanone and the second spectrum is that of 3-methyl-2-pentanone.

3-methyl-2-pentanone

$$m/z = 100 \longrightarrow m/z = 72$$

4-methyl-2-pentanone

$$m/z = 100 \longrightarrow m/z = 58$$

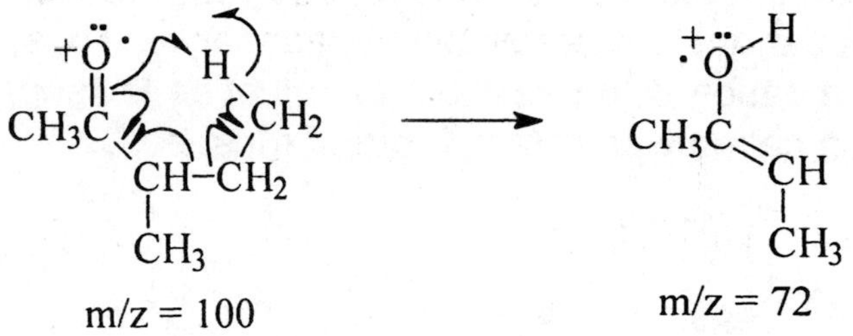

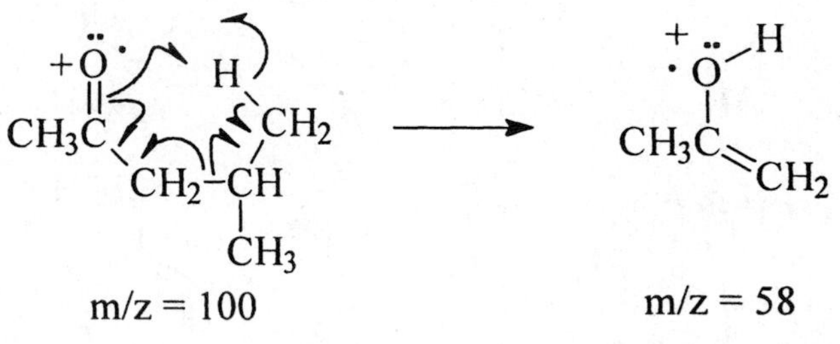

15.24 The DU for both isomers is 1. Based on their IR spectra, this must result from a carbonyl group in each. Therefore one must be propanal and other must be 2-propanone. Based on the position of the carbonyl absorption in the IR spectra, **A** must be propanal and **B** must be 2-propanone. This is confirmed by their mass spectra. The major fragmentation pathway in aldehyde and ketone is cleavage of the bond to the carbonyl carbon.

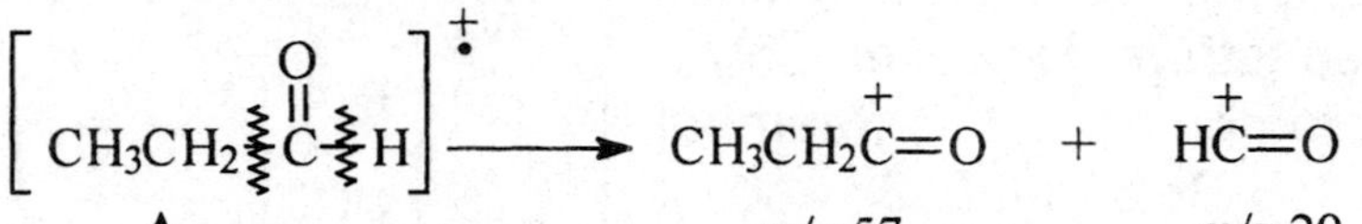

$$\left[CH_3CH_2 \overset{O}{\overset{||}{C}} H \right]^{+\bullet} \longrightarrow CH_3CH_2\overset{+}{C}=O \;+\; H\overset{+}{C}=O$$

A $\qquad\qquad\qquad$ m/z 57 $\qquad$ m/z 29

$$\left[CH_3 \overset{O}{\overset{||}{C}} -CH_3 \right]^{+\bullet} \longrightarrow CH_3\overset{+}{C}=O$$

B $\qquad\qquad\qquad$ m/z 43

15.25 The isomers have DU = 4. The ^{1}H-NMR peak near 7.25 δ (area 5) indicates the presence of a benzene ring with five H's (monosubstituted) in both isomers. Subtracting C_6H_5 from the formula leaves C_3H_7, so the remaining fragment must be either a propyl or an isopropyl group. The isomers are propylbenzene and isopropylbenzene. These isomers can be readily distinguished by mass spectrometry. The molecular ion of both compounds cleaves to form a carbocation with the positive charge on the carbon adjacent to the ring (a benzylic carbocation) because such carbocations are stabilized by resonance. Thus, propylbenzene cleaves to give a cation with m/z = 91 and must be **D**. Isopropylbenzene cleaves to give a cation with m/z = 105 and must be **C**.

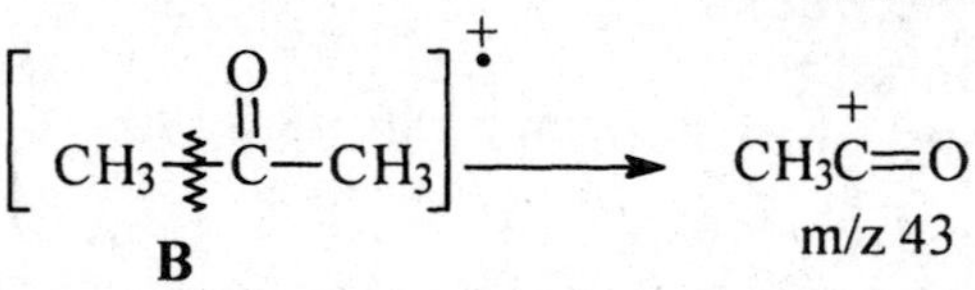

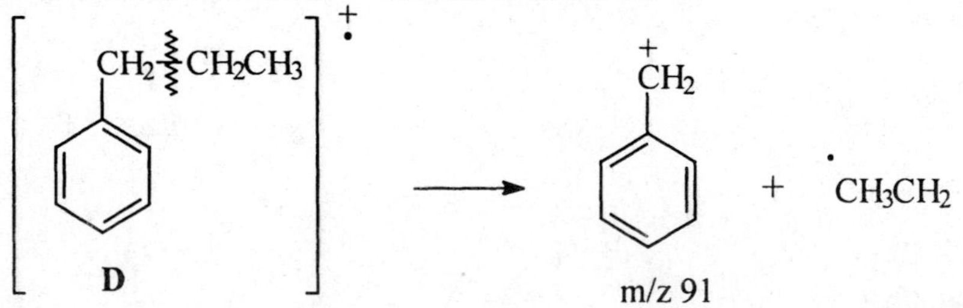

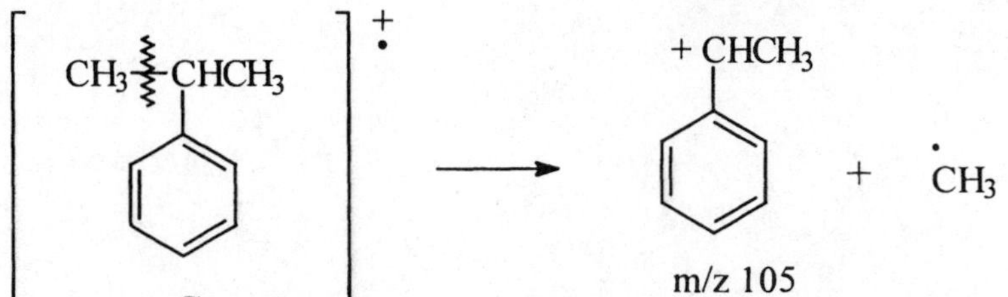

15.26 The DU of **E** and **F** is 3. Both produce **G**, with DU = 1, upon catalytic hydrogenation. **G** cannot have a pi bond because it is produced by hydrogenation, so its DU must be due to a ring. Therefore **E** and **F** must have two pi bonds and a ring. Because **G** shows only a single peak in its ^{13}C-NMR spectrum, all of its C's are identical. It must be cyclohexane. Thus **E** and **F** are isomers of cyclohexadiene. The λ_{max} at 259 nm in the UV spectrum of compound **F** suggests that its two double bonds are conjugated. The absence of any absorption maximum above 200 nm in the UV spectrum of **E** is consistent with non-conjugated double bonds.

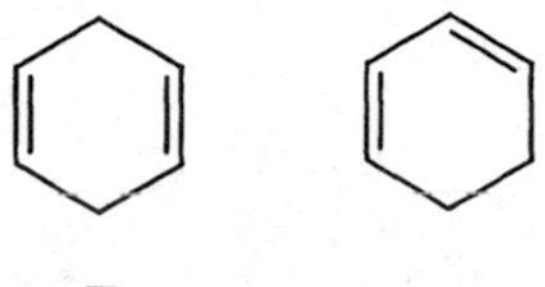

E F

15.27 The peaks of equal intensity at m/z 122 and 124 in the mass spectrum indicate the presence of a bromine atom in the unknown. Subtracting 79 from 122 leaves 43 as the mass of the rest of the molecule. The ^{1}H-NMR spectrum shows the presence of three different types of hydrogens in 2:2:3 ratio. The two H's at 3.4 δ appear as a triplet, so they must be coupled to two H's, the ones appearing at 1.9 δ. The three H's at 1.0 δ also appear as a triplet, so they are also coupled to the two H's at 1.9 δ. Therefore the two H's at 1.9 δ are coupled to five H's and appear as a sextet. This shows the presence of a propyl group, $CH_3CH_2CH_2$, in the compound. (The mass of a propyl group is 43.) Thus, the unknown is 1-bromopropane.

$CH_3CH_2CH_2Br$

15.28 To have an absorption at 280 nm, the amino acid must have conjugated double bonds in its side chain. Both phenylalanine and tryptophan have absorptions at 280 nm. Leucine, proline, and serine do not have any chromophores that absorb in this region.

15.29 As can be seen in the models, the double bonds of the compound on the left do not quite lie in the same plane, so its conjugation is decreased and its λ_{max} occurs at higher energy (shorter wavelength). The double bonds of the compound on the right lie in the same plane, so it is more conjugated and its λ_{max} occurs at lower energy (longer wavelength).

15.30 As can be seen in the models, the compound on the left (biphenyl) has its two benzene rings in the same plane. The methyl group in the ortho position of the compound on the right causes its benzene rings to lie in slightly different planes, decreasing their conjugation, and shifting the λ_{max} for this compound to higher energy (shorter wavelength).

15.31 The compound is 4-methyl-3-heptanone. It fragments in the mass spectrometer as follows:

$$CH_3CH_2-\overset{\overset{\displaystyle :\overset{\bullet+}{O}}{\|}}{C}-CHCH_2CH_2CH_3 \longrightarrow :\overset{+}{O}\equiv CCHCH_2CH_2CH_3 + CH_3CH_2C\equiv\overset{+}{O}:$$

with CH_3 groups below the first and second structures.

m/z 128   m/z 99   m/z 57

$$+ \quad CH_3CH_2\overset{\overset{\displaystyle :\overset{\bullet+}{O}H}{|}}{C}=CHCH_3 \quad \text{(from a McLafferty Rearrangement)}$$

m/z 86

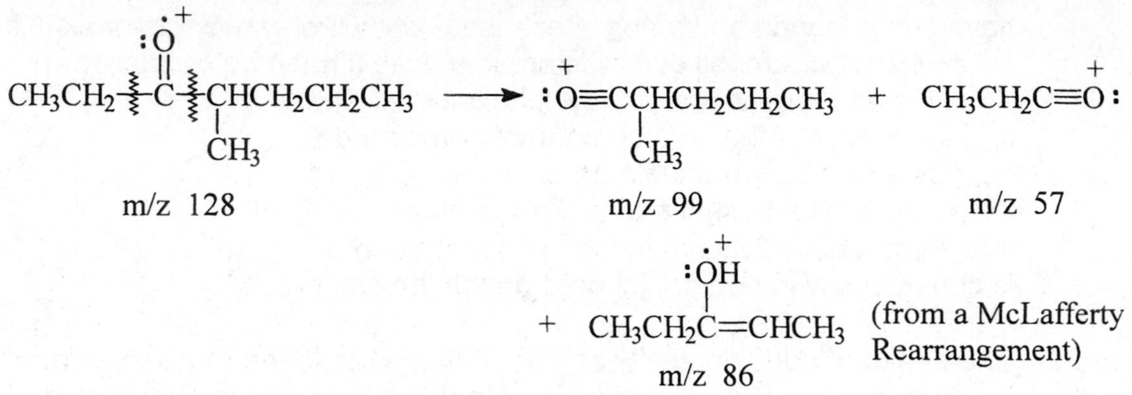

Review of Mastery Goals

After completing this chapter, you should be able to:

Determine whether a compound will absorb light in the ultraviolet or visible region.
(Problems 15.6, 15.17, 15.28, 15.29, and 15.30)

Identify the chromophore and type of transition responsible for absorption of UV-visible radiation.
(Problems 15.4, 15.5, 15.16, and 15.26)

Determine whether sulfur, chlorine, bromine, or nitrogen is present in a compound by examination of the M, M + 1, and M + 2 peaks in its mass spectrum.
(Problems 15.9, 15.10, 15.23, and 15.27)

Explain the major fragmentation pathways for compounds containing some of the simple functional groups.
(Problems 15.12, 15.13, 15.14, 15.18, 15.19, 15.20, 15.21, 15.22, 15.23, 15.24, 15.25, and 15.31)

Chapter 16
BENZENE AND AROMATIC COMPOUNDS

16.1

a) 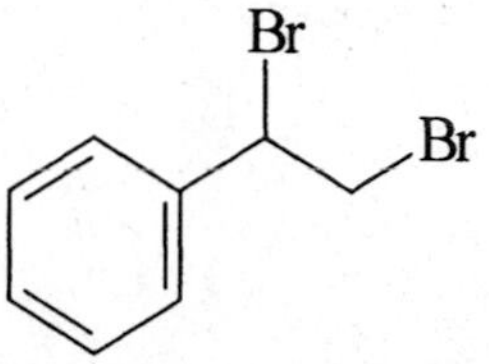 b) 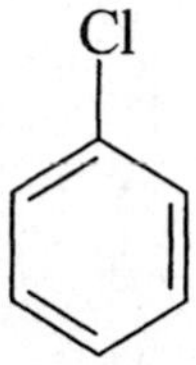 c) 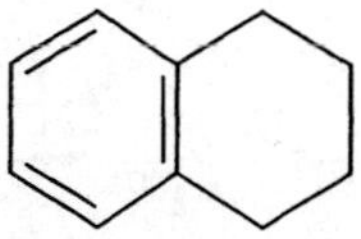 d) 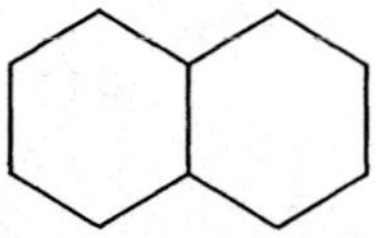

16.2 (5)(28.6 kcal/mol) - 80 kcal/mol = 63 kcal/mol.
(5)(120 kJ/mol) - 335 kJ/mol = 265 kJ/mol.
Yes, naphthalene should be termed aromatic because it has a very large resonance stabilization.

16.3

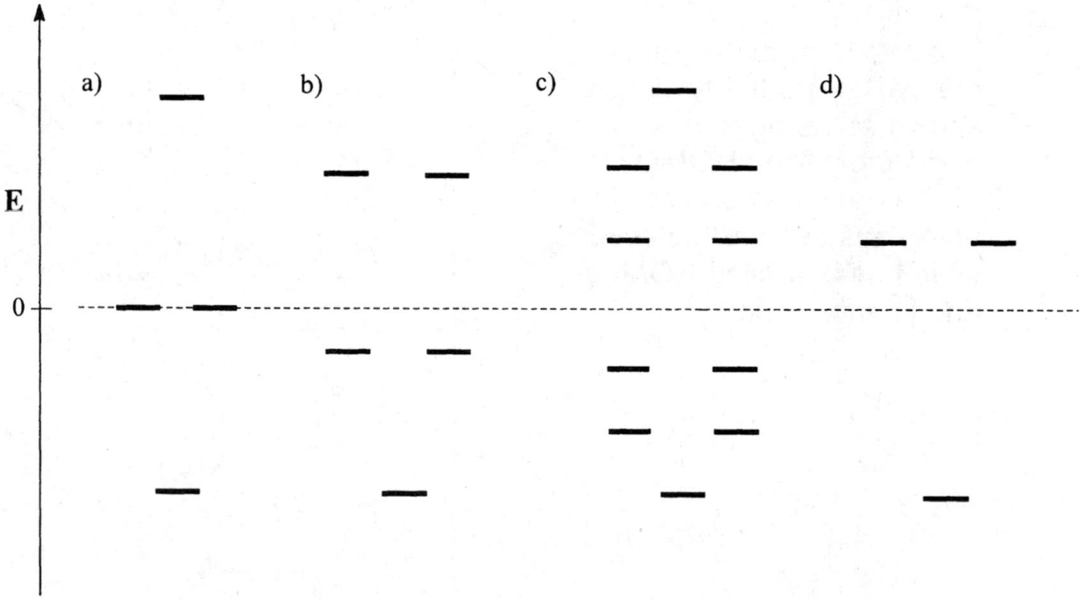

16.4 The HOMOs of this compound are half-filled, so it is antiaromatic.

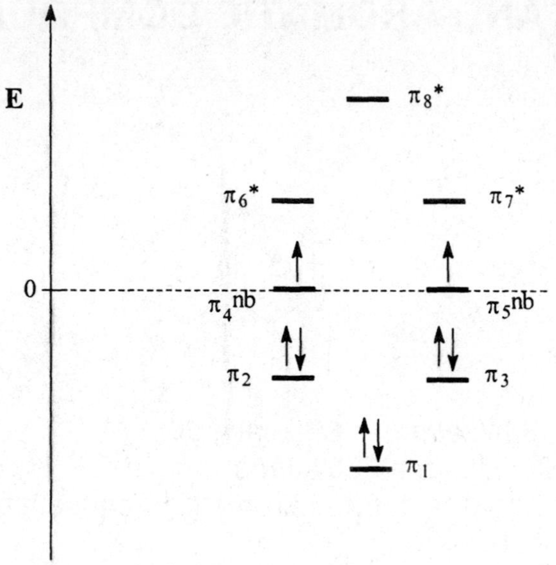

16.5 The MO energy diagram of the cycloheptatrienyl cation, shown at the right, has completely filled HOMOs. Therefore it is aromatic. The anion has two more electrons, so it has half-filled HOMOs and is antiaromatic.

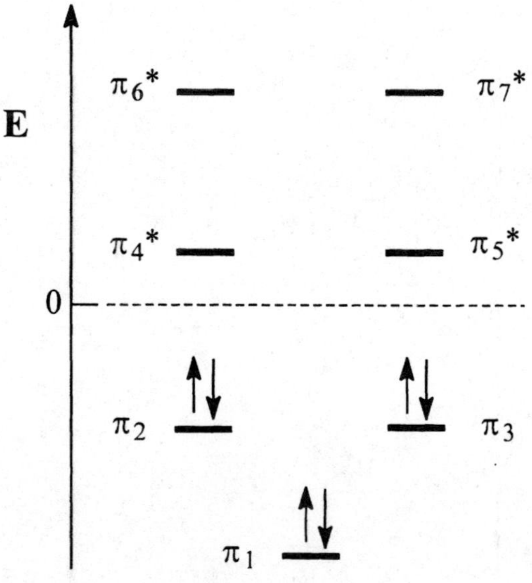

16.6 All of these compounds have filled HOMOs, so they are all aromatic.

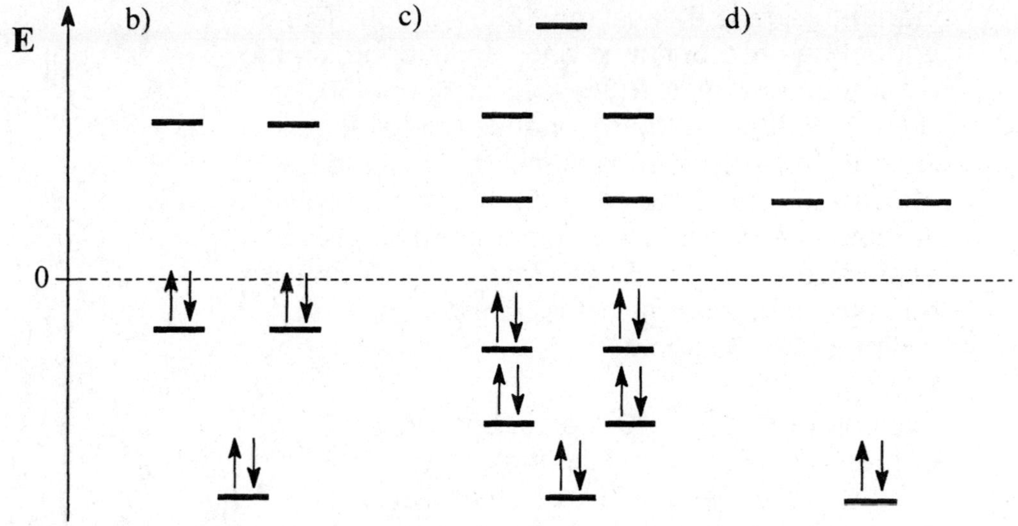

16.7 According to Huckel's rule, aromatic compounds are cyclic, fully conjugated, planar, and have (4n+2) π electrons. Antiaromatic compounds are cyclic, fully conjugated, planar, and have 4n π electrons.

a) 4 π electrons, antiaromatic
b) 6 π electrons, aromatic
c) 10 π electrons, aromatic if planar
d) 2 π electrons, aromatic

16.8 a) 6 π electrons, aromatic (do not count pairs on N's)
b) 6 π electrons, aromatic (count one pair on O)
c) 6 π electrons, aromatic (count one pair on S)
d) 6 π electrons, aromatic (count one pair on O)
e) 6 π electrons, aromatic (do not count pairs on N's)
f) 8 π electrons, antiaromatic (count pair on N)

16.9

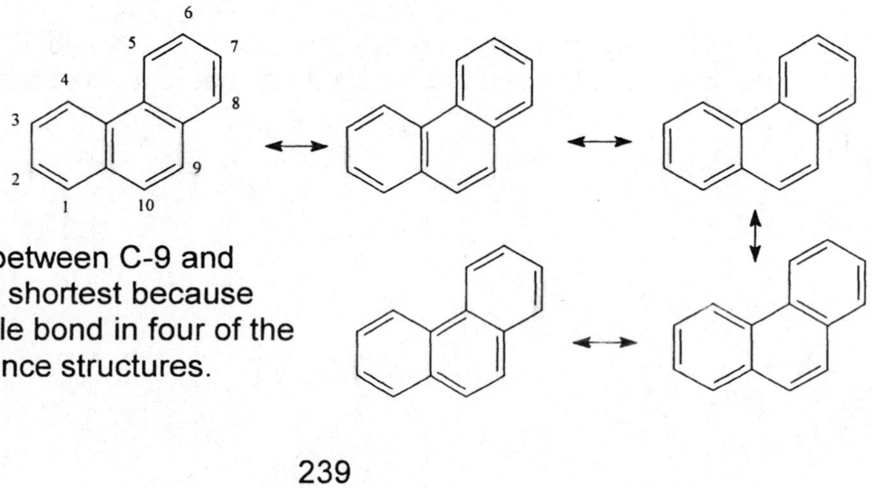

The bond between C-9 and C-10 is the shortest because it is a double bond in four of the five resonance structures.

16.10 The calculated resonance energy for phenanthrene is 92 kcal/mol (385 kJ/mol). The product has two benzene rings so its approximate resonance energy is 2(36) = 72 kcal/mol (301 kJ/mol). Therefore the resonance energy lost is equal to (92 - 72) = 20 kcal/mol (83.6 kJ/mol). Addition to one of the other double bonds leaves a naphthalene derivative, with an approximate resonance energy of 61 kcal/mol (255 kJ/mol), so the resonance energy lost is approximately 31 kcal/mol (129.6 kJ/mol).

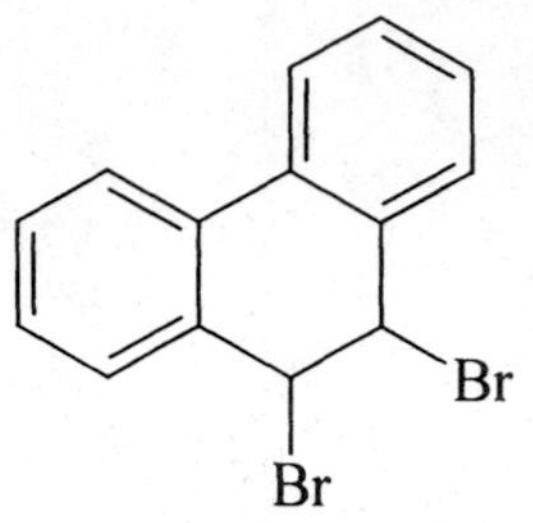

16.11 This compound has 18 π electrons, so it is aromatic. The diamagnetic ring current induces a magnetic field that is parallel to the external magnetic field on the outside of the ring and opposed in the center of the ring. This causes the NMR signals of hydrogens on the outside of the ring to appear downfield and the hydrogens inside the ring to appear upfield from the position of normal alkene hydrogens.

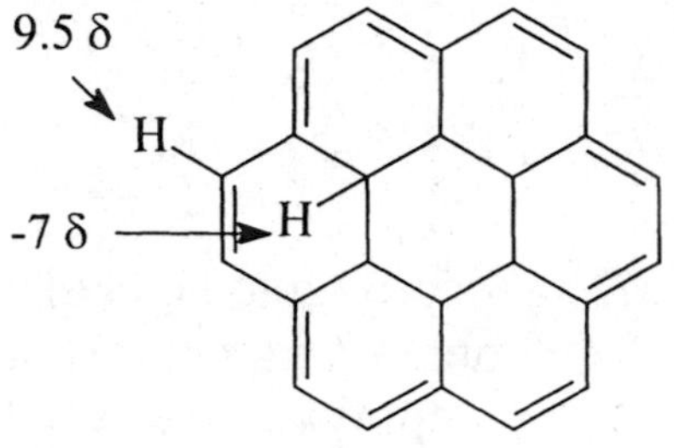

16.12 a) The compound on the left is a stronger acid because its conjugate base is aromatic.
b) The compound on the right is a stronger acid because conjugate base of the other compound is antiaromatic.
c) The compound on the right is a stronger acid because its conjugate base is aromatic.

16.13 The compound on the left has the faster rate of substitution by the S_N1 mechanism because the carbocation that is formed from it is aromatic.

16.14

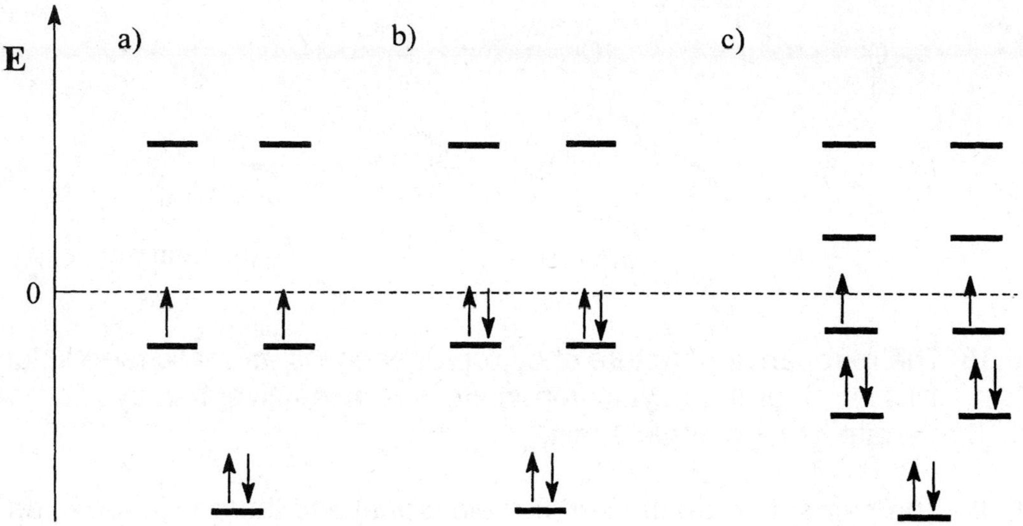

16.15 a) 8 π electrons, antiaromatic
b) 6 π electrons, aromatic (One electron pair on the O is part of the cycle of π electrons. The other pair on the O and the pair on the N are not part of the cycle.)
c) The cycle of p orbitals is interrupted by a sp^3-hybridized carbon, so the compound is neither aromatic nor antiaromatic.
d) 6 π electrons, aromatic (One electron pair on the O is part of the conjugated π system.)
e) Neither (The cycle of p orbitals is interrupted by the sp^3-hybridized nitrogen.)
f) 10 π electrons, aromatic (if it is planar)

16.16 Although cyclooctatetraene itself is neither aromatic nor antiaromatic because it is non-planar, the dianion is planar because it has 10 π electrons and is aromatic. It is much more stable than other dianions.

16.17 Recall that an important resonance structure for the carbonyl group moves the electrons of the carbon-oxygen double bond onto the oxygen, leaving an empty p-orbital on the carbon. This resonance structure for 2,4,6-cycloheptatrienone is aromatic, whereas the similar resonance structure of 2,4-cyclopentadienone is antiaromatic. Therefore 2,4,6-cycloheptatrienone is more stable than 2,4-cyclopentadienone.

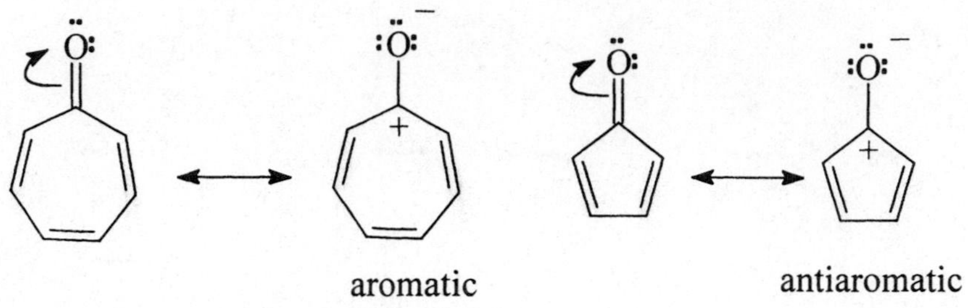

16.18 The resonance structure of cyclopropenone is aromatic (see solution 16.17). Therefore cyclopropenone is less reactive than cyclopropanone despite its high angle strain.

16.19 The compound on the left (8 π electrons) and the one in the middle (12 π electrons) are antiaromatic and are unstable. The compound on the right, known as azulene, has 10 π electrons and is very stable because it is aromatic.

16.20 In the resonance structure of coumarin shown below, the oxygen containing ring has 6 π electrons, giving it aromatic character.

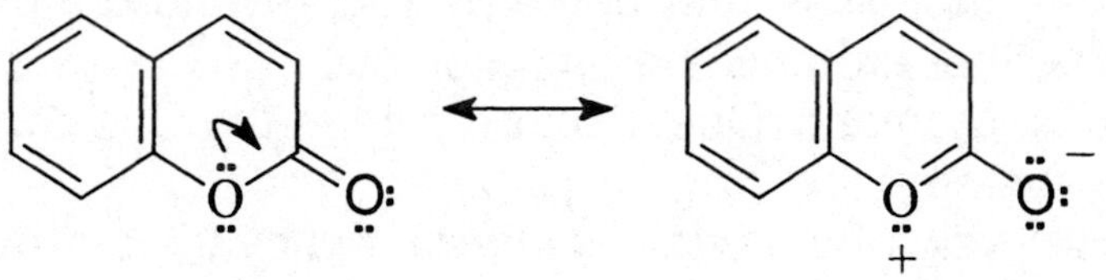

16.21 A resonance structure of the conjugate acid of this ketone is aromatic.

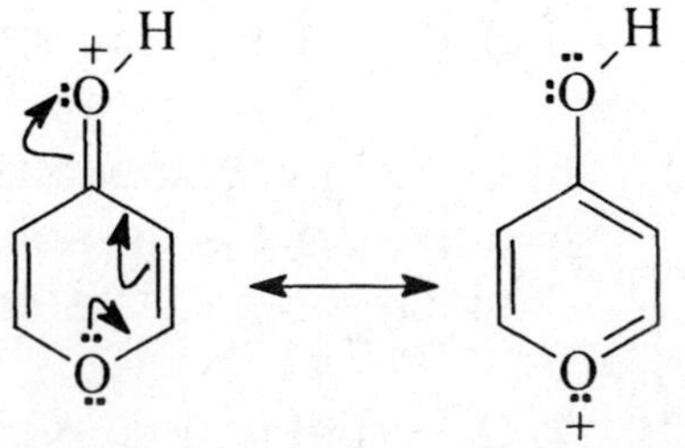

16.22 Benzocyclobutadiene is very unstable because it has an antiaromatic ring in addition to a considerable amount of angle strain.

16.23 a) Although this compound is not completely planar, the nitrogen bridge holds it nearly planar. There are 10 π electrons along the periphery of the ring, so the compound is aromatic.
b) This compound has 14 π electrons along the periphery of the ring, so it is aromatic.

16.24 This compound has 14 π electrons (only one pair of the triple bond is part of the π system), so it is aromatic. The hydrogens on the periphery of the ring should appear downfield due to the diamagnetic ring current. The hydrogens on the inside of the ring should appear further upfield.

16.25 The nitrogen with no hydrogen attached is more basic because the lone pair of electrons is in an sp^2 atomic orbital and is not part of the aromatic cycle of p orbitals. If the other nitrogen were to be protonated, the conjugate acid produced is no longer aromatic because it does not have a cycle of p orbitals completely around the ring.

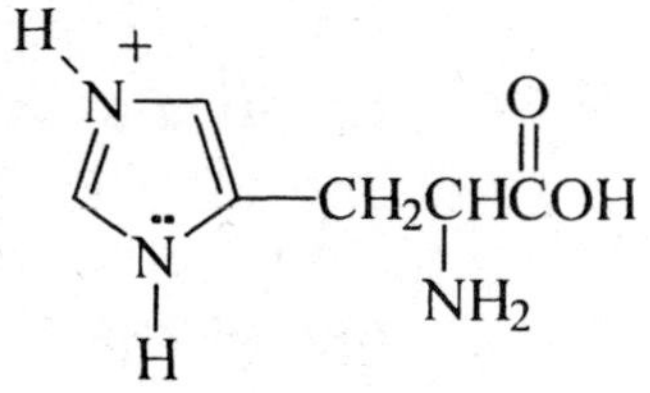

16.26 This compound is a very strong acid because its conjugate base is aromatic and is also stabilized by the electron withdrawing inductive effects of the five CF_3 groups.

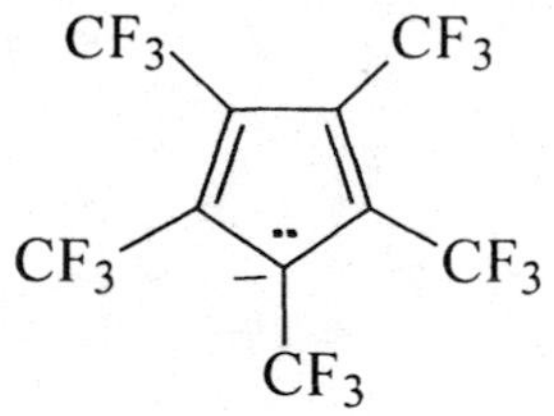

16.27 Using cyclopentene as a model, hydrogenation of the two double bonds of pyrrole should produce 2 x 26.6 kcal/mol (111 kJ/mol) = 53.2 kcal/mol (222 kJ/mol). But hydrogenation of pyrrole only produces 31.6 kcal/mol (132 kJ/mol). The difference, 53.2 - 31.6 = 21.6 kcal/mol (90 kJ/mol) is the resonance stabilization of pyrrole.

16.28 Butyllithium is a very strong base. Removal of two protons occurs readily because an aromatic dianion with 10 π electrons is formed.

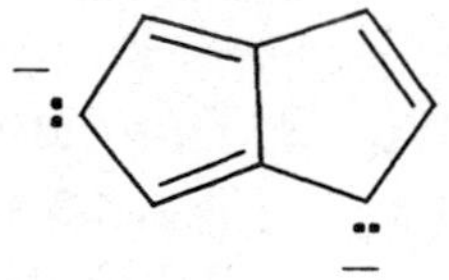

16.29 a) Indole can be viewed as resulting from the fusion of a benzene ring and a heterocyclic pyrrole ring, so its heterocyclic ring is aromatic.
b) The fused heterocyclic ring of benzimidazole is imidazole (see solution 16.8) which is aromatic.
c) The fused heterocyclic ring of quinoline is pyridine, which is aromatic.

16.30 Adenine can be viewed as resulting from the fusion of two aromatic rings, so it is aromatic.

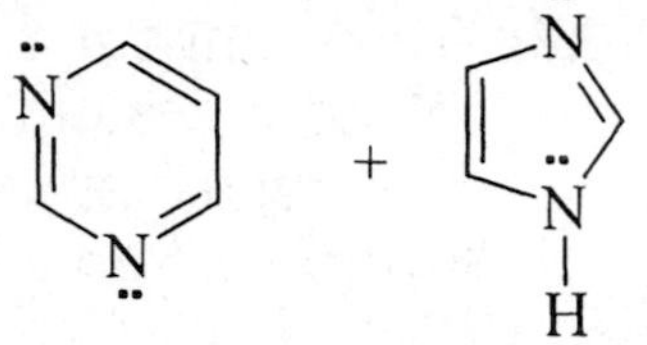

16.31 Both rings are aromatic in resonance structure **B**. Because of the importance of this resonance structure, electrons are transferred from the seven-membered ring to the five-membered ring creating a large dipole moment in the molecule.

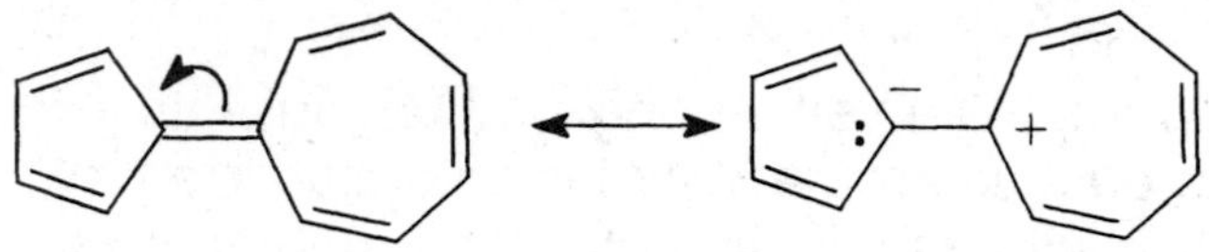

A **B**

16.32 With six π electrons, the cation formed in this reaction fulfills the criteria for aromaticity. All the hydrogens on the cation are on the periphery the ring and are equivalent, so they appear as a singlet. The signal is shifted downfield from the normal aromatic position because the lower electron density of the carbocation results in additional deshielding.

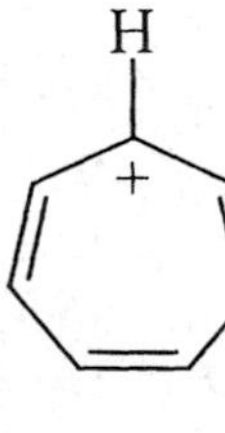

16.33 The dication species produced in this reaction is aromatic. The carbons of the methyl groups on the ring are all magnetically equivalent as are the four ring carbons. The ring carbons are very far downfield because they are part of an aromatic ring. They are deshielded even more due to the low electron density in the ring.

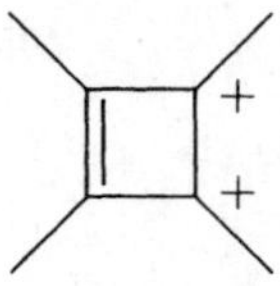

16.34 The benzylic cation rearranges to form cycloheptatrienyl cation, which is even more stable. Note that the cycloheptatrienyl carbocation is aromatic (6 π electrons) and has the positive charge delocalized around the entire ring.

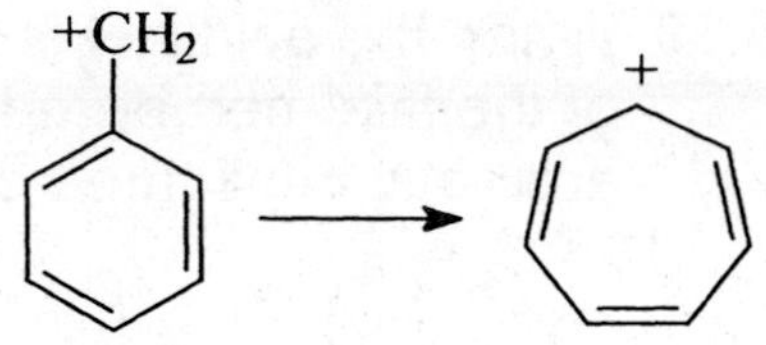

16.35 The two atoms of the double bond and the four atoms attached to them must lie in the same plane in order for the p-orbitals of the pi bond to be parallel. Examine the model to see for which adjacent carbons this is true. (When viewed down a C-C bond, the H's are eclipsed as are the C's.) These are the carbons that are connected by double bonds. When viewed down a C-C single bond, the H's are not eclipsed, nor are the C's.

16.36 Only the electron pair on the nitrogen bonded to the hydrogen is part of the pi system. The electrons on the nitrogens that are part of double bonds are perpendicular to the pi system. Both heterocyclic rings have six pi electrons and are aromatic, so purine is aromatic.

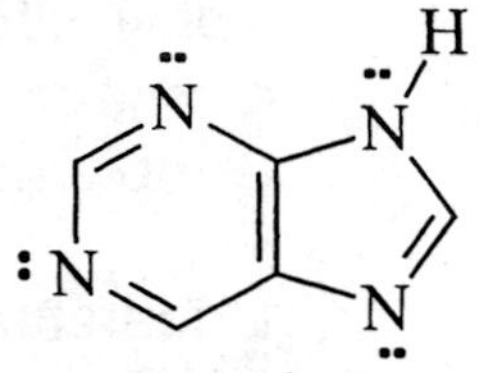

purine

16.37 For oxazole and isoxazole, one electron pair on the O is part of the pi system. The other electron pair on the O and the pair on the N is perpendicular to the pi system. Both compounds have six electron in their pi systems and are aromatic. For pyrazole, the pair of electrons on the N that is doubly bonded to the C is not part of the pi system. The pair of electrons on the N that is bonded to the H is part of the pi system. There are six electrons in the pi system, so pyrazole is aromatic. Like pyridine, the electrons on the N's of pyrimidine are not part of the pi system. There are six pi electrons, so pyrimidine is aromatic.

oxazole

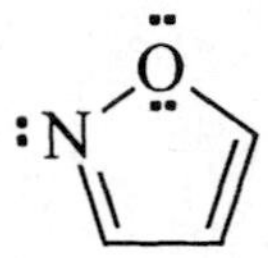

isoxazole

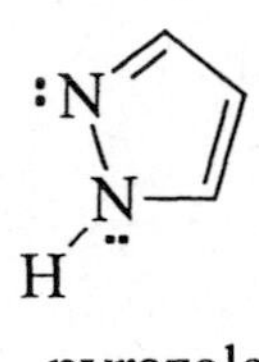

pyrazole

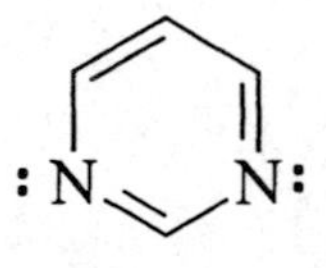

pyrimidine

16.38 When the acidic H is removed from the CH_2 group of the five membered ring, the resulting anion is aromatic in all three of its rings.

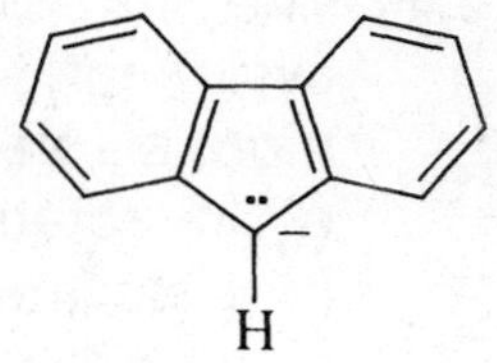

Review of Mastery Goals

After completing this chapter, you should be able to:

Show the MO energy levels for planar, cyclic, conjugated compounds. (Problems 16.3, 16.4, 16.5, 16.6, and 16.14)

Apply Hückel's rule and recognize whether a particular compound is aromatic, antiaromatic, or neither.
(Problems 16.6, 16.7, 16.8, 16.15, 16.19, 16.20, 16.23, 16.29, and 16.30)

Understand how aromaticity and antiaromaticity affect the chemistry and NMR spectra of compounds.
(Problems 16.1, 16.2, 16.9, 16.10, 16.11, 16.12, 16.13, 16.16, 16.17, 16.18, 16.21, 16.22, 16.24, 16.25, 16.26, 16.28, and 16.31)

Chapter 17
AROMATIC SUBSTITUTION REACTIONS

17.1 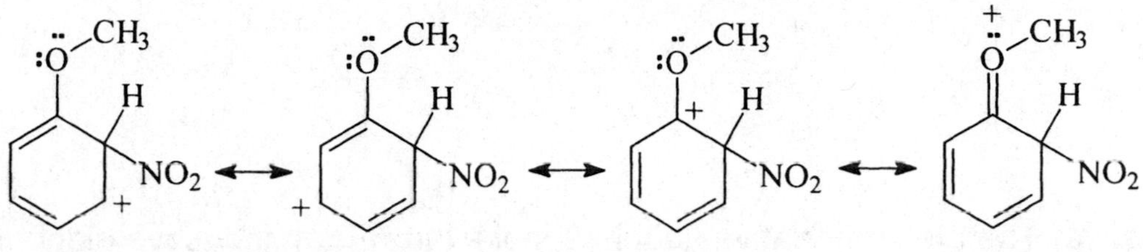

The last structure is especially stable because the octet rule is satisfied at all atoms.

17.2 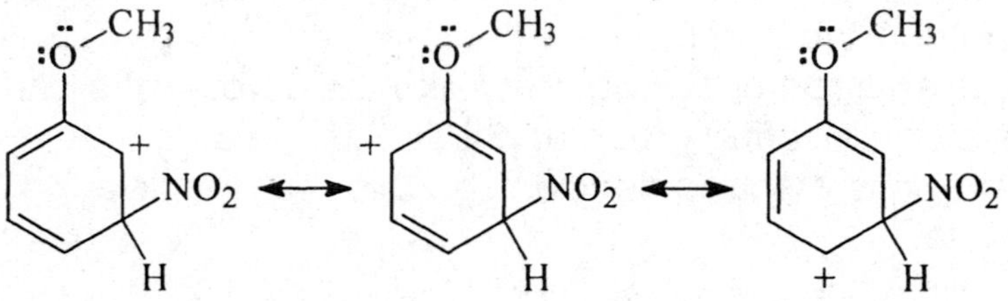

No especially stable resonance structure is formed because the positive charge is never located on the carbon that is bonded to the methoxy group.

17.3 a) The electron pair on the N can stabilize a positive charge on the adjacent carbon by resonance.

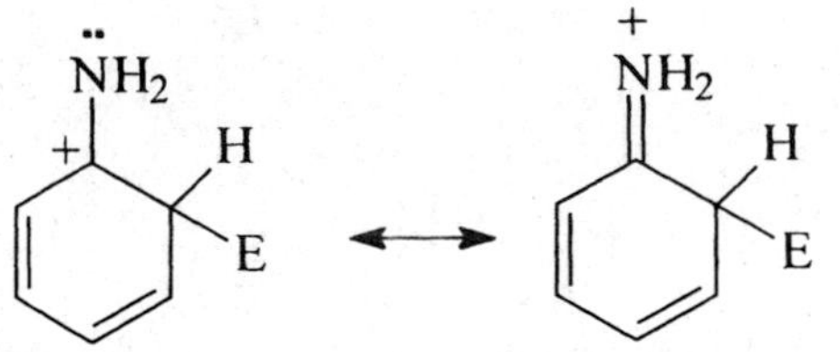

b) The ethyl group stabilizes a positive charge on the adjacent carbon by hyperconjugation.

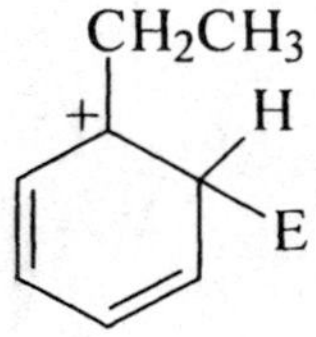

c) The phenyl group can stabilize a positive charge on the adjacent carbon by resonance.

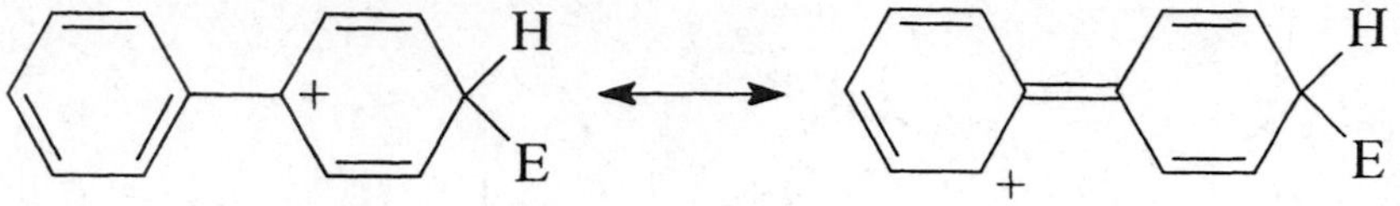

17.4 a) The electronegative fluorines make the CF_3 an inductive electron withdrawing group.
b) The positive nitrogen is an inductive electron withdrawing group.

17.5 a) Because of the unshared electrons on the S, this is an activating group and it directs ortho and para.
b) The S is electron deficient, much like the carbon of a carbonyl group, so this is a deactivating group and it directs meta.
c) The carbon is electron deficient, so this is a deactivating group and it directs meta.

17.6

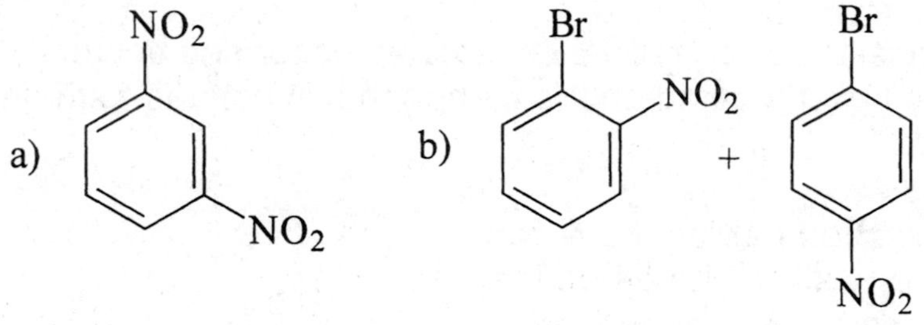

17.7 a) Both groups are directing to the same position, ortho to the ethyl group and meta to the nitro group.

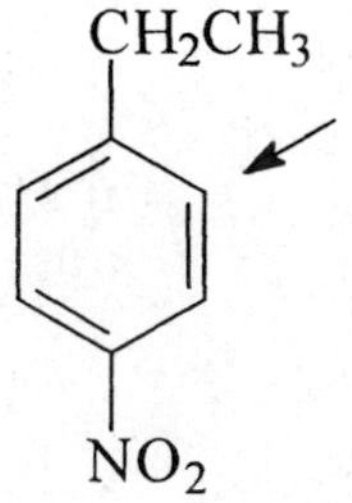

b) The groups are directing to different positions, so the more activating group will control the regiochemistry. The nitrogen attached to the ring is activating, whereas the carbonyl group is deactivating. Therefore the substitution occurs ortho and para to the nitrogen group, but not between the two groups as that position is sterically hindered.

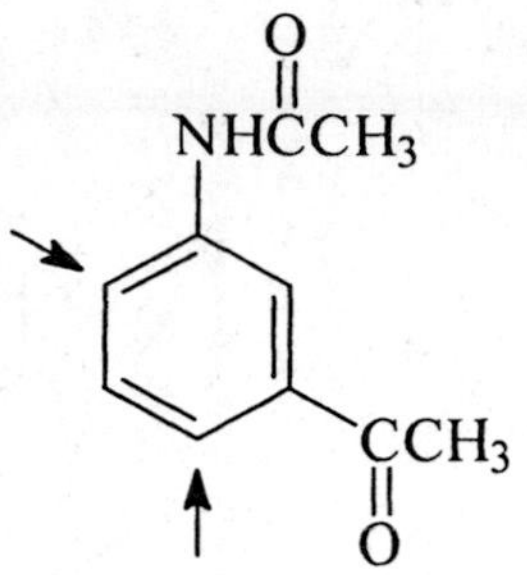

c) The methoxy group is a stronger activating group than the methyl group, so the substitution occurs ortho to the methoxy group.

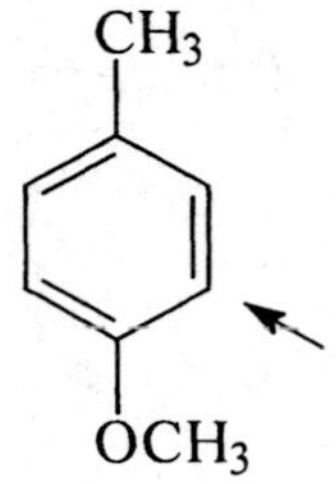

d) The groups are directing to the same positions, so the substitution occurs ortho to one and para to the other, but not at the position ortho to both because this position is too sterically hindered.

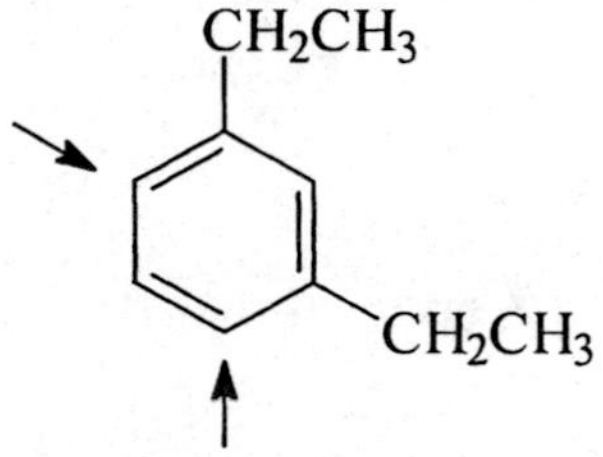

e) Both groups are directing to the same positions.

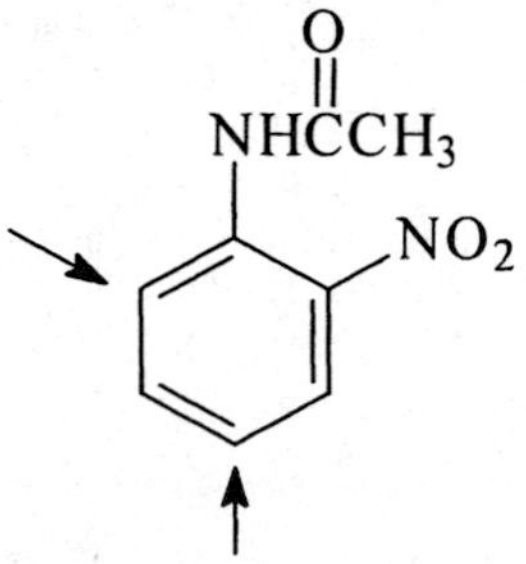

f) Substitution occurs ortho and para to the hydroxy group because it is more activating than chlorine.

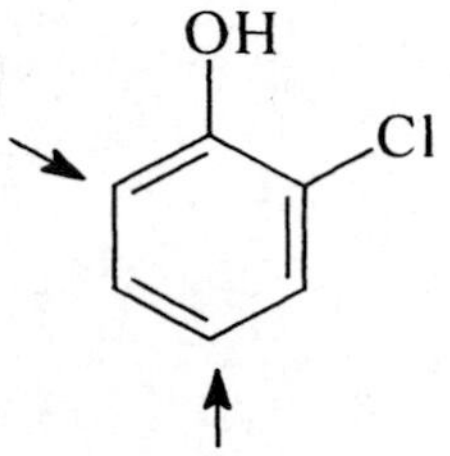

17.8

a) 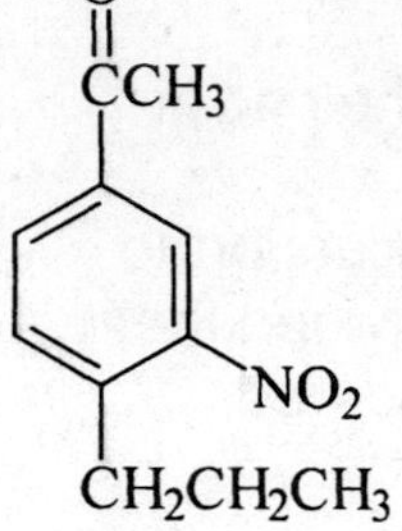

b) 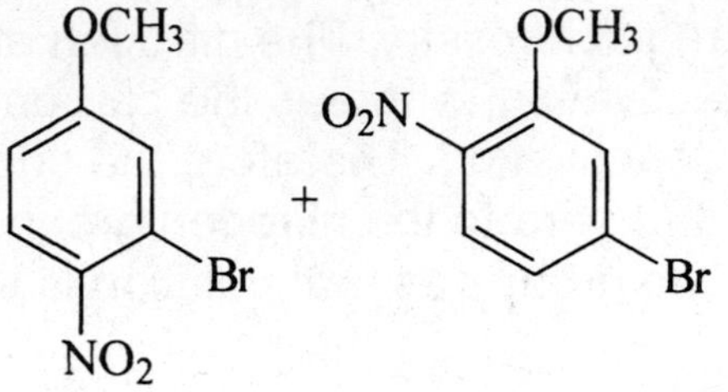

17.9

a) 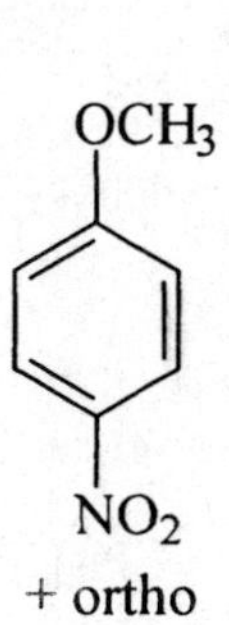

b) 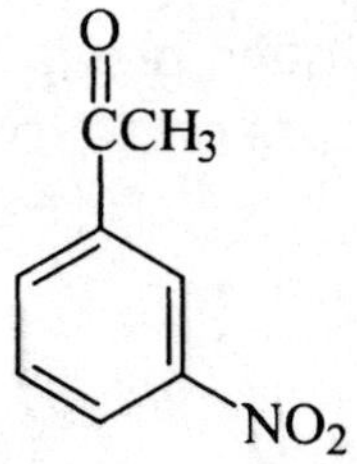

c) 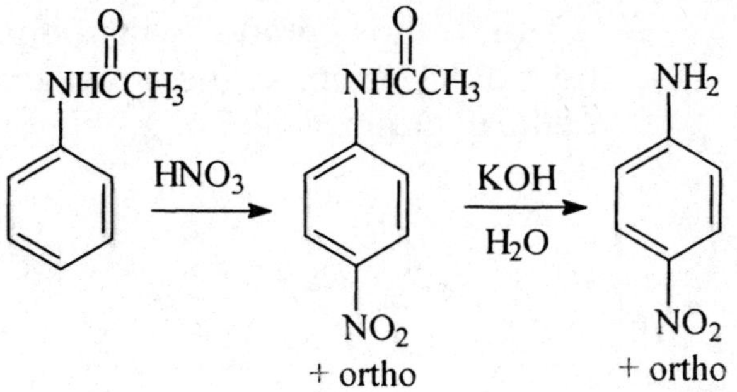

17.10

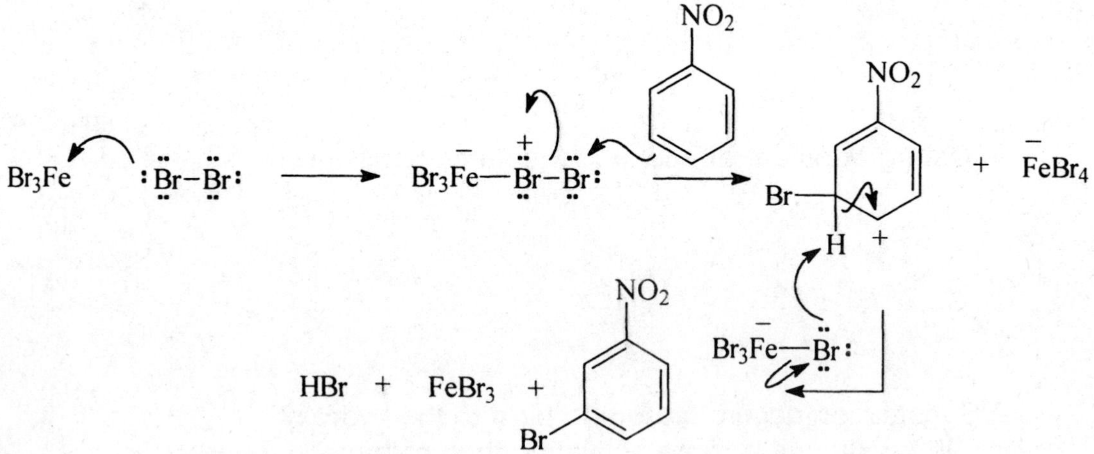

17.11

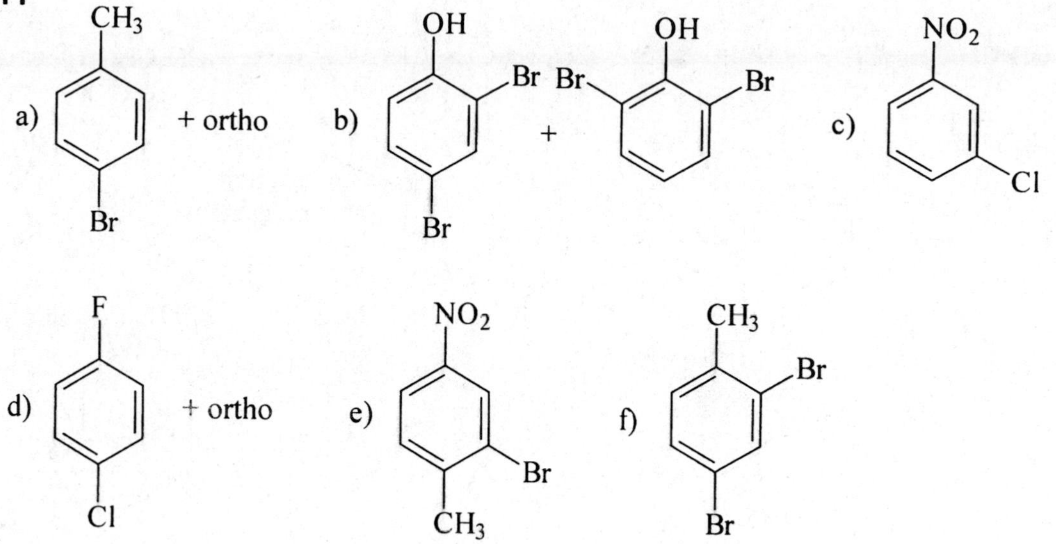

a)
CH_3
Br
+ ortho
b)
OH
Br
Br
+
OH
Br
Br
c)
NO_2
Cl
d)
F
Cl
+ ortho
e)
NO_2
Br
CH_3
f)
CH_3
Br
Br

17.12

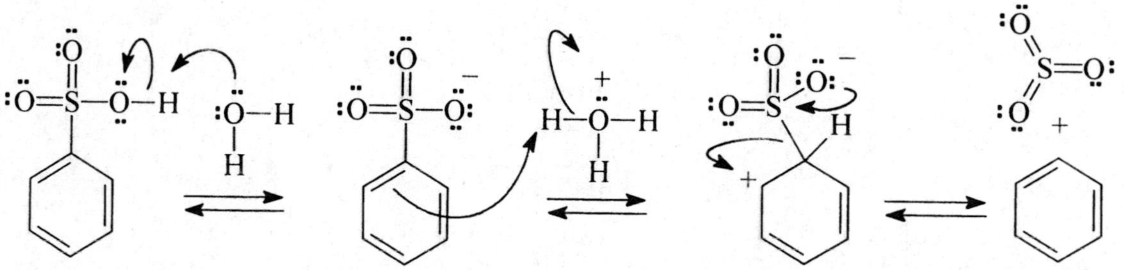

17.13

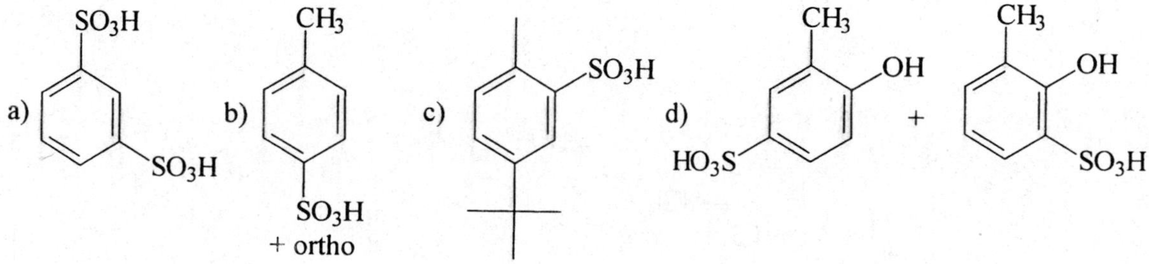

a)
SO_3H
SO_3H
b)
CH_3
SO_3H
+ ortho
c)
SO_3H
d)
CH_3
OH
HO_3S
+
CH_3
OH
SO_3H

17.14

$CH_3CH_2CH_2-\ddot{\underset{\cdot\cdot}{Cl}}: \ + \ AlCl_3 \ \rightleftharpoons \ CH_3CH_2CH_2-\overset{+}{\underset{\cdot\cdot}{Cl}}-\overset{-}{AlCl_3} \ \longrightarrow \ CH_3\overset{+}{CH}CH_2 \ + \ \overset{-}{AlCl_4}$

CH_3CHCH_3

$CH_3\overset{+}{CH}CH_3$

carbocation rearrangement

B:

$CH_2CH_2CH_3$

B:

17.15

a) OCH₃ (para isopropyl) + ortho   b) NHCCH₃ (with C=O), para tert-butyl + ortho   c) tert-butylbenzene

d) (2-methyl-2-phenylbutane type)   e) (tetrahydronaphthalene with methyl)   f) no reaction

17.16

a) benzene + (CH₃)₃C—Cl $\xrightarrow{AlCl_3}$ b) benzene + (CH₃)₂CHCl $\xrightarrow{AlCl_3}$

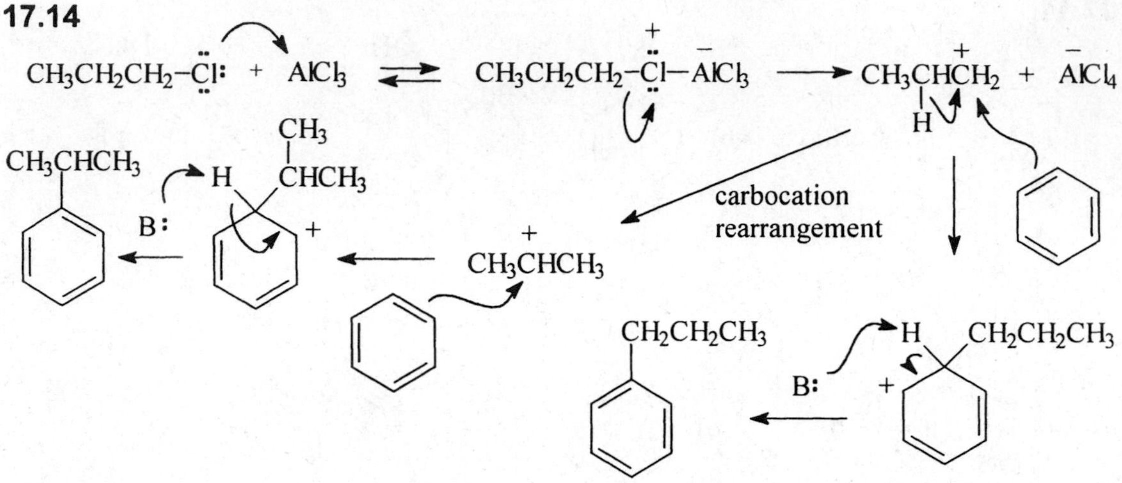

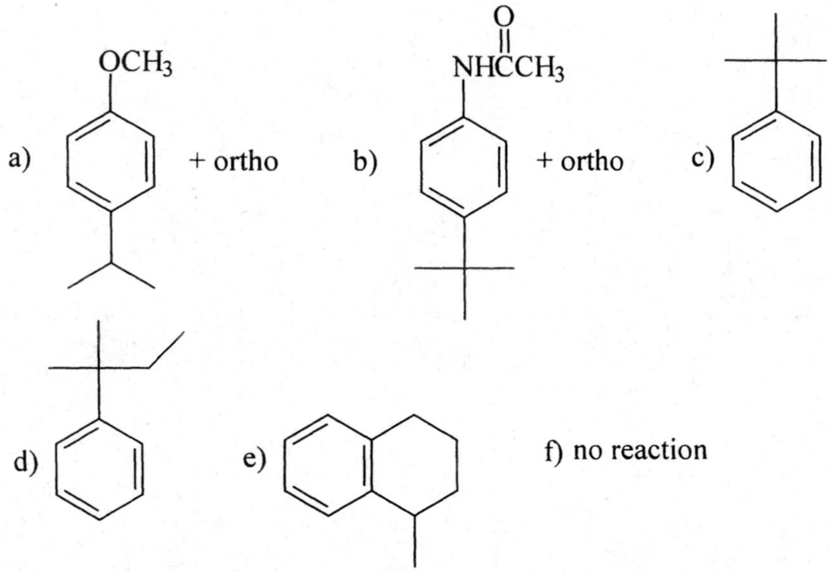

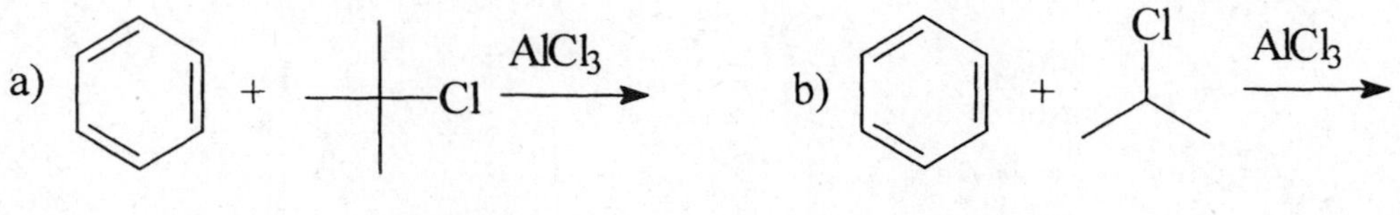

17.17

$$CH_3(CH_2)_7CH(CH_2)_5CH_3$$

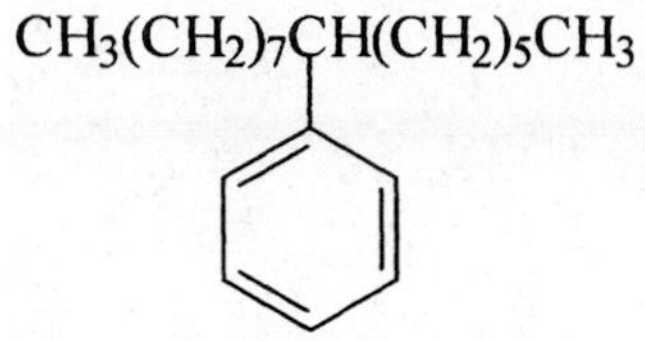

17.18 Both groups are weakly activating, but the position ortho to the methyl group is less sterically hindered than the position ortho to the isopropyl group. The Friedel-Crafts acylation reaction is very sensitive to steric effects.

17.19

a)

b) no reaction

c)

d)

e)

17.20

a) $+ CH_3CH_2CH_2CCl$ (with O double bond) $\xrightarrow{AlCl_3}$

b) $+ CH_3CCl$ (with O double bond) $\xrightarrow{AlCl_3}$

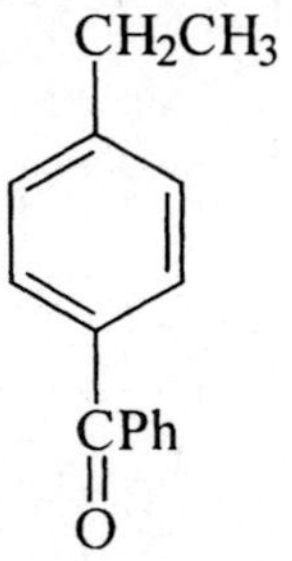

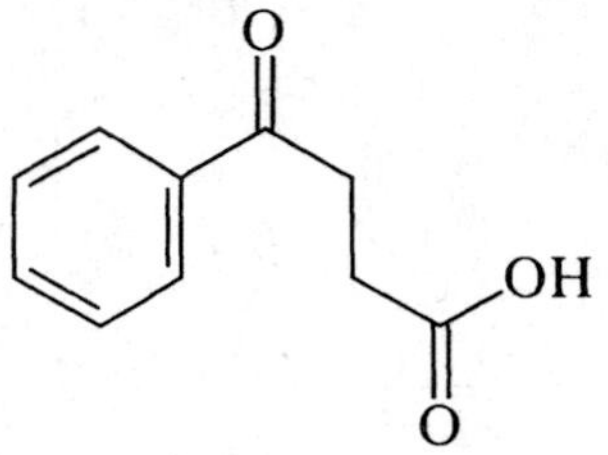

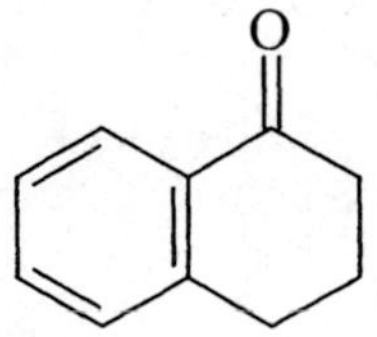

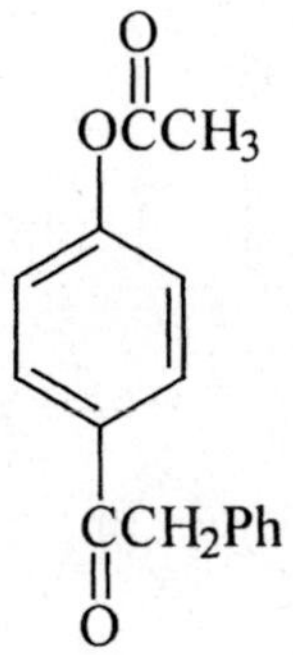

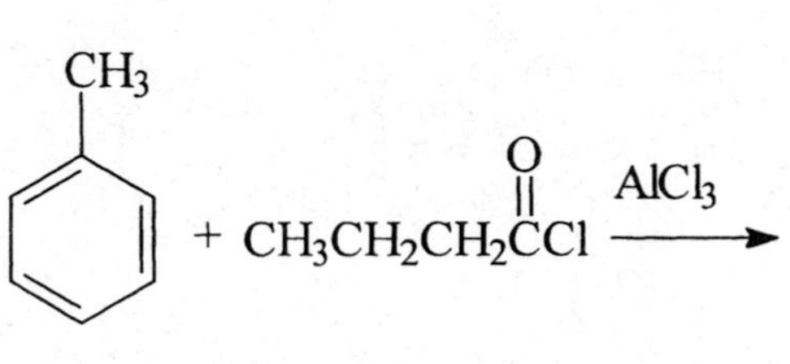

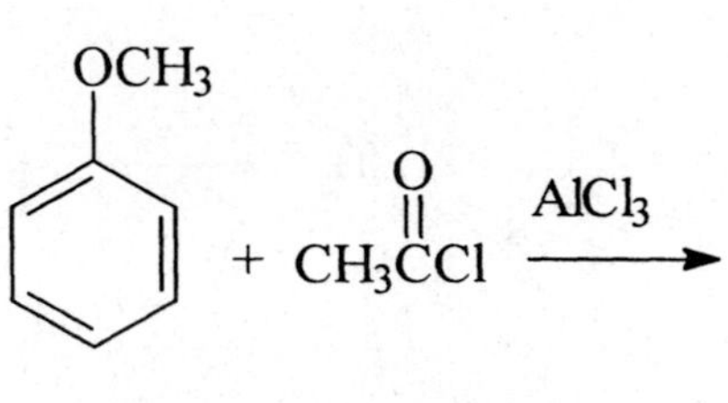

c) 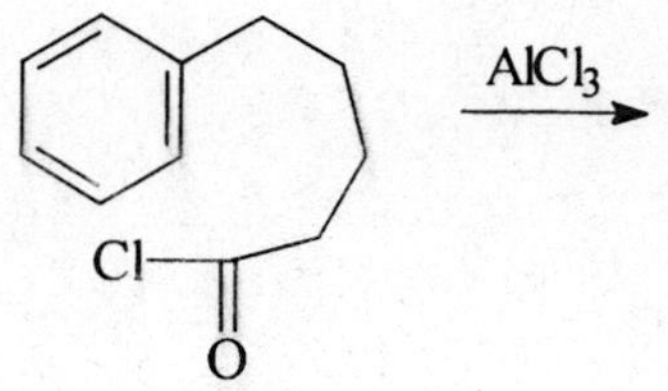

17.21

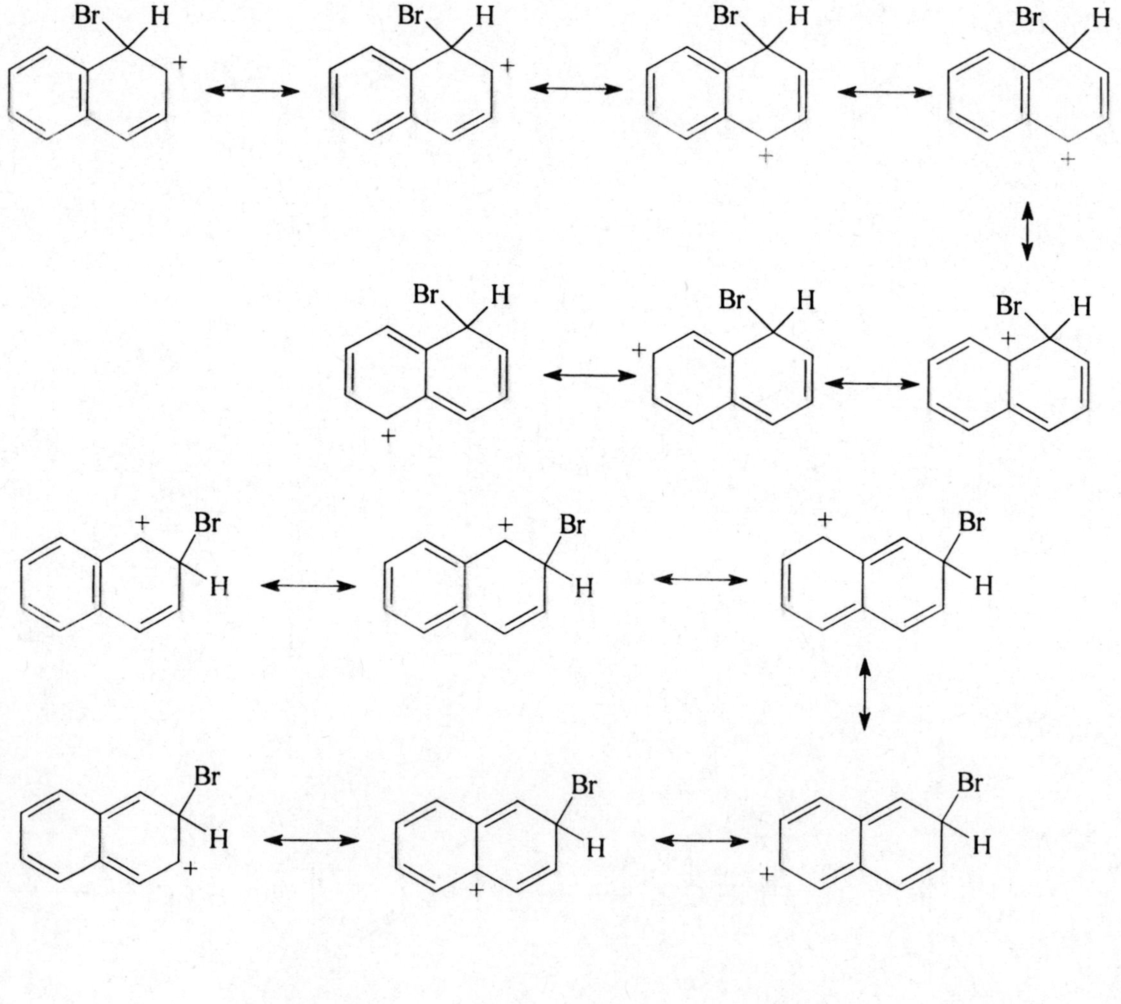

17.22

a) 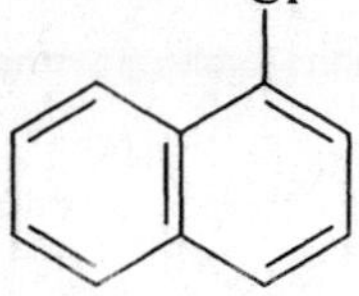

b) 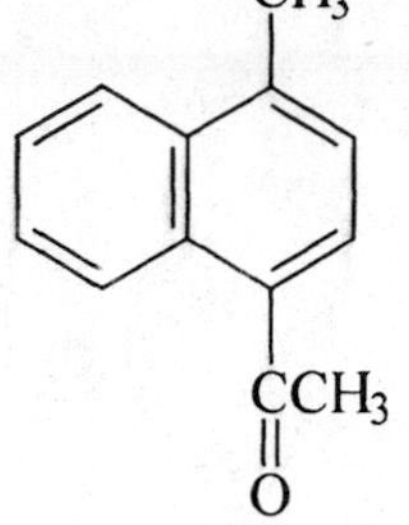

c) 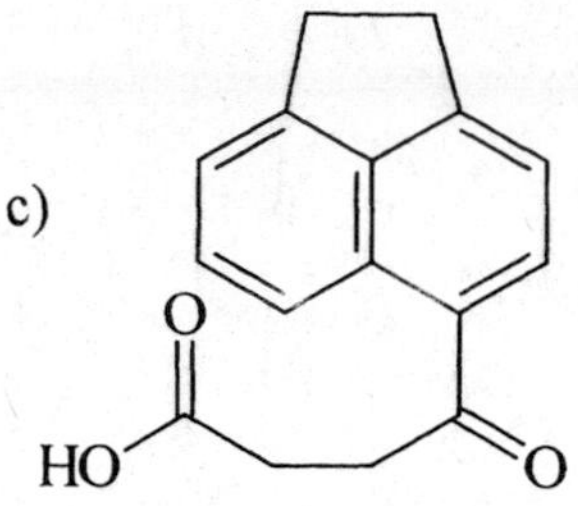

17.23

a) 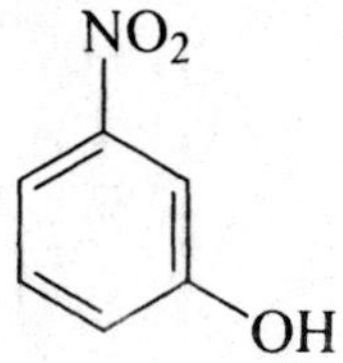

b) 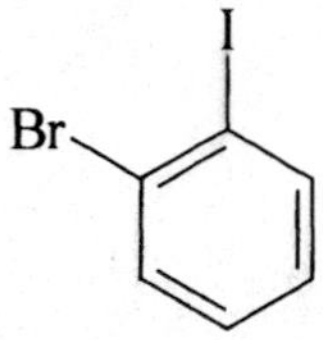

c) 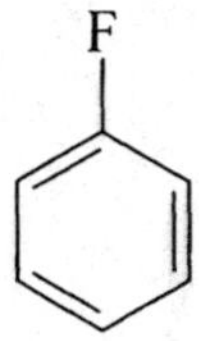

d) 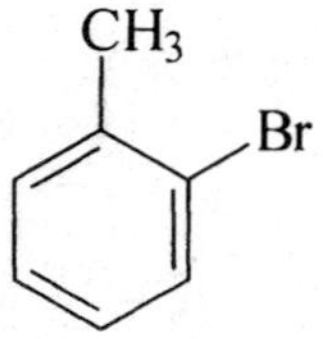

e) 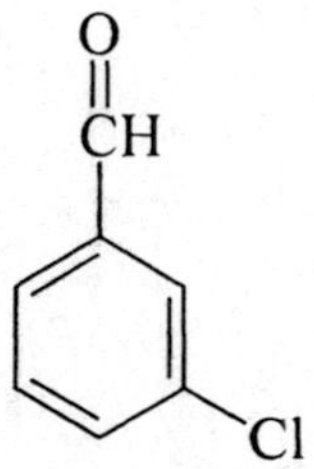

f) 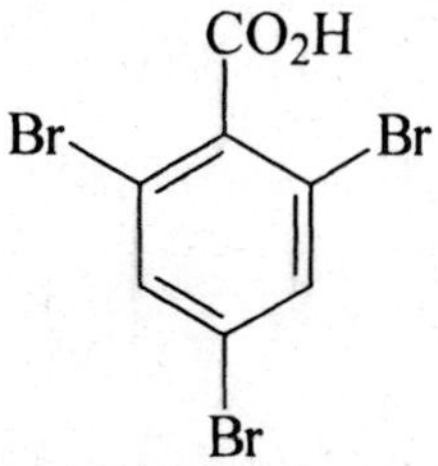

g) 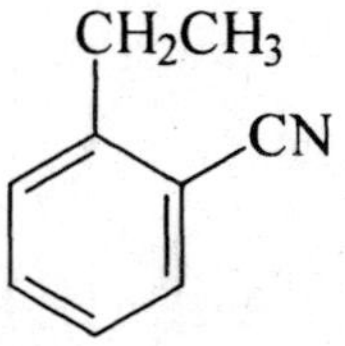

h) 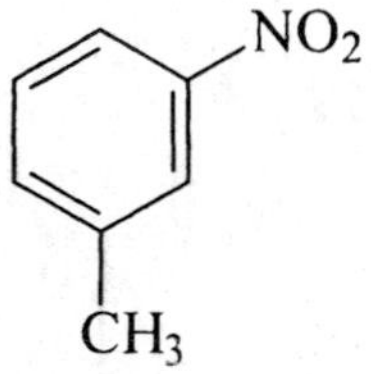

17.24 A nitro group in the para position is better at stabilizing the negative charge than is a nitro group in the meta position, but a nitro group in the meta position does help somewhat by its inductive effect.

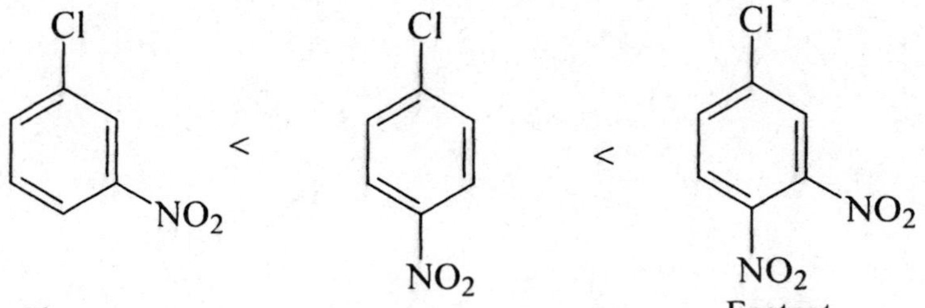

Slowest

Fastest

17.25

a) 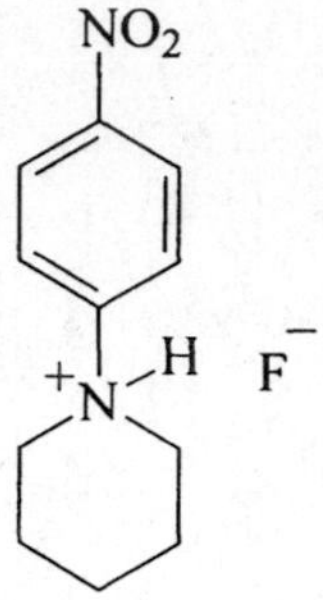
NO2
+N
H
F−

b) 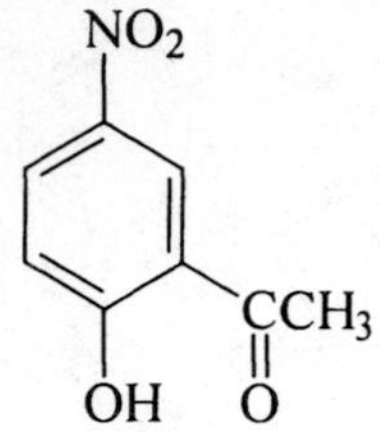
NO2
CCH3
OH
O

17.26

a) 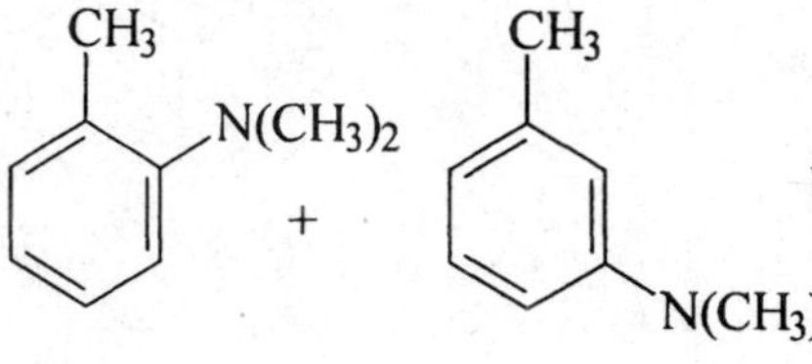
CH3
N(CH3)2
+
CH3
N(CH3)2

b) 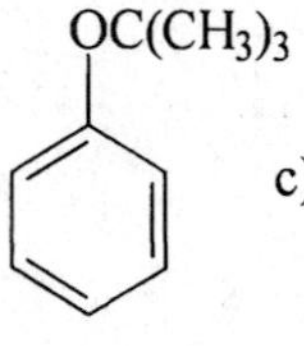
OC(CH3)3

c) 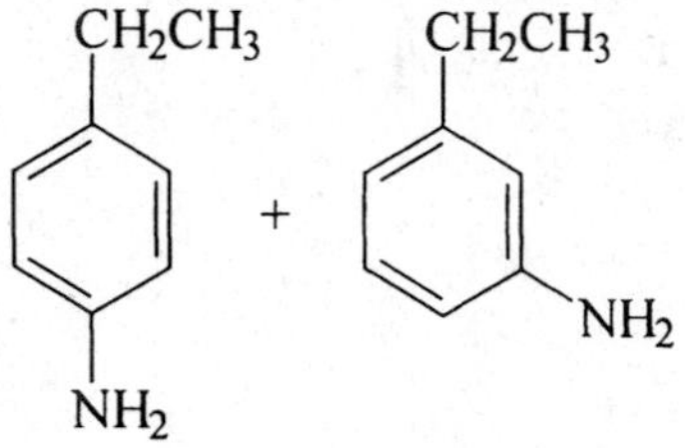
CH2CH3
NH2
+
CH2CH3
NH2

17.27

a) 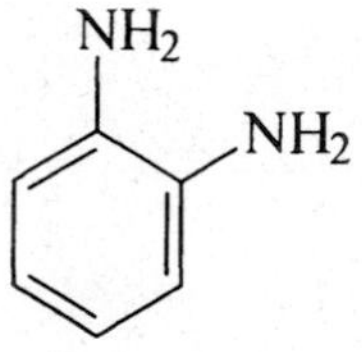
NH2
NH2

b) 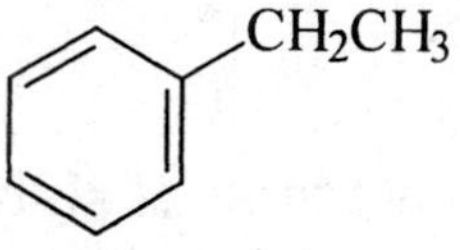
CH2CH3

c) 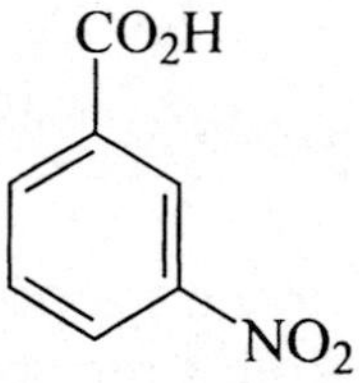
CO2H
NO2

d) 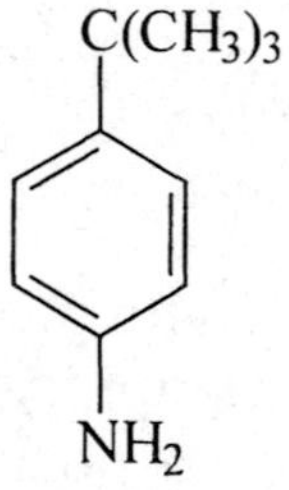
C(CH3)3
NH2

e) 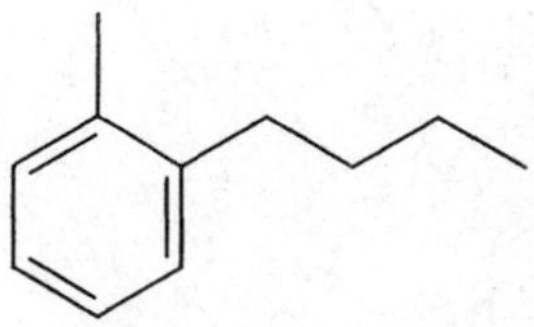

f) 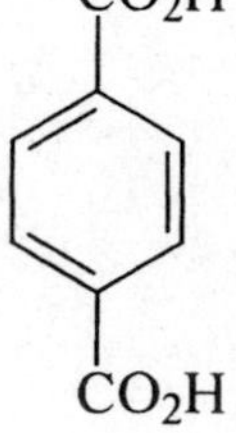
CO2H
CO2H

g) 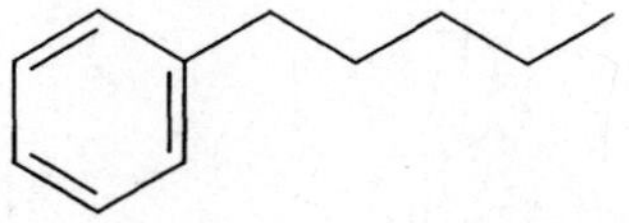

17.28

a) 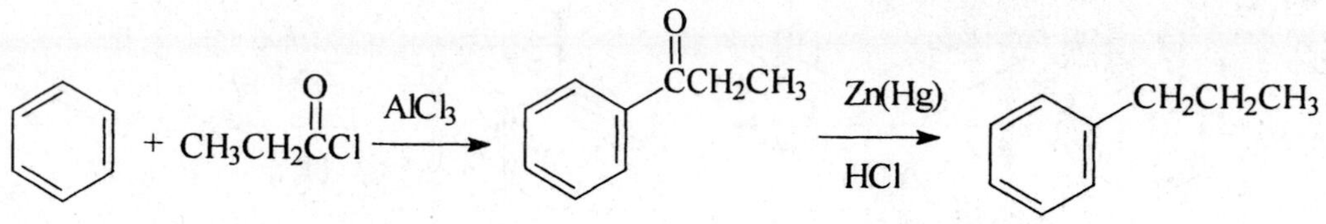

O
+ CH3CH2CCl
AlCl3
O
CCH2CH3
Zn(Hg)
HCl
CH2CH2CH3

b) 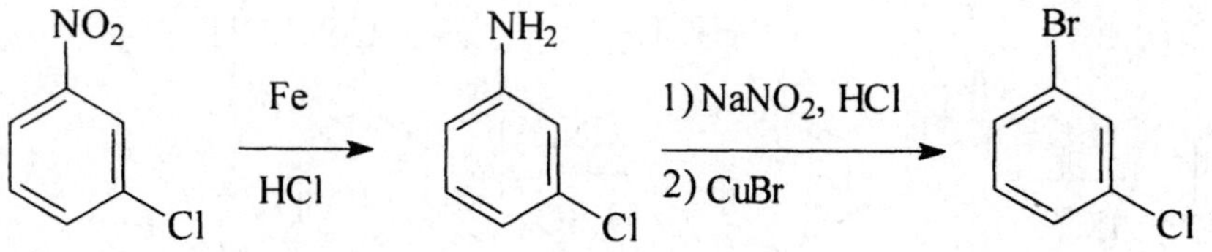

NO2
Cl
Fe
HCl
NH2
Cl
1) NaNO2, HCl
2) CuBr
Br
Cl

17.29

a) 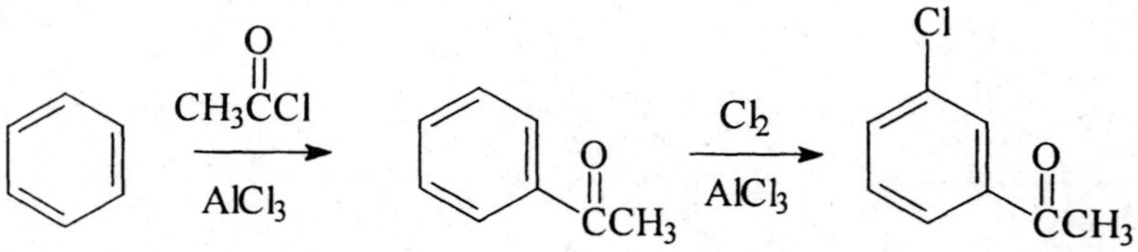

O
CH3CCl
AlCl3
O
CCH3
Cl2
AlCl3
Cl
O
CCH3

b) 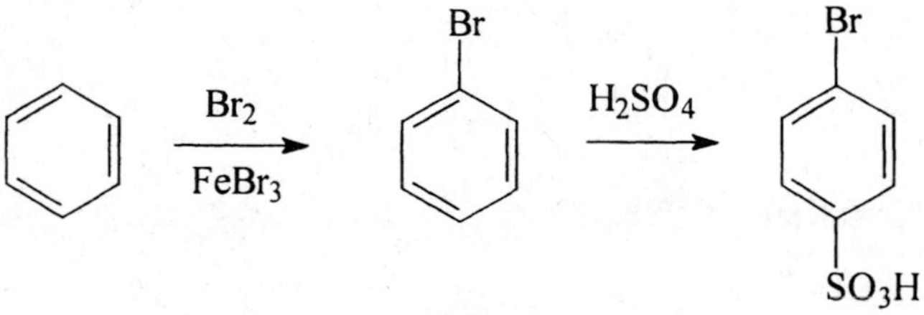

Br2
FeBr3
Br
H2SO4
Br
SO3H

c) 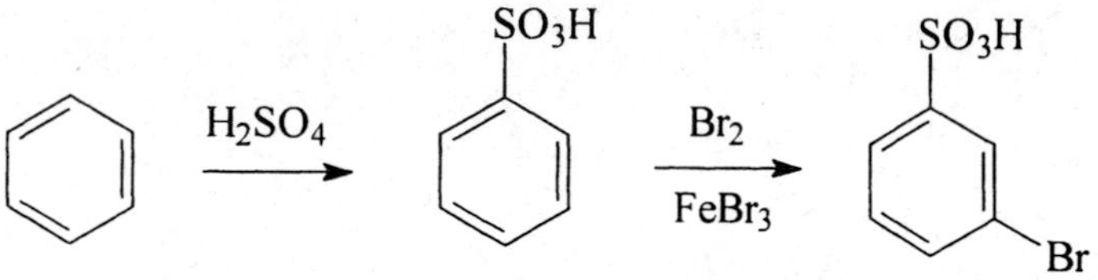

H2SO4
SO3H
Br2
FeBr3
SO3H
Br

d) 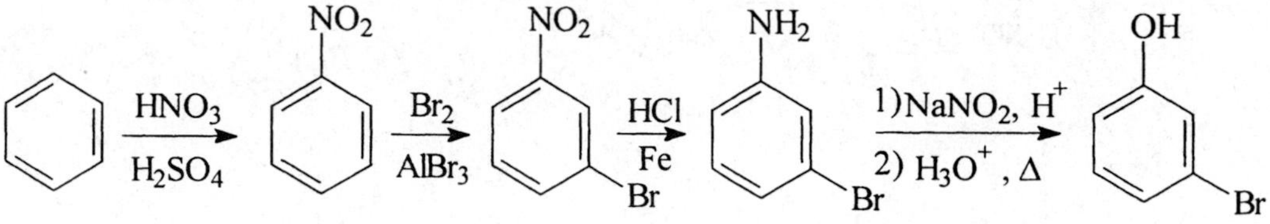

HNO3
H2SO4
NO2
Br2
AlBr3
NO2
Br
HCl
Fe
NH2
Br
1)NaNO2, H+
2) H3O+ , Δ
OH
Br

e) 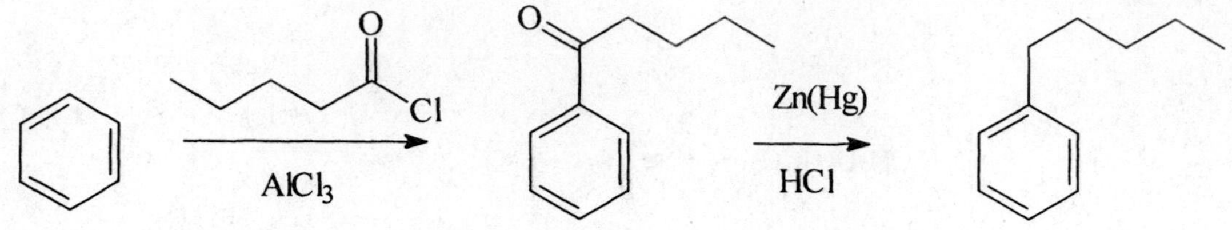

f) 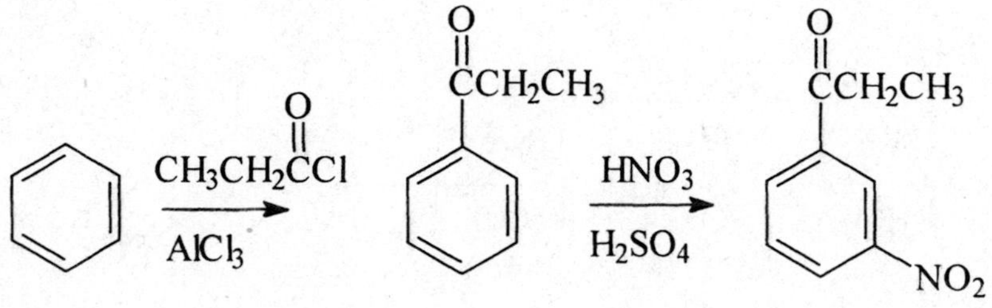

g) 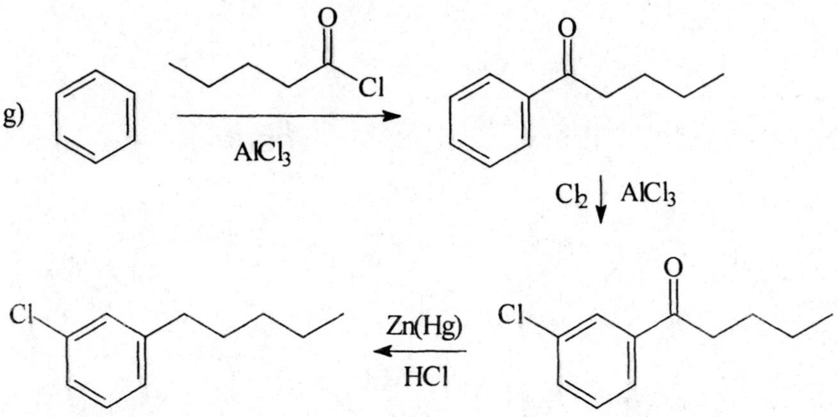

h) 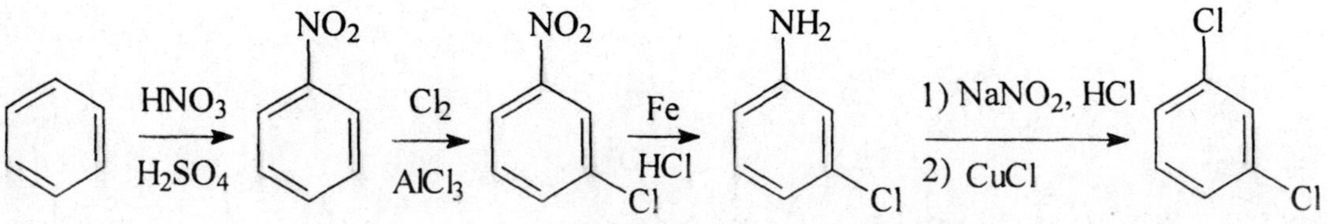

i)

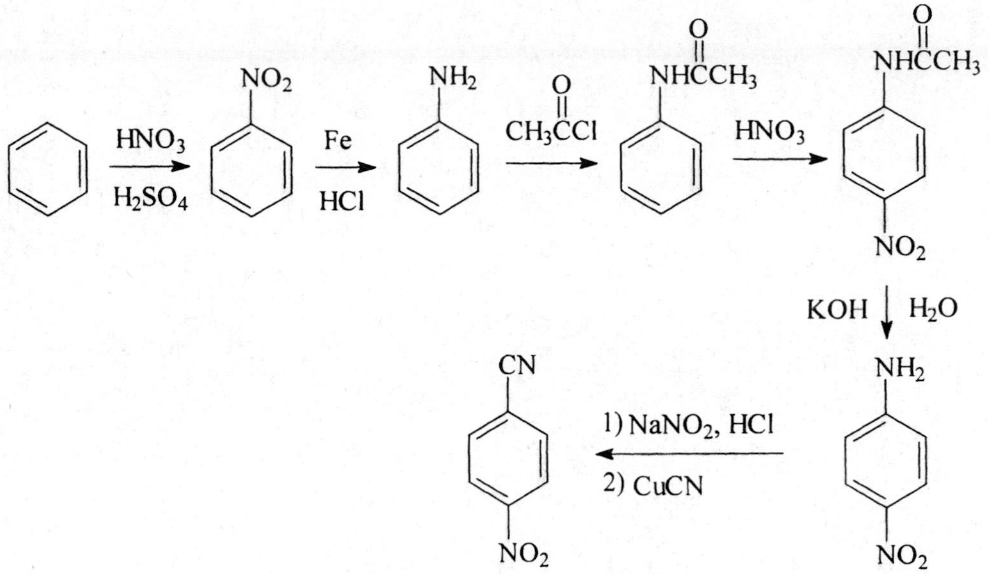

HNO3
H2SO4
NO2
Fe
HCl
NH2
CH3CCl
NHCCH3
HNO3
KOH
H2O
1) NaNO2, HCl
2) CuCN
CN

17.30

a)

Cl2
AlCl3
Cl
H2SO4
SO3H
HNO3
H2SO4
Δ
O2N
NO2
H2O
H2SO4
Δ

b)

AlCl3
HO
H2
Pd
H2SO4

c)

CH3I
AlCl3
CH3
HNO3
H2SO4
Δ
O2N
NO2
KMnO4
NaOH
CO2H

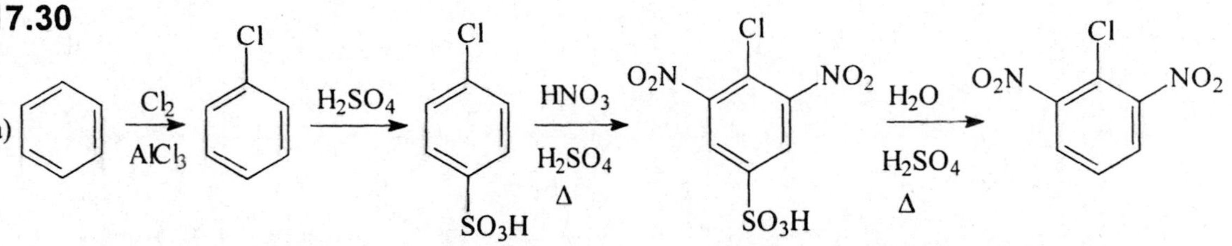

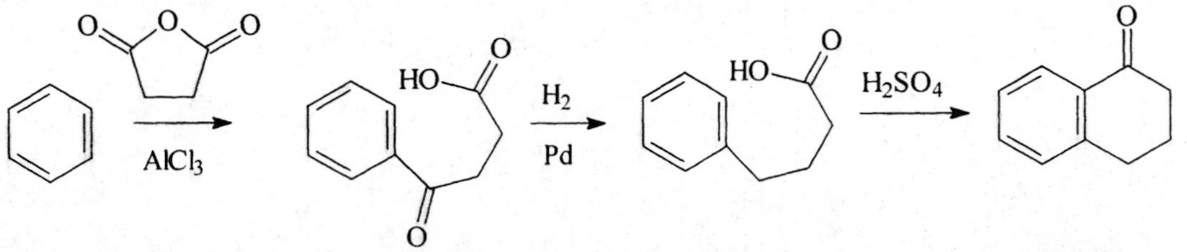

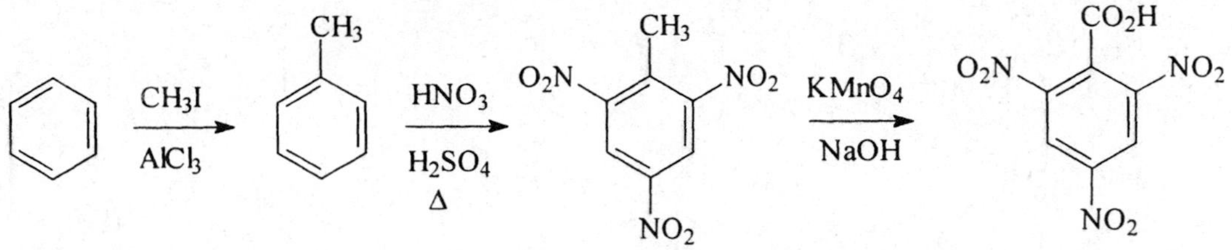

17.31

a) 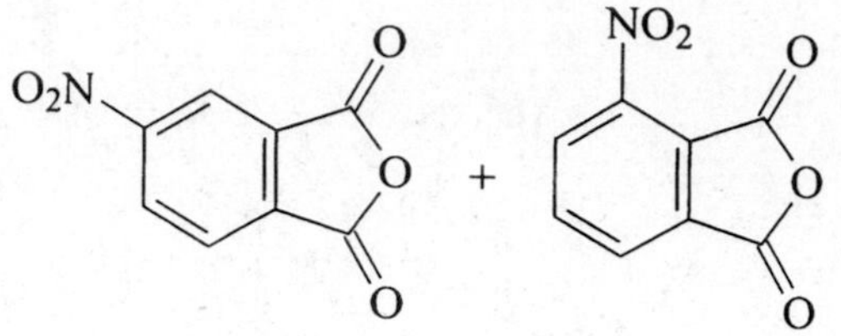

b) 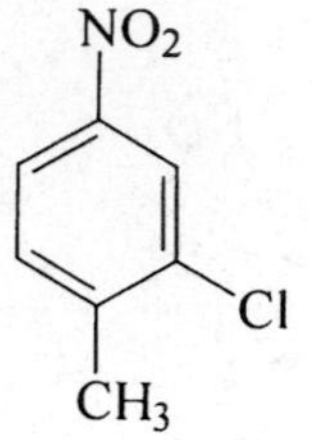

c) 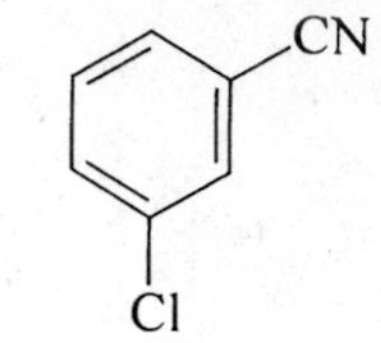

d) 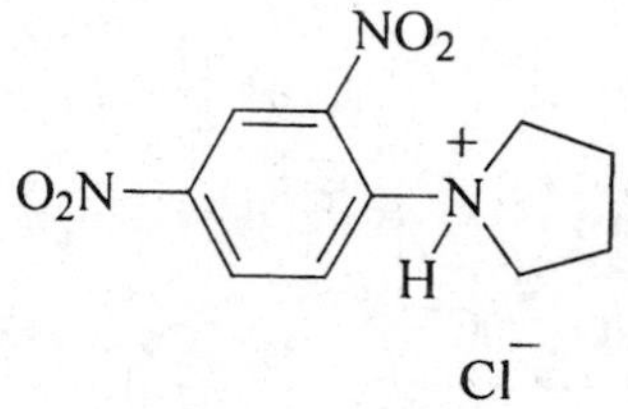

e) 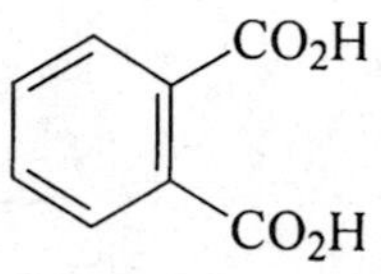

f) 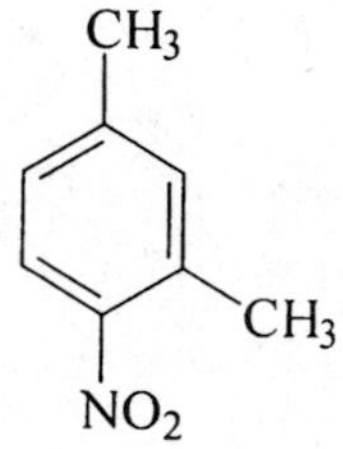

g) 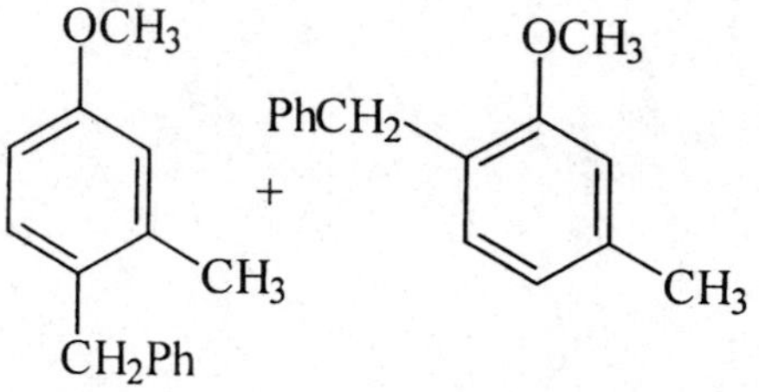

h) 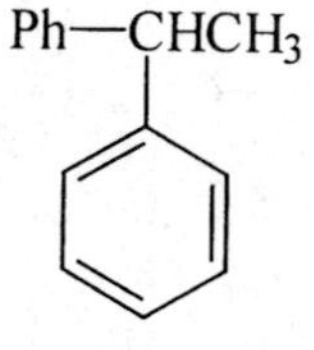

i) 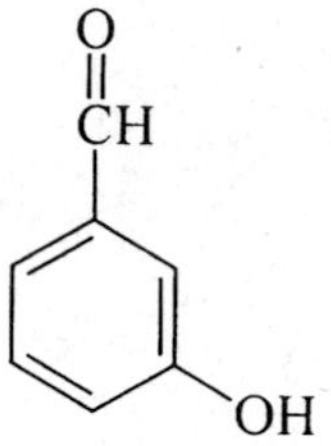

j) 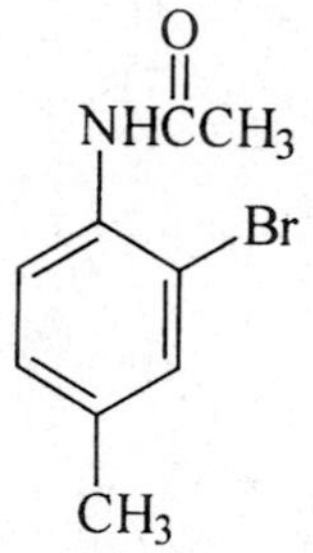

k) 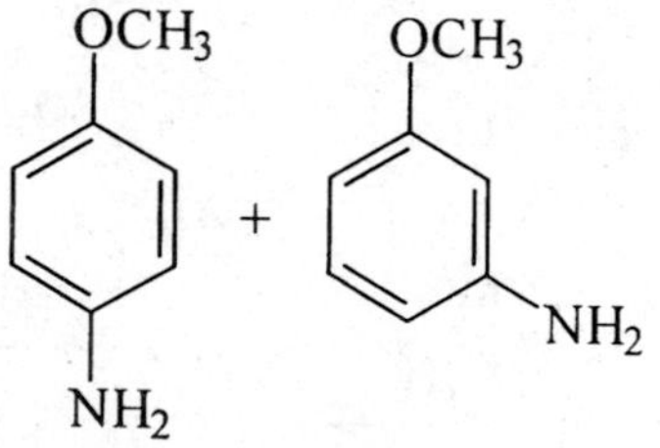

l) 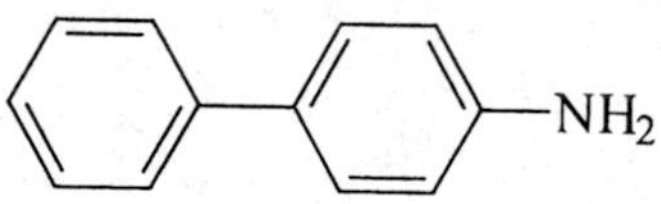

m) 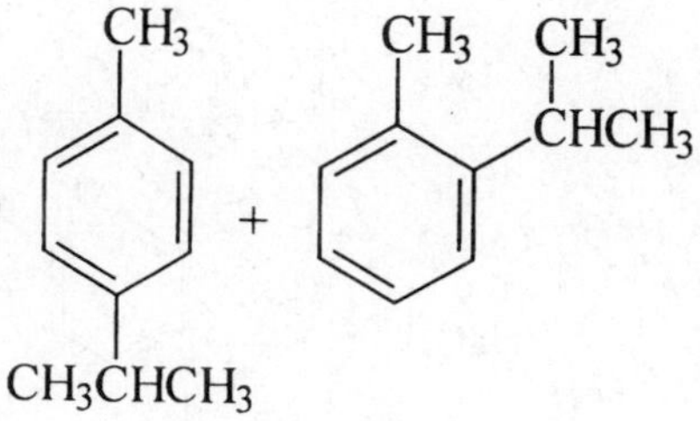

n) 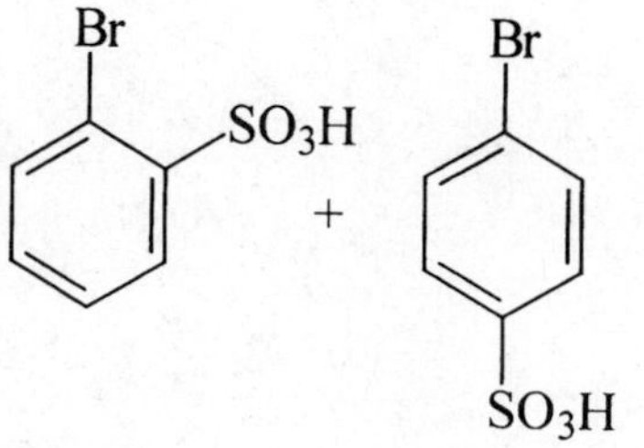

o) 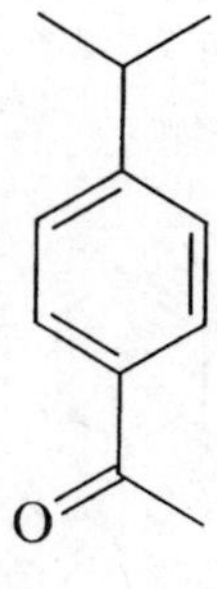

p) 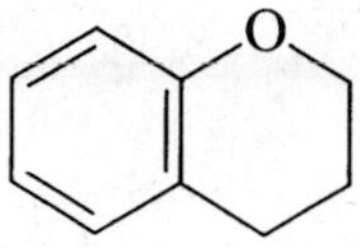 q) 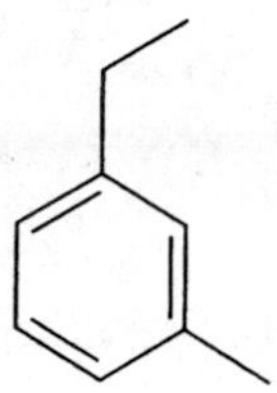 r) 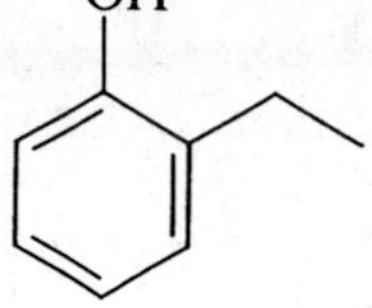

17.32

a) 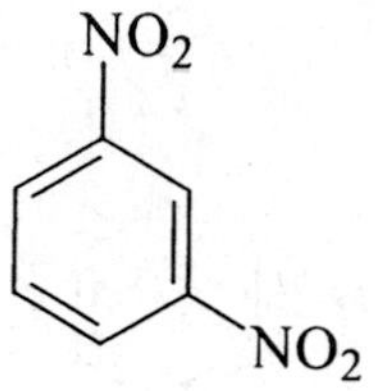 b) 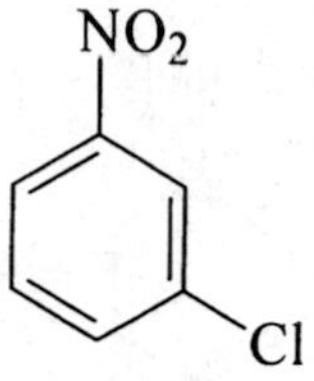 c) 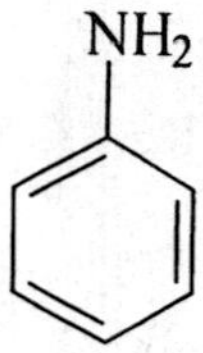

d) No reaction e) 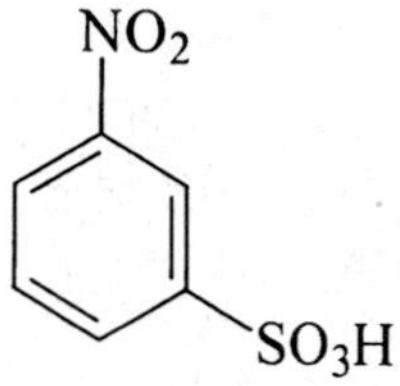 f) 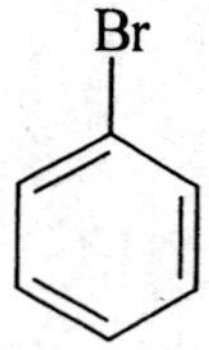

17.33

a) 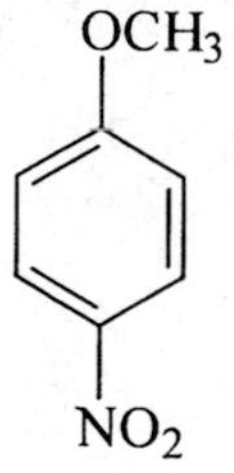
+ ortho

b) 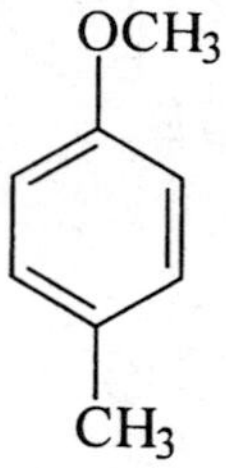
+ ortho

c) 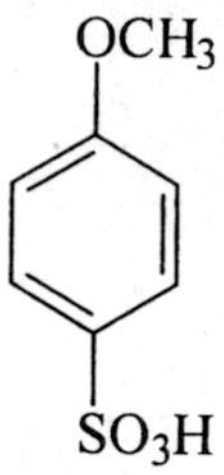
+ ortho

d) 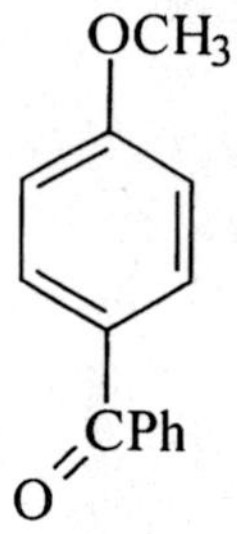

e) 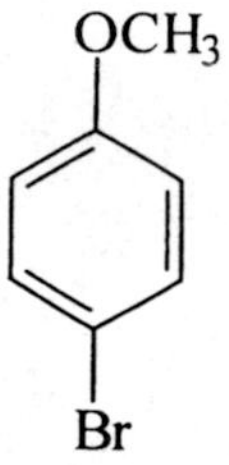
+ ortho

17.34

a) 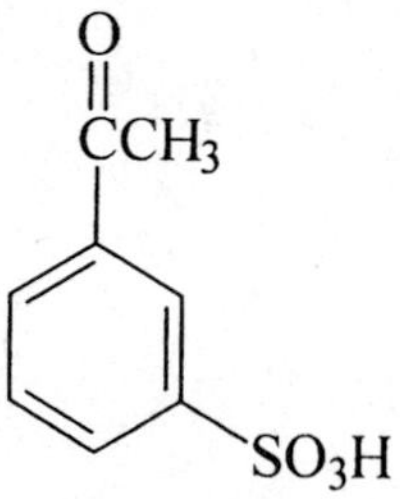 b) 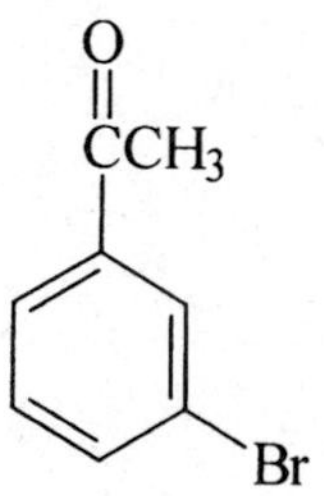 c) 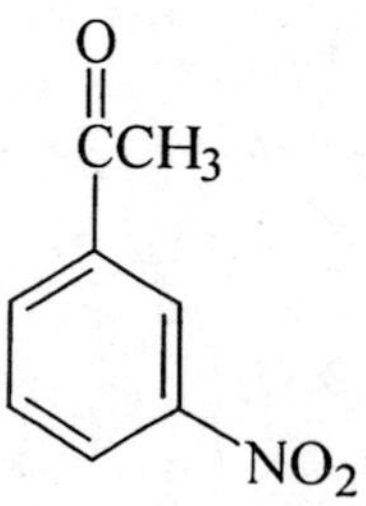 d), e), and f) 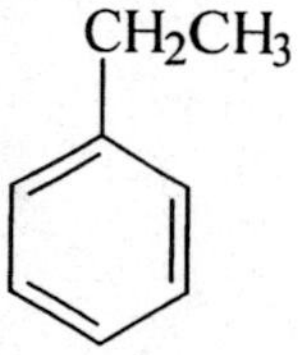

17.35

a) 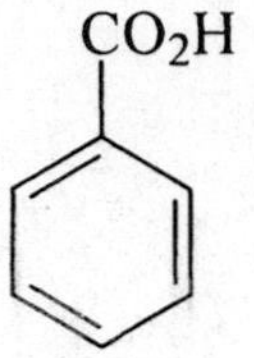

b) 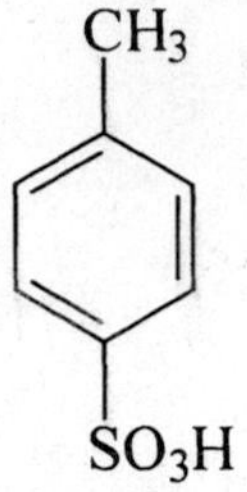 + ortho

c) 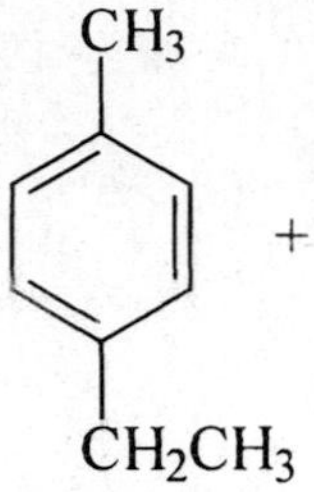 + ortho

d) 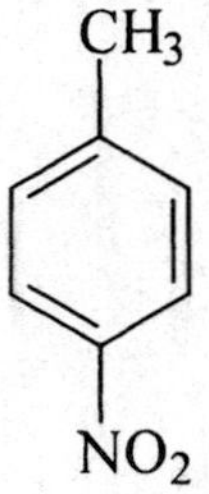 + ortho

e) 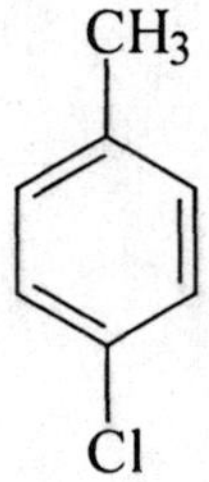 + ortho

f) 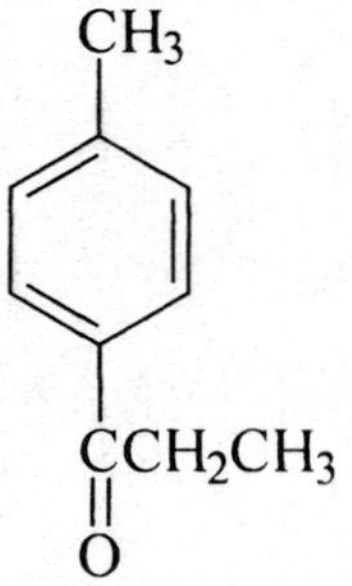

17.36

a) 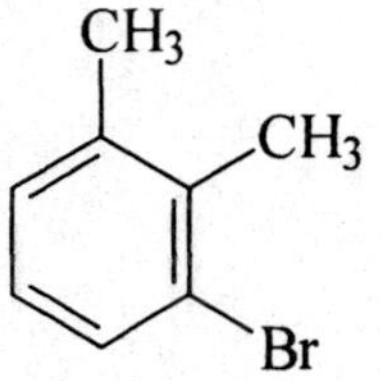 + 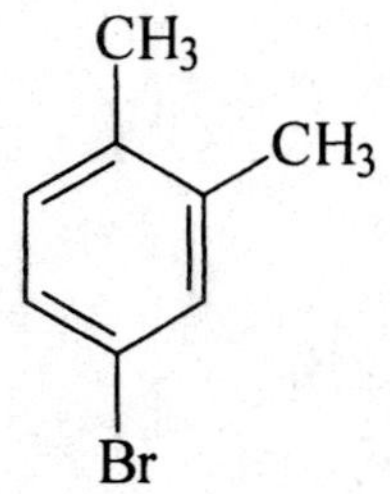

b) 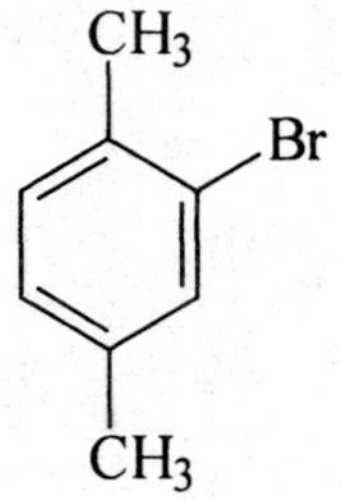

c) 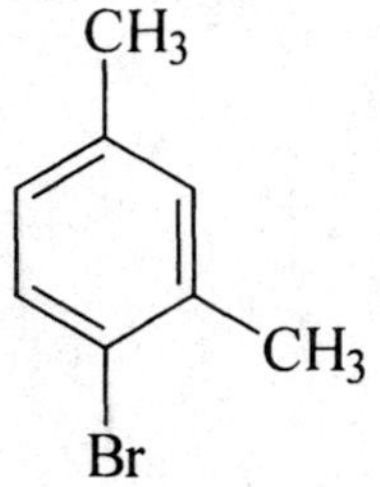

d) 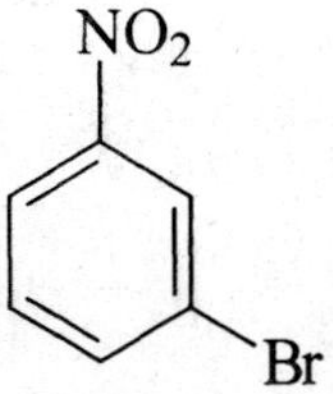

e) 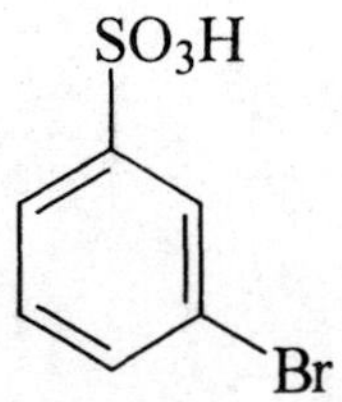

f) 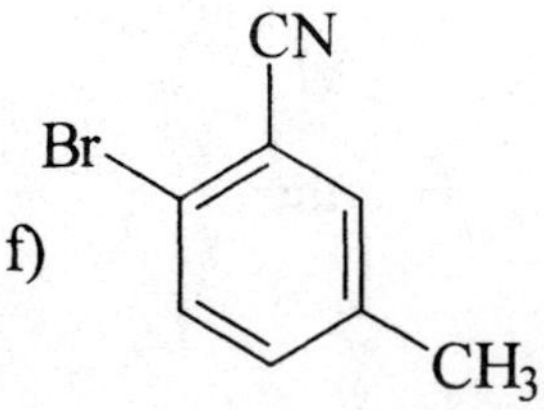 + 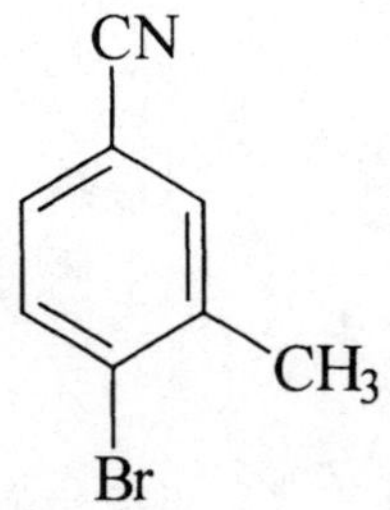

17.37

a) 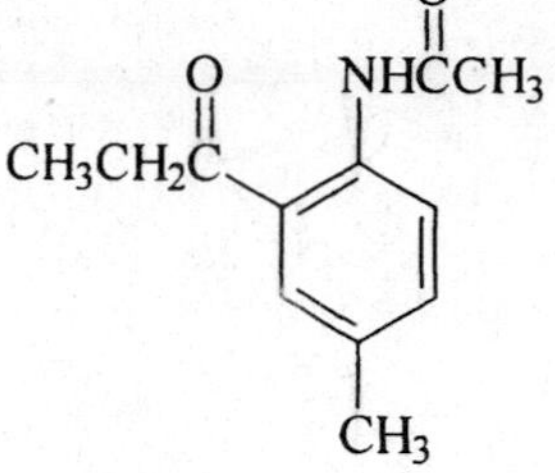

b) 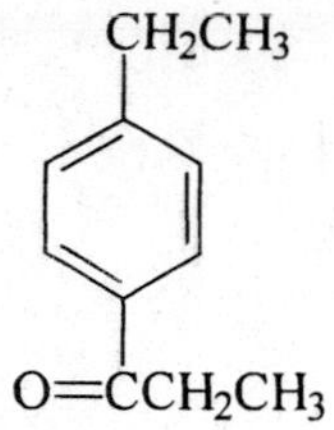

c) 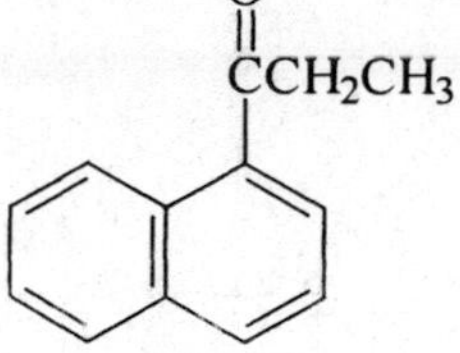

d) 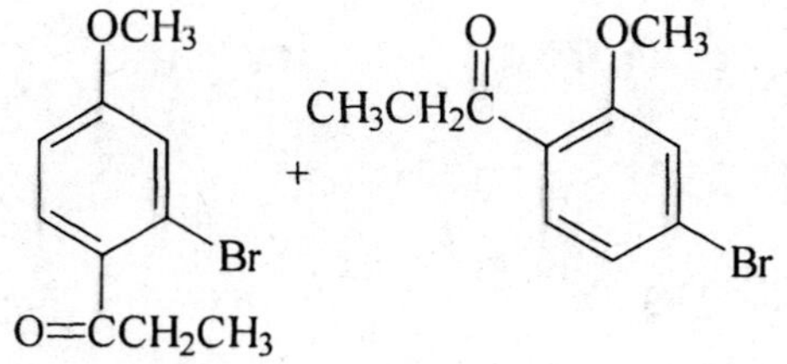

e) 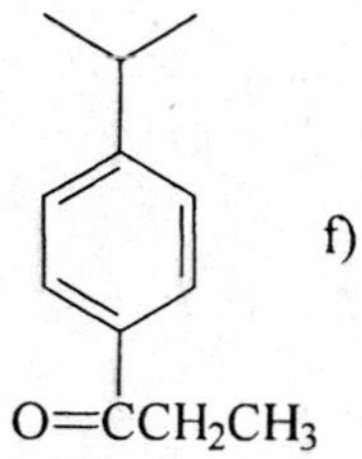

f) 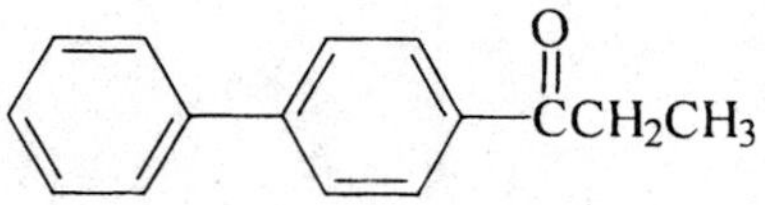

17.38

a) H_2SO_4 b) 1) HNO_3, H_2SO_4 2) Fe, HCl

c) 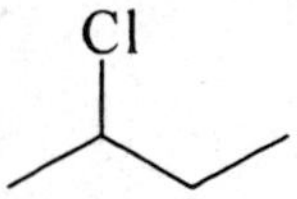

, $AlCl_3$ d) 1) 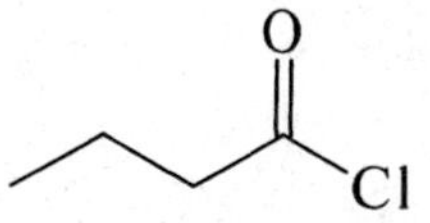

, $AlCl_3$ 2) Zn(Hg), HCl

e) 1) CH_3Cl, $AlCl_3$ 2) $KMnO_4$, NaOH, Δ 3) H_3O^+

f) 1) H_2, Pt 2) $NaNO_2$, H_3O^+ 3) CuCN g) Cl_2, $AlCl_3$ h) H_2SO_4, H_2O

17.39

a) 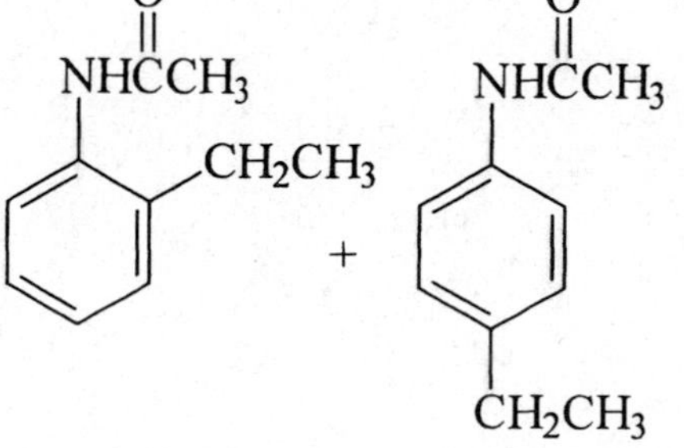

b) 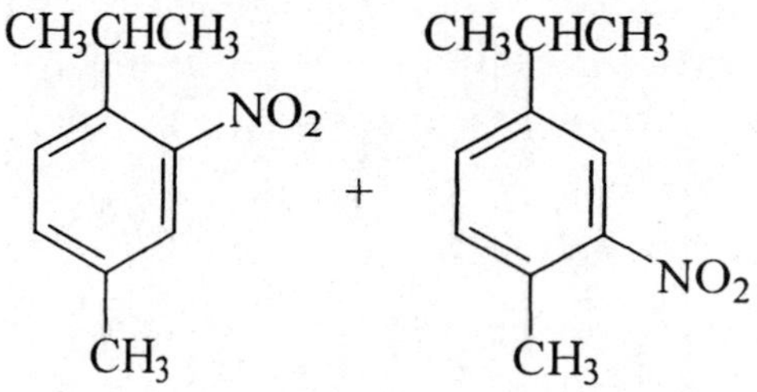

c) 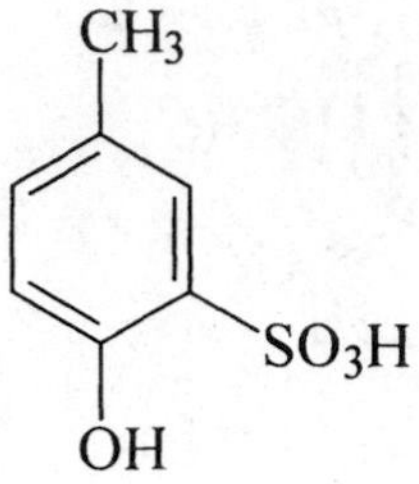
CH3
SO3H
OH
d) 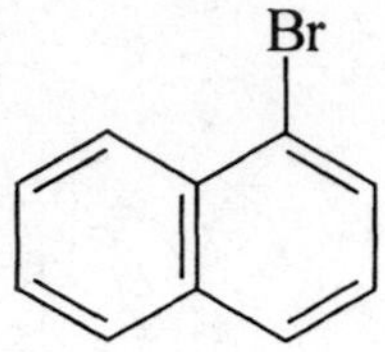
Br
e) 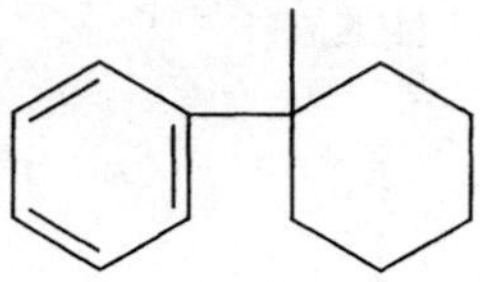

f) 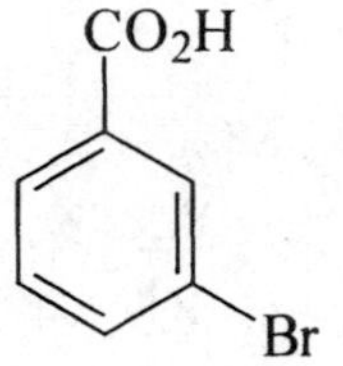
CO2H
Br
g) 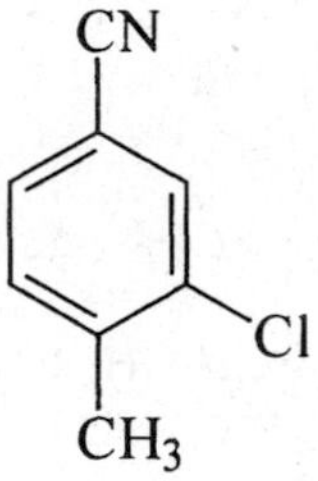
CN
Cl
CH3
h) 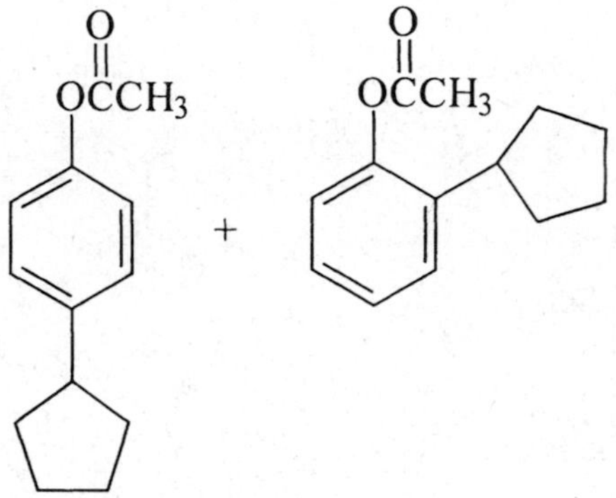
O
OCCH3
+
O
OCCH3

i) 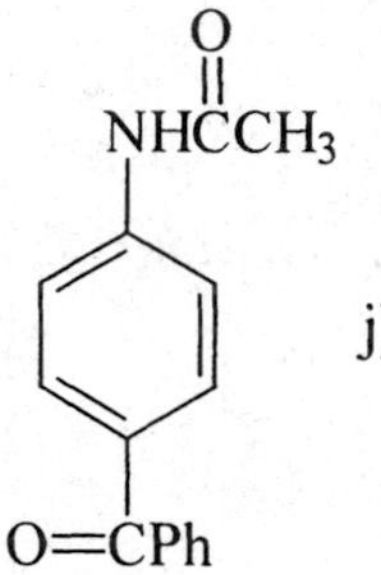
O
NHCCH3
O=CPh
j) 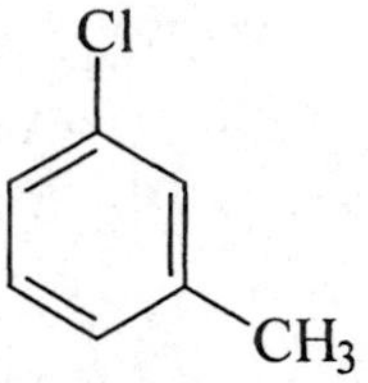
Cl
CH3
k) 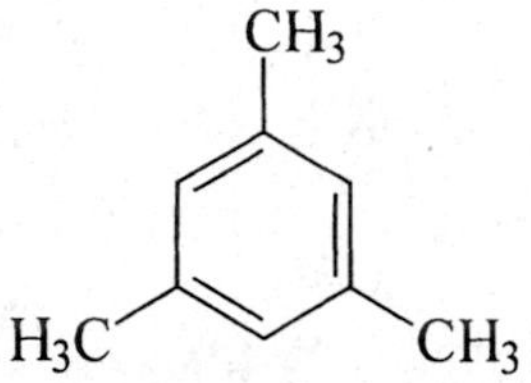
CH3
H3C
CH3
l) 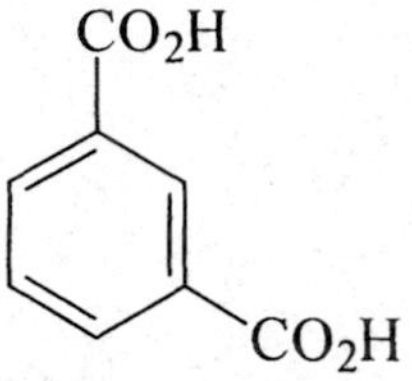
CO2H
CO2H

17.40

a) 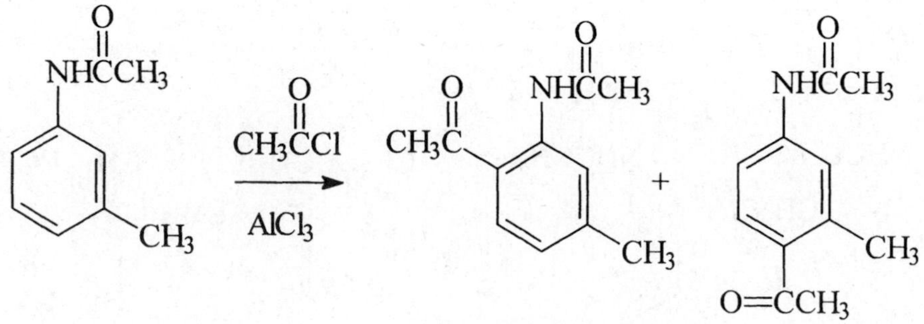
O
NHCCH3
CH3
O
CH3CCl
AlCl3
O
CH3C
O
NHCCH3
CH3
+
O
NHCCH3
CH3
O=CCH3

b) 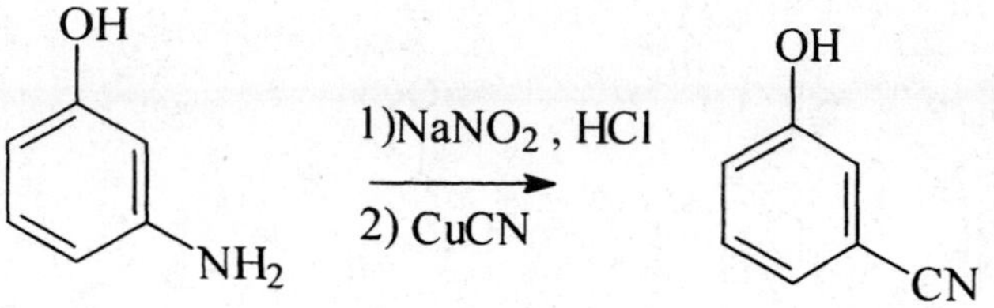

c) 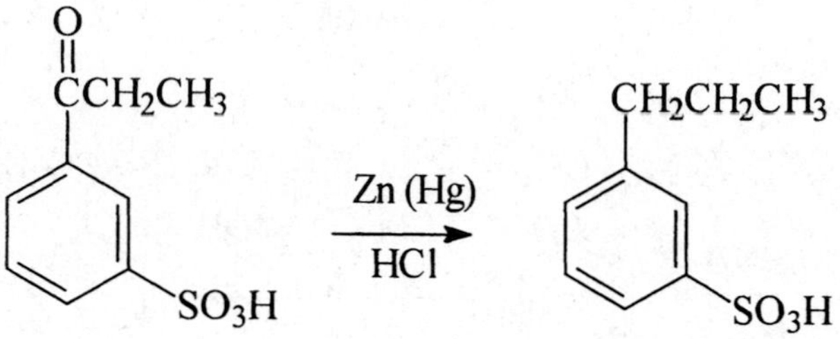

17.41

a) 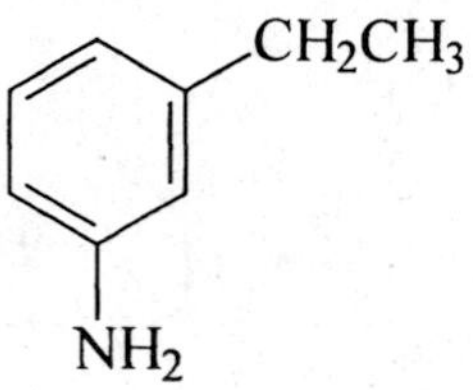

b) 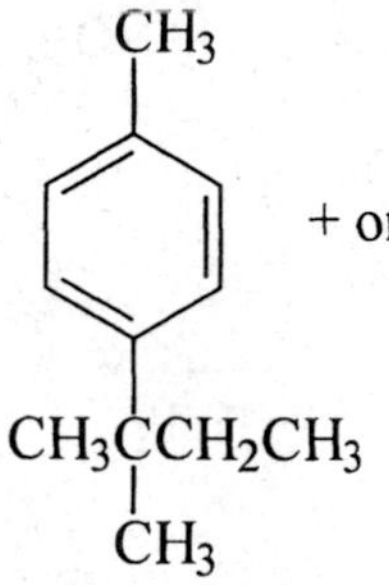 + ortho

c) 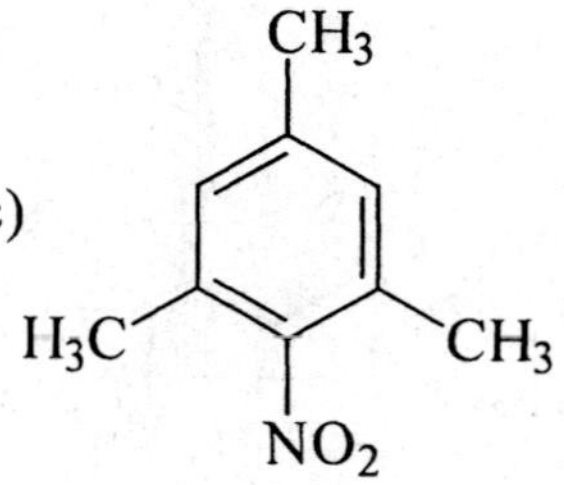

d) 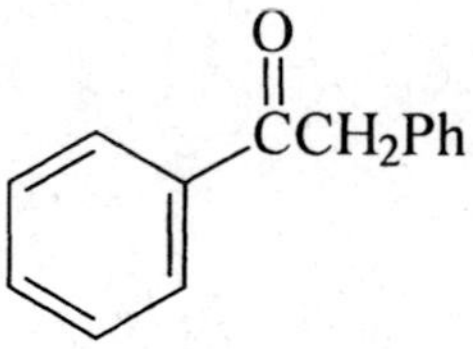

e) 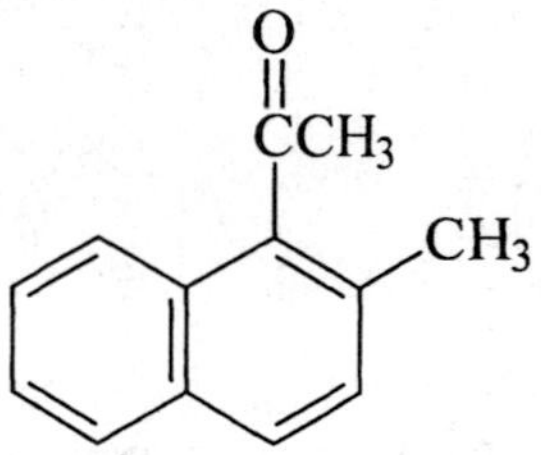

f) 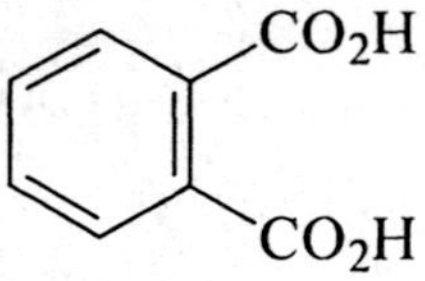

17.42

a) 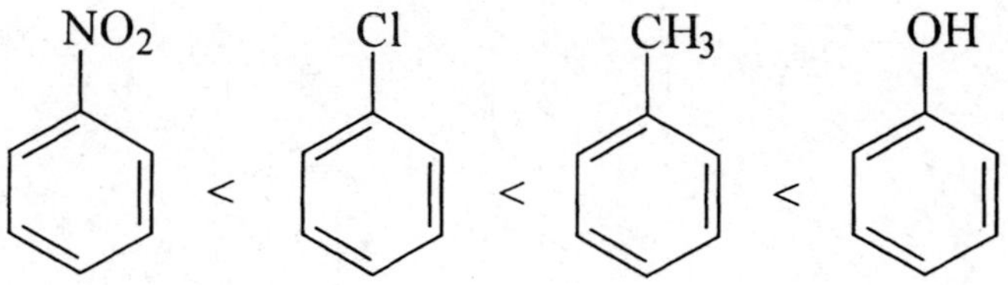

Slowest   Fastest

b) 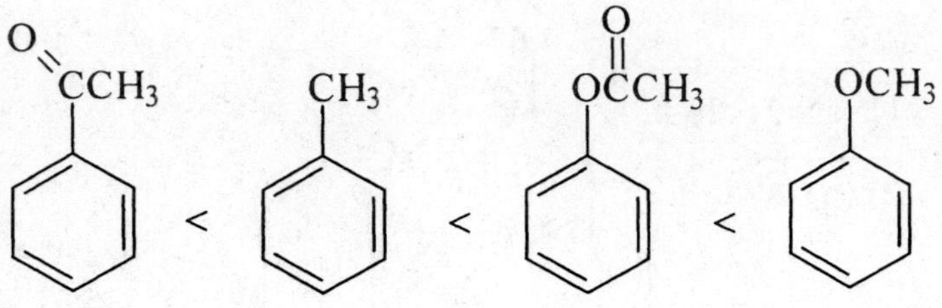

Slowest Fastest

17.43

a) 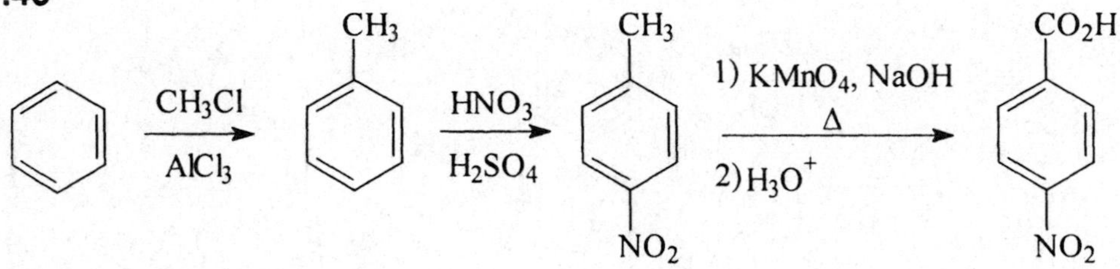

b) 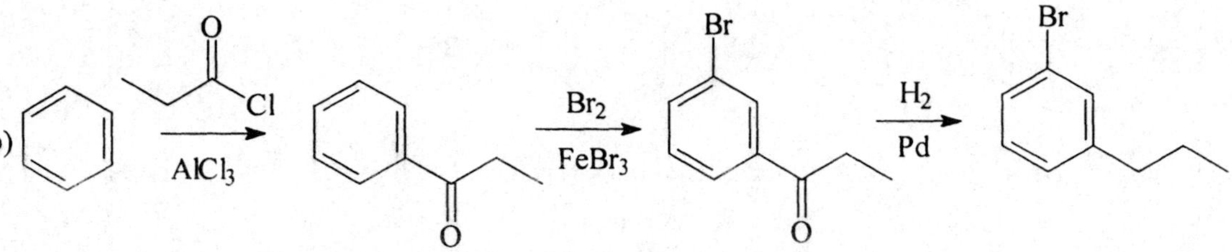

c) 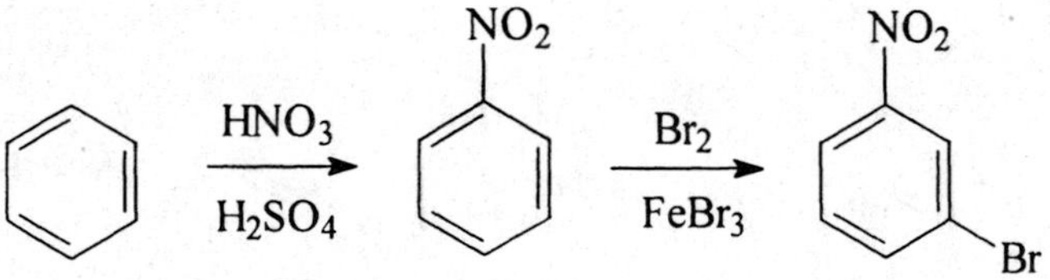

d) 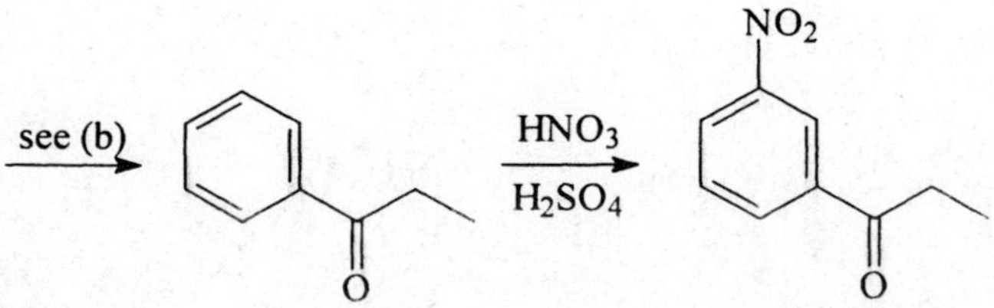

e) 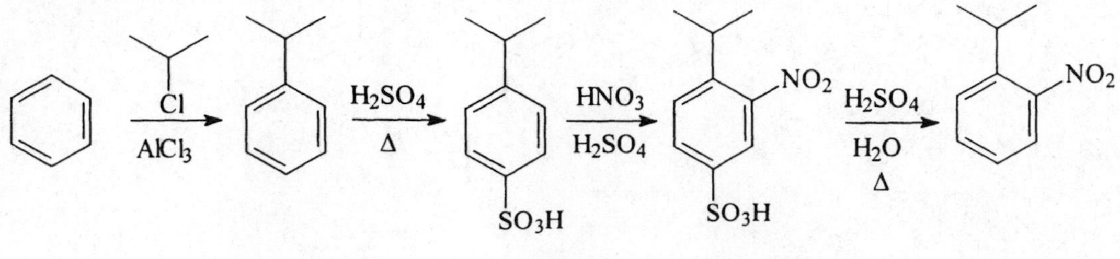

f) 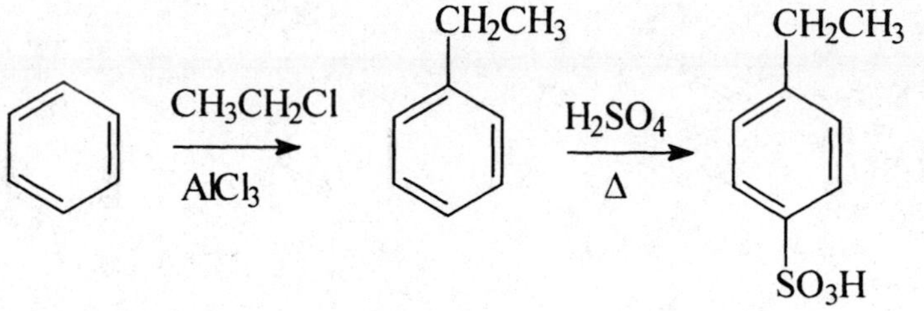

17.44

a) 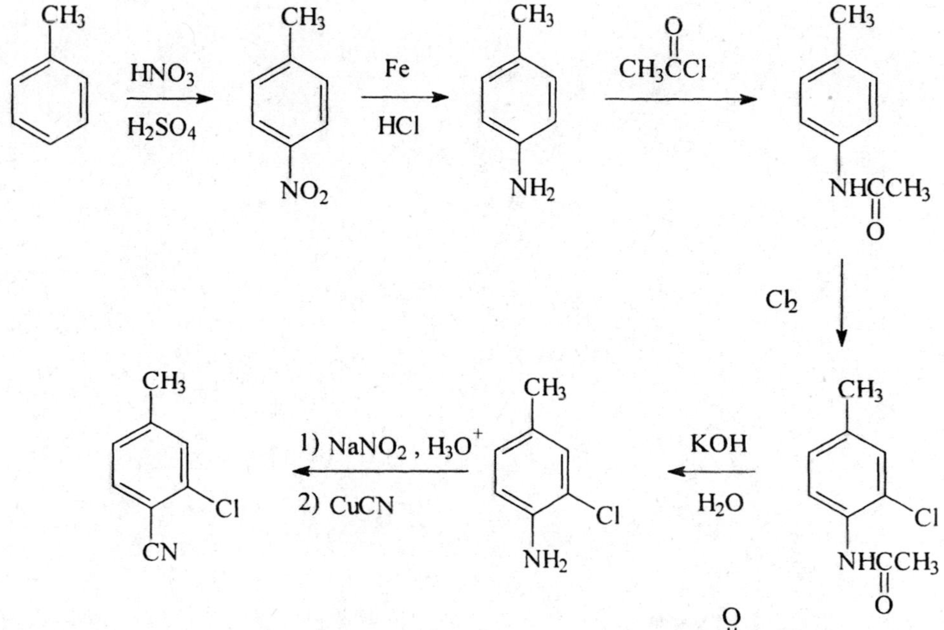

b) 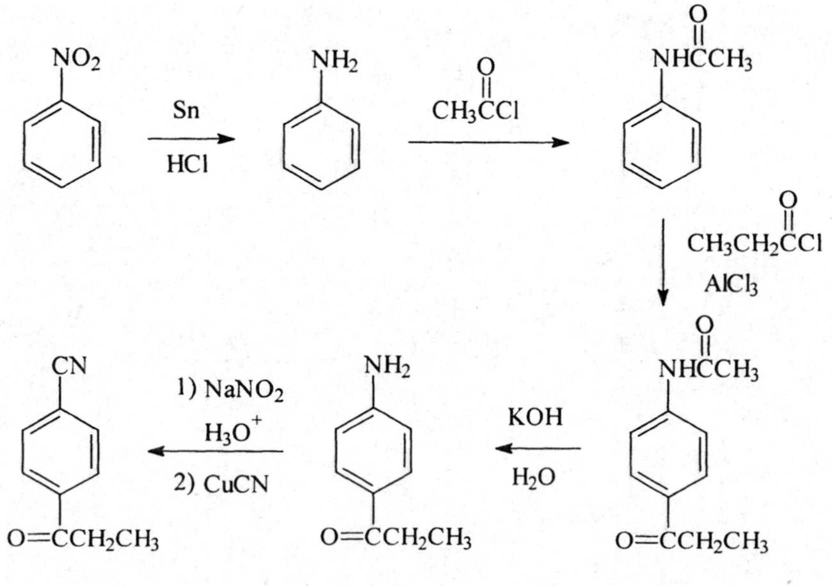

c)

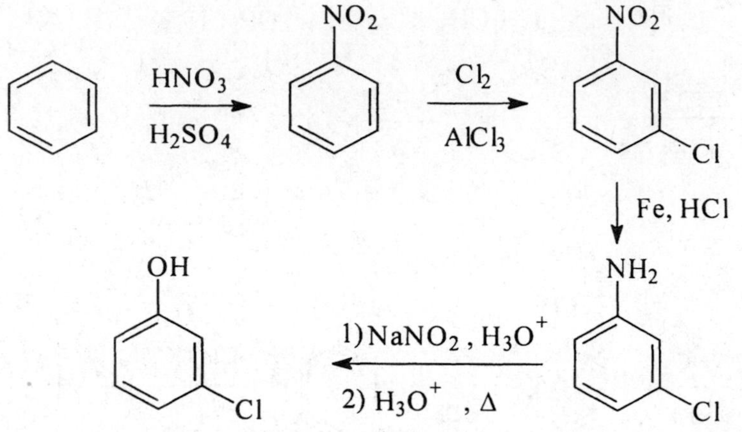

HNO3
H2SO4
NO2
Cl2
AlCl3
NO2
Cl
Fe, HCl
NH2
Cl
1) NaNO2 , H3O+
2) H3O+ , Δ
OH
Cl

d)

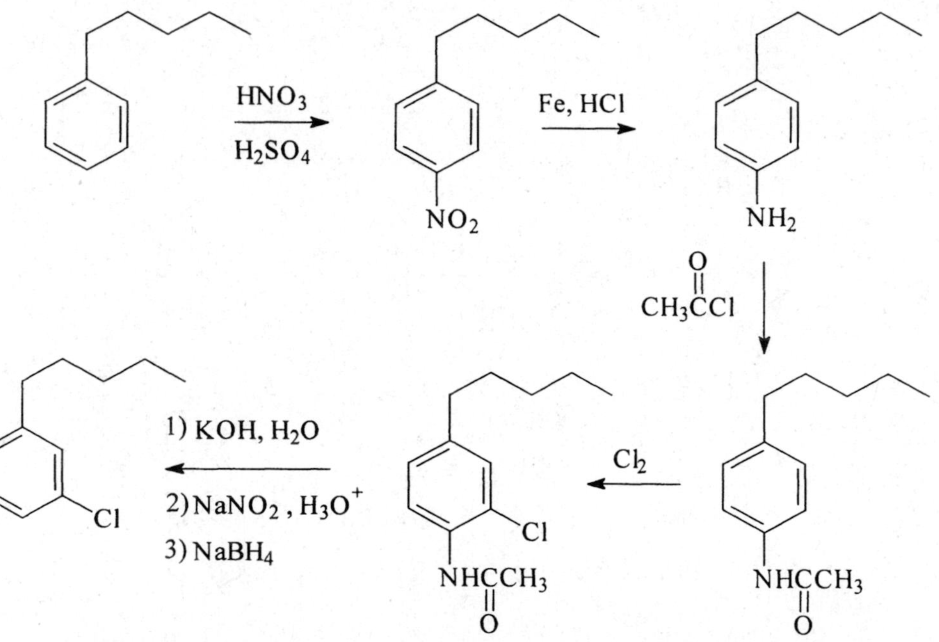

HNO3
H2SO4
NO2
Fe, HCl
NH2
O
CH3CCl
NHCCH3
O
Cl2
Cl
NHCCH3
O
1) KOH, H2O
2) NaNO2 , H3O+
3) NaBH4
Cl

e)

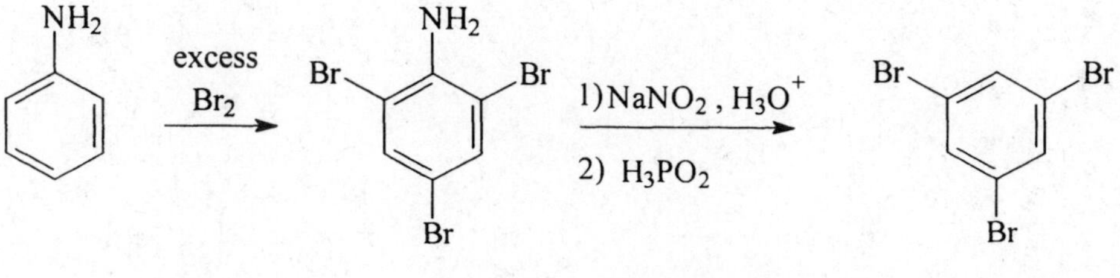

NH2
excess
Br2
NH2
Br
Br
Br
1) NaNO2 , H3O+
2) H3PO2
Br
Br
Br

f)

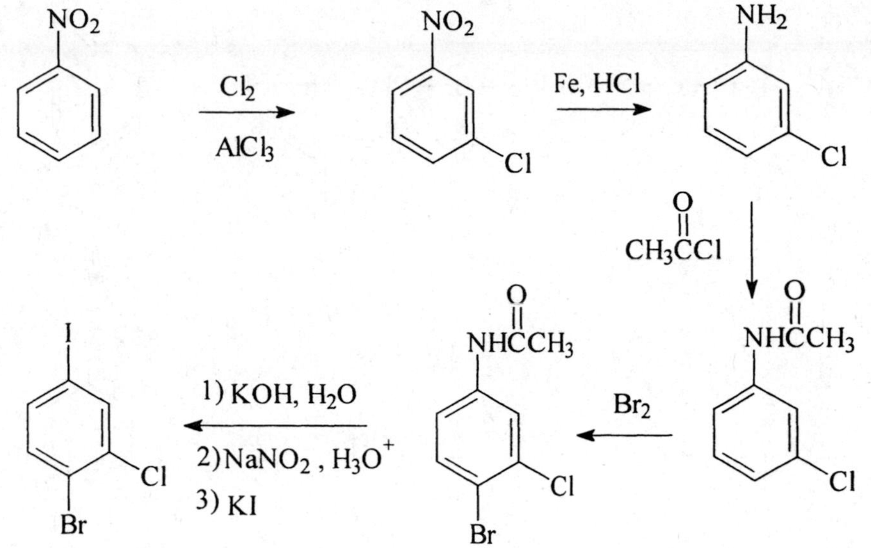

NO2
Cl2
AlCl3
NO2
Cl
Fe, HCl
NH2
Cl
O
CH3CCl
O
NHCCH3
Cl
Br2
O
NHCCH3
Cl
Br
1) KOH, H2O
2) NaNO2 , H3O+
3) KI
I
Cl
Br

g)

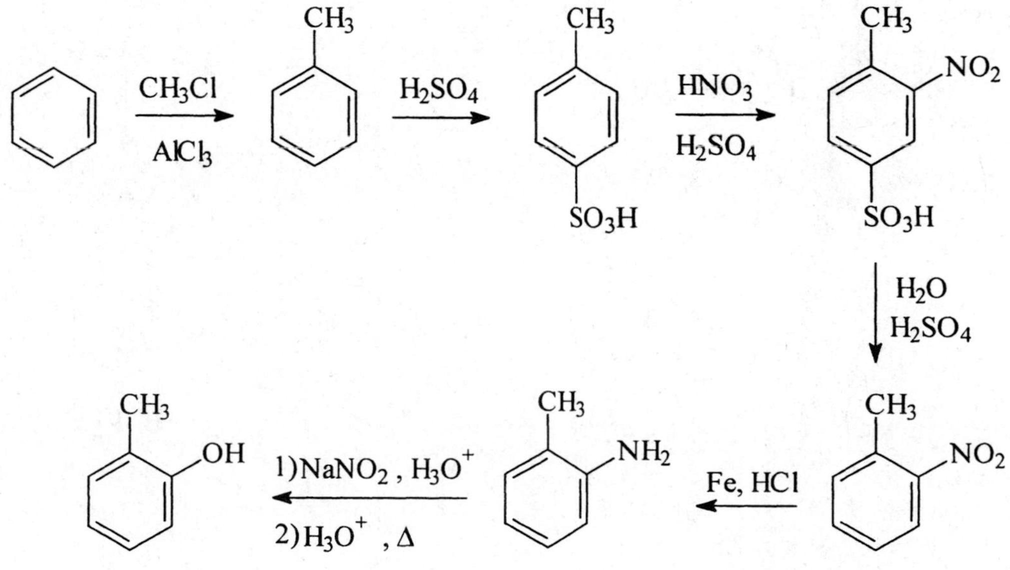

CH3Cl
AlCl3
CH3
H2SO4
CH3
SO3H
HNO3
H2SO4
CH3
NO2
SO3H
H2O
H2SO4
CH3
NO2
Fe, HCl
CH3
NH2
1) NaNO2 , H3O+
2) H3O+ , Δ
CH3
OH

17.45

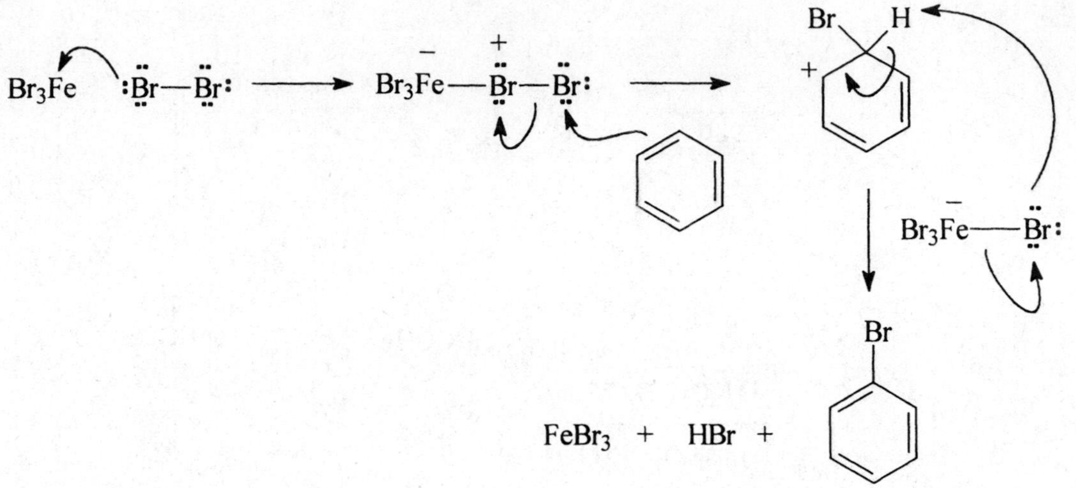

17.46

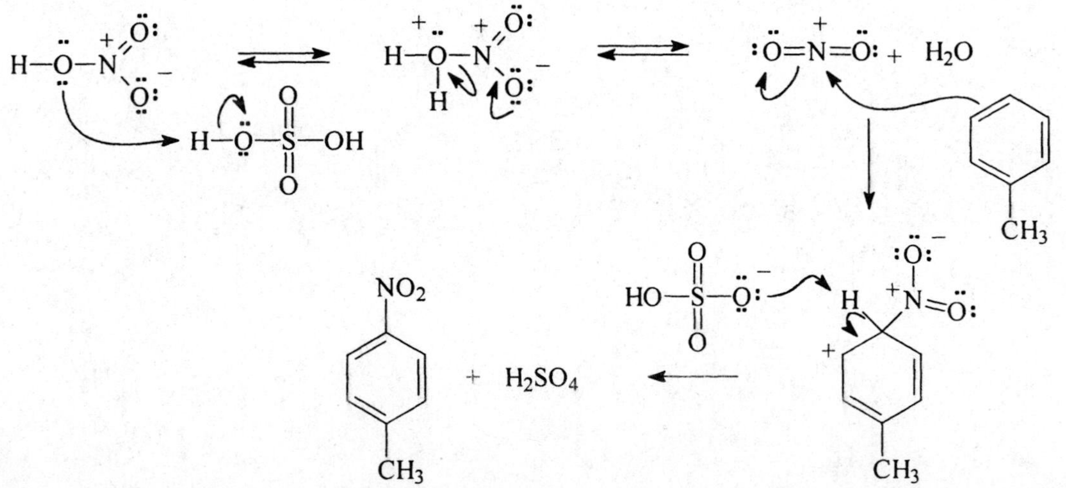

17.47

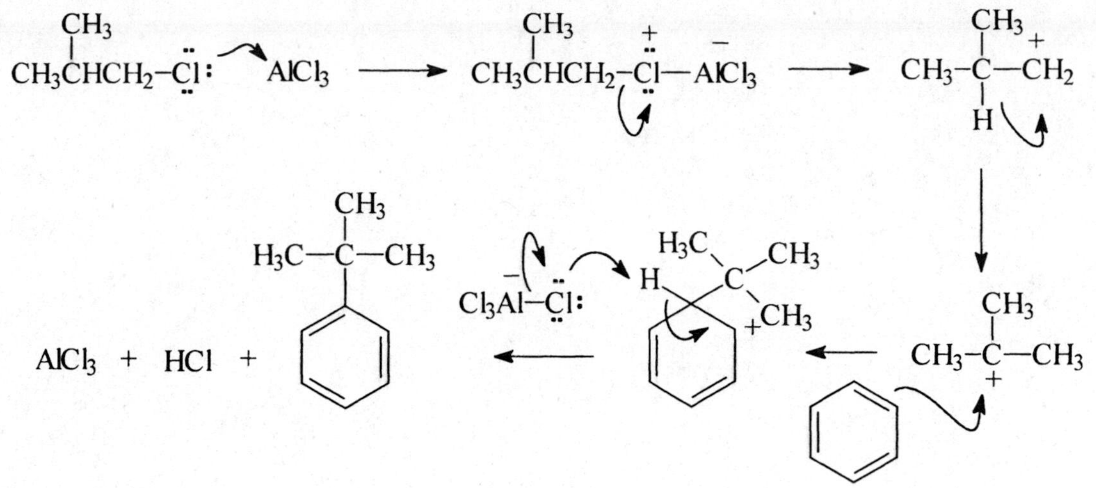

17.48

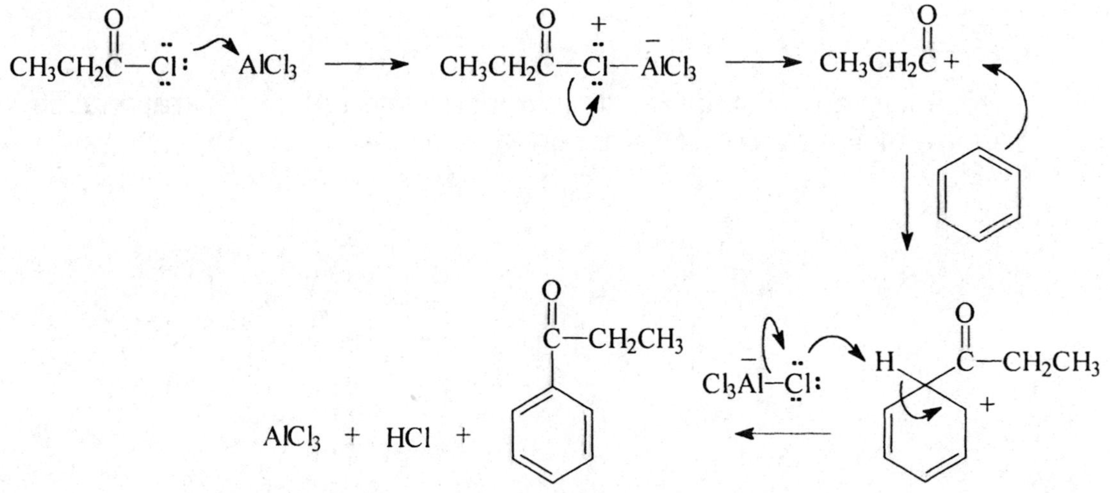

17.49

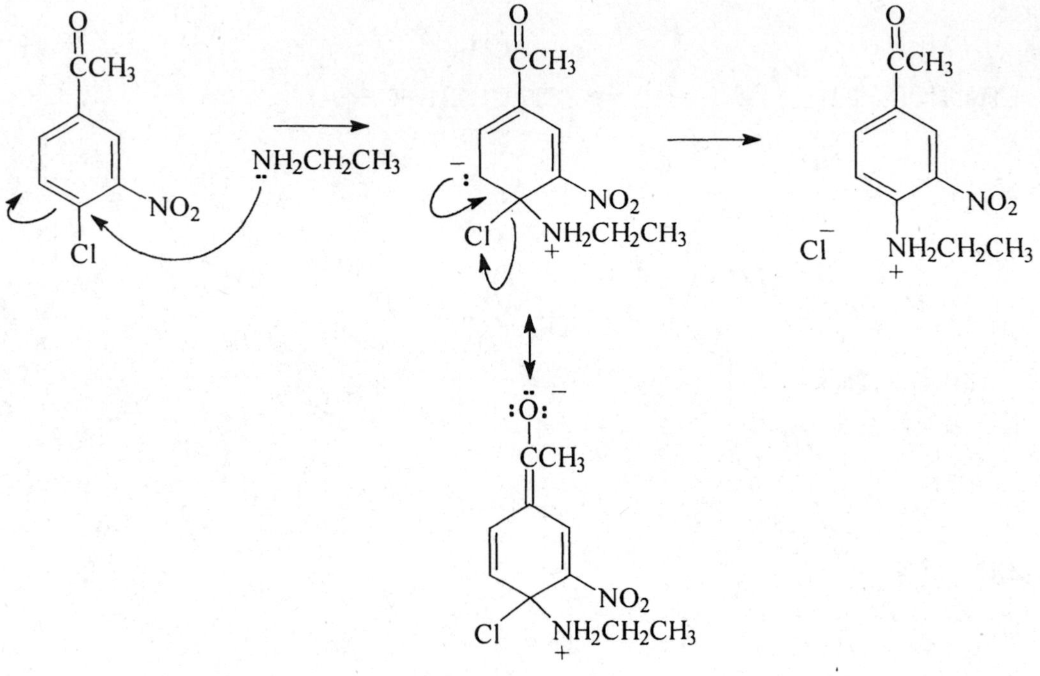

The acetyl group stabilizes the carbanion intermediate by resonance, as shown by the resonance structure above.

17.50

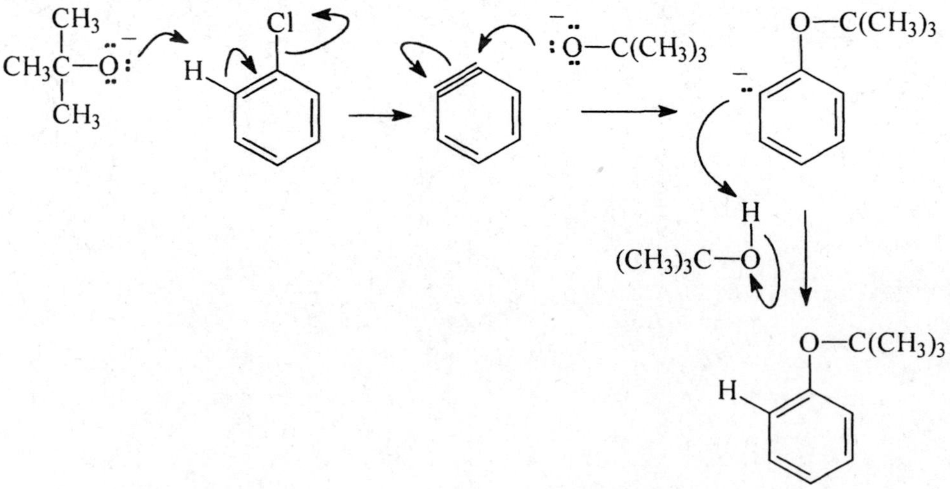

17.51

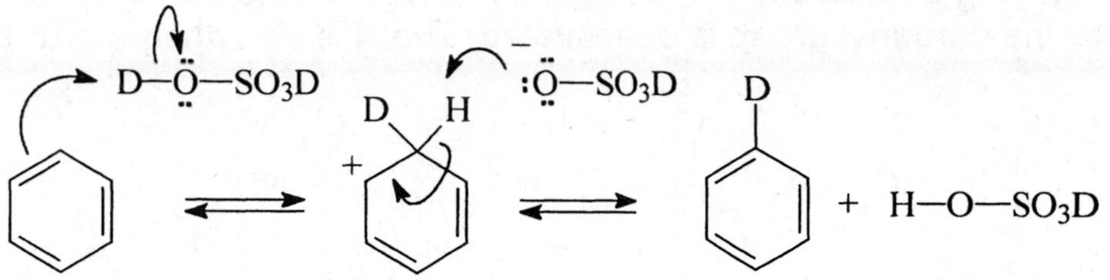

17.52 The amount of ortho product decreases as the size (steric bulk) of the alkyl group increases, indicating that the bromination reaction is somewhat subject to steric effects.

17.53 Both of these groups have a positively charged atom and thus should be inductive electron withdrawing groups. The S has an unshared pair of electrons, but, because of its positive charge, it does not want to donate them by resonance. Therefore, both groups should be deactivating and direct substitution to the meta position.

17.54

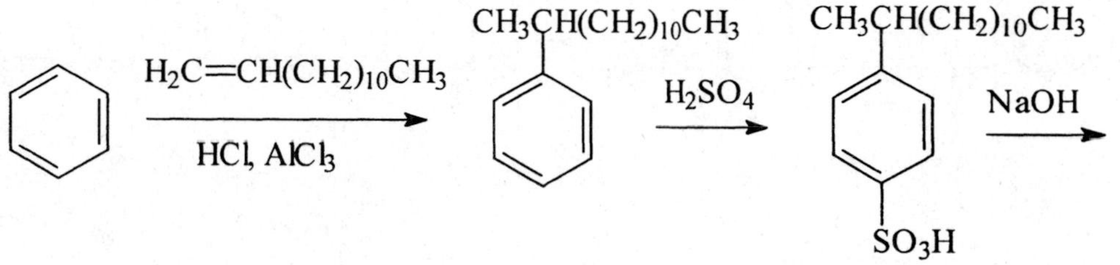

17.55 The nitroso group withdraws electrons by its inductive effect and donates electrons by its resonance effect. As was the case with the halogens, the inductive effect is controlling the rate of reaction and deactivating the compound, but the ortho and para positions are less deactivated due to donation of electrons back to those positions by resonance.

17.56 The ring attached to the nitrogen is activated, whereas the ring attached to the carbonyl group is deactivated. The N is an ortho-para director.

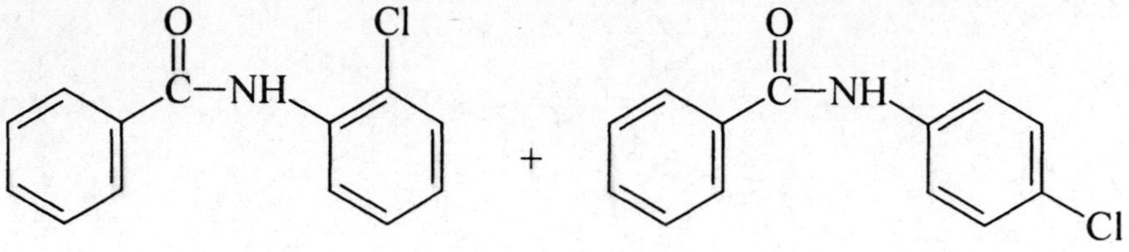

17.57

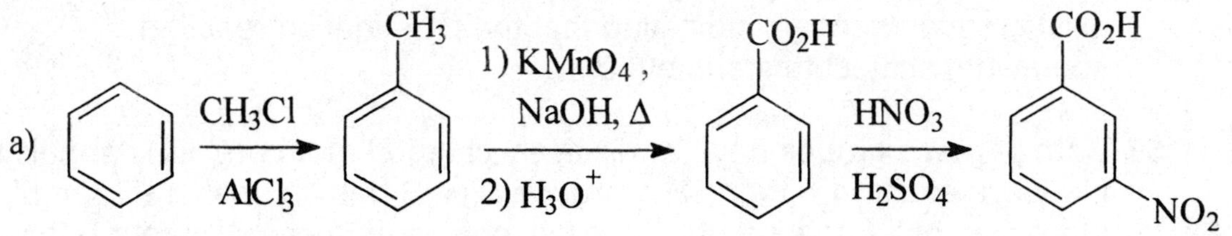

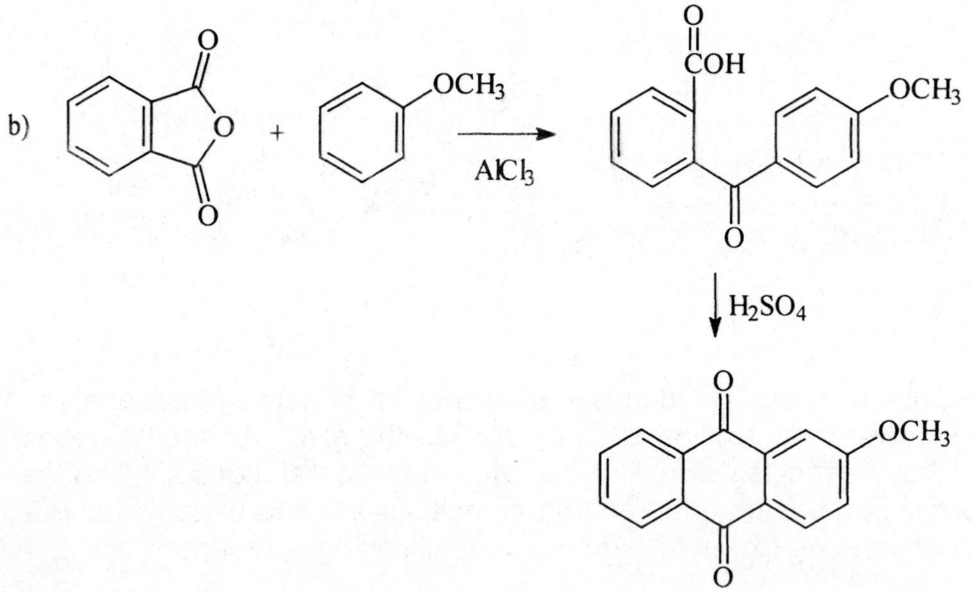

c)

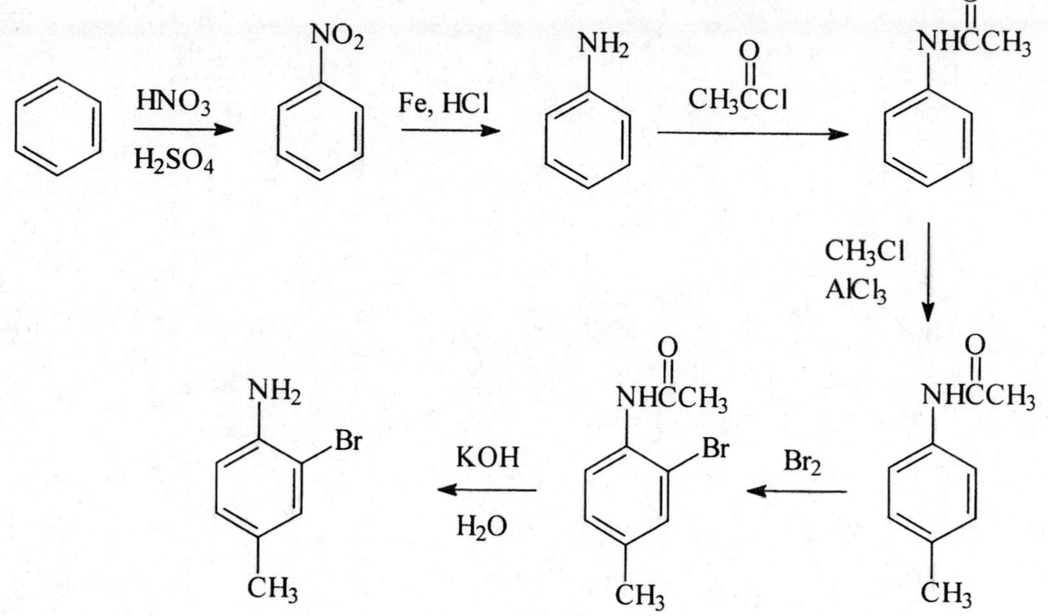
HNO_3
H_2SO_4
NO_2
Fe, HCl
NH_2
O
CH_3CCl
O
$NHCCH_3$
CH_3Cl
$AlCl_3$
O
$NHCCH_3$
CH_3
Br_2
O
$NHCCH_3$
Br
CH_3
KOH
H_2O
NH_2
Br
CH_3

d)

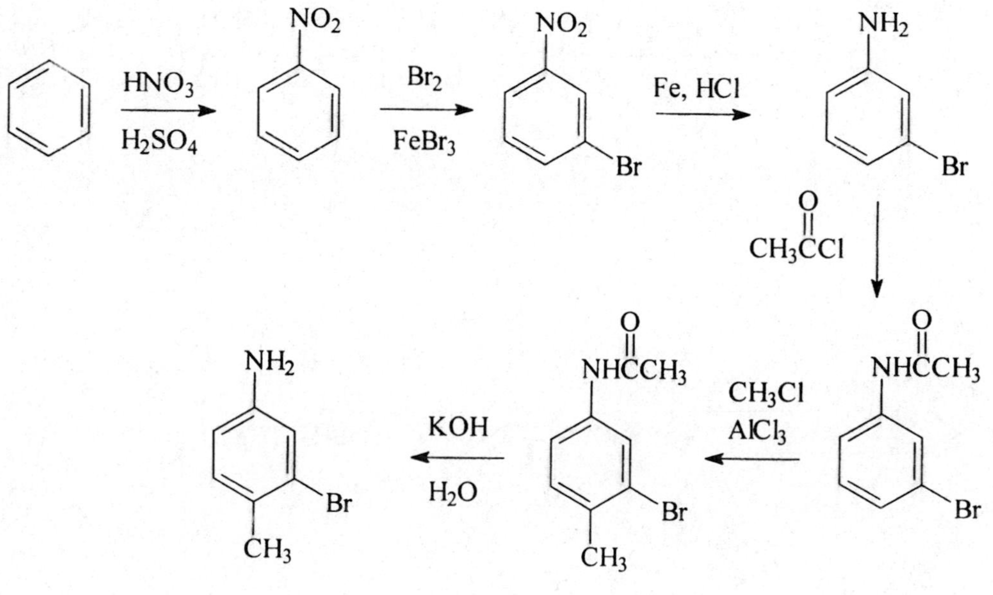
HNO_3
H_2SO_4
NO_2
Br_2
$FeBr_3$
NO_2
Br
Fe, HCl
NH_2
Br
O
CH_3CCl
O
$NHCCH_3$
Br
CH_3Cl
$AlCl_3$
O
$NHCCH_3$
Br
CH_3
KOH
H_2O
NH_2
Br
CH_3

e) 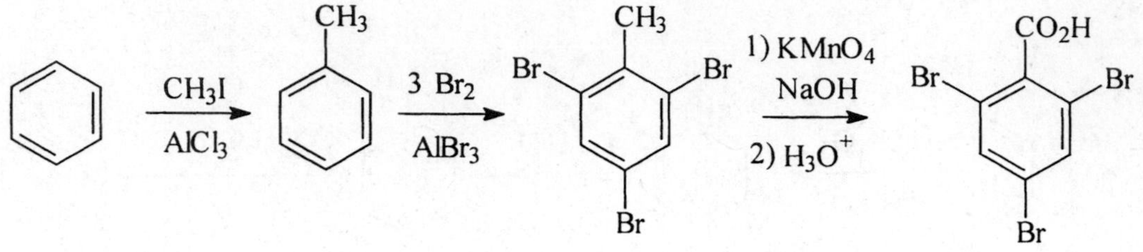
CH3
CH3I
AlCl3
3 Br2
AlBr3
CH3
Br
Br
Br
1) KMnO4
NaOH
2) H3O+
CO2H
Br
Br
Br

f) 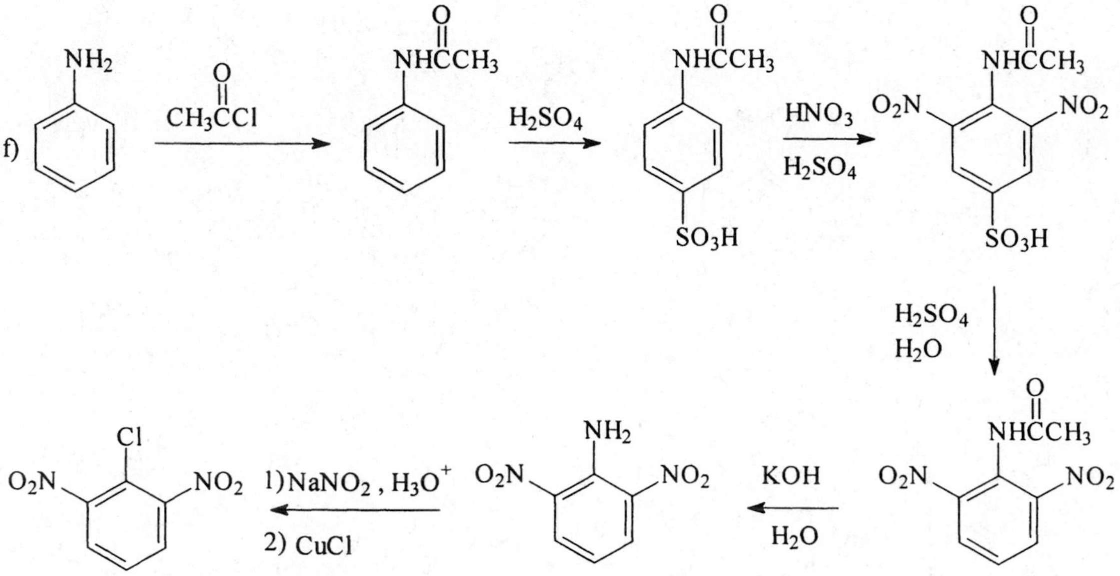
NH2
O
CH3CCl
O
NHCCH3
H2SO4
O
NHCCH3
SO3H
HNO3
H2SO4
O
NHCCH3
O2N
NO2
SO3H
H2SO4
H2O
O
NHCCH3
O2N
NO2
KOH
H2O
NH2
O2N
NO2
1)NaNO2 , H3O+
2) CuCl
Cl
O2N
NO2

17.58

a) 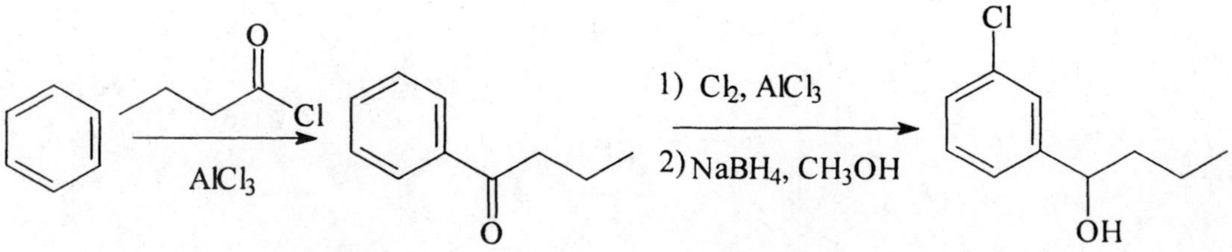
O
Cl
AlCl3
O
1) Cl2, AlCl3
2) NaBH4, CH3OH
Cl
OH

b)

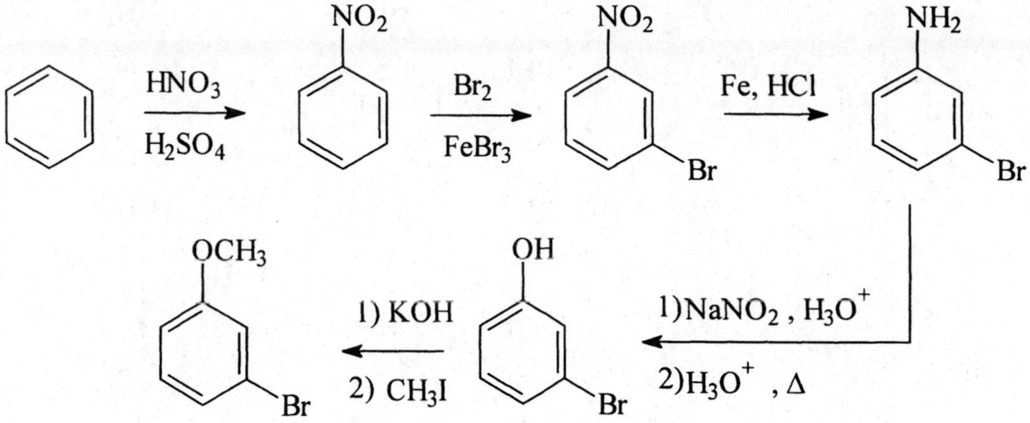
NO2
HNO3
H2SO4
NO2
Br2
FeBr3
Br
NH2
Fe, HCl
Br
OCH3
1) KOH
2) CH3I
Br
OH
Br
1)NaNO2 , H3O+
2)H3O+ , Δ

c)

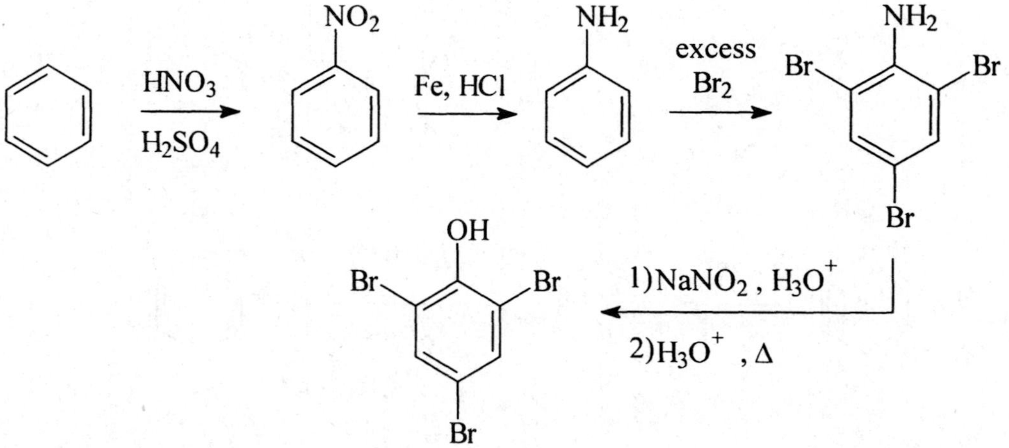
HNO3
H2SO4
NO2
Fe, HCl
NH2
excess
Br2
NH2
Br
Br
Br
OH
Br
Br
Br
1)NaNO2 , H3O+
2)H3O+ , Δ

d)

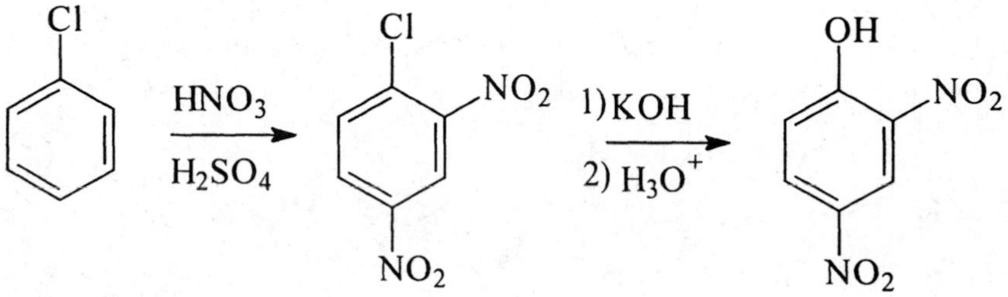
Cl
HNO3
H2SO4
Cl
NO2
NO2
1)KOH
2) H3O+
OH
NO2
NO2

e)

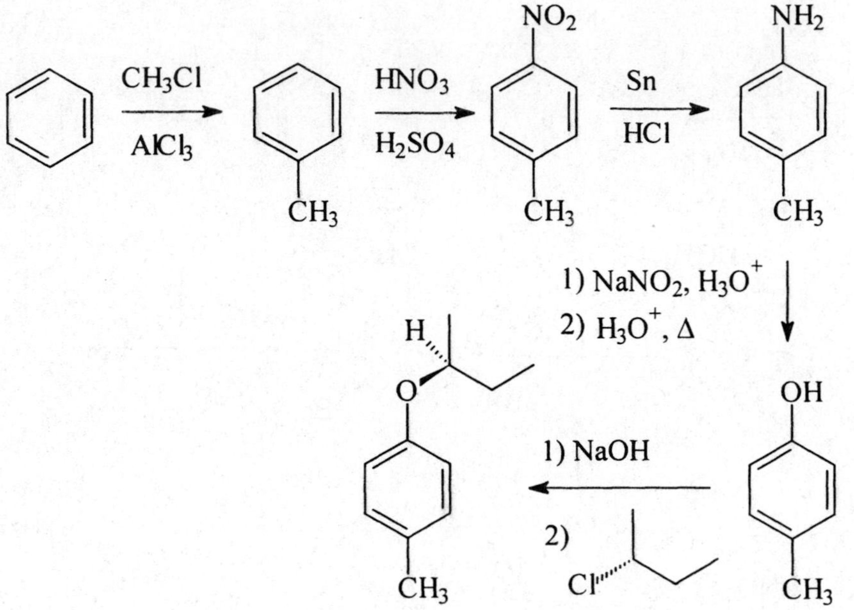
CH3Cl
AlCl3
HNO3
H2SO4
Sn
HCl
NO2
NH2
CH3
1) NaNO2, H3O+
2) H3O+, Δ
OH
H
O
1) NaOH
2)
Cl

f)

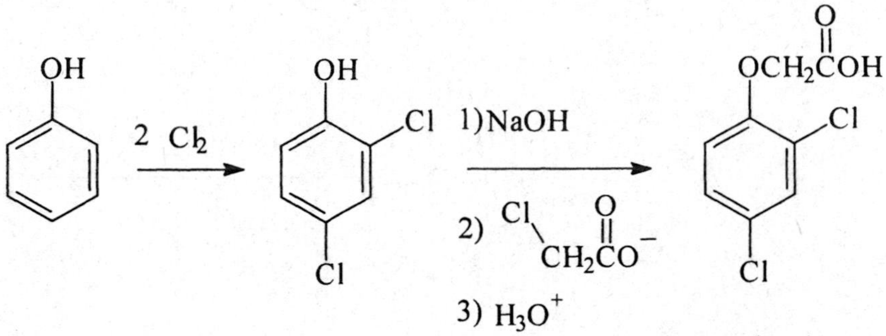
OH
2 Cl2
Cl
1)NaOH
2)
CH2CO−
3) H3O+
OCH2COH
O

17.59

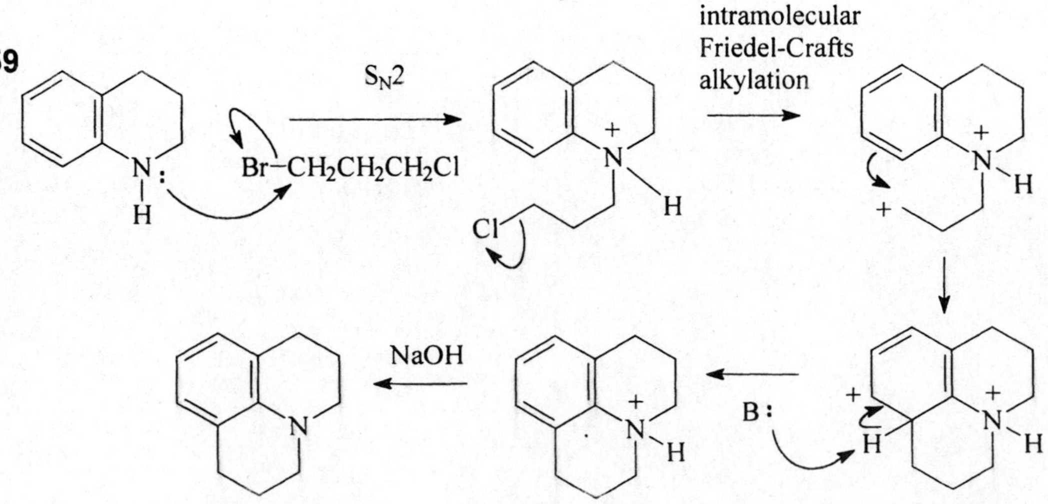
SN2
intramolecular
Friedel-Crafts
alkylation
N
H
Br−CH2CH2CH2Cl
Cl
NaOH
B :

17.60

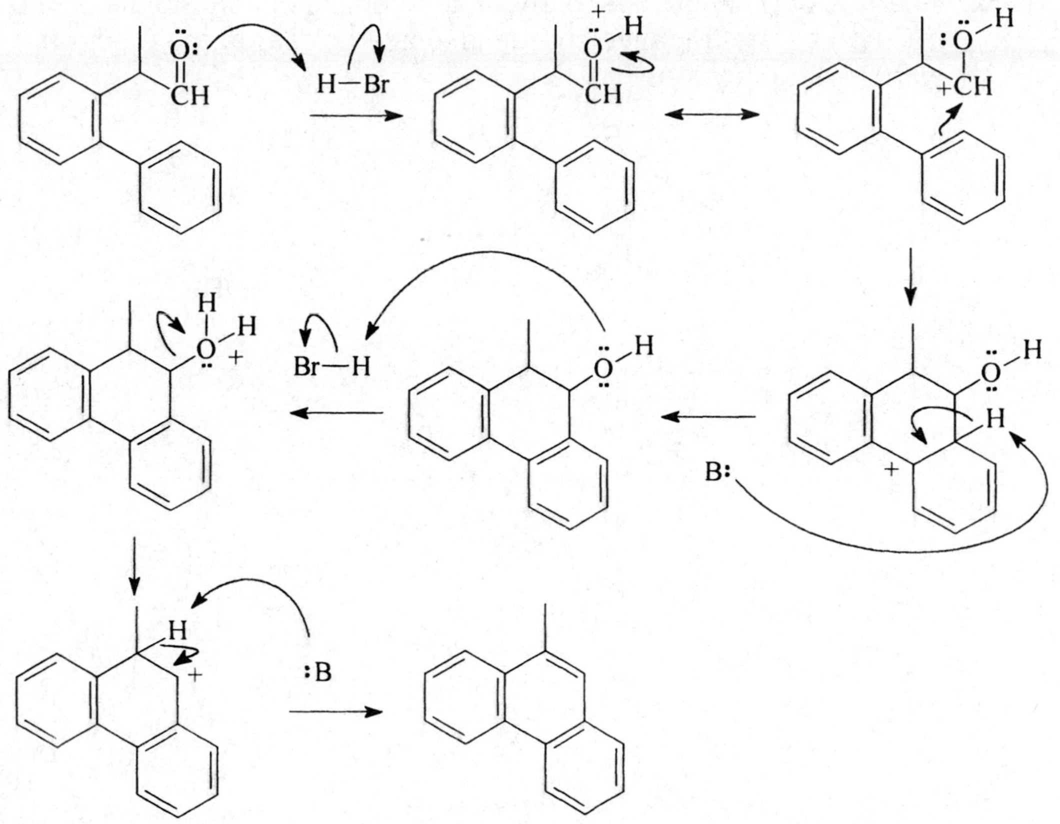

17.61 Nitrogen A is more nucleophilic than nitrogen B because the electron pair on nitrogen B is delocalized by resonance with the carbonyl group. The reaction is a nucleophilic aromatic substitution following the addition-elimination mechanism as described in Section 17.11. The N is the nucleophile and F is the leaving group.

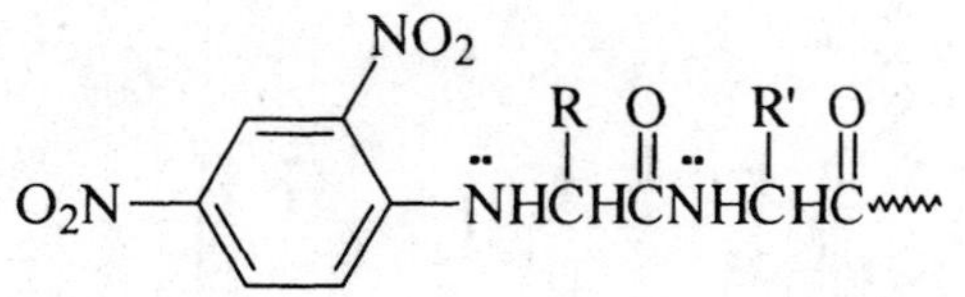

17.62 The nucleophile prefers to attack the benzyne intermediate at the meta position so the negative charge is located closer to the electron withdrawing oxygen.

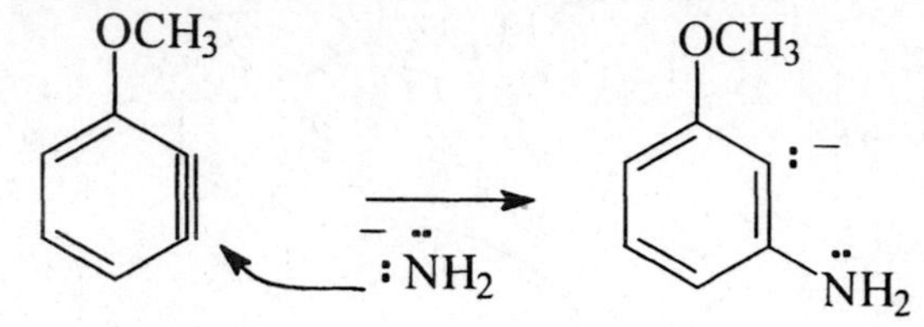

17.63

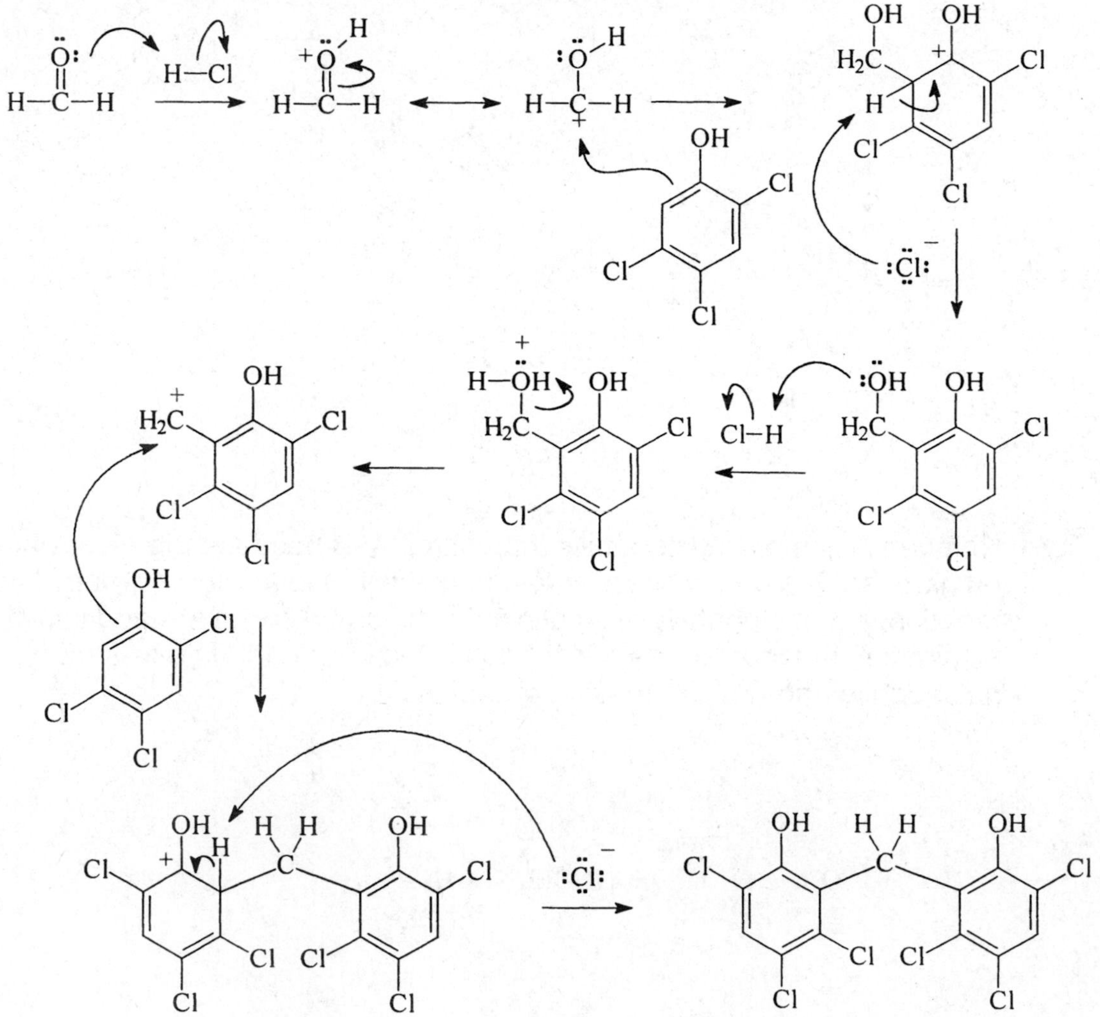

17.64

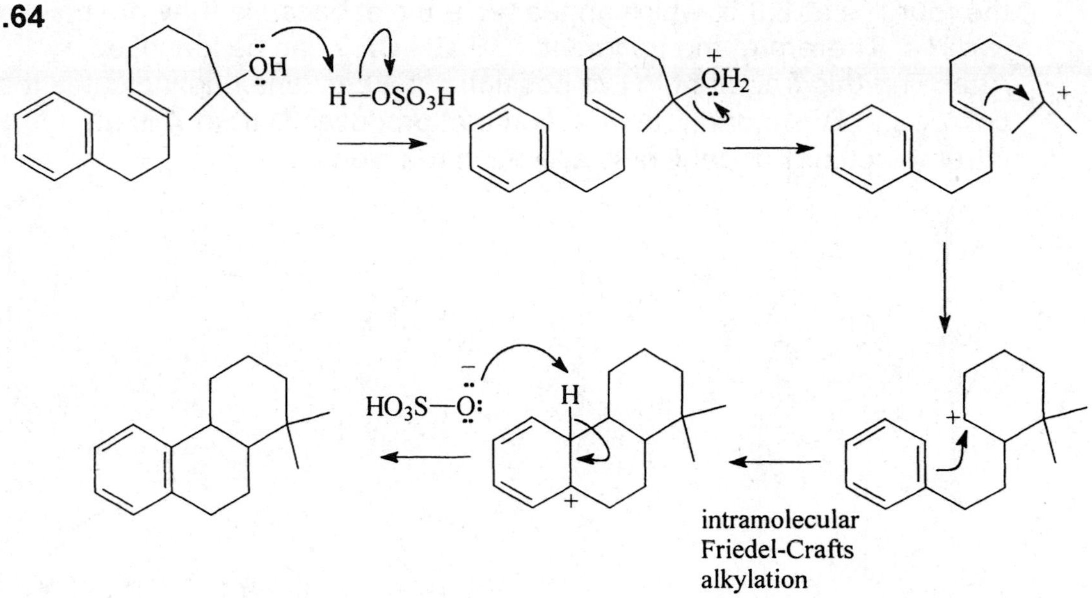

17.65 This is a Friedel-Crafts alkylation reaction followed by an S_N1 reaction.

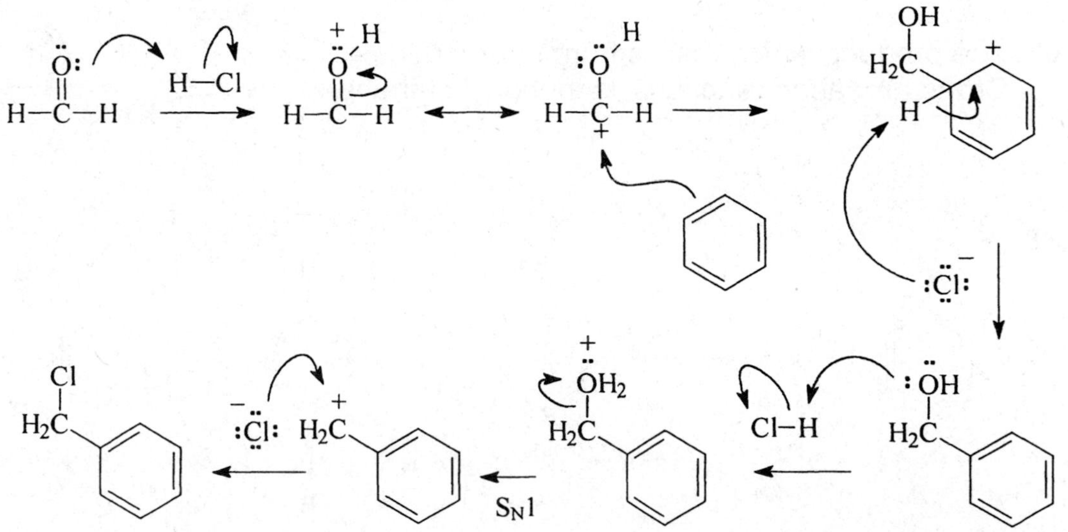

17.66 Compound **B** has a DU of 5. Because **B** has ten hydrogens, the ratio of the signals in the NMR spectrum must be 4:4:2. The signal for four H's near 7 δ indicates that **B** has a disubstituted aromatic ring. The quintuplet at 2 δ indicates that these two H's are coupled to four H's. These must be the four H's at 2.8 δ, which appear as a triplet because they are split by two H's. Therefore, the fragment $-CH_2CH_2CH_2-$ can be identified. Attaching this fragment to two positions of a benzene ring indicates that compound **B** is indan. The reaction that produces **B** from **A** is an intramolecular Friedel-Crafts alkylation reaction.

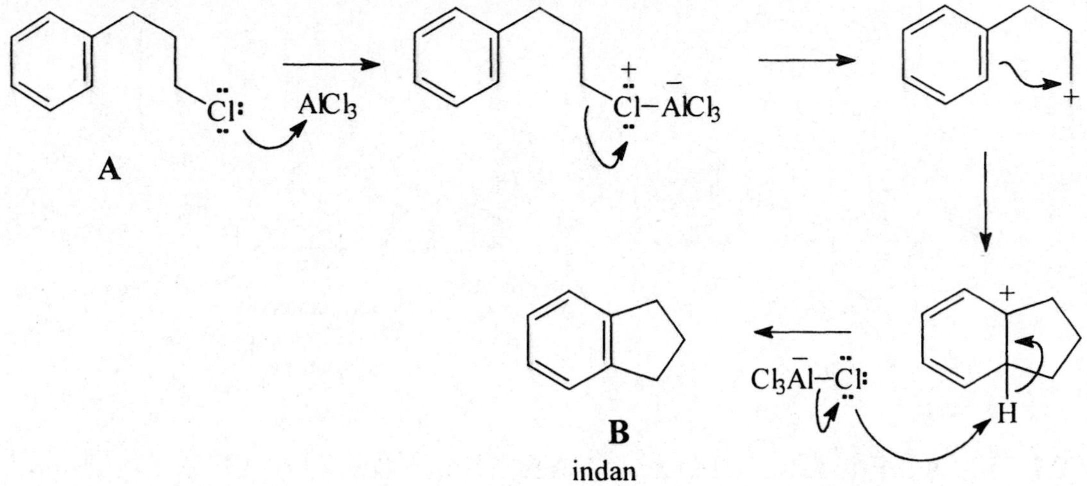

17.67 The product, which results from three Friedel-Crafts alkylation reactions, is triphenylmethane.

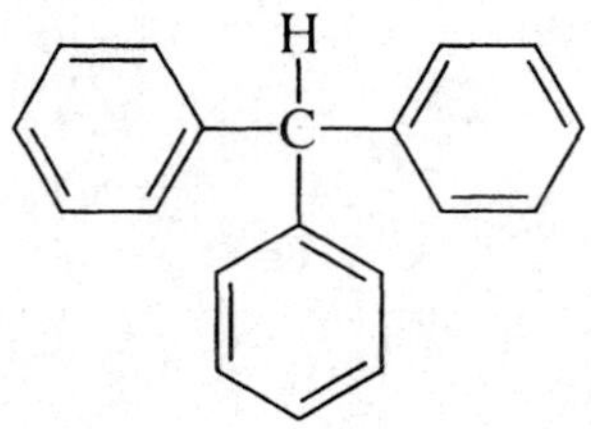

17.68

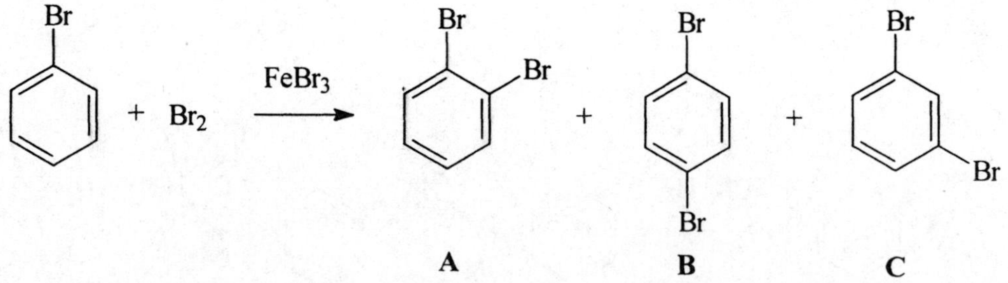

17.69 As the steric bulk of the alkyl group increases, less ortho product will be formed. The order is toluene > ethylbenzene > isopropylbenzene > *t*-butylbenzene.

17.70 Product 1 is the major product because the nitro group has been added ortho to the methyl group. Product 2, is formed in lesser amounts because the nitro group has been added ortho to the bulkier isopropyl group.

17.71 The nitrogen of *N,N*-dimethylaniline donates electrons to the ring by resonance, so it accelerates electrophilic aromatic substitution reactions. In contrast, steric hindrance between the methyl groups on the nitrogen and the ortho methyl groups causes the nitrogen of *N,N*,2,6-tetramethylaniline to rotate about its bond to the aromatic ring. As a result, the orbital containing the unshared electrons on the nitrogen is not parallel to the p orbitals of the pi system of the aromatic ring. This decreases the resonance donation of electrons and makes this compound less reactive in electrophilic aromatic substitution reactions.

Review of Mastery Goals

After completing this chapter, you should be able to:

Show the products of any of the reactions discussed in this chapter. (Problems 17.6, 17.8, 17.9, 17.11, 17.13, 17.15, 17.19, 17.22, 17.23, 17.26, 17.27, 17.31, 17.32, 17.33, 17.34, 17.35, 18.39, 17.40, 17.41, 17.66, and 17.67)

Show the mechanisms for the reactions whose mechanisms were discussed.
(Problems 17.10, 17.12, 17.14, 17.45, 17.46, 17.47, 17.48, 17.49, 17.50, 17.51, 17.59, 17.60, 17.61, 17.62, 17.63, 17.64, and 17.65)

Predict the effect of a substituent on the rate and the regiochemistry of an electrophilic aromatic substitution reaction.
(Problems 17.1, 17.2, 17.3, 17.4, 17.5, 17.7, 17.18, 17.24, 17.31, 17.33, 17.34, 17.35, 17.36, 17.37, 17.39, 17.40, 17.41, 17.42, 17.52, 17.53, 17.55, 17.56, 17.69, 17.70, and 17.71)

Use these reactions to synthesize aromatic compounds.
(Problems 17.16, 17.20, 17.28, 17.29, 17.30, 17.38, 17.43, 17.44, 17.54, 17.57, and 17.58)

Chapter 18
ADDITIONS TO THE CARBONYL GROUP
REACTIONS OF ALDEHYDES AND KETONES

18.1

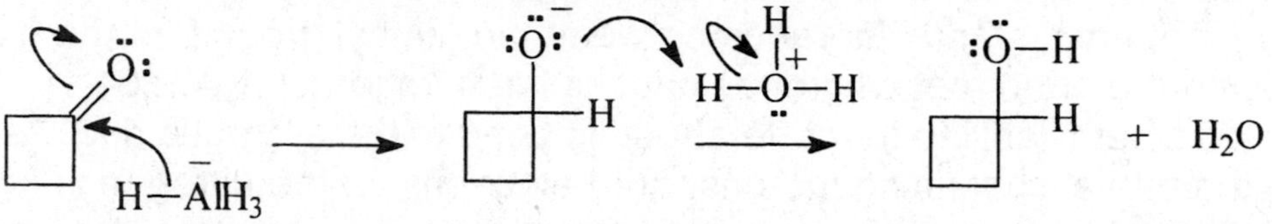

18.2

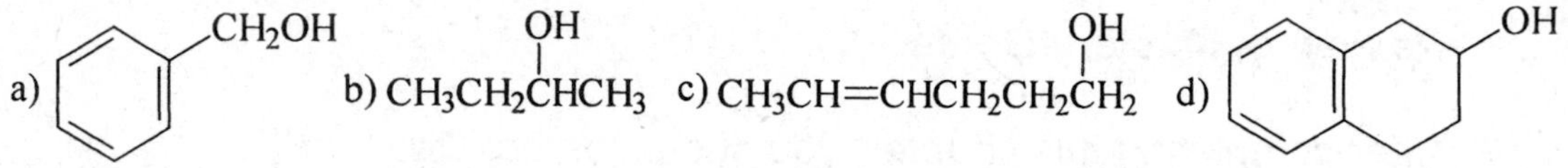

18.3

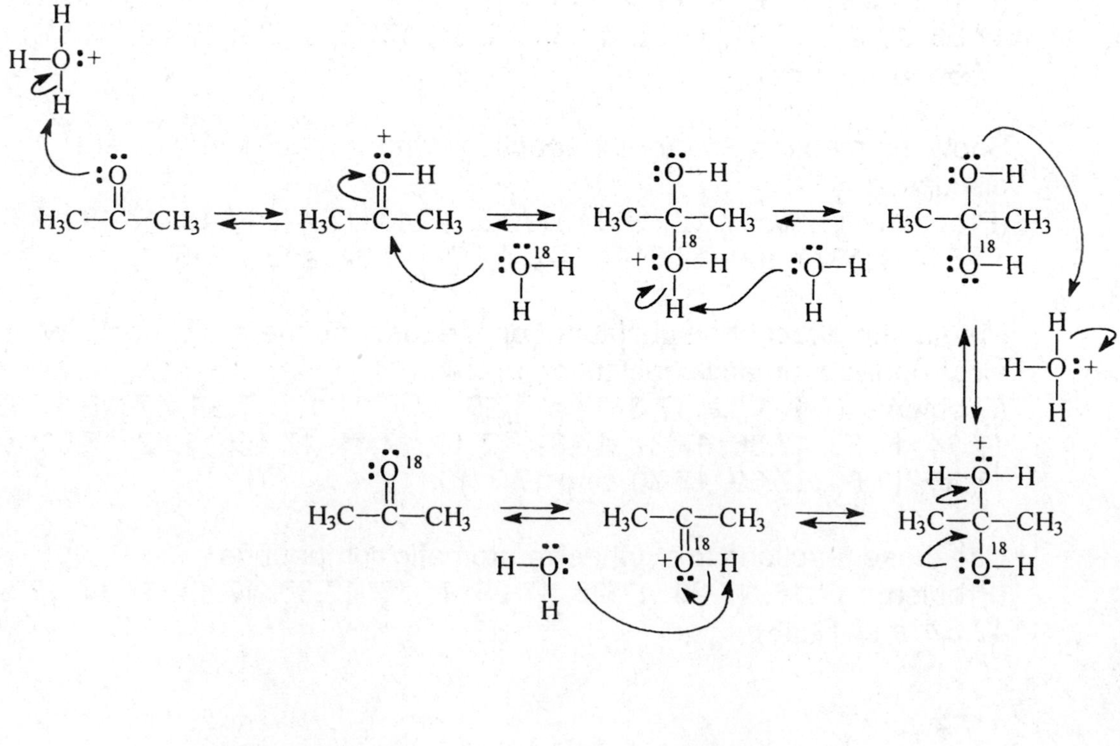

18.4 a) The compound on the right has a larger equilibrium constant because of the inductive electron withdrawing effect of the F's.
b) The aldehyde has a larger equilibrium constant because the inductive and steric effects of the methyl group make the ketone less reactive.
c) The compound on the left has a larger equilibrium constant because its carbonyl carbon is less sterically hindered.
d) The compound on the right has a larger equilibrium constant because of the inductive electron withdrawing effect of its Cl's.
e) There is not much difference in the equilibrium constants because the steric effect is too far from the reacting carbonyl carbon.

18.5

$CH_3CH_2CHO < ClCH_2CH_2CHO < CH_3CHClCHO < CH_3CF_2CHO$

smallest K ... largest K

18.6

a) 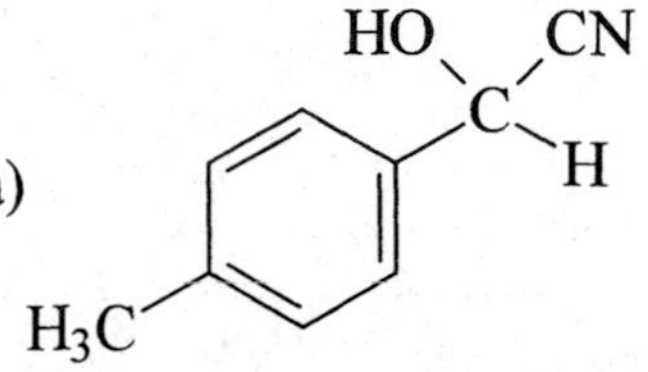

b) 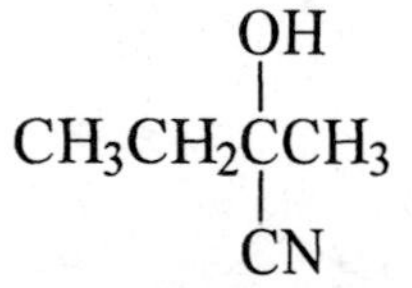

c) 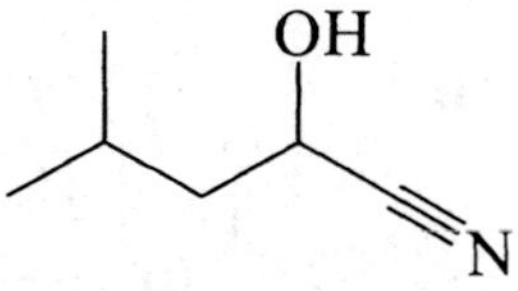

18.7 a) The compound on the right has a larger equilibrium constant because aldehydes are less sterically hindered than ketones.
b) The compound on the left has a larger equilibrium constant because resonance makes the carbonyl carbon of the compound on the right less electrophilic.
c) The compound on the left has a larger equilibrium constant because the ethoxy group of the compound on the right is a resonance electron donating group and makes the carbonyl carbon of this compound less electrophilic.
d) The compound on the right has a larger equilibrium constant because the electron withdrawing nitro group makes its carbonyl carbon more electrophilic.

18.8

a) $CH_3CH_2CH_2CH_3$ b) c) $CH_3CH_2C{\equiv}CMgBr + CH_3CH_3$

18.9

a) b) c)

d) $CH_3CH_2C(Ph){=}CHCH_3$ e) f) g) $CH_3C{\equiv}CCH(OH)CH_3$

18.10

a)

b) c)

d) e)

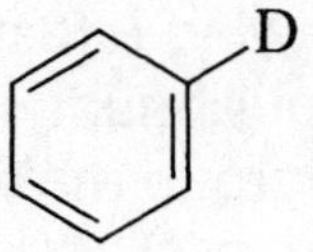

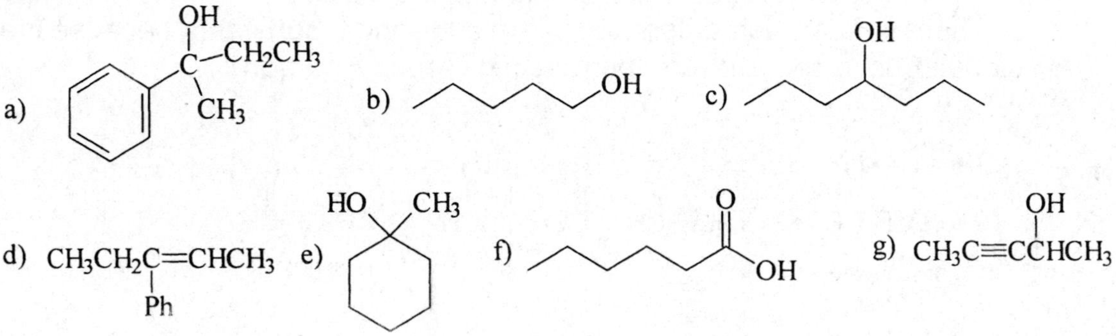

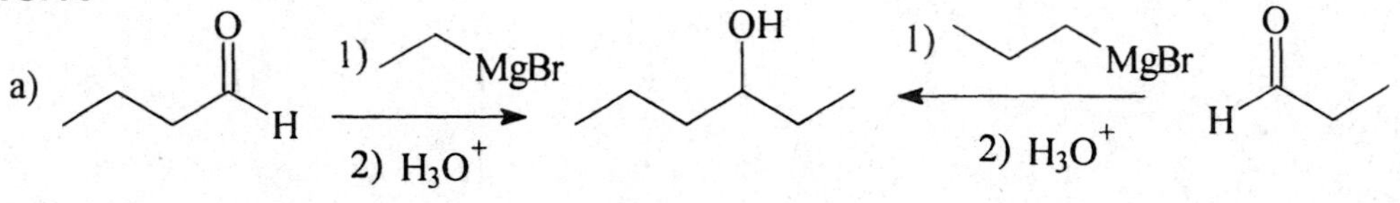

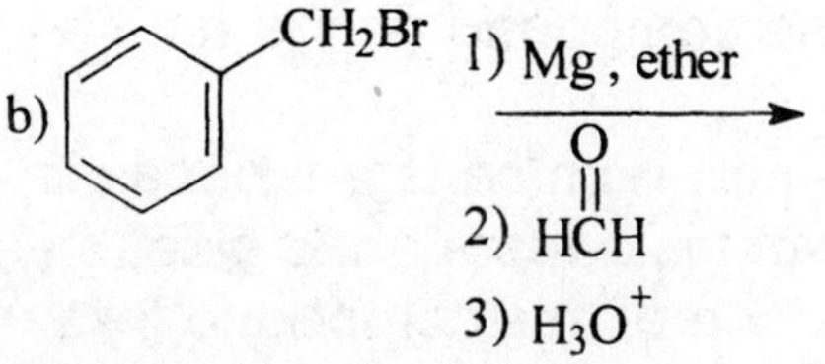

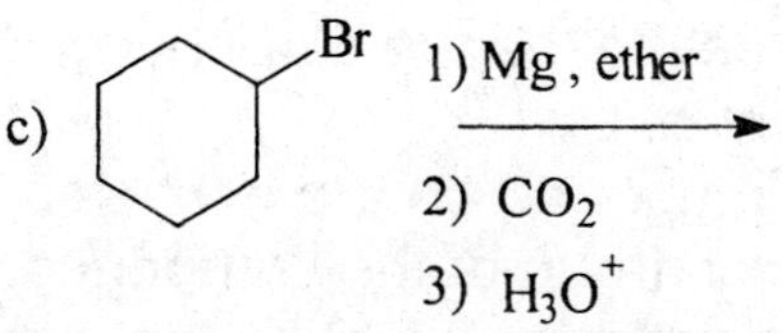

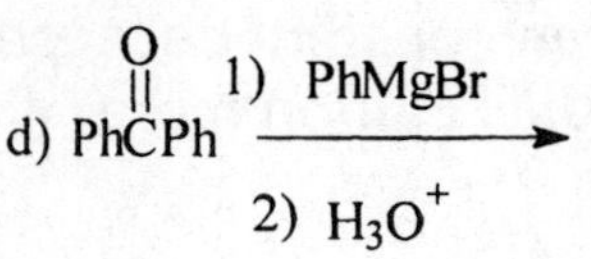

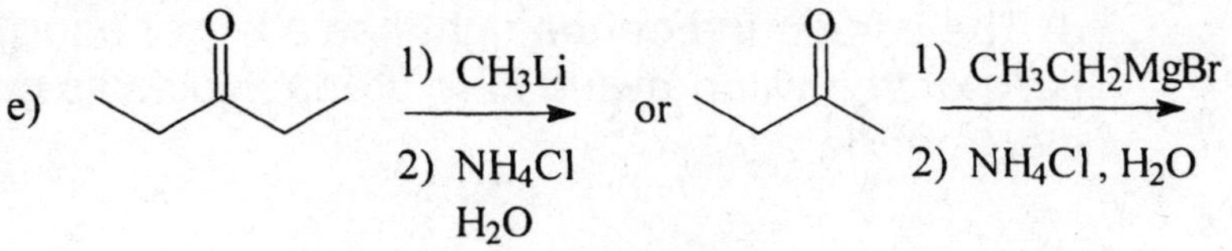

f) 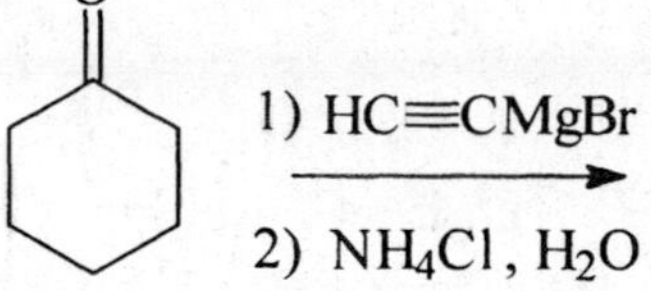

g) 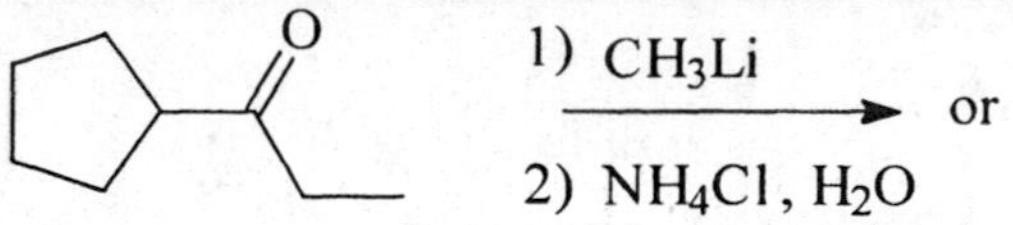

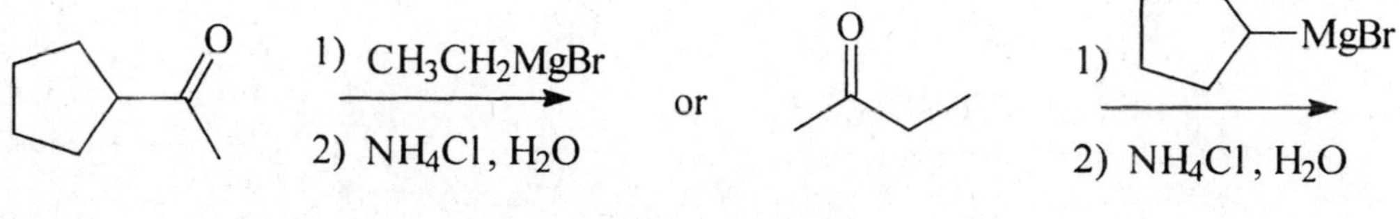

18.11 An acid-base reaction between the Grignard reagent (base) and the hydroxy group (acid) on the aldehyde destroys the Grignard reagent before it has a chance to react as a nucleophile.

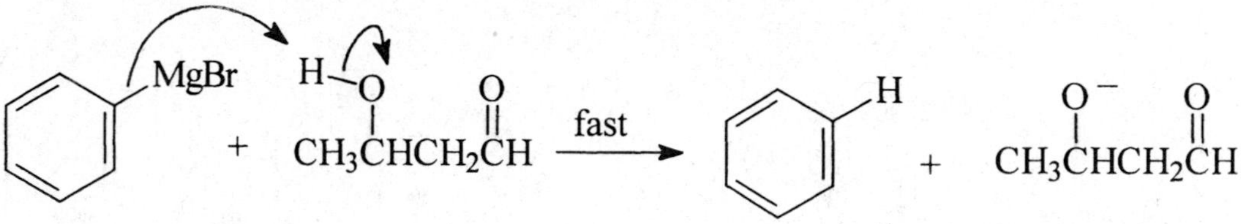

18.12 $Ph_3P + PhCH_2Cl \xrightarrow{S_N2} Ph_3\overset{+}{P}{-}CH_2Ph \quad Cl^-$

18.13

a) $PhCH{=}CHCH_2CH_3$ b) 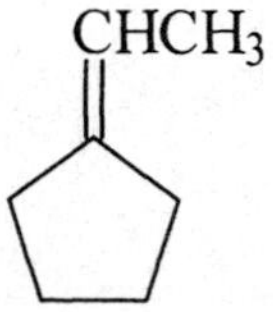 c) $HC{\equiv}CCH{=}CHCH_2CH_2CH_3$

d) 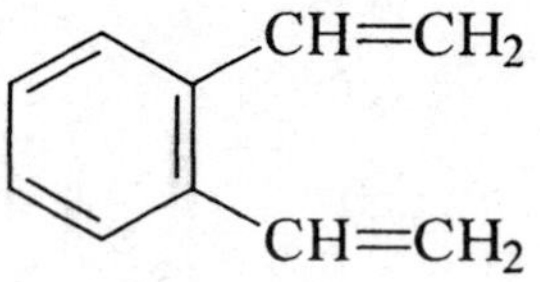 e) $PhCH{=}CH\overset{O}{\overset{\|}{C}}OEt$

18.14 The conjugate base (the ylide) is more stable because of resonance with the phenyl group.

18.15

a) 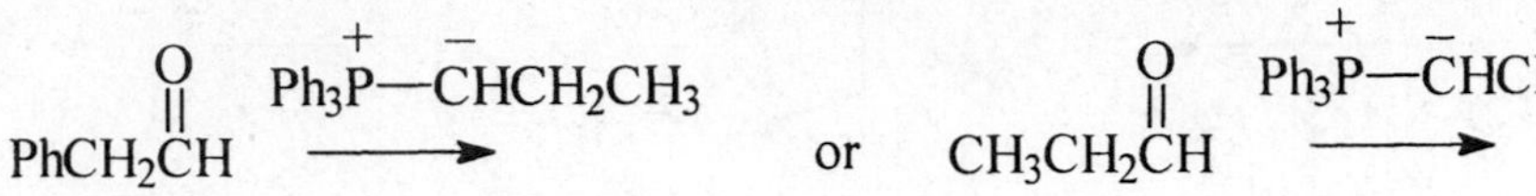

b) 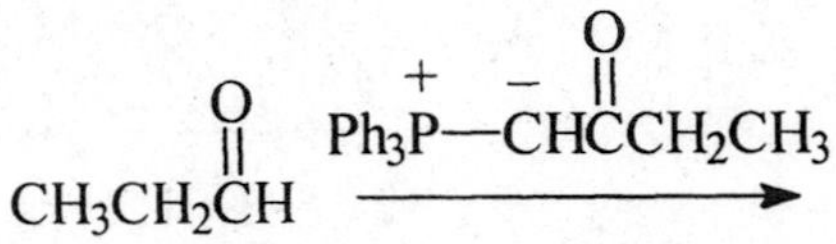

c) 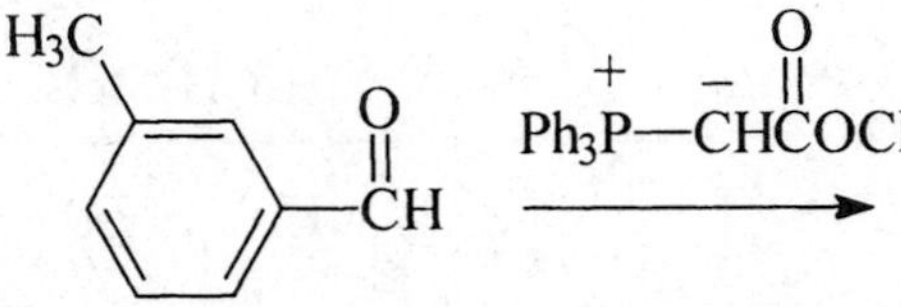

18.16

a) 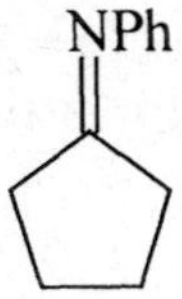

b) 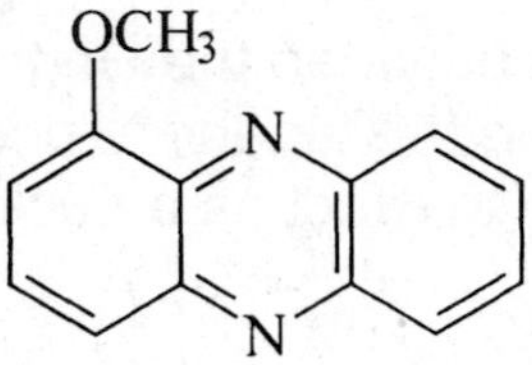

c) 

d) 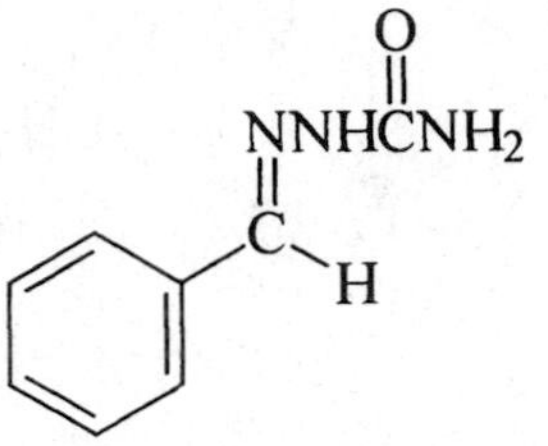

e) 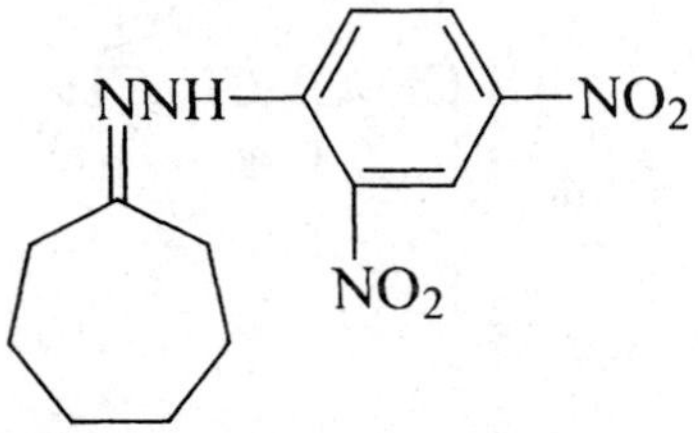

18.17

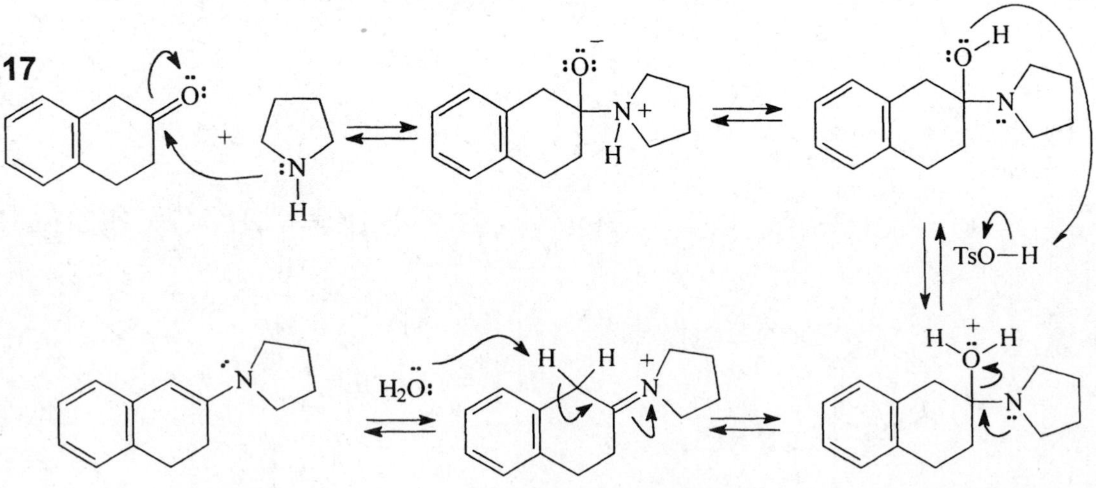

18.18 The enamine produced in problem 18.17 is conjugated whereas the enamine shown in this problem is not. Formation of the more stable conjugated enamine is preferred.

18.19

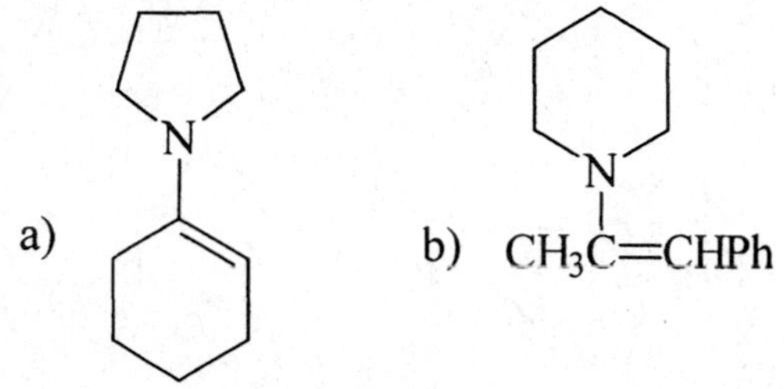

18.20

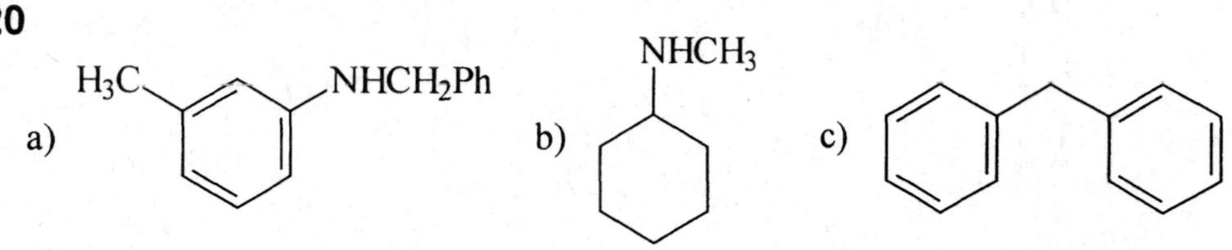

18.21 The compound on the right gives more cyclic hemiacetal at equilibrium because its five-membered ring product is more stable than the four-membered ring product of the compound on the left.

18.22

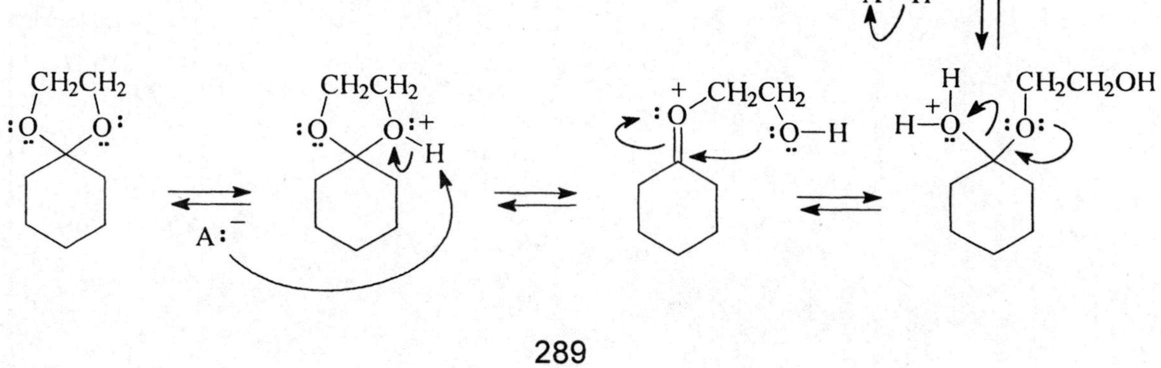

18.23

a)
$$CH_3CH_2O \quad OCH_2CH_3$$
$$CH_3CH_2CH_2CH$$

b)

c)

18.24

18.25

a)   b)   c)   d)   e)

18.26

a) $CH_3CH_2CH_2CH$ $\overset{\text{O}}{\parallel}$ $\xrightarrow[\text{2)}\ H_3O^+]{\text{1) }CH_3CH_2MgBr}$ $CH_3CH_2CH_2CHCH_2CH_3$ $\overset{\text{OH}}{|}$ $\xrightarrow[\text{H}_2SO_4]{\text{CrO}_3}$ $CH_3CH_2CH_2CCH_2CH_3$ $\overset{\text{O}}{\parallel}$

$\downarrow$ 1) CH_3MgI

$\quad$ 2) NH_4Cl, H_2O

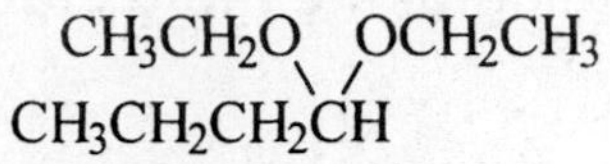

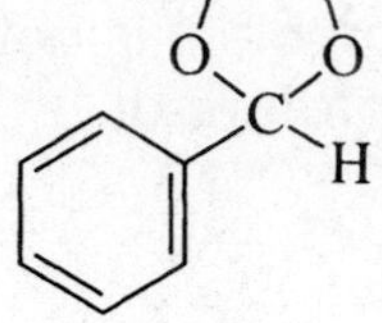

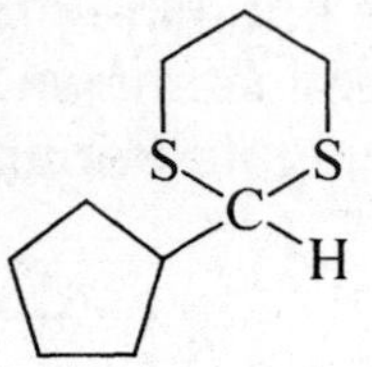

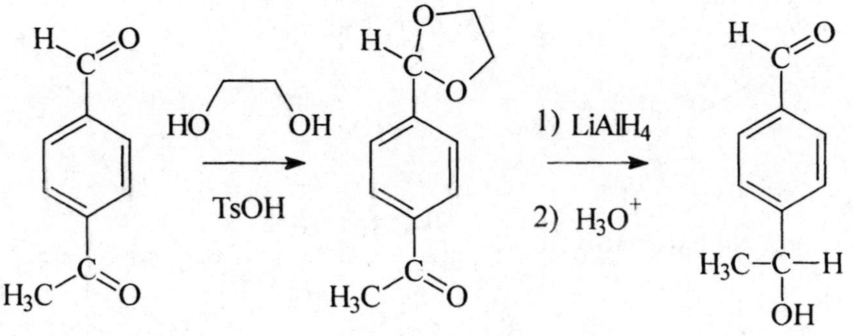

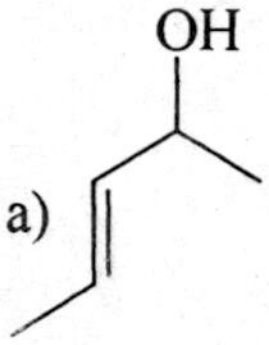

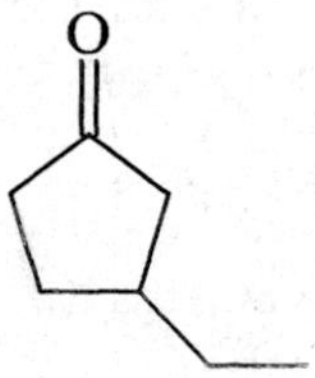

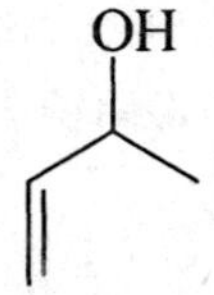

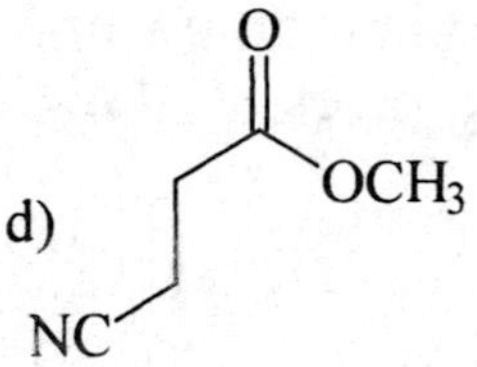

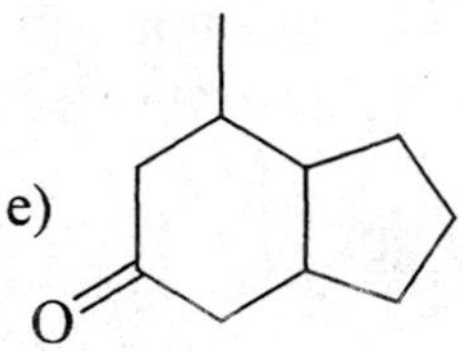

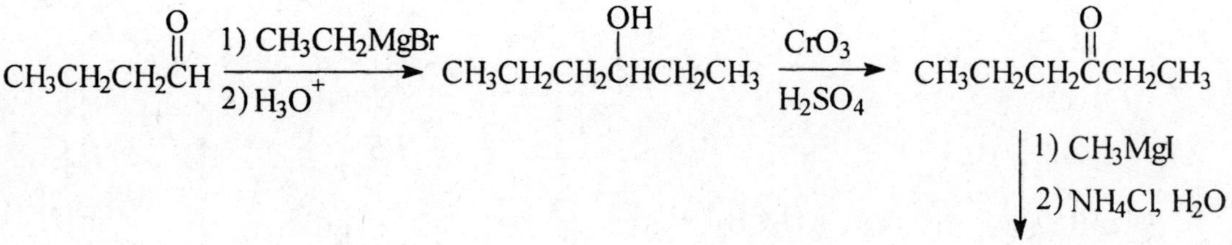

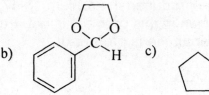

b)

1) $(CH_3CH_2)_2CuLi$
2) H_3O^+

$Ph_3\overset{+}{P}{-}\overset{-}{C}HCH_3$

CH_2CH_3

$CHCH_3$

CH_2CH_3

c) $\overset{O}{\overset{\|}{H}CCH{=}CH_2}$

1) $(CH_3CH_2)_2CuLi$
2) H_3O^+

$\overset{O}{\overset{\|}{H}CCH_2CH_2CH_2CH_3}$

1) CH_3MgI
2) H_3O^+

$\overset{OH}{\overset{|}{CH_3}CHCH_2CH_2CH_2CH_3}$

CrO_3
H_2SO_4

$\overset{CH_2}{\overset{\|}{CH_3}CCH_2CH_2CH_2CH_3}$

$Ph_3\overset{+}{P}{-}\overset{-}{C}H_2$

$\overset{O}{\overset{\|}{CH_3}CCH_2CH_2CH_2CH_3}$

d)

$SOCl_2$

1) Mg
2) $CH_3CH_2CH_2CH$ ($\overset{\|}{O}$ below)
3) H_3O^+

CrO_3
H_2SO_4

e)

$\overset{OH\ \ OH}{|\quad\ |}$
TsOH

1) Mg
2) CO_2
3) H_3O^+

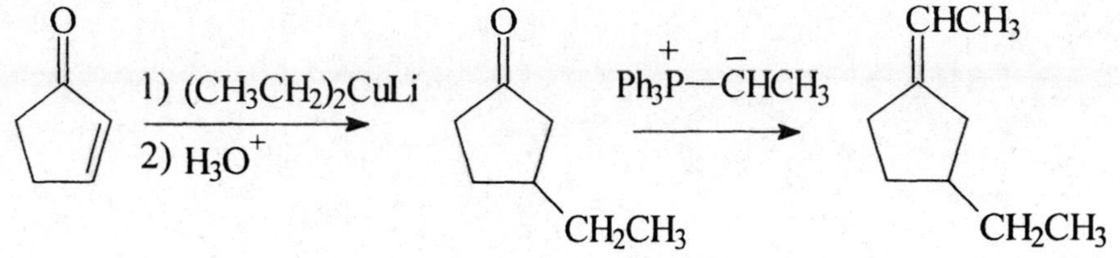

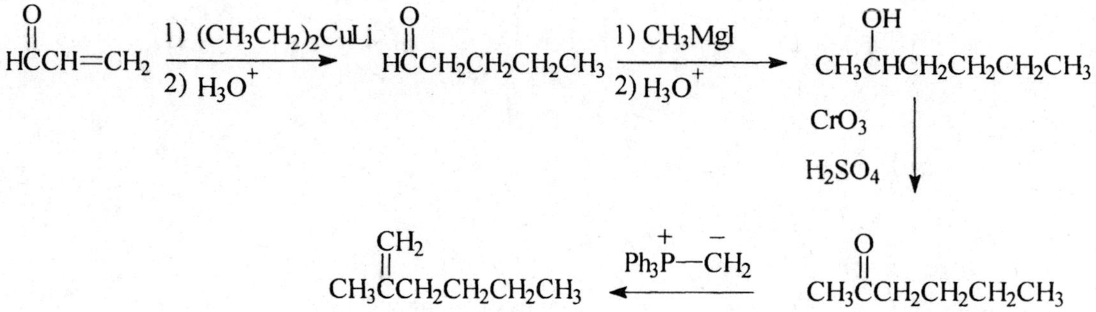

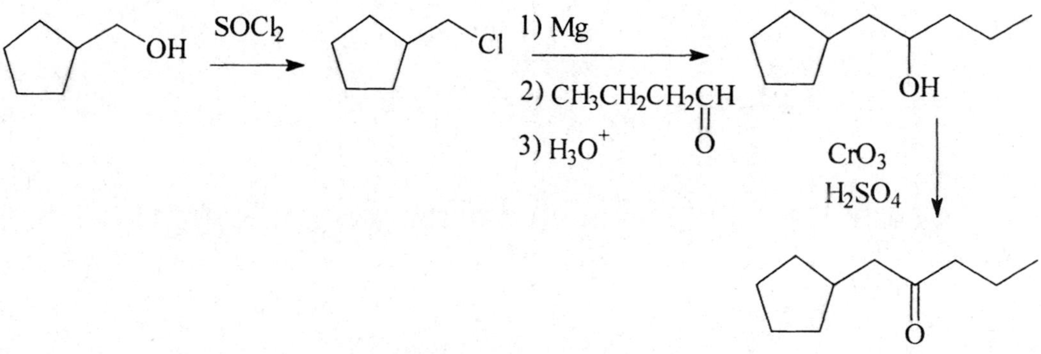

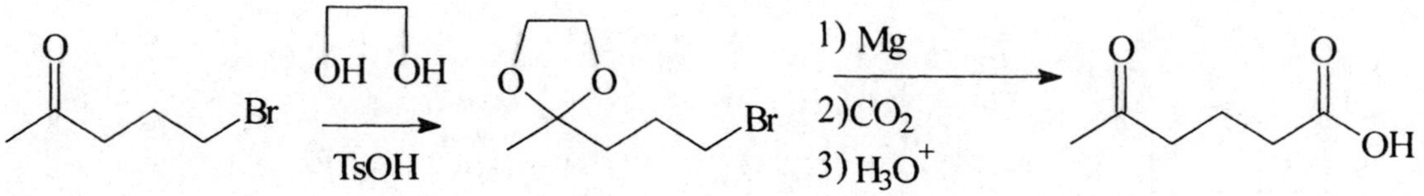

18.27

a) $PhCH_2CO_2H$

b)

c) $CH_3\overset{\displaystyle N-NHCNH_2}{\underset{\displaystyle }{C}}CH_3$ with O on the carbonyl

d)

e)

f)

g)

h)

i) $\underset{Ph-\overset{|}{C}H-\overset{|}{C}HCH_3}{CH_3NH \quad CH_3}$

j) $PhCH{=}CPh_2$

k) $CH_3CH_2CH_2CH_2\overset{OH}{\overset{|}{C}}HCN$

l) $CH_3CH_2CH_2CH_2\overset{OH}{\overset{|}{C}}H_2$

m)

n) $Ph\overset{\displaystyle N-OH}{\underset{\displaystyle }{C}}Ph$

o) $Ph\overset{}{\underset{\underset{CN}{|}}{C}H-CH_2\overset{O}{\overset{\|}{C}}Ph$

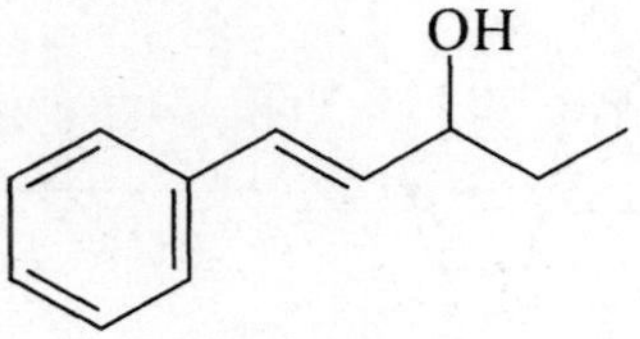

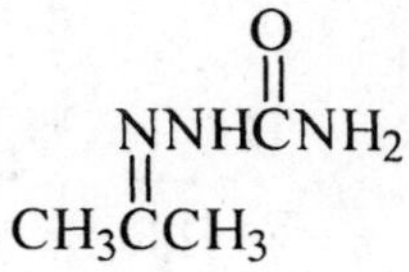

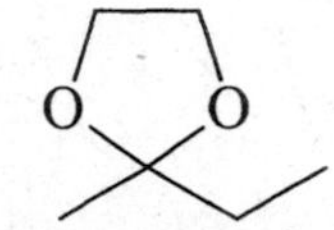

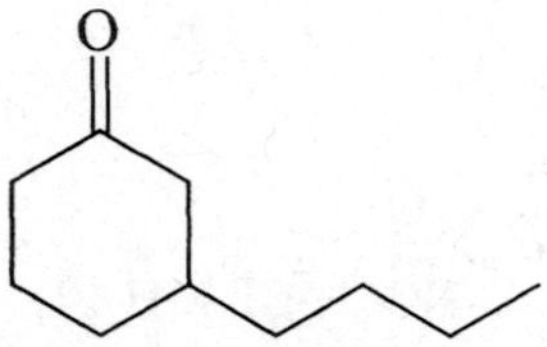

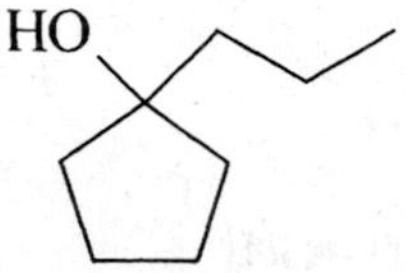

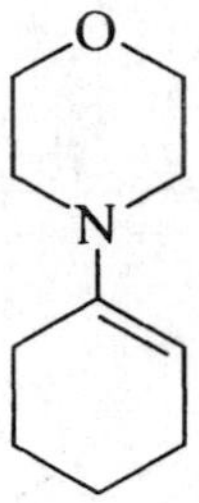

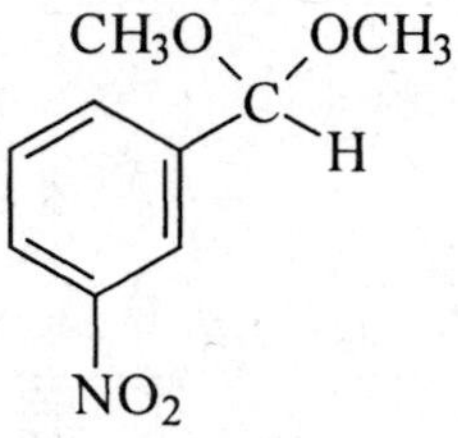

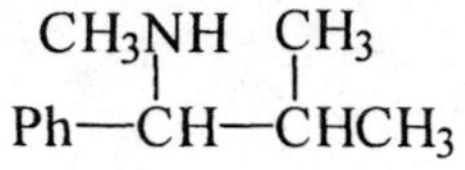

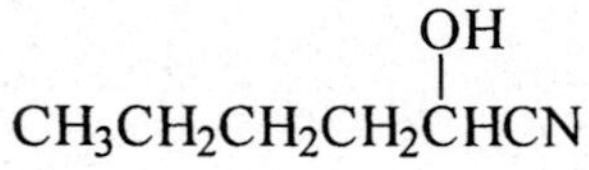

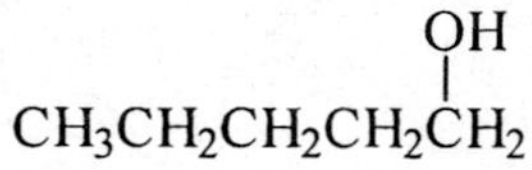

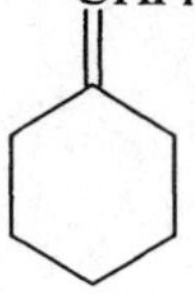

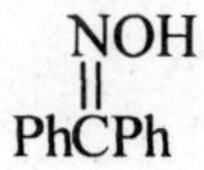

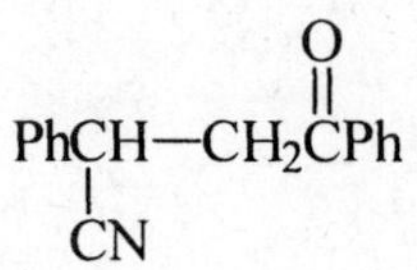

p)

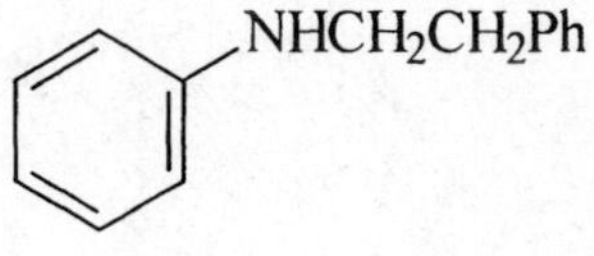

q) $CH_3CH_2CH_2CH_2CH{=}CH\overset{\overset{\displaystyle O}{\|}}{C}OCH_3$

r) 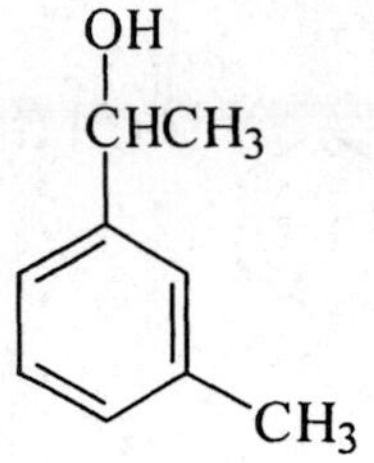

s) 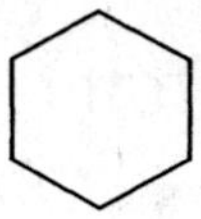

18.28

a) 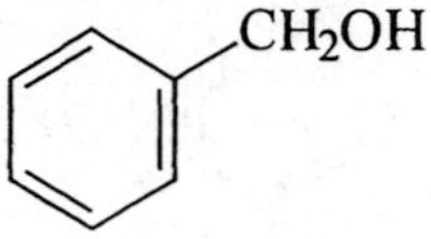

b) 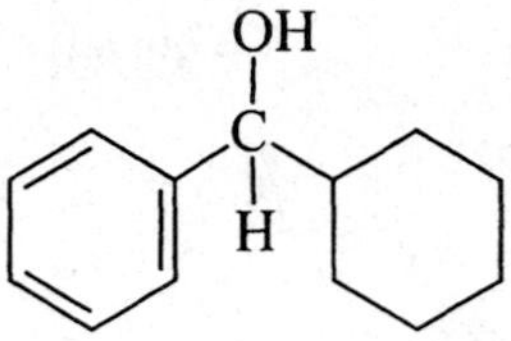

c) 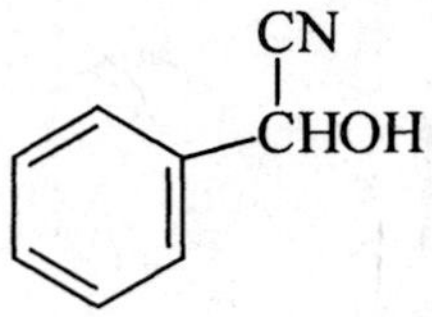

d) 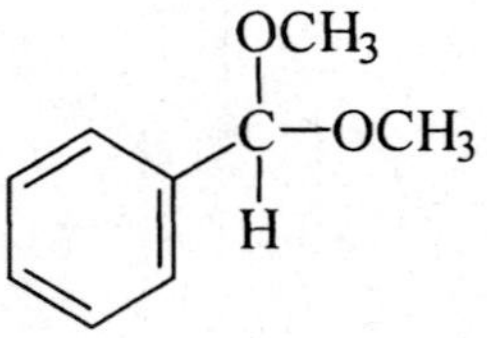

e) 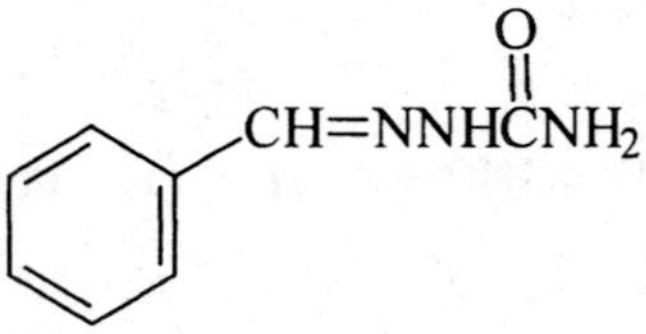

f) 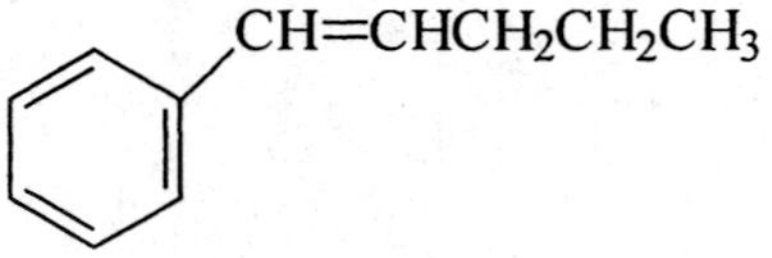

18.29

a) 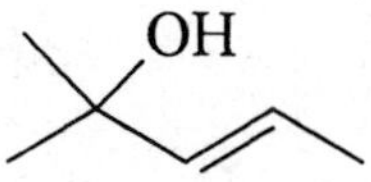

b) 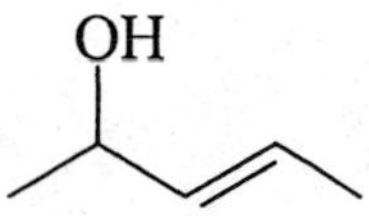

c) 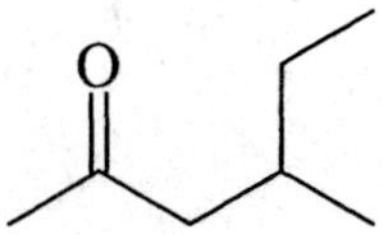

d) 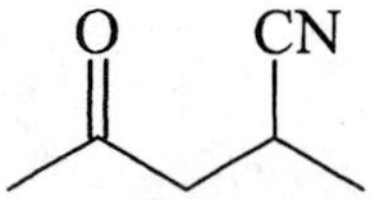

e) 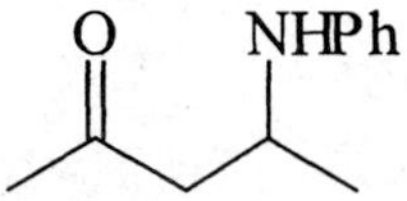

18.30 The reaction gives poor a yield because the carbonyl carbon is highly sterically hindered.

18.31

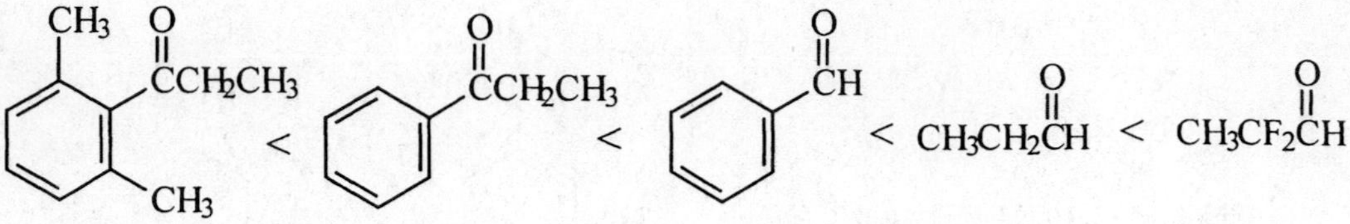

smallest *K* largest *K*

18.32

a) 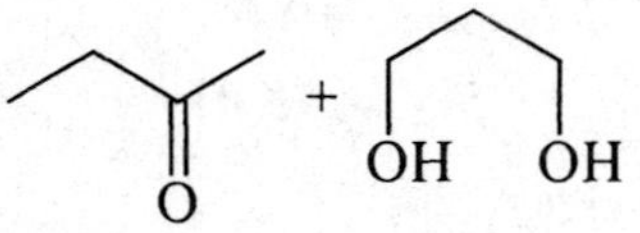 b) 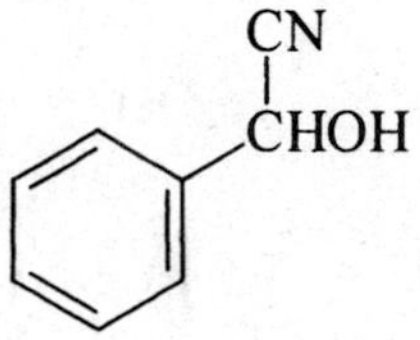 c) 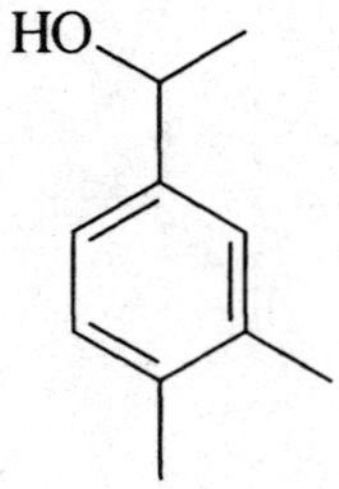

d) 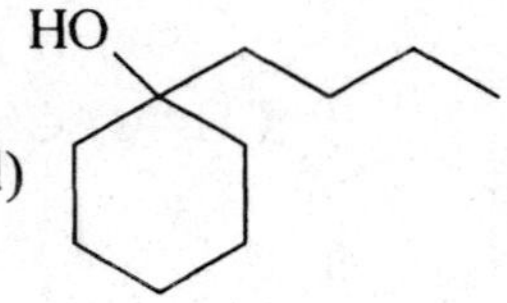 e) PhCH=CHCH=CH_2 f) 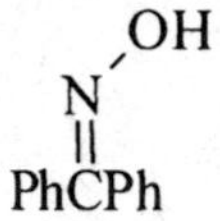 g) 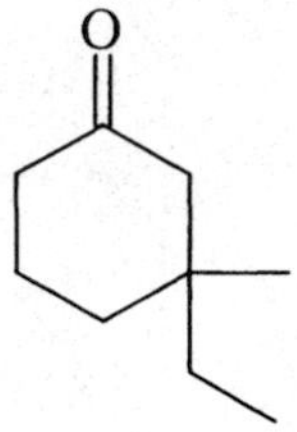

18.33

a) 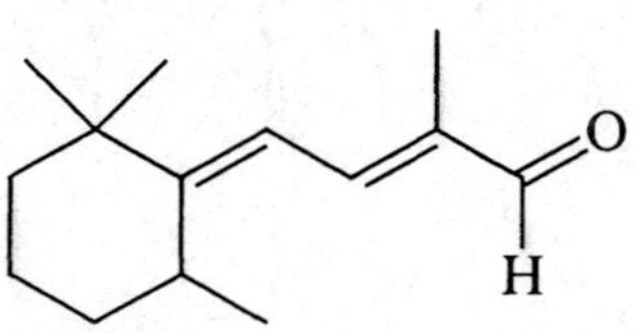 b) 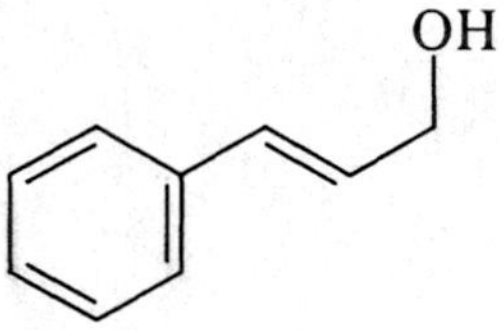 c) 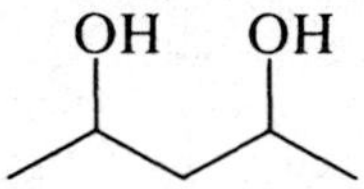

d) 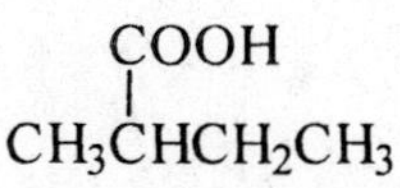 e) 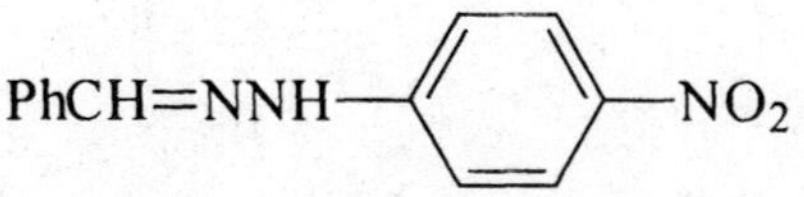 f) 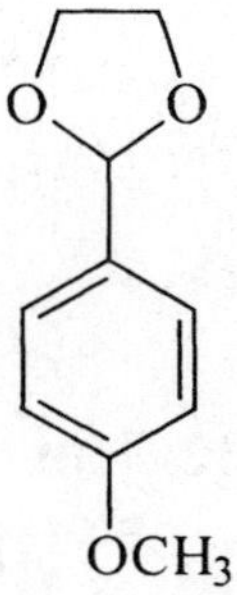

g) $PhNHCH_2CH_2—C{\equiv}N$

18.34 The amino group is strongly electron donating by resonance while the nitro group is strongly electron withdrawing by both its resonance and inductive effects. An electron withdrawing group in the para position makes the carbonyl carbon of the aldehyde group more reactive toward nucleophiles. Therefore the compound on the right has a higher equilibrium constant because its carbonyl carbon is more electrophilic.

18.35 a) The compound on the left has a larger equilibrium constant because its carbonyl carbon is less sterically hindered than the compound on the right.
b) The compound on the left has a larger equilibrium constant because its carbonyl group is less stabilized by resonance.
c) The compound on the right has a larger equilibrium constant because its carbonyl group is more electrophilic due to the electron withdrawing CCl_3 group.

18.36

a) 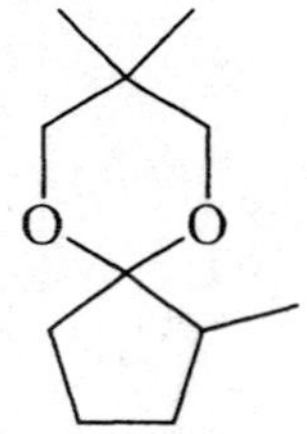

b) 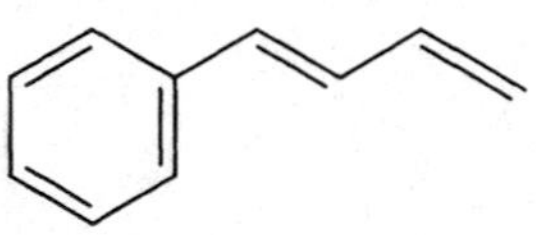

c) 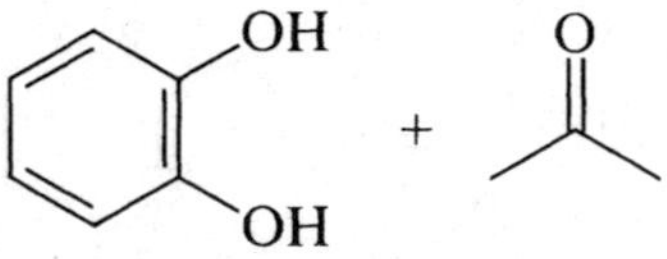

d) 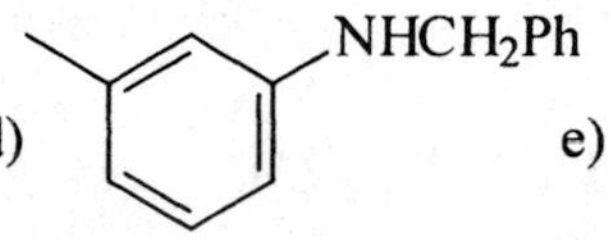

e) $CH_3CH_2CH_2CH{=}NPh$

f) 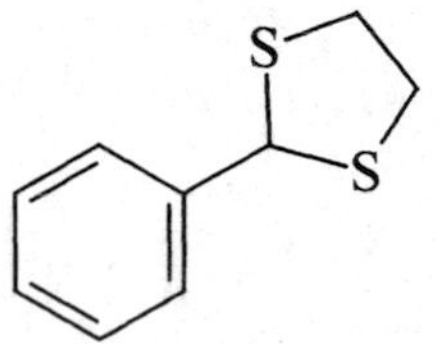

g) 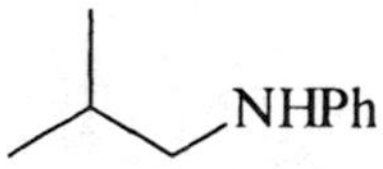

h) $PhCH(CN)-CH(CO_2Et)_2$

18.37

a) 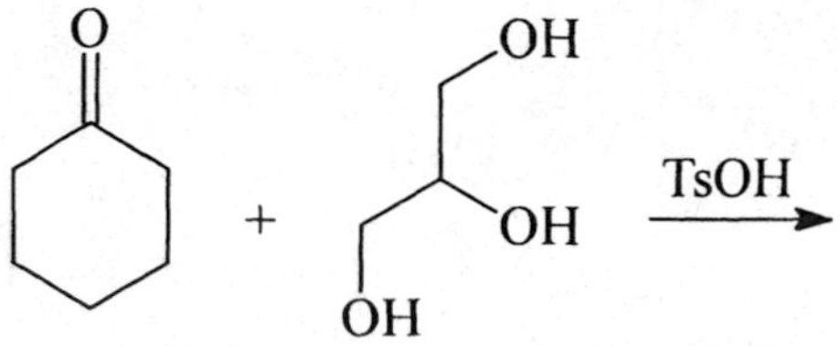

b) $CH_3CH(CH_3)CH_2CHO \xrightarrow[HCN]{^{-}CN}$

c) 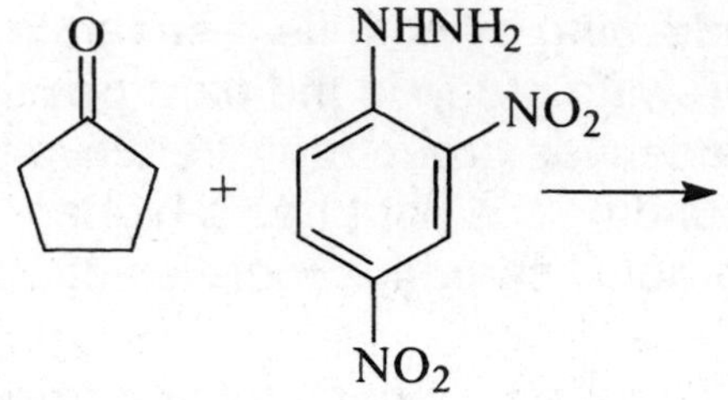
O
NHNH2
NO2
NO2
+

d) 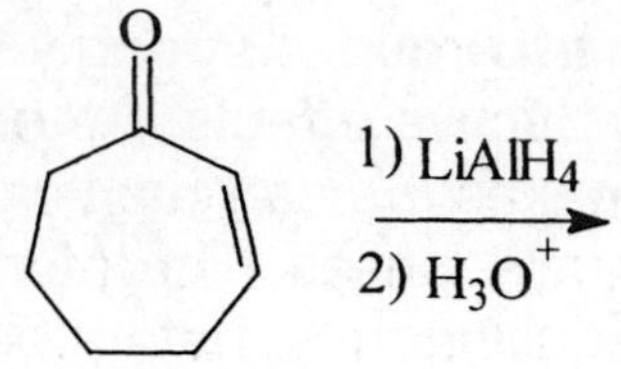
O
1) LiAlH4
2) H3O+

e) 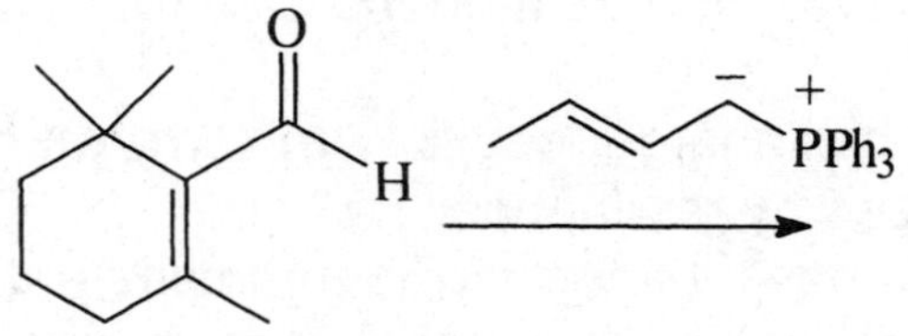
O
H
PPh3

f) 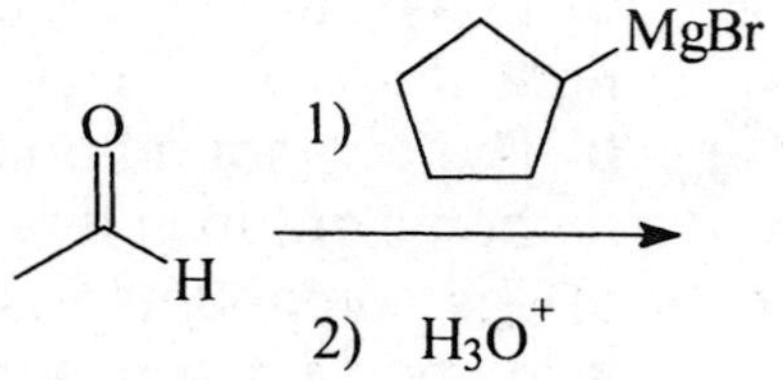
O
H
1)
MgBr
2) H3O+

18.38

a) 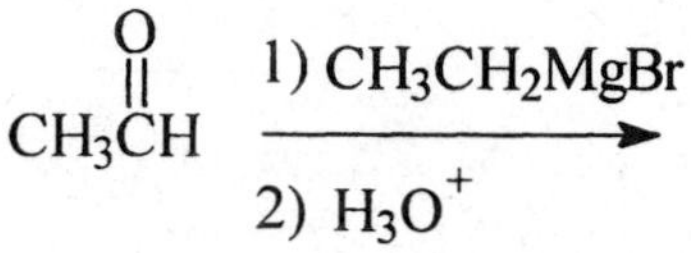
O
CH3CH
1) CH3CH2MgBr
2) H3O+

b) 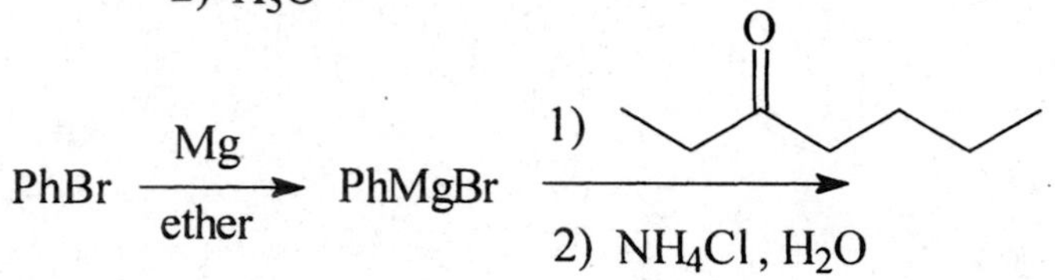
PhBr
Mg
ether
PhMgBr
1)
O
2) NH4Cl, H2O

c) 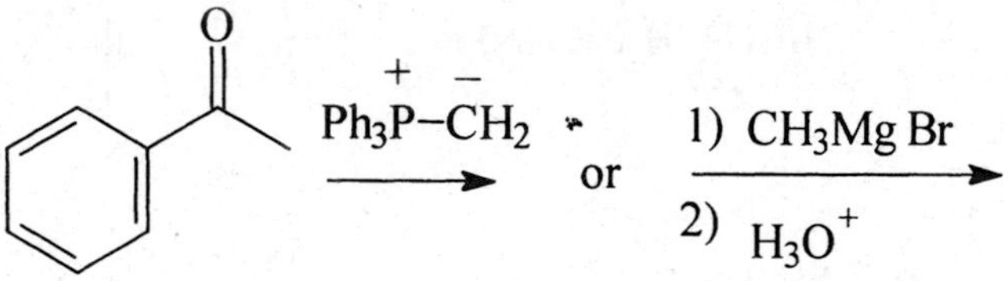
O
Ph3P–CH2
or
1) CH3Mg Br
2) H3O+

d) 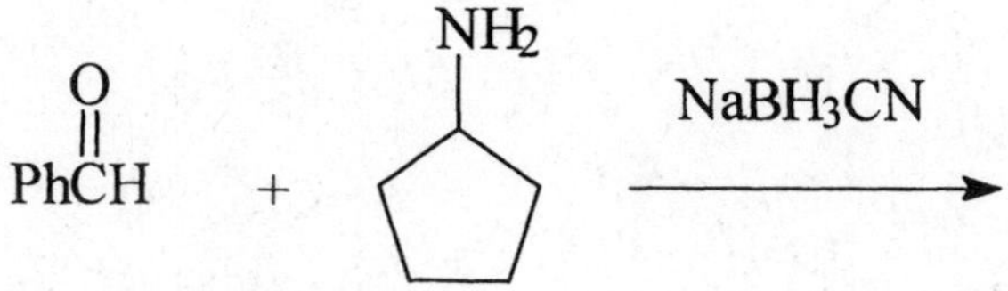
O
PhCH
+
NH2
NaBH3CN

e) 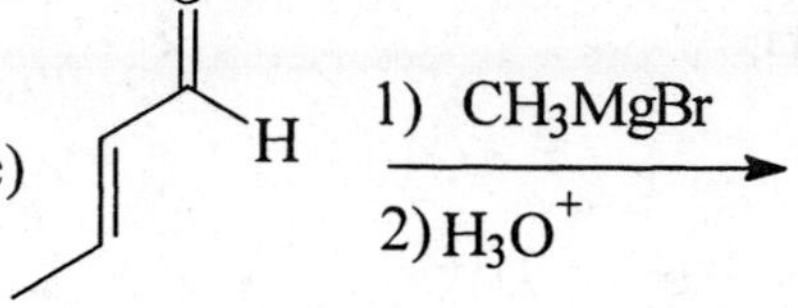

f) 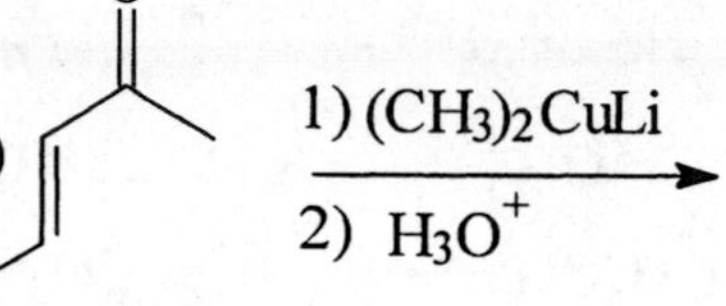

g) 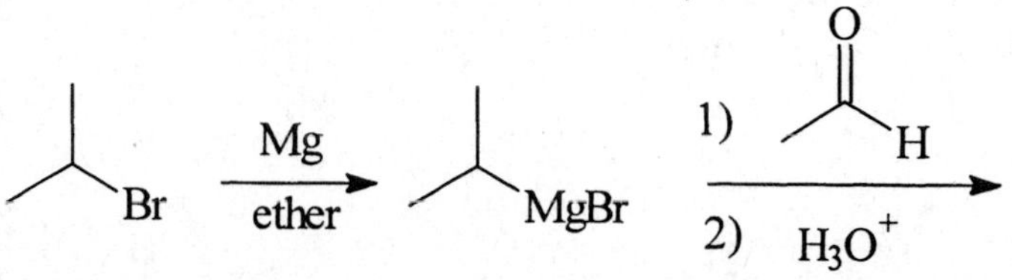

18.39

a) 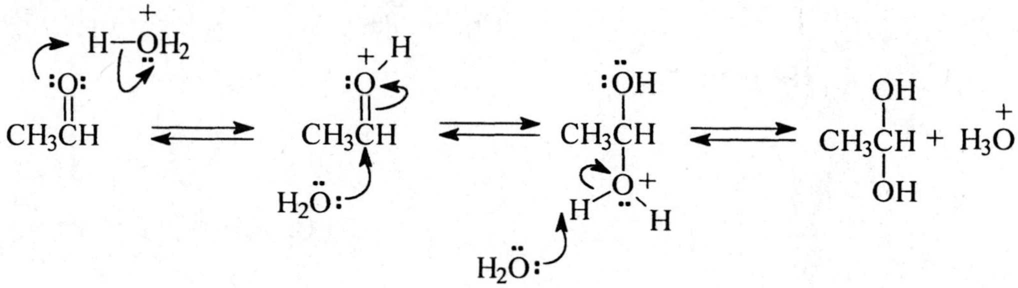

b) 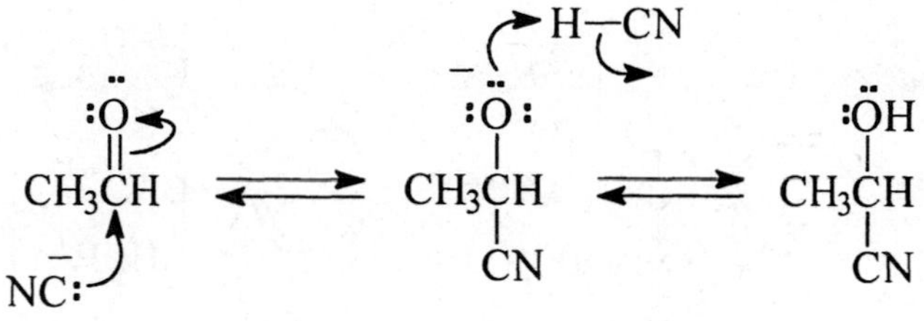

c) 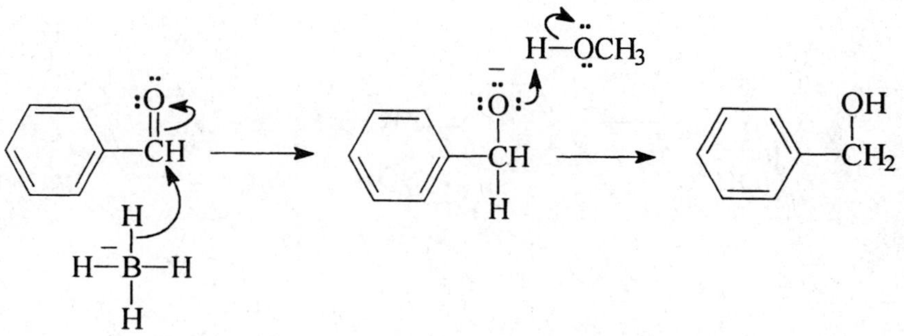

d)

e)

$$CH_3CH_2CH=CH_2 \ + \ Ph_3P=O$$

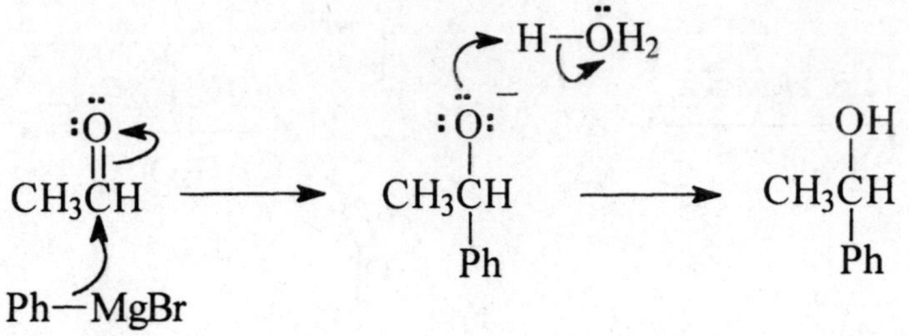

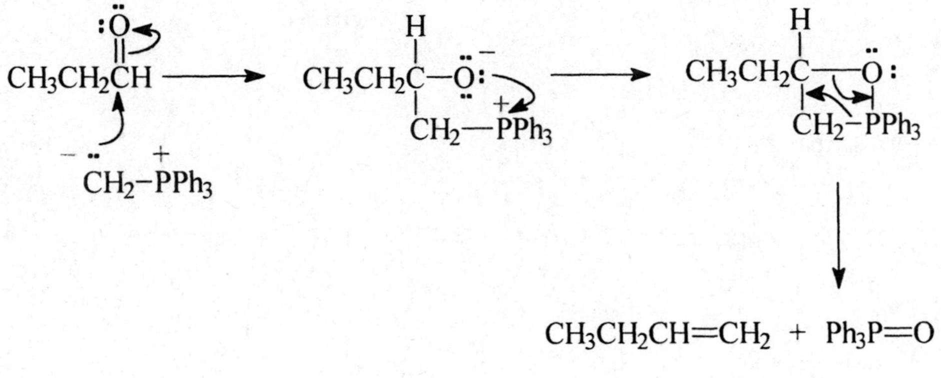

f)

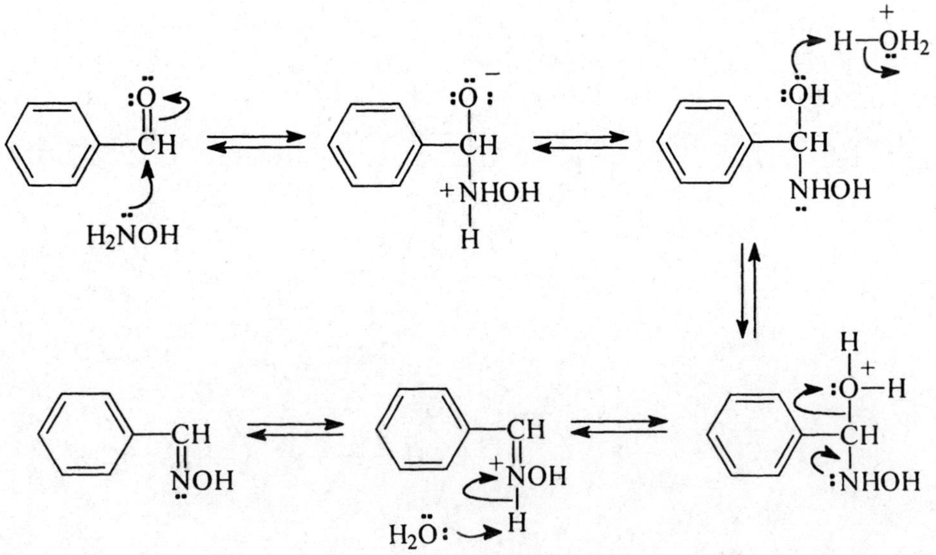

g)

$$CH_3\overset{\overset{\ddot{O}:}{\|}}{C}CH_3 \; \rightleftharpoons \; CH_3\overset{\overset{+}{\ddot{O}}-H}{C}CH_3 \; \rightleftharpoons \; CH_3\overset{\ddot{O}H}{\underset{}{C}}CH_3 \; \rightleftharpoons \; CH_3\overset{\ddot{O}H}{\underset{:\ddot{O}Et}{C}}CH_3$$

H—A   EtÖH   A:⁻ H—Ö⁺—Et   H—A

$$CH_3\overset{:\ddot{O}Et}{\underset{:\ddot{O}Et}{C}}CH_3 \; \rightleftharpoons \; CH_3\overset{\overset{+}{\ddot{O}}-Et}{\underset{:\ddot{O}Et}{C}}CH_3 \; \rightleftharpoons \; CH_3\overset{\|}{\underset{\overset{+}{\ddot{O}Et}}{C}}CH_3 \; \rightleftharpoons \; CH_3\overset{\overset{H-\overset{+}{\ddot{O}}-H}{|}}{\underset{:\ddot{O}Et}{C}}CH_3$$

A:⁻ H—Ö⁺—Et   EtÖH

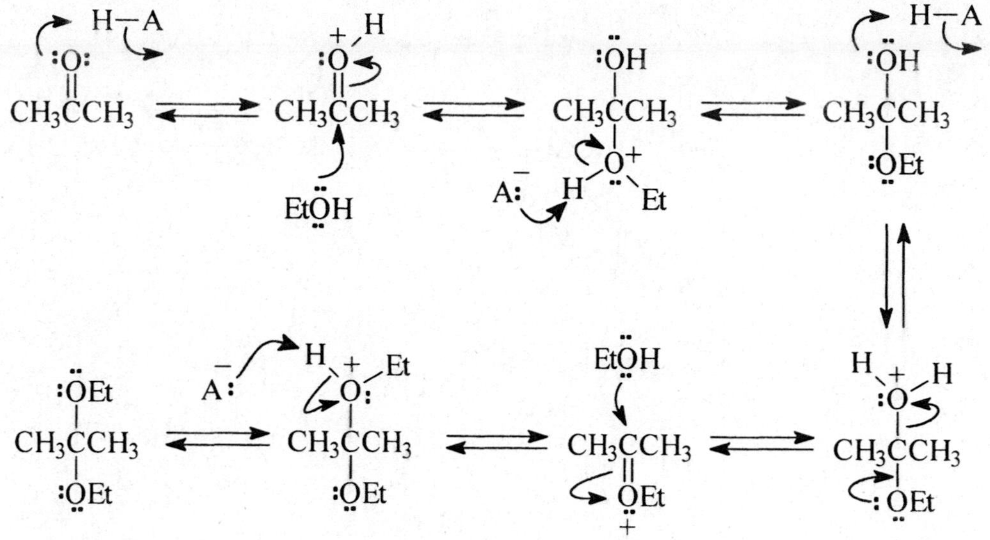

18.40 A conjugate addition reaction does not occur unless there is a group attached to the double bond that can help stabilize the carbanion intermediate by resonance. The first reaction does not occur because the methyl group will not stabilize the carbanion intermediate. The second reaction occurs because the nitro group can help stabilize the carbanion intermediate by resonance.

$$\overset{+}{PhNH_2}-CH_2-\overset{\ddot{..}}{\overset{-}{C}}H-CH_3$$

no resonance
stabilization

$$\overset{+}{PhNH_2}-CH_2-\overset{\ddot{..}}{\overset{-}{C}}H-NO_2$$

large resonance
stabilization

18.41

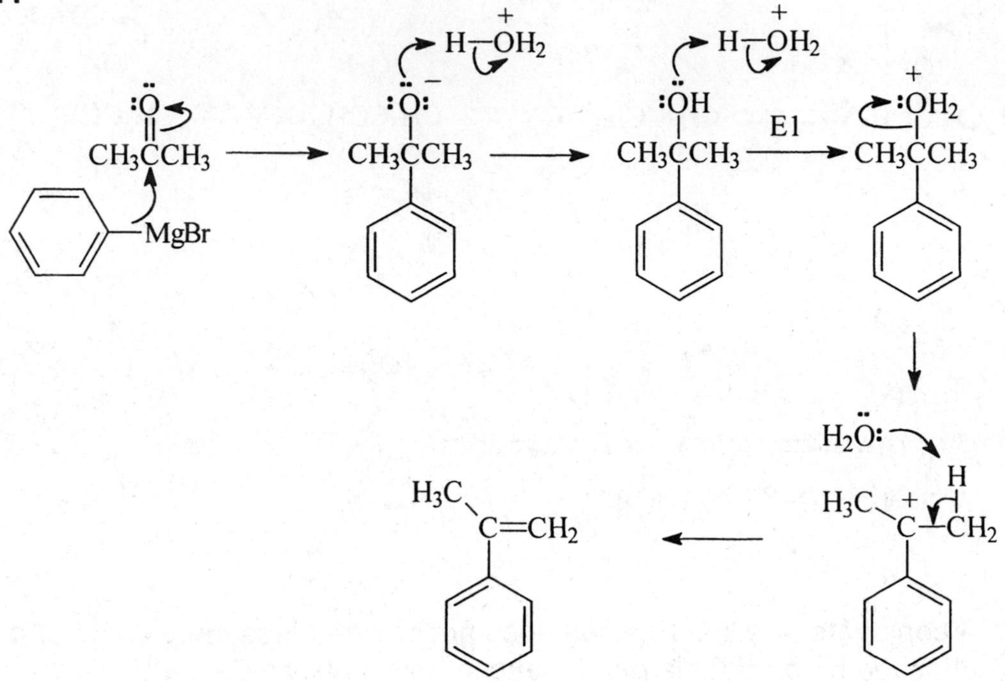

H—OH2
:O:
CH3CCH3
MgBr
:OH
E1
:OH2
H2O:
H
H3C
C
CH2
C=CH2

18.42

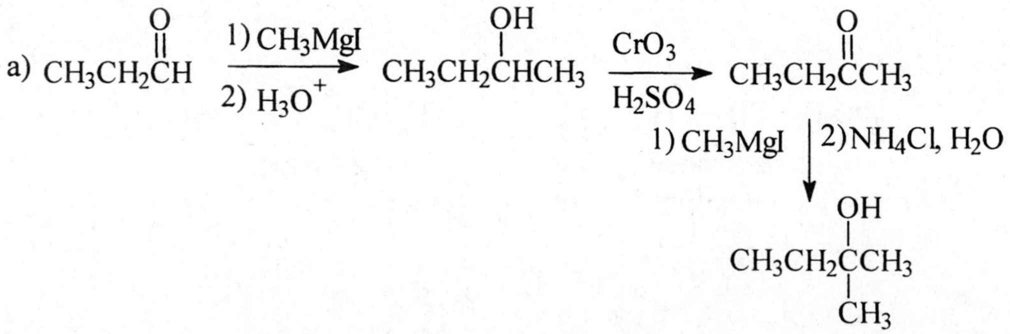

a) CH3CH2CH
1) CH3MgI
2) H3O+
CH3CH2CHCH3
CrO3
H2SO4
CH3CH2CCH3
1) CH3MgI
2)NH4Cl, H2O
OH
CH3

b) 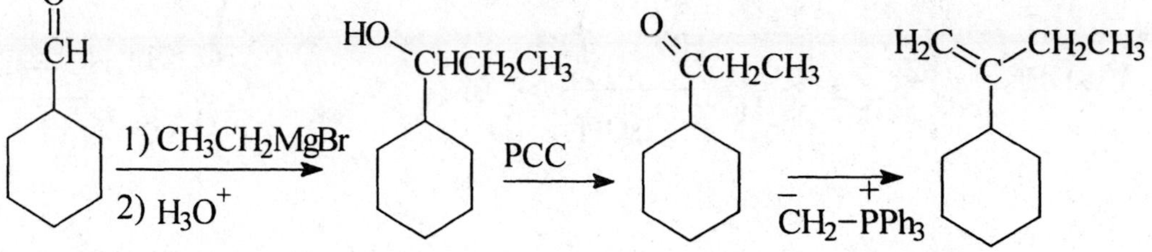
1) CH_3CH_2MgBr
2) H_3O^+
HO
$CHCH_2CH_3$
PCC
O
CCH_2CH_3
$\overset{-}{C}H_2-\overset{+}{P}Ph_3$
H_2C
C
CH_2CH_3

c) 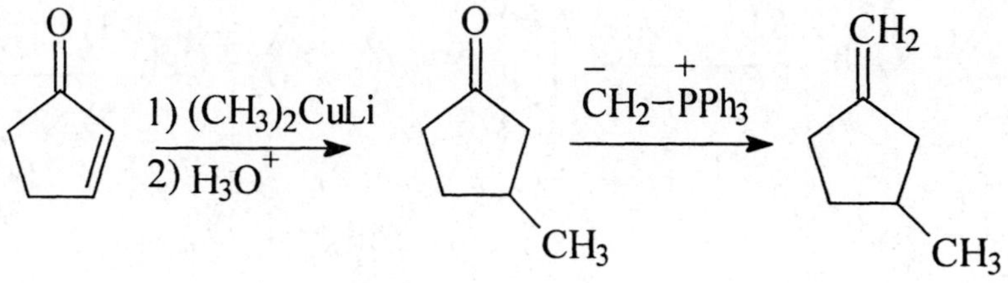
O
1) $(CH_3)_2CuLi$
2) H_3O^+
O
CH_3
$\overset{-}{C}H_2-\overset{+}{P}Ph_3$
CH_2
CH_3

d) 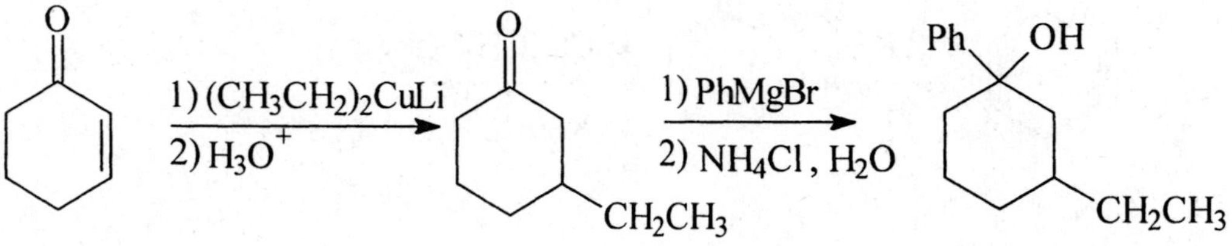
O
1) $(CH_3CH_2)_2CuLi$
2) H_3O^+
O
CH_2CH_3
1) PhMgBr
2) NH_4Cl, H_2O
Ph
OH
CH_2CH_3

e) 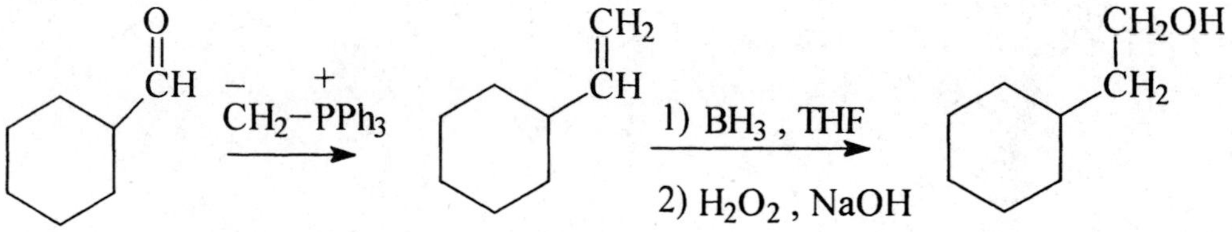
O
CH
$\overset{-}{C}H_2-\overset{+}{P}Ph_3$
CH_2
CH
1) BH_3, THF
2) H_2O_2, NaOH
CH_2OH
CH_2

f) 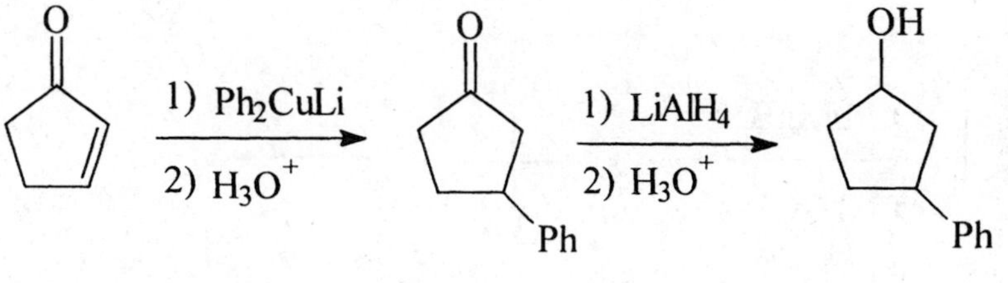
O
1) Ph_2CuLi
2) H_3O^+
O
Ph
1) $LiAlH_4$
2) H_3O^+
OH
Ph

18.43

a) 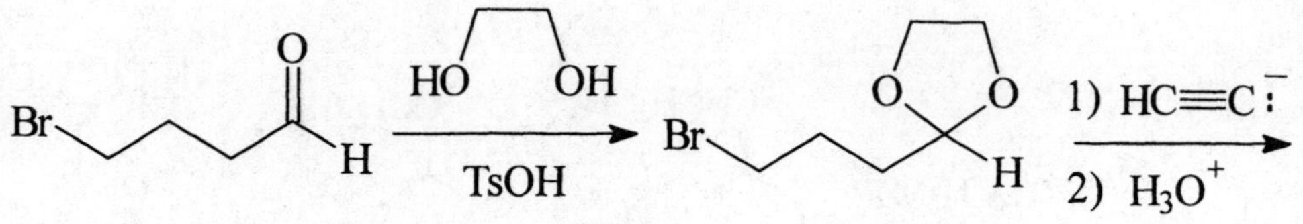

b) 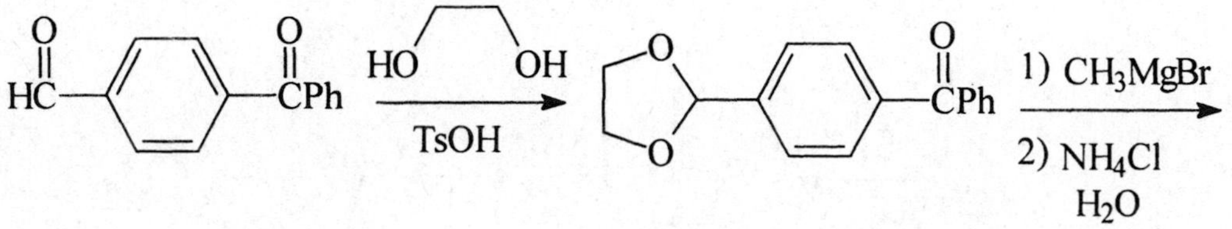

c) 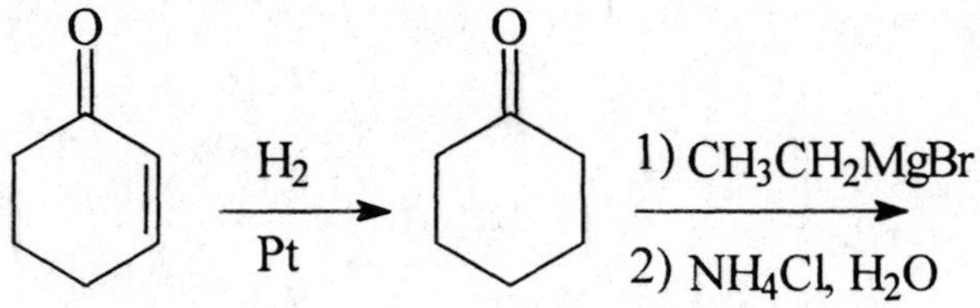

d) 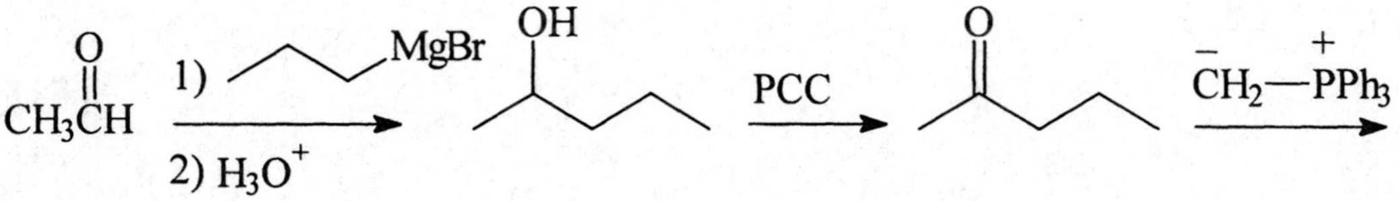

e) 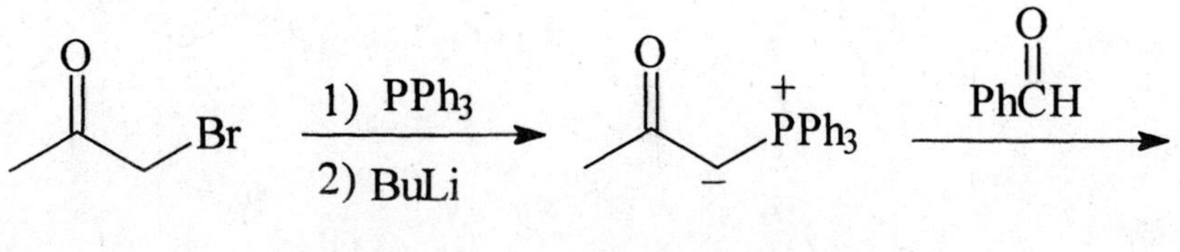

f) 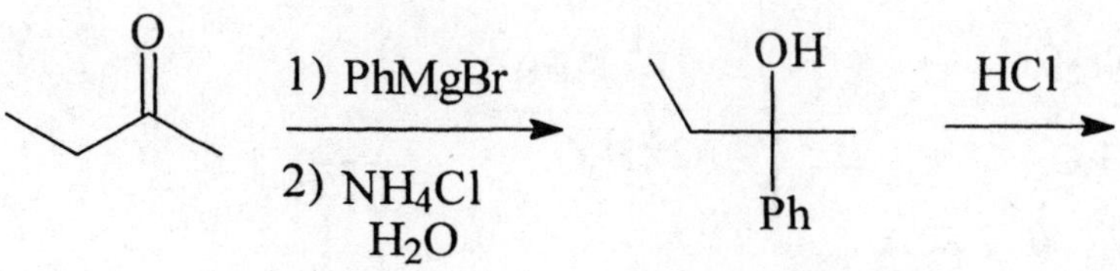

g) 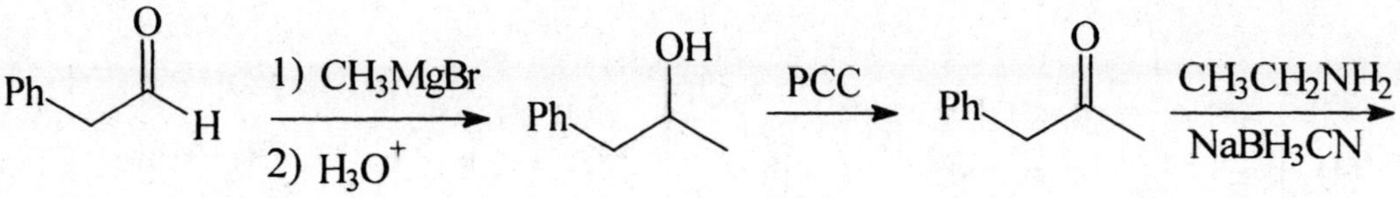
Ph
O
H
1) CH3MgBr
2) H3O+
OH
Ph
PCC
Ph
O
CH3CH2NH2
NaBH3CN

h) 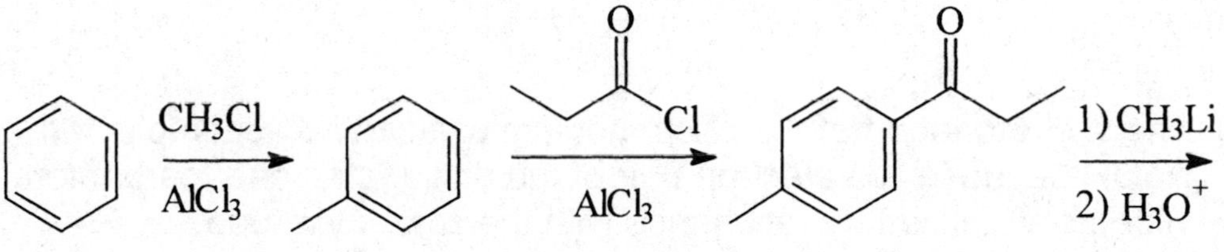
CH3Cl
AlCl3
O
Cl
AlCl3
O
1) CH3Li
2) H3O+

i) 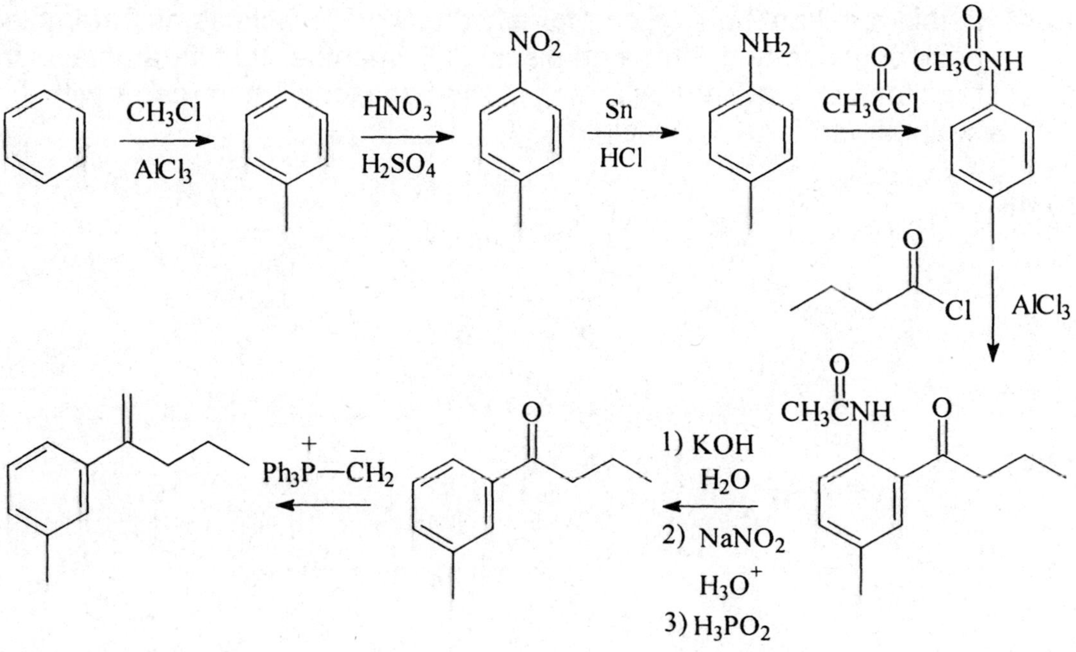
CH3Cl
AlCl3
HNO3
H2SO4
NO2
Sn
HCl
NH2
O
CH3CCl
O
CH3CNH
O
Cl
AlCl3
O
CH3CNH
O
1) KOH
H2O
2) NaNO2
H3O+
3) H3PO2
O
Ph3P—CH2

j) 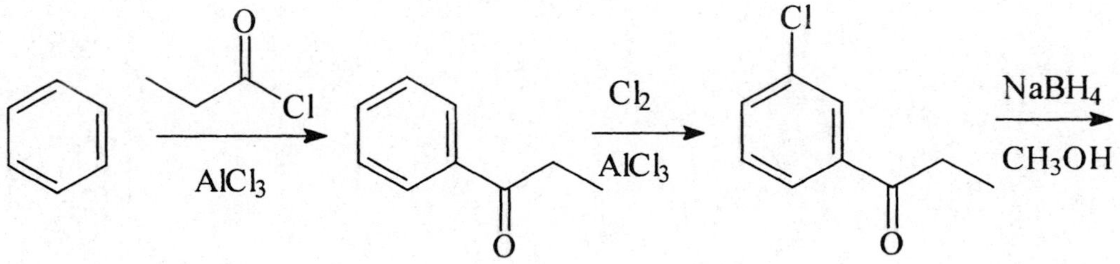
O
Cl
AlCl3
O
Cl2
AlCl3
Cl
O
NaBH4
CH3OH

k) 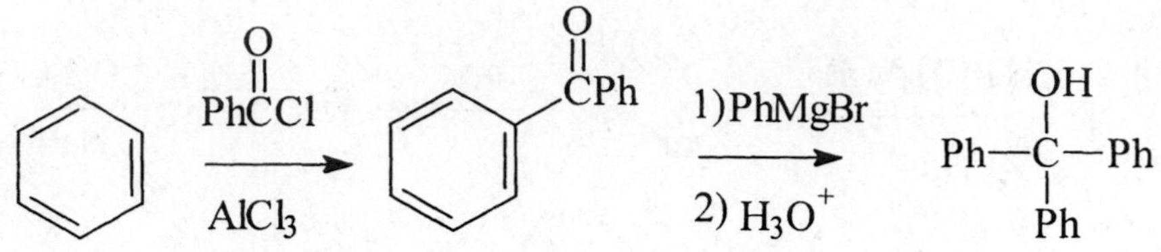

18.44 The deprotonation of the phosphonium salt can be accomplished with NaOH because the electron pair of the conjugate base (carbanion of the ylide) is stabilized by resonance with the carbonyl group.

18.45 In this reaction the oxygen of the hydroxy group acts as an intramolecular nucleophile to first form a hemiacetal. Remember that intramolecular reactions are favored by entropy. The hemiacetal then reacts with methanol to form the acetal.

18.46

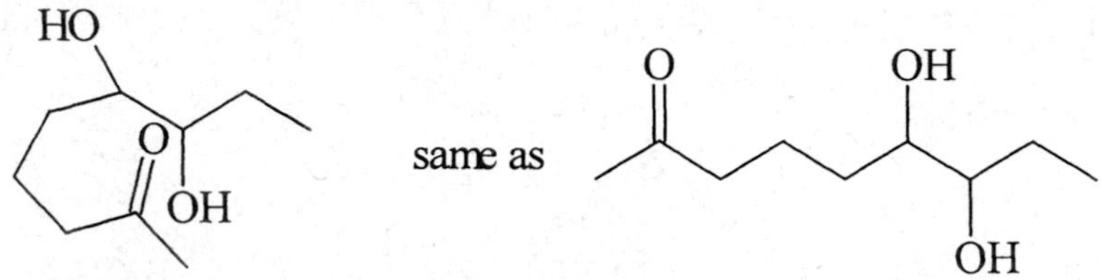

18.47

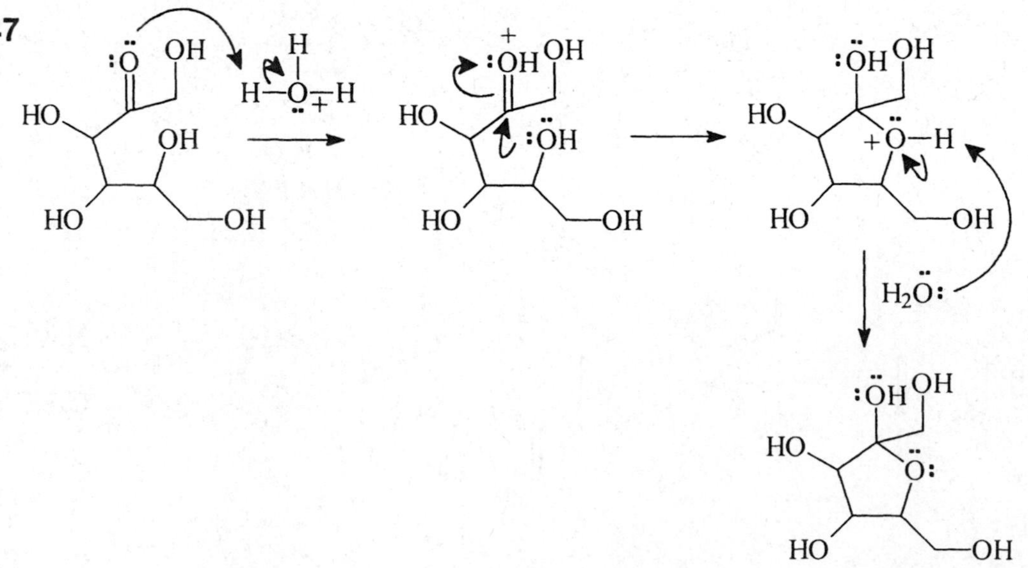

18.48

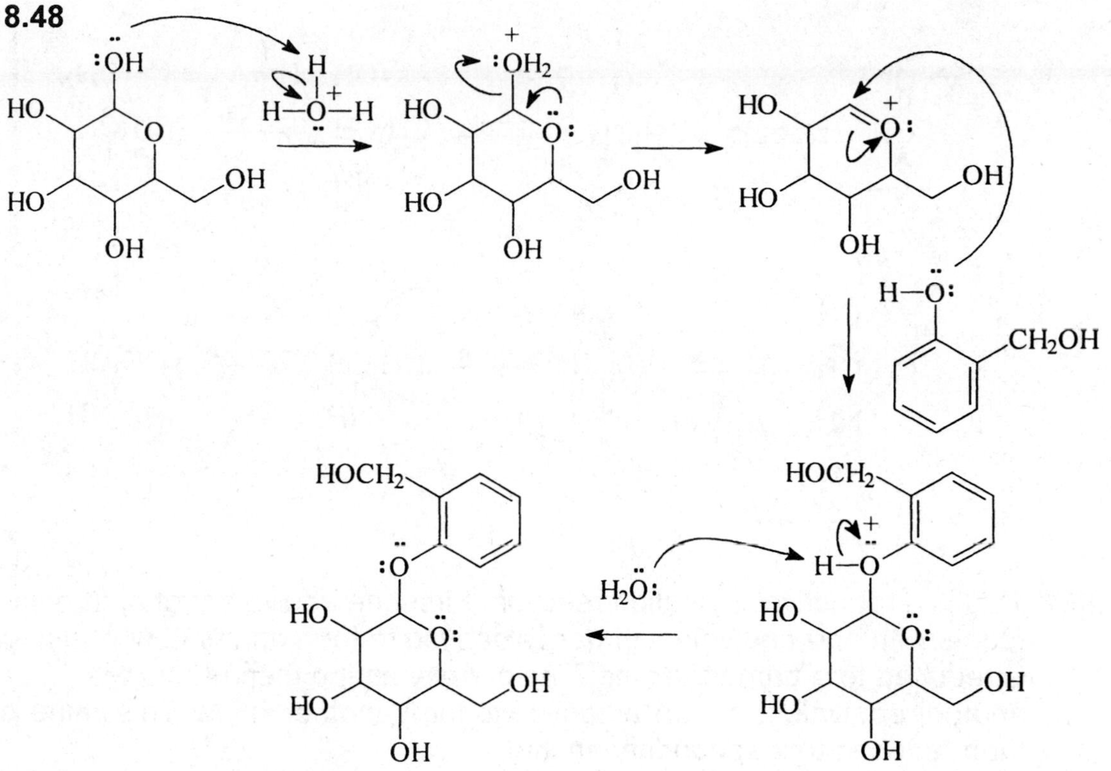

18.49 The first step forms a C=N, which reacts just like a C=O in the second step.

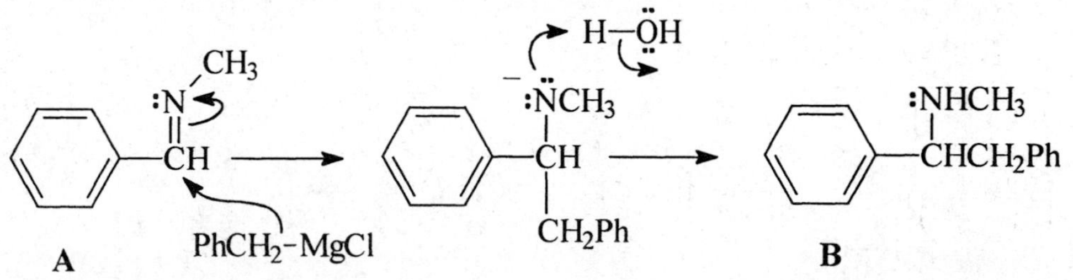

18.50 **A** is an imine (C=N), formed by the same mechanism as that shown in Figure 18.3. Hydrogen cyanide ion reacts with the C=N in the same manner as it reacts with a C=O.

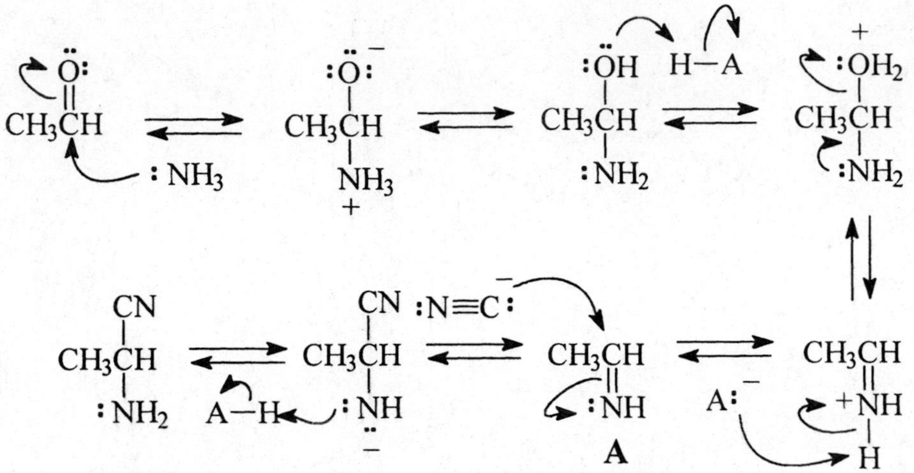

18.51 This is a reductive amination reaction. First one equivalent of ammonia reacts with one equivalent of benzaldehyde to form an imine which in turn is reduced to a primary amine. The primary amine then reacts with another equivalent of benzaldehyde to form another imine. This imine is then reduced to a secondary amine.

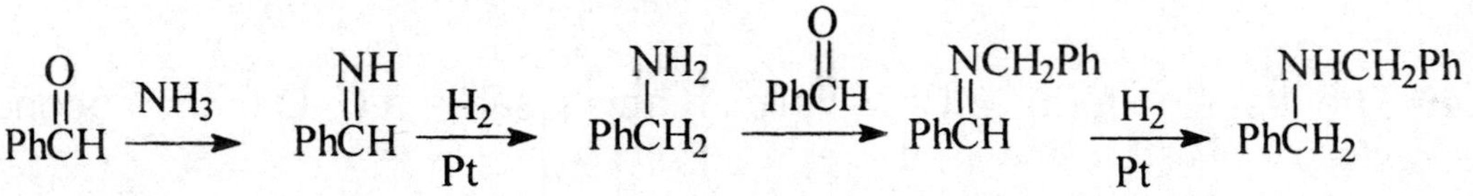

18.52

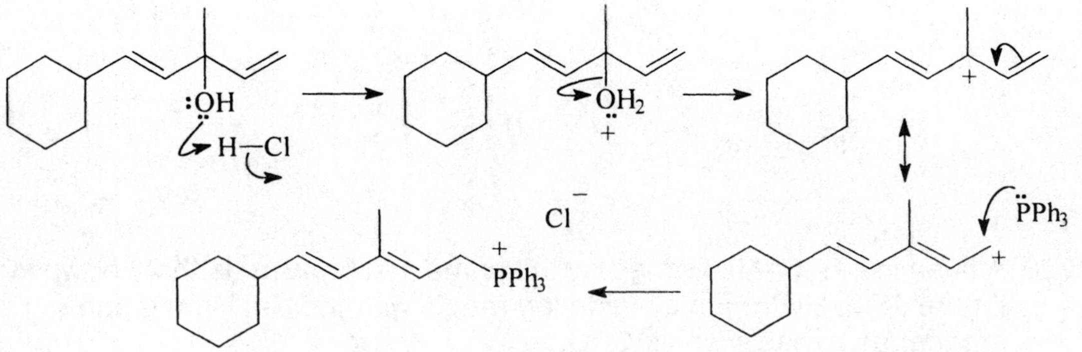

18.53 This is a conjugate addition reaction.

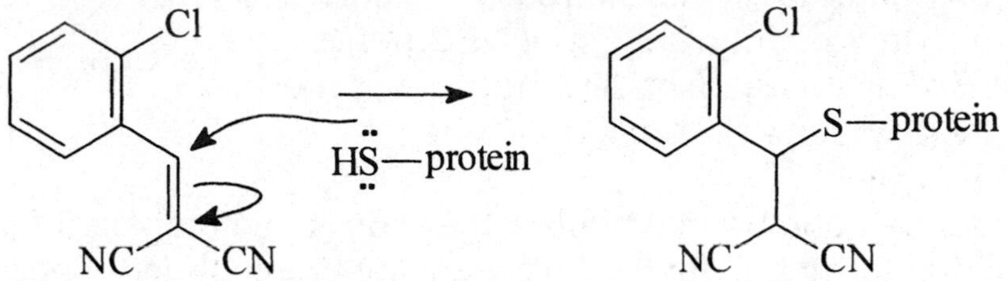

18.54 This compound can form a cyclic hemiacetal because it has an alcohol nucleophile and an aldehyde electrophile in the same molecule. Although hemiacetals are normally not favored at equilibrium, formation of a six-membered cyclic hemiacetal has a larger equilibrium constant than a comparable intermolecular reaction. In this case the equilibrium favors the cyclic hemiacetal, there is no carbonyl band in the IR spectrum.

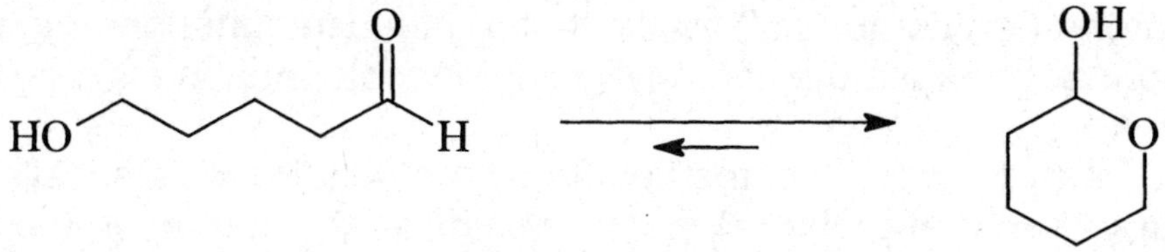

18.55

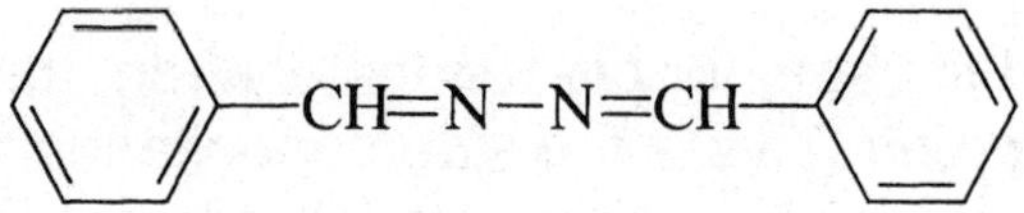

18.56 Grignard reactions with ketones give tertiary alcohols. These alcohols are very prone to undergo E1 elimination. In this reaction the initially formed alcohol undergoes an acid catalyzed elimination reaction to produce a stable conjugated alkene. This is the reason for not observing the OH absorption in the 3500-3200 cm^{-1} region in the IR spectrum.

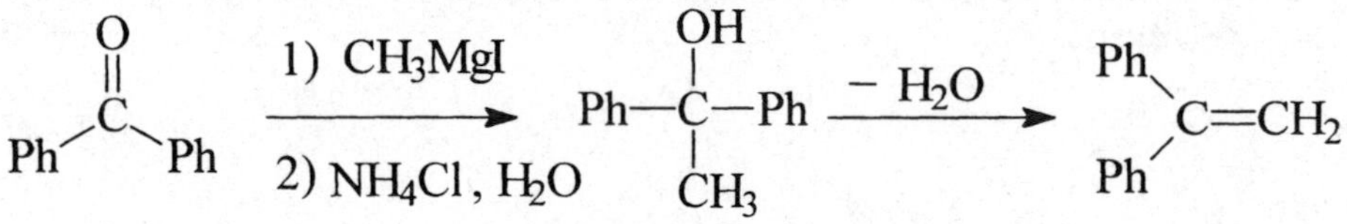

18.57 The absence of an absorption for an OH group in the IR spectrum and the presence of an absorption for a C=O group suggest that a conjugate addition occurred. The NMR spectrum confirms that the product is 2-pentanone, the result of a conjugate addition reaction rather than a 1,2-addition.

$$CH_3CH_2CH_2\overset{\overset{O}{\|}}{C}CH_3$$

18.58 a) Butanal (left compound) has a larger equilibrium constant for cyanohydrin formation than 2-butanone (right compound) because aldehydes are more reactive than ketones due to steric and inductive effects.
b) The aldehyde on the left has a larger equilibrium constant for cyanohydrin formation because it is less sterically hindered than its isomer on the right.
c) The aldehyde on the right has a larger equilibrium constant for cyanohydrin formation because it is less sterically hindered than the aldehyde on the left.

18.59 a) Acetone (left compound) reacts faster than acetophenone (right compound) because the carbonyl group of acetophenone is stabilized by resonance.
b) The carbonyl carbon of the ketone on the right is less sterically hindered than that of the ketone on the left, so the ketone on the right reacts faster.

18.60 The hydride is added to camphor predominately from the less hindered side of the carbonyl group. This is the side opposite the bridge with the two methyl groups and results in the formation of more isoborneol than borneol.

18.61 The different types of addition products result from differing amounts of steric hindrance at the carbonyl carbons and the β-carbons of these two ketones. The ketone on the left has one phenyl group on its β-carbon, so it reacts by conjugate addition (1,4-addition). The ketone on the right has two phenyl groups on its β-carbon. The additional steric hindrance at the β-carbon causes the reaction of this ketone to proceed by normal addition (1,2-addition).

Review of Mastery Goals

After completing this chapter, you should be able to:

Show the products resulting from the addition to aldehydes and ketones of all of the reagents discussed in this chapter.
(Problems 18.2, 18.6, 18.8, 18.9, 18.11, 18.13, 18.16, 18.19, 18.20, 18.23, 18.27, 18.28, 18.32, 18.33, and 18.36)

Show the products resulting from the addition of certain of these reagents to α,β-unsaturated compounds, noting whether 1,2- or 1,4-addition predominates.
(Problems 18.25, 18.29, 18.32, and 18.36)

Show the mechanisms for any of these additions.
(Problems 18.1, 18.3, 18.17, 18.18, 18.22, 18.39, 18.41, 18.45, 18.48, and 18.52)

Predict the effect of the structure of the aldehyde or ketone on the rate or the position of the equilibrium for these reactions.
(Problems 18.4, 18.5, 18.7, 18.14, 18.21, 18.30, 18.31, 18.34, 18.35, 18.58, 18.59)

Use these reactions, in combination with the reactions from previous chapters, to synthesize compounds.
(Problems 18.10, 18.12, 18.15, 18.26, 18.39, 18.42, 18.43)

Use acetals as protective groups in syntheses.
(Problems 18.24 and 18.43)

Chapter 19
SUBSTITUTIONS AT THE CARBONYL GROUP
REACTIONS OF CARBOXYLIC ACIDS AND DERIVATIVES

19.1 a) The ester on the left reacts faster because it is less sterically hindered.
b) The ester on the left reacts faster because it has a more electrophilic carbonyl carbon due to the electron withdrawing F's.
c) The ester on the left reacts faster because it has a more electrophilic carbonyl carbon due to the electron withdrawing nitro group.

19.2 The equilibrium favors the carboxylic acid derivative that is lower on the reactivity scale (less reactive).
a) The equilibrium favors the products because an anhydride is more reactive than an acid or an ester.
b) The equilibrium favors the reactants because an ester is more reactive than an amide.
c) The equilibrium favors the products because an acyl chloride is more reactive than an amide.

19.3 The reaction of the ester is faster because an ester is more reactive than an amide.

19.4 React propylamine as the nucleophile with a carboxylic acid derivative that is more reactive than the amide product, such as an acyl chloride.

$$CH_3CH_2CH_2\overset{\overset{O}{\|}}{C}Cl \; + \; NH_2CH_2CH_2CH_3 \longrightarrow$$

19.5

a) 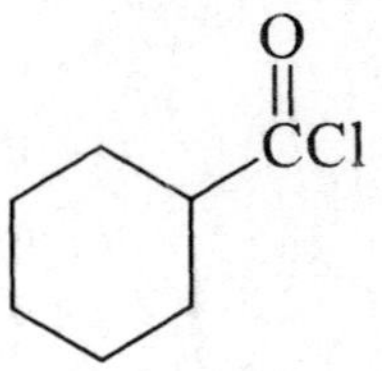

b) 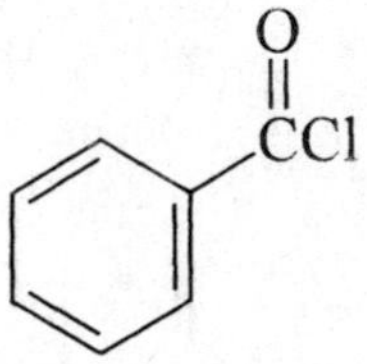

c) $CH_3CH_2CH_2\overset{\overset{O}{\|}}{C}Cl$

19.6

a) $CH_3CH_2\overset{\displaystyle O}{\overset{\|}{C}}O\overset{\displaystyle O}{\overset{\|}{C}}CH_2CH_3$ b)

$+$ $2\ CH_3\overset{\displaystyle O}{\overset{\|}{C}}OH$

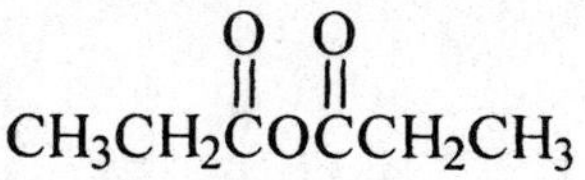

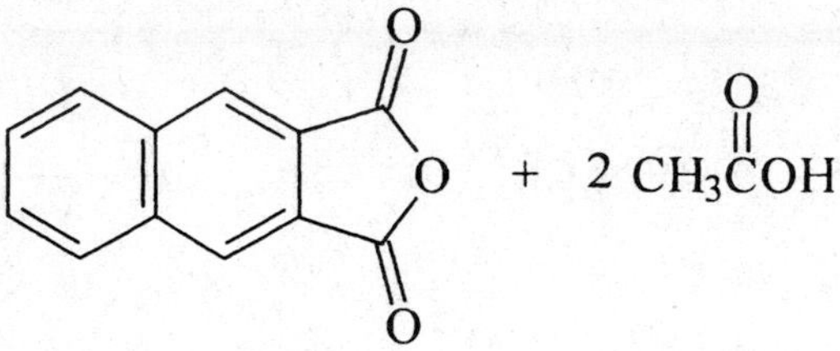

19.7

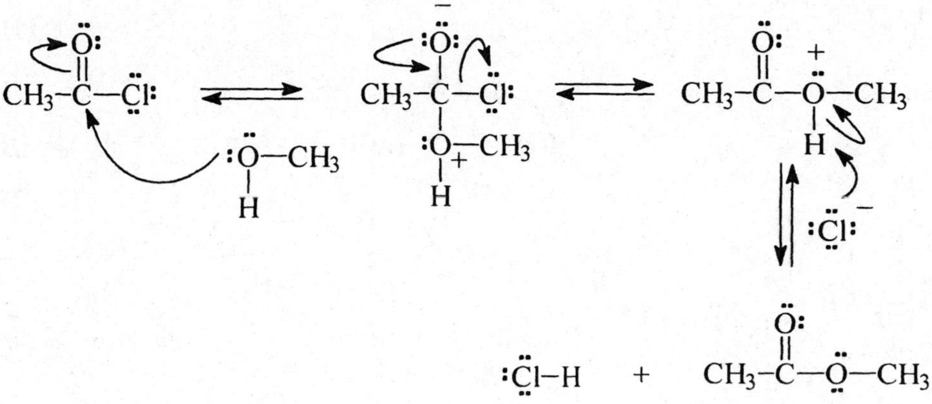

19.8 The reaction produces aspirin, an ester, rather than the more reactive anhydride.

19.9

a)

b)

c) $CH_3\overset{\displaystyle O}{\overset{\|}{C}}OCH_2\overset{\displaystyle CH_3}{\overset{|}{C}}HCH_3\ +\ CH_3\overset{\displaystyle O}{\overset{\|}{C}}OH$

d)

e) $\xrightarrow{\ CH_3OH\ }$

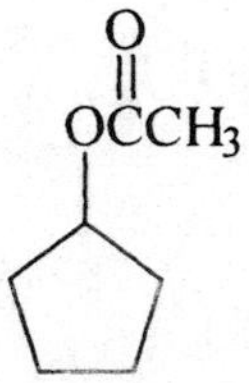

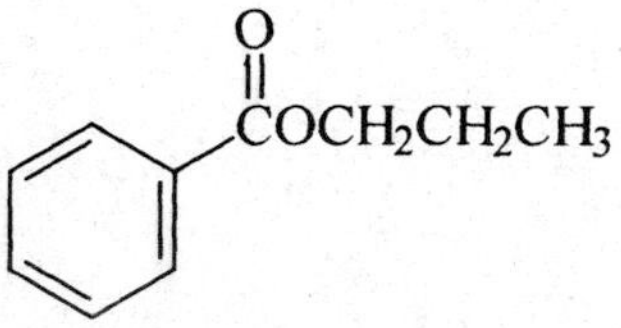

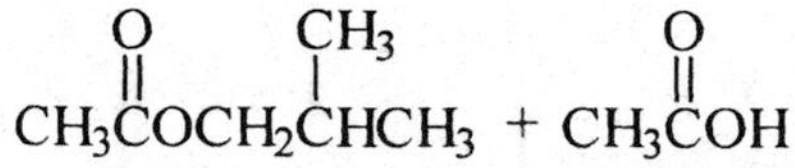

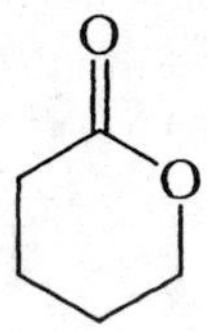

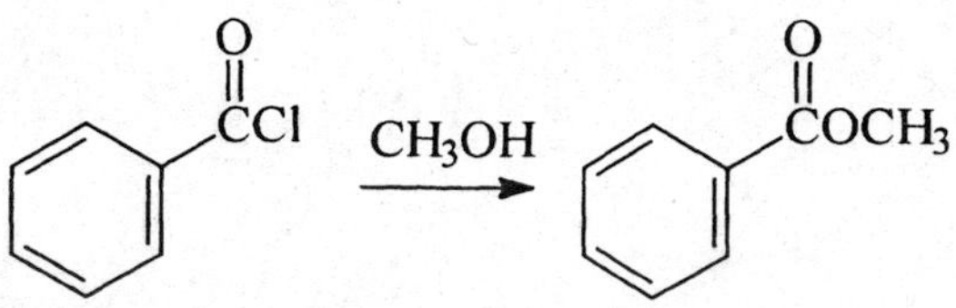

f)

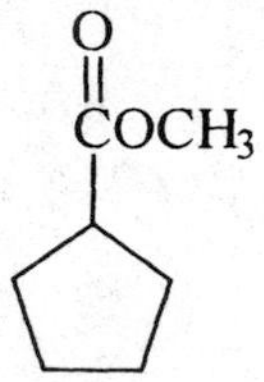

19.10

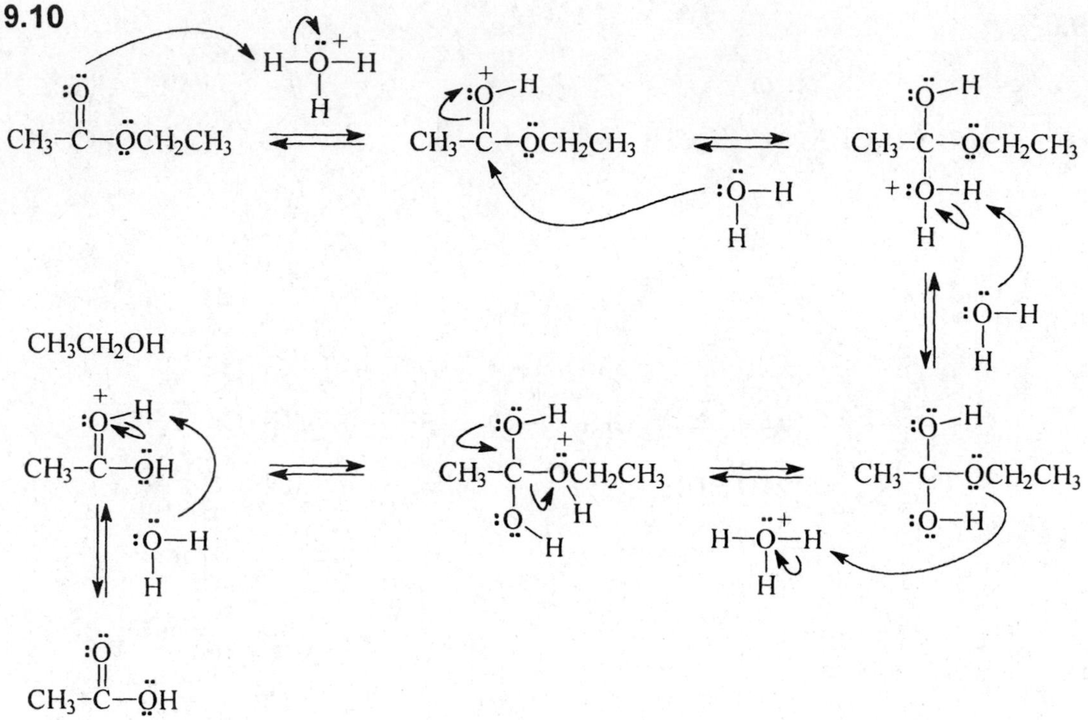

19.11

$$\begin{matrix} CH_2OH \\ | \\ CHOH \\ | \\ CH_2OH \end{matrix} \quad + \quad \begin{matrix} ^{-}OC(=O)(CH_2)_7CH{=}CH(CH_2)_7CH_3 \\ + \\ ^{-}OC(=O)(CH_2)_{14}CH_3 \\ + \\ ^{-}OC(=O)(CH_2)_{12}CH_3 \end{matrix}$$

19.12

a) $2\ CH_3C(=O)OH$ b) $PhC(=O)\text{-}O^{-}$ + $HOCH_2CH_3$ c) $CH_3C(=O)OH$ + $\overset{+}{N}H_3Ph\ Cl^{-}$

d) O_2N—⬡—$\overset{\overset{\displaystyle O}{\|}}{C}OH$ e) $CH_3CH_2\overset{\overset{\displaystyle O}{\|}}{C}OH$ + $HOCH_2CH_3$ f) Ph—$\overset{\overset{\displaystyle O}{\|}}{C}CH_2\overset{\overset{\displaystyle O}{\|}}{C}NH_2$

g) $CH_3CH_2\overset{\overset{\displaystyle OH}{|}}{C}HCH_3$ h) $CH_3O\overset{\overset{\displaystyle O}{\|}}{C}$—⬡—$\overset{\overset{\displaystyle O}{\|}}{C}OH$ i) $CH_3CH_2CH_2\overset{\overset{\displaystyle O}{\|}}{C}OH$

19.13

⬡—CH_2Br $\xrightarrow[\substack{\text{DMSO} \\ (S_N2, \text{ Chapter 10})}]{\text{NaCN}}$ ⬡—$CH_2C{\equiv}N$ $\xrightarrow[\substack{H_2O \\ \text{reflux}}]{H_2SO_4}$ ⬡—$CH_2\overset{\overset{\displaystyle O}{\|}}{C}OH$

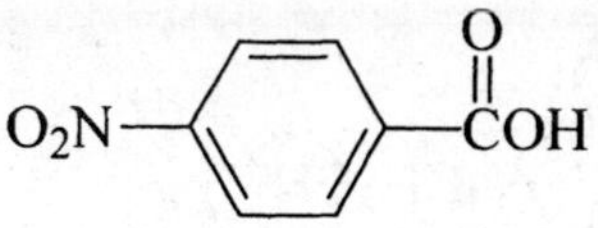

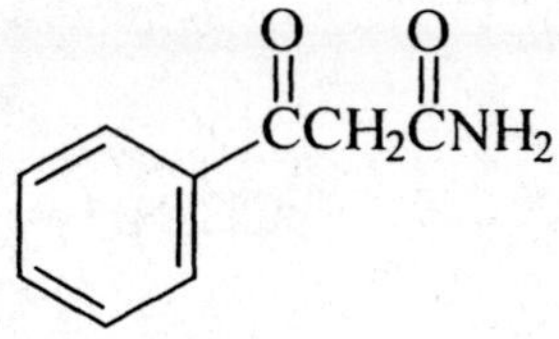

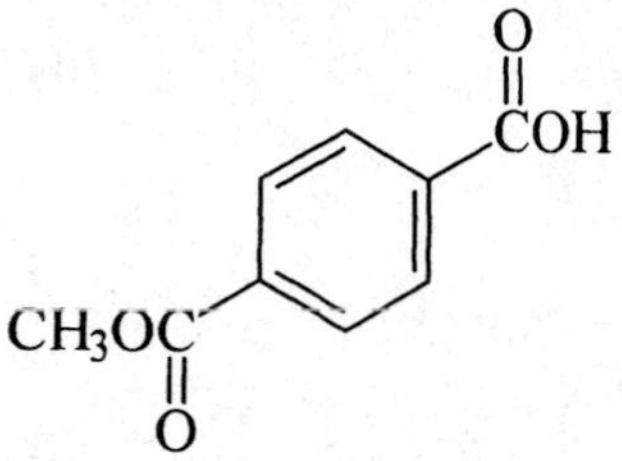

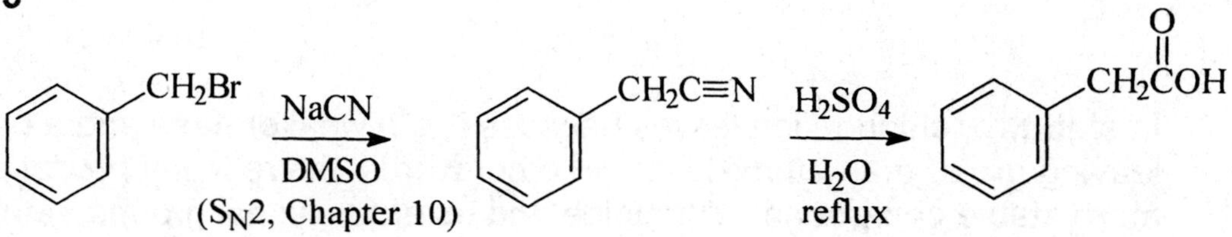

19.14

a)

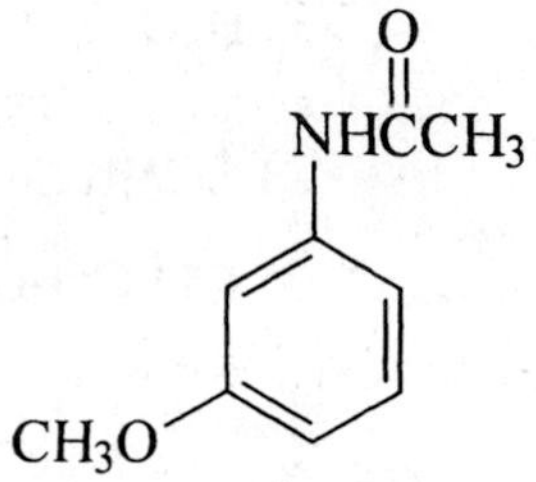

b) $CH_3CH_2CH_2\overset{\overset{\displaystyle O}{\|}}{C}N(CH_2CH_3)_2$

c)

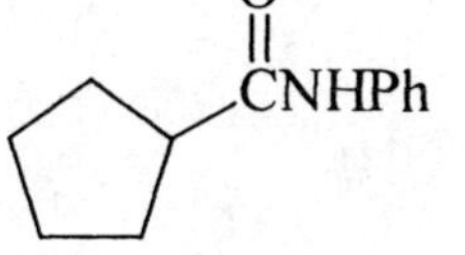

d)   + CH_3OH

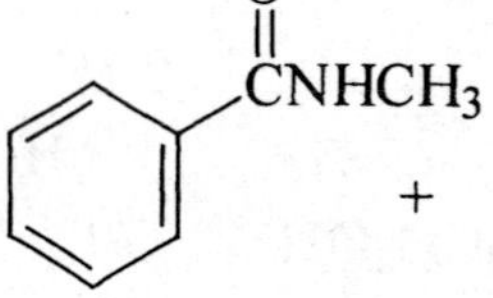

19.15

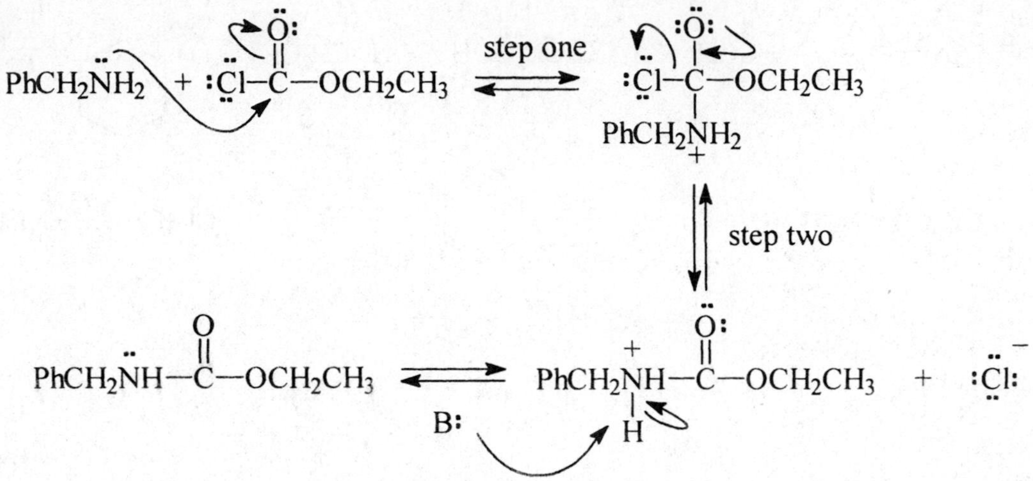

In step two, chloride ion leaves because it is a weaker base and a better leaving group than ethoxide ion. In other words, the reaction produces the more stable compound, with amide and ester functional groups, rather than the less stable compound, with amide and acyl chloride functional groups.

19.16

a)
1) $SOCl_2$
2) OH

b)
1) $SOCl_2$
2) excess
$CH_3CH_2CH_2NH_2$

c)
1) H_3O^+, Δ
2) $SOCl_2$
3) CH_3CH_2OH

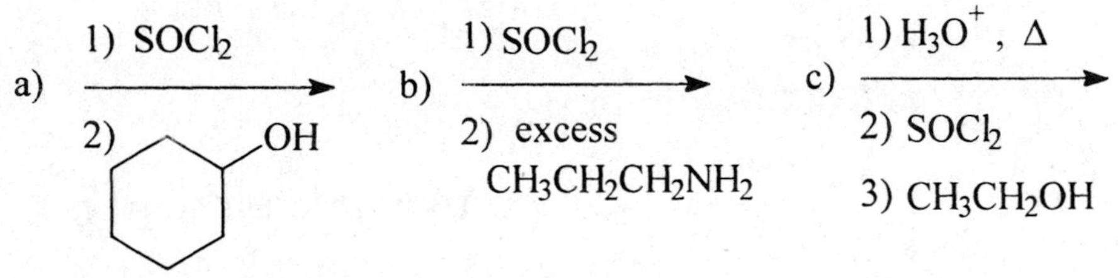

19.17

a) $CH_3CH_2CH_2CH_2OH$
+ CH_3OH

b)

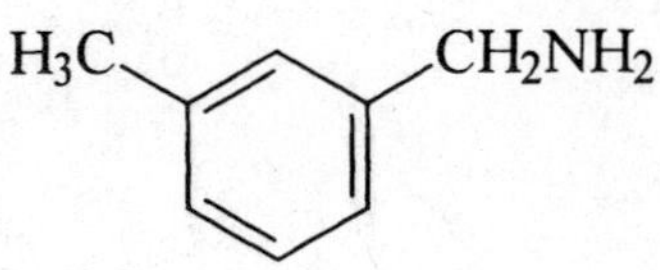

c)
+ CH_3CH_2OH

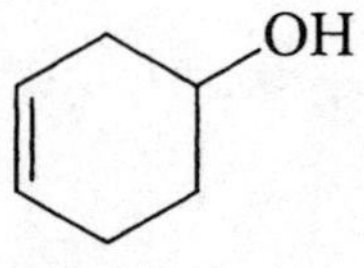

d)

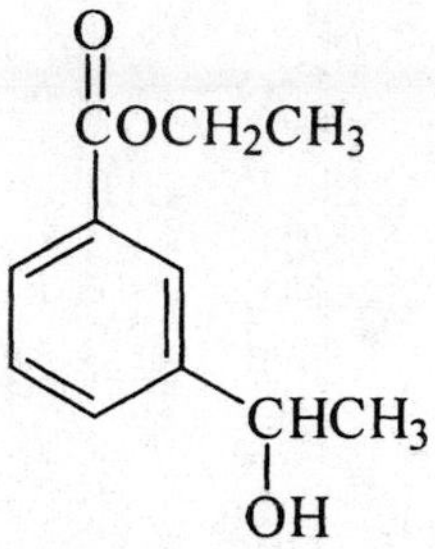

e)

19.18 If acid were used, the product, an amine, would also be protonated.

19.19

a)
$$CH_3\overset{\underset{\displaystyle CH_3}{|}}{CH}-\overset{\underset{\displaystyle }{\overset{\displaystyle O}{\|}}}{CH}$$

b)

c)

19.20

a) $\xrightarrow{\begin{array}{l}1)\ SOCl_2 \\ 2)\ LiAlH(Ot\text{-}Bu)_3 \\ \quad -78^\circ\ C\end{array}}$

b) $\xrightarrow{\begin{array}{l}1)\ SOCl_2 \\ 2)\ PhNH_2 \\ 3)\ LiAlH_4 \\ 4)\ H_2O\end{array}}$

c) $\xrightarrow{\begin{array}{l}1)\ H_3O^+,\ \Delta \\ 2)\ SOCl_2 \\ 3)\ LiAlH(Ot\text{-}Bu)_3 \\ \quad -78\ ^\circ C\end{array}}$

d) $\xrightarrow{\begin{array}{l}1)\ LiAlH_4 \\ 2)\ H_2O \\ 3)\ CH_3\overset{O}{\overset{\|}{C}}Cl \\ 4)\ LiAlH_4 \\ 5)\ H_2O\end{array}}$

19.21

a)

b)
$$CH_3CH_2CH_2\overset{\underset{\displaystyle CH_3}{|}}{\overset{\underset{\displaystyle }{\overset{\displaystyle OH}{|}}}{C}}CH_3$$

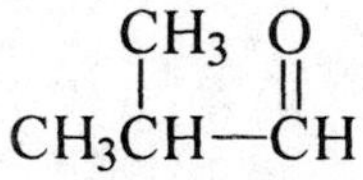

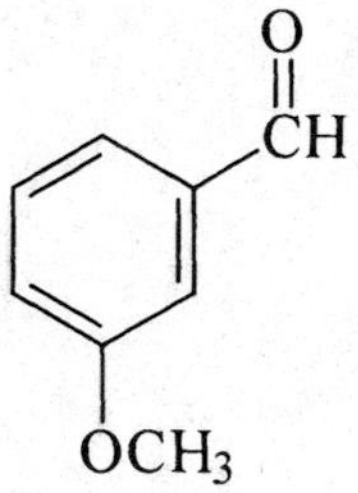

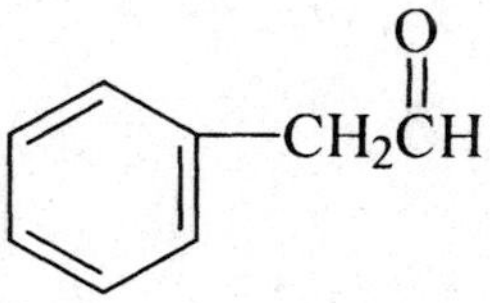

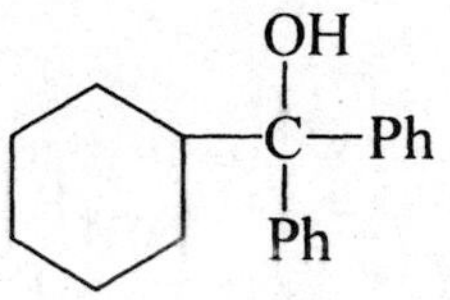

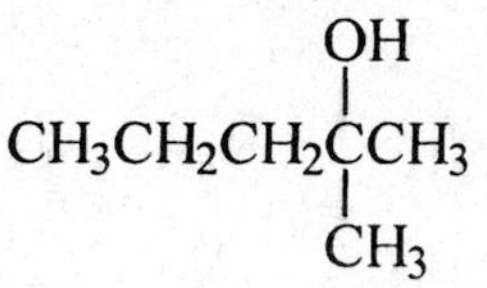

c)

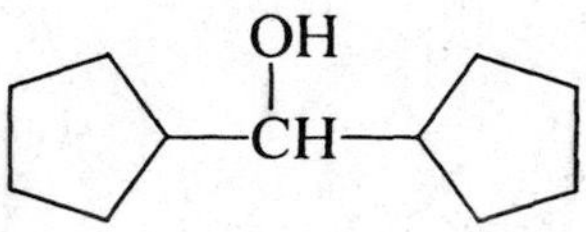

19.22

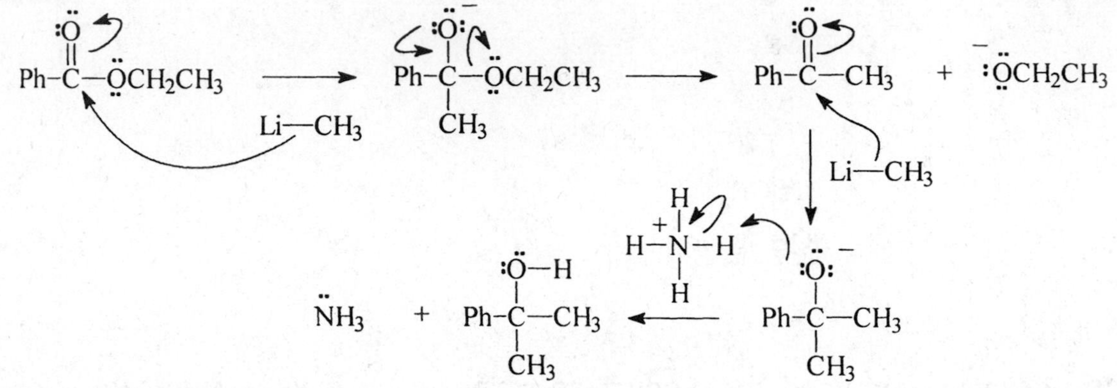

19.23

a) 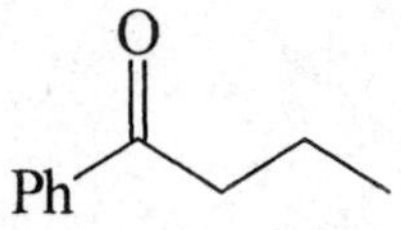

b) 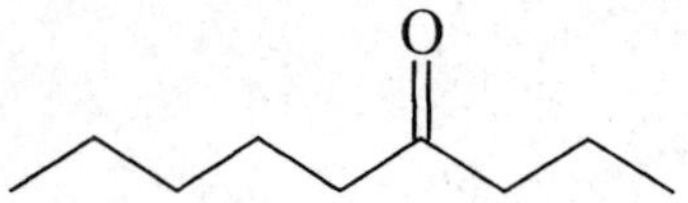

19.24

a) 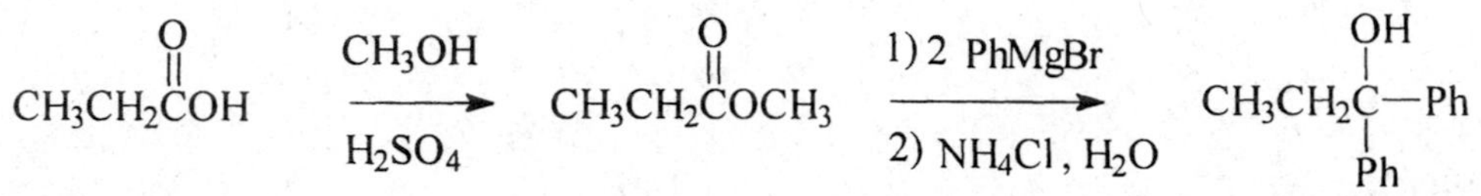

b) 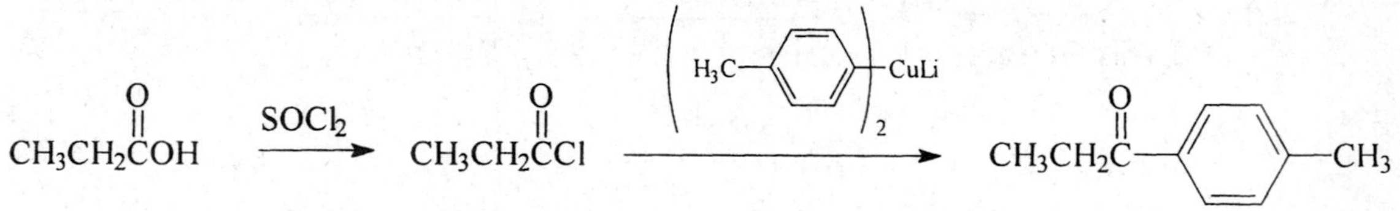

c) 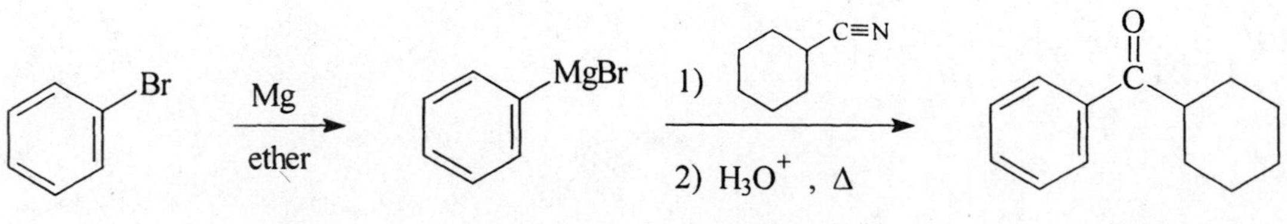

d) 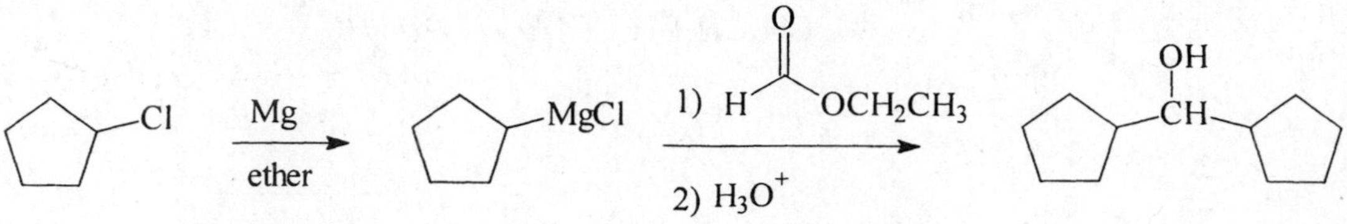

19.25

a) $H_3C\text{—}$ (p-tolyl)$\text{—}SO_2\text{—}O\text{—}$cyclohexyl

b) Ph–NH–S(=O)$_2$–CH$_3$

c) $CH_3CH_2OP(=O)(OCH_2CH_3)(OCH_2CH_3)$

19.26

a) $CH_3CH_2O\text{—}\overset{O}{\overset{\|}{C}}\text{—}OCH_2CH_3$

b) $CH_3CH_2\overset{O}{\overset{\|}{C}}Cl$

c) $PhCH_2\overset{O}{\overset{\|}{C}}Cl \quad \xrightarrow{\;PhCH_2\overset{O}{\overset{\|}{C}}O^-\;} \quad PhCH_2\overset{O}{\overset{\|}{C}}O\overset{O}{\overset{\|}{C}}CH_2Ph$

d) Ph–$CO_2CH_2CH_2CH_2CH_3$

e) o-(CH$_2$OH)(CH$_3$)C$_6$H$_4$

f) $CH_3CH_2CH_2CH_2\overset{O}{\overset{\|}{C}}CH_2CH_2CH_2CH_3$

g) $CH_3CH_2CH_2CH_2NH_2$

h) $CH_2\text{=}CH\overset{O}{\overset{\|}{C}}OCH_2CH_3$

i) $CH_3\overset{CH_3}{\overset{|}{C}H}\overset{O}{\overset{\|}{C}}OCl \quad \xrightarrow[H_2O]{NH_3} \quad CH_3\overset{CH_3}{\overset{|}{C}H}\overset{O}{\overset{\|}{C}}ONH_2$

j) $CH_3(CH_2)_{10}CH_2OSO_2$–(p-tolyl)

k) $HO\overset{O}{\overset{\|}{C}}CH_2CH_2CH_2\overset{O}{\overset{\|}{C}}OH$

l) m-(H$_3$C)(CH$_2$CH$_2$OH)C$_6$H$_4$

m) p-(CHO)(OCH$_3$)C$_6$H$_4$

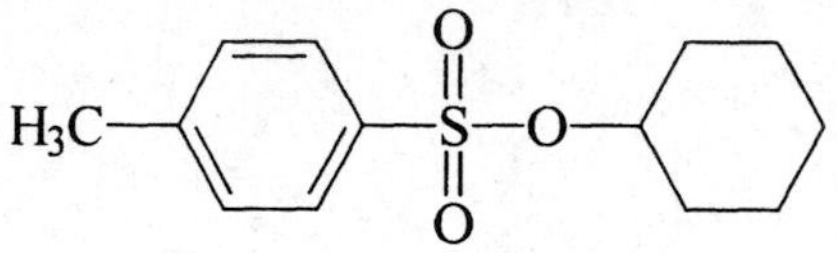

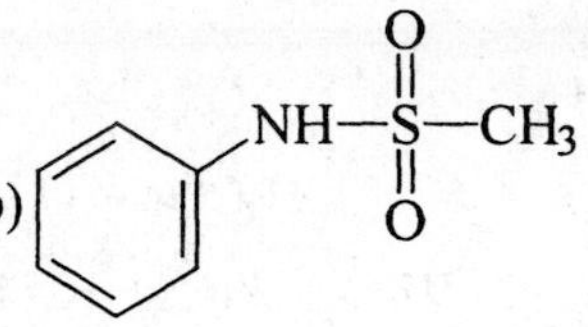

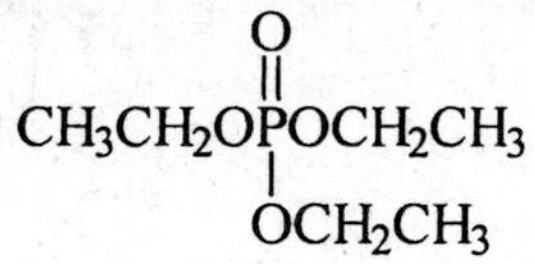

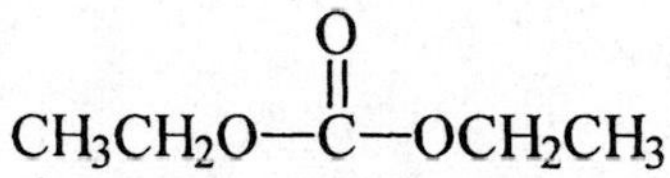

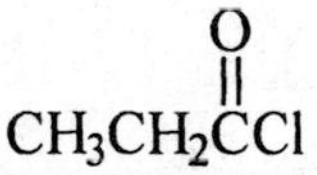

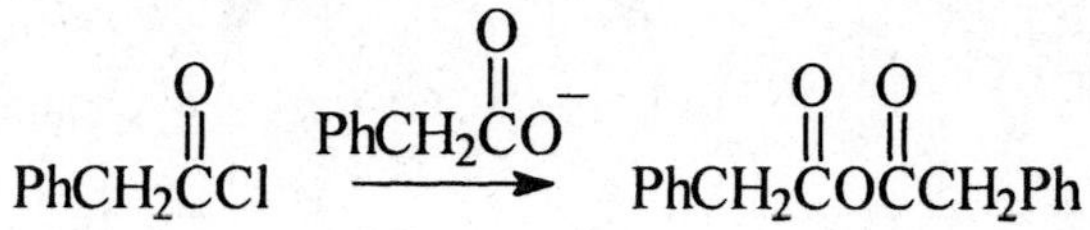

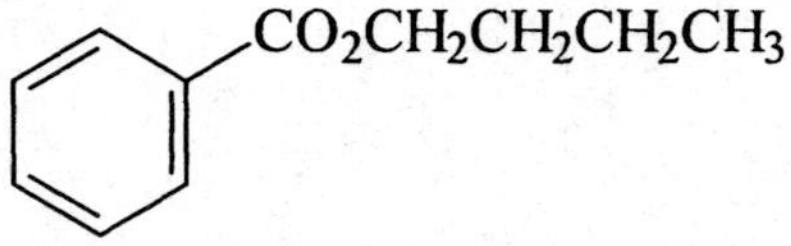

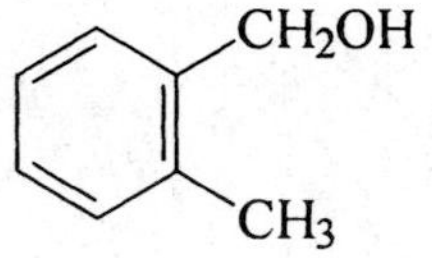

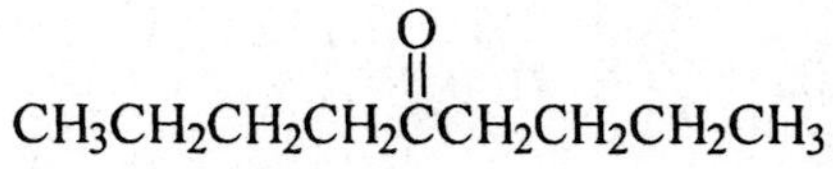

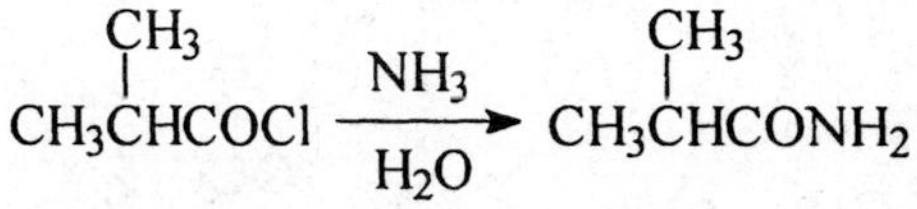

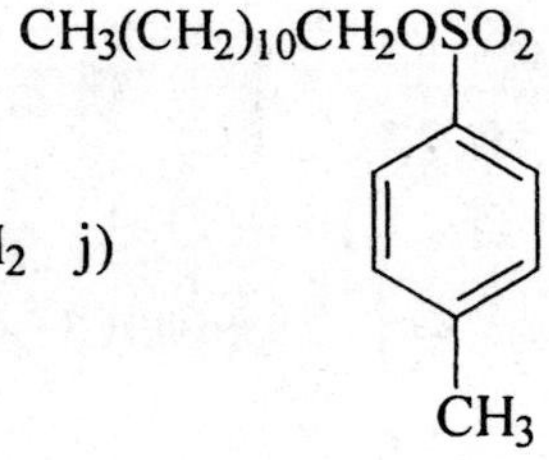

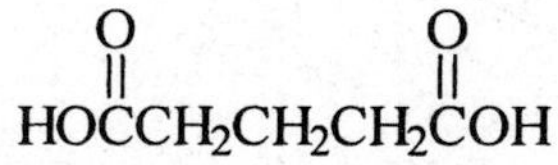

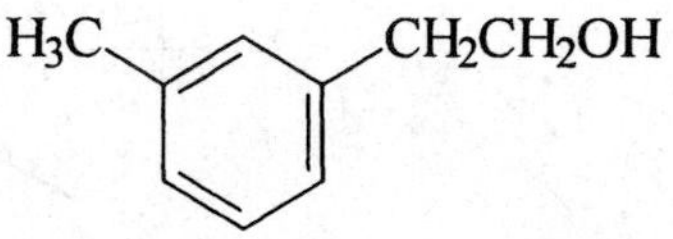

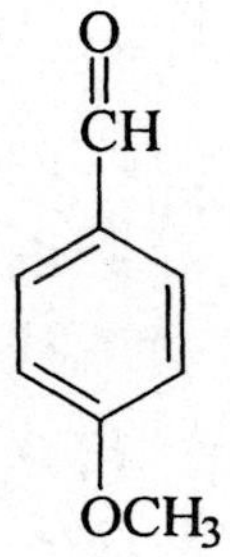

n) 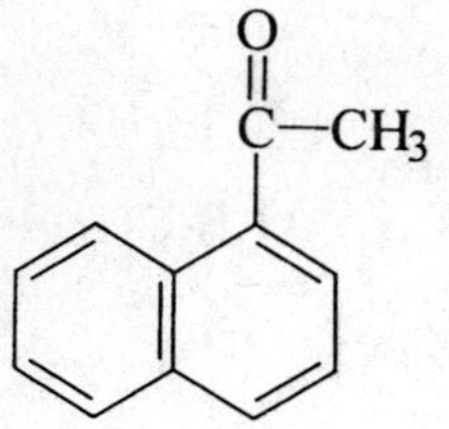

o) 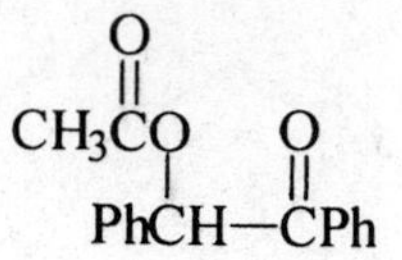

p) 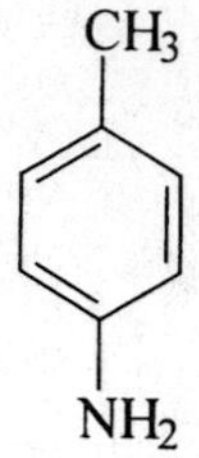

q) 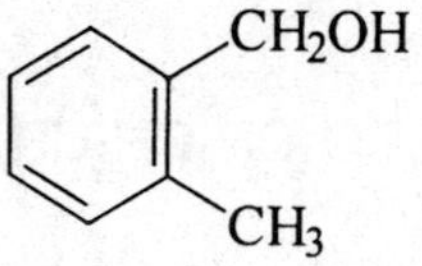

r) 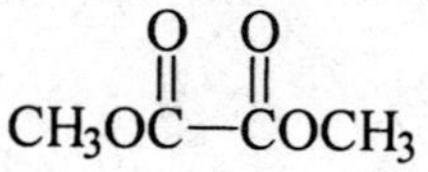

s) 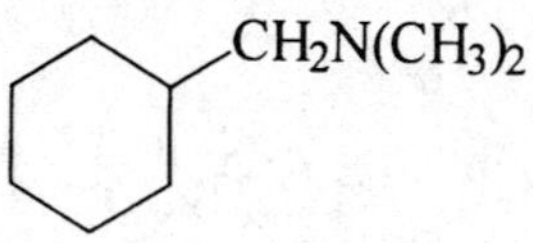

t) 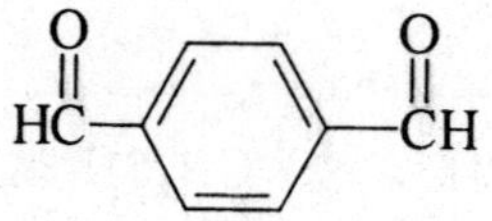

19.27

a) 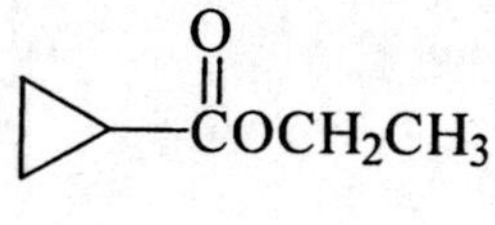

b) 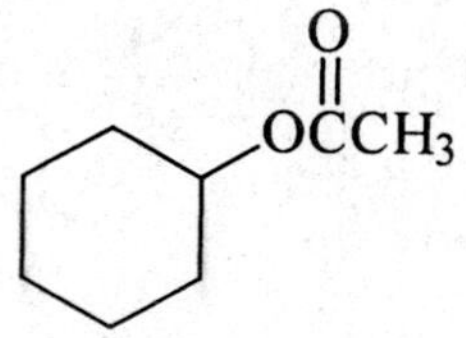

c) 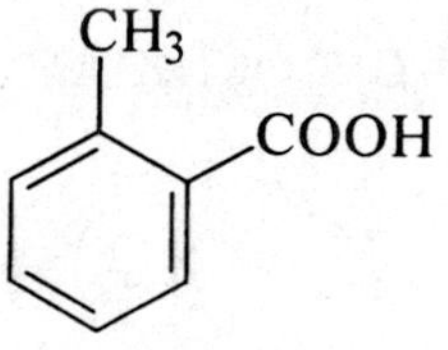

d) 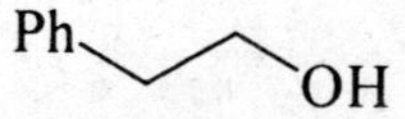

+ CH_3OH

e) 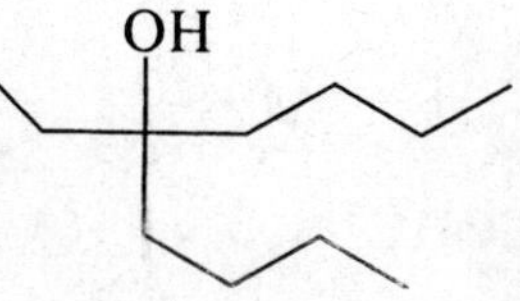

19.28

a) 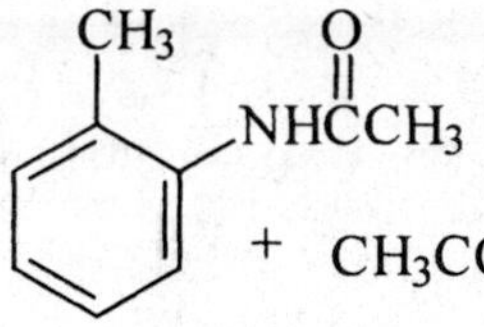 + CH_3COOH

b) 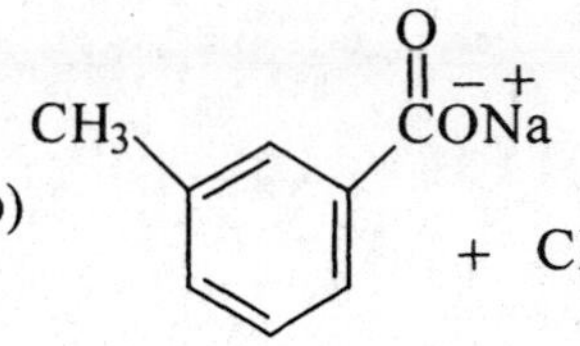 + CH_3CH_2OH

c) 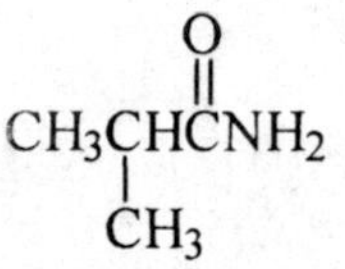

d) 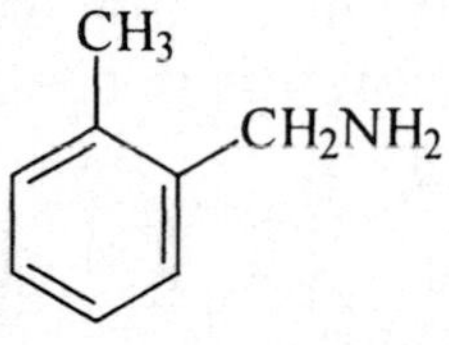

e) 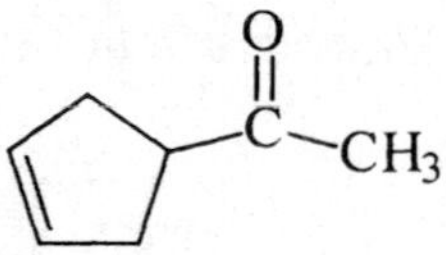

19.29

a) 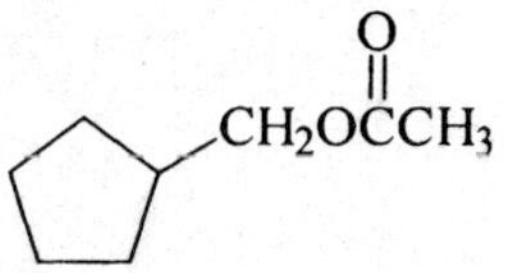

b) 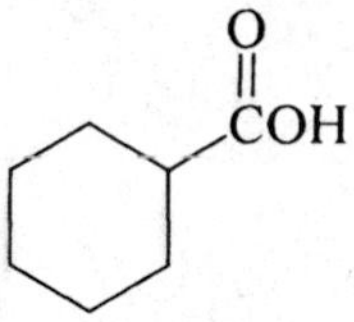

c) $CH_3CH_2CH_2NH_2$

d) 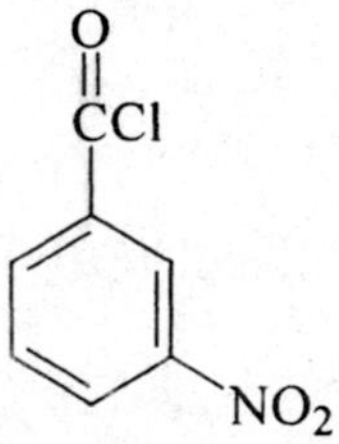

e) 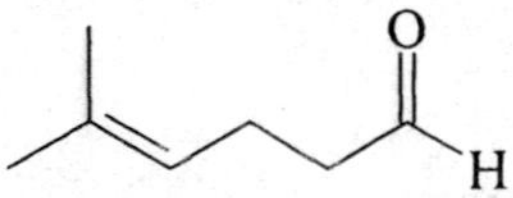

f) 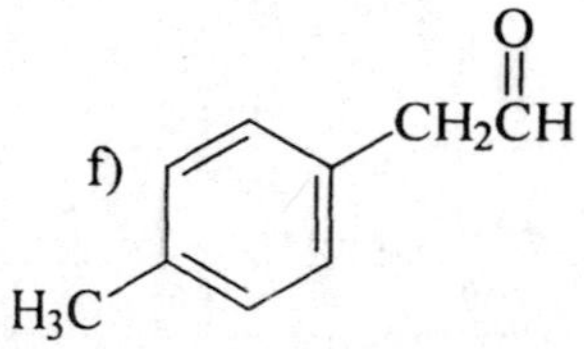

19.30

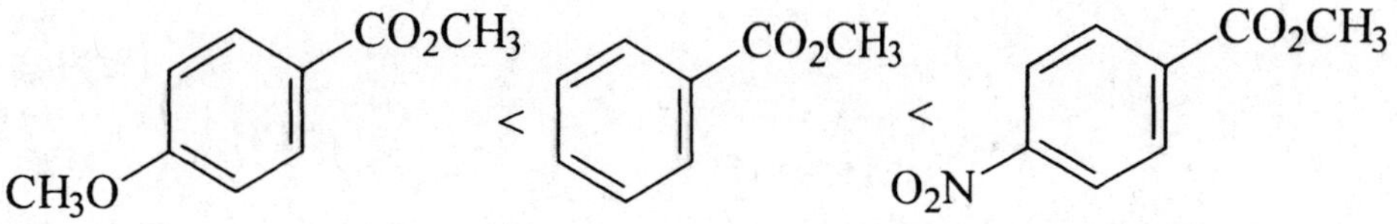

Slowest

Fastest

The methoxy group donates electrons by resonance. It makes the carbonyl carbon less electrophilic and slows the reaction. The nitro group is an electron withdrawing group by both its resonance and inductive effects. It makes the carbonyl carbon more electrophilic and accelerates the reaction.

19.31

$$\underset{\underset{\text{Slowest}}{CH_3}}{\overset{CH_3}{\underset{|}{\overset{|}{CH_3-C-CO_2CH_2CH_3}}}} < \underset{\underset{H}{\overset{CH_3}{\underset{|}{\overset{|}{CH_3-C-CO_2CH_2CH_3}}}}}{} < \underset{\underset{H}{\overset{H}{\underset{|}{\overset{|}{CH_3-C-CO_2CH_2CH_3}}}}}{\text{Fastest}}$$

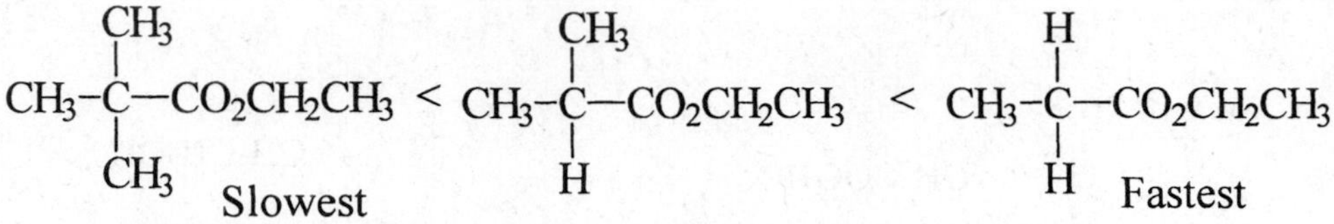

Steric hindrance slows the rate of saponification. Thus, the rate increases as the amount of steric hindrance decreases.

19.32

a)

EtOH
H_2SO_4

1) 2 CH_3MgBr
2) NH_4Cl, H_2O

b)

1) $LiAlH_4$
2) H_3O^+

c)

$SOCl_2$

$LiAlH(OtBu)_3$
$-78°C$

d)

1) $SOCl_2$
2) $(CH_3)_2NH$
   excess

1) $LiAlH_4$
2) H_2O

e)

1) $SOCl_2$
2) $(CH_3CH_2)_2CuLi$

f) as (e)

1) $PhMgBr$
2) NH_4Cl, H_2O

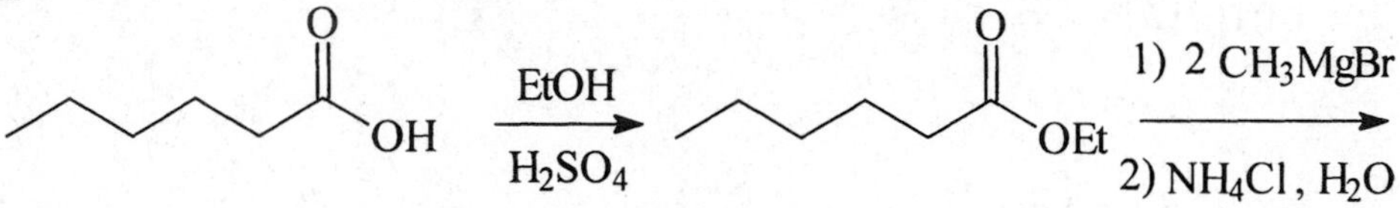

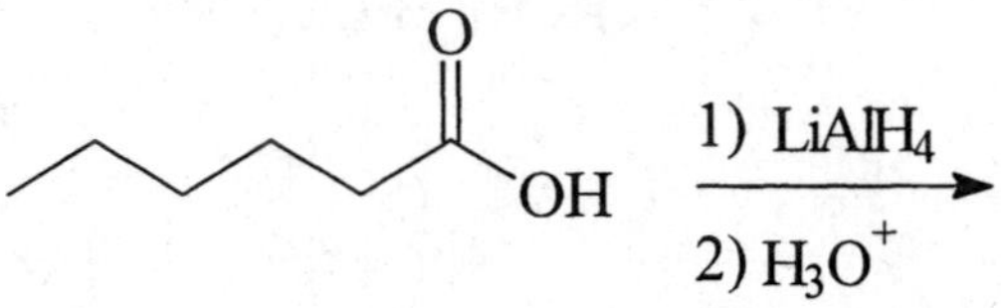

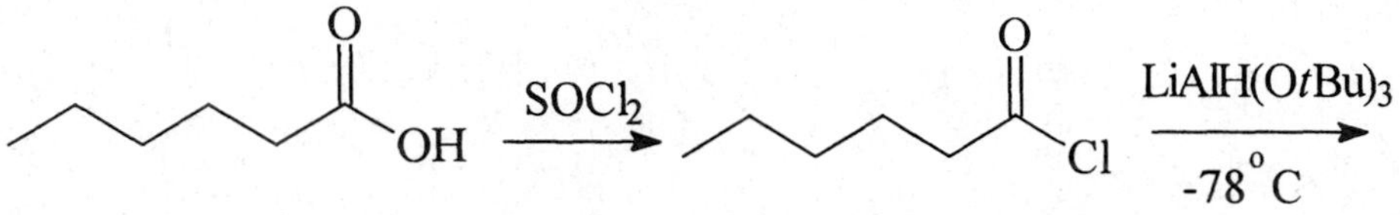

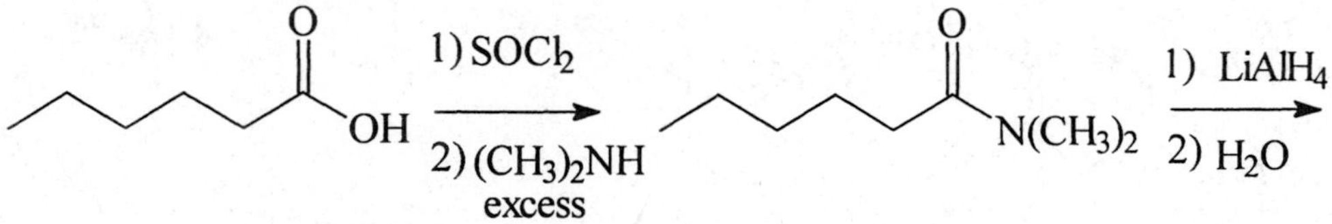

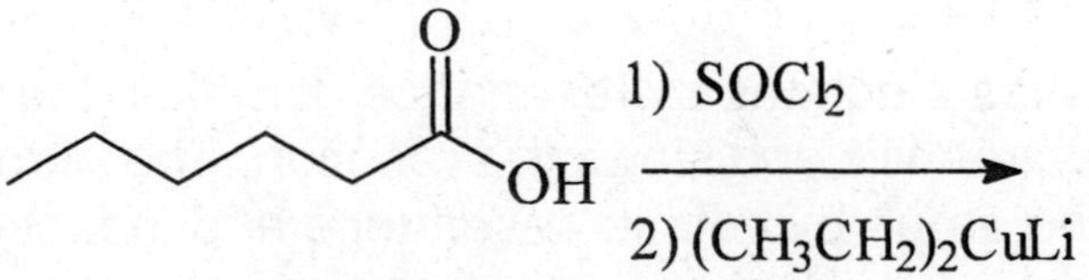

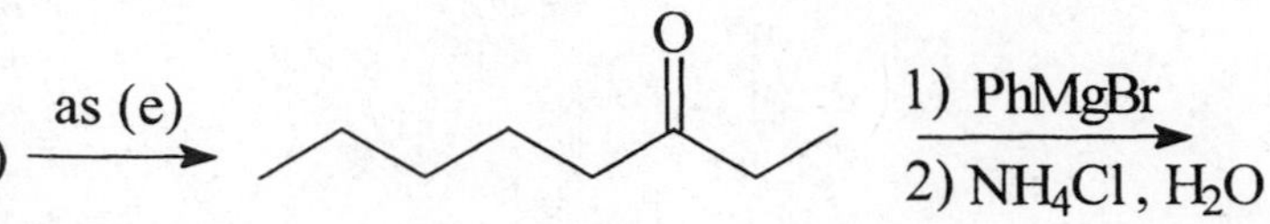

19.33

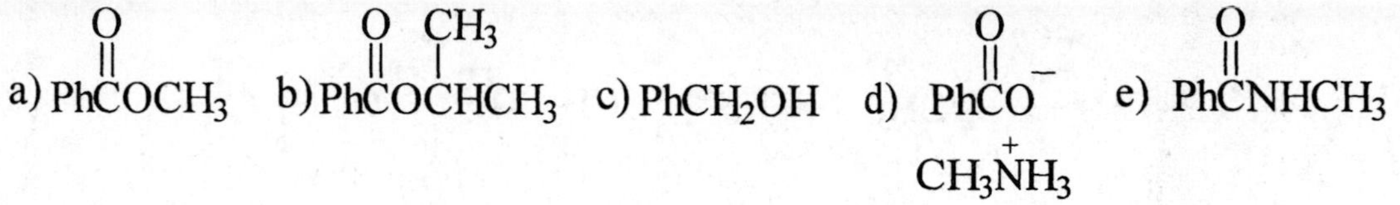

a) $Ph\overset{O}{\overset{\|}{C}}OCH_3$ b) $Ph\overset{O}{\overset{\|}{C}}O\overset{CH_3}{\overset{|}{C}}HCH_3$ c) $PhCH_2OH$ d) $Ph\overset{O}{\overset{\|}{C}}O^-$ $CH_3\overset{+}{N}H_3$ e) $Ph\overset{O}{\overset{\|}{C}}NHCH_3$

19.34

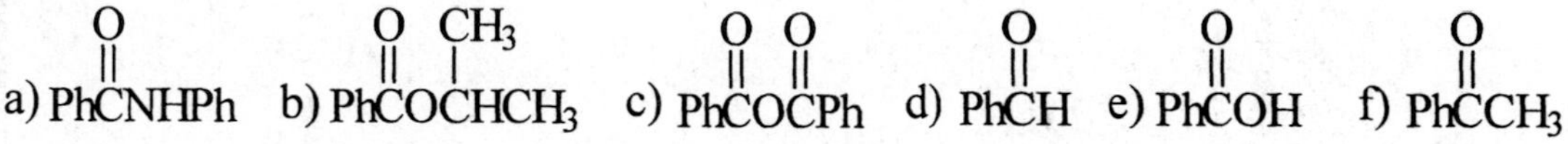

a) $Ph\overset{O}{\overset{\|}{C}}NHPh$ b) $Ph\overset{O}{\overset{\|}{C}}O\overset{CH_3}{\overset{|}{C}}HCH_3$ c) $Ph\overset{O}{\overset{\|}{C}}O\overset{O}{\overset{\|}{C}}Ph$ d) $Ph\overset{O}{\overset{\|}{C}}H$ e) $Ph\overset{O}{\overset{\|}{C}}OH$ f) $Ph\overset{O}{\overset{\|}{C}}CH_3$

19.35

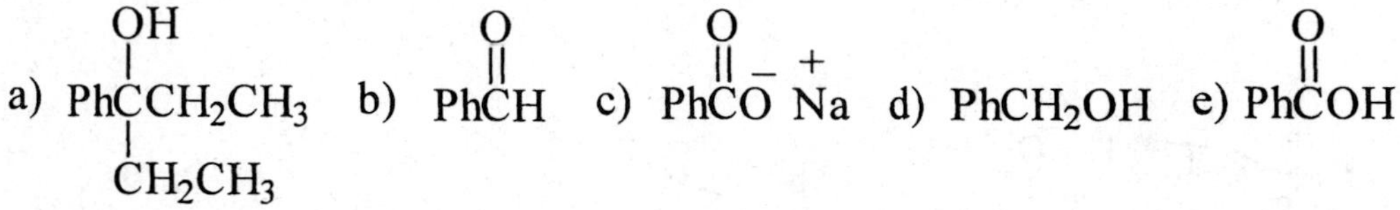

a) $Ph\overset{OH}{\overset{|}{\underset{CH_2CH_3}{\underset{|}{C}}}}CH_2CH_3$ b) $Ph\overset{O}{\overset{\|}{C}}H$ c) $Ph\overset{O}{\overset{\|}{C}}O^-\ Na^+$ d) $PhCH_2OH$ e) $Ph\overset{O}{\overset{\|}{C}}OH$

19.36

a) $Ph\overset{O}{\overset{\|}{C}}OH + CH_3\overset{+}{N}H_3$ b) $PhCH_2NHCH_3$

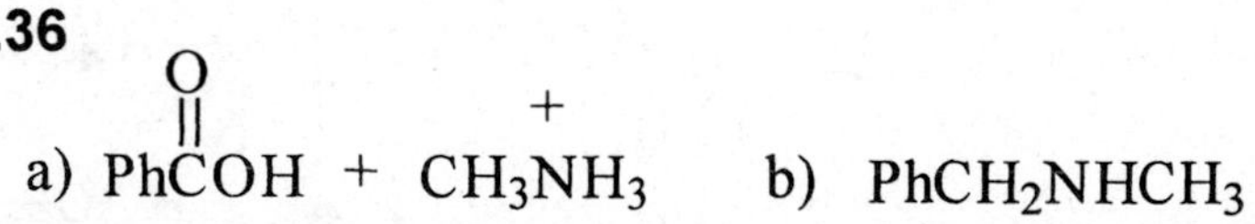

19.37

a)

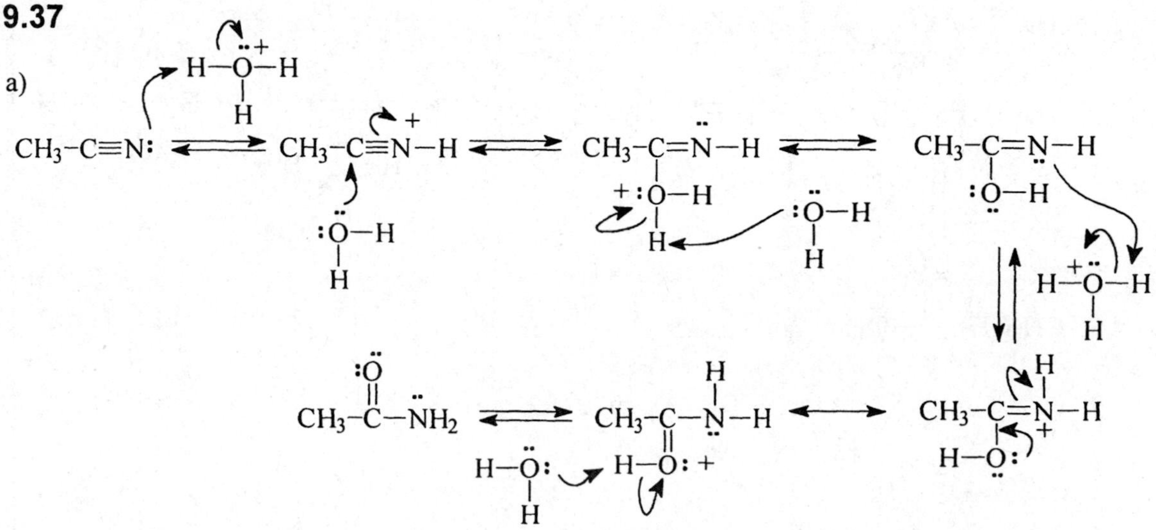

b)

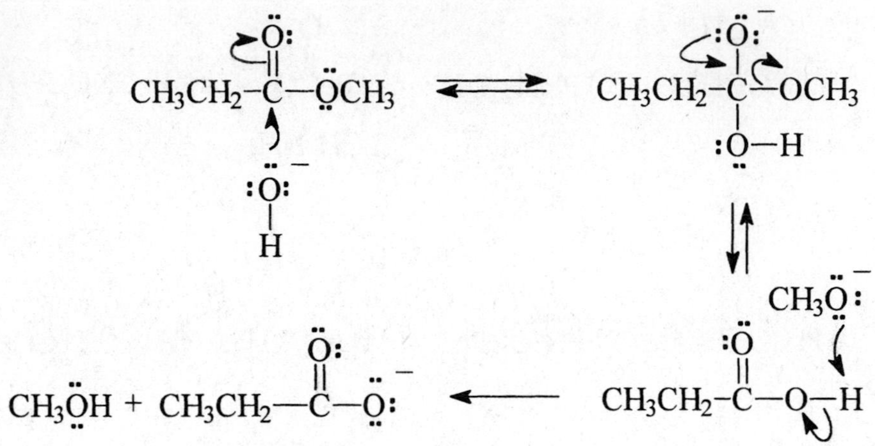

c)

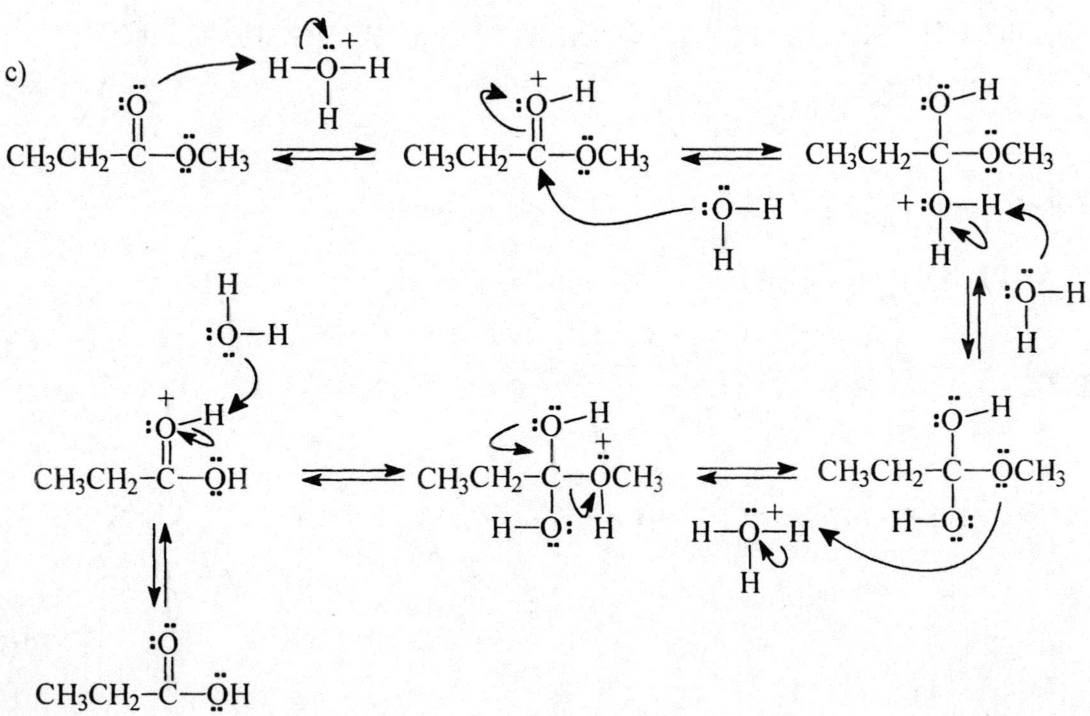

d)

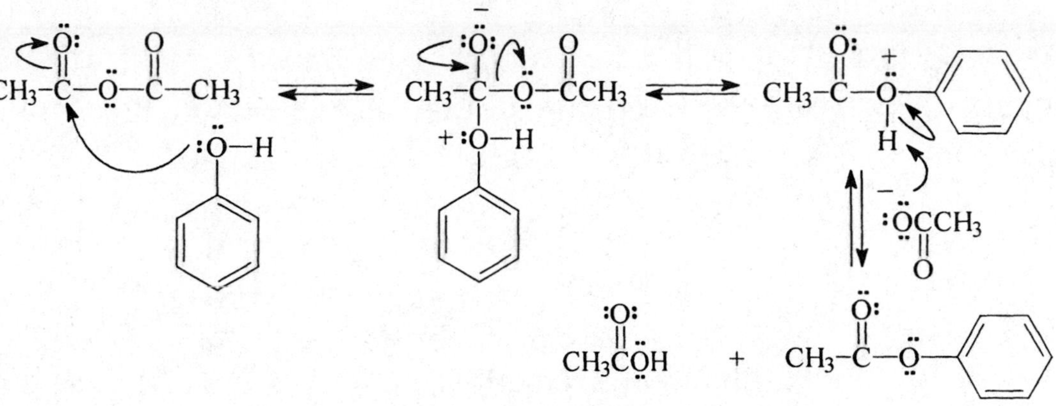

e)

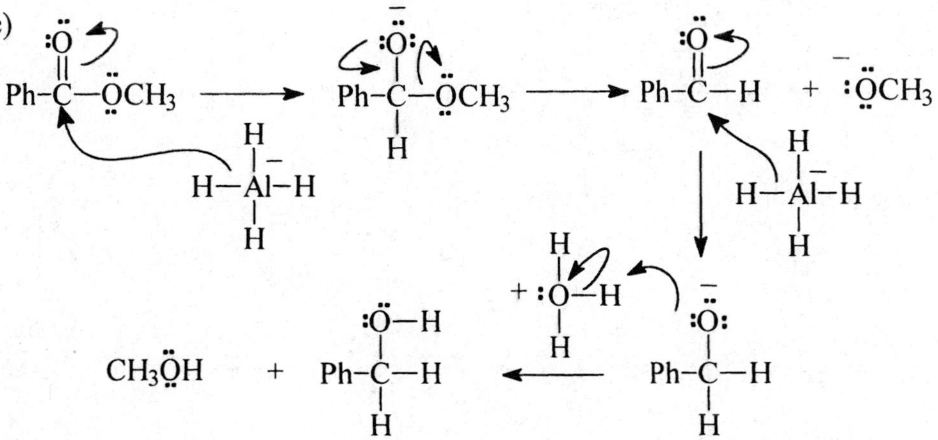

f)

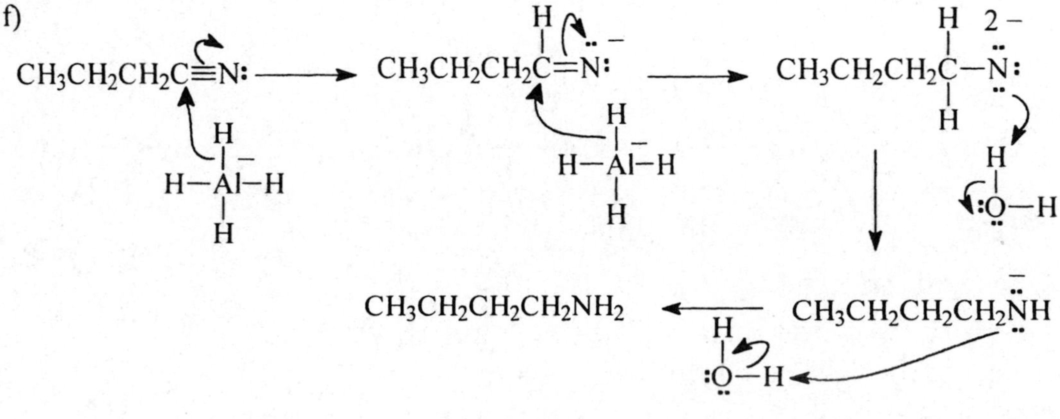

g)

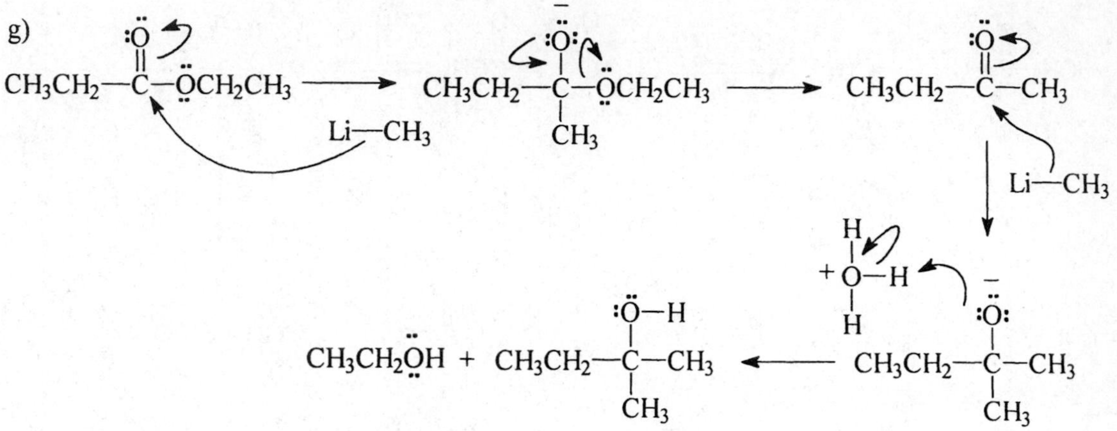

h)

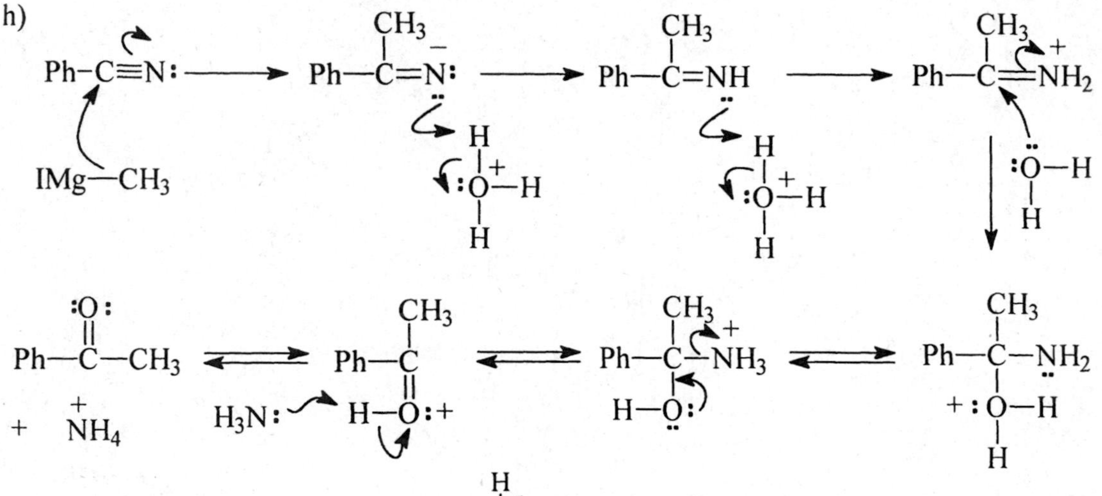

i)

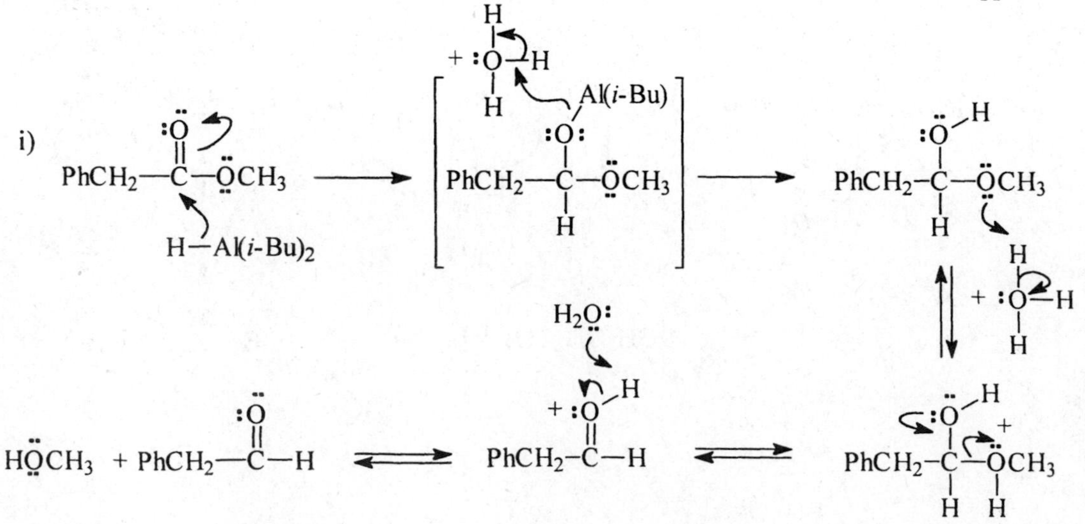

j)

19.38

a)

b)

$$PhCOH \xrightarrow[H_2SO_4]{CH_3OH} PhCOCH_3 \xrightarrow[\text{2) } H_3O^+, \Delta]{\text{1) 2 CH}_3\text{MgI}}$$

c)

1) CH$_3$MgBr
2) NH$_4$Cl, H$_2$O
3) HCl

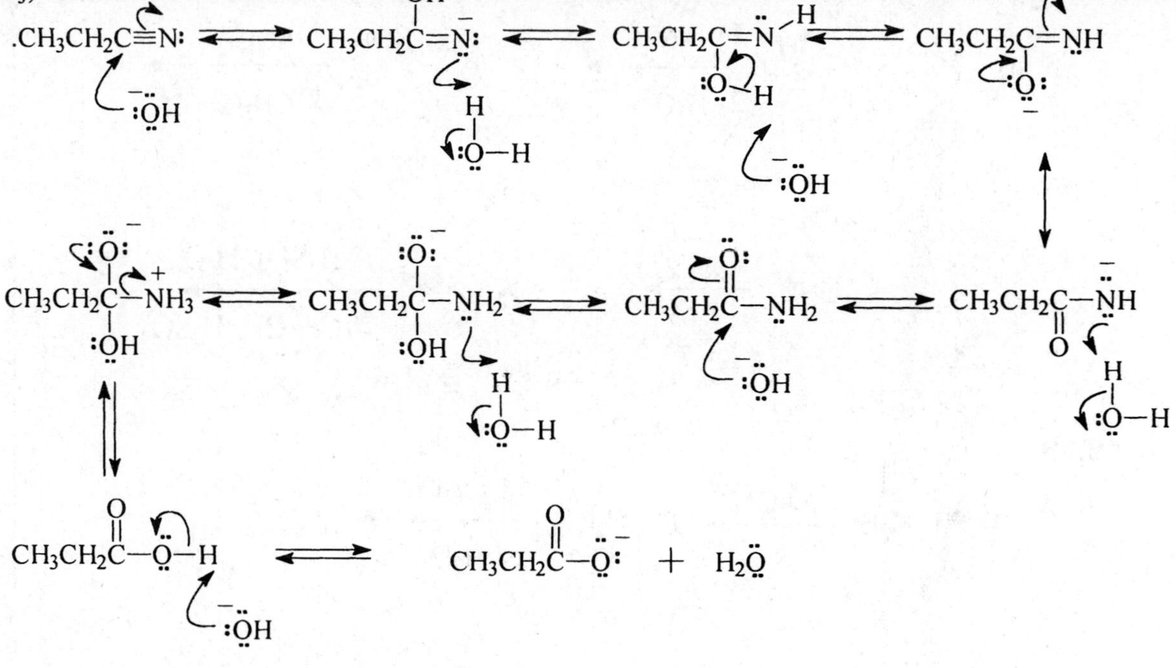

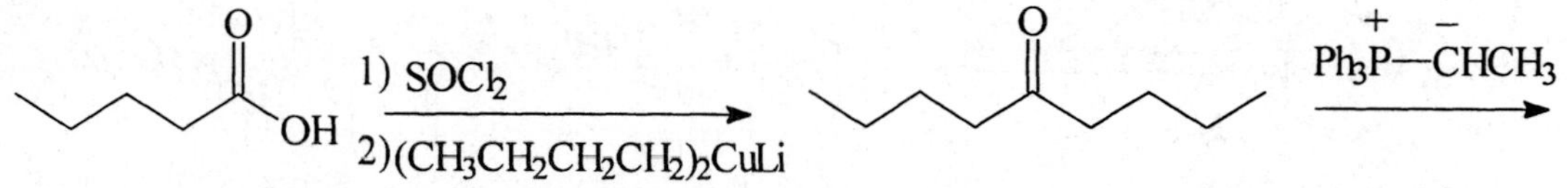

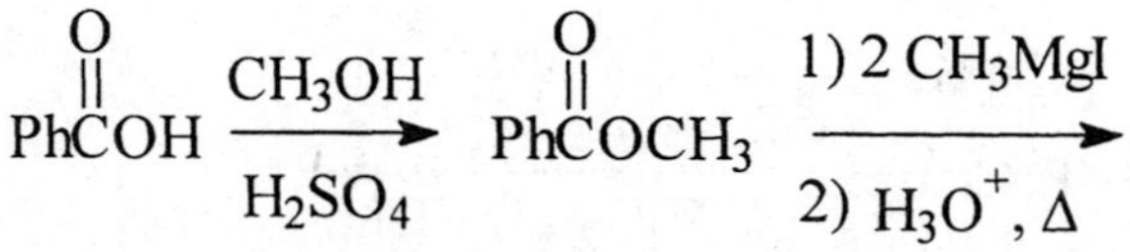

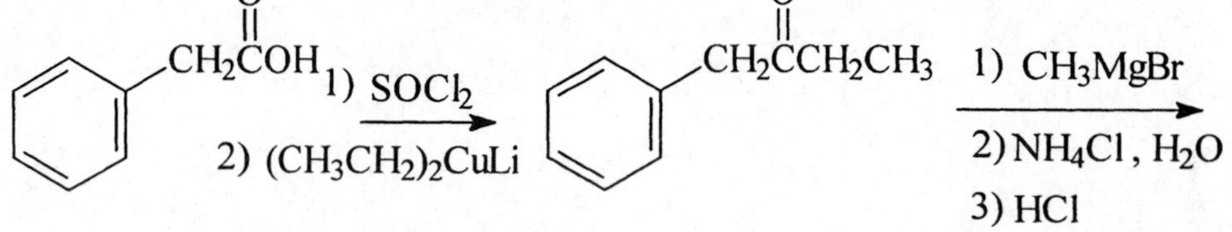

d)

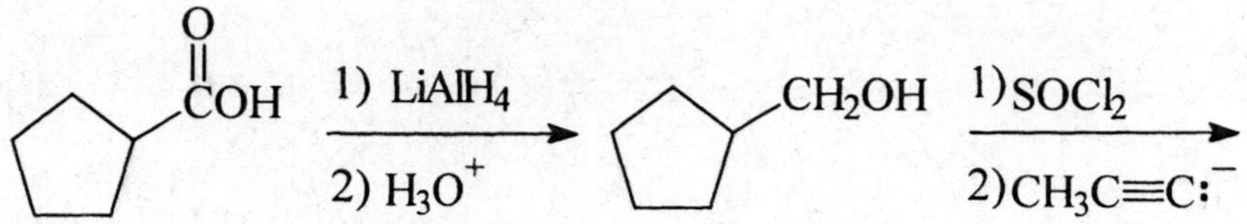

cyclopentane-COOH $\xrightarrow[\text{2) }H_3O^+]{\text{1) LiAlH}_4}$ cyclopentane-CH$_2$OH $\xrightarrow[\text{2)}CH_3C\equiv C:^-]{\text{1)}SOCl_2}$

e)

Br $\xrightarrow{\text{NaCN}}$ CN $\xrightarrow[\text{2)}CH_3OH, H_2SO_4]{\text{1) }H_2SO_4, H_2O}$

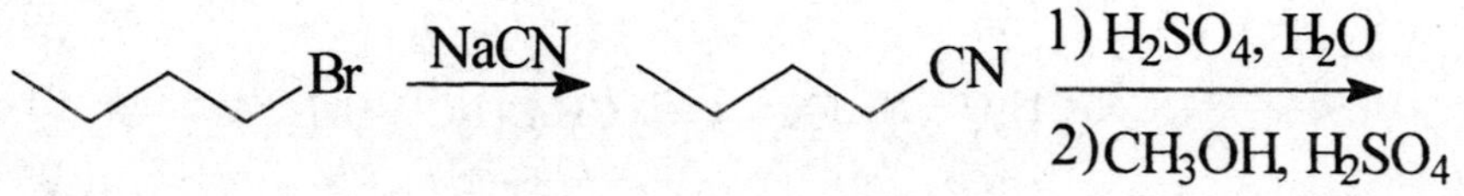

19.39

a) PhCH$_2$Br $\xrightarrow{\text{NaCN}}$ PhCH$_2$CN $\xrightarrow[\text{2) }H_2O]{\text{1) LiAlH}_4}$

b)

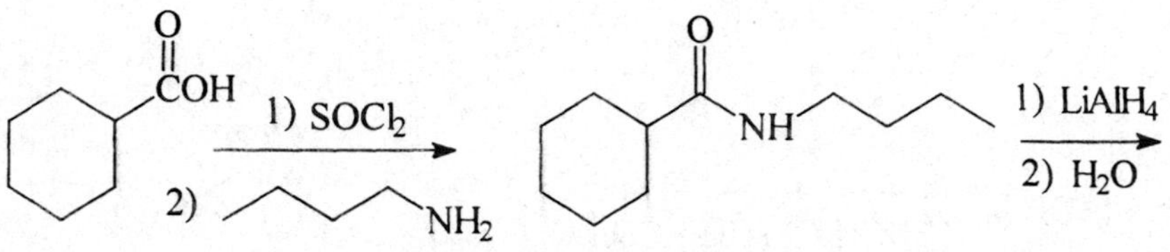

cyclohexane-COOH $\xrightarrow[\text{2) } \overset{}{\text{NH}_2}]{\text{1) SOCl}_2}$ cyclohexane-C(=O)-NH-butyl $\xrightarrow[\text{2) }H_2O]{\text{1) LiAlH}_4}$

c)

PhCOH (O) $\xrightarrow[\substack{\text{2) 2 CH}_3\text{MgBr} \\ \text{3) }H_3O^+}]{\text{1) CH}_3\text{OH, H}_2\text{SO}_4}$ PhC(CH$_3$)=CH$_2$ $\xrightarrow[\text{2)}H_2O_2, \text{NaOH}]{\text{1) BH}_3 , \text{THF}}$

d)

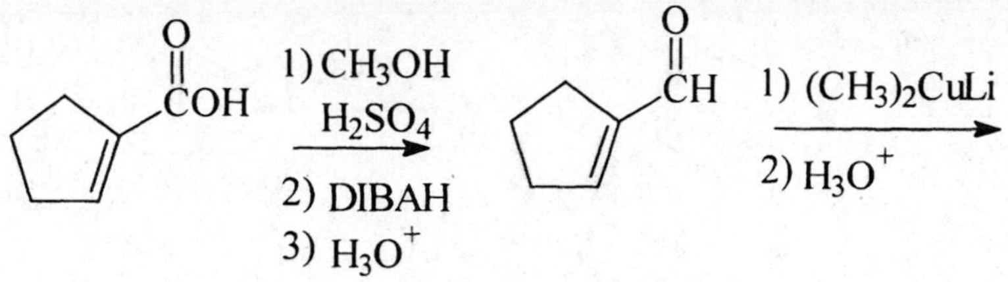
O
COH
1) CH_3OH
H_2SO_4
2) DIBAH
3) H_3O^+
O
CH
1) $(CH_3)_2CuLi$
2) H_3O^+

e)

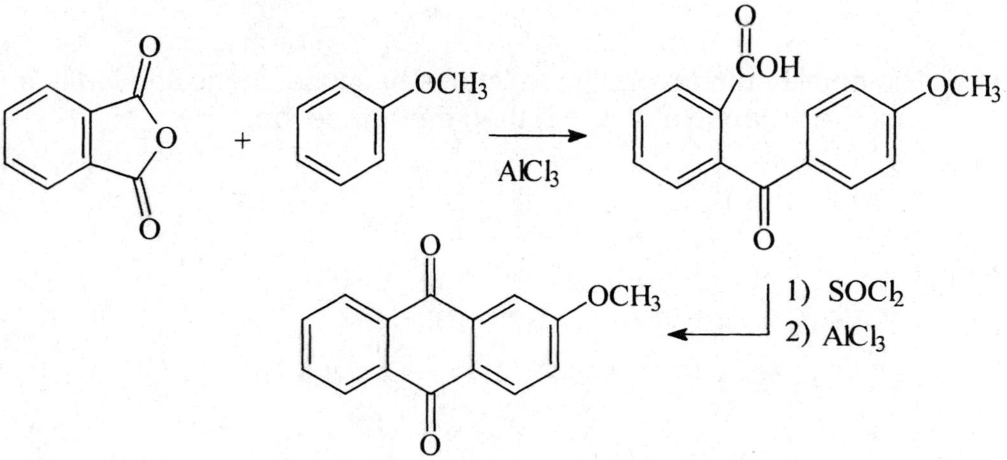
O
O
O
OCH_3
+
$AlCl_3$
O
COH
OCH_3
O
1) $SOCl_2$
2) $AlCl_3$
O
OCH_3
O

f)

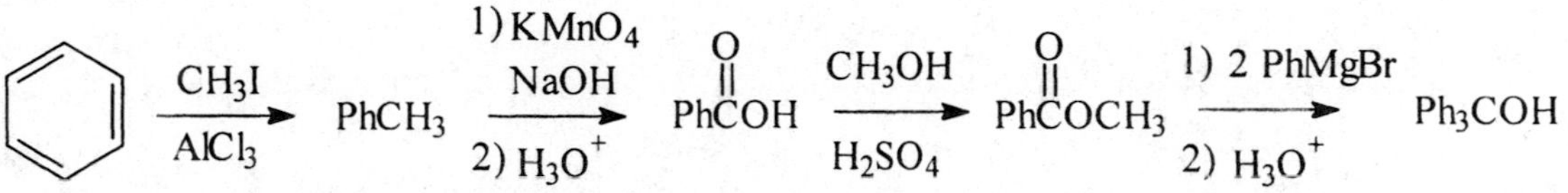
CH_3I
$AlCl_3$
$PhCH_3$
1) $KMnO_4$
NaOH
2) H_3O^+
O
PhCOH
CH_3OH
H_2SO_4
O
$PhCOCH_3$
1) 2 PhMgBr
2) H_3O^+
Ph_3COH

g)

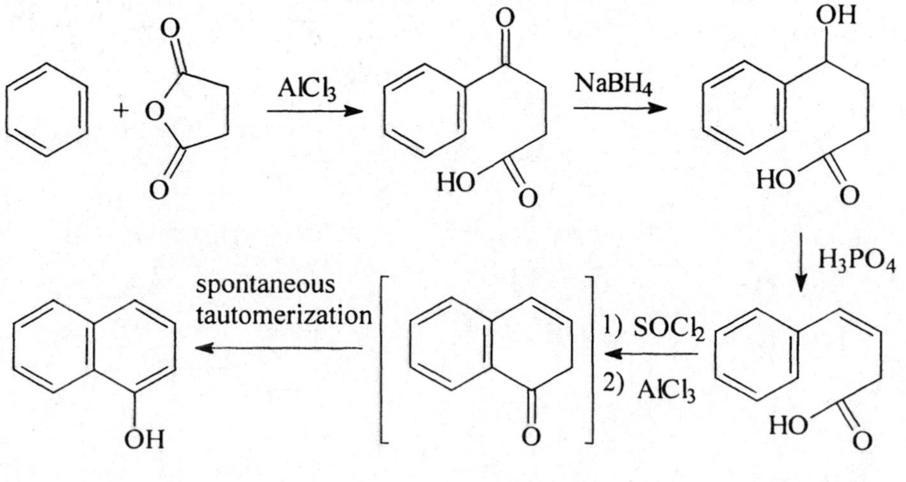
O
+ O
O
$AlCl_3$
O
HO
O
$NaBH_4$
OH
HO
O
H_3PO_4
HO
O
1) $SOCl_2$
2) $AlCl_3$
O
spontaneous
tautomerization
OH

h) 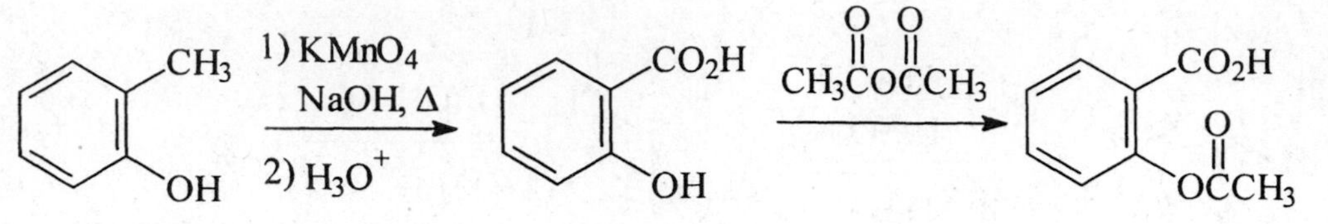

19.40 The equilibrium favors the reactants because phenoxide ion is a weaker base (resonance stabilized) than methoxide ion.

19.41

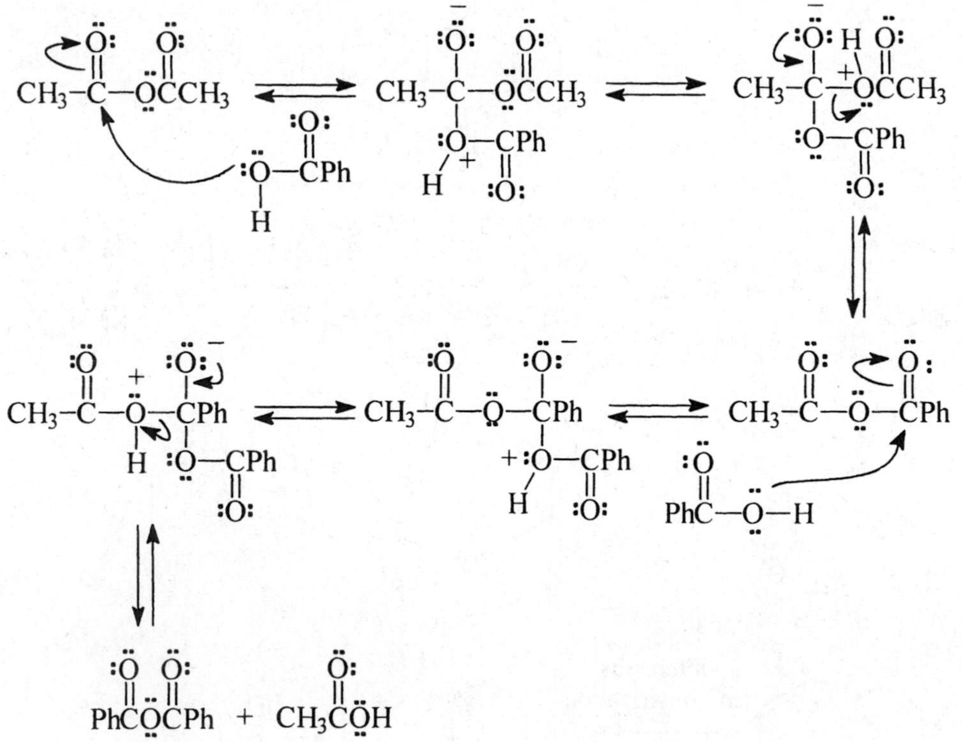

19.42

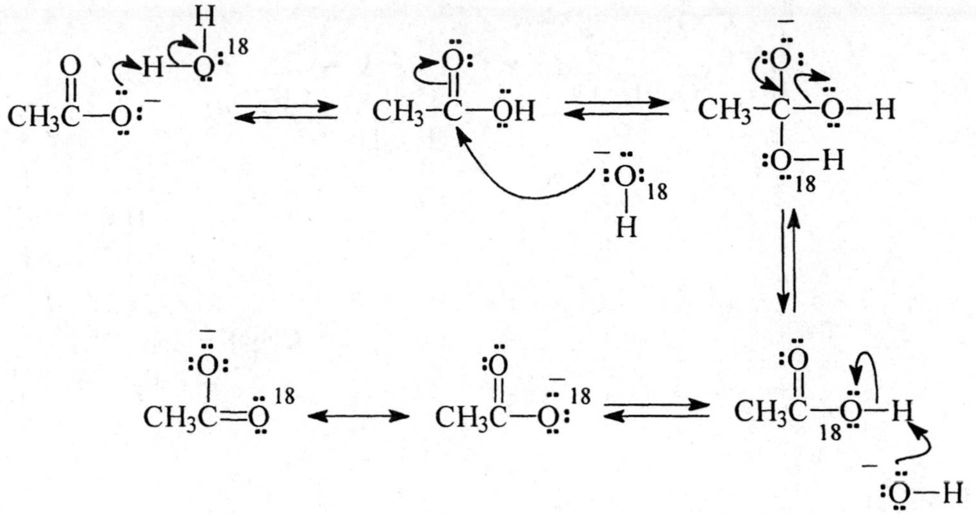

19.43

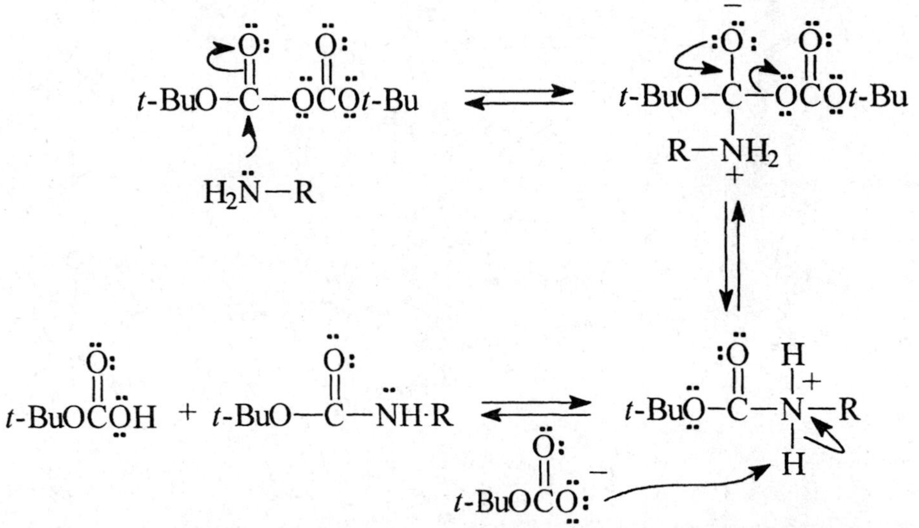

The products of this reaction are determined by the stability of the leaving group. The *t*-BuOCOO$^-$ group is the conjugate base of a carboxylic acid and is a weaker base than the *t*-BuO$^-$ group, so it is a better leaving group.

19.44

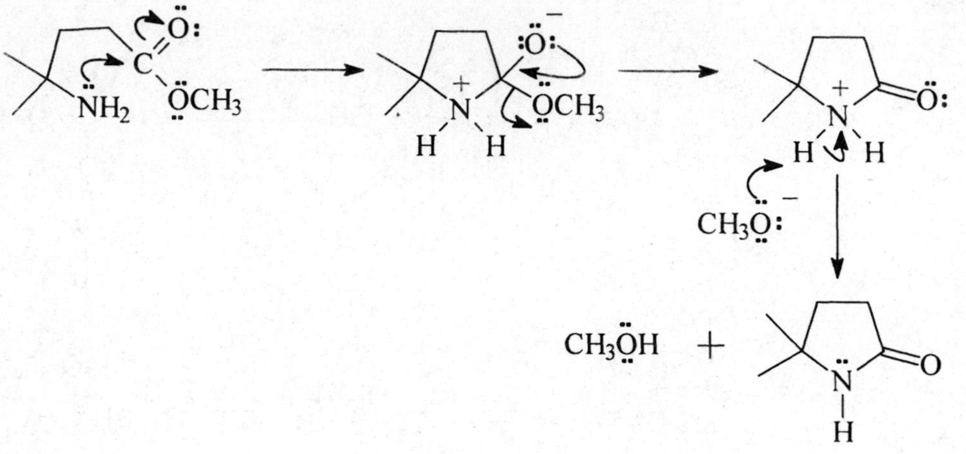

19.45 a)

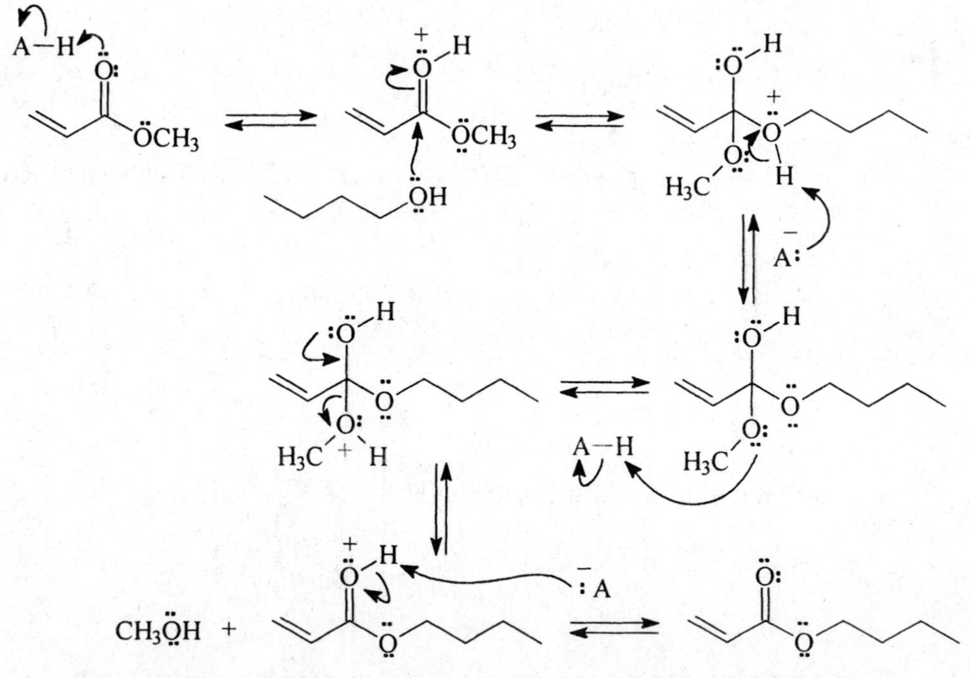

b) The equilibrium constant for this reaction should be about one because the reactants and the products have approximately equal stabilities.

c) The equilibrium can be driven toward the product either by using an excess of butanol in the reaction or by removing methanol by distillation during the reaction.

19.46

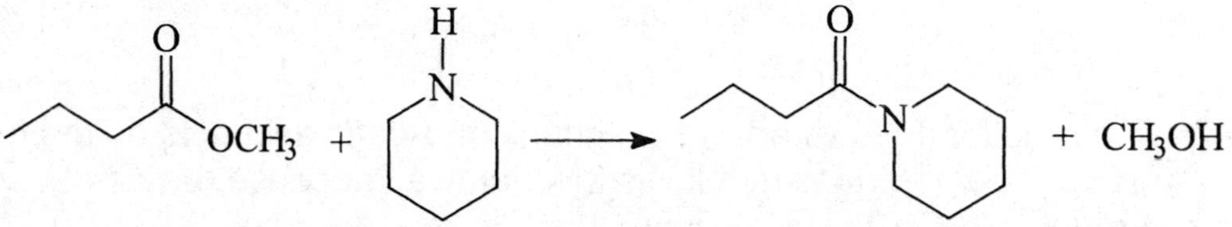

The product, an amide, is more stable than the reactant, an ester, so it is not necessary to drive the equilibrium towards the products.

19.47 The equilibrium favors the product because the strain of the four-membered ring of the reactant is relieved during this reaction.

19.48

$$Ph\!-\!\overset{\overset{\displaystyle OH}{|}}{\underset{\underset{\displaystyle Ph}{|}}{C}}\!-\!Ph \quad + \quad 2\ CH_3CH_2OH$$

19.49

a)

$+ \ H_2NCH_2CHPh$

b)

c)

d)

e) $CH_3O\overset{\overset{\displaystyle O}{\|}}{C}\!-\!\overset{\overset{\displaystyle O}{\|}}{C}OCH_3$

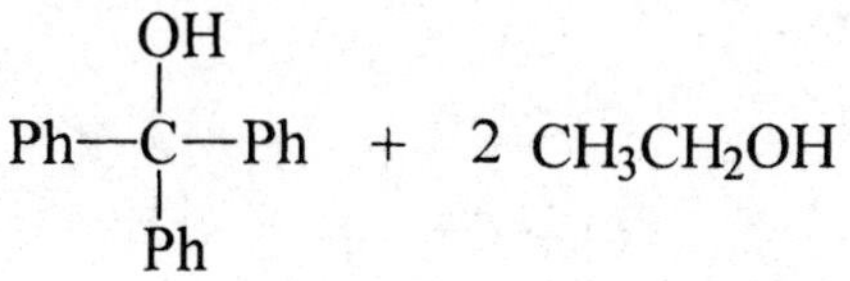

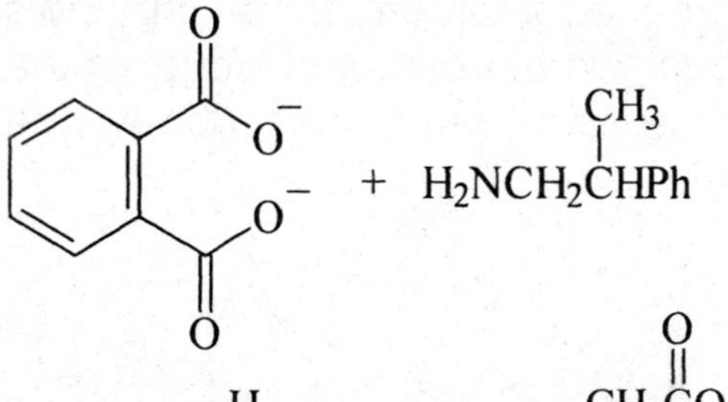

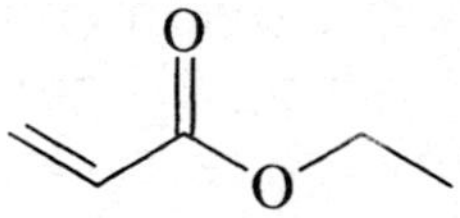

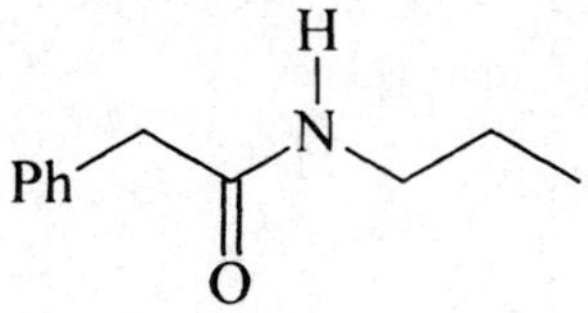

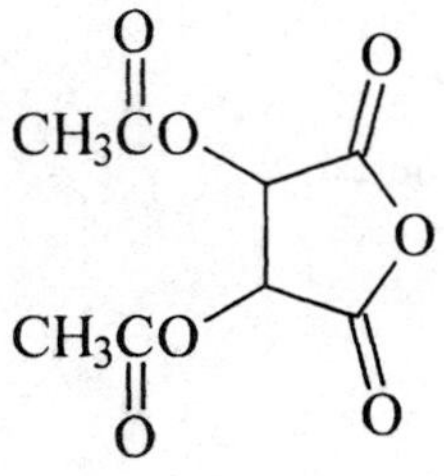

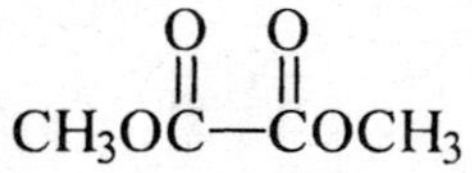

f) 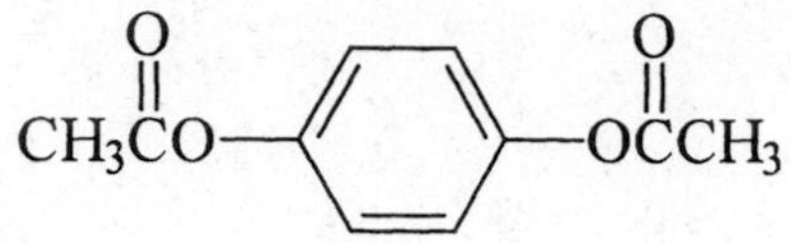

g) 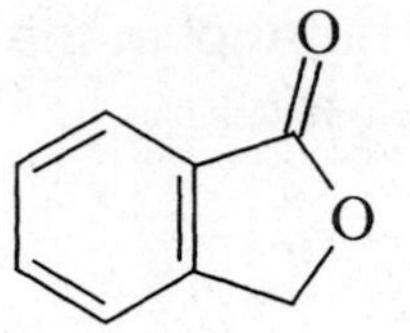

19.50 The reaction follows an S_N1 mechanism. Acetic acid acts as the leaving group. The mechanism follows this pathway because the carbocation is tertiary.

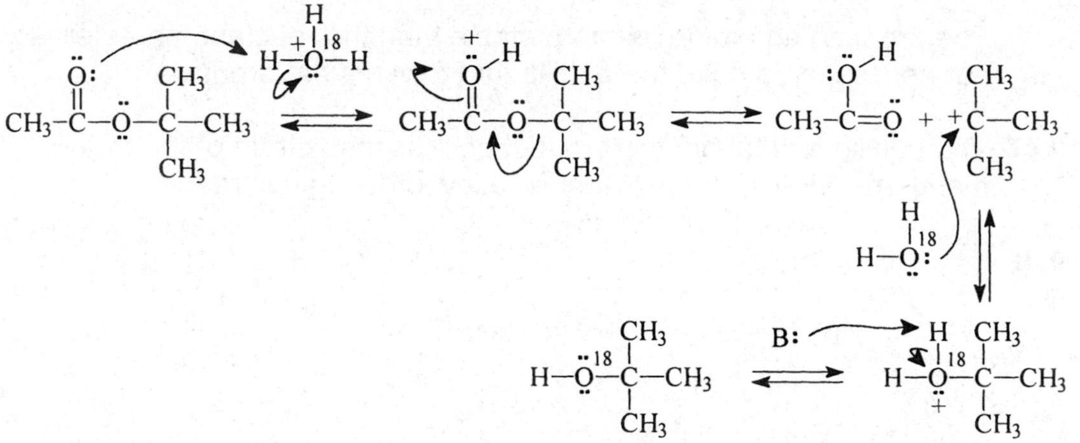

19.51 Only the *cis*-isomer forms the lactone because the boat conformation of the *cis*-isomer provides the required geometry for an intramolecular reaction between the OH group and the carbonyl group. In the *trans*-isomer, The OH group cannot possibly reach the carbonyl carbon, so lactone formation is impossible.

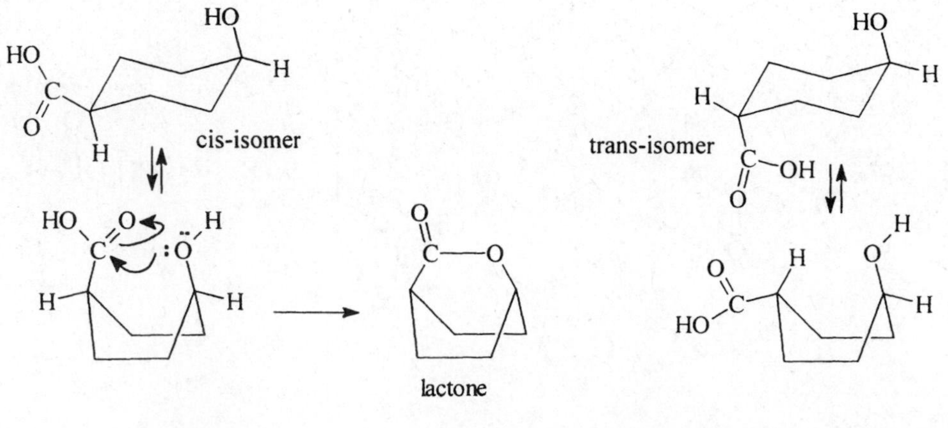

19.52 The lone pair of electrons on the nitrogen of an amide is delocalized by resonance. If this pair of electrons reacted as a base, the resonance stabilization would be lost. As a result, the electron pair on the oxygen, which is not involved in resonance, is more basic.

19.53

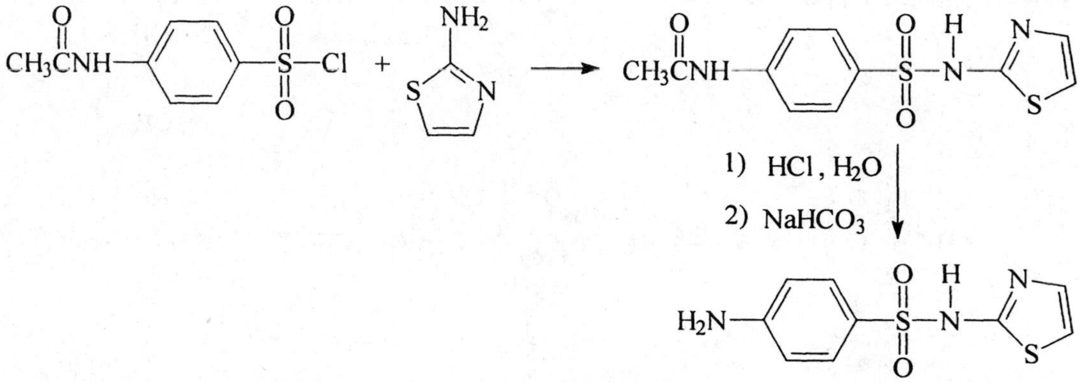

19.54 The electron withdrawing effect of the positive N makes the carbonyl carbon of the reactant more electrophilic than that of an amide. In addition, trimethylamine is a relatively good leaving group.

19.55

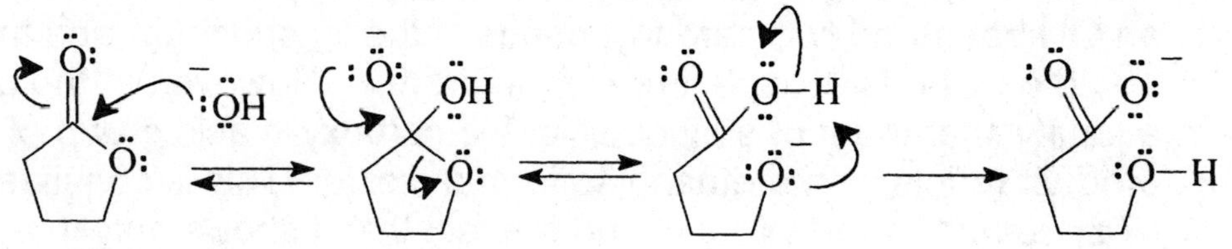

GBL

GHB

The last step in the mechanism, the formation of the carboxylate anion, a weaker base than hydroxide ion, drives the equilibrium to GHB. When acid is added, GHB is protonated. The hydroxy group and the carboxylic acid group undergo spontaneous ester (lactone) formation because an ester and a carboxylic acid are of similar reactivities and intramolecular reactions are more favorable.

19.56 This reaction is just acid-catalyzed hydrolysis of an amide. The mechanism is the same as that shown in the solution for Practice Problem 19.2 in the text.

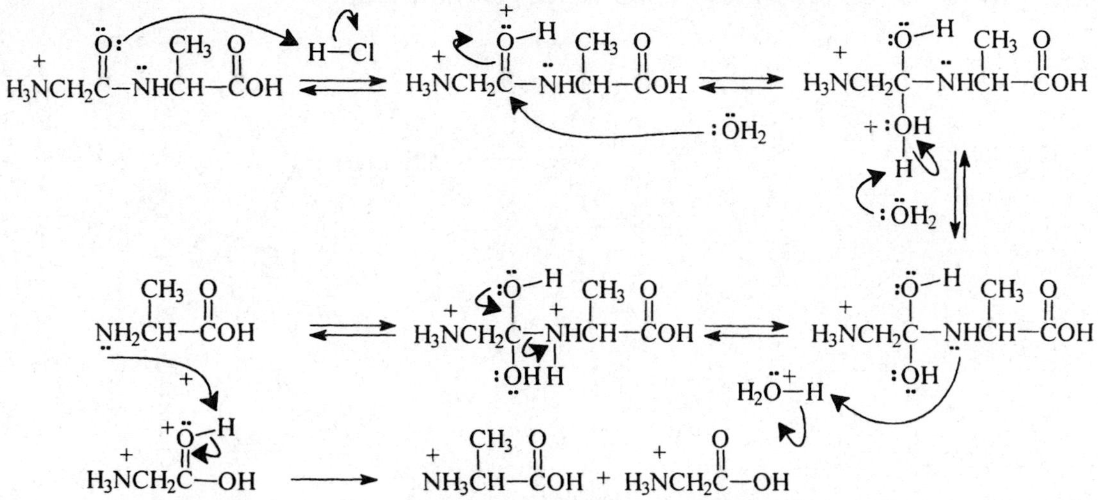

19.57 An anhydride reacts with an alcohol to produce an ester and a carboxylic acid. So when succinic anhydride is heated in methanol it produces the monomethyl ester of succinic acid. This is apparent from the presence of an OH band and two carbonyl bands in the IR spectrum and the signal at 10.25 δ in the NMR spectrum of the product. However, in the presence of a catalytic amount of sulfuric acid, the carboxylic acid group of the initial product reacts with methanol to form a diester. This is consistent with the IR spectrum of this product and the fact that it shows only three absorptions in its ^{13}C-NMR spectrum.

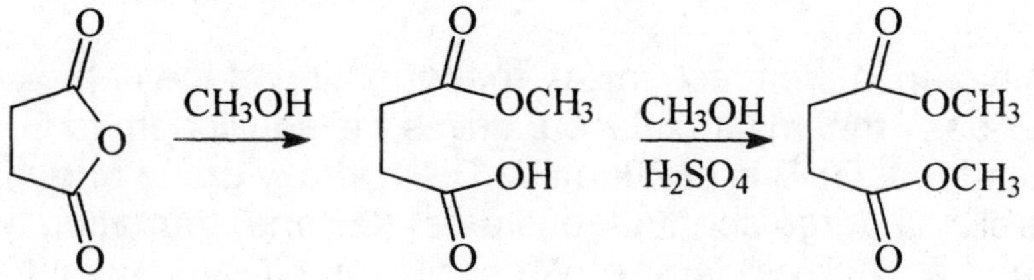

19.58 Sodium borohydride selectively reduces the aldehyde group of this compound. The negatively charged O which is generated then attacks the carbonyl group of the ester, resulting in an intramolecular esterification reaction to form a lactone. This is apparent from the absence of an OH band in the IR spectrum of the product. The carbonyl absorption of the lactone occurs at higher wavenumber because it is part of a five membered ring (1740 + 30 = 1780 cm^{-1}). The NMR spectrum shows two H's at 4.3 δ split by the two H's at 2.3 δ, and two H's (at 2.5 δ) also split by the two H's at 2.3 δ. The two H's at 2.3 δ are split into 5 lines by the four nearby H's. (The coupling is actually a little more complicated than this.) This corresponds to the $CH_2CH_2CH_2$ group of the lactone.

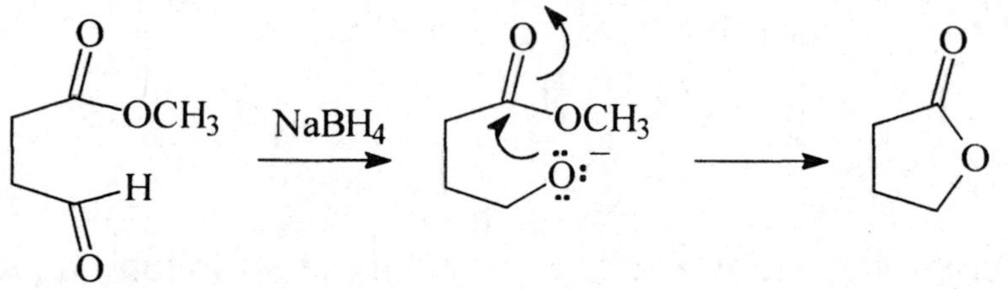

19.59 The rate of esterification is slowed by steric hindrance at the carbonyl carbon. Therefore, the carboxylic acid on the left (2,2-dimethylpropanoic acid) reacts slowest, followed by the one on the right (2-methylpropanoic acid). The acid in the middle (propanoic acid) is least hindered and reacts fastest.

19.60 The two ortho methyl groups on the benzene ring dramatically hinder the approach of the nucleophile to the carbonyl carbon.

19.61 The *tert*-butyl group locks each isomer in the conformation shown. A lactone can form as long as the OH group can reach the carbonyl carbon. This is possible as long as at least one of these groups are equatorial. The only isomer where the OH is too far from the carbonyl carbon is the one at the upper right, where both the OH and the carboxyl group are axial.

19.62 The cis double bond of oleic acid causes it to have a bent shape. Stearic acid, without this double bond, has a more linear shape, so it packs better into a crystal lattice. As a result, stearic acid has a higher melting point than oleic acid.

19.63 When the carboxylate anions are protonated, the one on the left has an OH group ortho to the CH_2CO_2H group and is capable of forming a lactone (intramolecular ester). Lactone formation is favored by entropy and occurs readily. The lactone has no carboxylic acid group and is not very acidic. The isomer on the right, however, has the OH para to the CH_2CO_2H group and is not capable of forming a lactone. It has a carboxylic acid group that is relatively acidic, with a pK_a near 5.

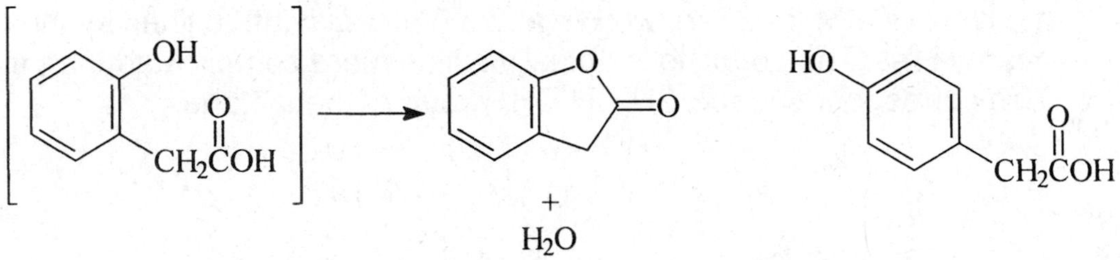

19.64 These compounds are the cis (on the left) and trans (on the right) isomers of 3-hydroxycyclohexanecarboxylic acid. In order to form a lactone, the O of the OH group must be able to bond to the C of the carbonyl group. This requires both the OH and the carboxylic acid group to be axial. Only the *cis*-isomer can attain this conformation, so only the *cis*-isomer can form a lactone. The *trans*-isomer always has one group equatorial, pointed away from the ring, so the groups are too far apart to form a bond.

19.65 The carbonyl group of the *cis*-isomer is axial and that of the *trans*-isomer is equatorial. The axial H's in the *cis*-isomer hinder the approach of the nucleophile, thus slowing the reaction.

Review of Mastery Goals

After completing this chapter, you should be able to:

Show the products of any of the reactions discussed in this chapter. (Problems 19.5, 19.6, 19.9, 19.11, 19.12, 19.14, 19.17, 19.19, 19.21, 19.23, 19.25, 19.26, 19.27, 19.28, 19.29, 19.33, 19.34, 19.35, 19.36, and 19.50)

Show the mechanism for these reactions. (Problems 19.7, 19.10, 19.15, 19.22, 19.37, 19.41, 19.42, 19.43, 19.44, 19.45, 19.55, and 19.56)

Predict the effect of a change in structure on the rate and equilibrium of a reaction.
(Problems 19.1, 19.2, 19.3, 19.8, 19.30, 19.31, 19.40, 19.47, 19.59, and 19.65)

Use these reactions to interconvert any of the carboxylic acid derivatives and to prepare aldehydes, ketones, alcohols and amines.
(Problems 19.4, 19.16, 19.20, 19.24, 19.32, 19.46, and 19.54)

Use these reactions in combination with reactions from previous chapters to synthesize compounds.
(Problems 19.13, 19.38, and 19.39)

Chapter 20
ENOLATE AND OTHER CARBON NUCLEOPHILES

20.1

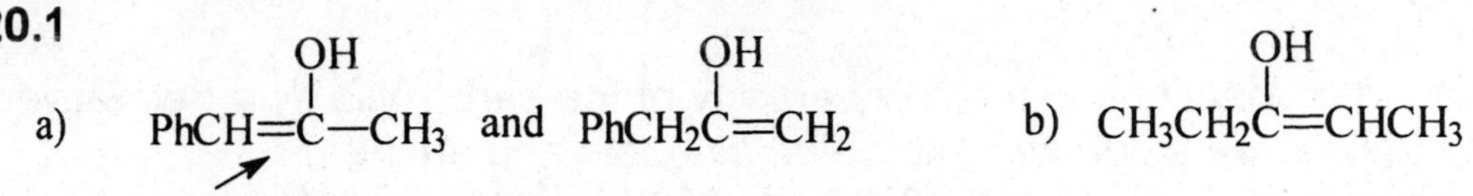

a)

OH
|
PhCH=C—CH₃   and   PhCH₂C=CH₂

OH
|

b)   CH₃CH₂C=CHCH₃

OH
|

This one is more stable because it is conjugated.

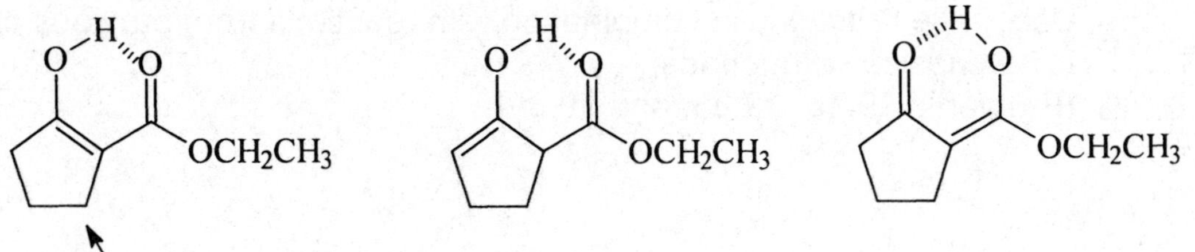

c)

This one is more stable due to conjugation, hydrogen bonding and loss of the ketone rather than the ester carbonyl group.

20.2

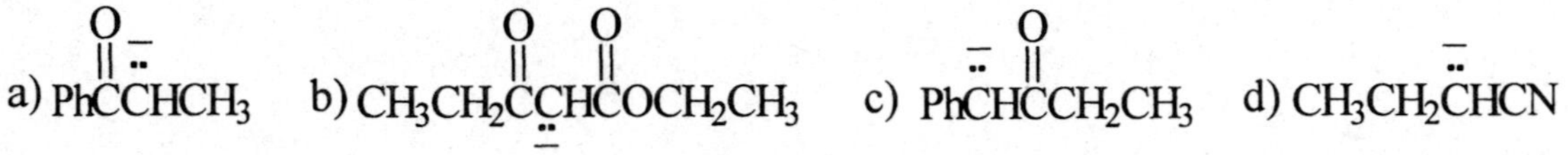

a) PhCCHCH₃   b) CH₃CH₂CCHCOCH₂CH₃   c) PhCHCCH₂CH₃   d) CH₃CH₂CHCN

20.3

a)

b)

$$H_3C-\overset{\underset{\textstyle CH_3}{|}}{\underset{\textstyle CH_3}{C}}-\overset{\textstyle O}{\overset{\|}{C}}-OH \;\; + \;\; CHBr_3$$

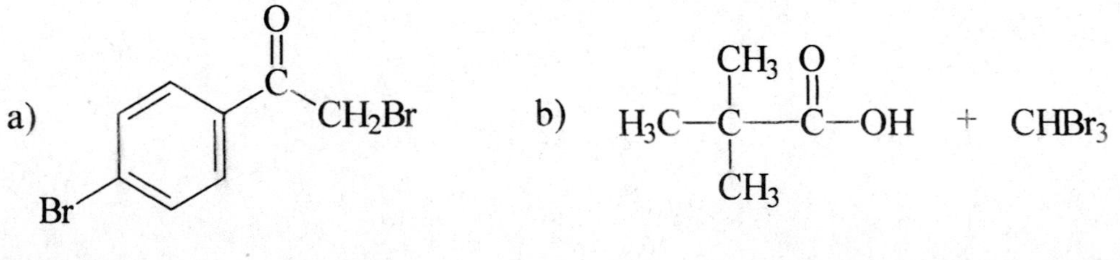

20.4 Butyllithium reacts with cyclohexanone as a nucleophile rather than as a base.

Li—CH₂CH₂CH₂CH₃   ⟶

Li⁺   :Ö: CH₂CH₂CH₂CH₃

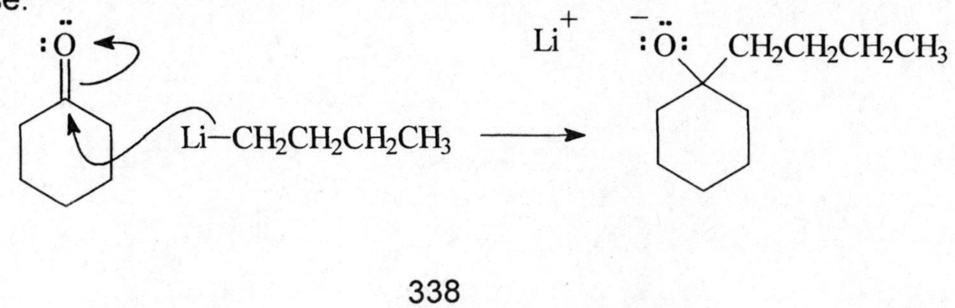

20.5

a) [structure: cyclohexane with CH(CN)(CHPh) substituent]

b) [structure: ethyl 2-ethylhexanoate]

c) PhCCHCH$_2$CH$_3$ (with C=O, and CH$_2$CH$_3$ substituent)

d) [cyclopropane]—CN

f) [structure: ethyl 2-propyl-4-pentenoate]

20.6

a) CH$_3$CCHCOEt (with two C=O groups, CH$_2$Ph substituent) $\xrightarrow{\text{1) NaOH, H}_2\text{O} \quad \text{2) H}_3\text{O}^+, \Delta}$ CH$_3$CCH$_2$CH$_2$Ph (C=O)

b) EtOCCHCOEt (with two C=O groups, CH$_2$CH$_2$CH$_3$ substituent) $\xrightarrow{\text{1) NaOH, H}_2\text{O} \quad \text{2) H}_3\text{O}^+, \Delta}$ CH$_3$CH$_2$CH$_2$CH$_2$COH (C=O)

c) [cyclohexanone ring with ethyl and COOEt substituents at C1; C=O on ring]

d) CH$_3$CCCOEt (with two C=O, H$_3$C and CH$_2$CH$_2$CH$_2$CH$_3$ substituents) $\xrightarrow{\text{1) NaOH, H}_2\text{O} \quad \text{2) H}_3\text{O}^+, \Delta}$ CH$_3$C—CHCH$_2$CH$_2$CH$_2$CH$_3$ (C=O, CH$_3$ substituent)

e) N≡CCHCCH$_3$ (C=O, CH$_2$CH$_2$CH$_3$ substituent)

20.7

a) EtOCCH$_2$COEt (two C=O) $\xrightarrow[\text{2) CH}_3\text{CHCH}_3 \, \text{Br}]{\text{1) NaOEt, EtOH}}$ EtOCCHCOEt (two C=O, CHCH$_3$ with CH$_3$ substituent) $\xrightarrow{\text{1) NaOH, H}_2\text{O} \quad \text{2) H}_3\text{O}^+, \Delta}$

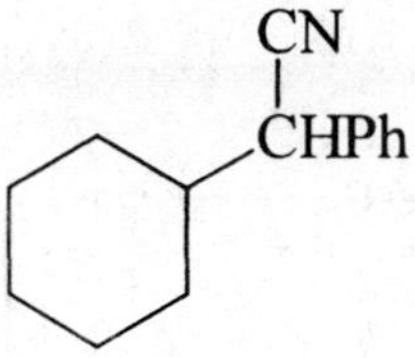

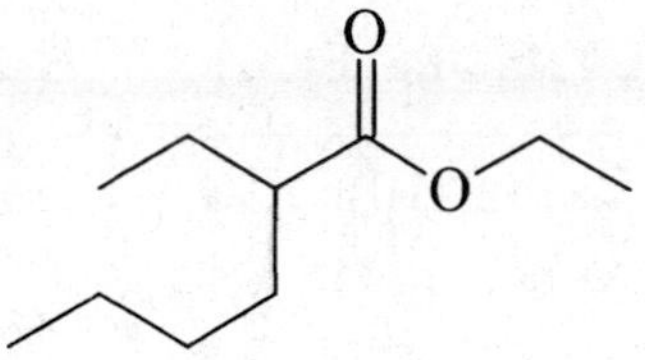

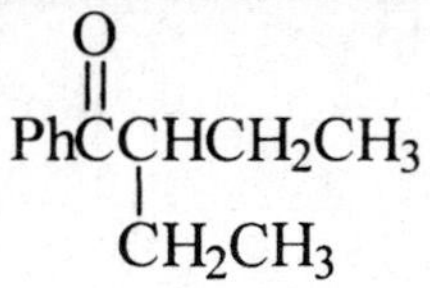

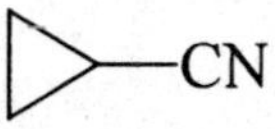

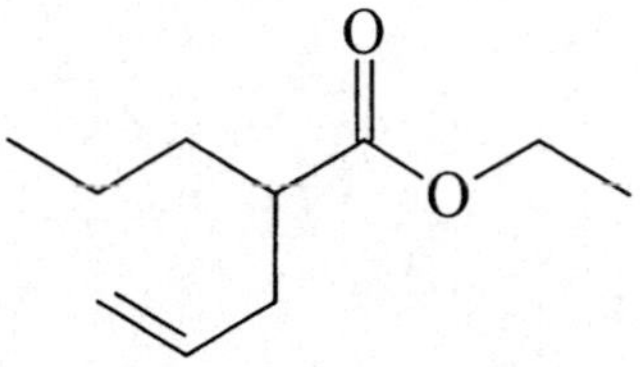

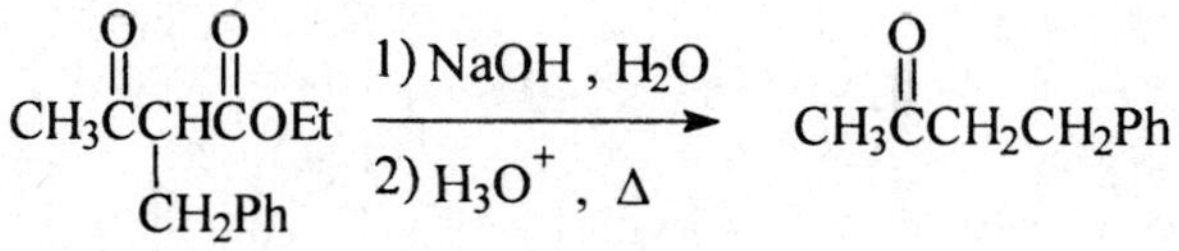

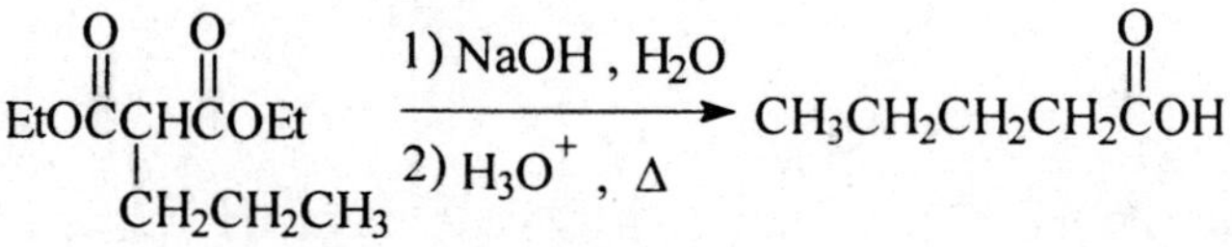

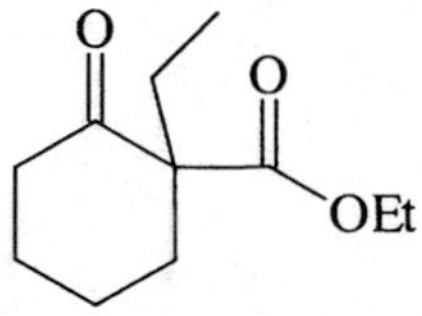

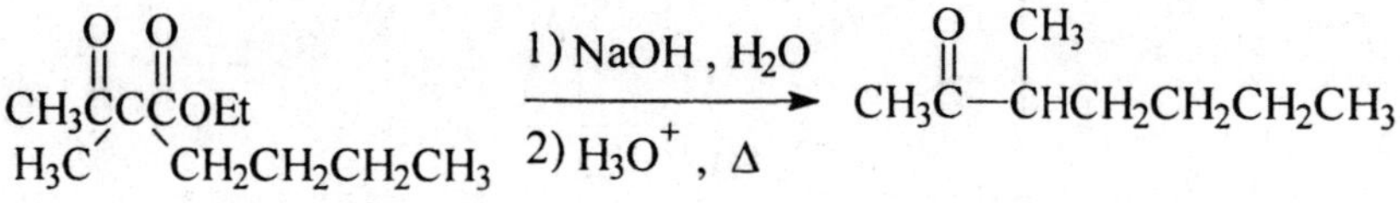

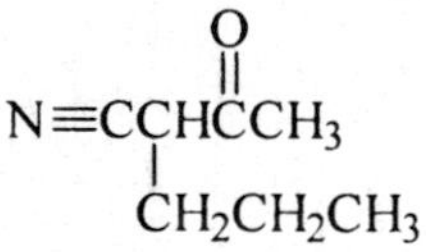

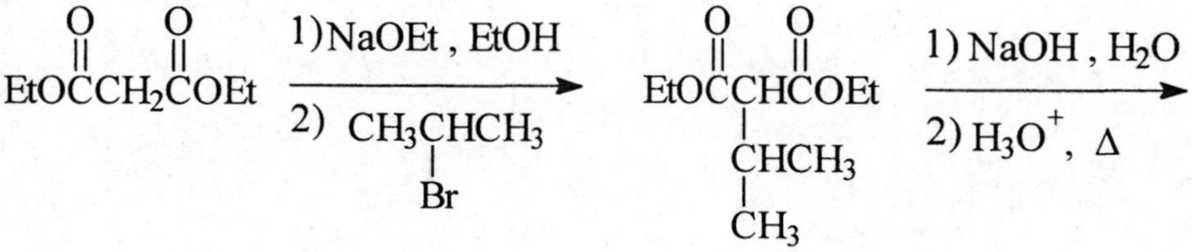

b) 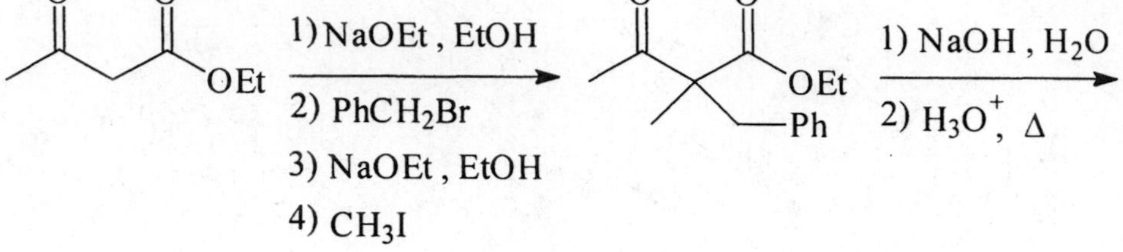
O
O
OEt
1) NaOEt, EtOH
2) PhCH2Br
3) NaOEt, EtOH
4) CH3I
O
O
OEt
Ph
1) NaOH, H2O
2) H3O+, Δ

c) 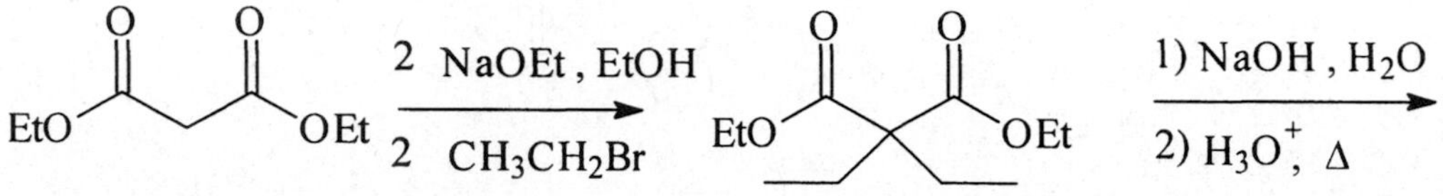
O
O
EtO
OEt
2 NaOEt, EtOH
2 CH3CH2Br
O
O
EtO
OEt
1) NaOH, H2O
2) H3O+, Δ

d) 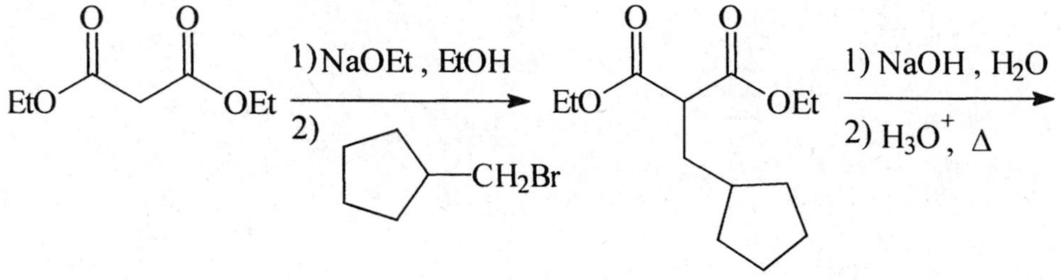
O
O
EtO
OEt
1) NaOEt, EtOH
2)
CH2Br
O
O
EtO
OEt
1) NaOH, H2O
2) H3O+, Δ

e) 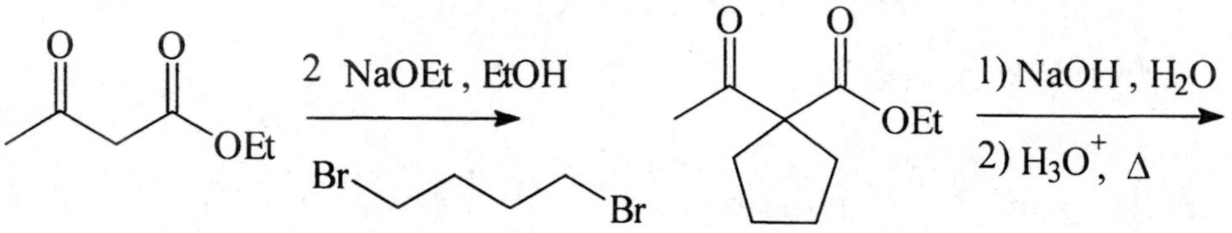
O
O
OEt
2 NaOEt, EtOH
Br
Br
O
O
OEt
1) NaOH, H2O
2) H3O+, Δ

20.8

a) 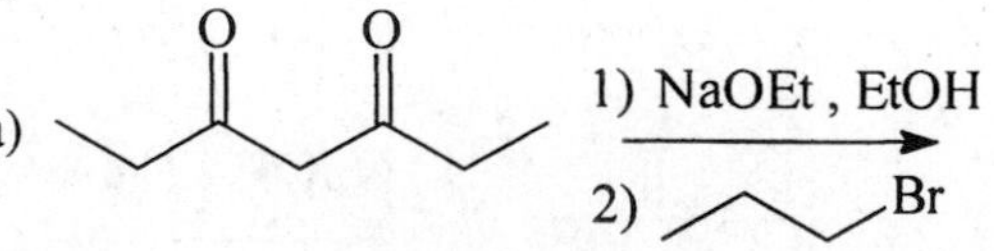
O
O
1) NaOEt, EtOH
2)
Br

b) 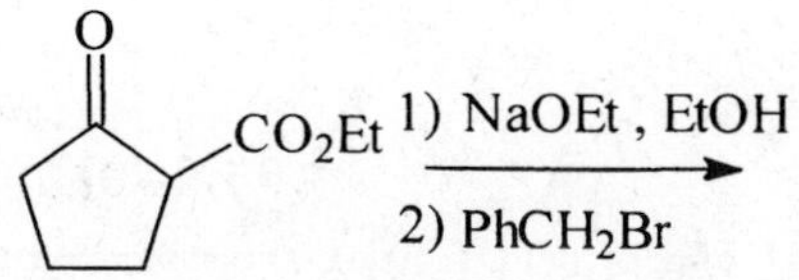
O
CO2Et
1) NaOEt, EtOH
2) PhCH2Br

c) 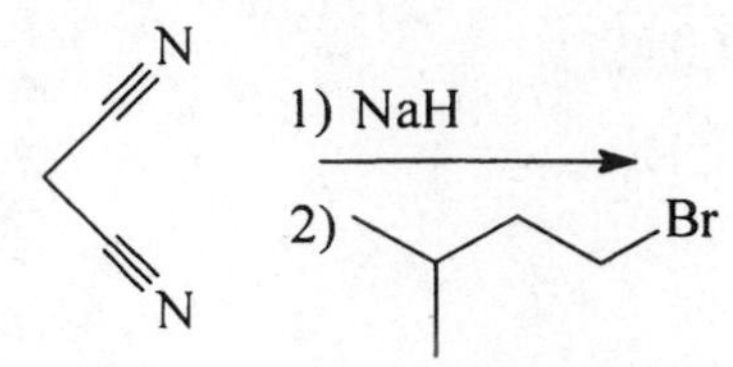
N
N
1) NaH
2)
Br

d) 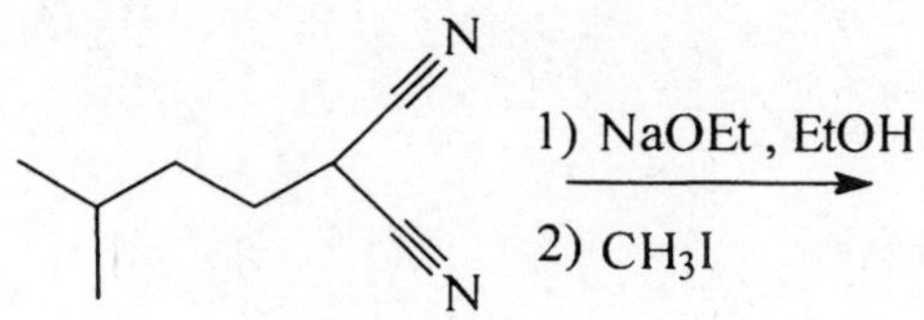
N
N
1) NaOEt, EtOH
2) CH3I

20.9

a) 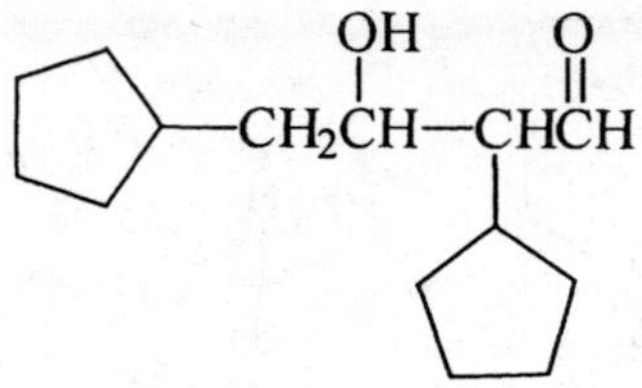

b) 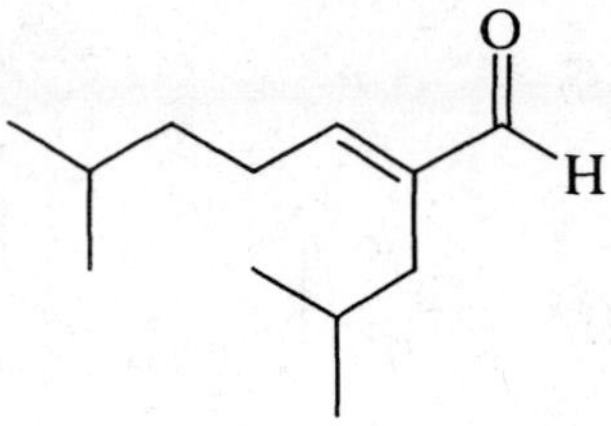

c) 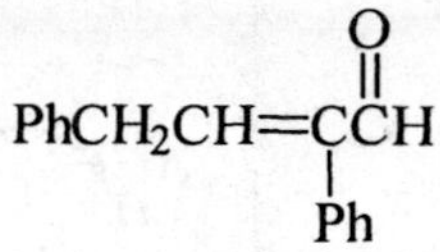

20.10

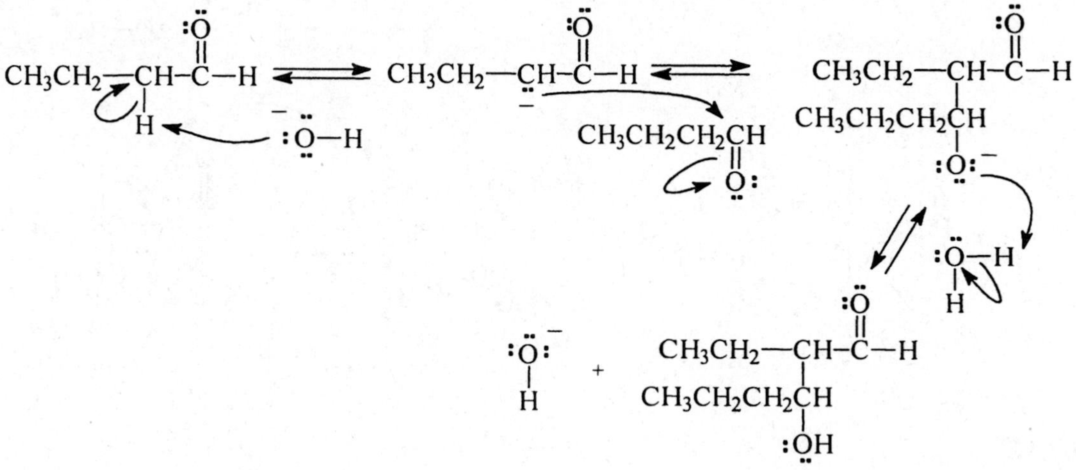

20.11

a) 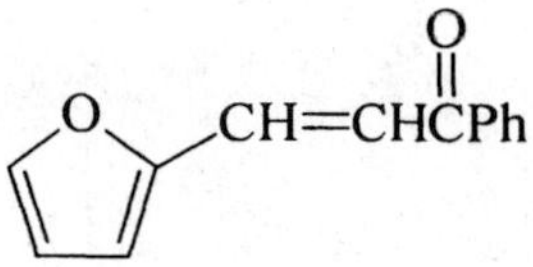

b) 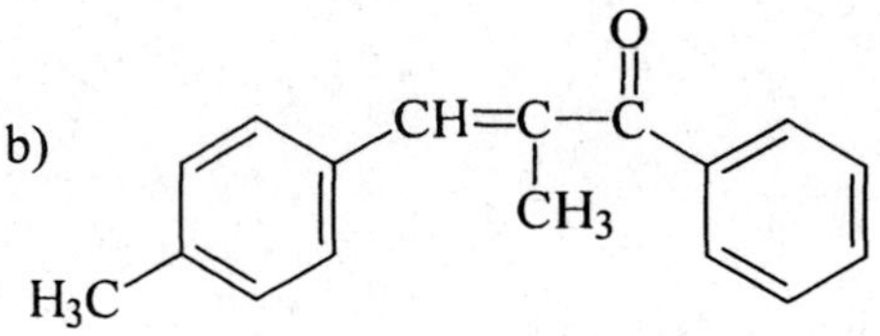

c) 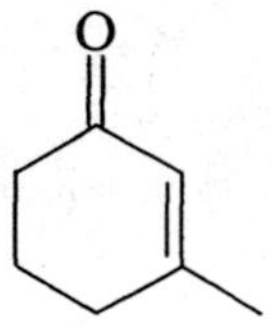

d) 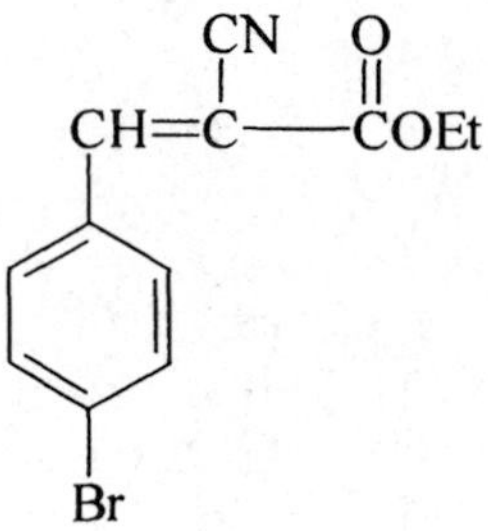

e) 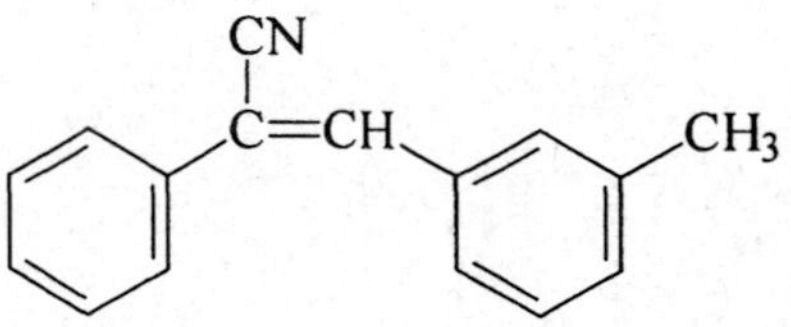

20.12

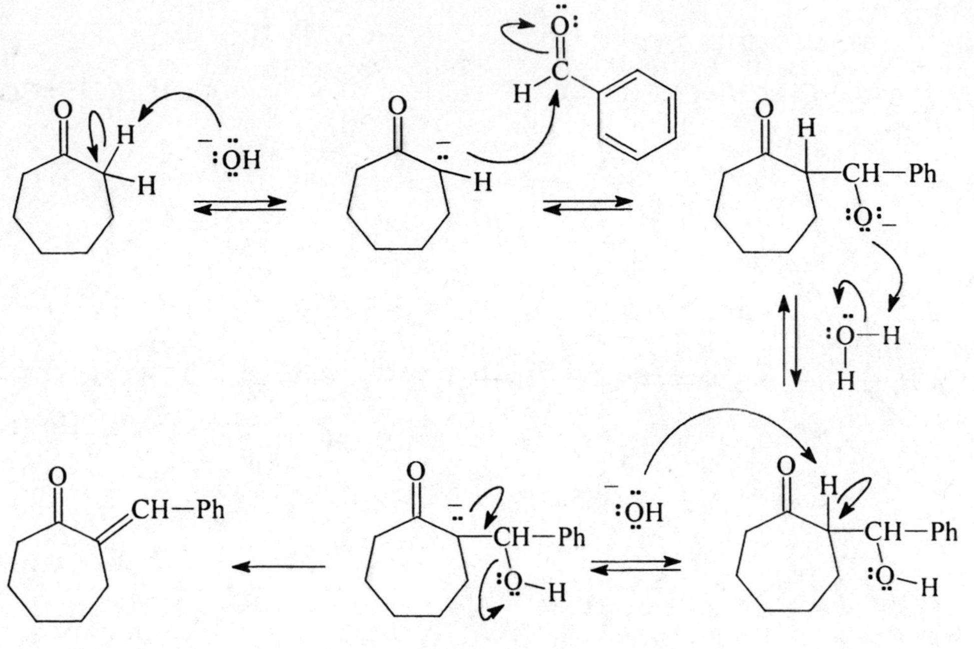

20.13

a) 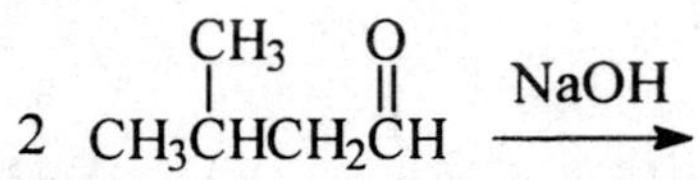

b) 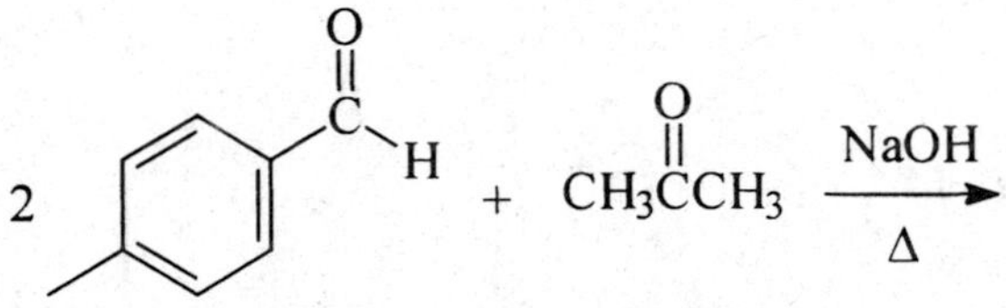

c) 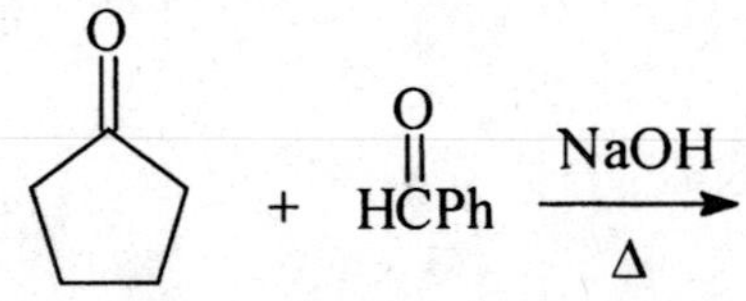

d) 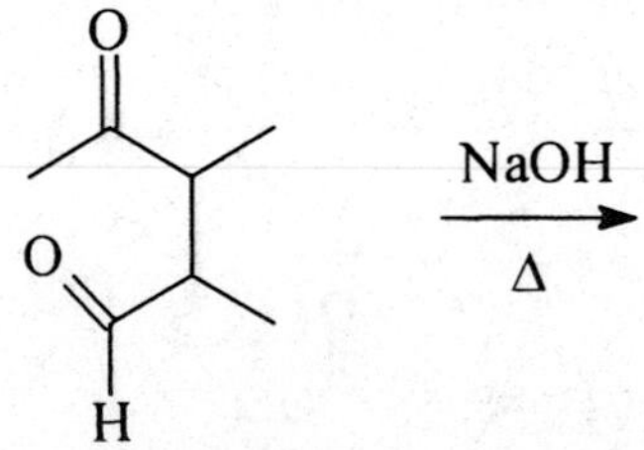

20.14

a) 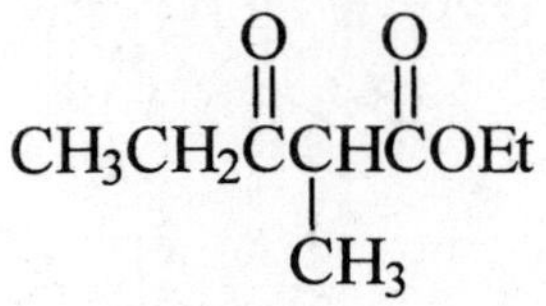

b) 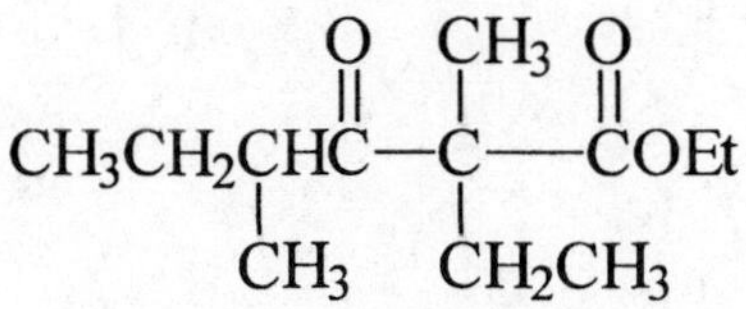

c) 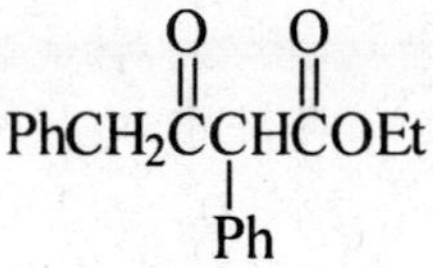

d) 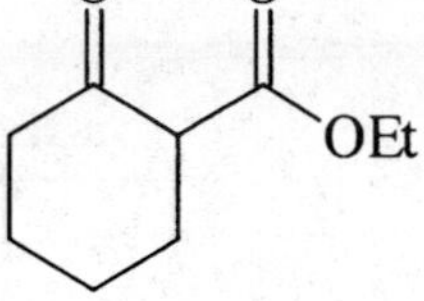

20.15 The other possible product would result from the enolate at the other α-carbon attacking the other carbonyl carbon. This product does not form because there is no acidic H between the two carbonyl groups so the equilibrium driving step cannot occur.

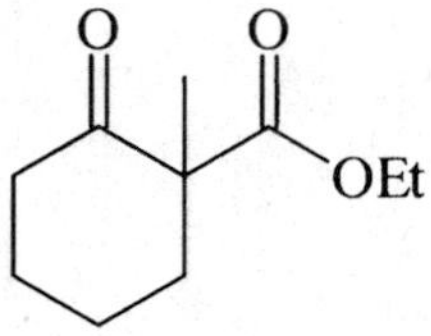

20.16

a) 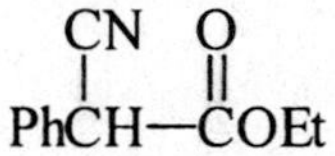

b) 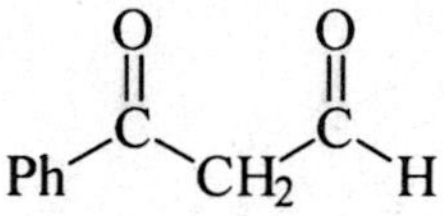

c) 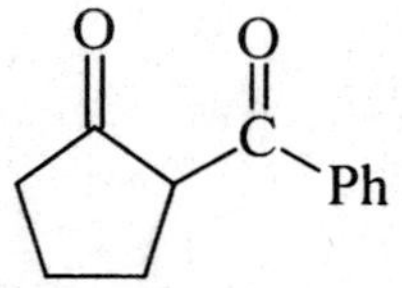

d) 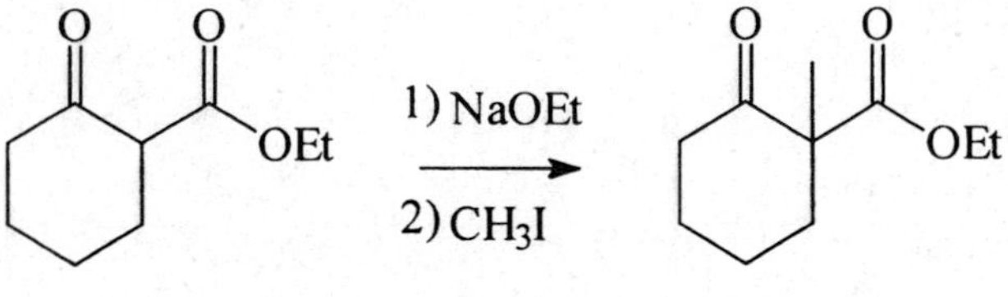

20.17

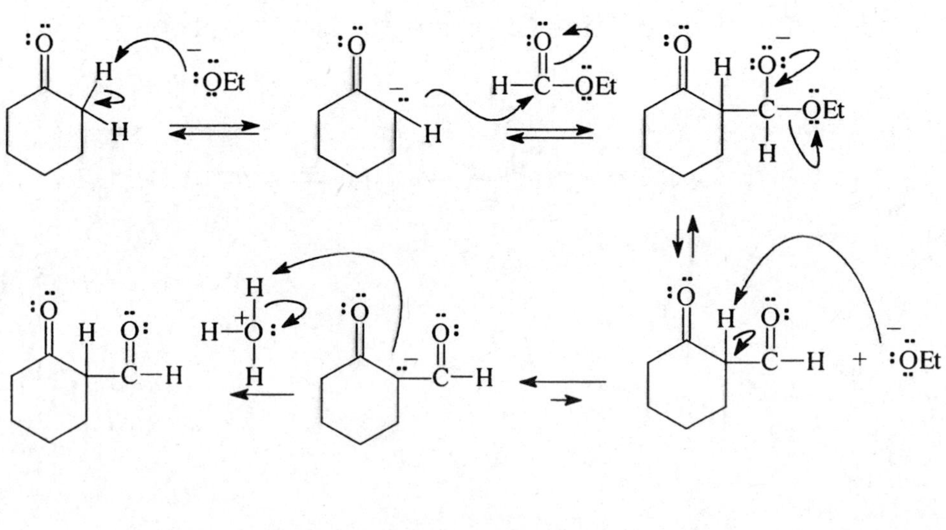

20.18

a) 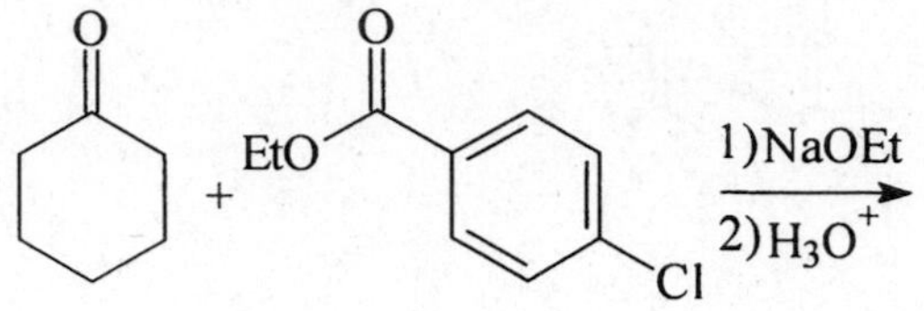

b) 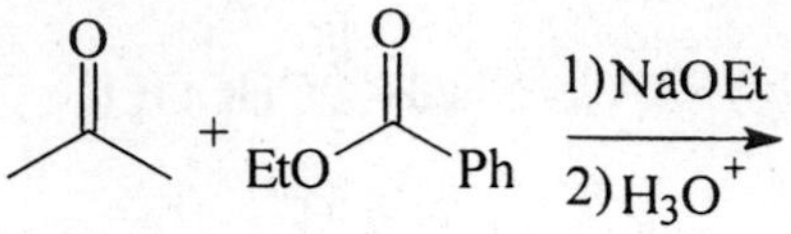

c) 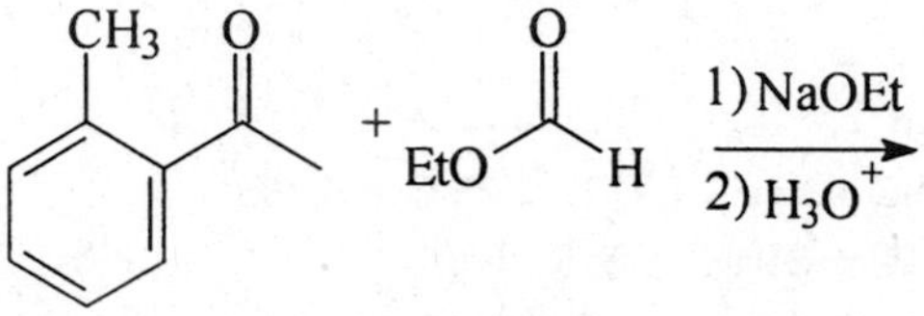

d) 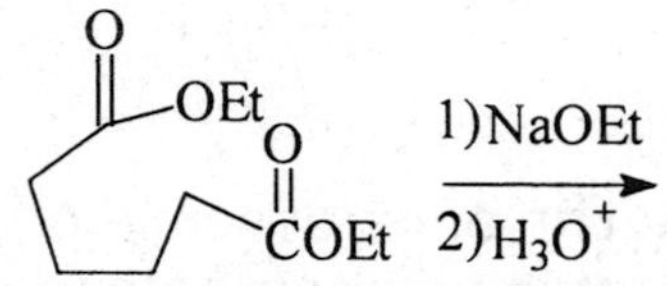

e) 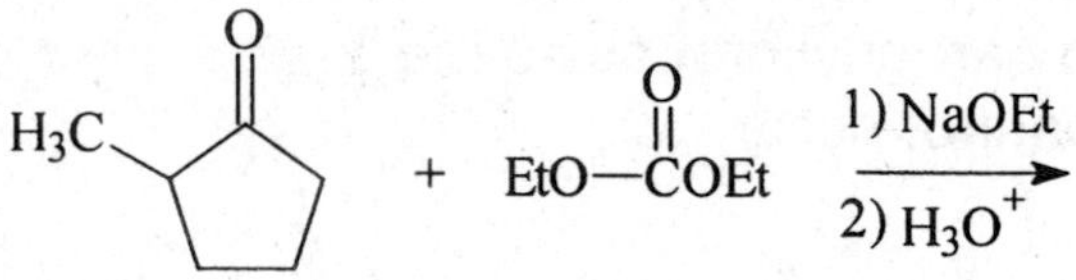

20.19 Use excess benzaldehyde to minimize the amount of this product.

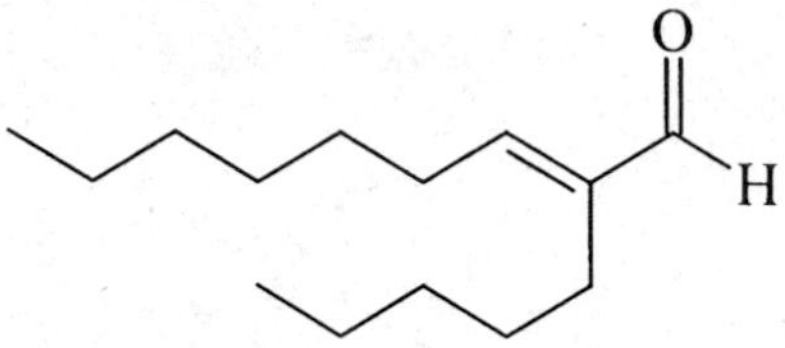

20.20

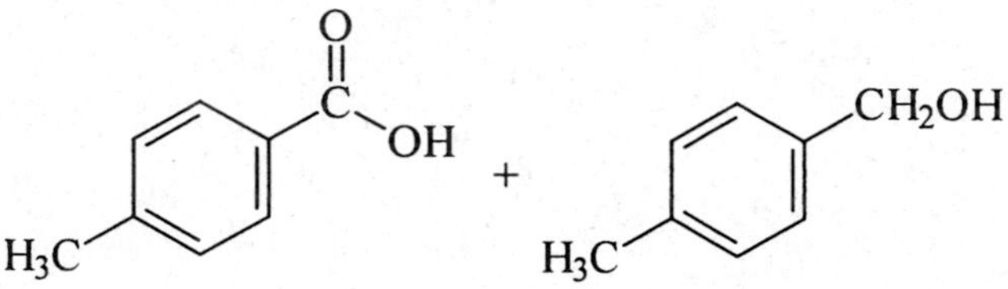

20.21 This is the reverse of the Claisen ester condensation.

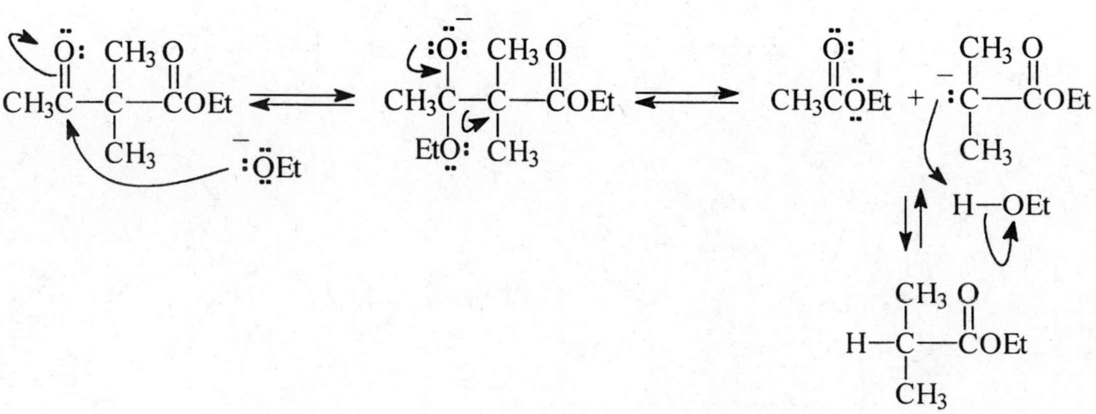

20.22

a)

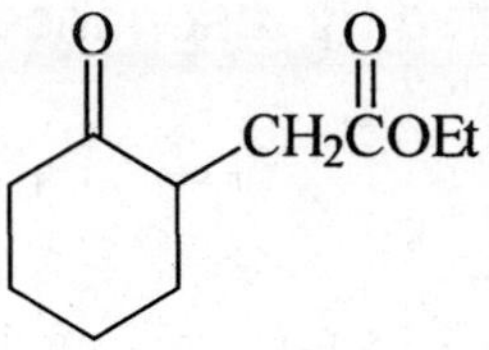

b)

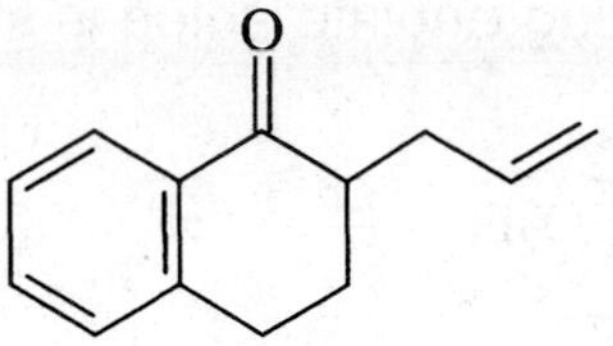

c)

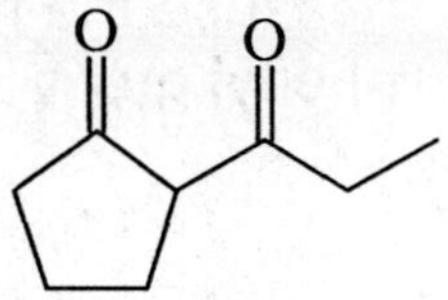

20.23

a)

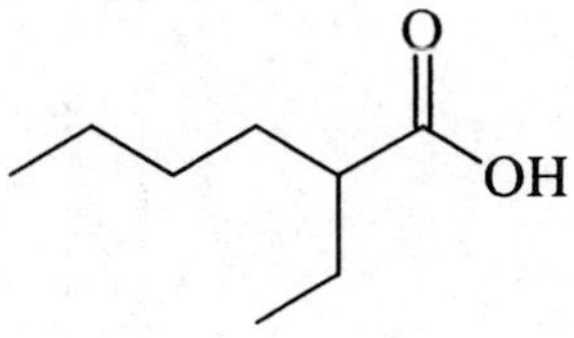

b)

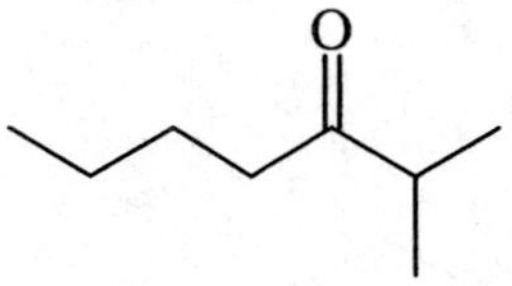

c)

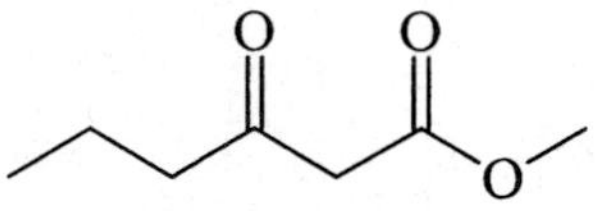

d) Ph

20.24

a)

b)

1) SH   SH , BF$_3$
2) BuLi
3) PhCH$_2$Br
4) Hg^{2+}, H$_2$O

c) PhCH$_2$CO$_2$H

1) 2 LDA
2) CH$_3$I
3) H$_3$O$^+$

20.25

a) EtOCCH—CH$_2$CH$_2$CCH$_3$
         CO$_2$Et

b) PhCHCH$_2$CPh
    EtO$_2$CCHCN

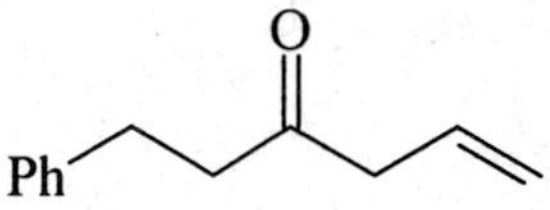

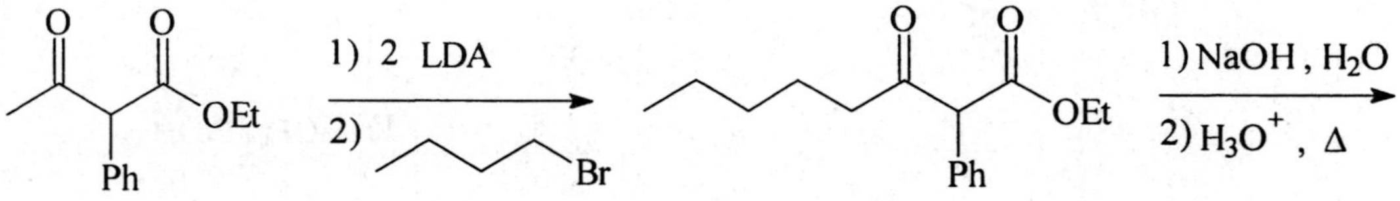

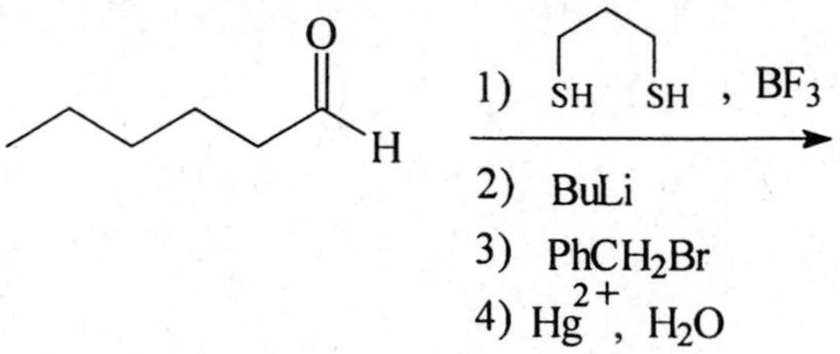

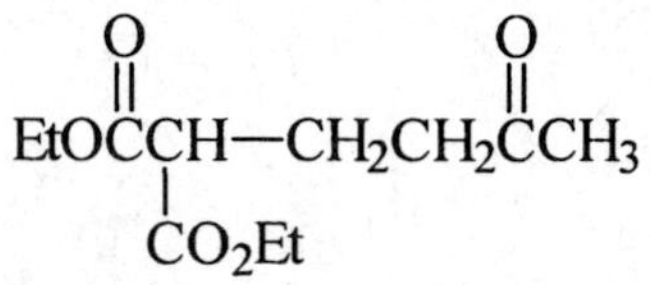

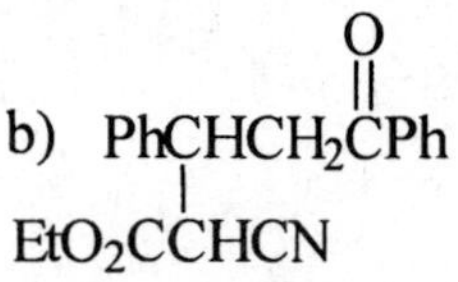

c)

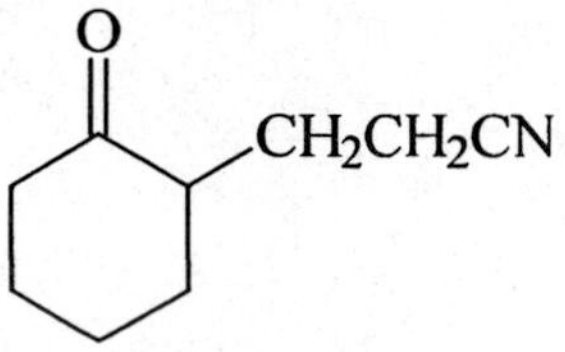

20.26 The hydrogen on the α-carbon bonded to the phenyl group is more acidic because the resulting enolate anion is stabilized by resonance with the phenyl group.

20.27

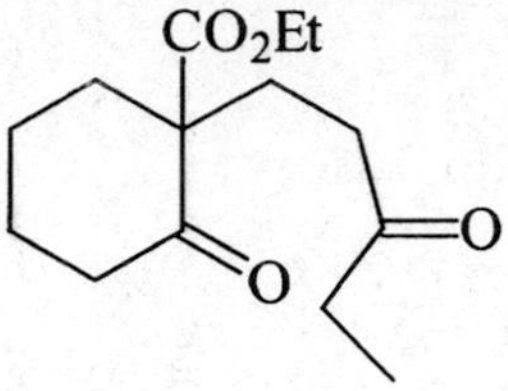

20.28

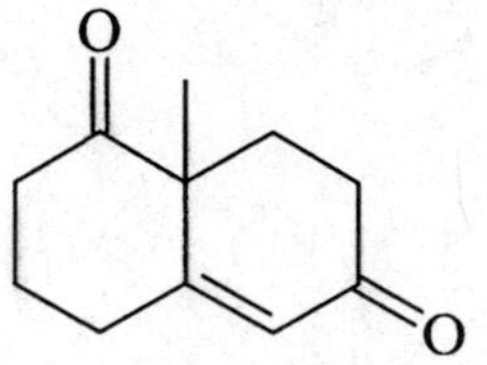

20.29

a)

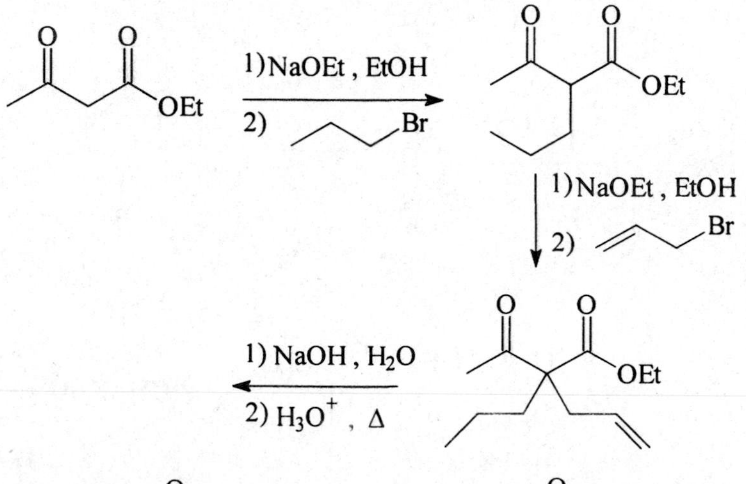

b) 2

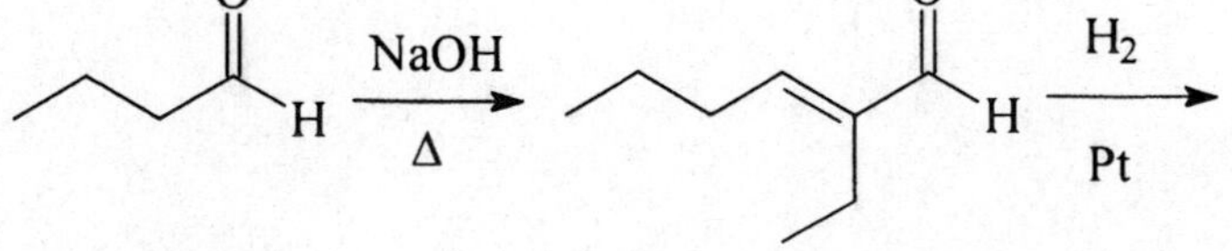

c) 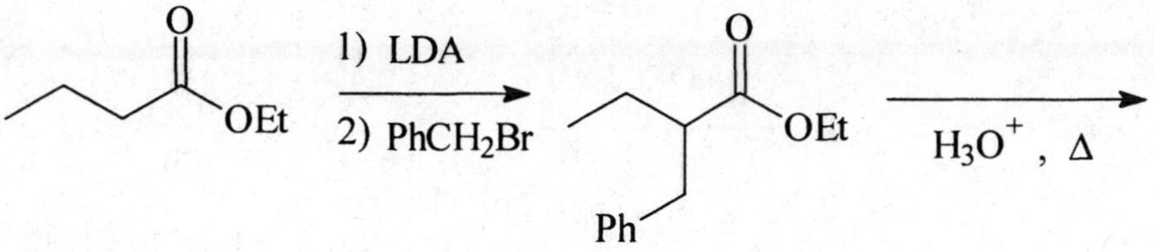
O
OEt
1) LDA
2) PhCH2Br
O
OEt
Ph
H3O+ , Δ

d) 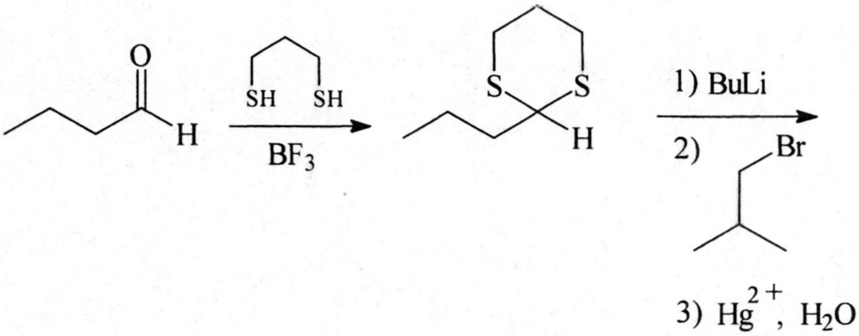
O
H
SH SH
BF3
S S
H
1) BuLi
2) Br
3) Hg2+, H2O

e) 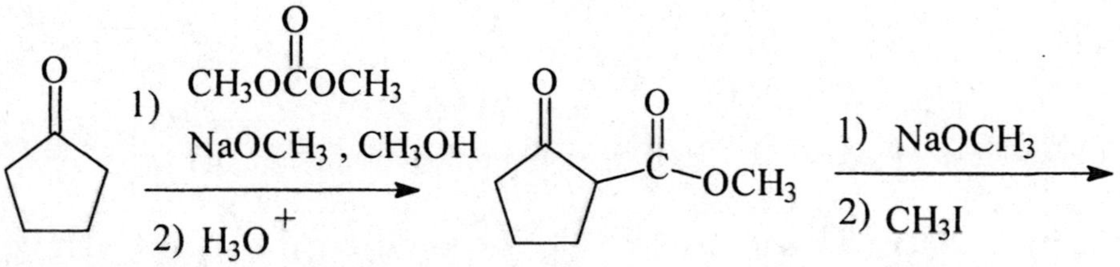
O
1) CH3OCOCH3
NaOCH3 , CH3OH
2) H3O+
O
O
C
OCH3
1) NaOCH3
2) CH3I

f) 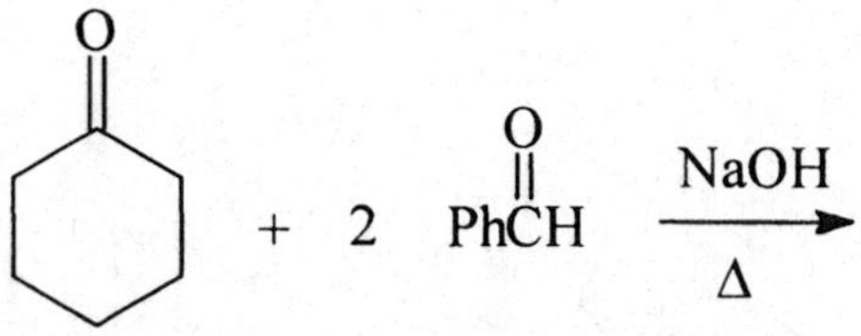
O
+ 2 PhCH
NaOH
Δ

g) 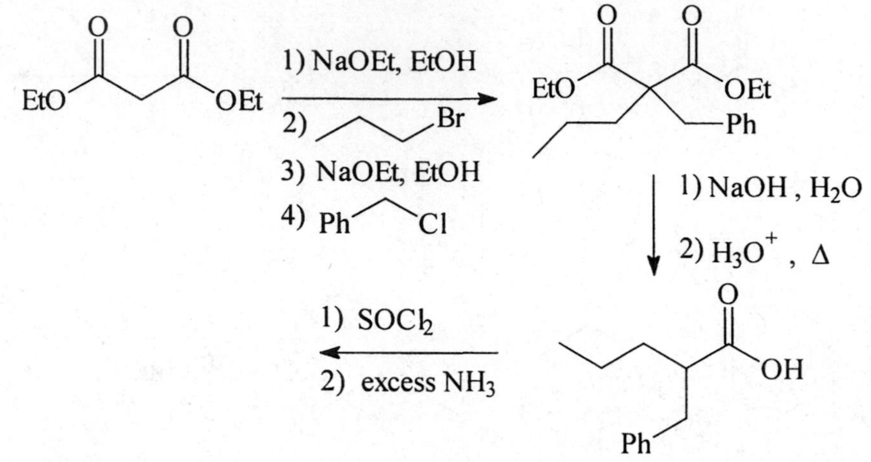

h) 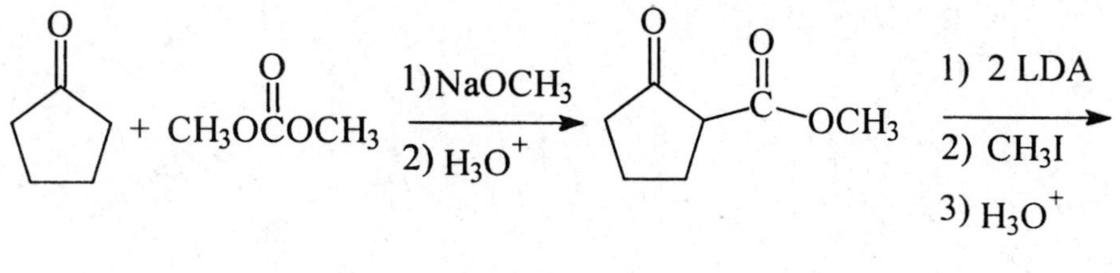

20.30

a) 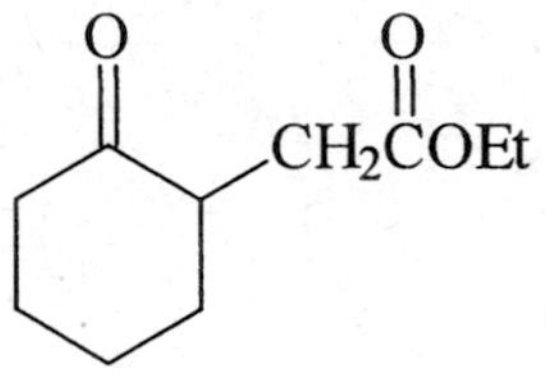

b) 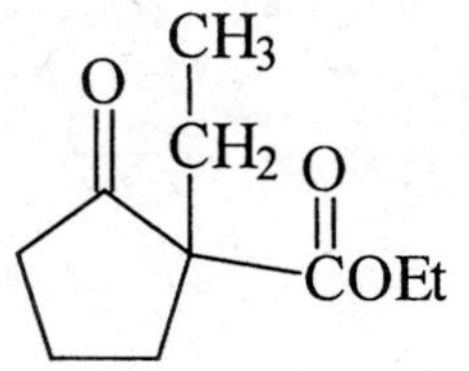

c) 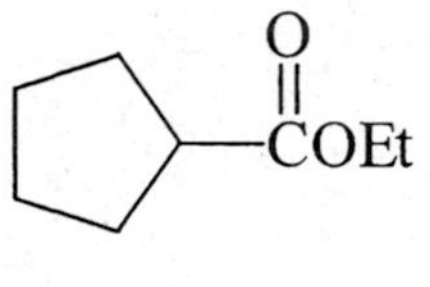

d) 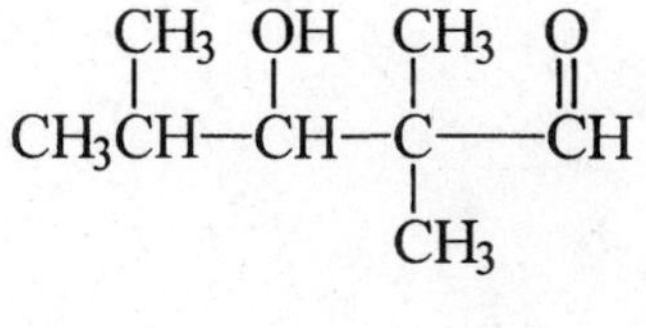

e) 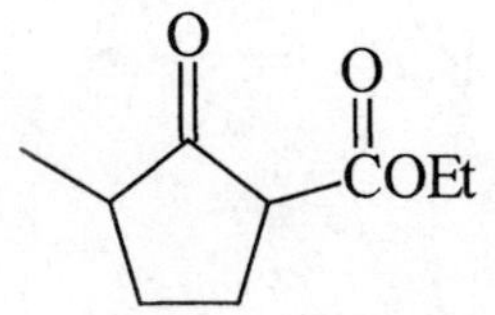

f) $CH_3CH_2\overset{O}{\overset{\|}{C}}CH_2CH_2CH_2Ph$

g) 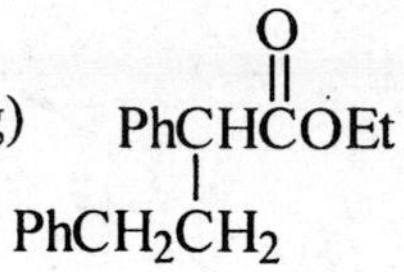

h) 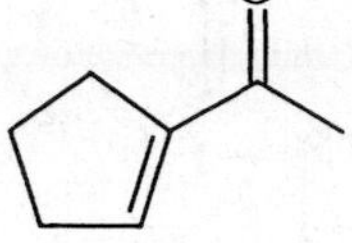

i) 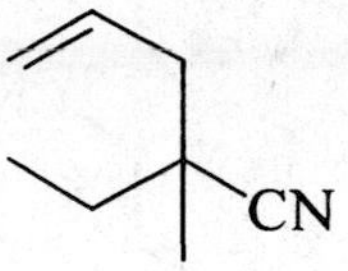

j) 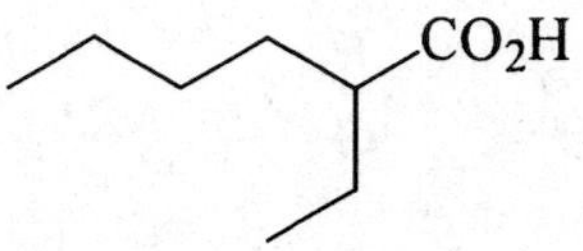

k) 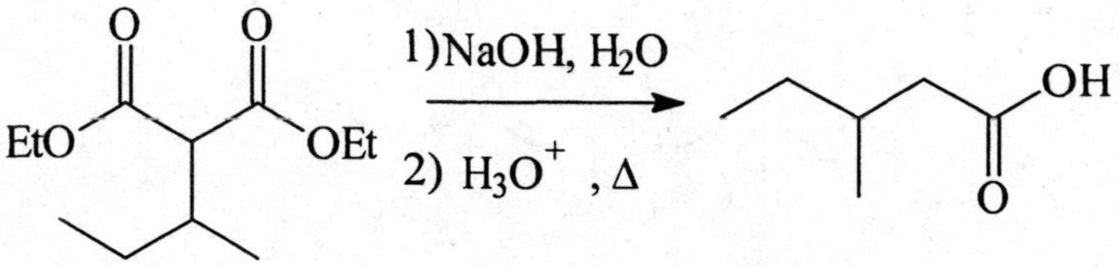

l) 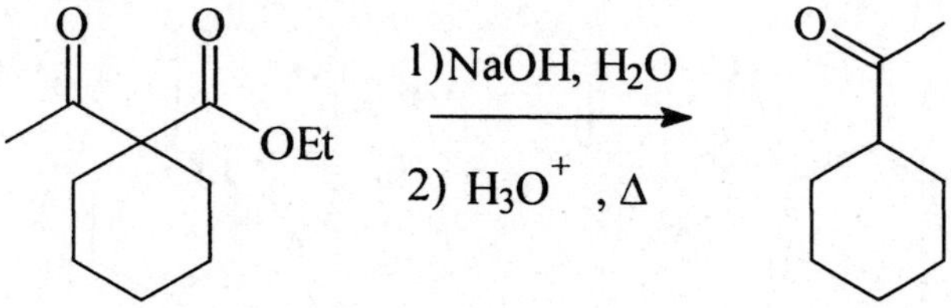

m) 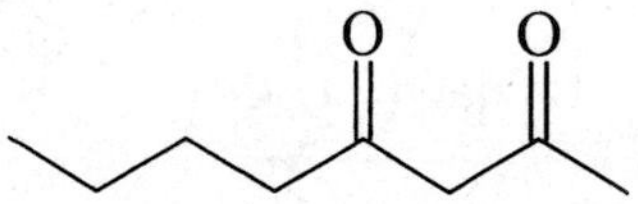

n) 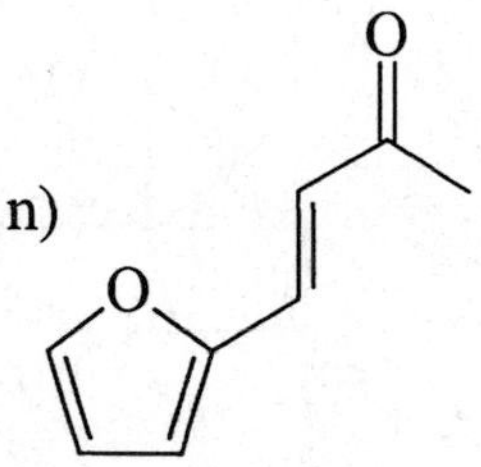

o) 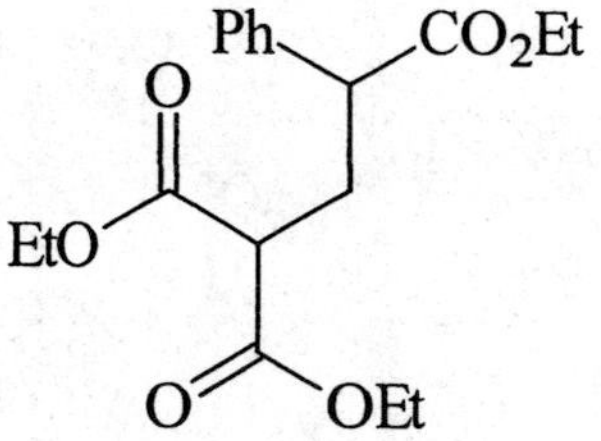

p) 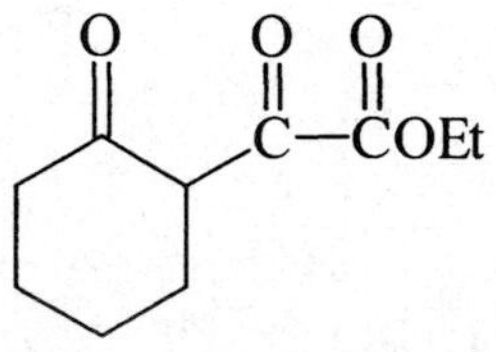

q) 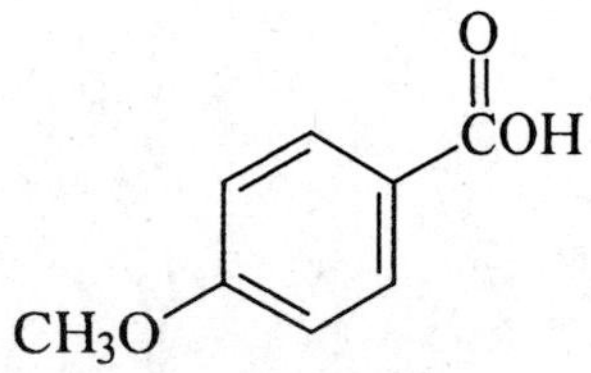

r) 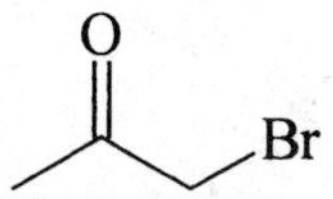

s) 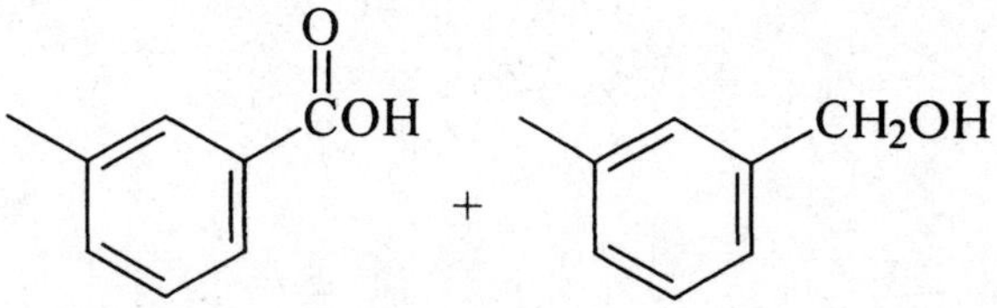

20.31

a)

b) $\xrightarrow[\text{2) } H_3O^+]{\text{1) NaOH , } H_2O}$

c)

d)

e)

f) Ph

g)

h)

i) $\xrightarrow[\text{2) } H_3O^+, \Delta]{\text{1) NaOH, } H_2O}$

20.32

a)

b) $\xrightarrow[\text{2) } H_3O^+]{\text{1) NaOH, } H_2O}$

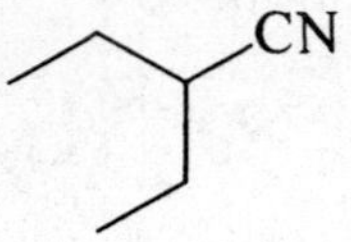

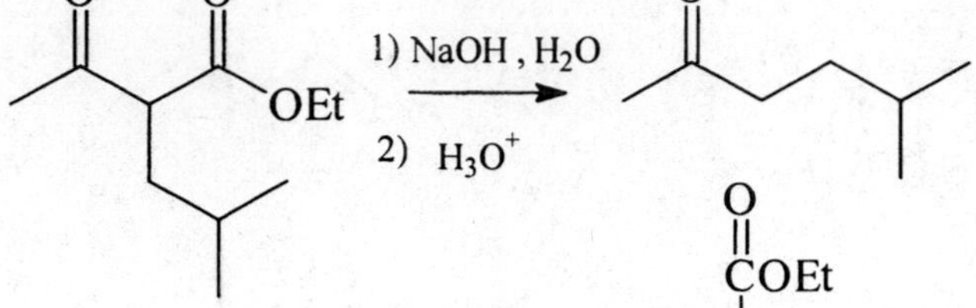

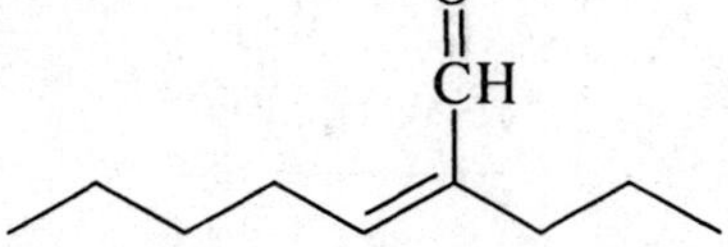

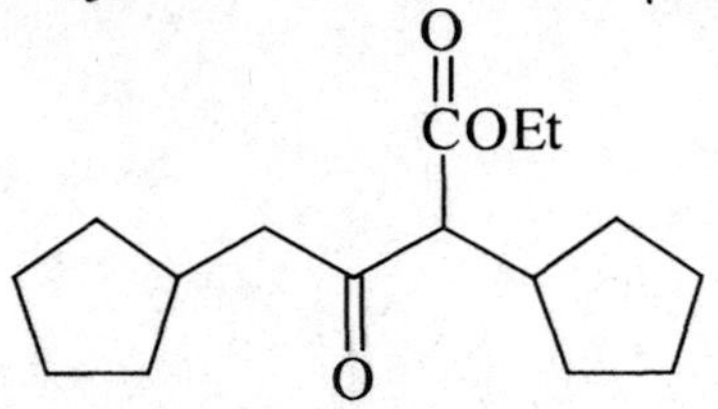

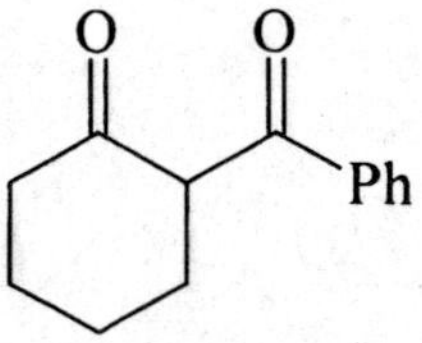

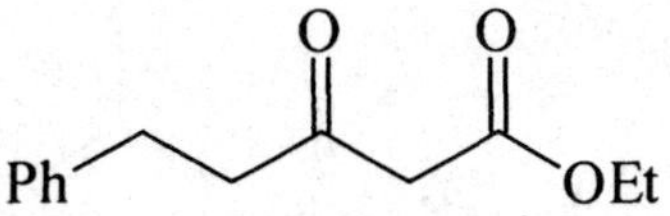

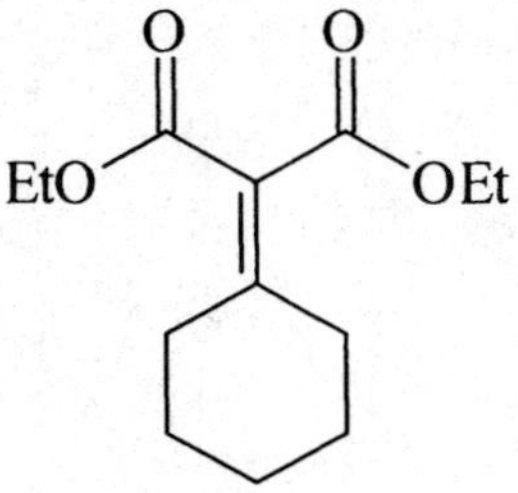

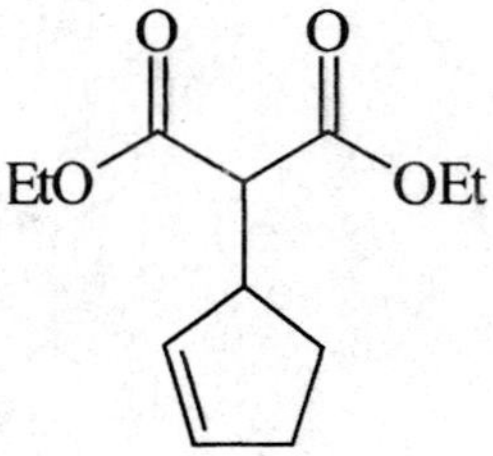

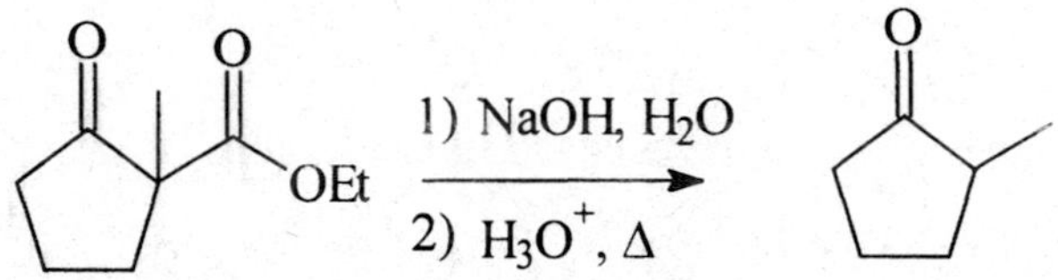

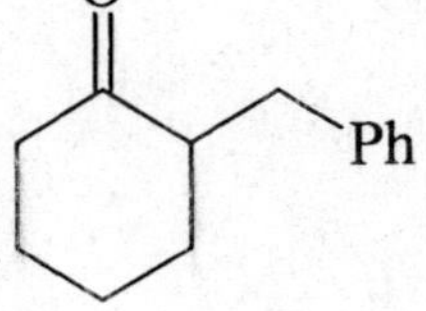

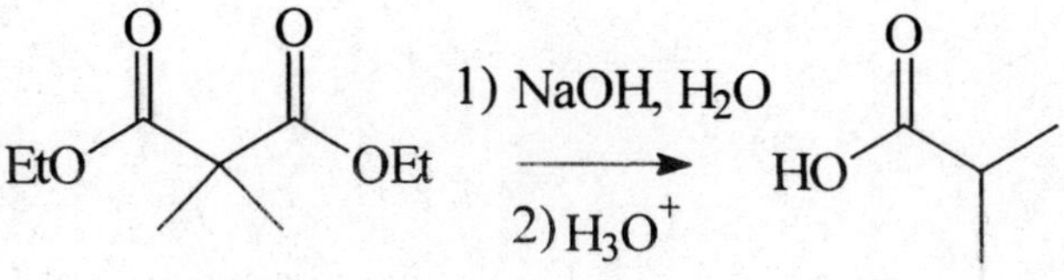

c) 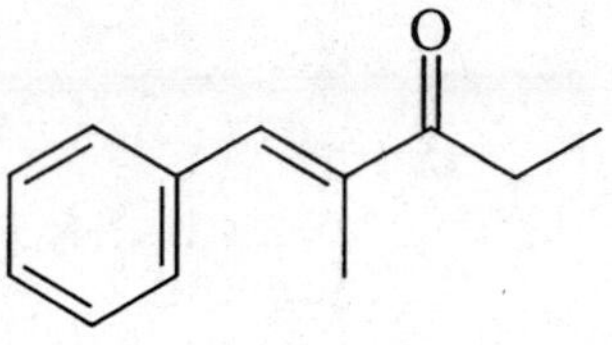 d) 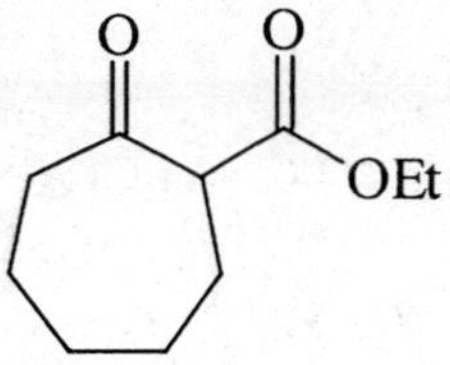

e) 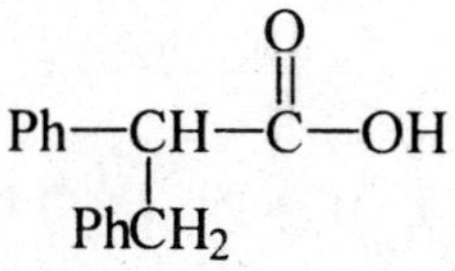 f) 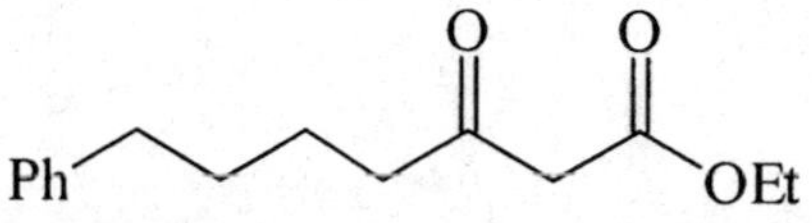

20.33

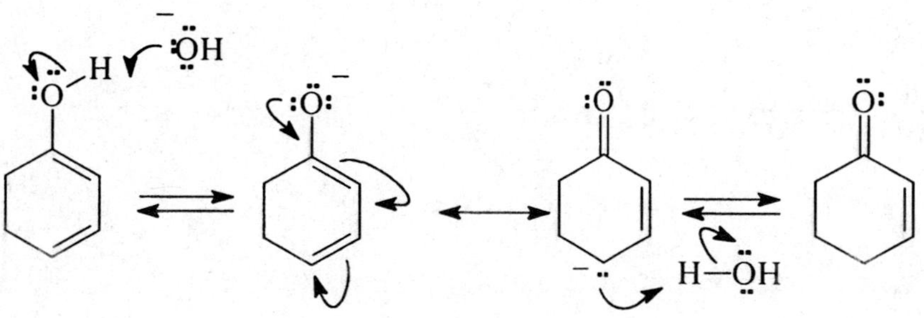

20.34

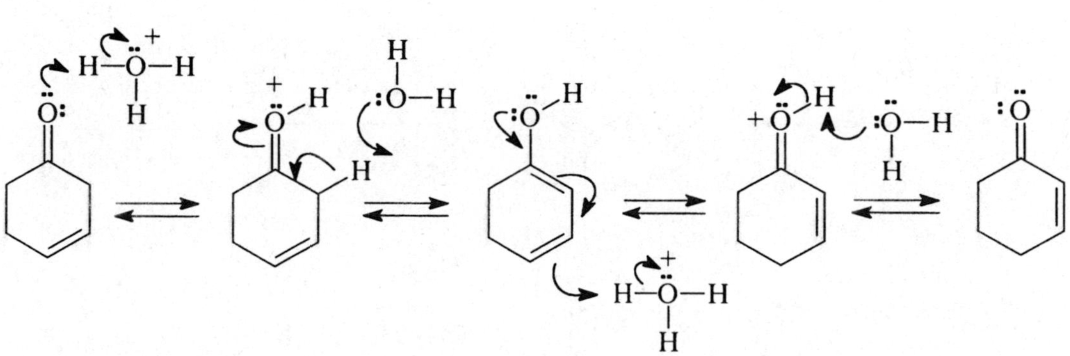

20.35

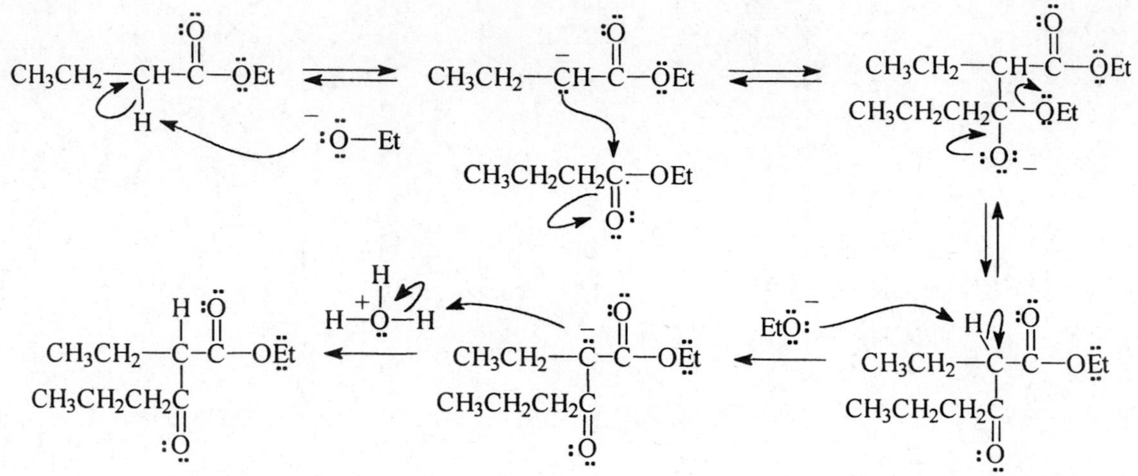

20.36

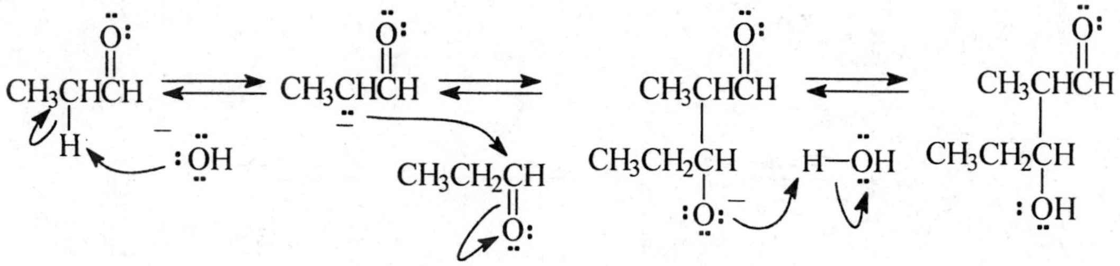

20.37

a) 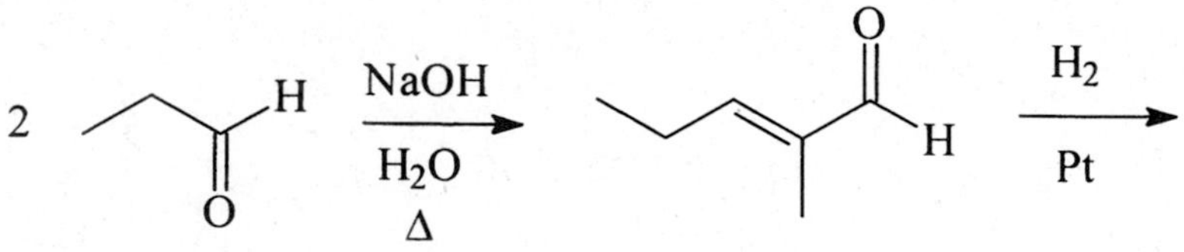
NaOH
H2O
Δ
H2
Pt

b) as (a) 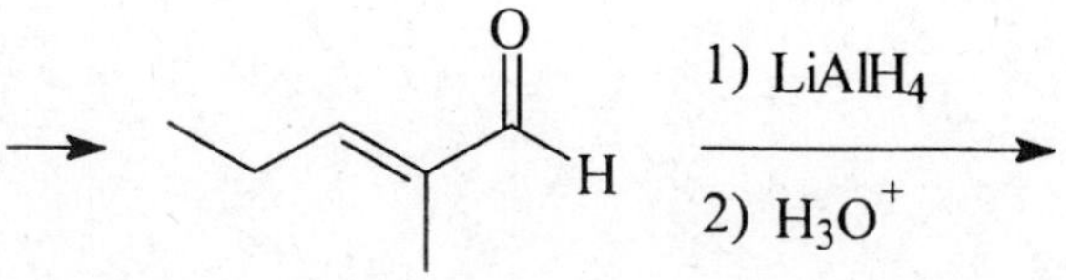
1) LiAlH4
2) H3O+

c) as (a) 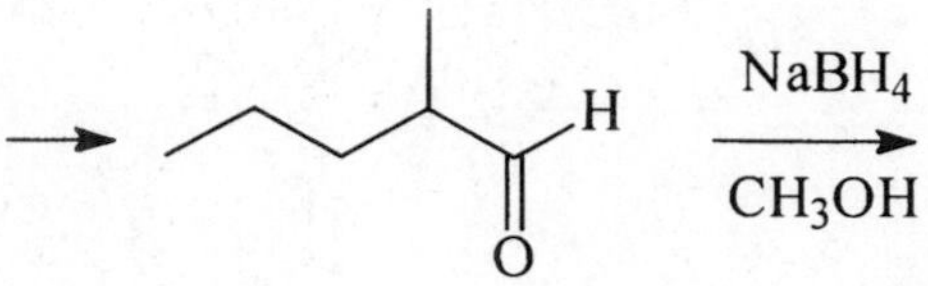
NaBH4
CH3OH

d) $\underset{PhCH}{\overset{O}{\parallel}}$ + $\underset{H}{\overset{O}{\parallel}}$ $\xrightarrow[\underset{\Delta}{H_2O}]{NaOH}$ $Ph\diagup\diagdown\underset{H}{\overset{O}{\parallel}}$ $\xrightarrow[Pd]{H_2}$

20.38

a) 2 $\underset{OEt}{\overset{O}{\parallel}}$ $\xrightarrow[2)\ H_3O^+]{1)\ NaOEt,\ EtOH}$

b) as (a) $\longrightarrow$ $\underset{OEt}{\overset{O\ \ \ O}{\parallel\ \ \ \parallel}}$ $\xrightarrow[CH_3OH]{NaBH_4}$

c) as (a) $\longrightarrow$ $\underset{OEt}{\overset{O\ \ \ O}{\parallel\ \ \ \parallel}}$ $\xrightarrow[2)\ CH_3Br]{1)\ NaOEt,\ EtOH}$

d) as (a) $\longrightarrow$ $\underset{OEt}{\overset{O\ \ \ O}{\parallel\ \ \ \parallel}}$

1) NaOEt, EtOH
2) CH_3CH_2Cl
3) KOH, H_2O
4) H_3O^+, Δ

e) as (a) $\longrightarrow$ $\underset{OEt}{\overset{O\ \ \ O}{\parallel\ \ \ \parallel}}$ $\xrightarrow[2)\ PhCH_2Cl]{1)\ NaOEt,\ EtOH}$ $\underset{Ph}{\underset{OEt}{\overset{O\ \ \ O}{\parallel\ \ \ \parallel}}}$ $\xrightarrow[2)\ H_3O^+]{1)\ LiAlH_4}$

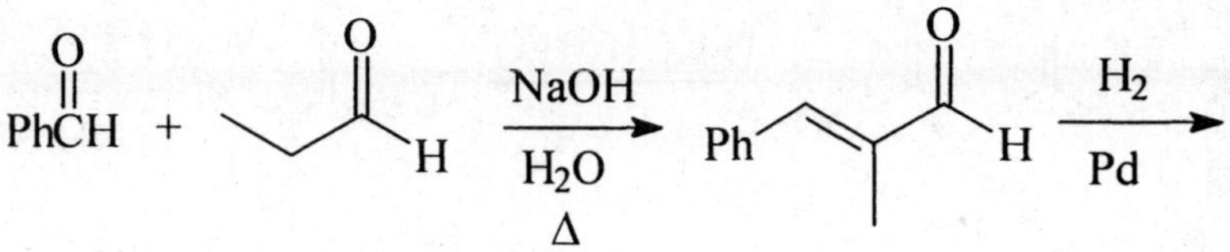

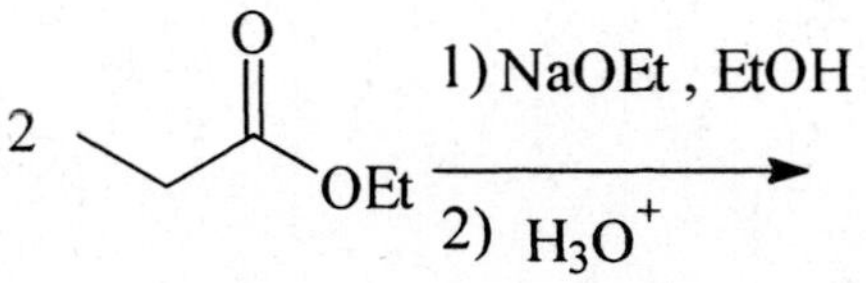

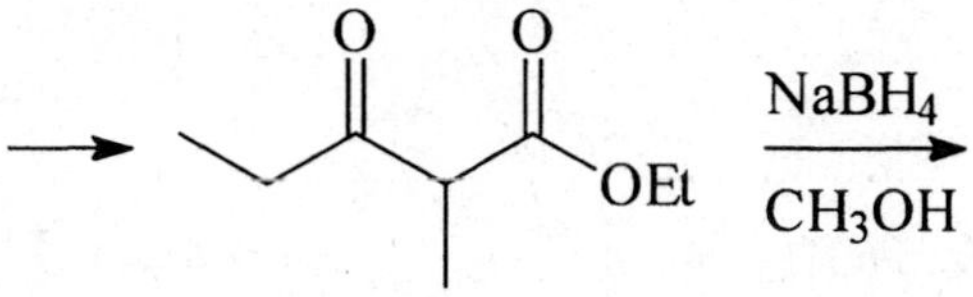

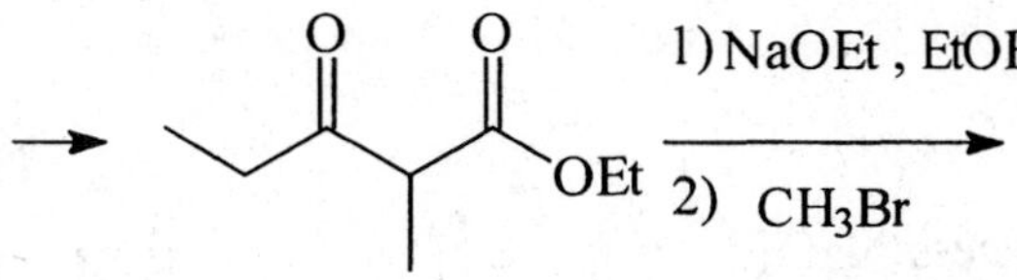

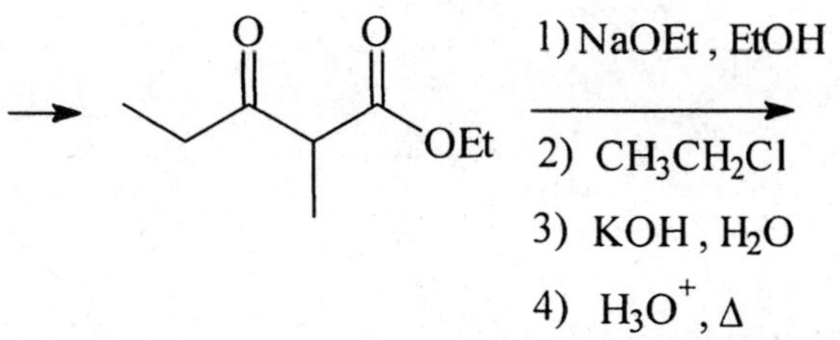

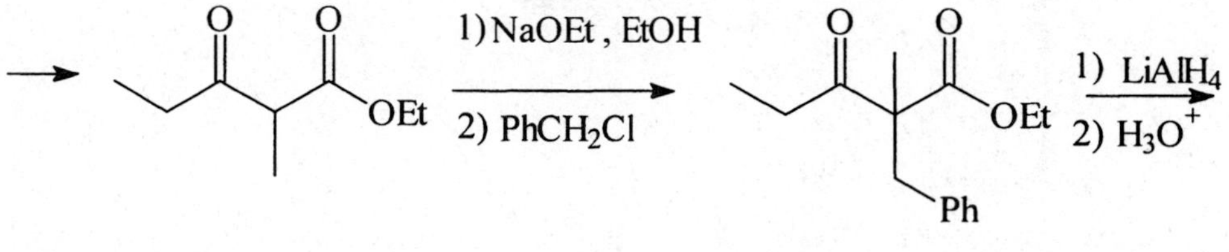

f) as (c)

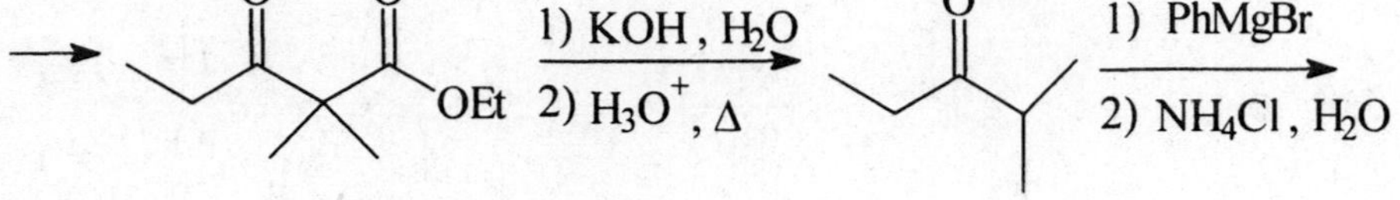
O
O
OEt
1) KOH, H2O
2) H3O+, Δ
O
1) PhMgBr
2) NH4Cl, H2O

g)

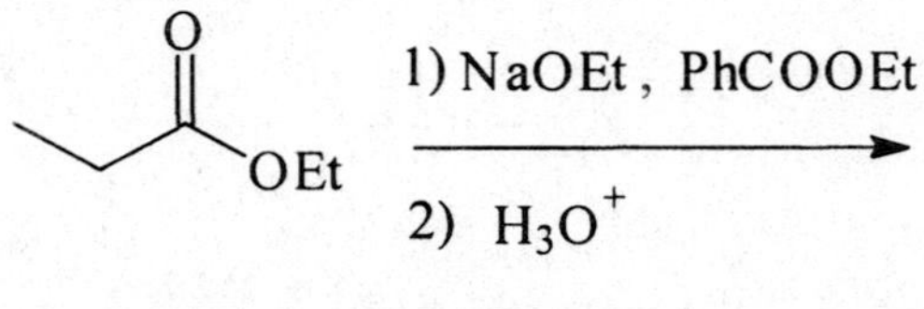
O
OEt
1) NaOEt, PhCOOEt
2) H3O+

20.39

a)

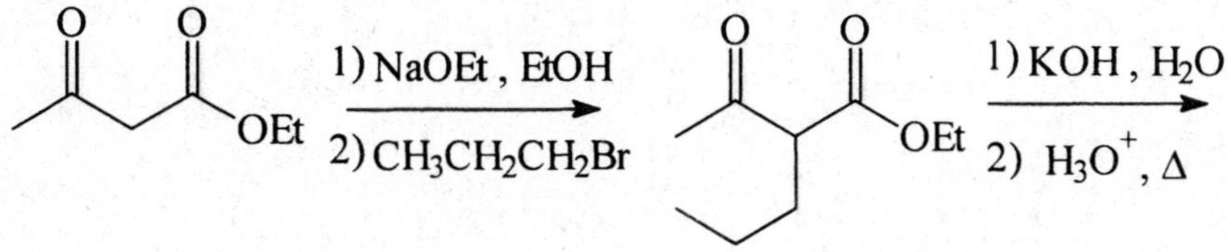
O
O
OEt
1) NaOEt, EtOH
2) CH3CH2CH2Br
O
O
OEt
1) KOH, H2O
2) H3O+, Δ

b)

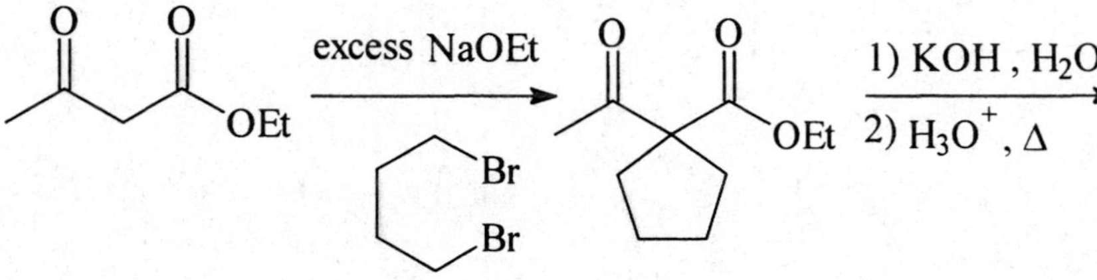
O
O
OEt
excess NaOEt
Br
Br
O
O
OEt
1) KOH, H2O
2) H3O+, Δ

c)

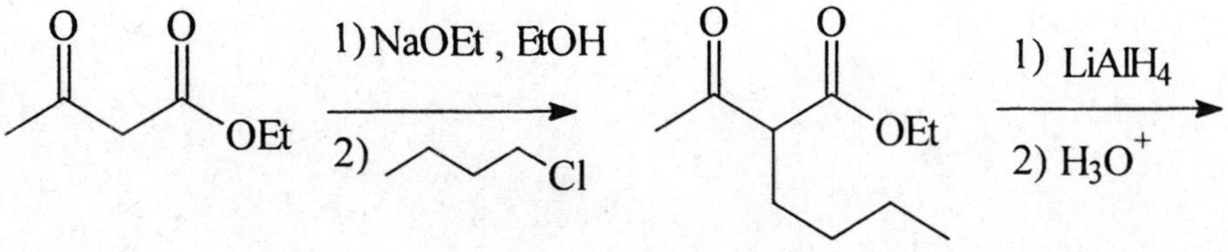
O
O
OEt
1) NaOEt, EtOH
2)
Cl
O
O
OEt
1) LiAlH4
2) H3O+

d) 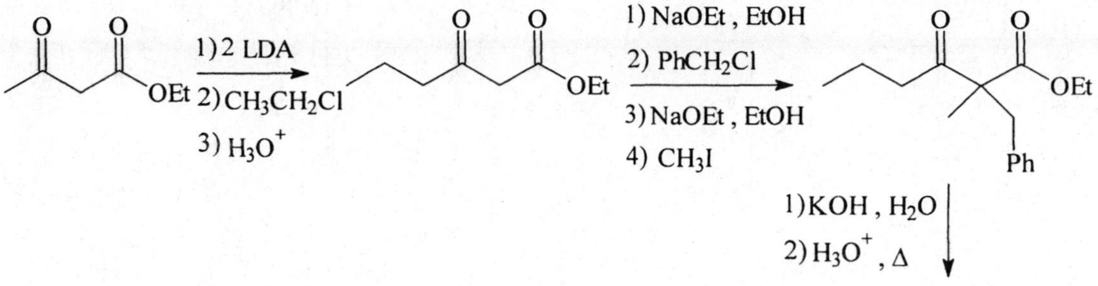

20.40 The planar enolate ion formed from **A** can be protonated from either side, resulting in racemization. **B** does not racemize because the chiral center is not the α-carbon.

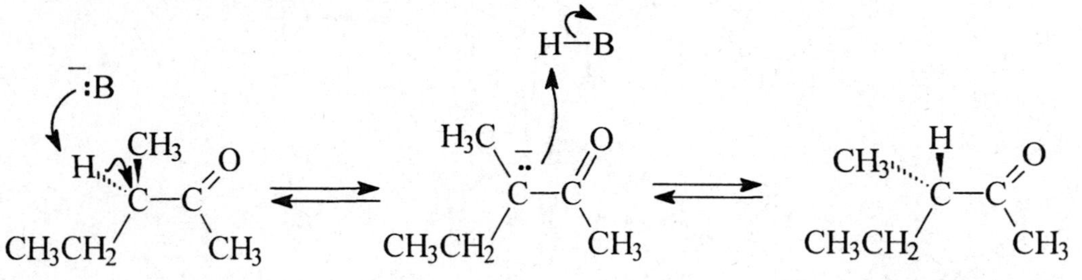

A enolate anion enantiomer

20.41

a) 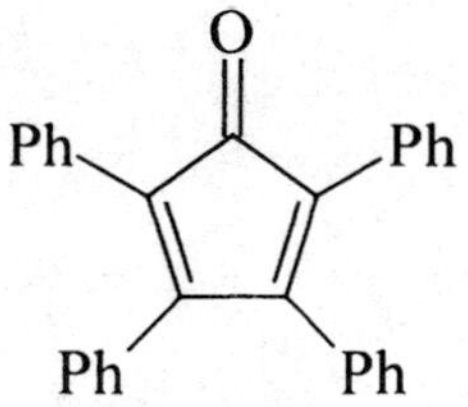

b) 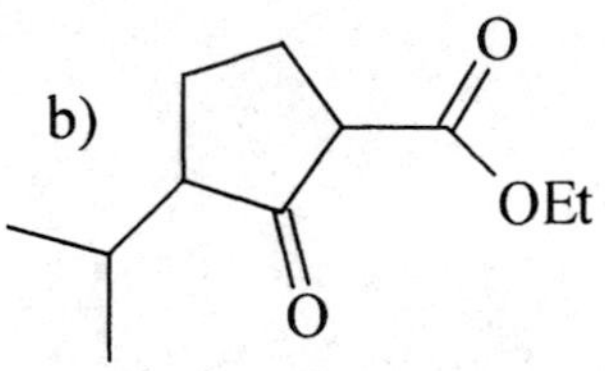

c) 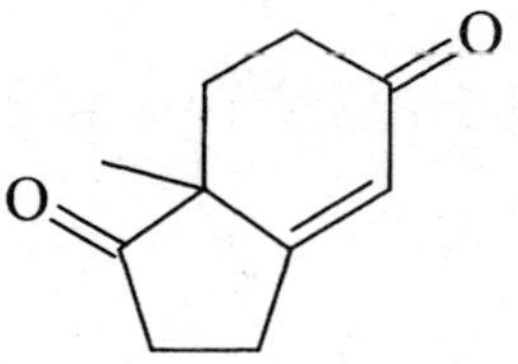

d) 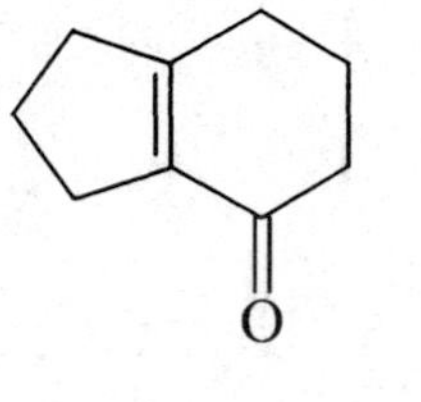

e) 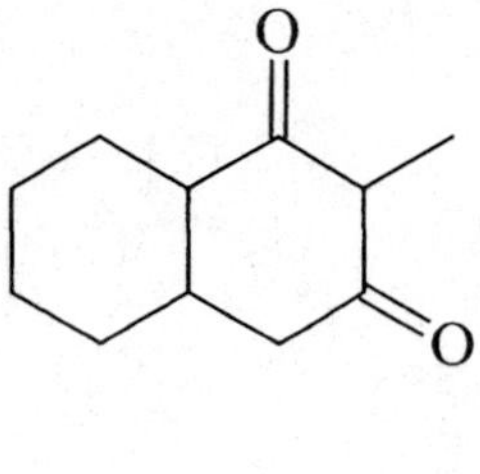

f) 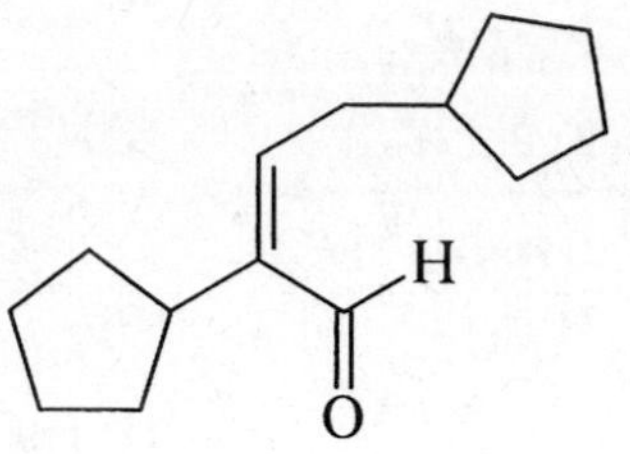

g) 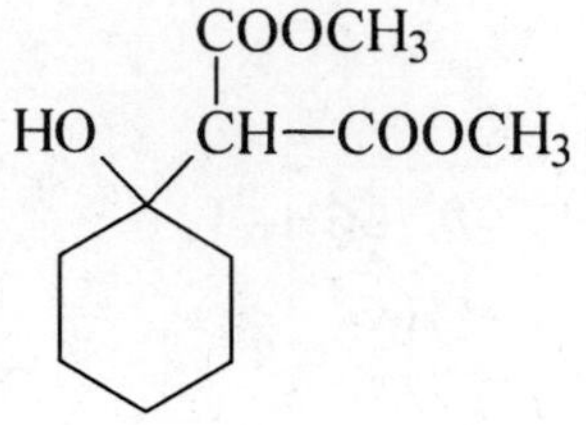
COOCH$_3$
HO CH—COOCH$_3$

20.42

a) 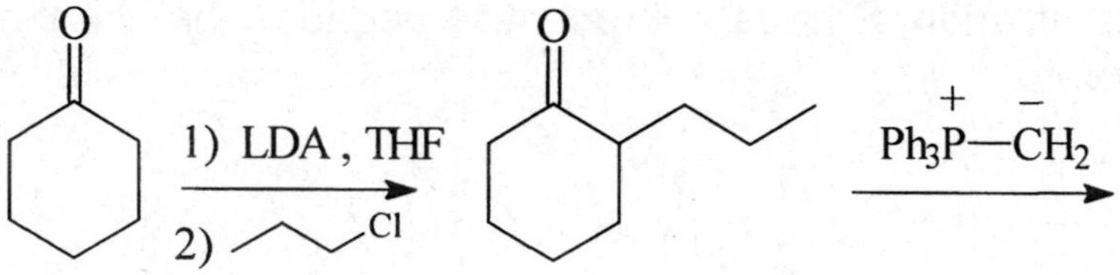
O
1) LDA, THF
2) Cl
O
$Ph_3\overset{+}{P}—\overset{-}{C}H_2$

b) 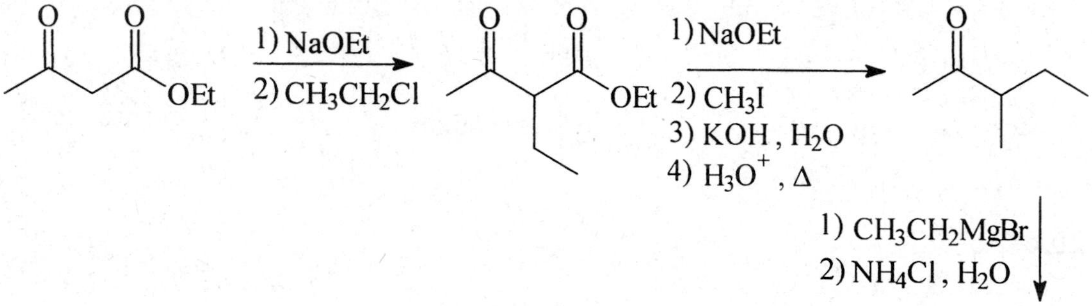
O O
OEt
1) NaOEt
2) CH$_3$CH$_2$Cl
O O
OEt
1) NaOEt
2) CH$_3$I
3) KOH, H$_2$O
4) H$_3$O$^+$, Δ
O
1) CH$_3$CH$_2$MgBr
2) NH$_4$Cl, H$_2$O

c) 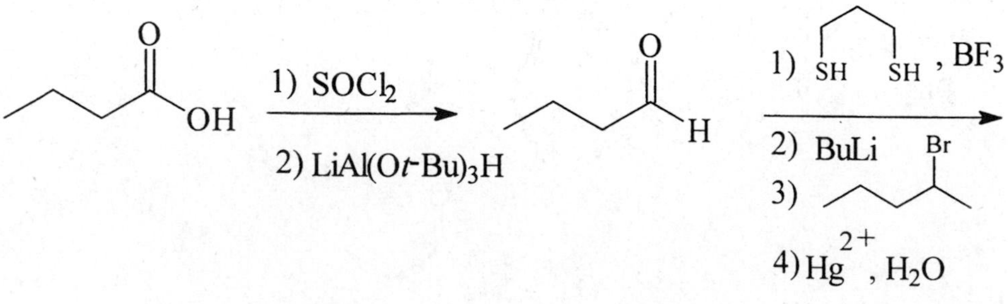
O
OH
1) SOCl$_2$
2) LiAl(O*t*-Bu)$_3$H
O
H
1) SH SH, BF$_3$
2) BuLi
3) Br
4) Hg^{2+}, H$_2$O

d)

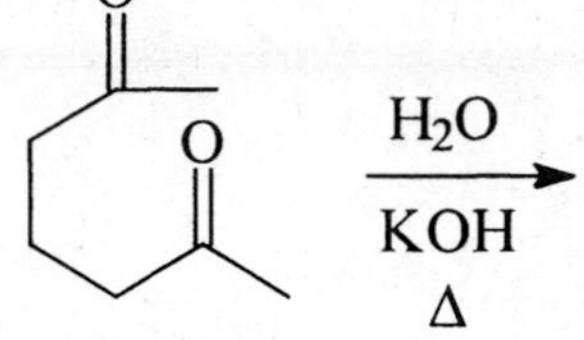
H2O
KOH
Δ

e)

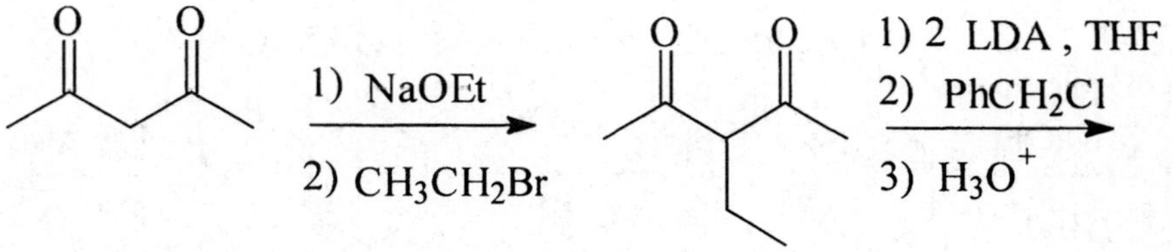
1) NaOEt
2) CH3CH2Br
1) 2 LDA, THF
2) PhCH2Cl
3) H3O+

f)

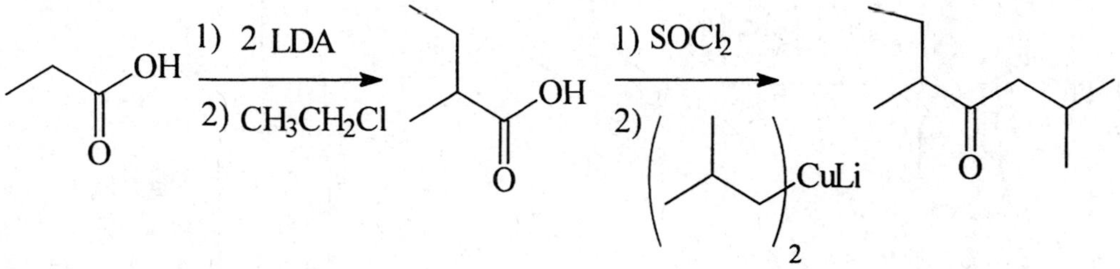
OH
1) 2 LDA
2) CH3CH2Cl
OH
1) SOCl2
2)
CuLi
2

g)

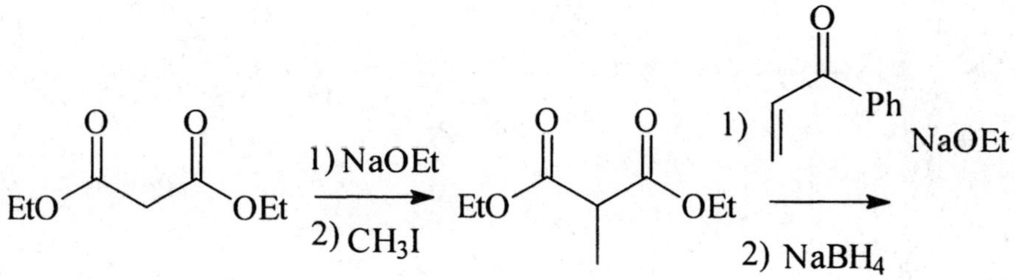
EtO
OEt
1) NaOEt
2) CH3I
EtO
OEt
1)
Ph
NaOEt
2) NaBH4

20.43

a)

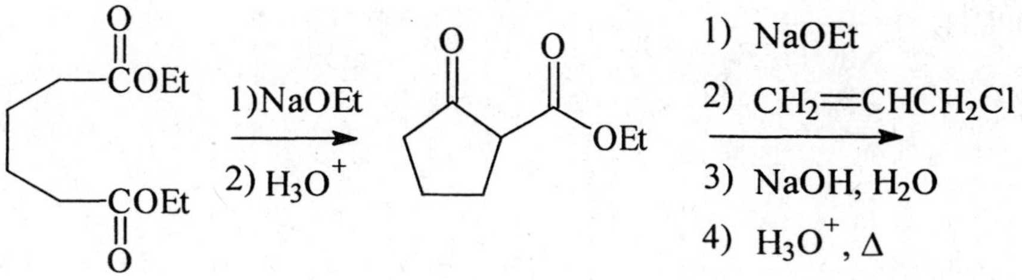
COEt
COEt
1)NaOEt
2) H3O+
OEt
1) NaOEt
2) CH2=CHCH2Cl
3) NaOH, H2O
4) H3O+, Δ

b) 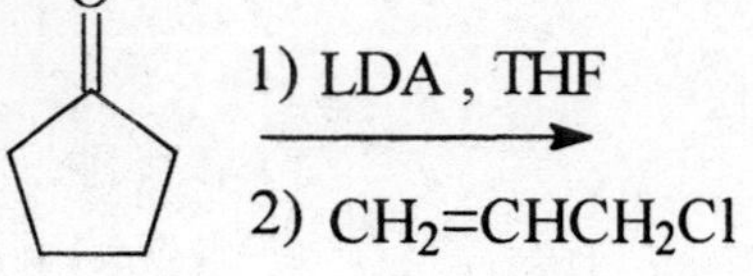

c) 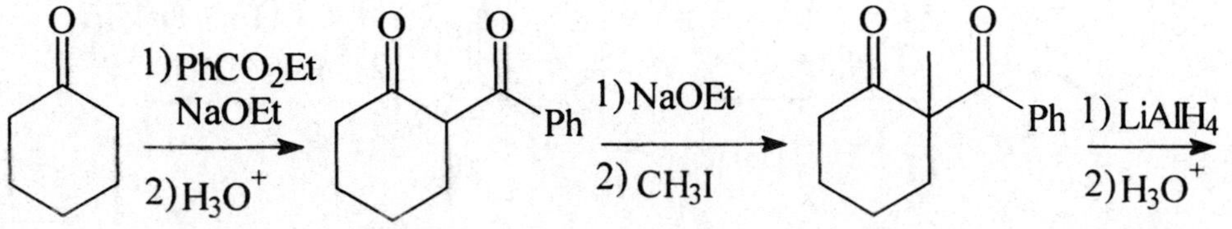

d) 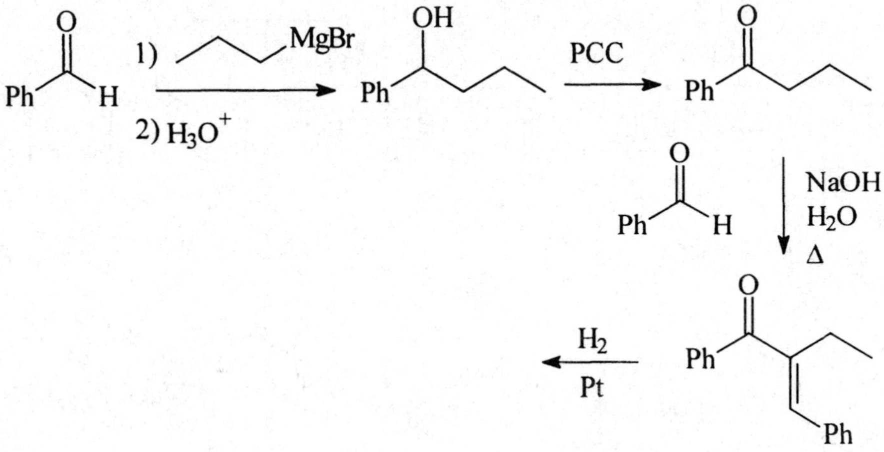

e) 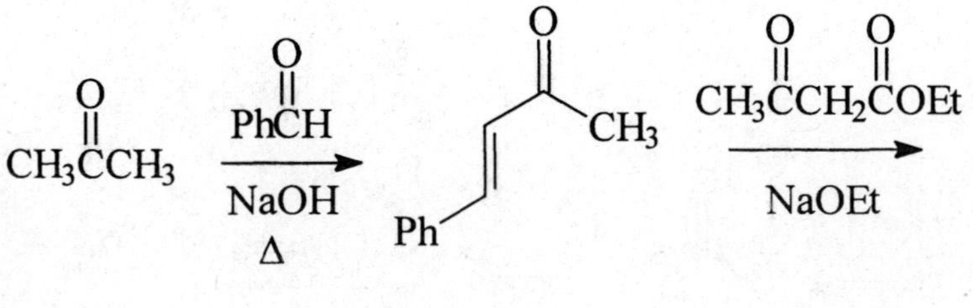

20.44

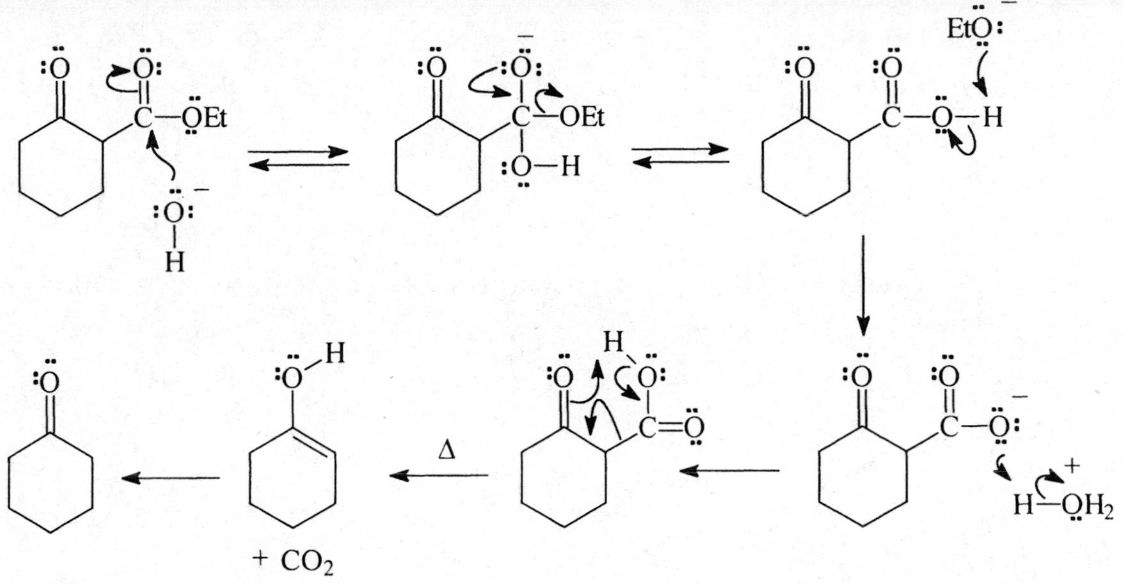

20.45

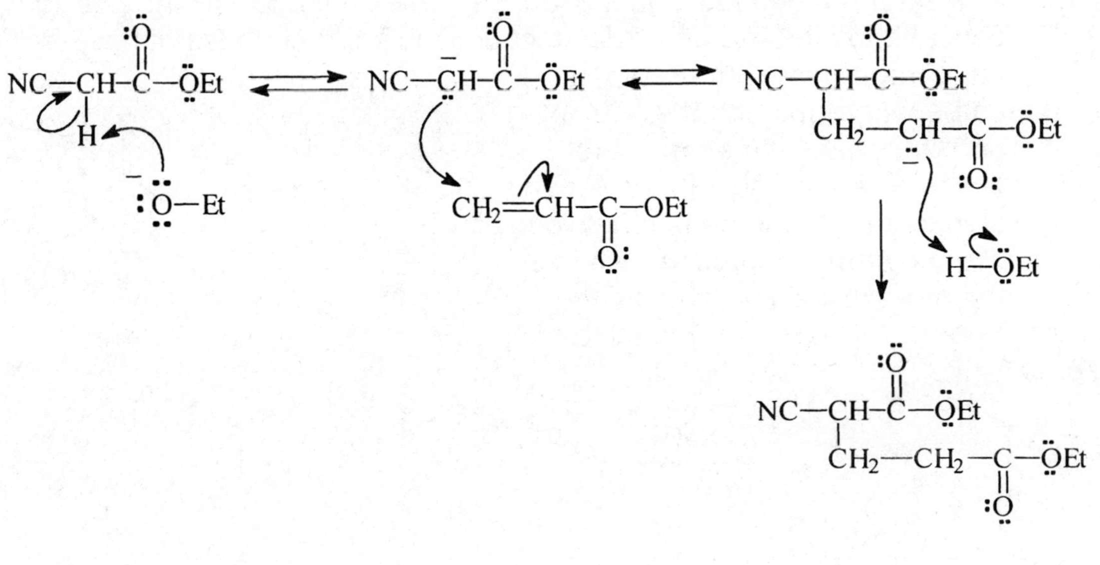

20.46

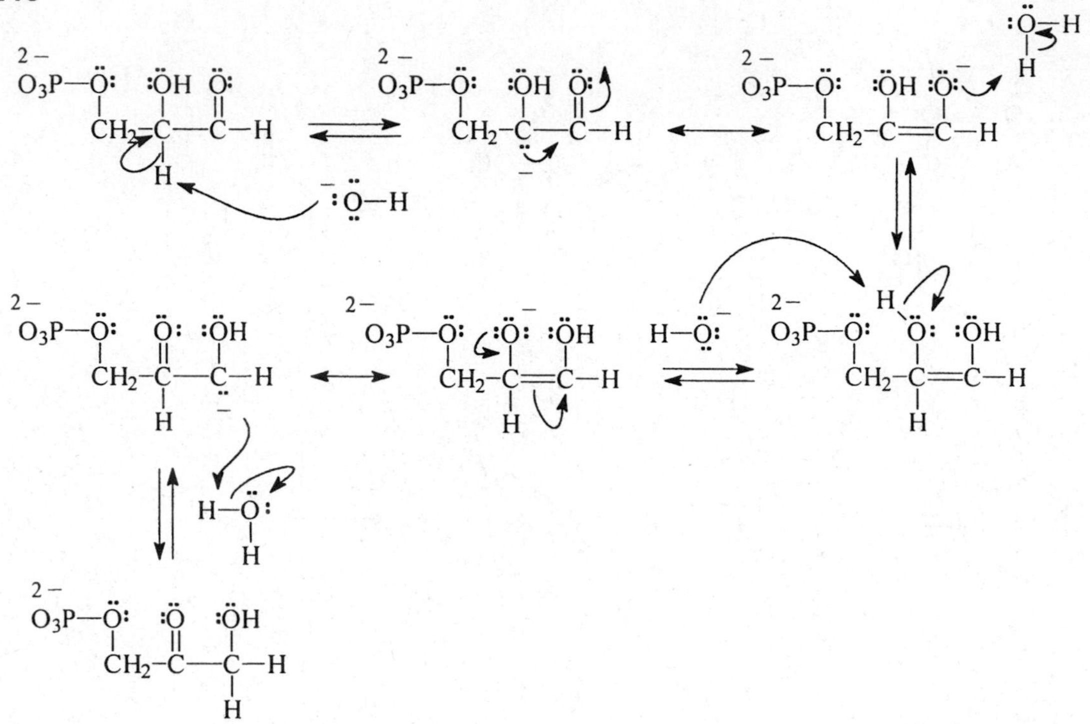

20.47 The ester condensation in the first step does not occur at the position occupied by the methyl group because the product does not have a H on the carbon between the two carbonyl groups and the equilibrium driving step cannot occur. The alkylation in the second step occurs at the carbon between the two carbonyl groups because the most acidic H is located there.

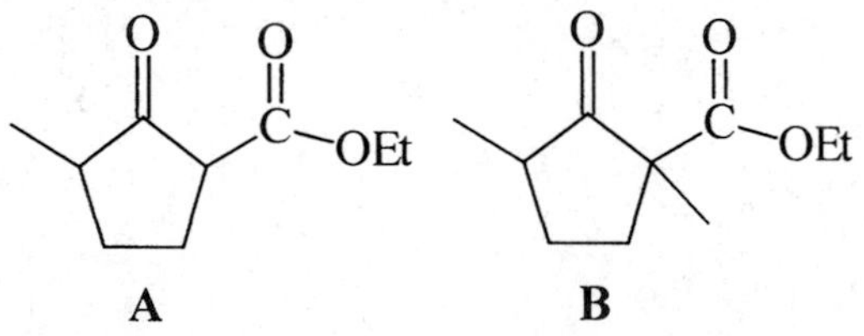

20.48

a) 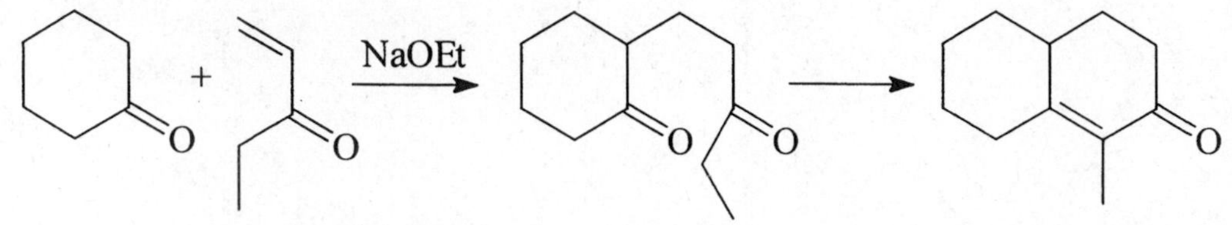

b)

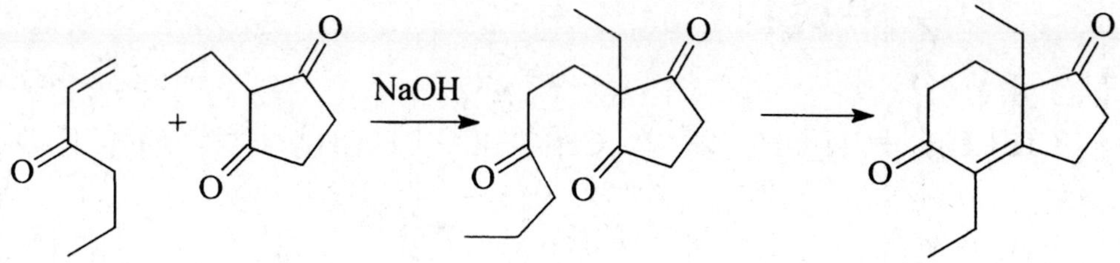
NaOH

20.49

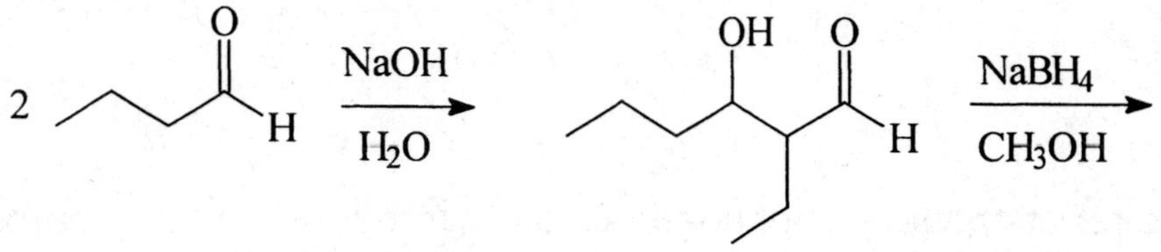
2
NaOH
H2O
NaBH4
CH3OH

20.50

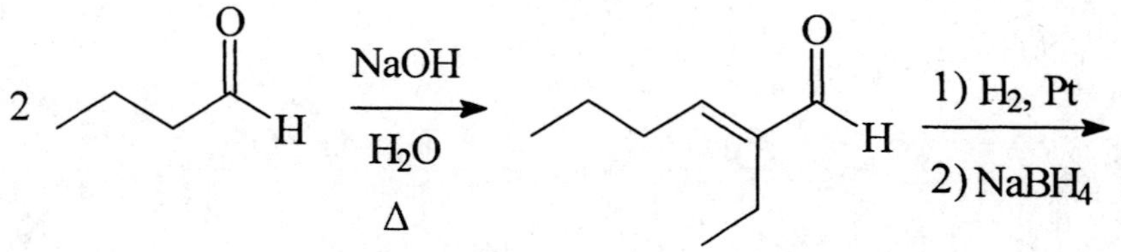
2
NaOH
H2O
Δ
1) H2, Pt
2) NaBH4

20.51

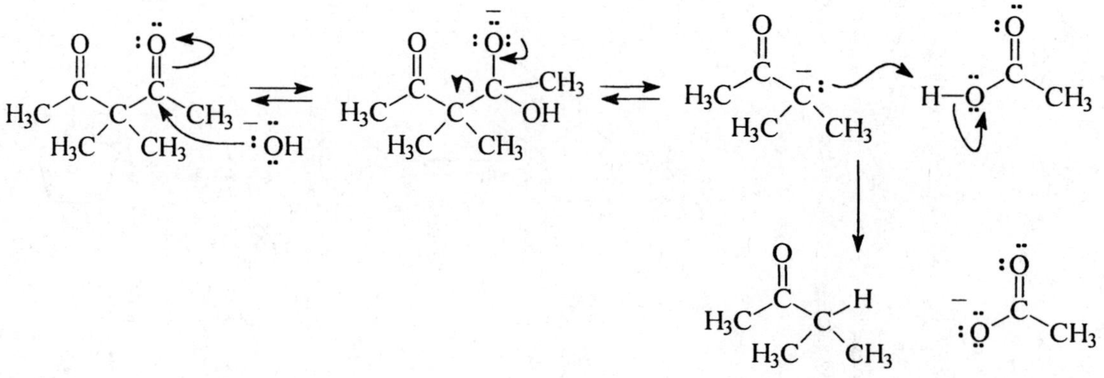

20.52

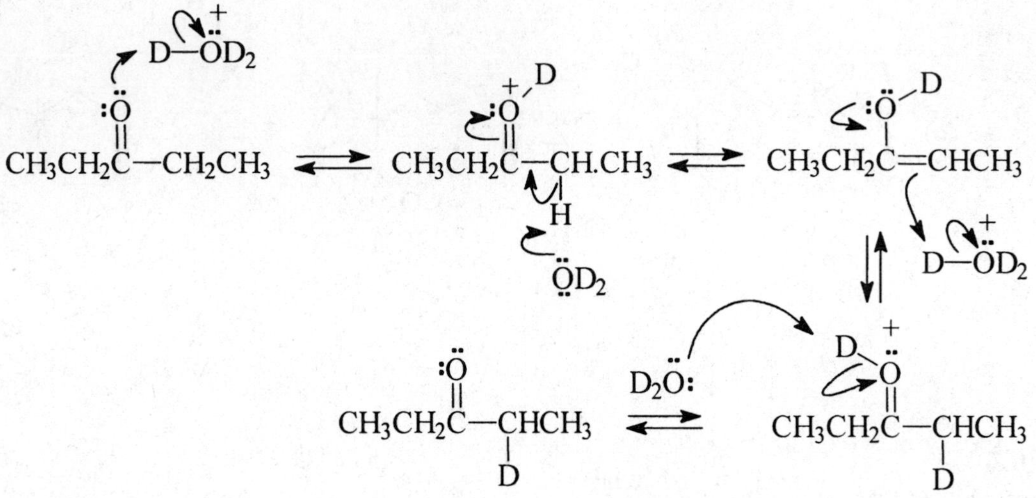

If this reaction were continued, all four hydrogens on the α-carbons on either side of the carbonyl group would be exchanged for deuterium.

20.53

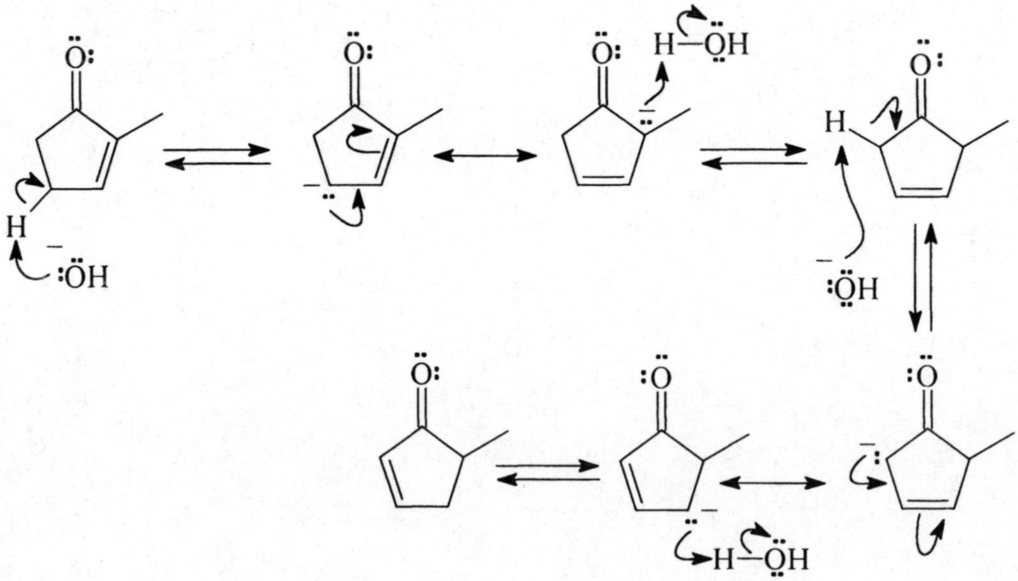

20.54 a) There are two electrophilic carbonyl carbons and three nucleophilic α-carbons in this molecule. Aldehyde **A** would result from the α-carbon of the aldehyde bonding to the carbonyl carbon of the ketone. This product is not favored because the ketone carbonyl carbon is not as electrophilic as that of the aldehyde. Ketone **B** and the observed product result from one of the α-carbons of the ketone reacting with the carbonyl carbon of the aldehyde. The five-membered ring product is formed rather than the seven-membered ring as a result of more favorable entropy.

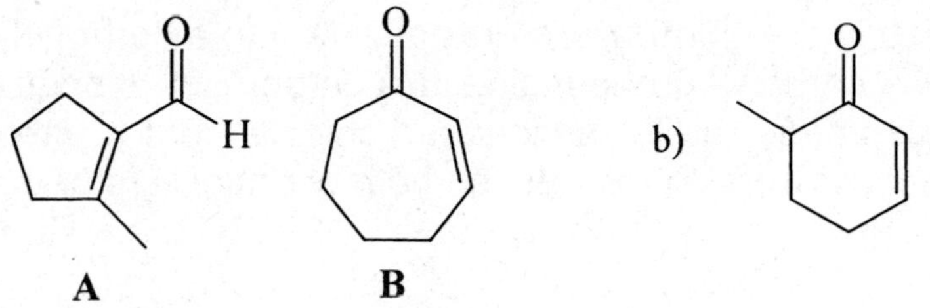

20.55 **A** results from an ester condensation. **B** results from a conjugate addition. The final product is formed from **B** by an aldol condensation. The conversion of **A** to the final product is a Robinson annulation.

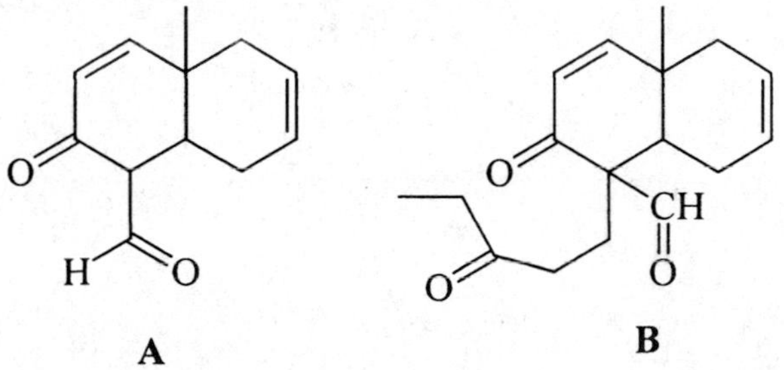

20.56 This is an intramolecular Cannizzaro reaction.

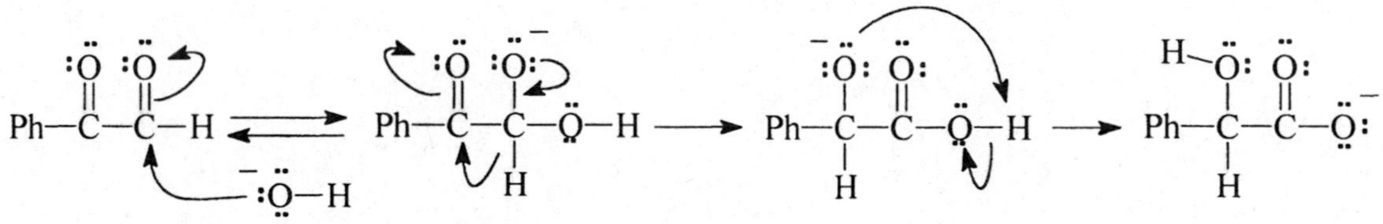

20.57

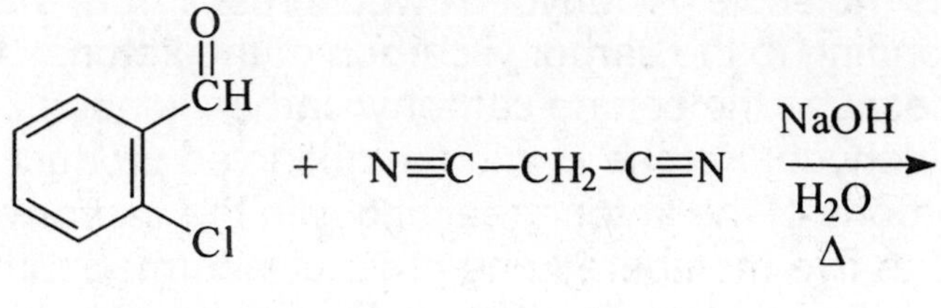

20.58 The carbon which is the chirality center is adjacent to a carbonyl group. Enolization of this carbonyl group makes this carbon sp^2 hybridized with trigonal planar geometry, so it is no longer a chirality center. Protonation of the enol can occur from either side, so both enantiomers of thalidomide are formed.

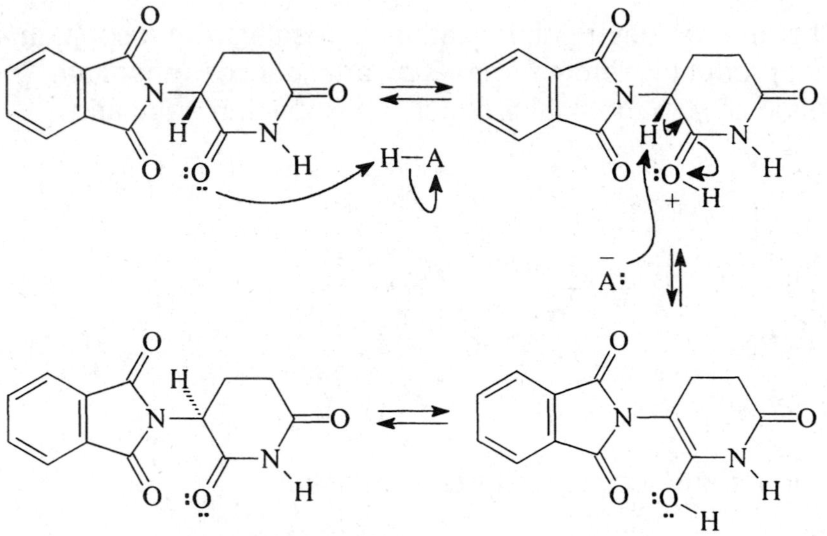

20.59

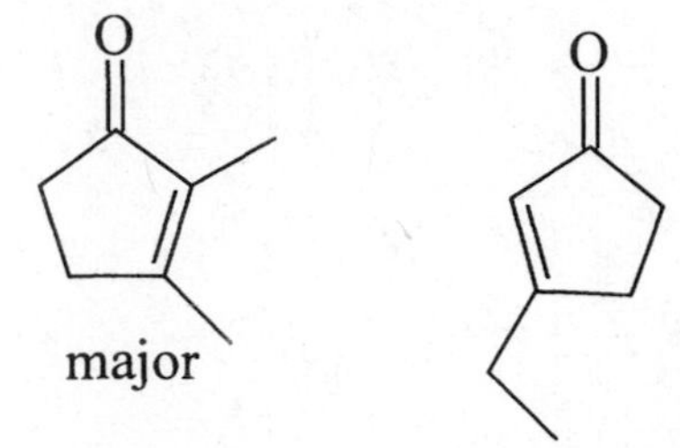

20.60 The H on the C between the carbonyl group and the phenyl group is more acidic because the conjugate base is stabilized by resonance with both the carbonyl group and the phenyl group.

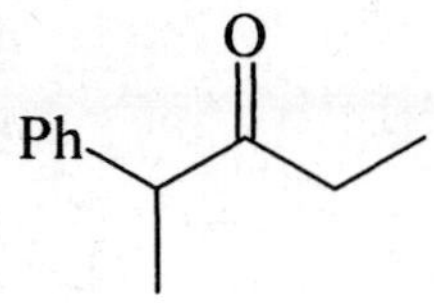

20.61

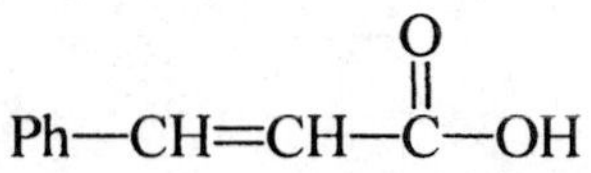

20.62 When the acidic H on the carbon alpha to the carbonyl group is removed from either of these ketones, the same planar enolate anion is formed. When this enolate anion is reprotonated, the proton can add from either side, so an equilibrium mixture of the two ketones is formed. The stereoisomer on the left, with the *tert*-butyl group and the ethyl group cis, is the major component of the equilibrium mixture because it has both groups equatorial. The *trans*-isomer is less stable because it has the ethyl group in the axial position.

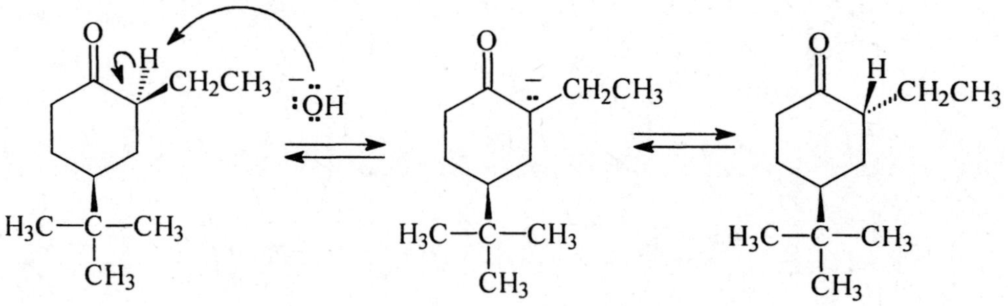

20.63 These products all result from an enolate anion nucleophile generated by removal of an acidic hydrogen from the starting ketone acting as an intramolecular nucleophile and displacing the bromine. Product **A** is a result of the enolate anion formed from the alpha CH_2 group acting as the nucleophile. Product **B** is the result of the enolate anion formed on the other side of the carbonyl group acting as the nucleophile. And Product **C** is the result of the oxygen of this enolate anion acting as the nucleophile.

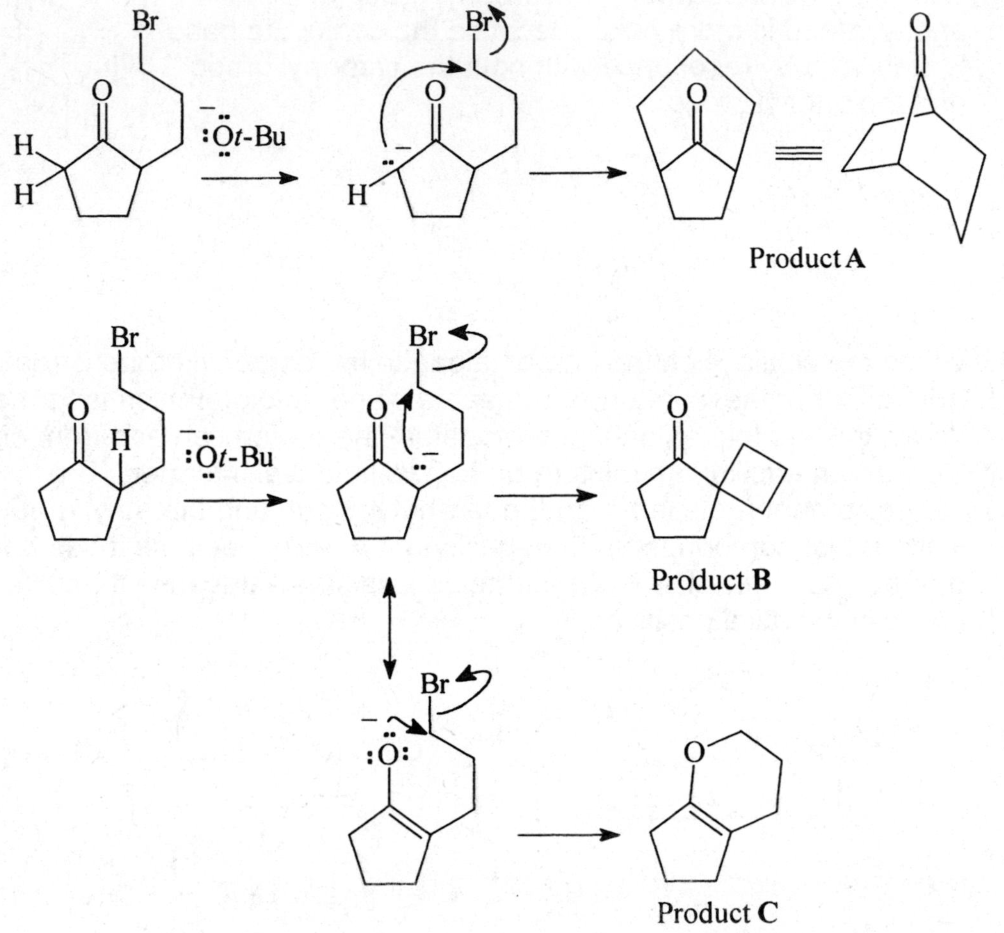

20.64 Both of the carbon-carbon double bonds in the product are *E* or trans because this product is more stable due to the larger groups being on opposite sides of the double bonds.

Review of Mastery Goals

After completing this chapter, you should be able to:

Show the products of any of the reactions discussed in this chapter. (Problems 20.1, 20.2, 20.3, 20.4, 20.5, 20.6, 20.9, 20.11, 20.14, 20.16, 20.19, 20.20, 20.22, 20.23, 20.25, 20.27, 20.28, 20.30, 20.31, 20.32 20.41, 20.47, 20.54, and 20.55.)

Show the mechanism for any of these reactions.
(Problems 20.10, 20.12, 20.17, 20.21, 20,26, 20.33, 20.34, 20.35, 20.36, 20.40, 20.44, 20.45, 20.46, 20.51, 20.52, 20.53, 20.56, 20.58, 20.62, and 20.63)

Use these reactions in combination with reactions from previous chapters to synthesize compounds.
(Problems 20.7, 20.8, 20.13, 20.15, 20.18, 20.24, 20.29, 20.37, 20.38, 20.39, 20.42, 20.43, 20.48, 20.49, 20.50, and 20.57)

Chapter 21
THE CHEMISTRY OF RADICALS

21.1   a) $:\overset{\cdot}{N}=\overset{\cdot\cdot}{O}:$ b)

$$:\overset{\cdot\cdot}{O}\diagdown\!\!\underset{\underset{+}{\overset{\cdot}{N}}}{}\!\!\diagup\overset{\cdot\cdot}{O}:^{-}$$

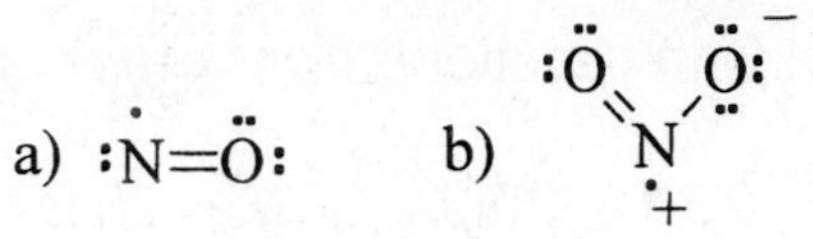

21.2   Radicals are stabilized by the same factors that stabilize carbocations.

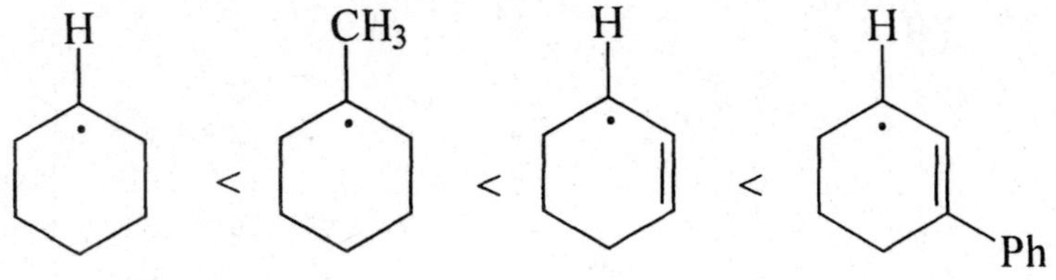

Least stable   Most stable

21.3

a) 2 (cyclohexyl-CN radical) $+ \ N_2$ b) $2 :\overset{\cdot\cdot}{\underset{\cdot\cdot}{Br}}\cdot$ c) $2 \ CH_3\overset{CH_3}{\overset{|}{CH}}-\overset{\cdot\cdot}{\underset{\cdot\cdot}{O}}\cdot$

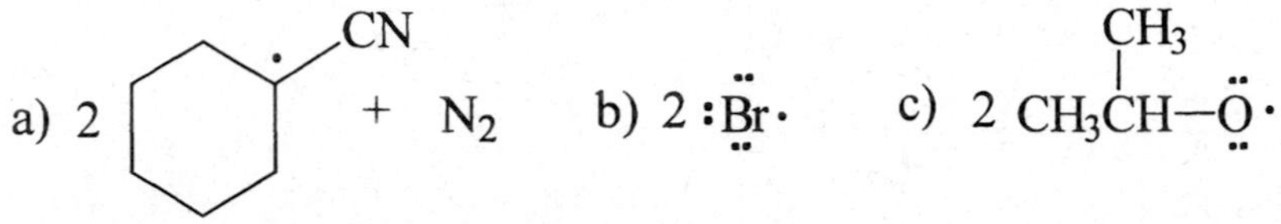

21.4   a) Abstraction of a tertiary hydrogen occurs more readily than abstraction of a secondary hydrogen, so the second reaction is faster.
b) Formation of the resonance stabilized benzylic radical occurs more readily, so the first reaction is faster.

21.5

a) $CH_3\overset{\cdot}{C}HCH_2Br$ b) $2 \ CH_3\overset{\overset{O}{\|}}{\underset{\cdot\cdot}{C}}\overset{\cdot\cdot}{O}\cdot$ c) $CH_3(CH_2)_6CH_3 \ + \ CH_3CH_2CH_2CH_3$

$+ \ CH_3CH_2CH=CH_2$

d)

$$\overset{:\overset{\cdot\cdot}{O}}{\underset{-:\overset{\cdot\cdot}{O}:}{}}\!\!\diagdown\!\!\underset{+N-N+}{}\!\!\diagup\overset{\overset{\cdot\cdot}{O}:}{\underset{:\overset{\cdot\cdot}{O}:^{-}}{}}$$

e) $2 \ CH_3\overset{CH_3}{\underset{CH_3}{\overset{|}{\underset{|}{C}}}}\cdot \ + \ N_2$

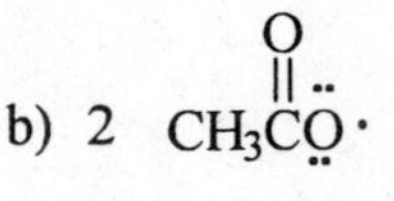

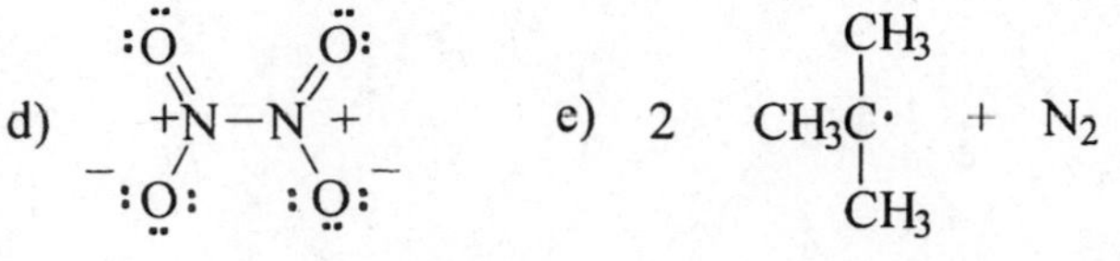

21.6

$$:\ddot{C}l\text{---}\ddot{C}l: \xrightarrow{h\nu} 2 \;\; :\ddot{C}l\cdot$$

$$:\ddot{C}l\cdot \; + \; H\text{---}CH_2Cl \longrightarrow :\ddot{C}l\text{--}H \; + \; \cdot CH_2Cl$$

$$ClH_2C\cdot \; + \; :\ddot{C}l\text{--}\ddot{C}l: \longrightarrow CH_2Cl_2 \; + \; :\ddot{C}l\cdot$$

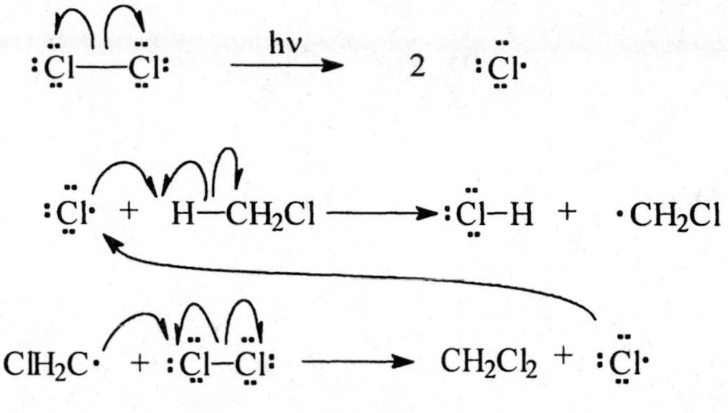

+ various terminations

21.7 If the strength of a typical CH bond is taken as 98 kcal/mol (410 kJ/mol) (see Table 2.1), then the abstraction of the hydrogen atom by a fluorine atom is exothermic by 135 - 98 = 37 kcal/mol (155 kJ/mol) whereas a similar abstraction by an iodine atom is endothermic by 98 - 71 = 27 kcal/mol (113 kJ/mol).

21.8

a) CH_2Br (on benzene ring)

b) $CHBr_2$ (on benzene ring with Br)

c) CH_2Br, CH_3 (on benzene ring)

d) Br, $C\text{---}CH_2CH_3$, CH_3 (on benzene ring) — racemic

e) CH_2Br (on naphthalene)

f) (allylic bromide) + (allylic bromide) Br $(Z$ and $E)$

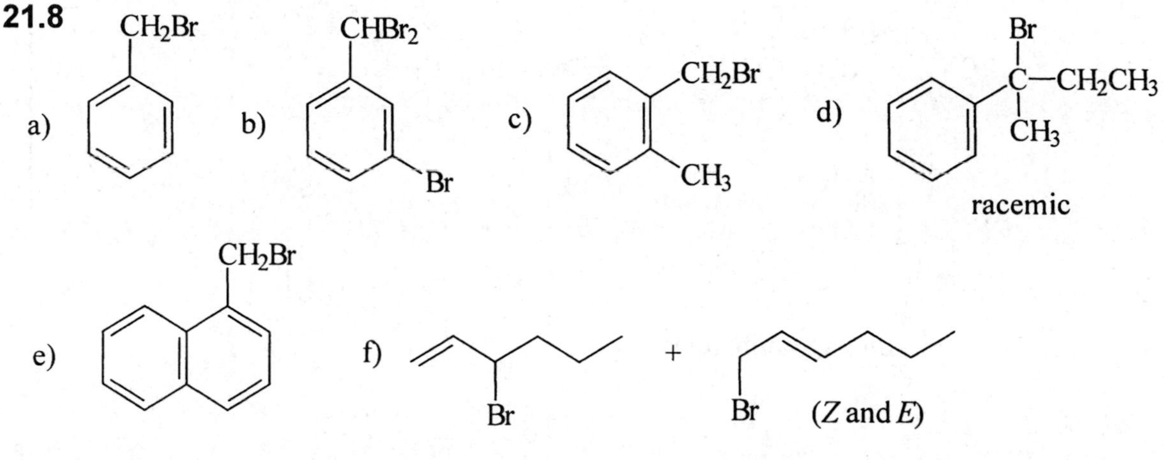

21.9 The radical intermediate in this reaction is stabilized by resonance and provides two sites for forming a bond to the bromine. The conjugated product is more stable, so it is the major product.

$$PhCH{=}CH\overset{\cdot}{C}H_2 \longleftrightarrow Ph\overset{\cdot}{C}HCH{=}CH_2 \longrightarrow \underset{\text{major}}{PhCH{=}CH\overset{\overset{\displaystyle Br}{|}}{C}H_2} + \underset{\text{minor}}{Ph\overset{\overset{\displaystyle Br}{|}}{C}HCH{=}CH_2}$$

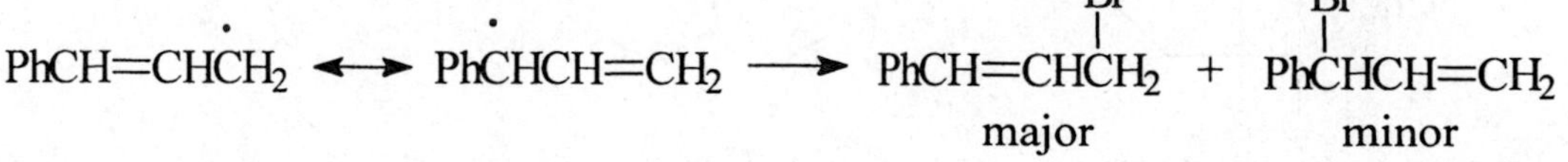

21.10

a)

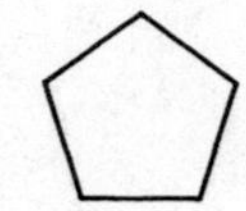

b) Ph

21.11

a)

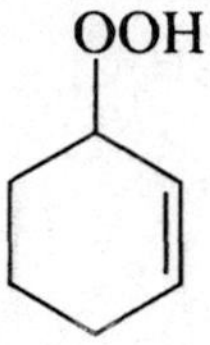

b)

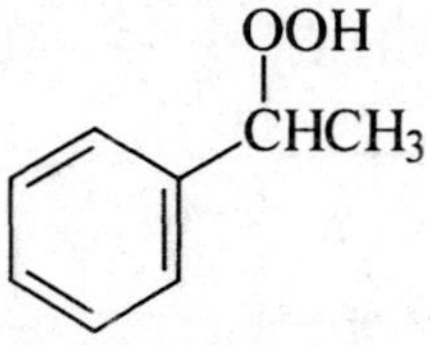

21.12 The radical produced from vitamin E is stabilized by resonance and its radical center is sterically hindered. Therefore it is not very reactive.

21.13

$$Cl_3C-\overset{..}{\underset{..}{Cl}}: \quad \xrightarrow{h\nu} \quad Cl_3C\cdot \ + \ \cdot\overset{..}{\underset{..}{Cl}}:$$

$$Cl_3C\cdot \ + \ CH_2=\overset{\overset{\textstyle CH_3}{|}}{C}CH_3 \quad \longrightarrow \quad Cl_3CCH_2-\overset{\overset{\textstyle CH_3}{|}}{\underset{\cdot}{C}}CH_3$$

$$Cl_3CCH_2-\overset{\overset{\textstyle CH_3}{|}}{\underset{\cdot}{C}}CH_3 \ + \ :\overset{..}{\underset{..}{Cl}}-CCl_3 \quad \longrightarrow \quad Cl_3CCH_2-\overset{\overset{\textstyle CH_3}{|}}{\underset{\underset{\textstyle :\overset{..}{\underset{..}{Cl}}:}{|}}{C}}CH_3 \ + \ Cl_3C\cdot$$

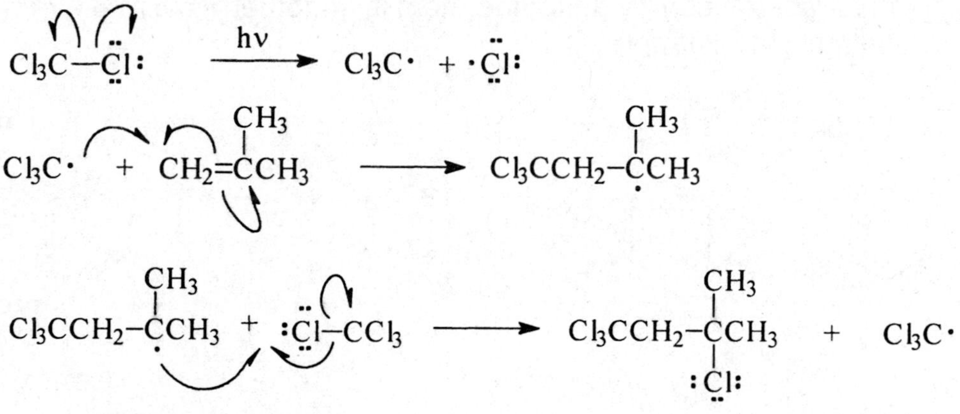

+ various terminations

21.14

a)

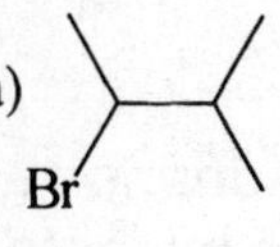

b)

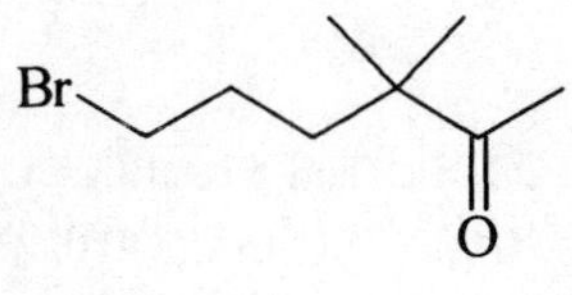

c)

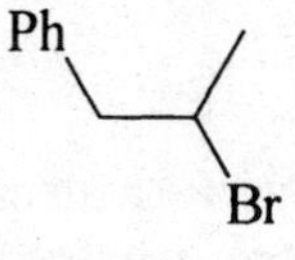

d)

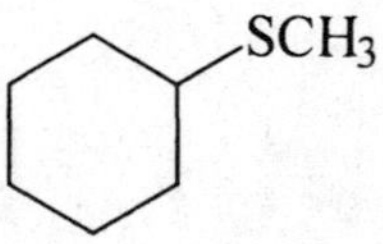

e)

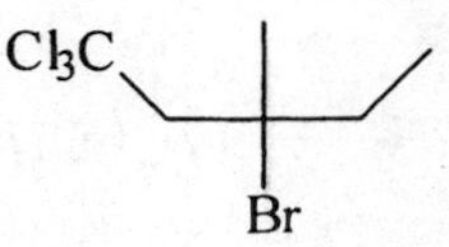

f)

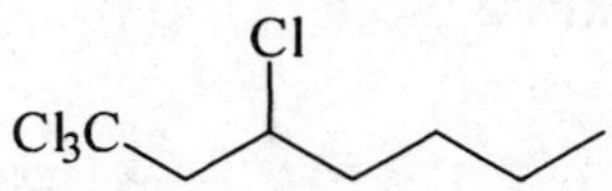

21.15

a) 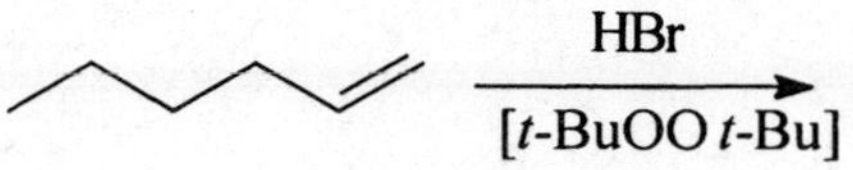

b) 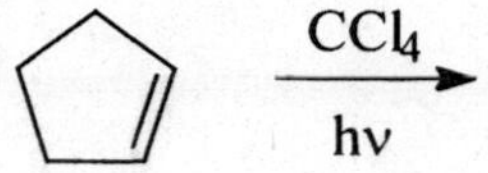

c) 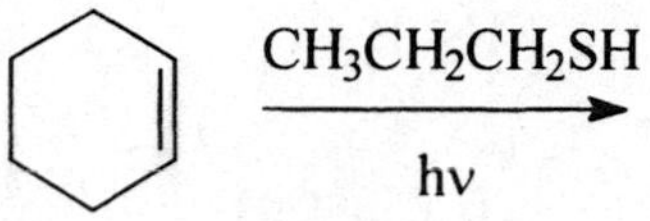

d) 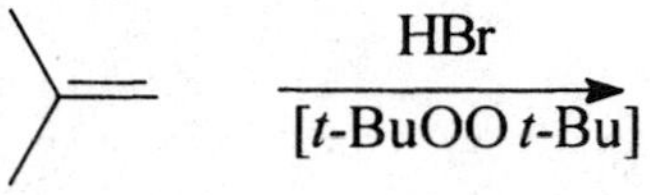

e) 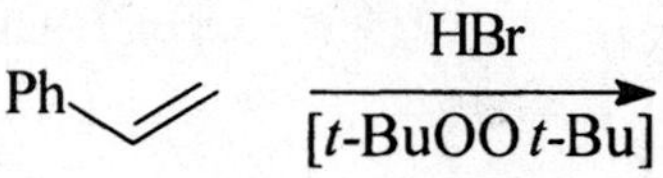

f) 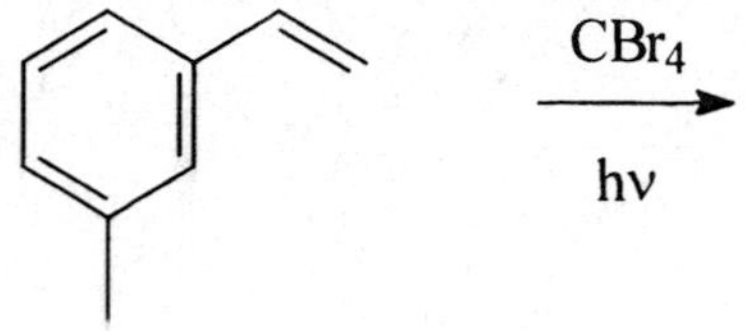

g) 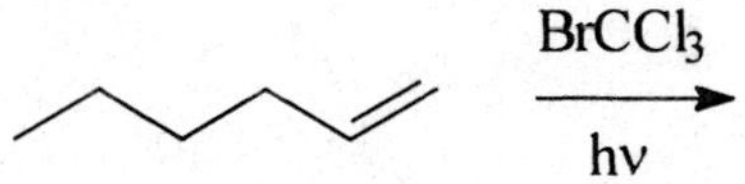

21.16

a) 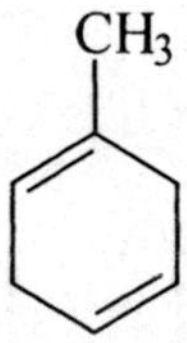

b) 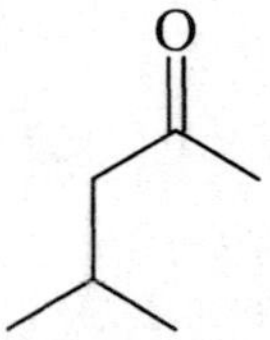

c) 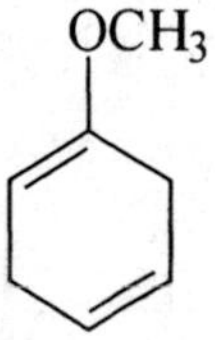

d) 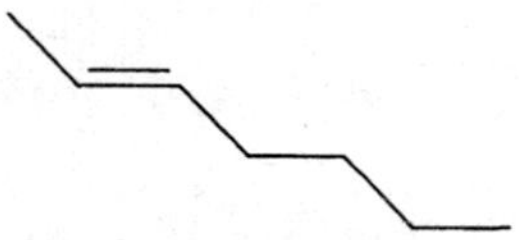

e) 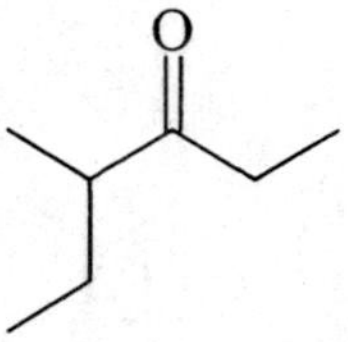

f) 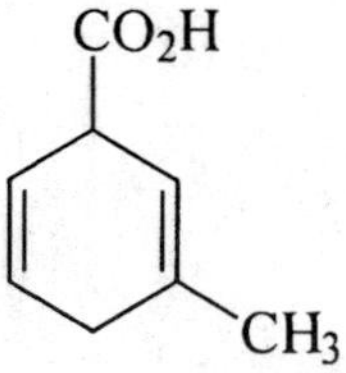

g) 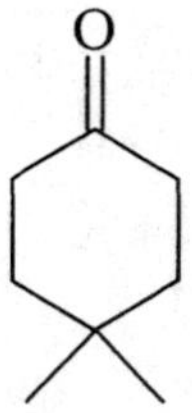

h) 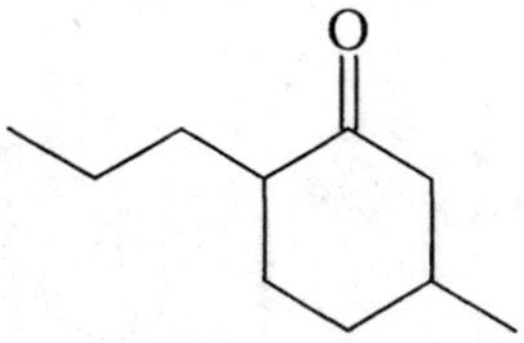

21.17 a) 1) Li, NH_3 (*l*); 2) $PhCH_2Br$ b) 1) Li, NH_3 (*l*); 2) H_3O^+

21.18

a) 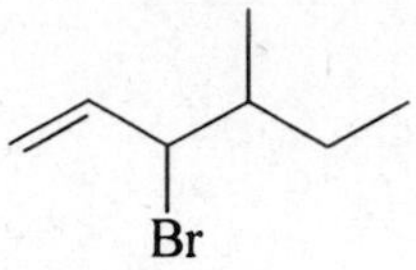 + 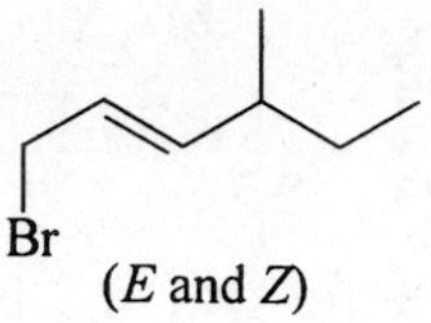

(*E* and *Z*)

b) 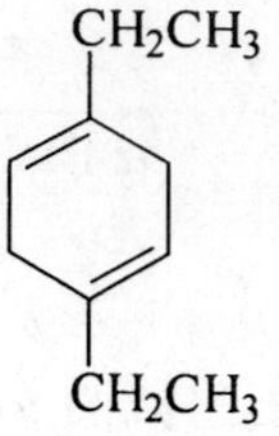

c) 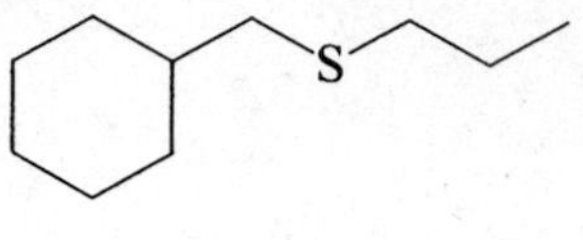

d) 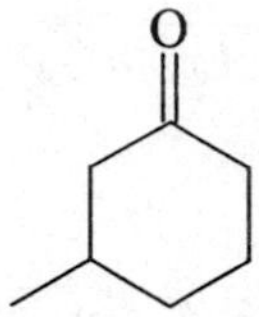

e) 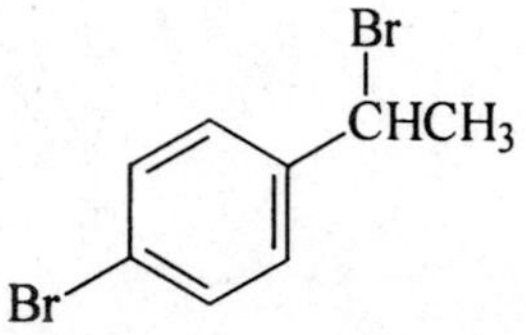

f) 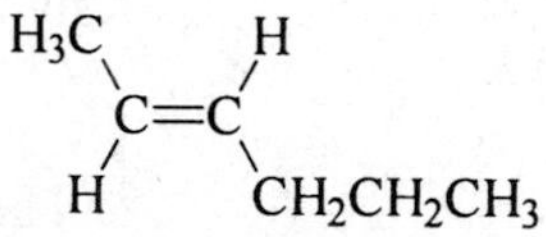

g) 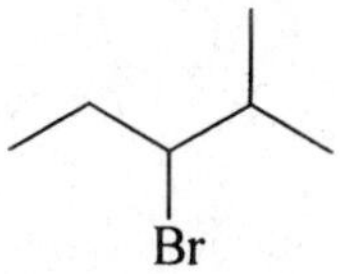

h) 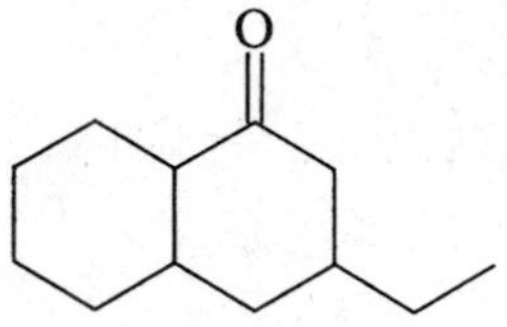

i) 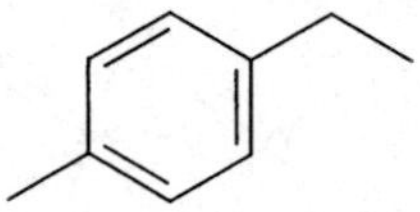

j) 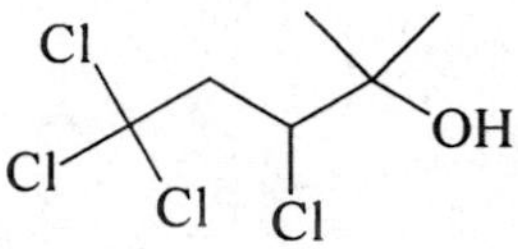

k) 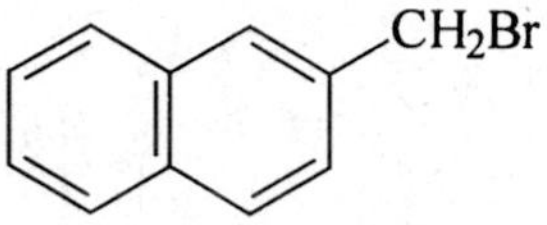

l) 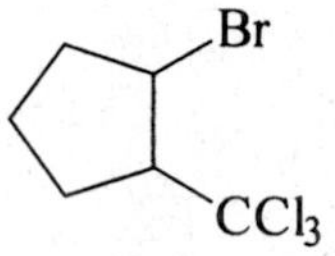

21.19

a) 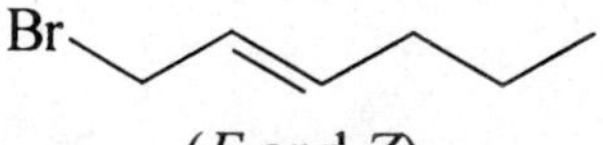 + 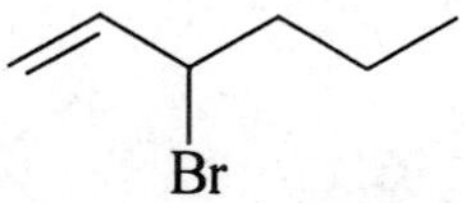

(*E* and *Z*)

b) 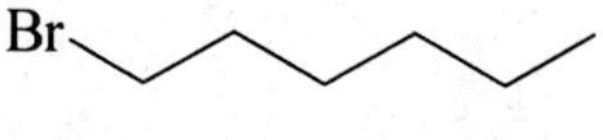

c) 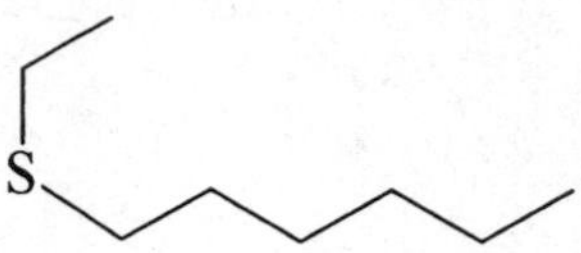

d) 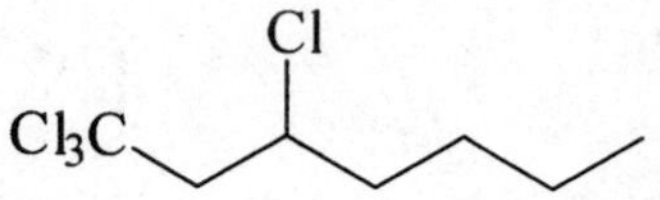

e) 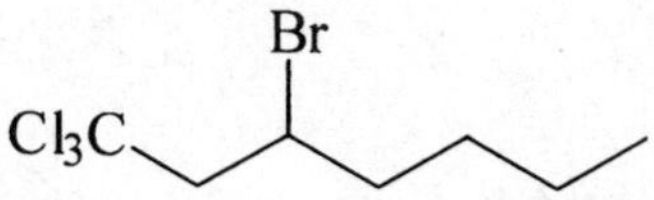

21.20

a) 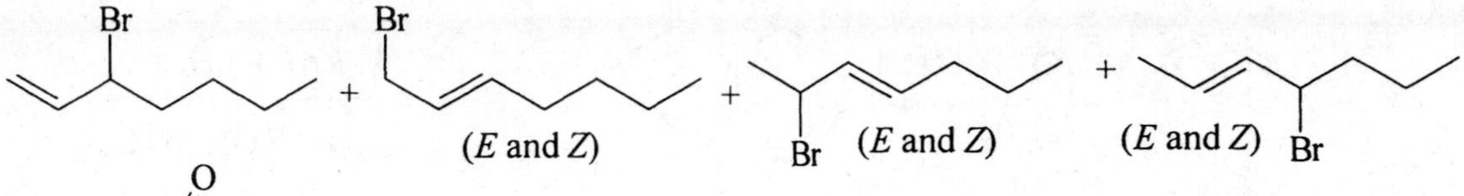

b) 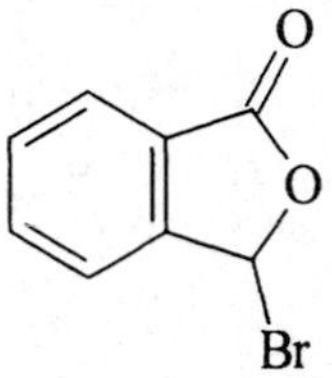 c) 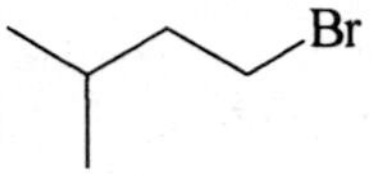 d) 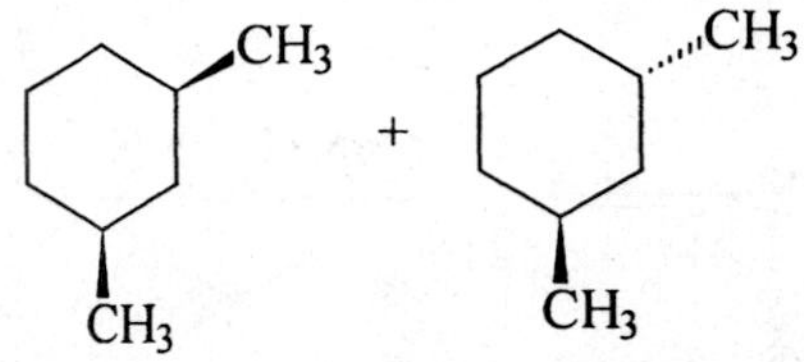

e) 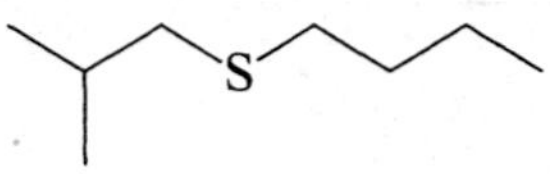 f) 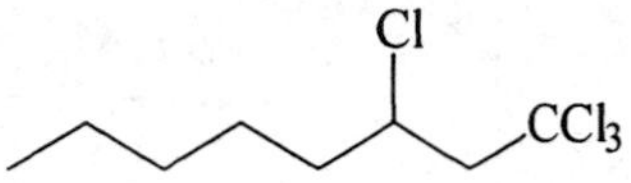 g) 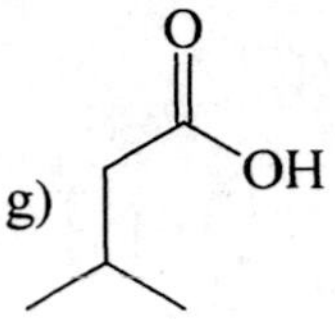 h) 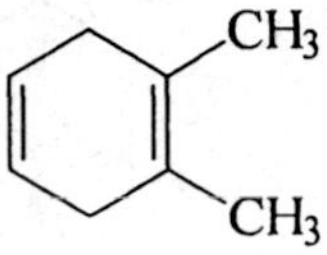

21.21

a) 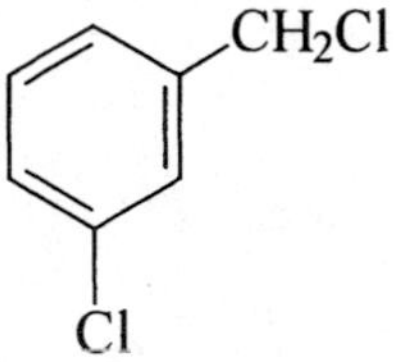 b) 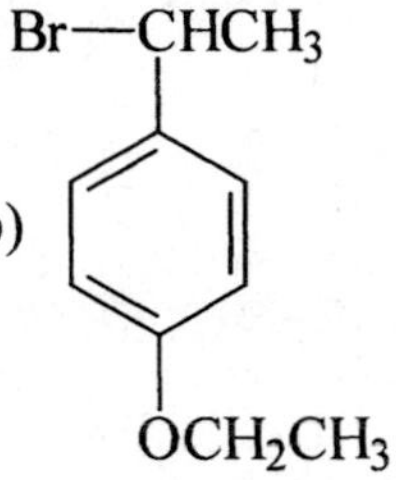 c) 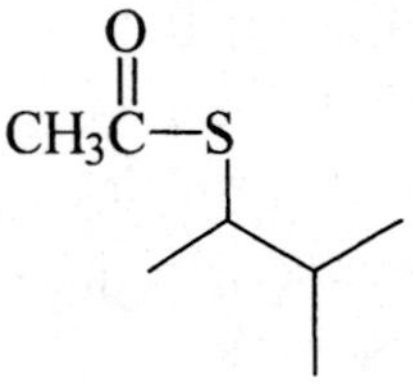 d) 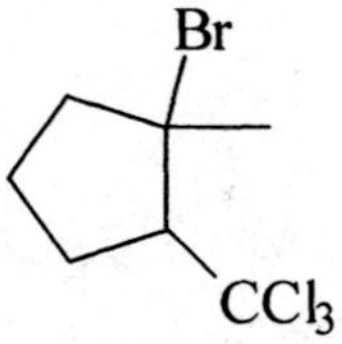

e) 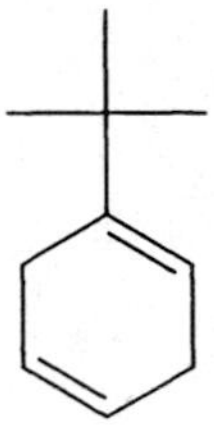 f) 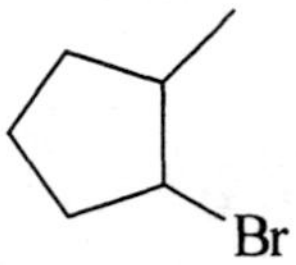 g) 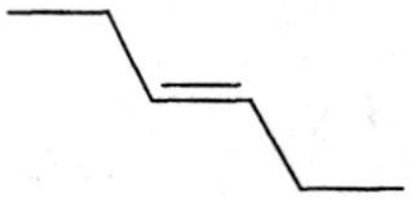 h) 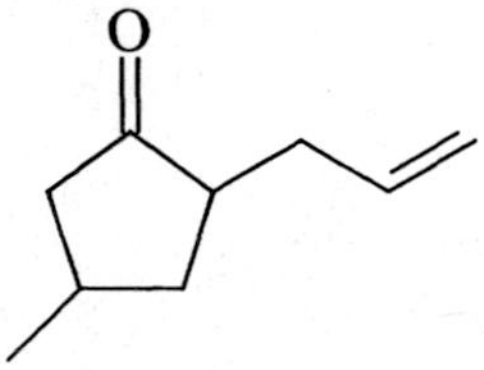

21.22

a) 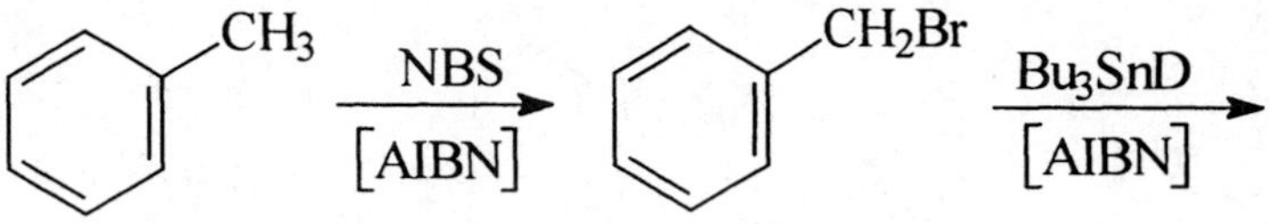

b) 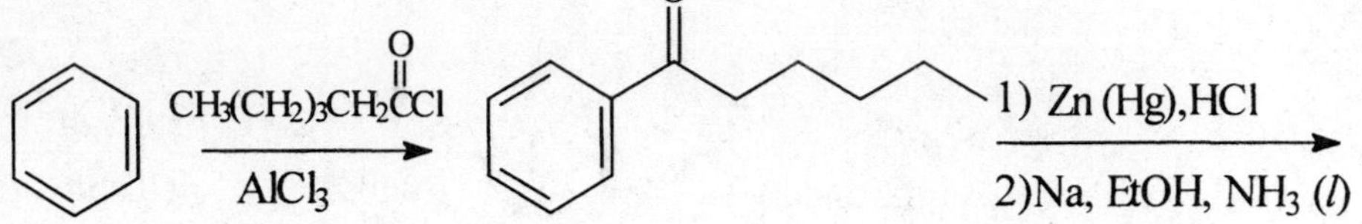
$CH_3(CH_2)_3CH_2CCl$
$AlCl_3$
1) Zn (Hg),HCl
2)Na, EtOH, NH_3 (*l*)

c) 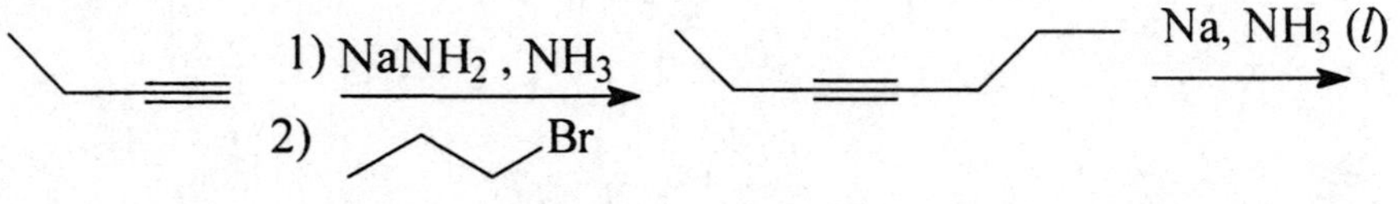
1) $NaNH_2$, NH_3
2)
Br
Na, NH_3 (*l*)

d) 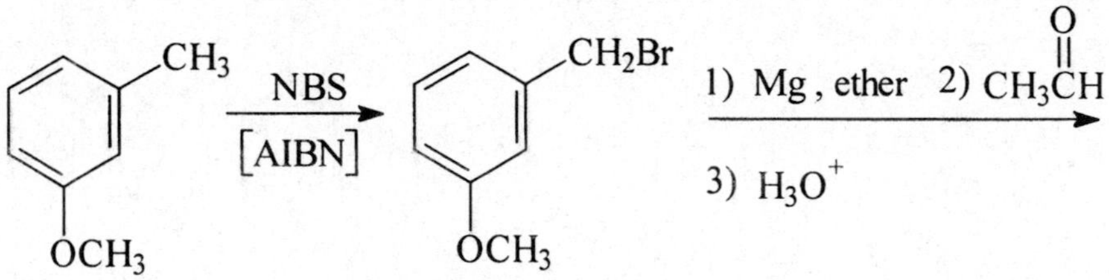
CH_3
OCH_3
NBS
[AIBN]
CH_2Br
OCH_3
1) Mg , ether 2) CH_3CH
3) H_3O^+

e) 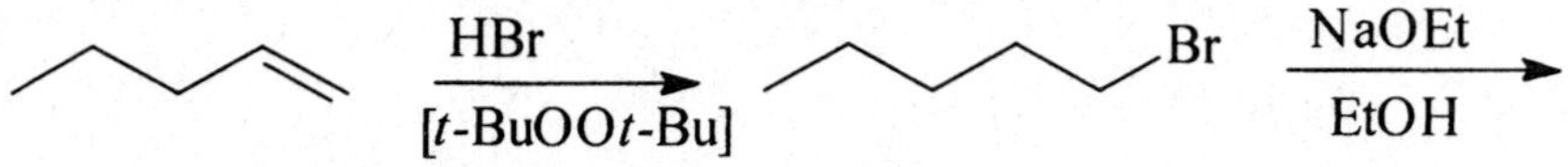
HBr
[*t*-BuOO*t*-Bu]
Br
NaOEt
EtOH

f) 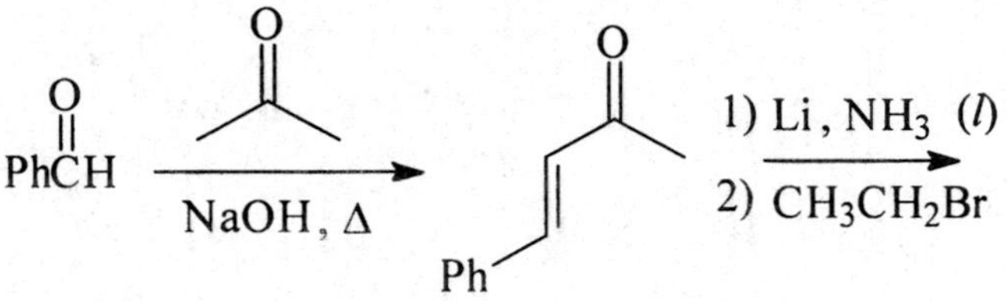
PhCH
NaOH, Δ
Ph
1) Li , NH_3 (*l*)
2) CH_3CH_2Br

21.23

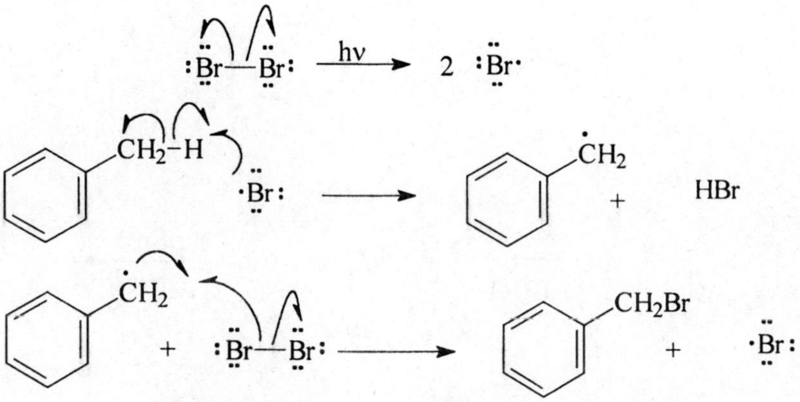
hν
2
CH_2–H
HBr
CH_2
CH_2Br

and various termination steps

21.24

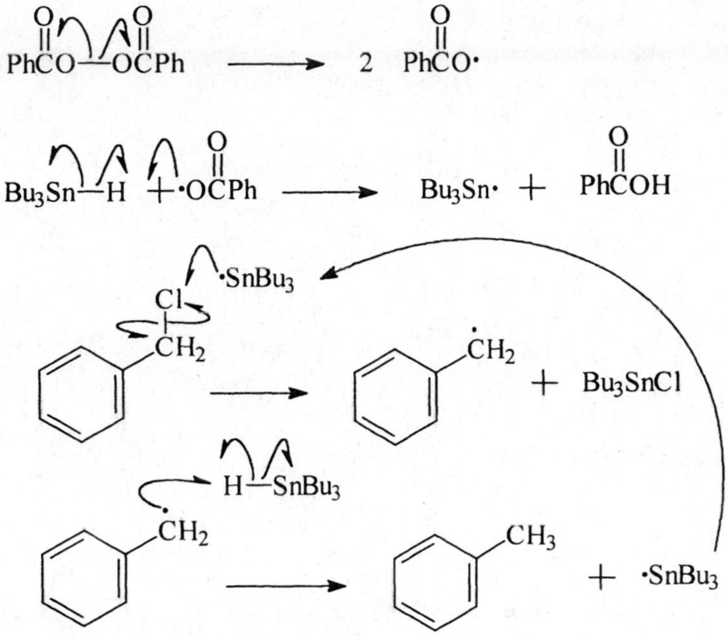

and various termination steps

21.25

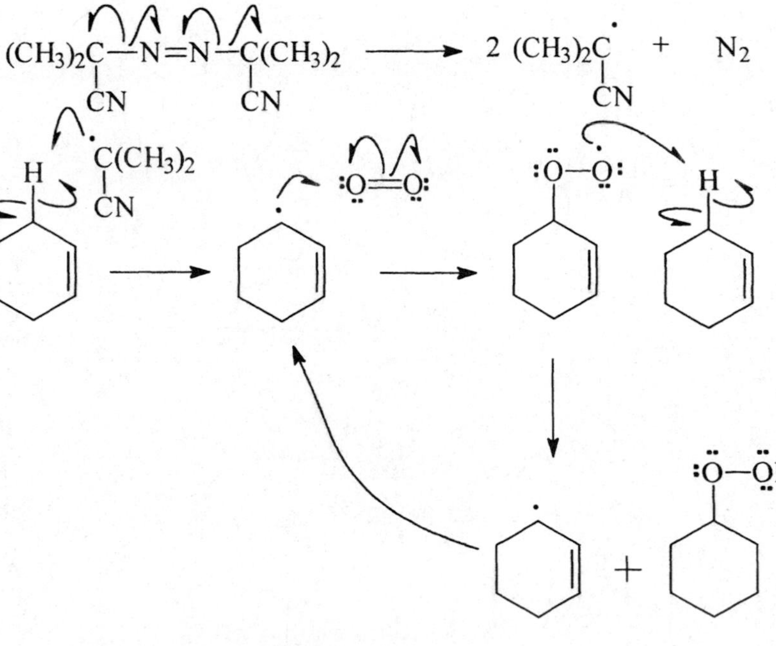

and various termination steps

21.26

$$CH_3CH_2S{-}H \xrightarrow{h\nu} CH_3CH_2S\cdot \ + \ \cdot H$$

$$CH_3CH_2S\cdot \ + \ CH_2{=}\overset{CH_3}{\underset{}{C}}CH_3 \longrightarrow CH_3CH_2SCH_2{-}\overset{CH_3}{\underset{\cdot}{C}}CH_3$$

$$CH_3CH_2SCH_2{-}\overset{CH_3}{\underset{\cdot}{C}}CH_3 \ + \ H{-}SCH_2CH_3 \longrightarrow CH_3CH_2SCH_2{-}\overset{CH_3}{\underset{H}{C}}CH_3 \ + \ CH_3CH_2S\cdot$$

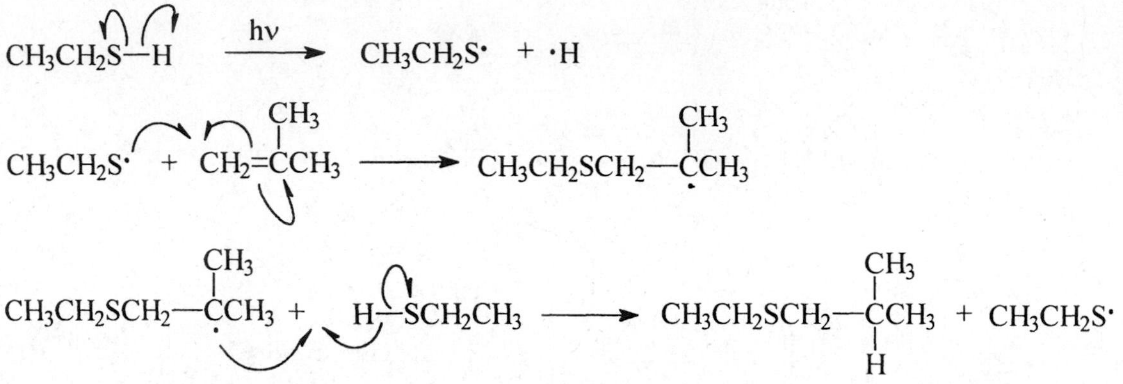

+ various terminations

21.27

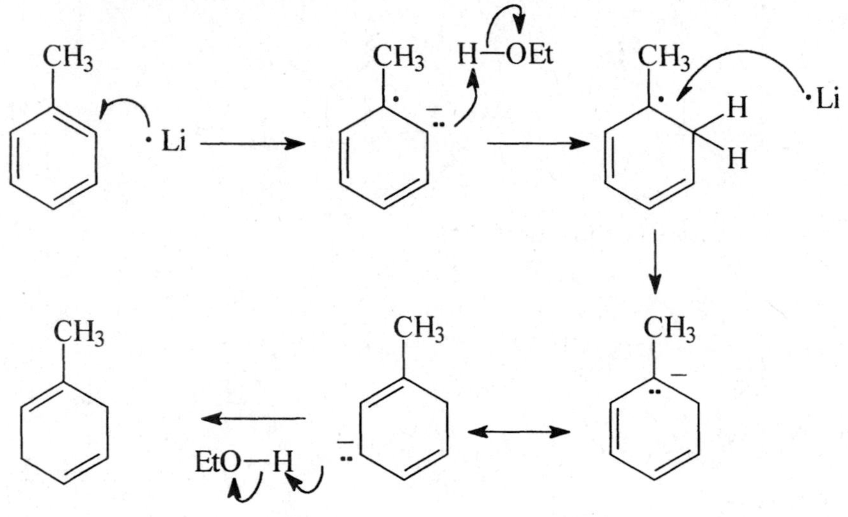

21.28

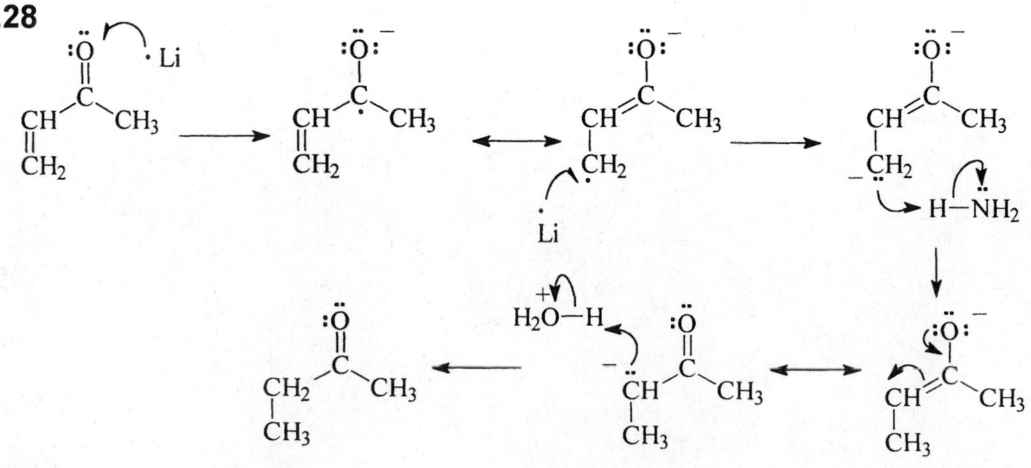

21.29 There are two types of allylic hydrogens available to be abstracted in this compound. Each of the resulting allylic radicals (**A** and **C**) provide two radical sites that can abstract a bromine atom. The radical sites of resonance structures **A** and **B** are identical, so only one product is formed. However, the radical sites of the resonance structures **C** and **D** are different, so two products are formed.

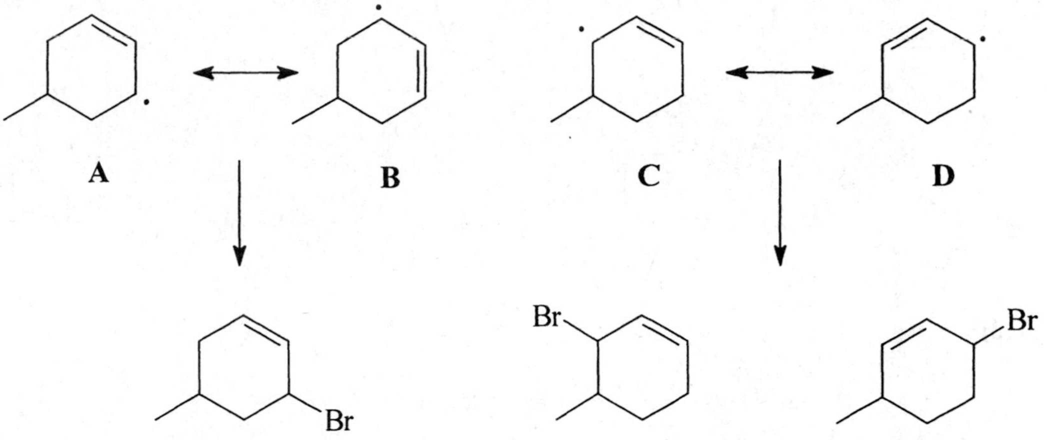

21.30 The difference in bond dissociation energies to produce a secondary radical (95 kcal/mol, 397 kJ/mol) and to produce a tertiary radical (92 kcal/mol, 385 kJ/mol) is equal to 3 kcal/mol (12 kJ/mol). This should be a reasonable estimate of the heat given off when a secondary radical rearranges to a tertiary radical.

21.31 Azobenzene is not a useful radical initiator because the phenyl radicals produced are not very stable. Therefore dissociation is quite slow.

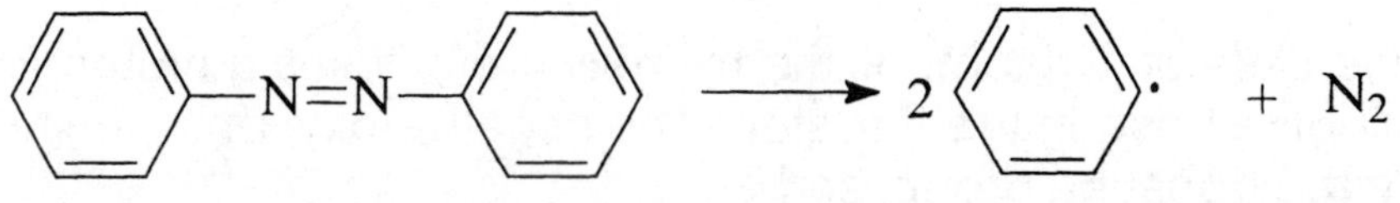

21.32 The triphenylmethyl radicals add to O_2 to form peroxide radicals.

21.33 Because it is not conjugated, the π antibonding MO of a 1,4-cyclohexadiene is too high in energy for an electron from Li atom to be readily added to it.

21.34

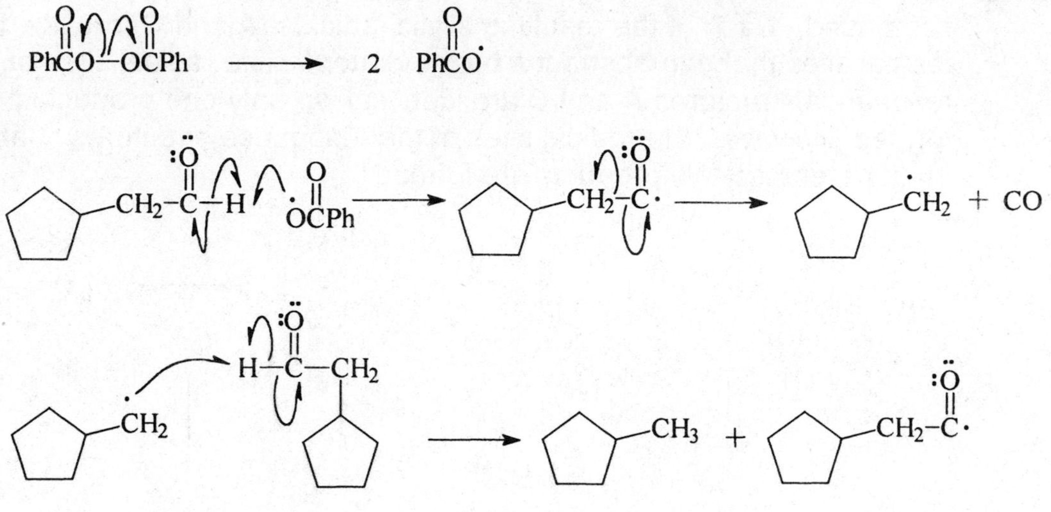

21.35

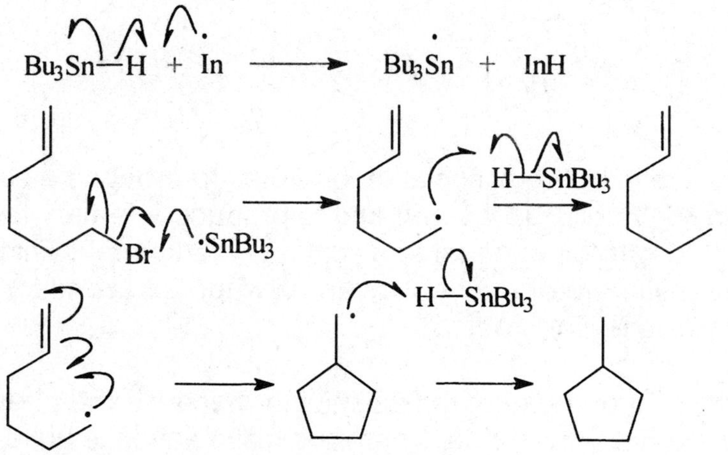

21.36 In the case of 1-naphthol, the hydroxy group loses a proton form its conjugate base in the first step. The negative oxygen is an inductive electron donating group, so it directs the reduction to the other ring. In contrast, the ethoxy group of 1-ethoxynaphthalene is an inductive electron withdrawing group, so it directs the reduction to the ring to which it is attached.

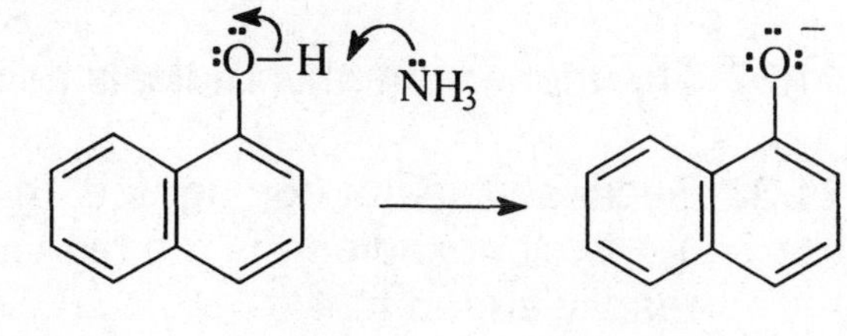

21.37

PhC(=O)—O—O—C(=O)Ph $\xrightarrow{\text{heat}}$ 2 PhC(=O)—O·

PhC(=O)—O· + CH₂=CH₂ ⟶ PhCOCH₂—·CH₂

PhCOCH₂—·CH₂ → (+ CH₂=CH₂) → PhCOCH₂—CH₂—CH₂—·CH₂ (+ CH₂=CH₂)

PhCOCH₂CH₂CH₂CH₂—CH₂—·CH₂ ⟵ PhCOCH₂—CH₂—CH₂—·CH₂

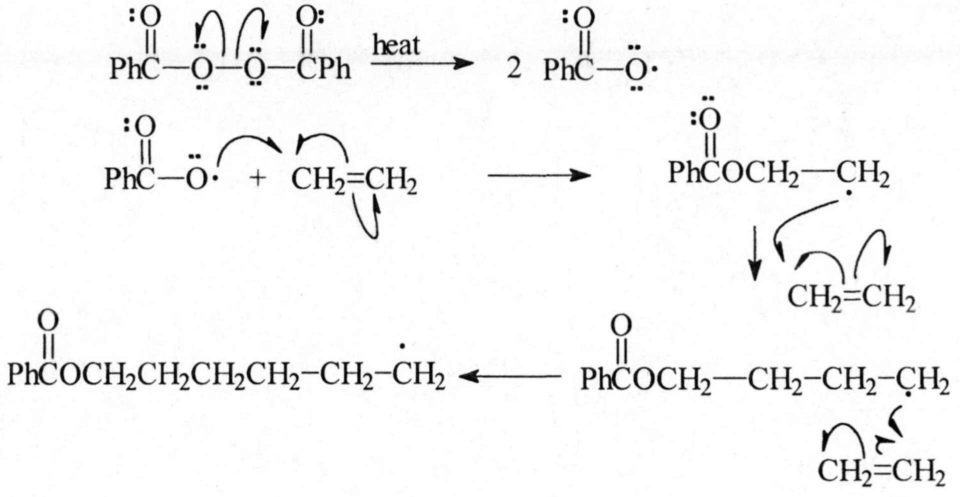

21.38

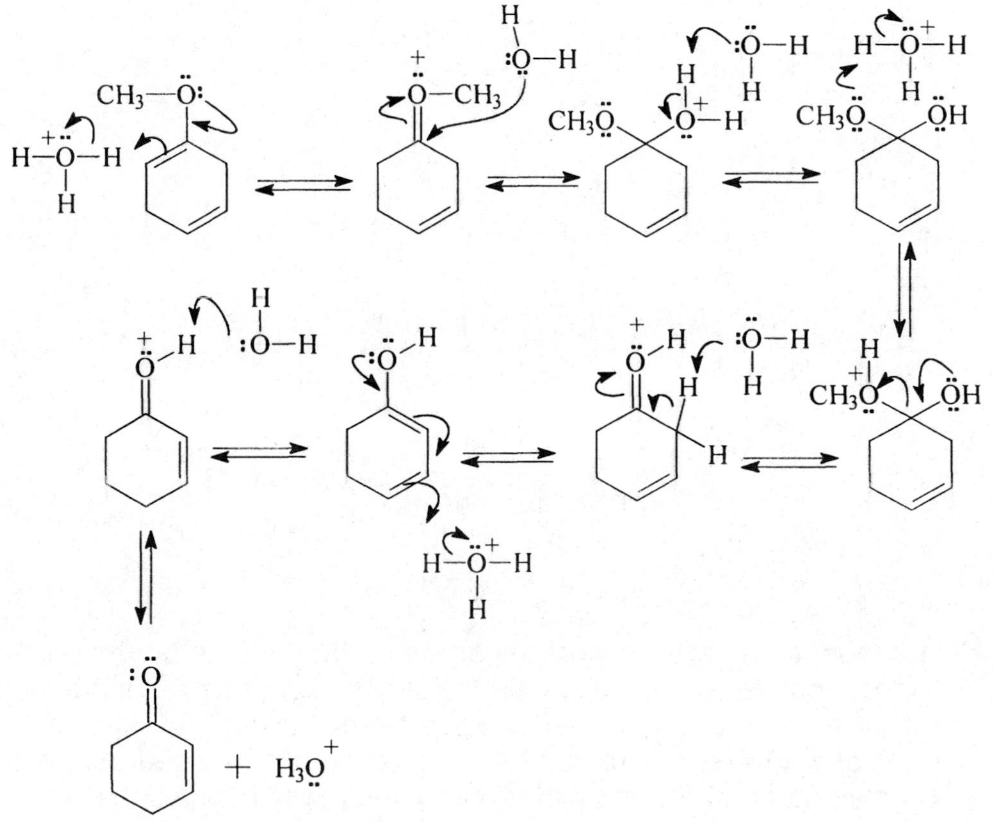

21.39

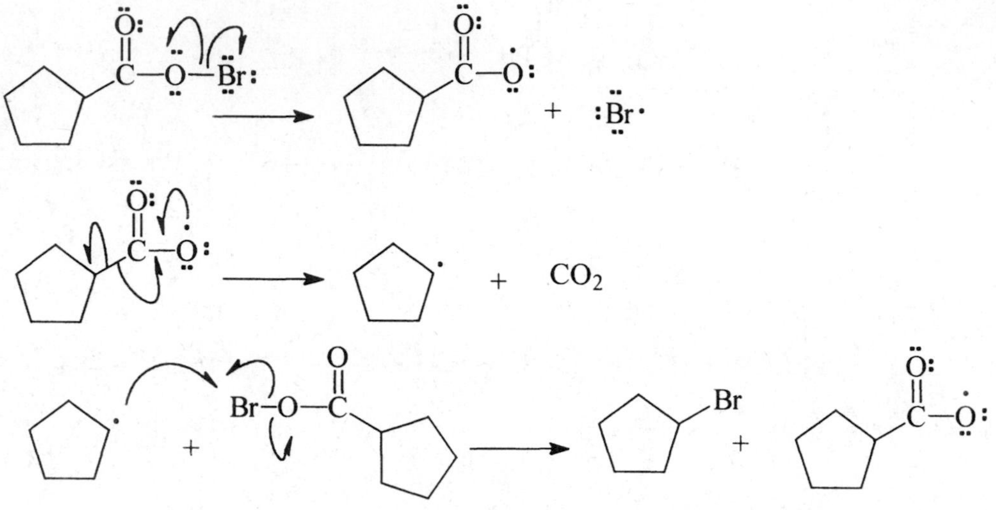

21.40

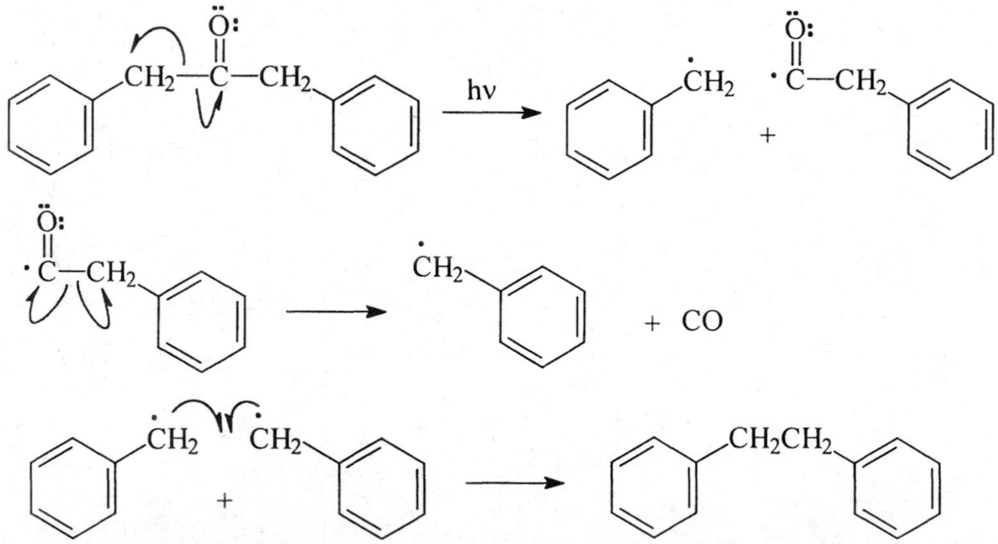

21.41 Hydrogens on carbons next to carbon-carbon double bonds (allylic hydrogens) are most readily abstracted because the resulting radicals are stabilized by resonance. An oleate ester has two allylic sites where a hydrogen can be abstracted (positions 8 and 11 in the structure below). Oxygen can then add to either end of the allylic radicals that are produced, so there are four possible products, neglecting cis-trans isomers. One of these products is shown below. The OOH group could also be bonded to C-11. Another isomer that would be formed has the double bond between C-8 and C-9, with the OOH group on C-10. The final isomer has the double bond between C-10 and C-11 and the OOH group on C-9.

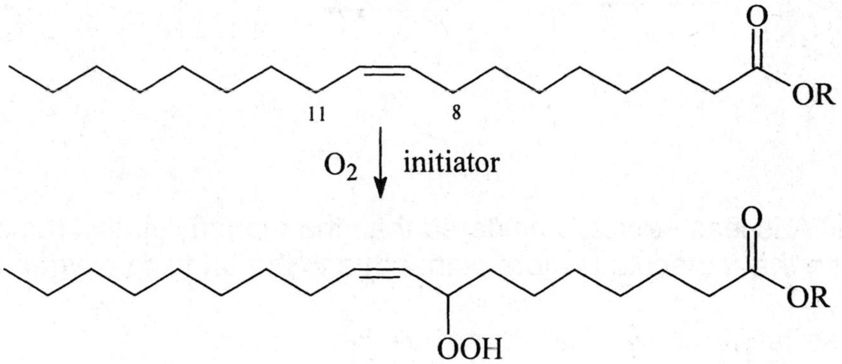

21.42

$$Cl{-}CF_3 \xrightarrow{h\nu} :\ddot{Cl}\cdot + \cdot CF_3$$

$$:\ddot{Cl}\cdot + O_3 \longrightarrow :\ddot{Cl}{-}\ddot{O}\cdot + O_2$$

$$:\ddot{Cl}{-}\ddot{O}\cdot + \cdot\ddot{O}\cdot \longrightarrow O_2 + :\ddot{Cl}\cdot$$

21.43

$$CH_3(CH_2)_6{-}C(=\ddot{O}:){-}\ddot{O}:^- \xrightarrow{-e^-} CH_3(CH_2)_6{-}C(=\ddot{O}:){-}\ddot{O}\cdot \longrightarrow CH_3(CH_2)_5\dot{C}H_2 + CO_2$$

$$2\ CH_3(CH_2)_5\dot{C}H_2 \longrightarrow CH_3(CH_2)_5CH_2{-}CH_2(CH_2)_5CH_3$$

21.44

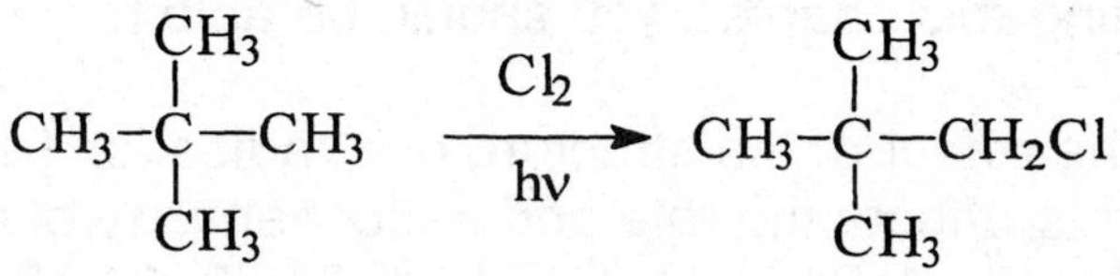

21.45

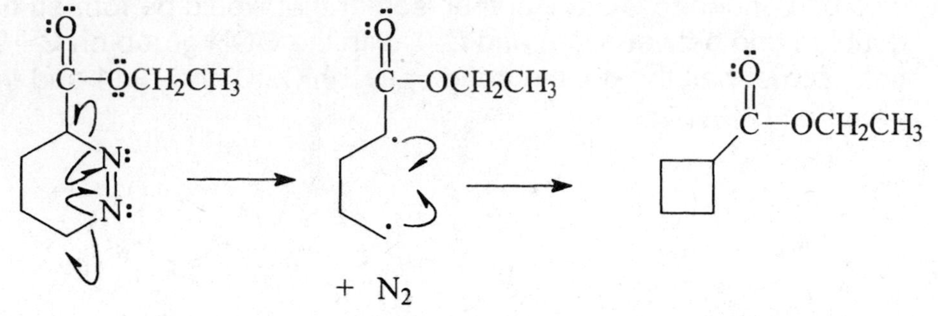

21.46 Radical A is less sterically hindered than the triphenylmethyl radical because the molecule is more planar due to the bridging oxygen atoms.

21.47 In the structure shown, galvinoxyl has the radical on the lower oxygen. It has another resonance structure, equivalent to this one, which has the radical located on the upper oxygen. In addition, it has a number of resonance structures which have the radical located on the carbons ortho and para to the two oxygens, and on the carbon connecting the two rings. Because it has considerable resonance stabilization, and the positions where the radical electron density is located are sterically hindered, the radical is relatively unreactive.

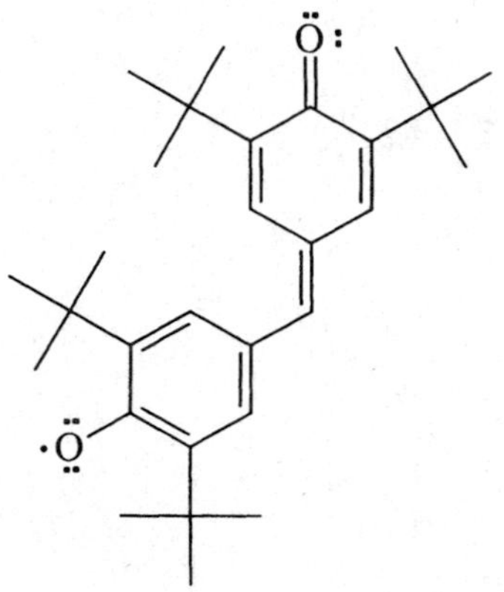

21.48 Dichloride 1 and dichloride 2 are enantiomers and are formed by abstraction of the tertiary hydrogen at the chirality center of the reactant. The resulting radical is no longer chiral, so both enantiomers of the product must be formed in equal amounts.

Review of Mastery Goals

After completing this chapter, you should be able to:

Understand the effect of the structure of a radical on its stability and explain how this affects the rate and regiochemistry of radical reactions. (Problems 21.2, 21.4, 21.7, 21.30, 21.31, 21.33, 21.36, and 21.41)

Show the products of the reactions discussed in this chapter.
(Problems 21.3, 21.5, 21.8, 21.10, 21.11, 21.14, 21.16, 21.18, 21.19, 21.20, and 21.12)

Show the mechanisms of these reactions.
(Problems 21.6, 21.9, 21.13, 21.23, 21.24, 21.25, 21.26, 21.27, 21.28, 21.29, 21.32, 21.34, 21.35, 21.37, 21.39, 21.41, 21.42, and 21.43)

Use these reactions in synthesis.
(Problems 21.15, 21.17, and 21.22)

Chapter 22
PERICYCLIC REACTIONS

22.1 a) conrotation b) disrotation

21.2

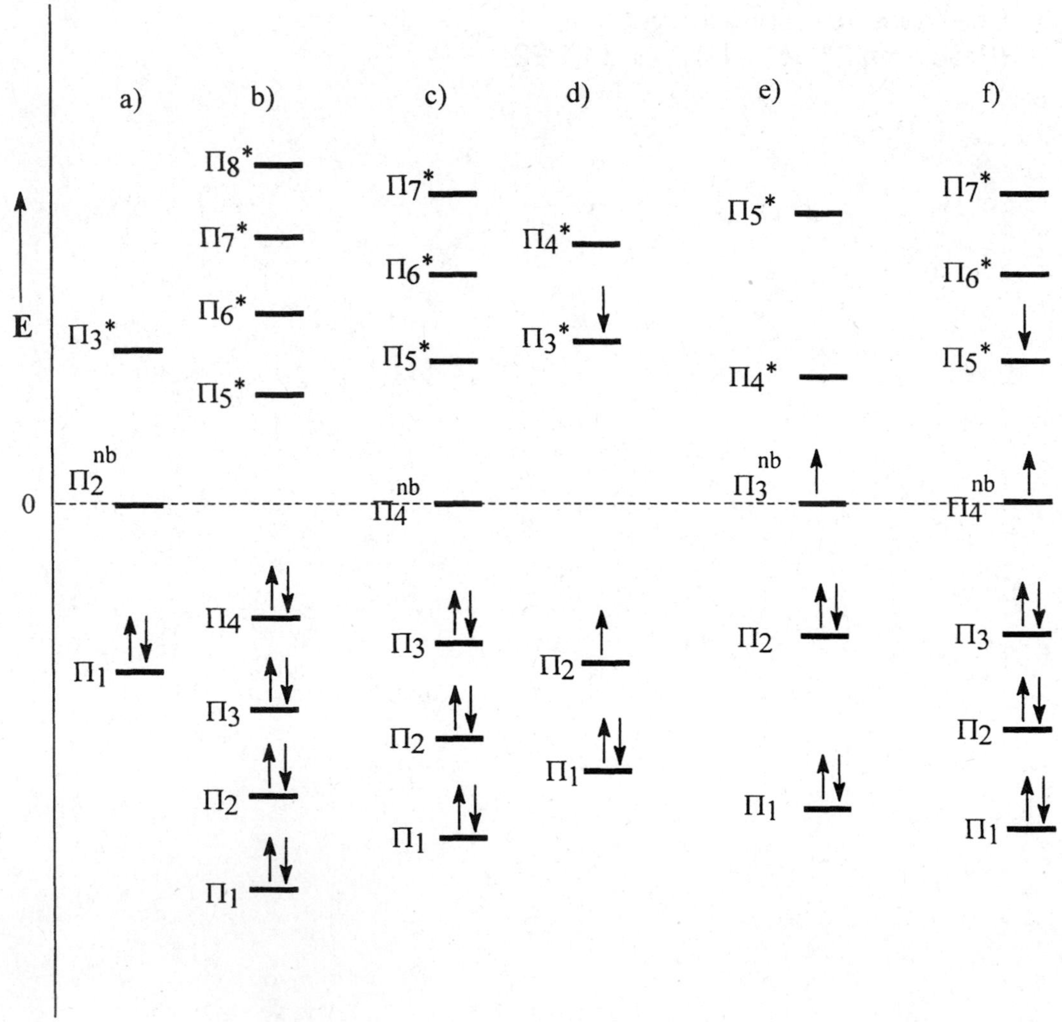

22.3

a) $\Pi 3^{nb}$
2 nodes

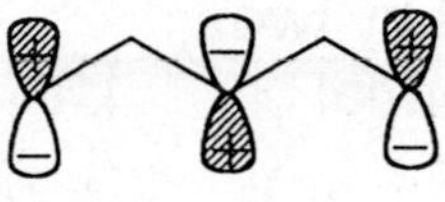

b) $\Pi 2$
1 node

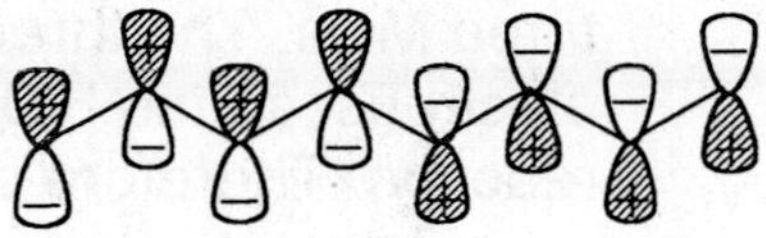

c) $\Pi 4^*$
3 nodes

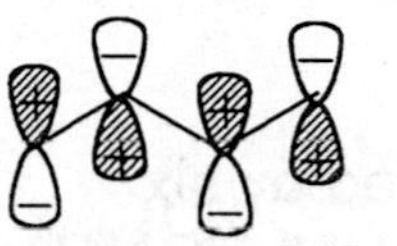

d) $\Pi 4^{nb}$
3 nodes

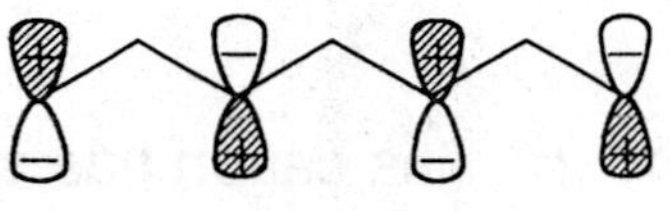

e) $\Pi 3$
2 nodes

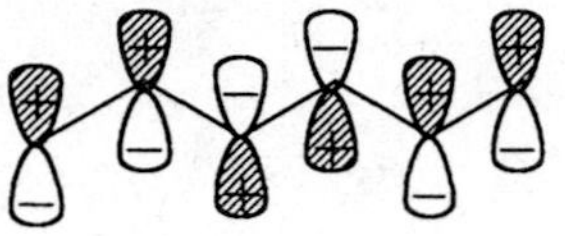

22.4 For the photochemical reaction, the HOMO of the diene is π_3^* (two nodes), so conrotation is forbidden.

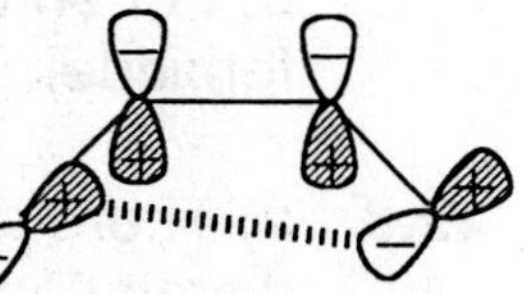

22.5 For the thermal reaction, the HOMO of triene is π_3 (two nodes), so conrotation is forbidden.

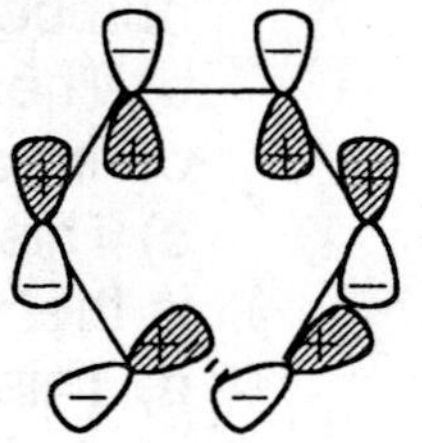

For the photochemical reaction, the HOMO of triene is π_4^* (three nodes), so disrotation is forbidden.

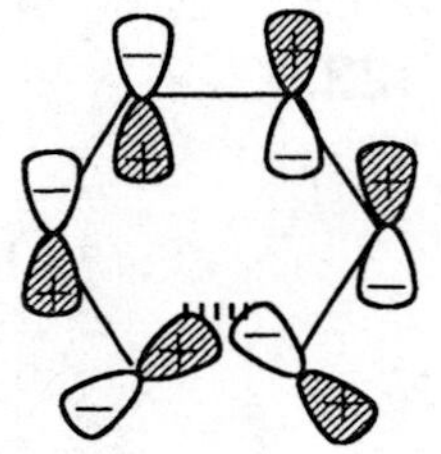

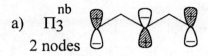

22.6 a) This cation has three p orbitals, so there are three MOs. The three MOs are occupied by two electrons, so the HOMO is π_1 for the thermal reaction. Therefore disrotation is allowed.

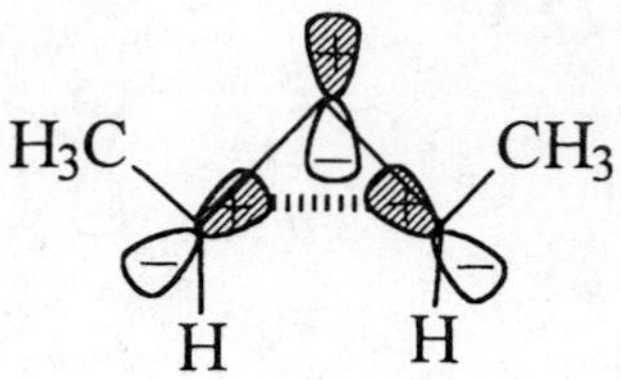

b) This cation has seven p orbitals occupied by six electrons. For the photochemical reaction, the HOMO is π_4^{nb}. The disrotatory closure here is forbidden.

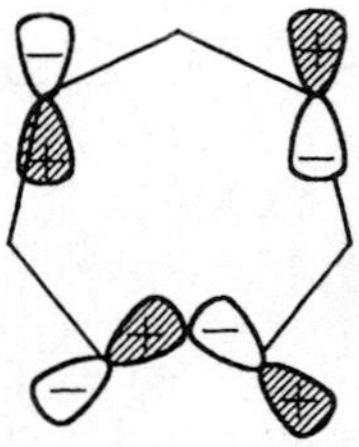

22.7 a) For an odd number of electron pairs (1), disrotation is thermally allowed.
b) For an odd number of electron pairs (3), disrotation is photochemically forbidden.

22.8 a) There are two electron pairs involved and the reaction occurs by disrotation. According to the chart, a disrotation involving an even number of electron pairs is photochemically allowed.
b) There are three electron pairs involved and the reaction occurs by conrotation. According to the chart this reaction is thermally forbidden.
c) This reaction involves two electron pairs and occurs by disrotation, so it is thermally forbidden.
d) This reaction involves four electron pairs and occurs by disrotation, so it is photochemically allowed.

22.9

a) 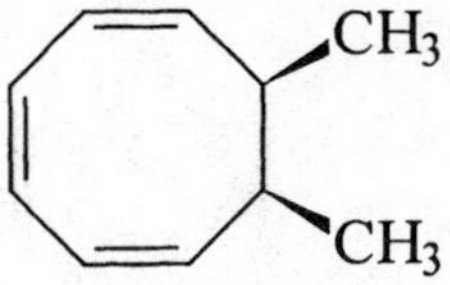

b) 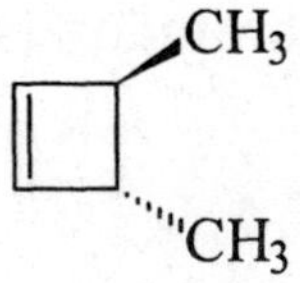

c) 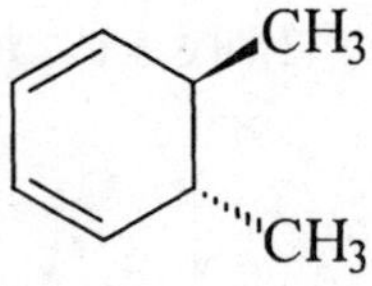

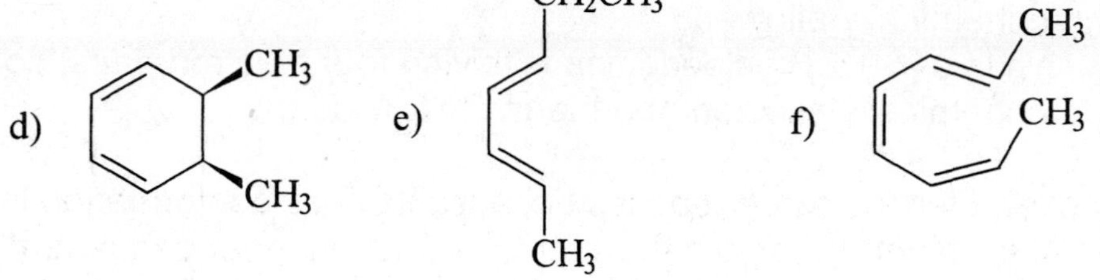

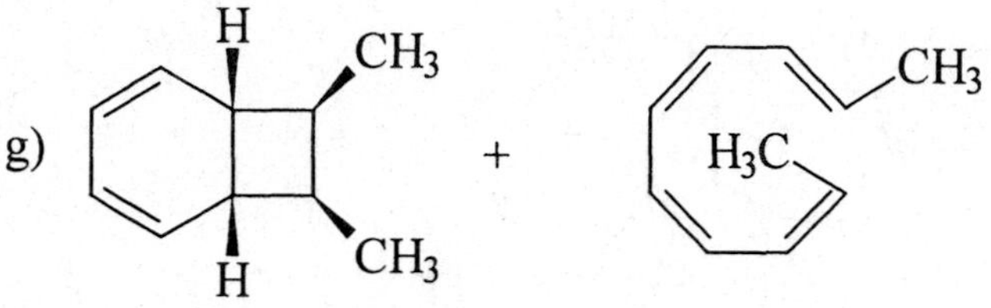

22.10 For a cycloaddition reaction, if the overlaps of the orbitals of the HOMO of one component and LUMO of the other component are bonding, then the reaction is allowed; but if one of these overlaps is antibonding, the reaction is forbidden. In this orbital drawing of the photochemical [4 + 2] cycloaddition, one overlap is bonding and the other is antibonding, so the reaction is forbidden.

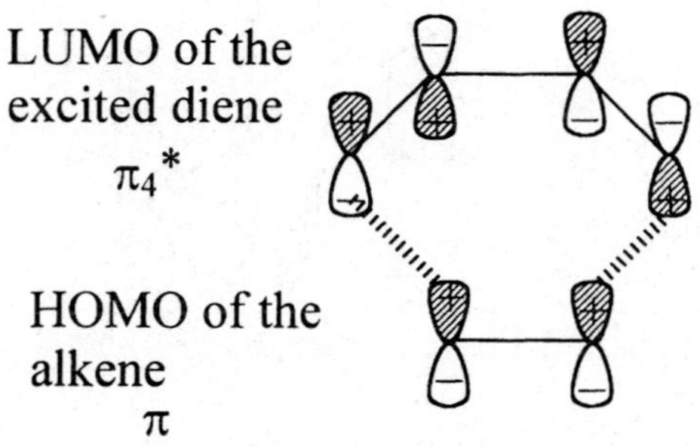

22.11 In this orbital drawing of the thermal [4 + 4] cycloaddition, one overlap is bonding and the other is antibonding so the reaction is forbidden.

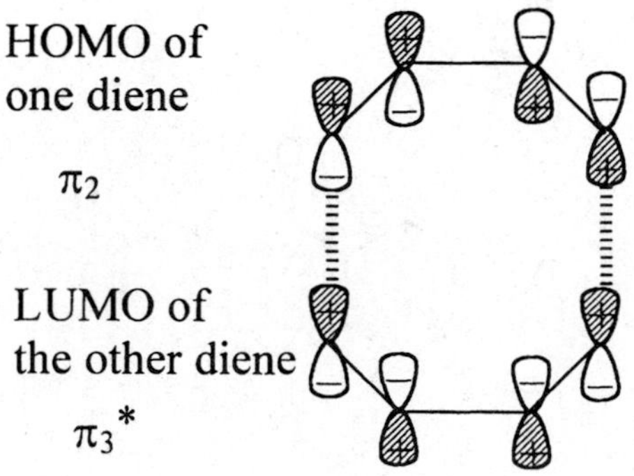

22.12 a) This is a [4 + 4] cycloaddition involving four electron pairs, so it is photochemically allowed.
b) This is a [6 + 2] cycloaddition involving four electron pairs, so it is photochemically allowed and thermally forbidden.

22.13 The left diene is more reactive because its *s*-cis conformation is less sterically hindered and is therefore present in higher concentration.

22.14

a) 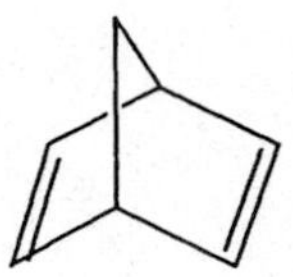

b) 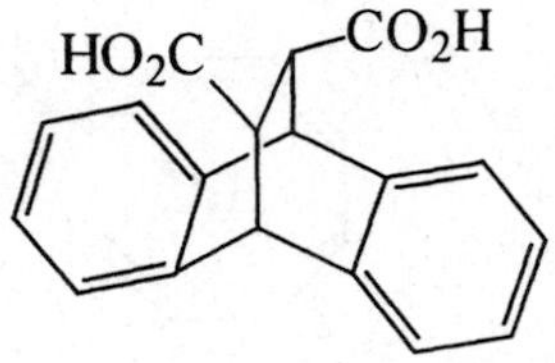

c) 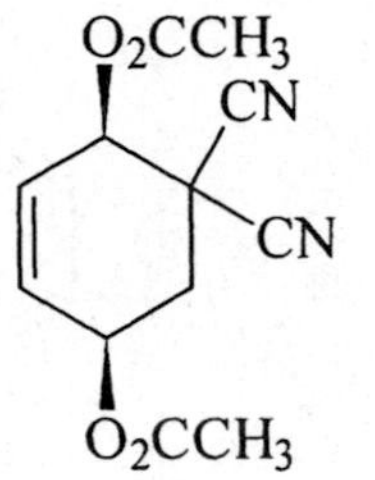

d) 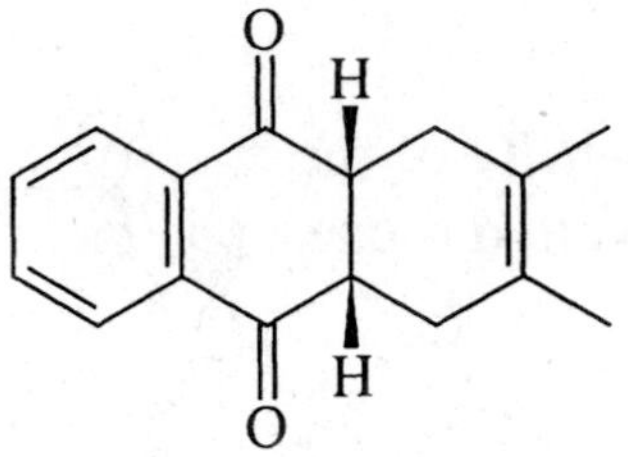

e) 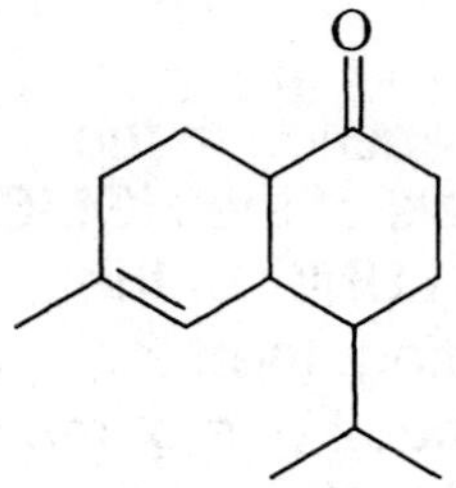

f) 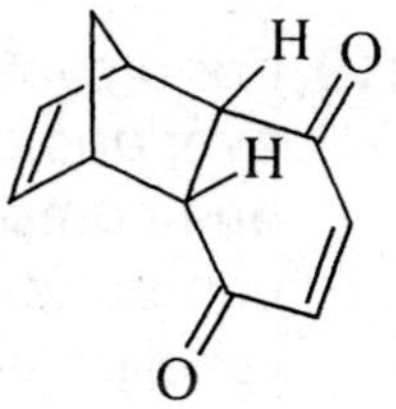

g) 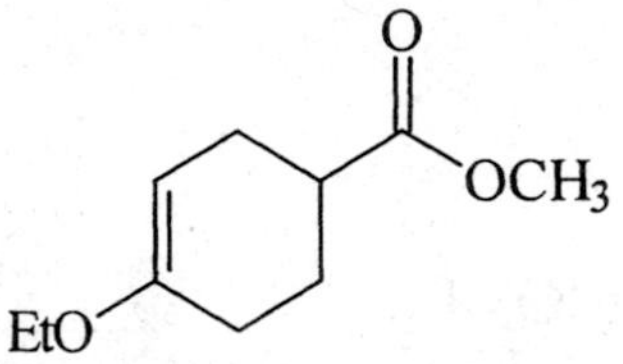

h) 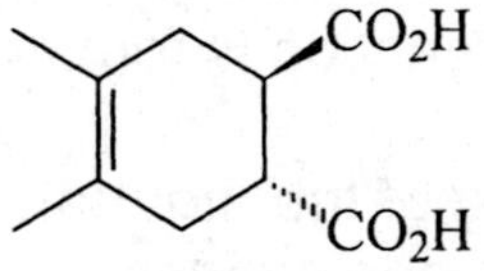

i) 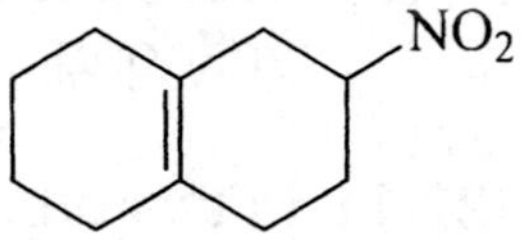

j) 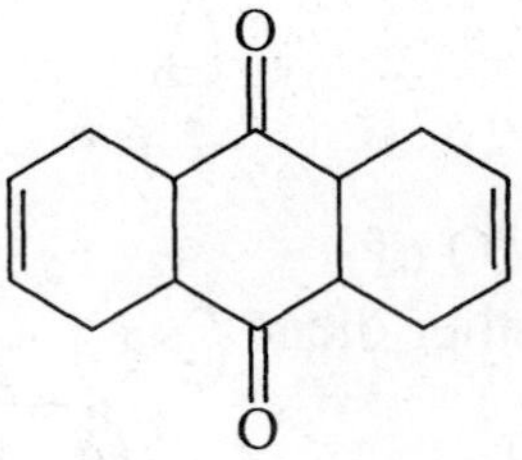

k) 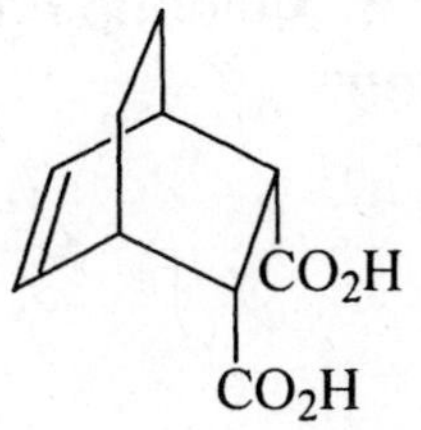

22.15 This is a [10 + 2] cycloaddition. It involves six electron pairs, so it is photochemically allowed.

22.16

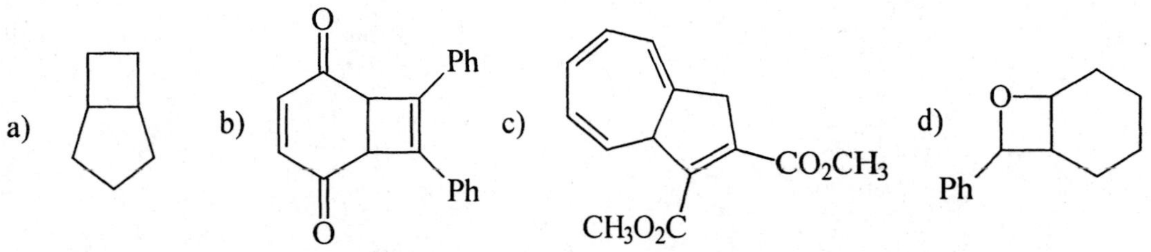

22.17 A sigmatropic rearrangement is an intramolecular reaction involving migration of a σ-bonded group over a conjugated π system. The *i* and *j* values in the notation of sigmatropic rearrangements refer to the number of atoms in each migrating component. Find the σ bond that is broken in the reaction and give number 1 to both atoms of this bond. Then number the atoms of each migrating component up to the atoms where the new σ bond is formed. The numbers of the atoms where the new σ bond is formed in the product are used to designate the rearrangement.

a) The new σ bond is formed between atom 3 of one component and atom 3 of the other, so this is a [3,3] sigmatropic rearrangement.

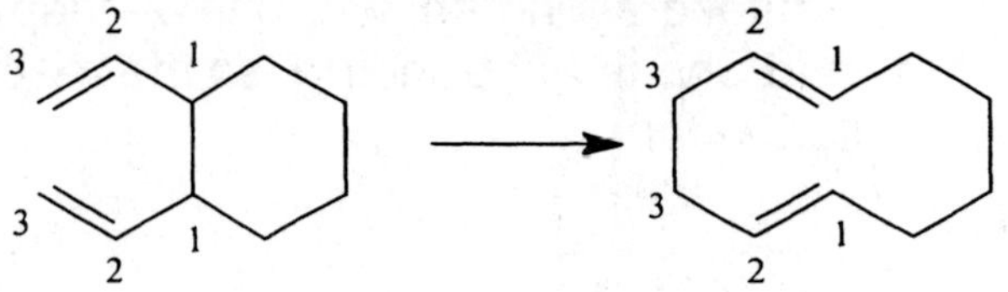

b) The new σ bond is formed between atom 1 (the H) of one component and atom 7 of the other, so this is a [1,7] sigmatropic rearrangement.

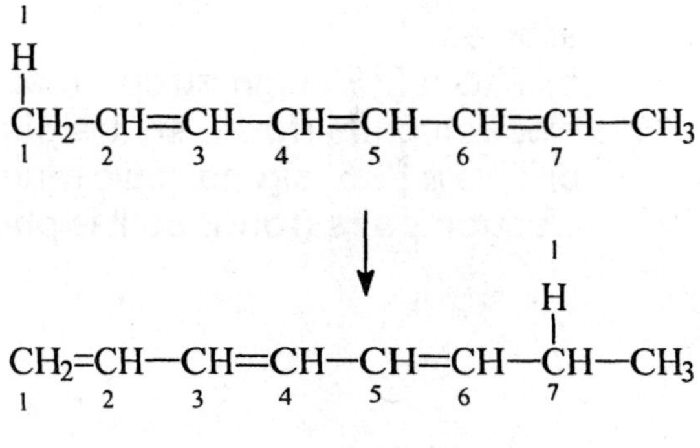

c) The new σ bond is formed between atom 3 of one component and atom 5 of the other, so this is a [3,5] sigmatropic rearrangement.

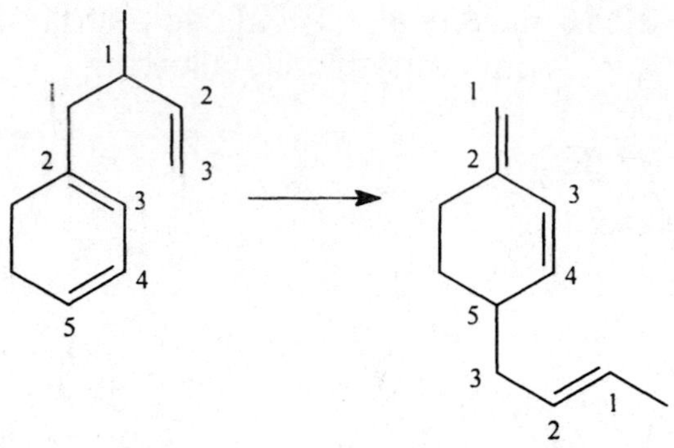

22.18 The two fragments of a [1,5] sigmatropic rearrangement are a hydrogen atom and a pentadienyl radical: The HOMO of the pentadienyl radical for the photochemical reaction is π_4^* (three nodes). According to the orbital drawing one overlap is bonding and one is antibonding, so the reaction is photochemically forbidden.

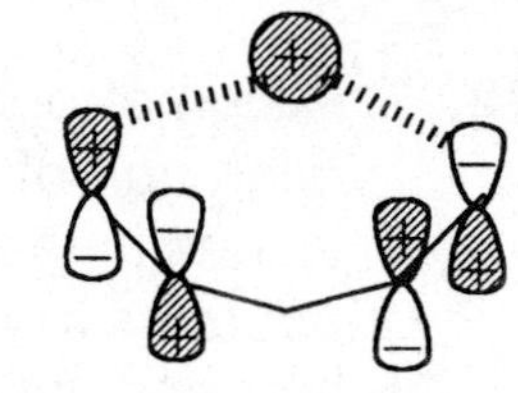

22.19 For the thermal reaction the HOMO of the hexatrienyl radical is π_4^{nb} (three nodes). In the orbital drawing shown, one overlap is bonding and one is antibonding, so the reaction is forbidden.

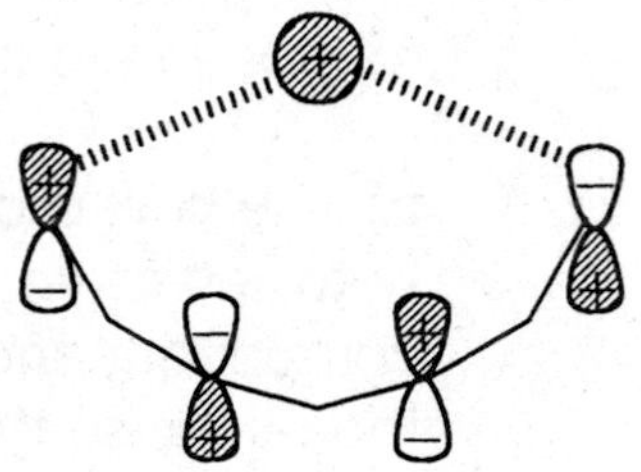

22.20 a) This a [3,3] sigmatropic rearrangement involving an odd number of electron pairs (three), so it is thermally allowed.

b) This a [1,7] sigmatropic rearrangement involving an even number of electron pairs (four), so it is photochemically allowed.

c) This a [3,5] sigmatropic rearrangement involving an even number of electron pairs (four), so it is photochemically allowed.

22.21

a)

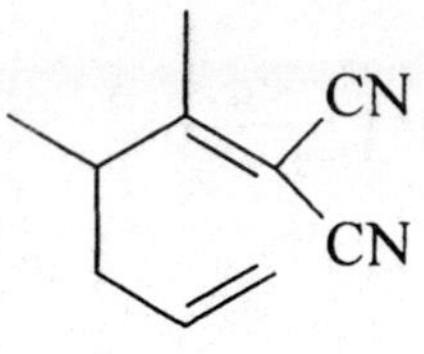

b)

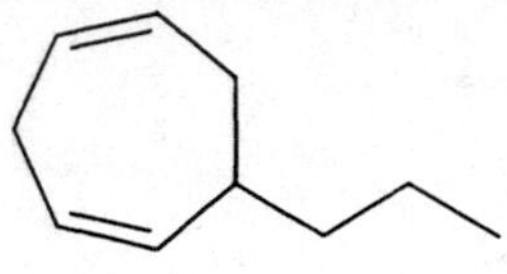

c)

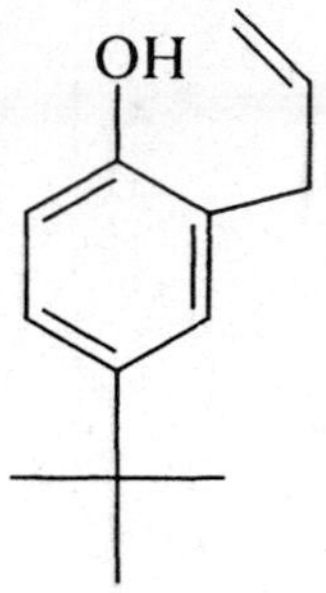

d)

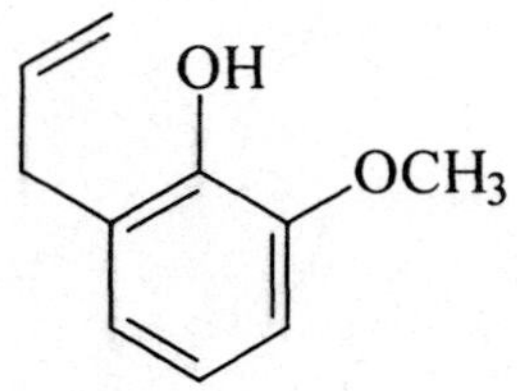

e)

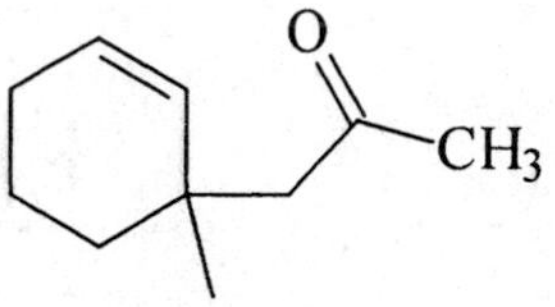

22.22 This reaction is a [3,3] sigmatropic rearrangement involving an odd number of electron pairs, so it is thermally allowed. The reactant is favored at equilibrium because it has less ring strain.

22.23

a) $Et-\underset{\underset{Et}{|}}{\overset{\overset{Et}{|}}{C}}-\overset{\overset{O}{\|}}{C}-Et$ b) $PhNH\overset{\overset{O}{\|}}{C}Ph$ c)   d)

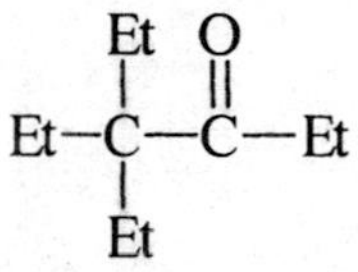

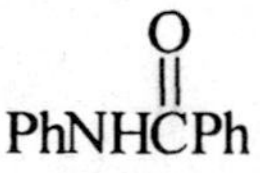

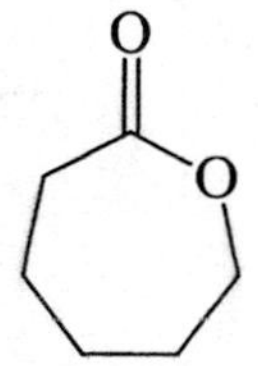

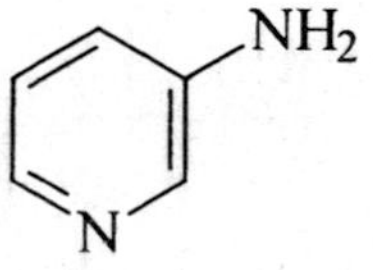

22.24 In the Curtius rearrangement, N_2 acts as the leaving group and rearrangement occurs to form an isocyanate. The isocyanate is converted to a carbamic acid which loses CO_2 in exactly the same manner as shown in Figure 22.7 for the Hofmann rearrangement.

$+ N_2$

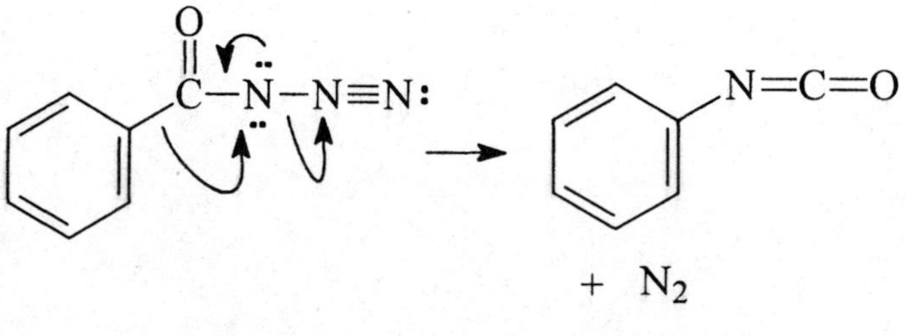

22.25

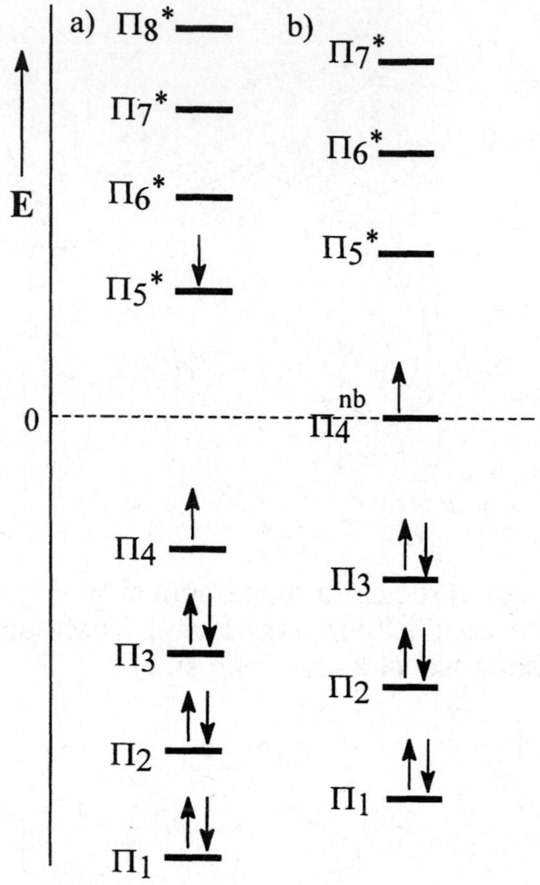

E
a)
Π8*
Π7*
Π6*
Π5*
0
Π4
Π3
Π2
Π1
b)
Π7*
Π6*
Π5*
nb
Π4
Π3
Π2
Π1

22.26

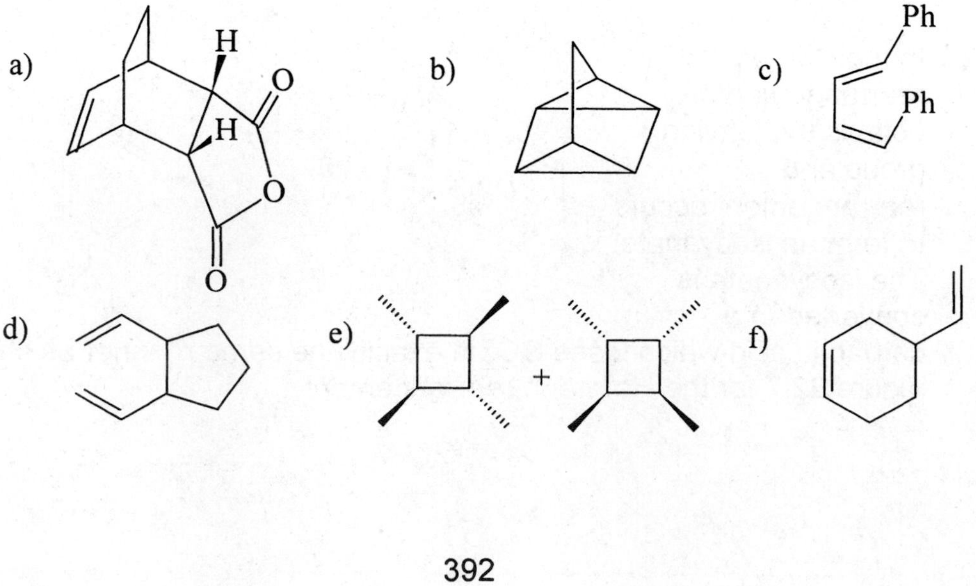

a)
H
O
H
O
O
b)
c)
Ph
Ph
d)
e)
+
f)

g) 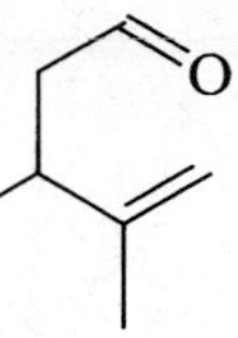 h) 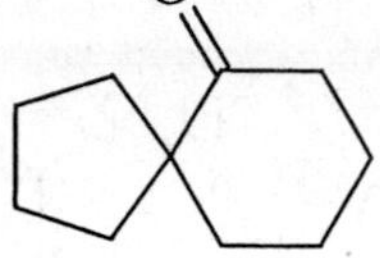 i) 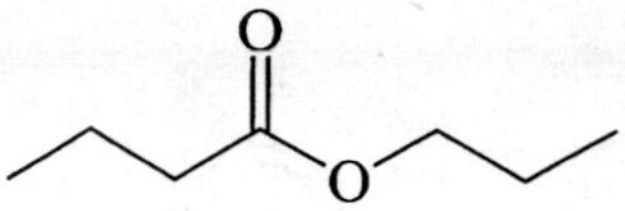

22.27

a) 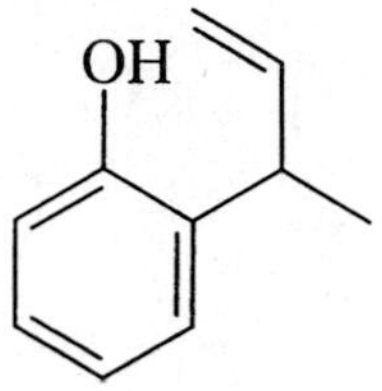
OH

b) 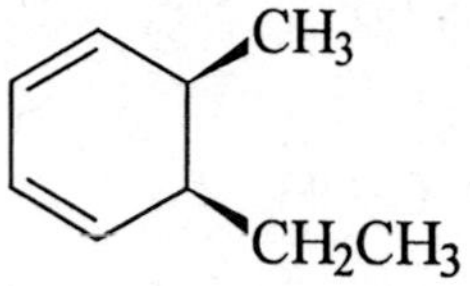
CH3
CH2CH3

c) 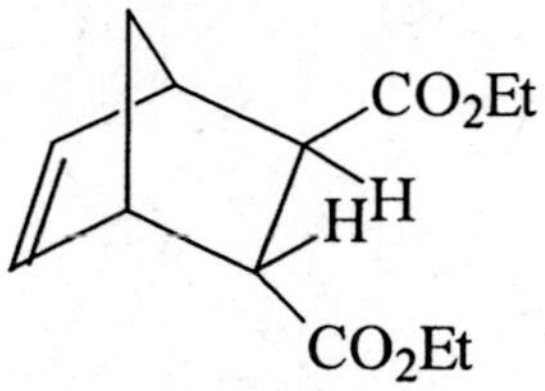
CO2Et
H
H
CO2Et

d) 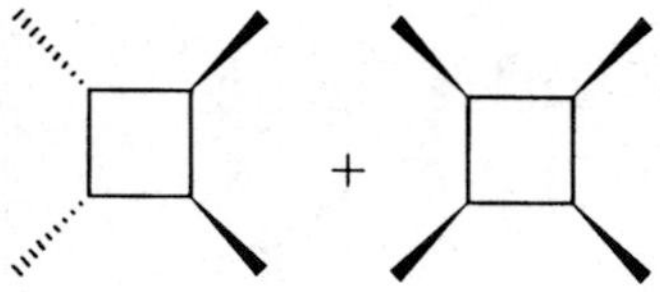

e) 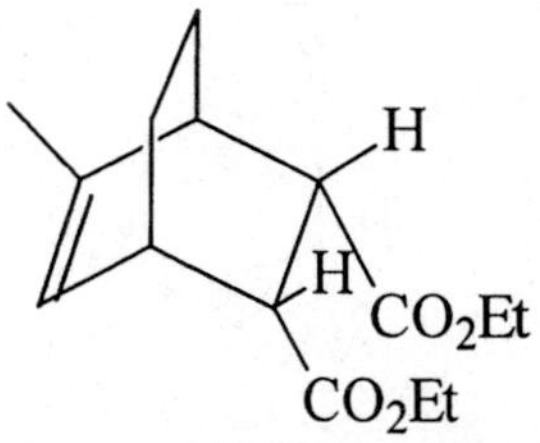
H
H
CO2Et
CO2Et

f) 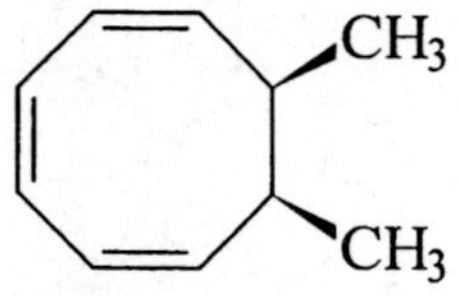
CH3
CH3

g) 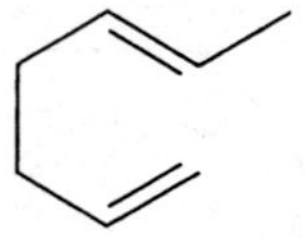

h) 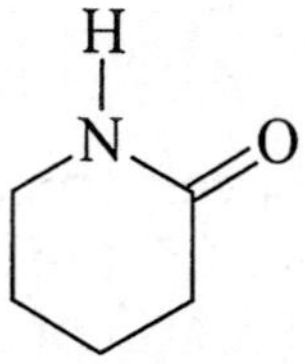
H
N
O

i) 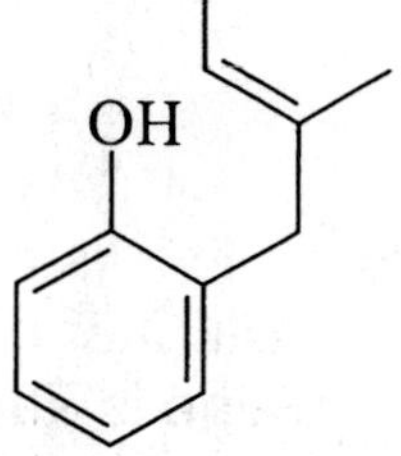
OH

j) 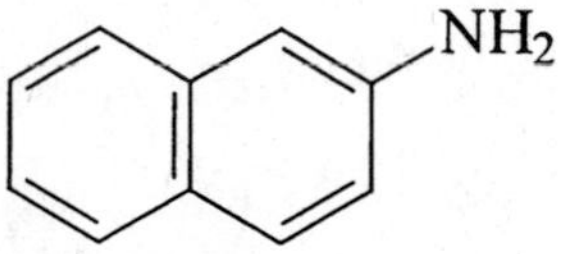
NH2

22.28

a) 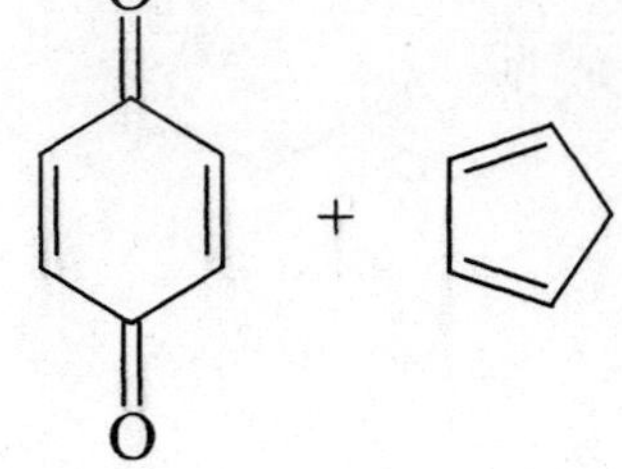

b) 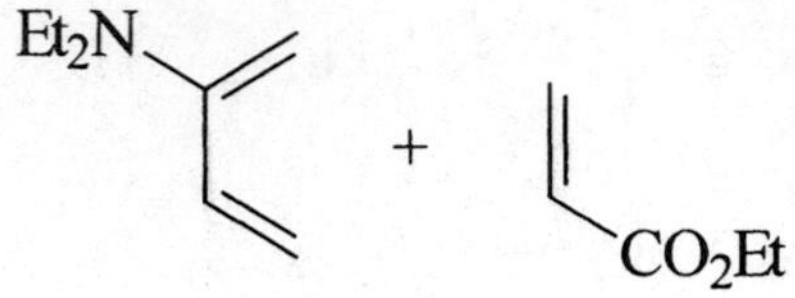

c) 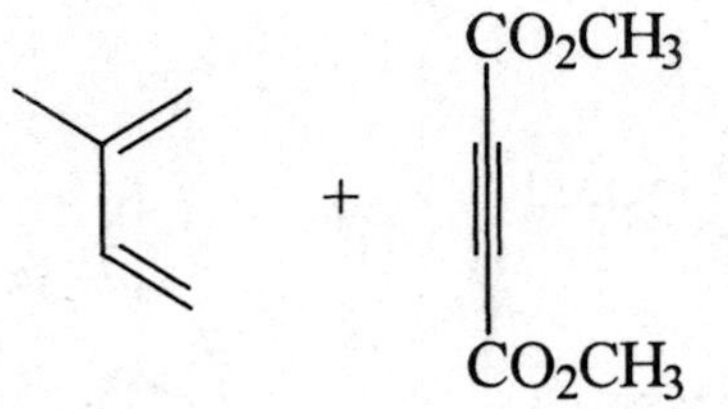

d) 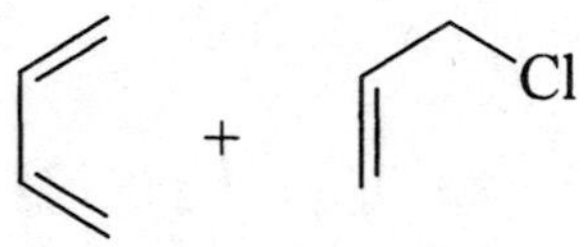

e) 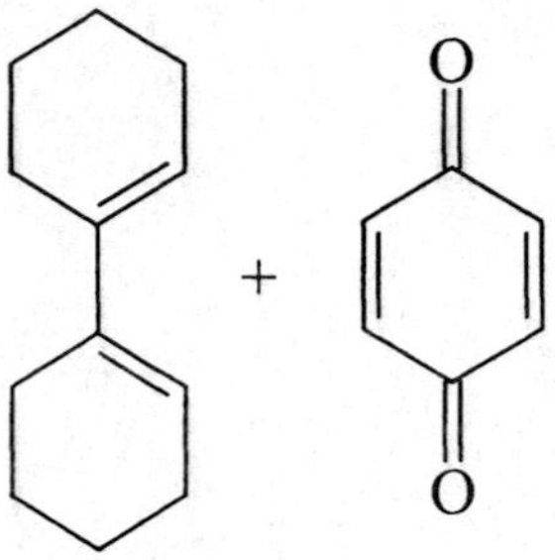

f) 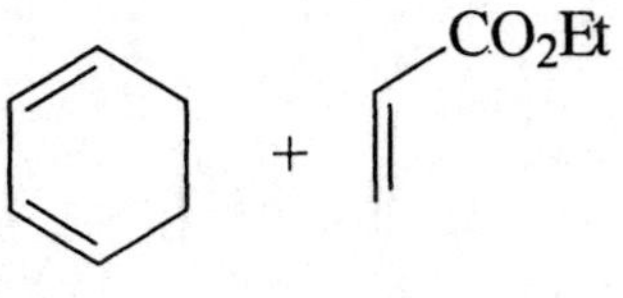

g) 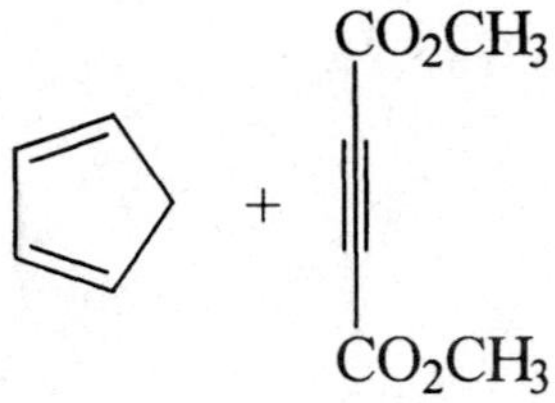

22.29 a) This is a thermal [6+2] cycloaddition. Use the LUMO of one component (π_4*, the LUMO of the triene is shown) and the HOMO of the other component (π_1, the HOMO of ethene is shown). One overlap is bonding and the other is antibonding, so the reaction is forbidden.

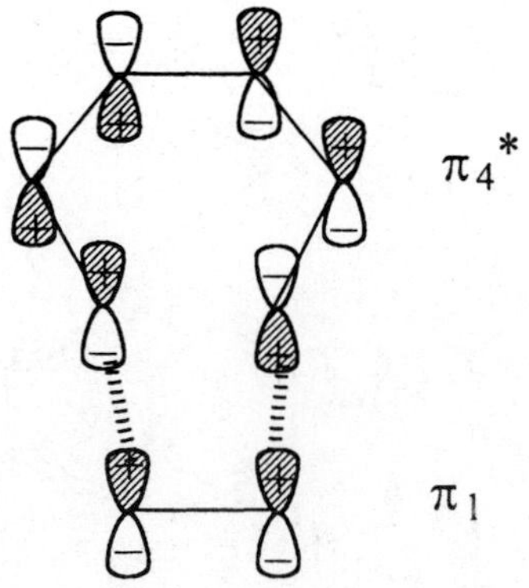

b) This is a photochemical [6+2] cycloaddition. The LUMO of the excited state triene (π_5^*) and the HOMO of ethene (π_1) are shown. Both overlaps are bonding, so the reaction is allowed.

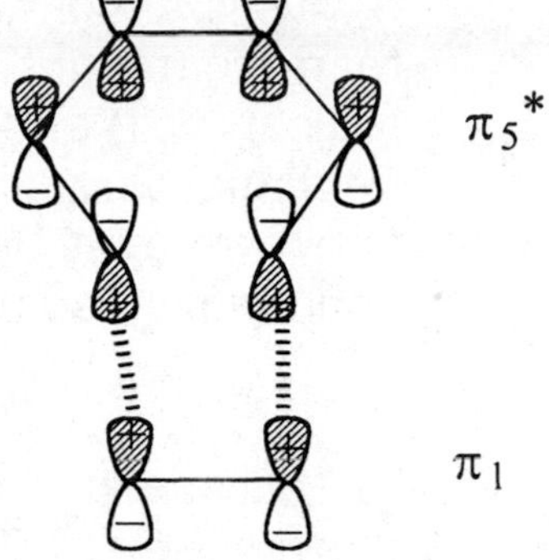

c) For an electrocyclic reaction, examine the new overlap in the HOMO when the proper rotation occurs. This is conrotation of a triene. The overlap in the HOMO (π_3) is antibonding, so conrotation is forbidden.

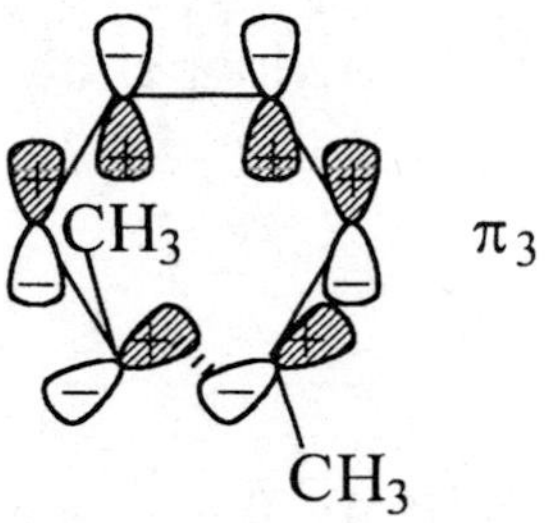

d) This is a photochemical [4+4] cycloaddition. The LUMO of the excited state diene (π_4^*) and the HOMO of the ground state diene (π_2) are shown. Both overlaps are bonding, so the reaction is allowed.

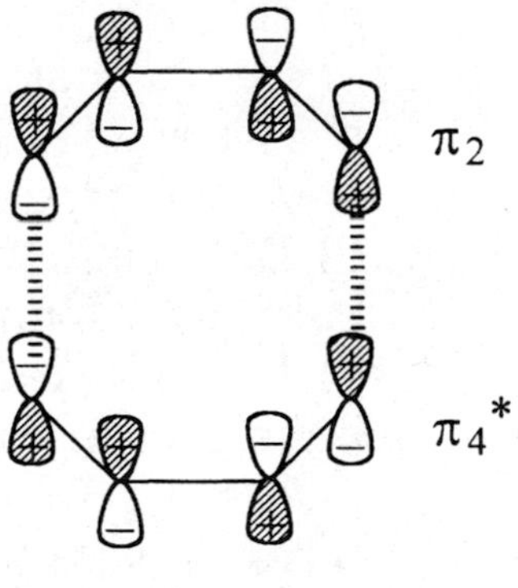

e) This is a conrotatory electrocyclic reaction. The HOMO for the ground state is π_4^{nb}. The new overlap is bonding, so conrotation is allowed.

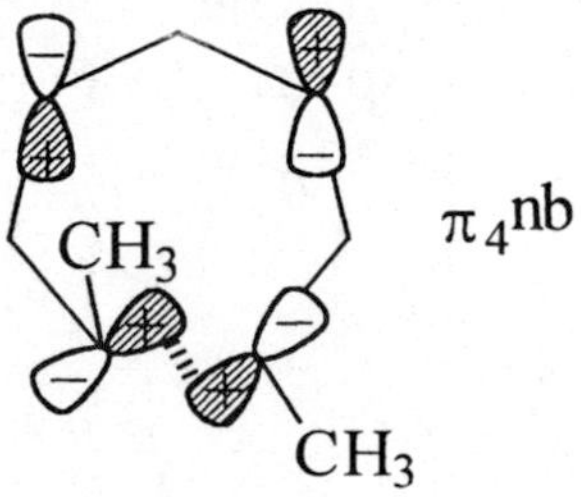

f) This is a thermal [3,5] sigmatropic rearrangement. The HOMO of the allylic radical (π_2^{nb}) and the HOMO of the pentadienyl radical (π_3^{nb}) are shown. One overlap is bonding and the other is antibonding, so the reaction is forbidden.

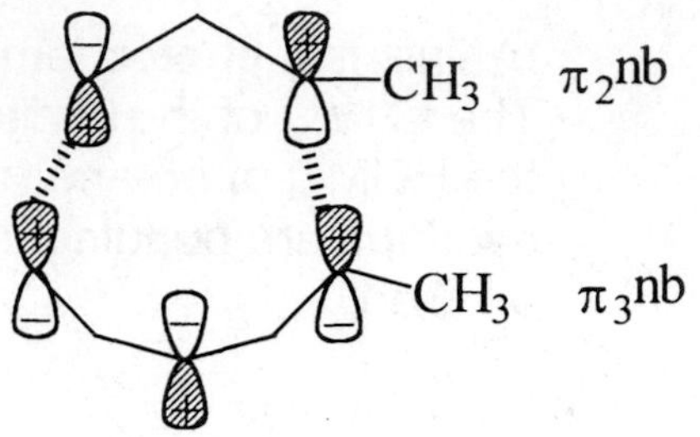

g) This is a thermal [1,5] sigmatropic rearrangement. The HOMO of the methyl radical (p orbital) and the HOMO of the pentadienyl radical (π_3^{nb}) are shown. Both overlaps are bonding, so the reaction is allowed.

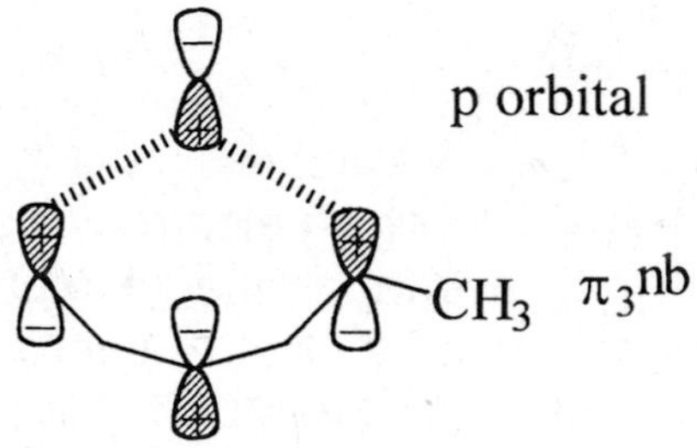

22.30 a) This is a [6+2] cycloaddition involving an even number (4) of electron pairs, so it is photochemically allowed.
b) This is a [1,5] sigmatropic rearrangement involving an odd number (3) of electron pairs, so it is thermally allowed.
c) This is a [1,7] sigmatropic rearrangement involving an even number (4) of electron pairs, so it is photochemically allowed.
d) This is a [6+2] cycloaddition involving an even number (4) of electron pairs, so it is photochemically allowed.
e) This is a [2+2] cycloaddition involving an even number (2) of electron pairs, so it is photochemically allowed.
f) This is a [6+4] cycloaddition involving an odd number (5) of electron pairs, so it is thermally allowed.
g) This is a [3,7] sigmatropic rearrangement involving an odd number (5) of electron pairs, so it is thermally allowed.
h) This is a [6+4] cycloaddition involving an odd number (5) of electron pairs, so it is thermally allowed.

22.31

a) 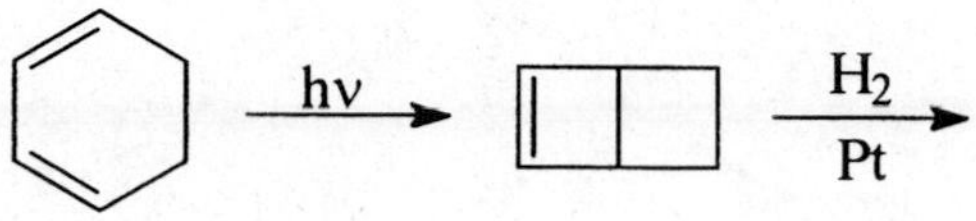
hν
H2
Pt

b) 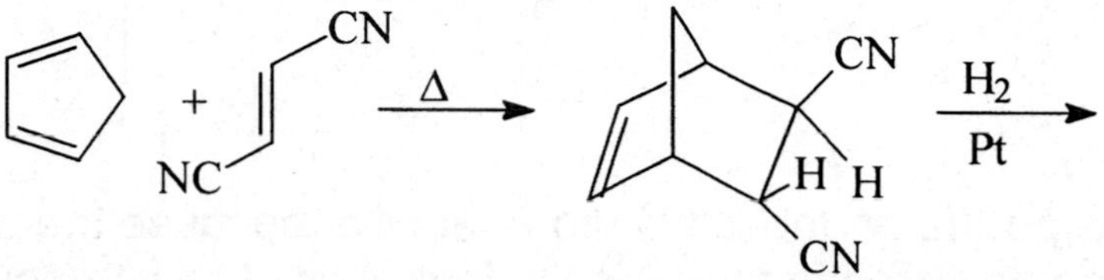
CN
NC
Δ
CN
H H
CN
H2
Pt

c) 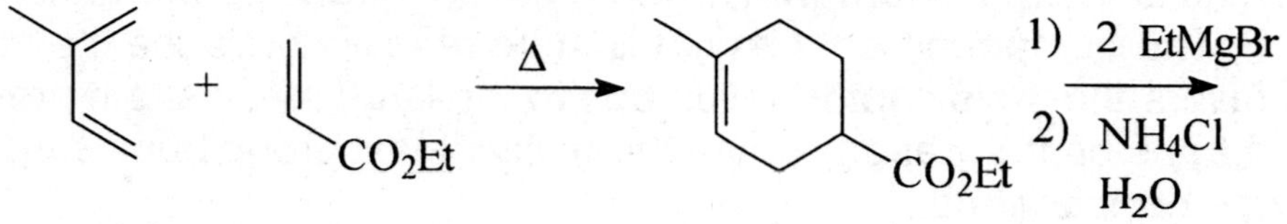
CO2Et
Δ
CO2Et
1) 2 EtMgBr
2) NH4Cl
H2O

d) 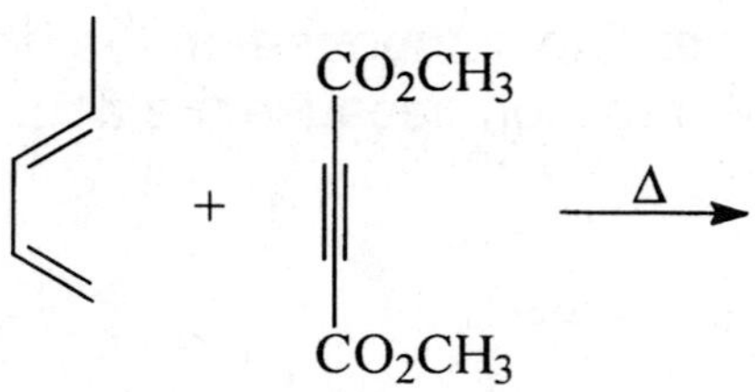
CO2CH3
CO2CH3
Δ

e) 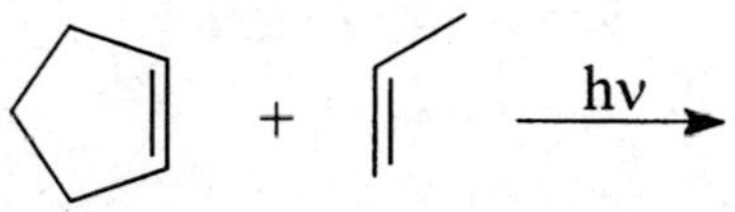
hν

f) 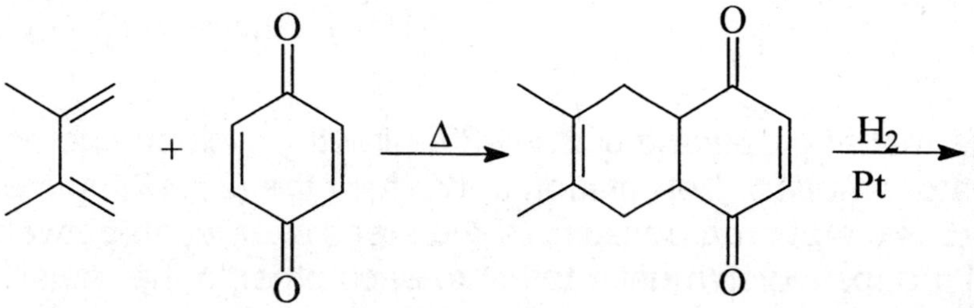
O
O
Δ
O
O
H2
Pt

g) 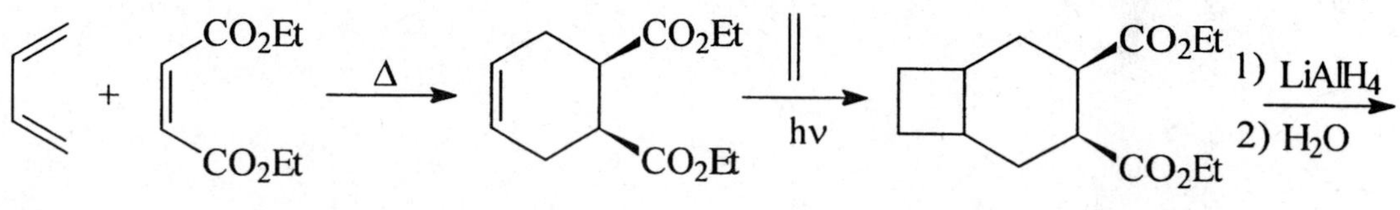
CO2Et
CO2Et
Δ
CO2Et
CO2Et
hν
CO2Et
CO2Et
1) LiAlH4
2) H2O

22.32

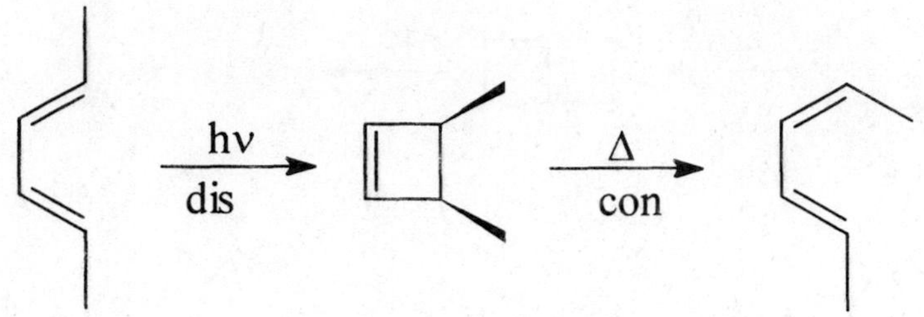

22.33 a) The compound on the left is more reactive because the double bonds of the diene are fixed in the *s*-cis conformation. The compound on the right is in equilibrium with its more stable *s*-trans conformation.
b) The compound on the right is more reactive because the equilibrium favors the *s*-cis conformation due to the large steric strain present in the *s*-trans conformation, where the bulky *t*-butyl group bumps into the CH_2 of the double bond.
c) The compound on the left is more reactive because the double bonds are fixed in an *s*-cis conformation. The compound on the right cannot react as a diene in a Diels-Alder reaction because the double bonds are held *s*-trans.

22.34

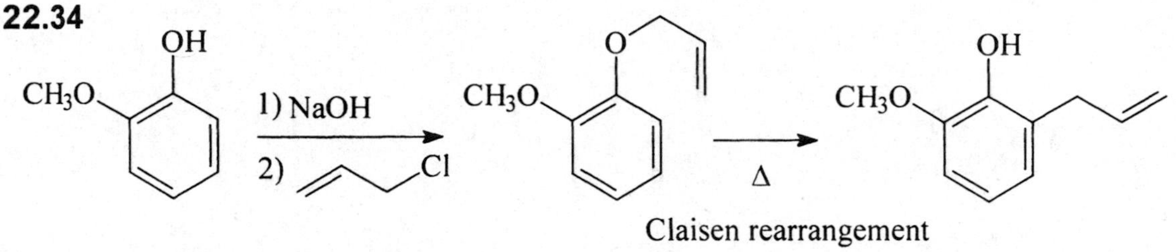

22.35 The thermal ring opening of *trans*-3,4-dimethylcyclobutene occurs by a conrotatory motion. Conrotation to produce the (2*Z*,4*Z*)-isomer, although allowed, is disfavored because of the steric interaction between the two methyl groups which must rotate into each other in the transition state.

22.36 The protonated OH on the carbon bonded to the phenyl groups preferentially leaves because the carbocation formed is more stable than the primary carbocation that would be formed if the other OH were to leave. This leads to the formation of the aldehyde rather than the ketone.

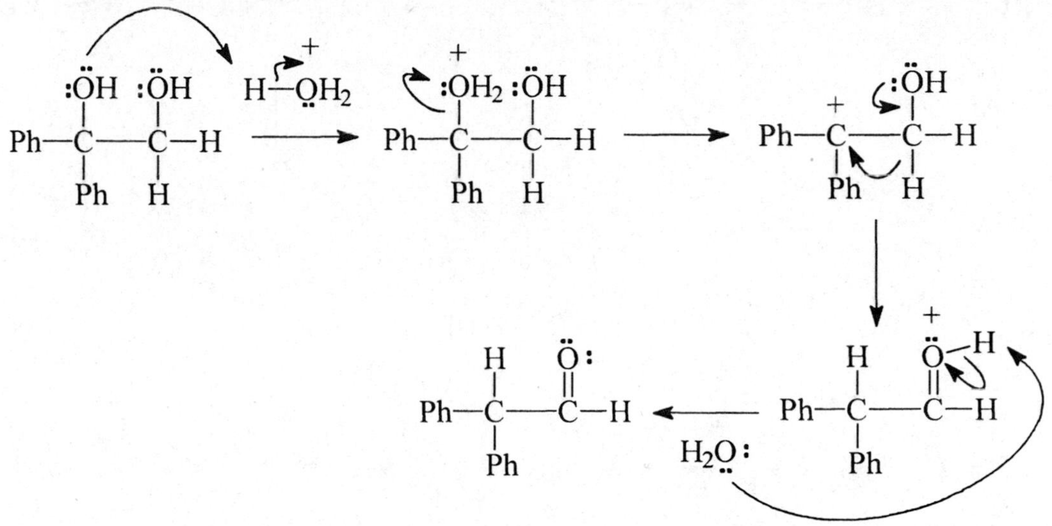

22.37

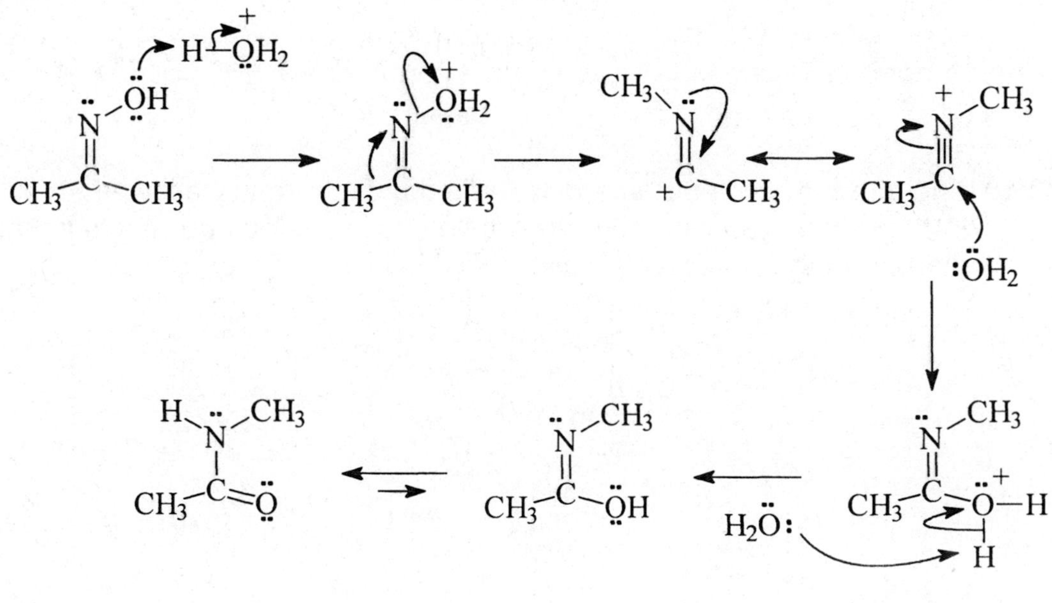

22.38

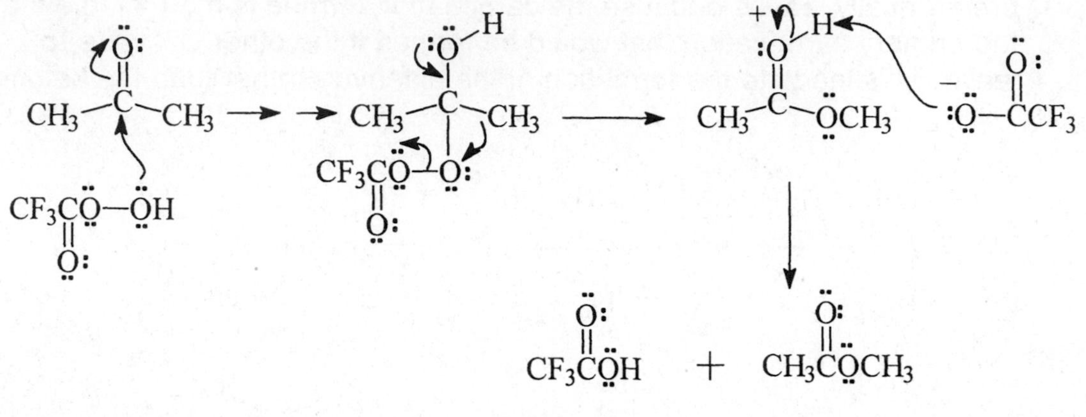

22.39

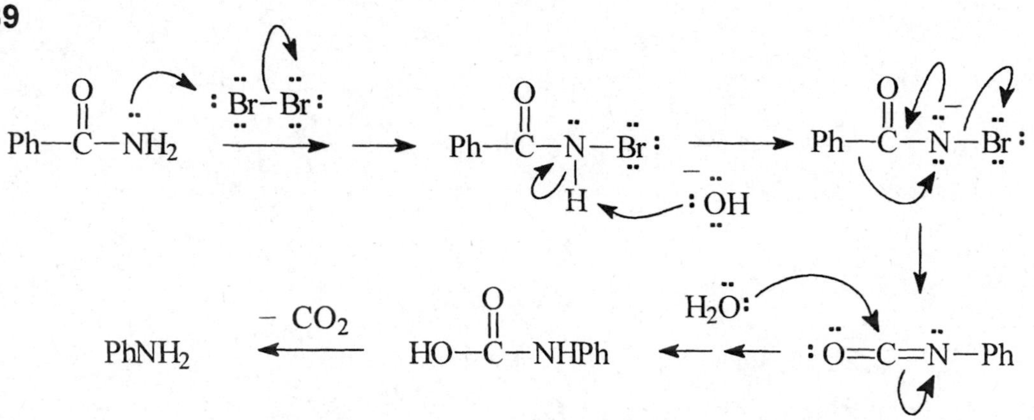

22.40 The dimer of cyclobutadiene is formed by a thermally allowed [4+2] cycloaddition reaction. One molecule of cyclobutadiene reacts as the diene and the other as the dienophile in a Diels-Alder reaction.

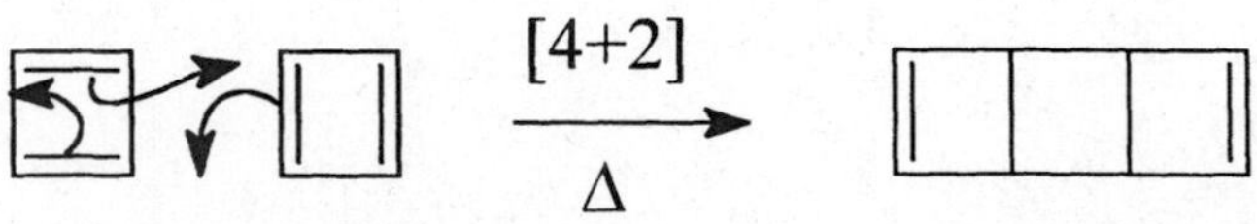

22.41 The first step of this reaction is a thermally allowed [4+2] cycloaddition (Diels-Alder reaction) to produce an intermediate bicyclic compound. This intermediate rapidly undergoes a thermally allowed reverse Diels-Alder reaction to produce the stable aromatic diester and carbon dioxide.

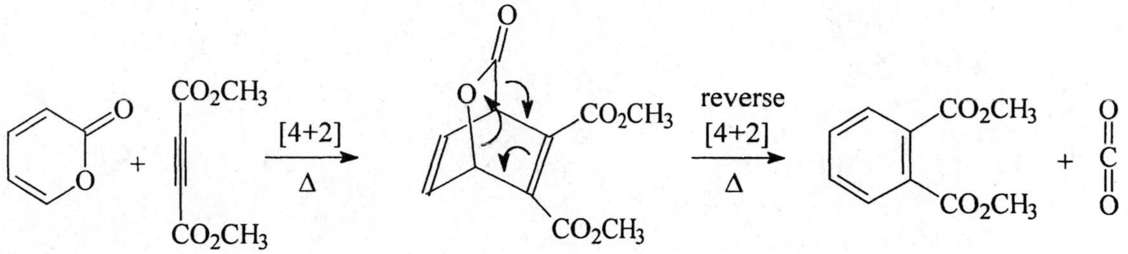

22.42 This reaction is related to the pinacol rearrangement.

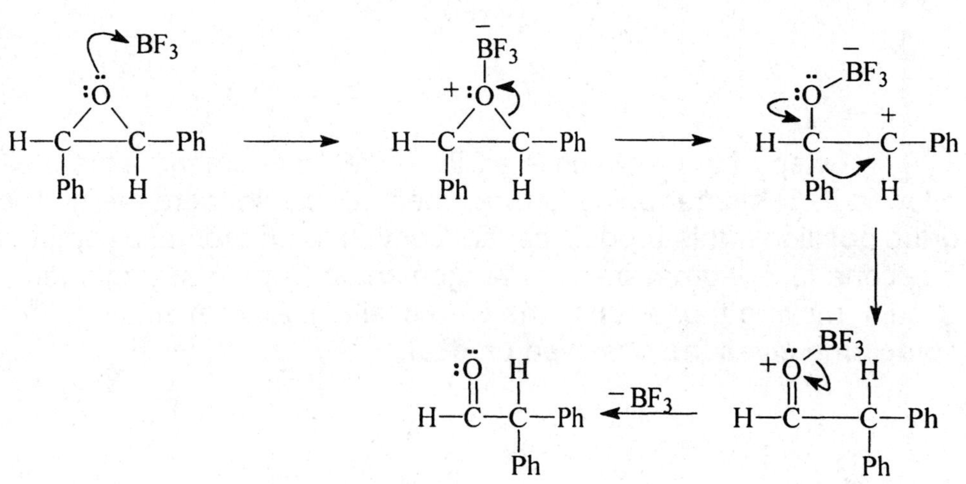

22.43

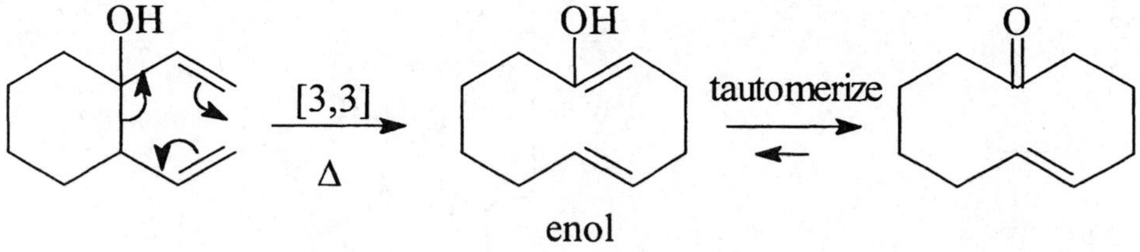

22.44 This is a thermally allowed [1,5] sigmatropic rearrangement involving the migration of a hydrogen atom. Because there is free rotation at the migration origin, there can be two possible transitions states producing two products. The stereochemistry at the new double bond and the stereochemistry at the new chirality center are interrelated.

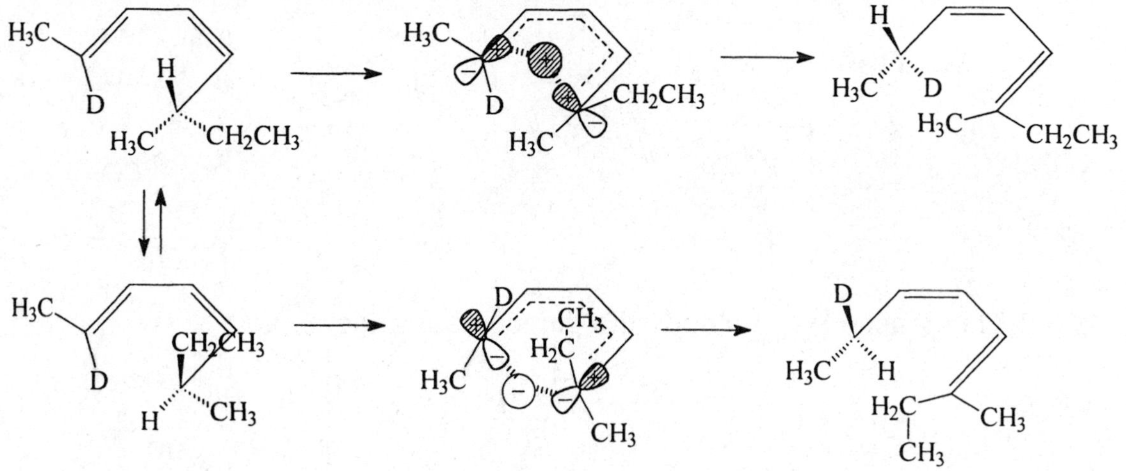

22.45 The first step in the reaction is a Claisen rearrangement, a thermally allowed [3,3] sigmatropic rearrangement. Because there are no H's on the ortho positions, this product cannot convert to an aromatic compound. So a second [3,3] sigmatropic rearrangement, a Cope rearrangement, occurs, moving the group to the para position. Aromatization of this compound gives the observed product.

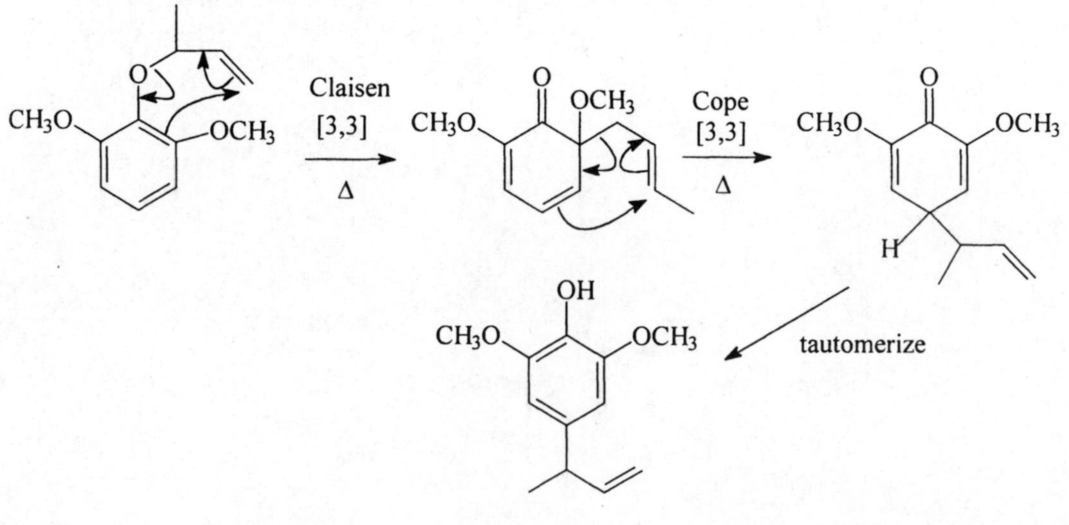

22.46

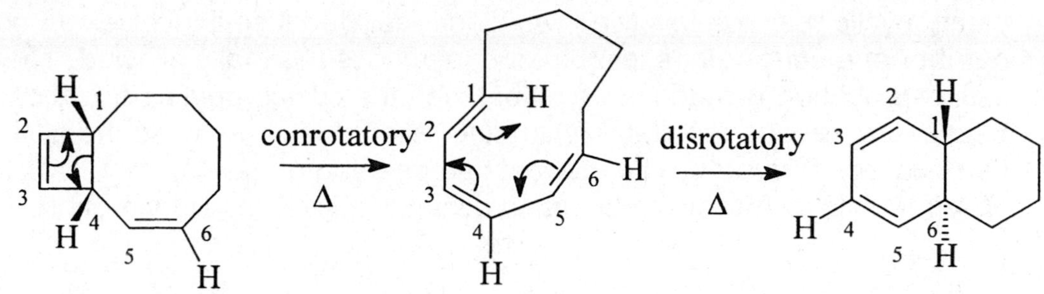

22.47 The reaction is very exothermic because it produces two very stable aromatic compounds. But it is a reverse [2+2] cycloaddition reaction and, as such, it is thermally forbidden, so the reactant is stable at room temperature.

22.48 The allowed thermal ring opening of a cyclobutene occurs with a conrotatory motion. The ring opening of the *cis*-isomer would produce a highly strained six-membered ring with a trans double bond. Therefore ring opening by a pericyclic process is disfavored for the *cis*-isomer. The formation of the observed product from the *cis*-isomer must be due to some other high energy reaction pathway, perhaps a nonconcerted reaction.

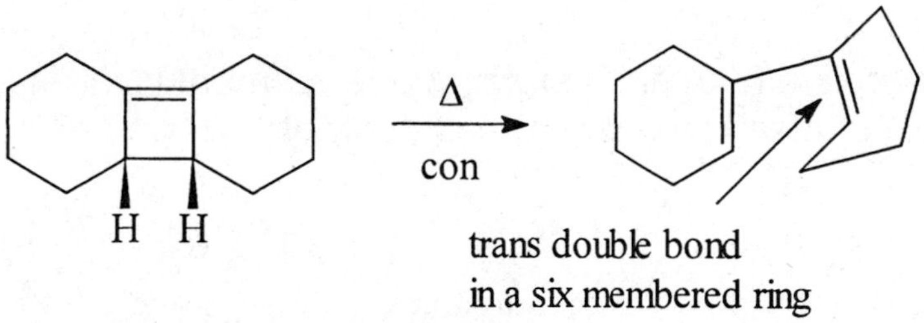

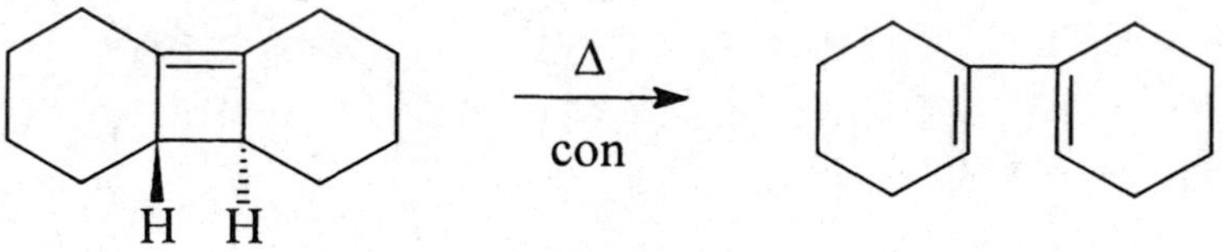

22.49 The opening of a cyclopropyl carbocation to form an allyl carbocation is an electrocyclic reaction involving two electrons. Because it involves an odd number of electron pairs, disrotatory opening is thermally allowed. The allyl carbocation is much more stable than the cyclopropyl carbocation because of resonance stabilization and relief of ring strain, so the allyl carbocation is favored. The thermal ring opening of the *cis*-dimethylcyclopropyl carbocation occurs with a disrotatory motion.

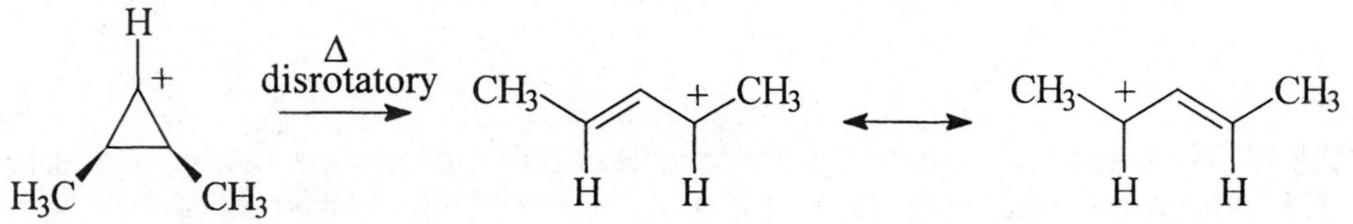

22.50 The dimer results from a [2+2] cycloaddition, which is photochemically allowed.

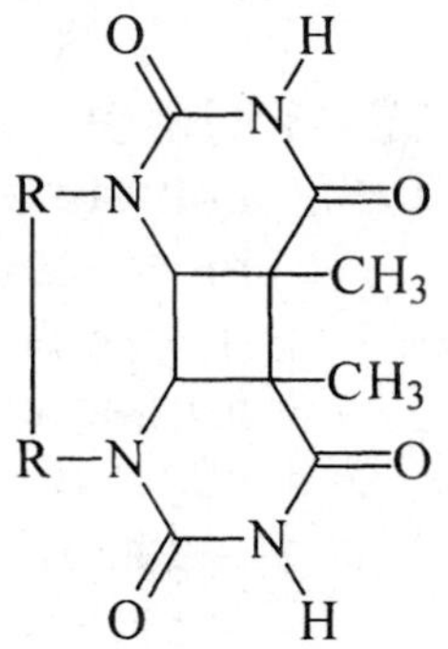

22.51 The reaction is a [3,3] sigmatropic rearrangement involving an odd number (3) of electron pairs, so it is thermally allowed.

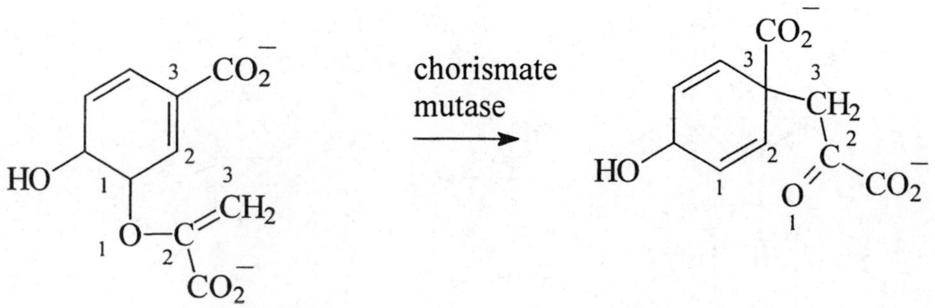

22.52 Benzyne is very reactive as a dienophile in the Diels-Alder reaction. The central ring of anthracene reacts like a diene because the adduct still has the resonance stabilization of two benzene rings. A very symmetrical product, with only four types of carbons, results.

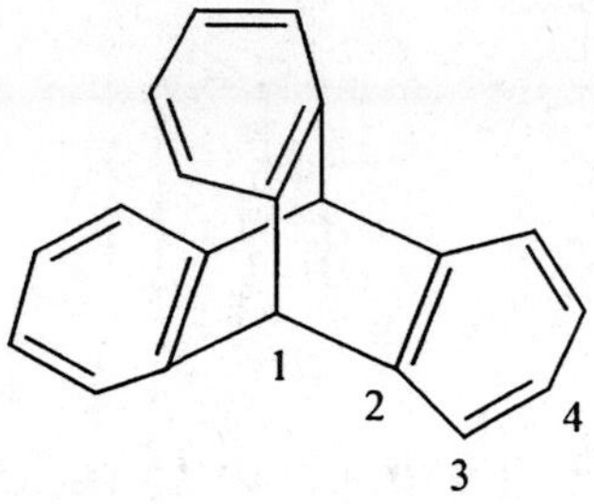

22.53

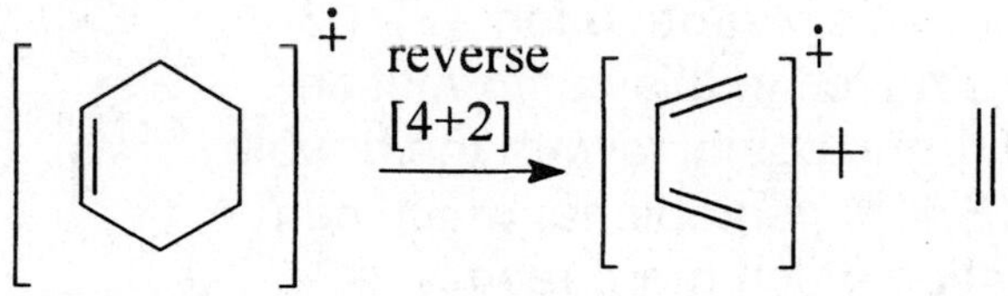

m/z 82   m/z 54

22.54

$$CH_3-\overset{\overset{\displaystyle O}{\|}}{C}-O\underset{\underset{\displaystyle CH_3}{|}}{CH}CH_3$$

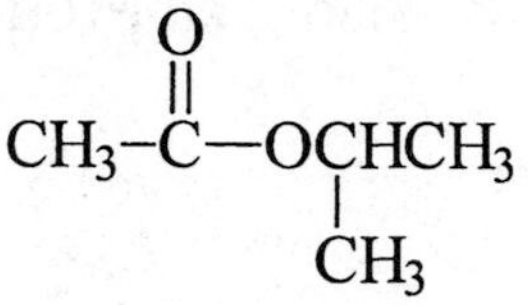

22.55 a) disrotation   b) conrotation   c) disrotation

22.56 a) This electrocyclic reaction involved two electron pairs and occurs by conrotation, so it is thermally allowed.
b) This electrocyclic reaction involves three electron pairs and occurs by conrotation, so it is photochemically allowed.

22.57 The octatriene has three electron pairs. When heated, disrotation occurs to give Product A, with the methyl groups cis. Upon irradiation with ultraviolet light, conrotation occurs to give Product B, with the methyl groups trans.

22.58

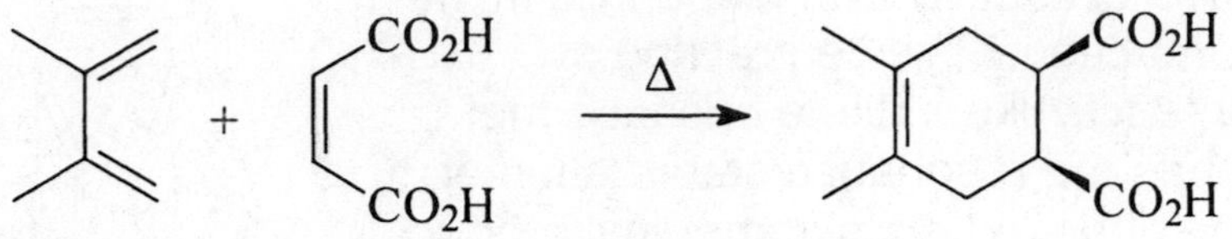

22.59 In order to react in the Diels-Alder reaction, the diene must be in the *s*-cis conformation. (*Z*)-1,3-Pentadiene is very strained in this conformation because its methyl group points toward the double bond. As a result, (*E*)-1,3-pentadiene, which can attain this conformation much more readily, is more reactive.

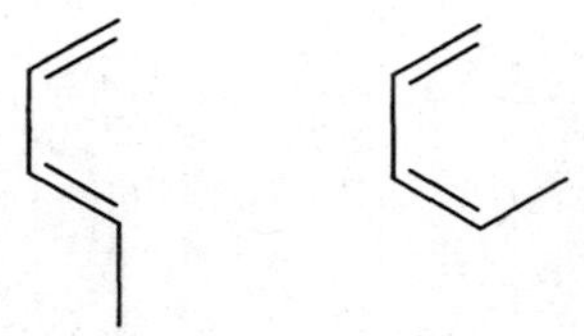

E *Z*

22.60 The Diels-Alder reaction occurs with endo addition, so Product B is formed.

Review of Mastery Goals

After completing this chapter, you should be able to:

Show the energies and nodal properties of the pi MOs of a small conjugated system, whether it is composed of an even or odd number of orbitals.
(Problems 22.2, 22.3, and 22.25)

Classify reactions as electrocyclic reactions, [*x*+*y*] cycloadditions, or [*i,j*] sigmatropic rearrangements.
(Problems 22.17 and 22.25)

Use the pi MOs to explain whether these reactions are allowed or forbidden.
(Problems 22.4, 22.5, 22.6, 22.10, 22.11, 22.18, 22.19, and 20.29)

Show the products, including stereochemistry, of any of these reactions.
(Problems 22.9, 22.14, 22.16, 22.21, 22.23, 22.26 and 22.27)

Show the mechanisms of the pinacol, Beckmann, Hofmann, and Baeyer-Villiger rearrangements.
(Problems 22.24, 22.36, 22.37, 22.38, and 22.39)

Use the principles of pericyclic reactions to explain the mechanism, selectivity, and stereochemistry of a concerted reaction.
(Problems 22.7, 22.8, 22.12, 22.15, 22.20, 22.22, 22.36, 22.40, 22.41, 22.42, 22.43, 22.44, 22.45, 22.46, 22.47, 22.48, and 22.49)

Use these reactions in synthesis.
(Problems 22.28, 22.31, 22.32, and 22.34)

Chapter 23
THE SYNTHESIS OF ORGANIC COMPOUNDS

23.1

a) 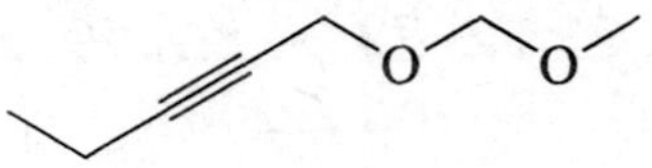

b) $PhCH_3$ + 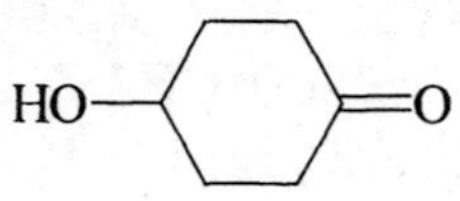

c) 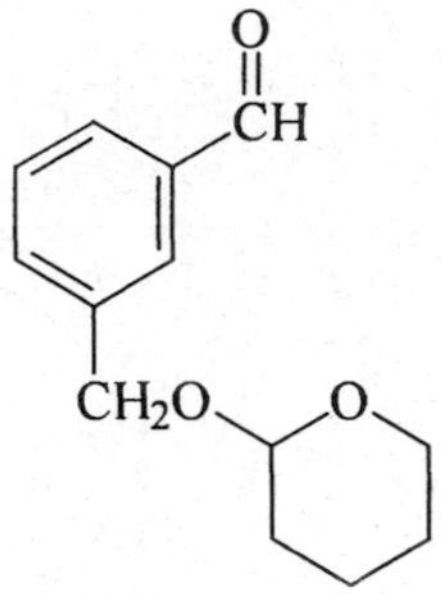

d) 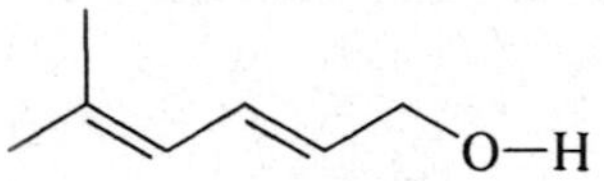

e) 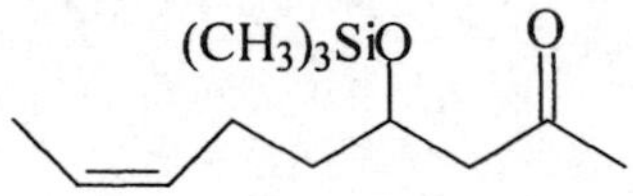

f) 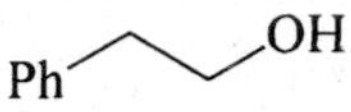

g) 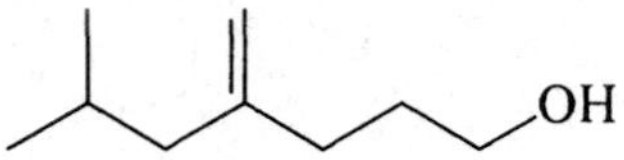

23.2

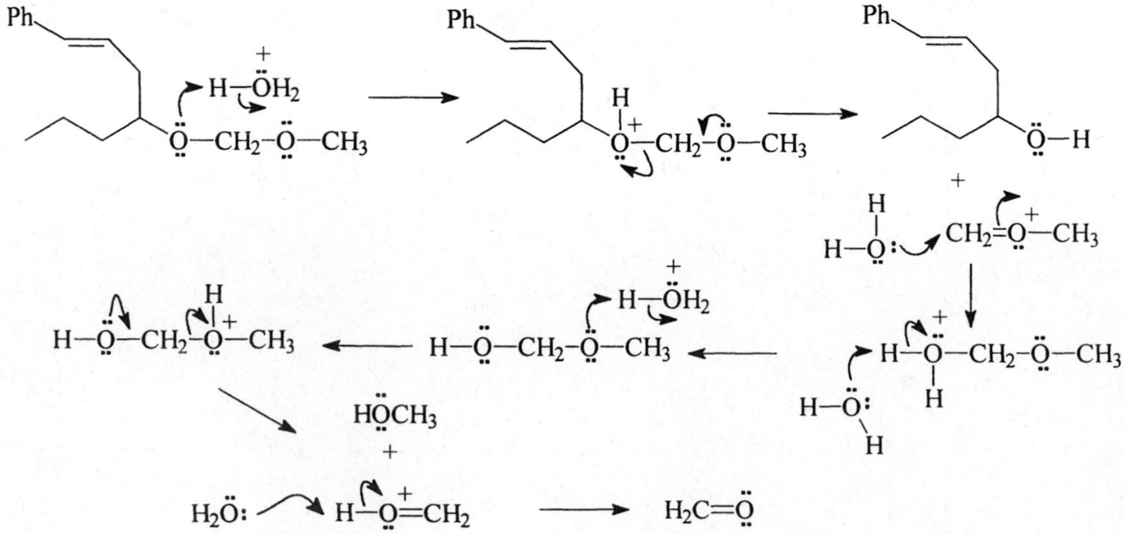

23.3

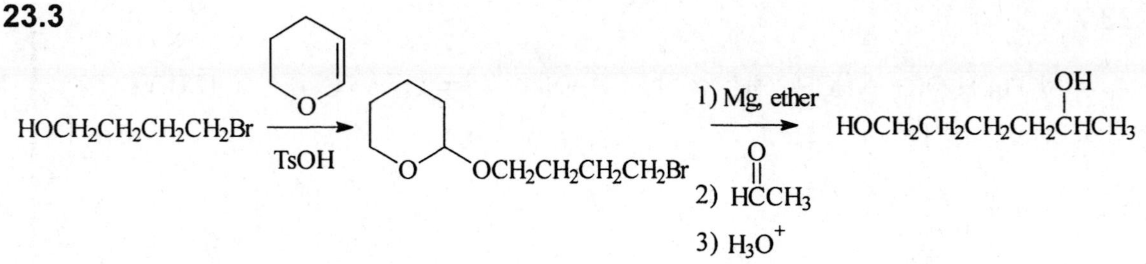

23.4

a) 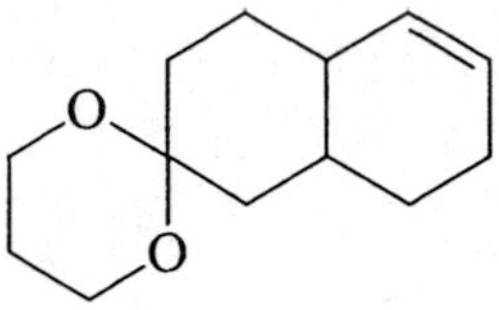

b) 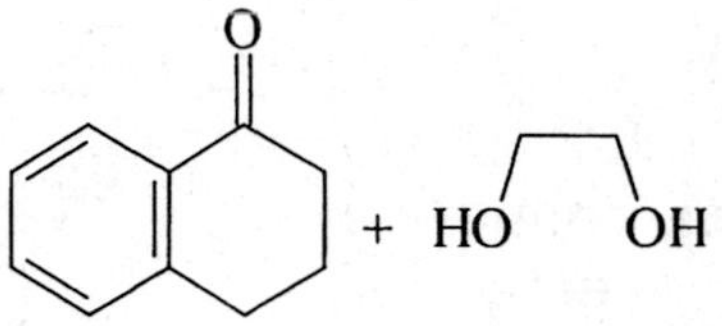

23.5

a) 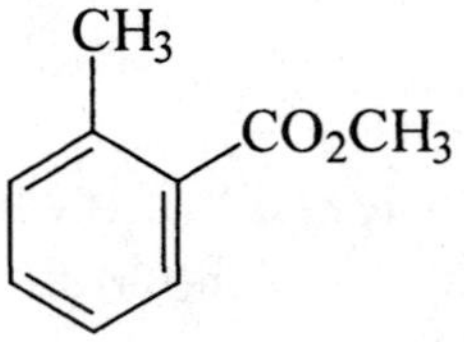

b) 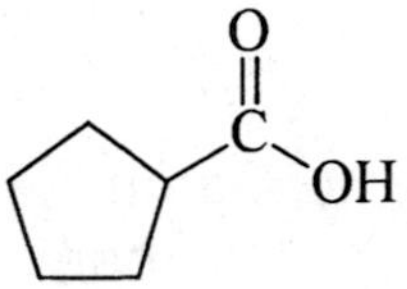

c) 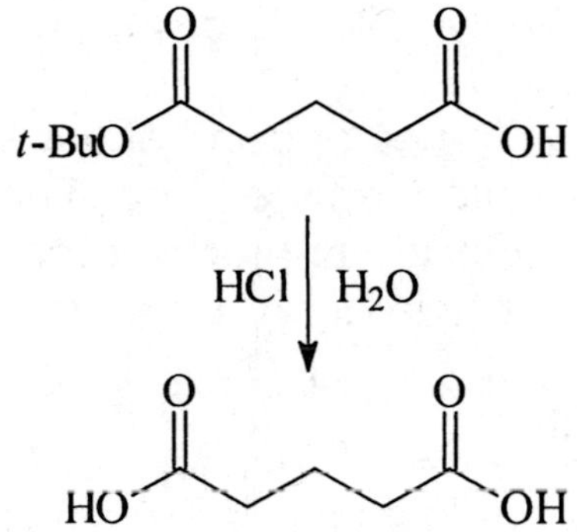

d) 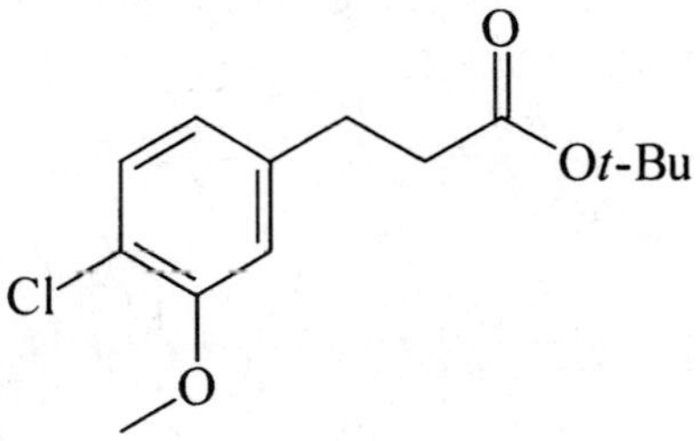

23.6

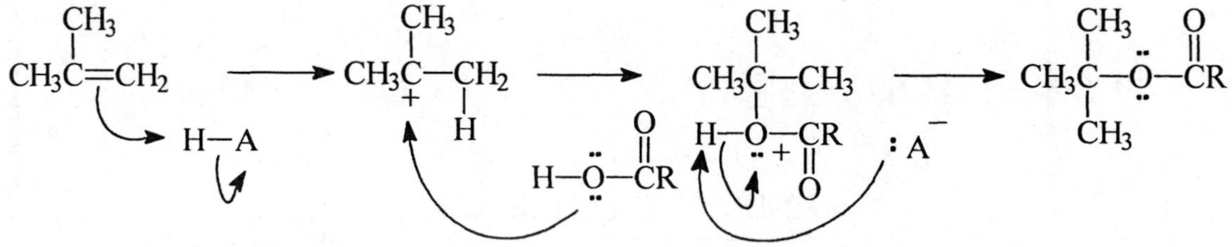

23.7

a)
$$PhCH_2O\overset{O}{\overset{\|}{C}}-NHCH\overset{O}{\overset{\|}{C}}OH$$
with CH_3 on the CH

b)
$$H-N\text{(cyclic)}=O + CO_2 + (CH_3)_3COH$$

c) $PhCH_3 + NH_2CH_2CH_2Ph + CO_2$

d)
$$(CH_3)_3CO\overset{O}{\overset{\|}{C}}NHCH_2CO_2H$$

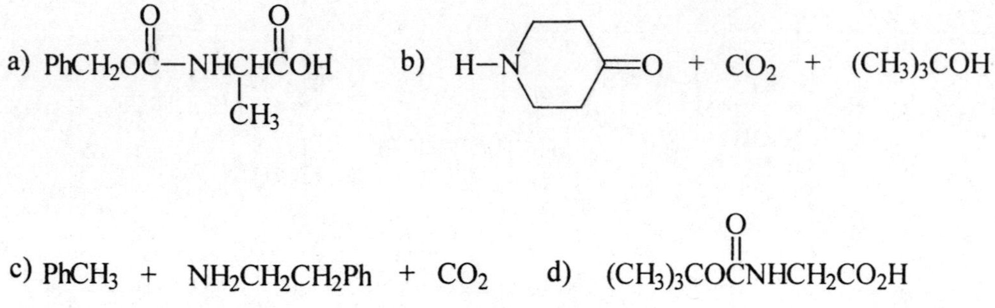

23.8 The amide cannot be prepared directly because the unprotected amino group will react with the acyl chloride group.

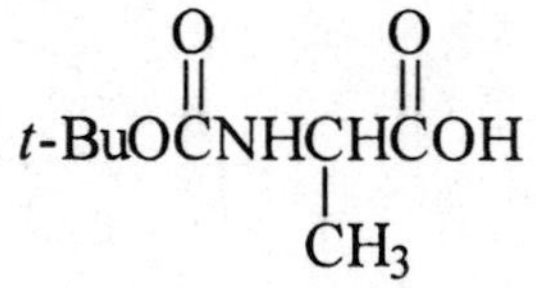

$$t\text{-BuO}\overset{O}{\overset{\|}{C}}NHCH\overset{O}{\overset{\|}{C}}OH$$
CH_3

A

$$t\text{-BuO}\overset{O}{\overset{\|}{C}}NHCH\overset{O}{\overset{\|}{C}}N(CH_3)_2$$
CH_3

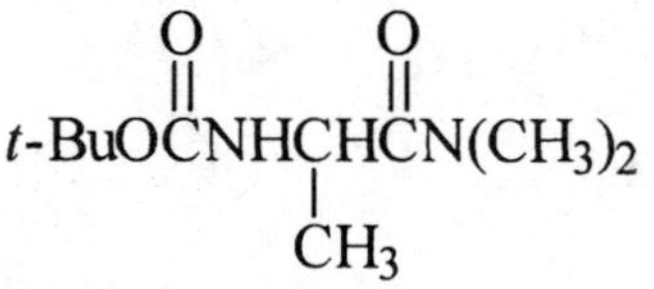

B

23.9 The ester bond of the carbamate group is hydrolyzed more rapidly than either amide bond. The resulting carbamic acid then eliminates carbon dioxide to produce the amine. Conditions drastic enough to hydrolyze an amide bond are never employed in this process.

23.10

a)
$$CH_3CH_2CH_2 \overset{OH}{\underset{CH_3}{\overset{|}{{-}CHCH_2CHCH_3}}} \Longrightarrow CH_3CH_2CH_2 \quad {}^+\overset{OH}{\underset{CH_3}{\overset{|}{CHCH_2CHCH_3}}}$$

1) $CH_3CH_2CH_2MgBr$ 2) H_3O^+

$$H\overset{O}{\overset{\|}{C}}CH_2\underset{CH_3}{\overset{|}{CHCH_3}}$$

b)
$$PhC{\equiv}C{-}CH_2CH_2CH_2CH_3 \Longrightarrow PhC{\equiv}\overset{-}{C} \quad {}^+CH_2CH_2CH_2CH_3$$

1) $NaNH_2$ 2) $CH_3CH_2CH_2CH_2Br$

$$PhC{\equiv}C{-}H$$

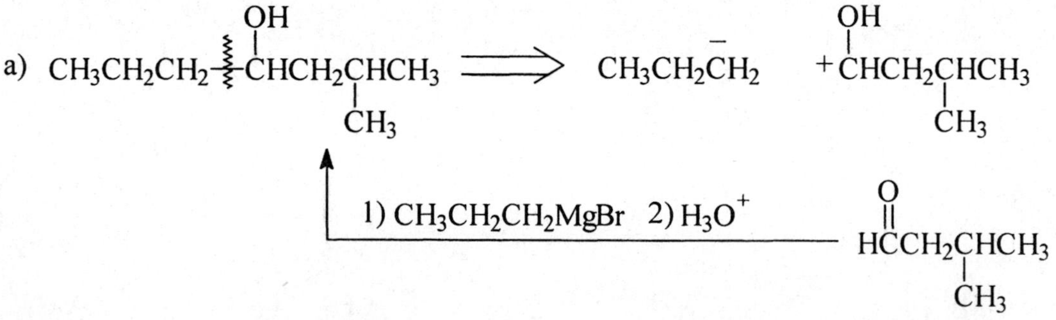

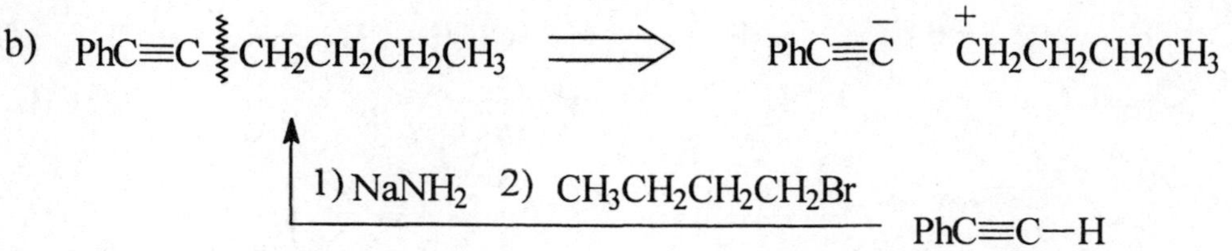

c)

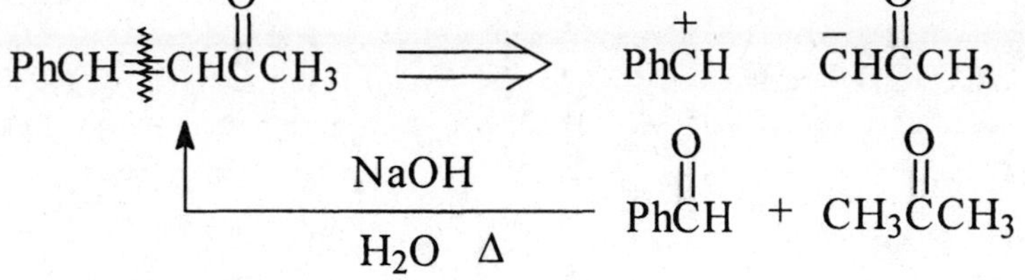
PhCH—CHCCH₃ (O)
PhCH (+)
CHCCH₃ (−, O)
NaOH
H₂O Δ
PhCH (O) + CH₃CCH₃ (O)

23.11

a)

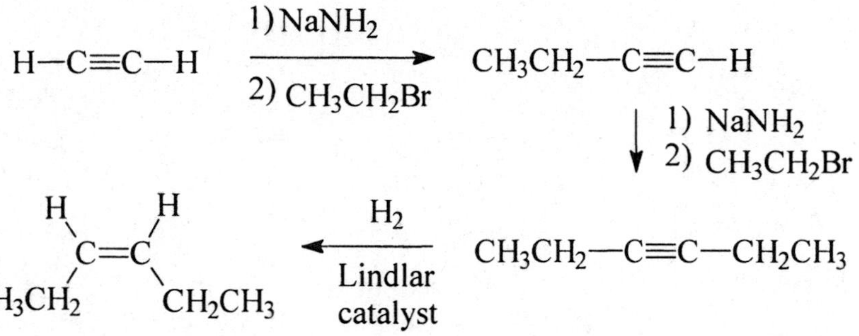
H—C≡C—H
1) NaNH₂
2) CH₃CH₂Br
CH₃CH₂—C≡C—H
1) NaNH₂
2) CH₃CH₂Br
CH₃CH₂—C≡C—CH₂CH₃
H₂
Lindlar
catalyst
H H
C=C
CH₃CH₂ CH₂CH₃

b)

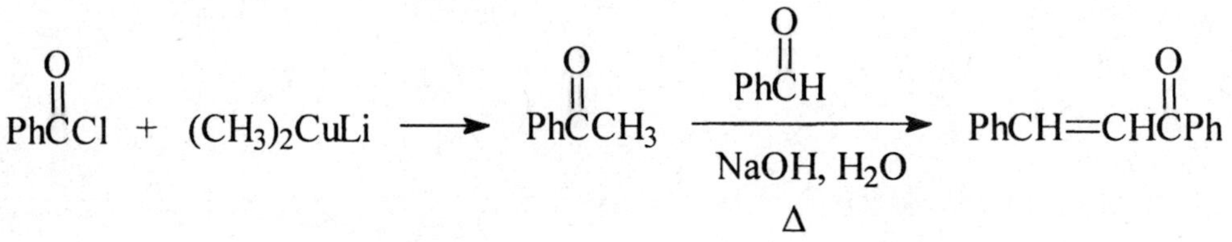
PhCCl (O) + (CH₃)₂CuLi
PhCCH₃ (O)
PhCH (O)
NaOH, H₂O
Δ
PhCH=CHCPh (O)

c)

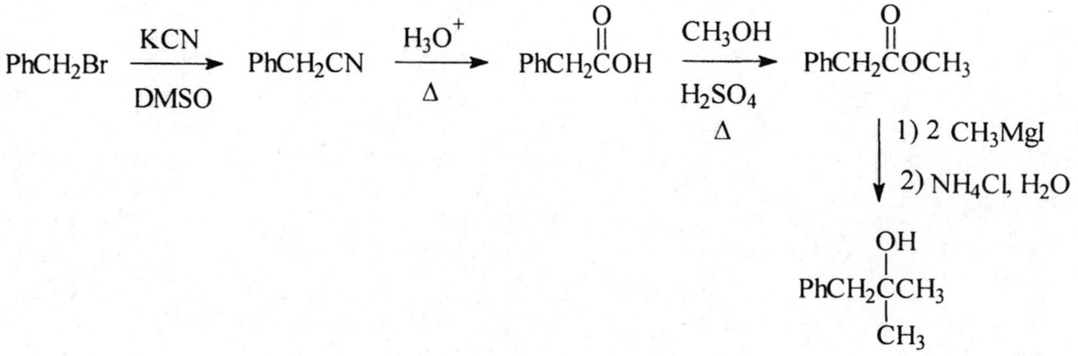
PhCH₂Br
KCN
DMSO
PhCH₂CN
H₃O⁺
Δ
PhCH₂COH (O)
CH₃OH
H₂SO₄
Δ
PhCH₂COCH₃ (O)
1) 2 CH₃MgI
2) NH₄Cl, H₂O
OH
PhCH₂CCH₃
CH₃

d)

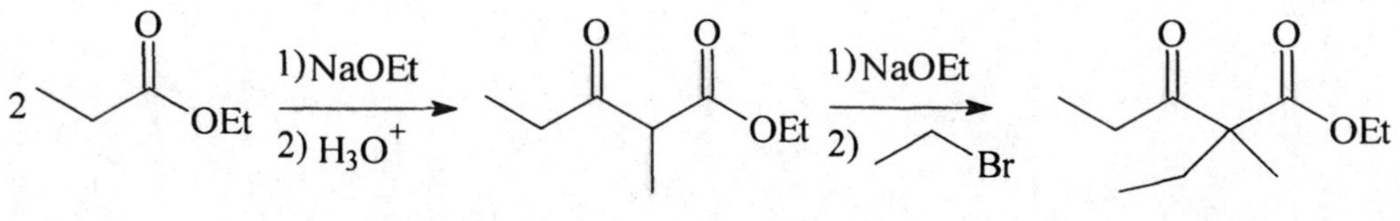
2
O
OEt
1) NaOEt
2) H₃O⁺
O O
OEt
1) NaOEt
2)
Br
O O
OEt

e)

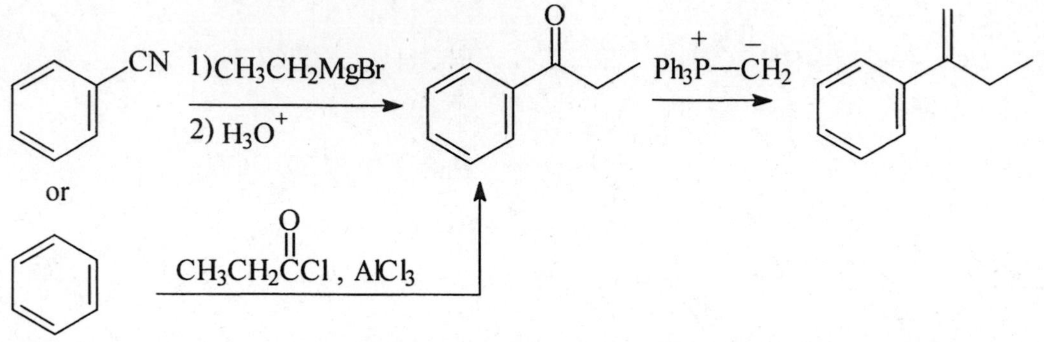

CN
1)CH3CH2MgBr
2) H3O+
O
+ −
Ph3P—CH2
or
O
CH3CH2CCl , AlCl3

f)

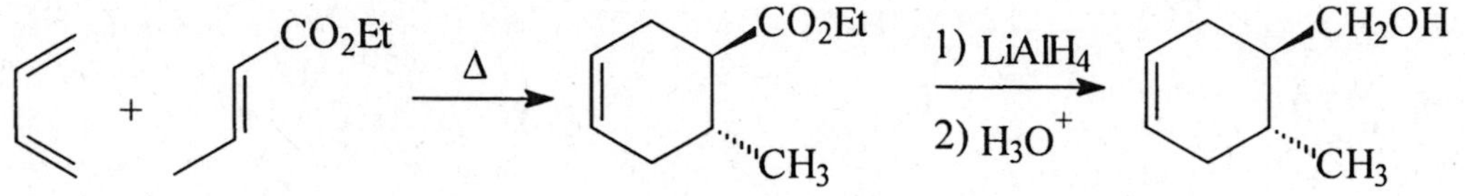

+
CO2Et
Δ
CO2Et
CH3
1) LiAlH4
2) H3O+
CH2OH
CH3

g)

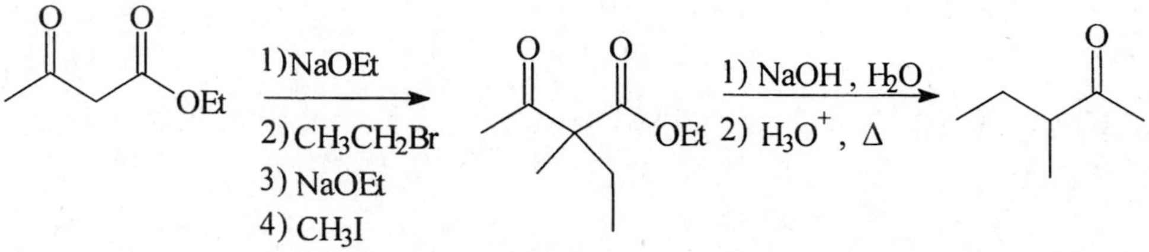

O O
OEt
1)NaOEt
2) CH3CH2Br
3) NaOEt
4) CH3I
O O
OEt
1) NaOH , H2O
2) H3O+ , Δ
O

h)

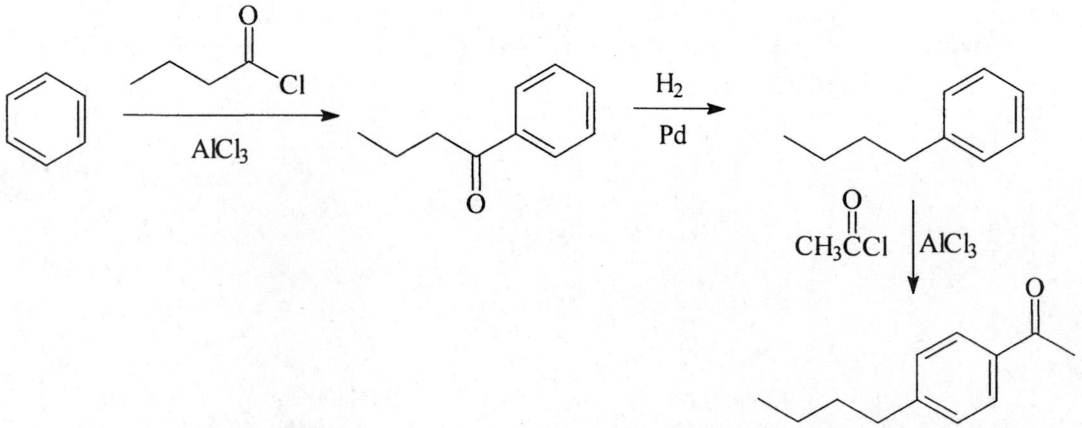

O
Cl
AlCl3
O
H2
Pd
O
CH3CCl
AlCl3
O

23.12

a) $CH_3OH \xrightarrow{HI} CH_3I$

$CH_3CH_2OH \xrightarrow{Na} CH_3CH_2ONa$

$CH_3I + CH_3CH_2ONa \longrightarrow CH_3OCH_2CH_3$

b) $CH_3CH_2OH \xrightarrow[H_2SO_4]{Na_2Cr_2O_7} CH_3CO_2H \xrightarrow[H_2SO_4]{CH_3OH} CH_3CO_2CH_3$

c) $CH_3CH_2OH \xrightarrow{PCC} CH_3\overset{O}{\overset{\|}{C}}H \xrightarrow[\Delta]{NaOH} CH_3CH{=}CH\overset{O}{\overset{\|}{C}}H \xrightarrow[Pt]{H_2} CH_3CH_2CH_2\overset{O}{\overset{\|}{C}}H$

d)

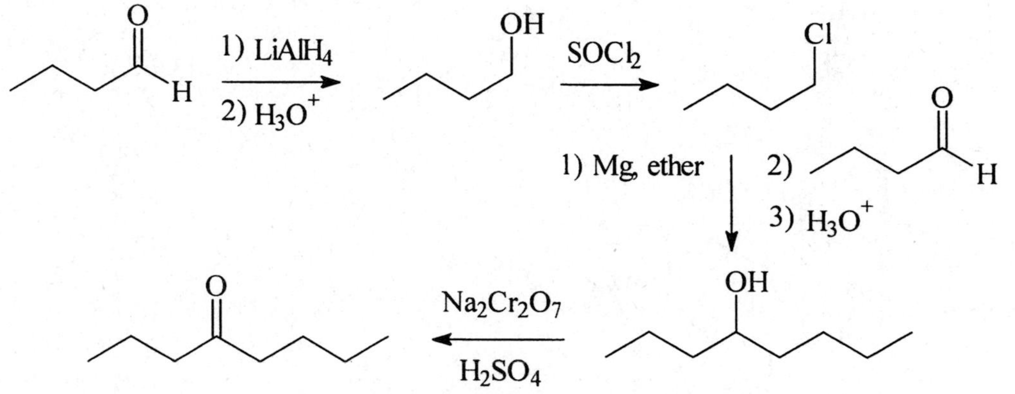

e)

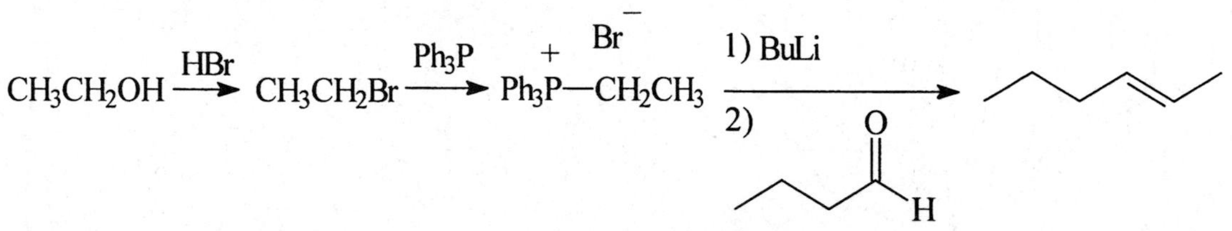

f)

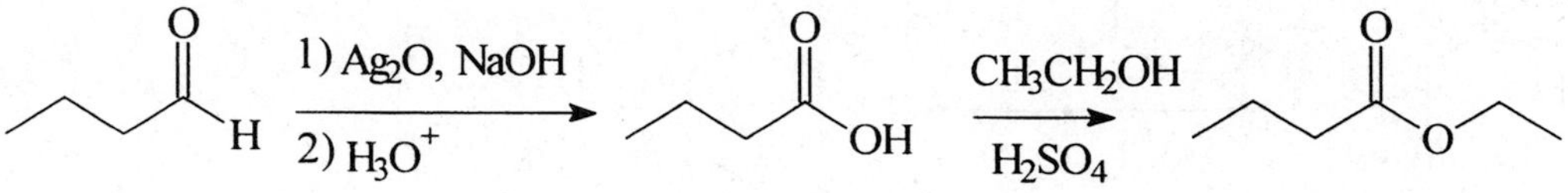

g) 2

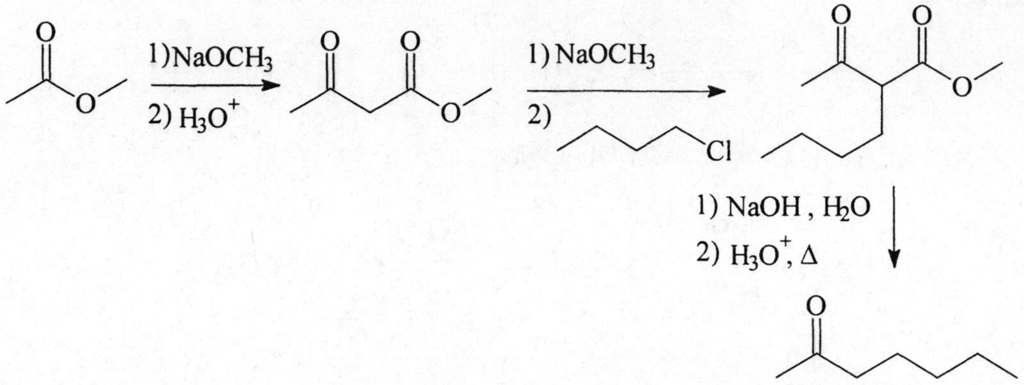
1)NaOCH3
2) H3O+
1) NaOCH3
2)
Cl
1) NaOH , H2O
2) H3O+, Δ

h)

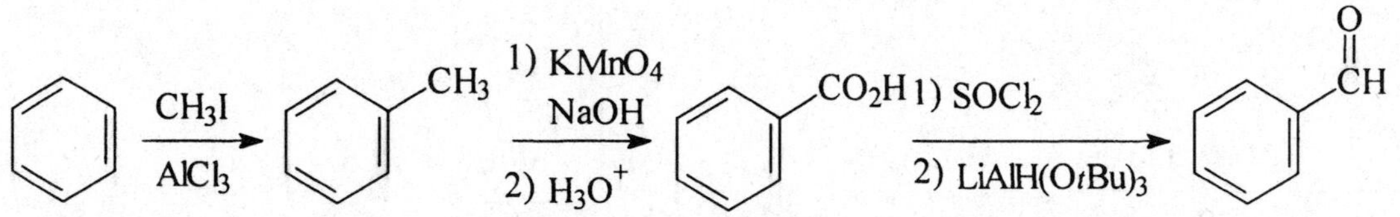
CH3I
AlCl3
CH3
1) KMnO4
NaOH
2) H3O+
CO2H
1) SOCl2
2) LiAlH(OtBu)3
CH

i)

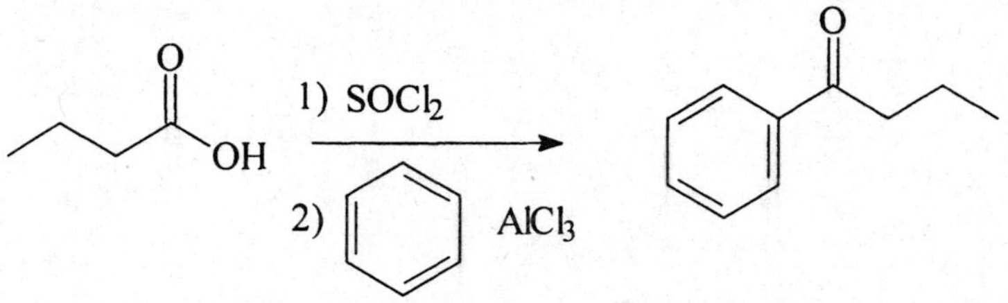
OH
1) SOCl2
2)
AlCl3

j)

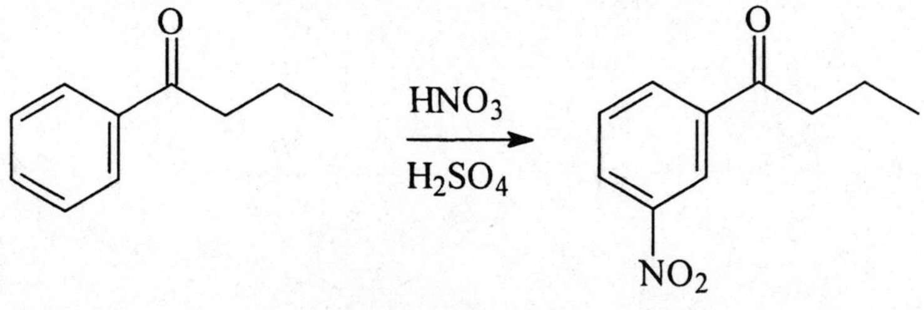
HNO3
H2SO4
NO2

k)

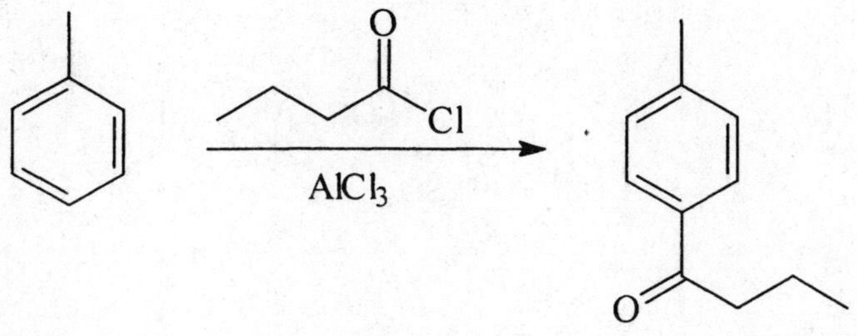
Cl
AlCl3

l) 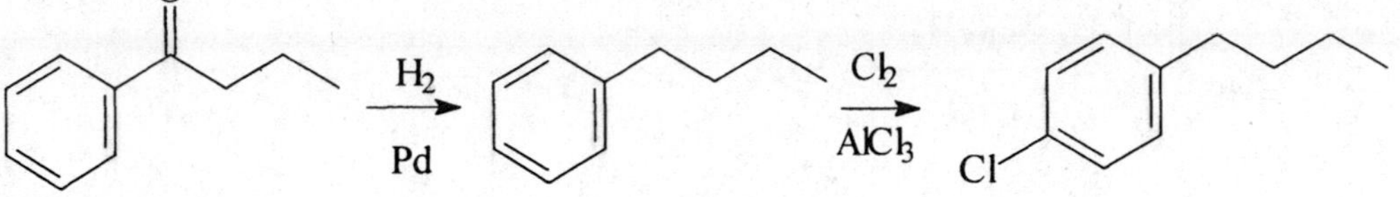

O
H2
Pd
Cl2
AlCl3
Cl

m) 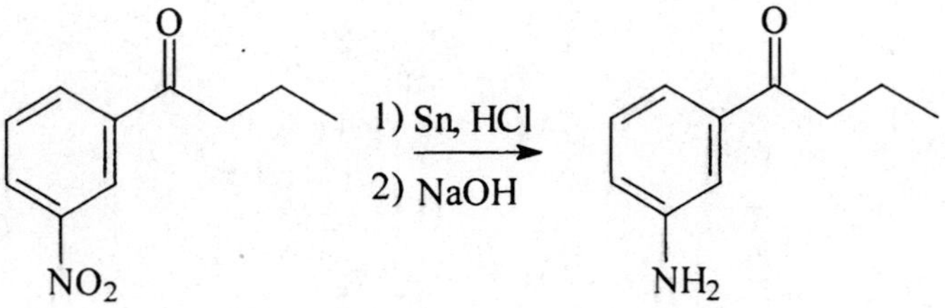

O
NO2
1) Sn, HCl
2) NaOH
O
NH2

n) 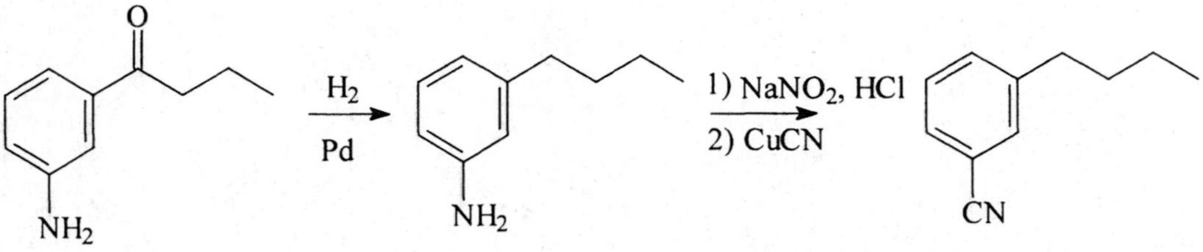

O
NH2
H2
Pd
NH2
1) NaNO2, HCl
2) CuCN
CN

o) 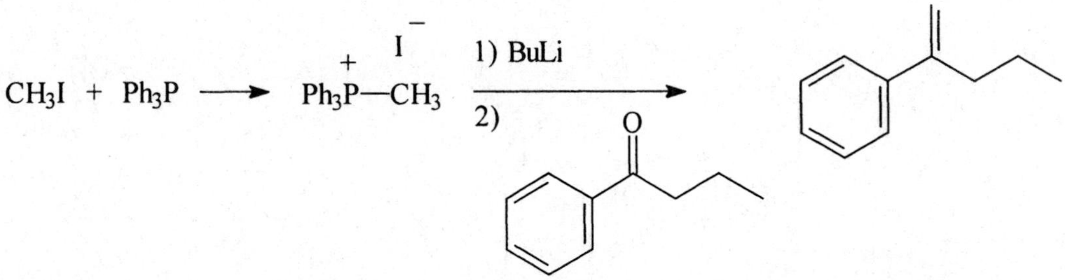

CH3I + Ph3P
+
I
Ph3P—CH3
1) BuLi
2)
O

p) see (g)

q) 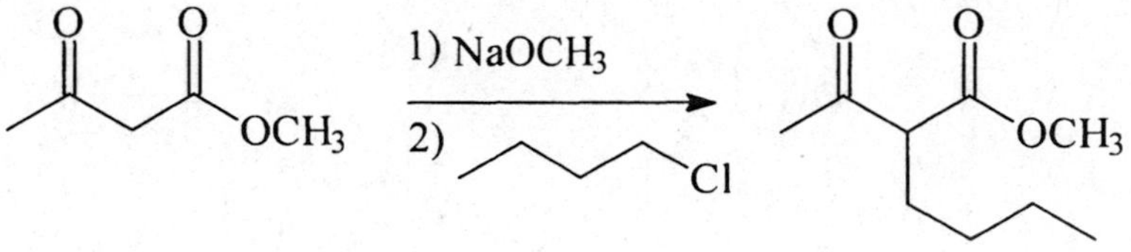

O
O
OCH3
1) NaOCH3
2)
Cl
O
O
OCH3

r) 2 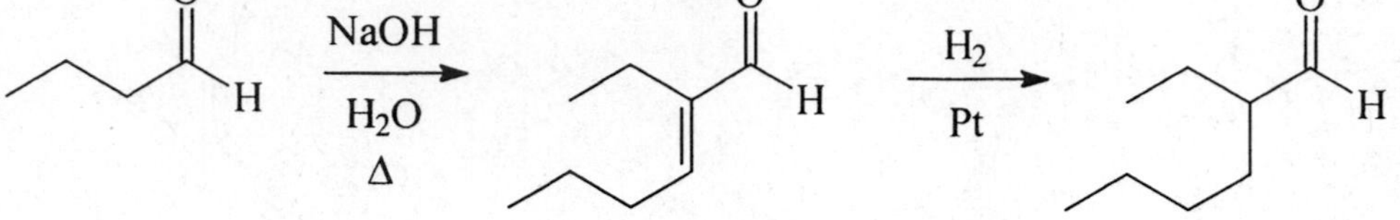
O
H
NaOH
H_2O
Δ
O
H
H_2
Pt
O
H

s) 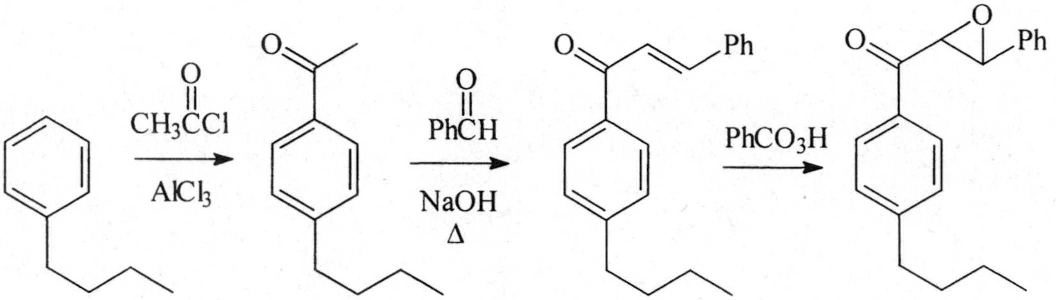
O
$CH_3\overset{||}{C}Cl$
$AlCl_3$
O
O
$Ph\overset{||}{C}H$
NaOH
Δ
O
Ph
$PhCO_3H$
O
O
Ph

t) 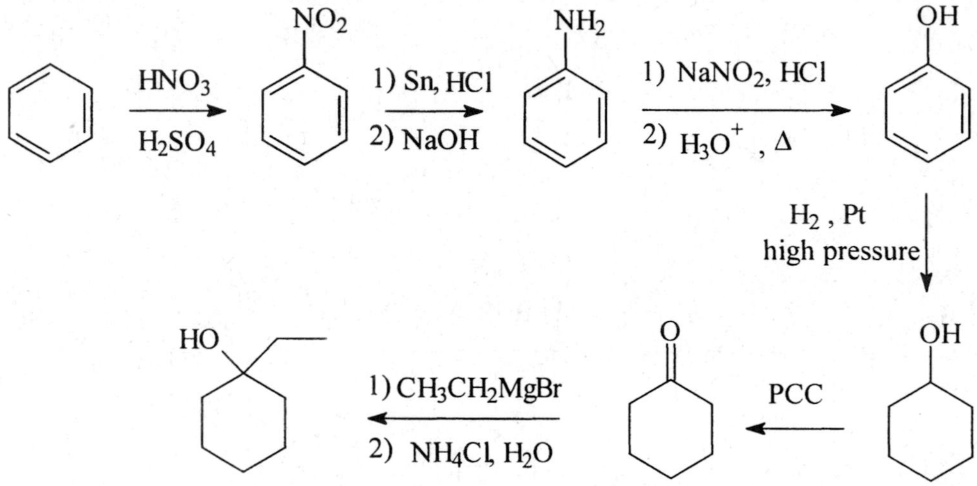
HNO_3
H_2SO_4
NO_2
1) Sn, HCl
2) NaOH
NH_2
1) $NaNO_2$, HCl
2) H_3O^+ , Δ
OH
H_2 , Pt
high pressure
OH
PCC
O
1) CH_3CH_2MgBr
2) NH_4Cl, H_2O
HO

23.13

a) 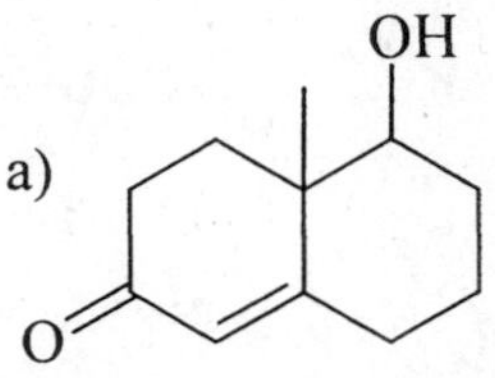
OH
O

b) 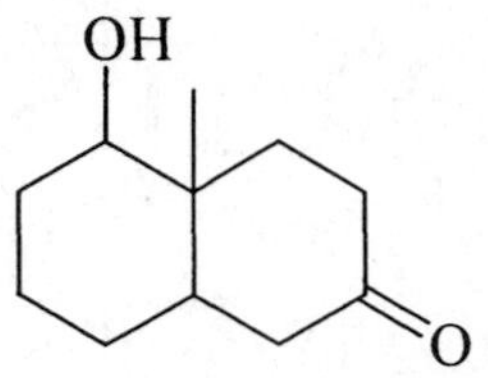
OH
O

c) 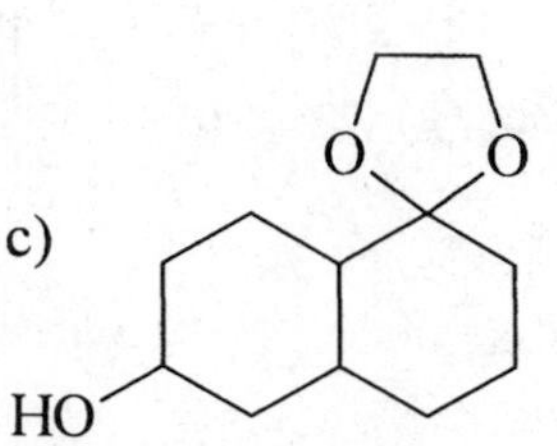
O
O
HO

d) 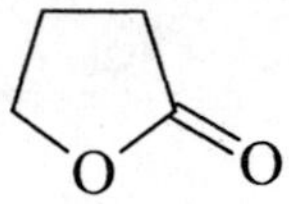 e) 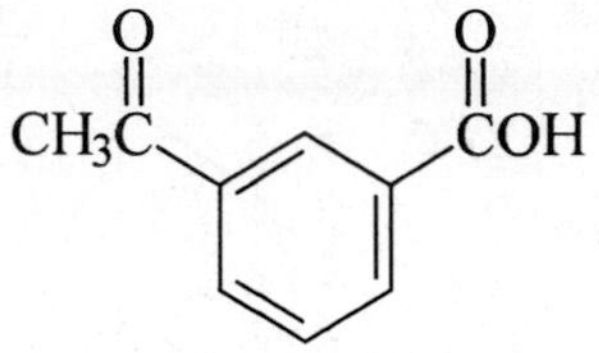 f) 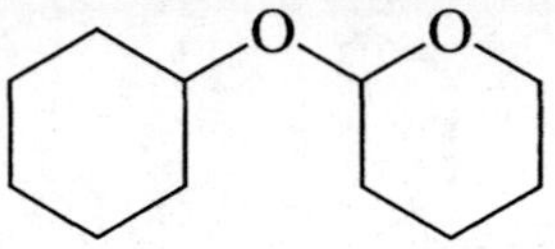

g) 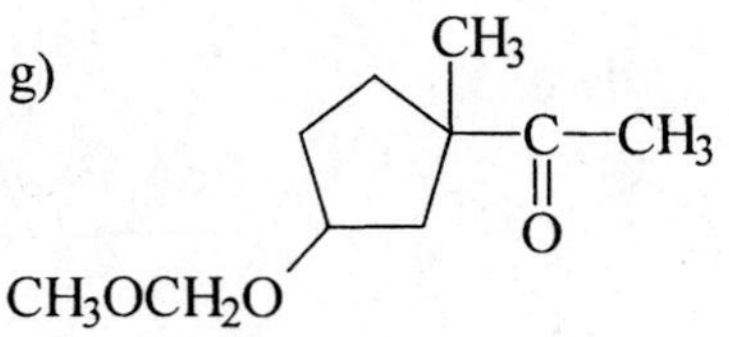 h) 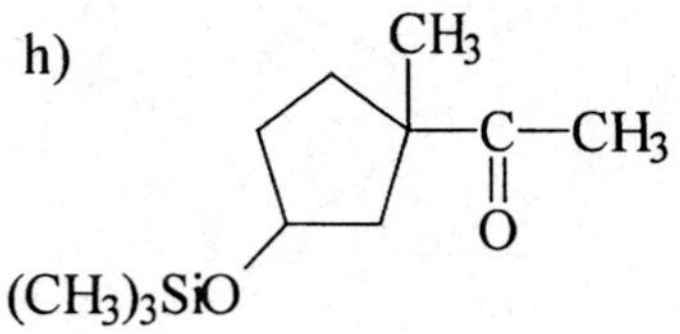 i) 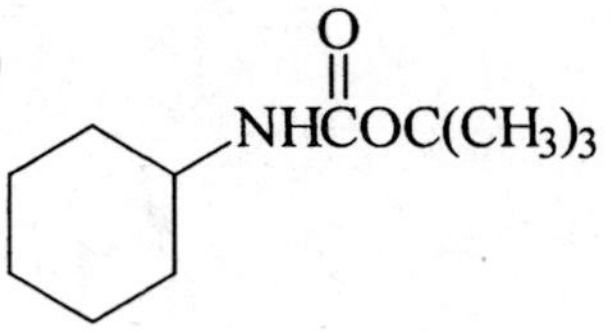

j) 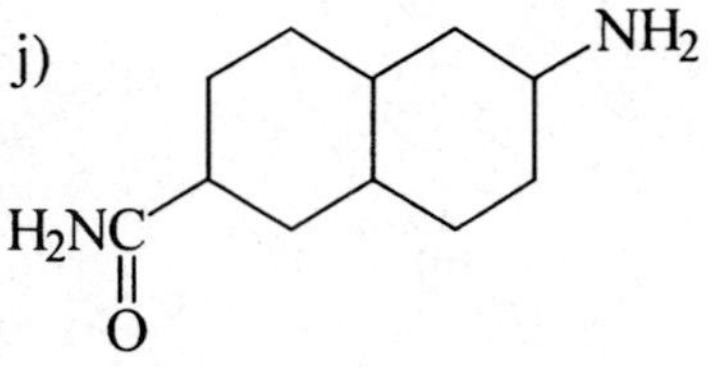 k) 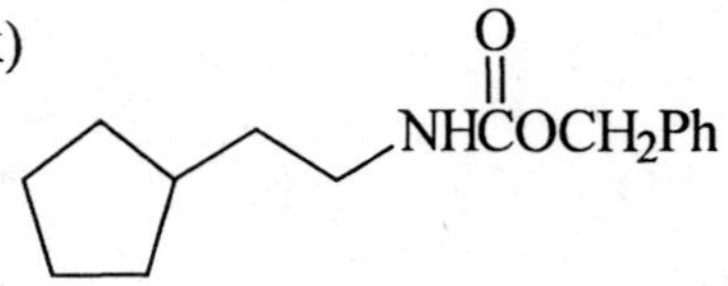 l) 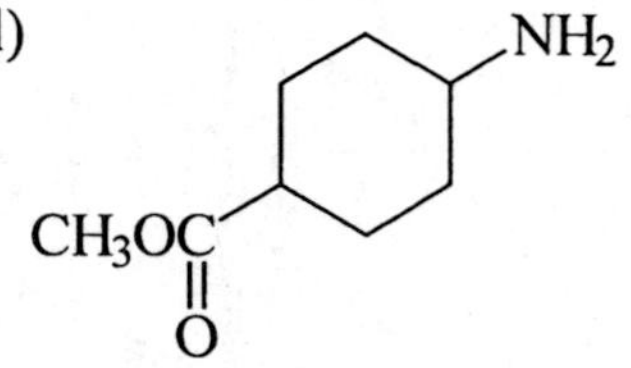

21.14

a) 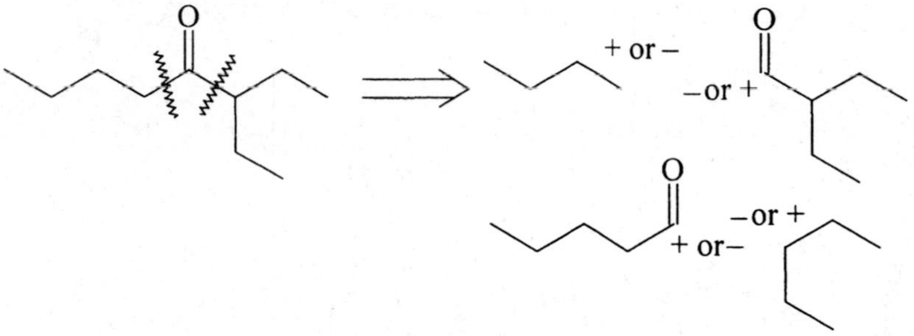

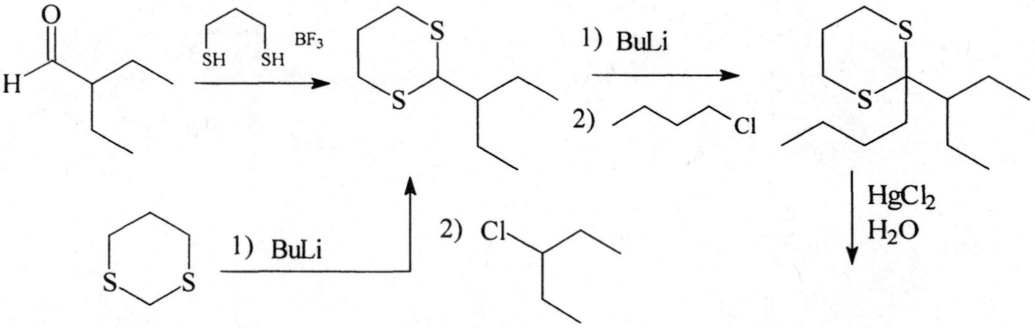

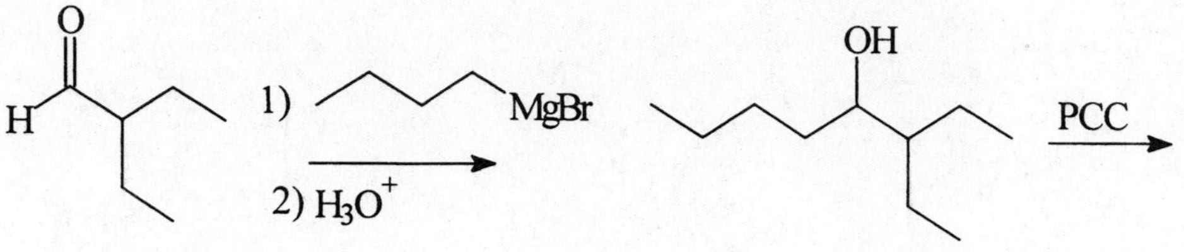
O
H
1)
MgBr
2) H3O+
OH
PCC

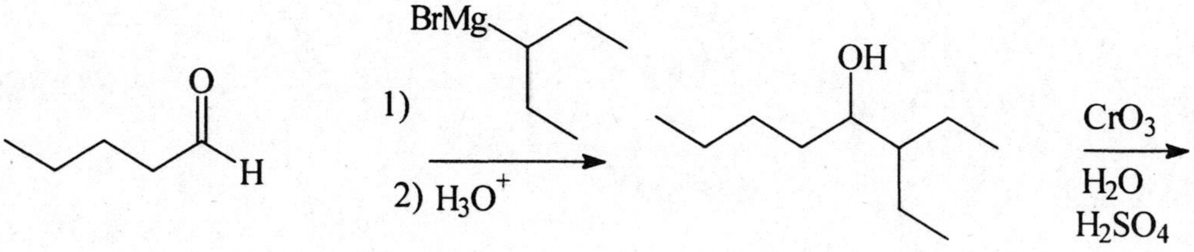
BrMg
O
H
1)
2) H3O+
OH
CrO3
H2O
H2SO4

b)

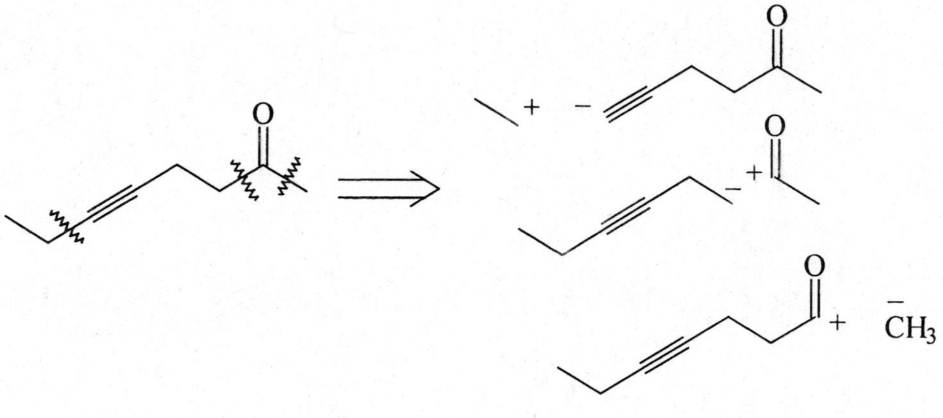
O
+
−
CH3

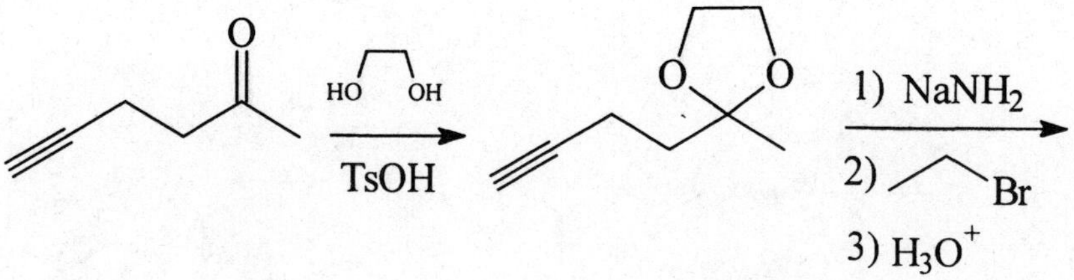
O
HO
OH
TsOH
1) NaNH2
2)
Br
3) H3O+

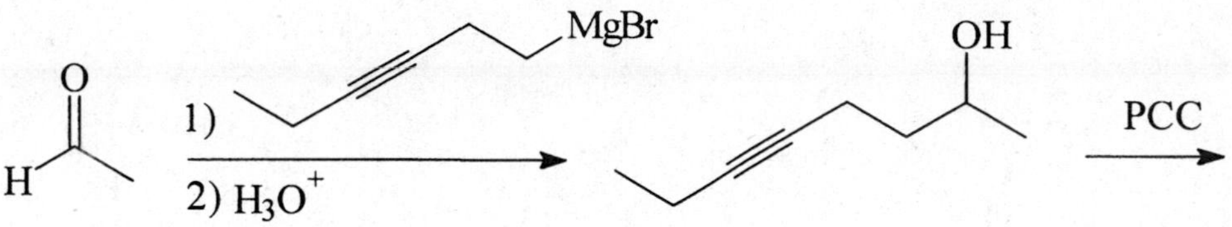
MgBr
1)
2) H3O+
OH
PCC

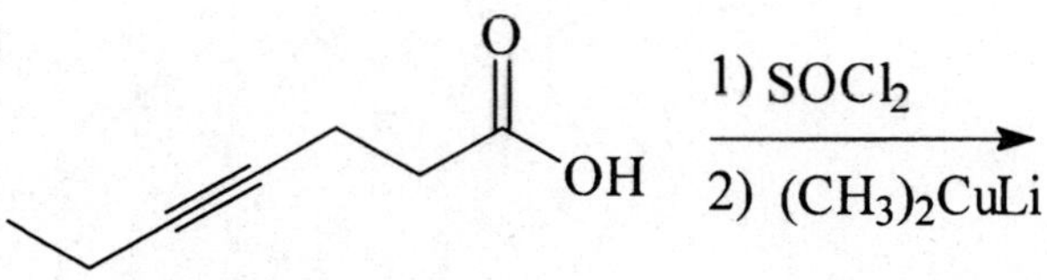
OH
1) SOCl2
2) (CH3)2CuLi

c)

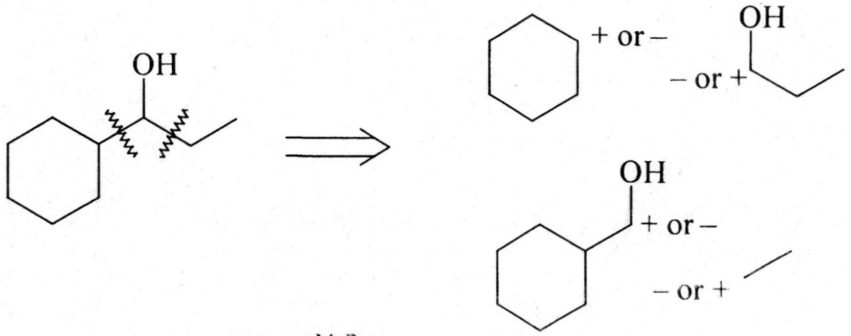
OH
+ or –
– or +
OH
+ or –
– or +

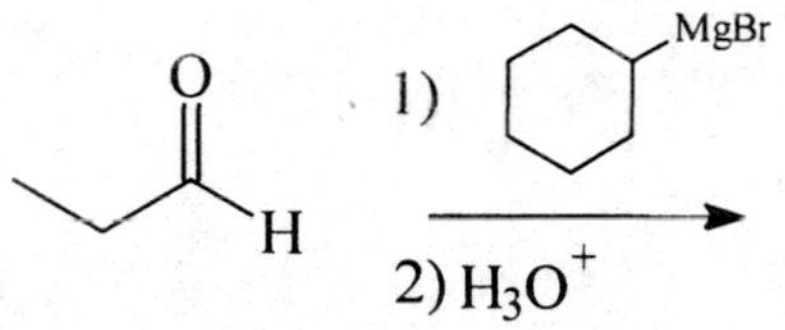
H
1)
MgBr
2) H3O+

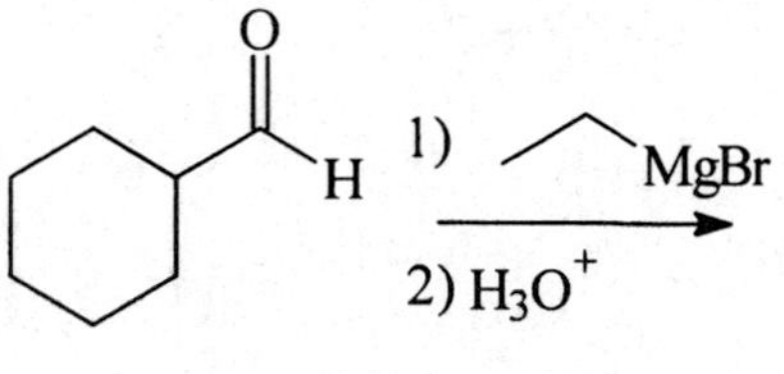
H
1)
MgBr
2) H3O+

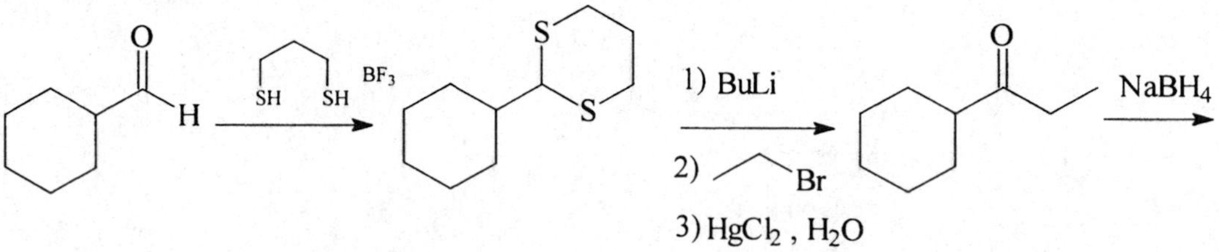
H
SH
SH
BF3
S
S
1) BuLi
2)
Br
3) HgCl2 , H2O
NaBH4

d)

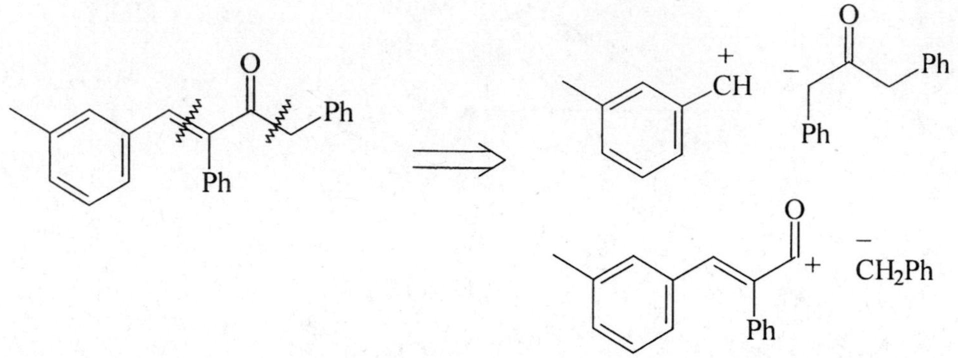
Ph
Ph
CH
Ph
Ph
Ph
CH2Ph

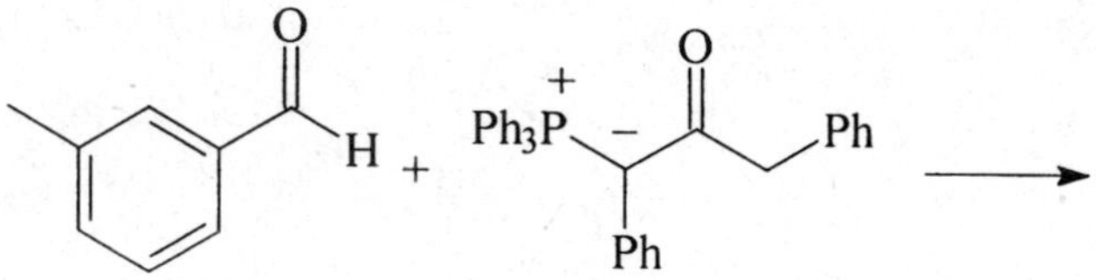
H
Ph3P
Ph
Ph

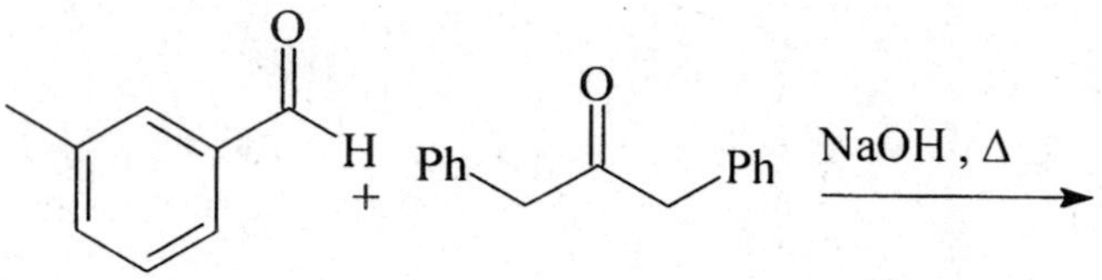
H
Ph
Ph
NaOH , Δ

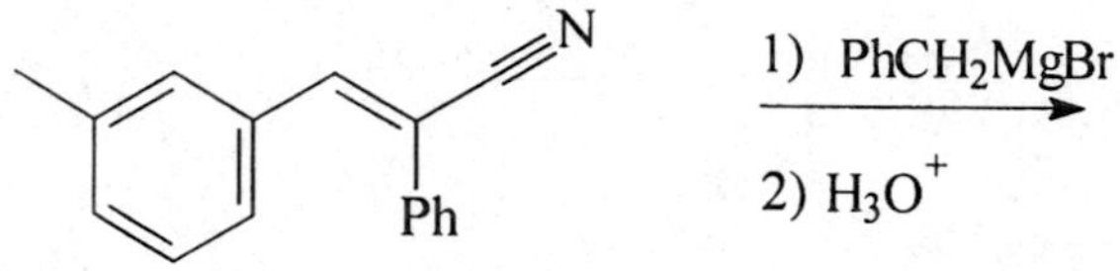
N
Ph
1) PhCH2MgBr
2) H3O+

23.15

a)

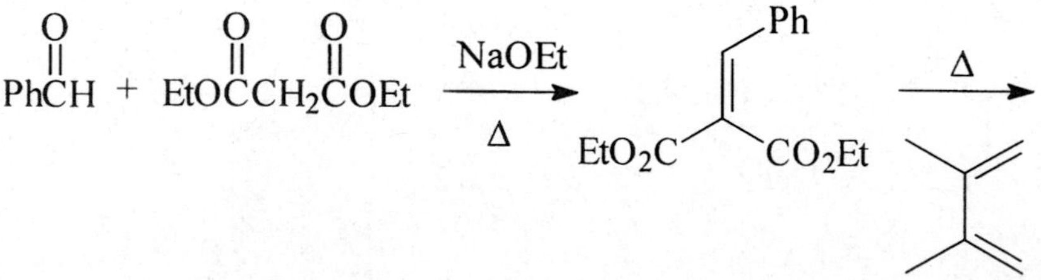
PhCH
EtOCCH2COEt
NaOEt
Δ
Ph
EtO2C
CO2Et
Δ

b)

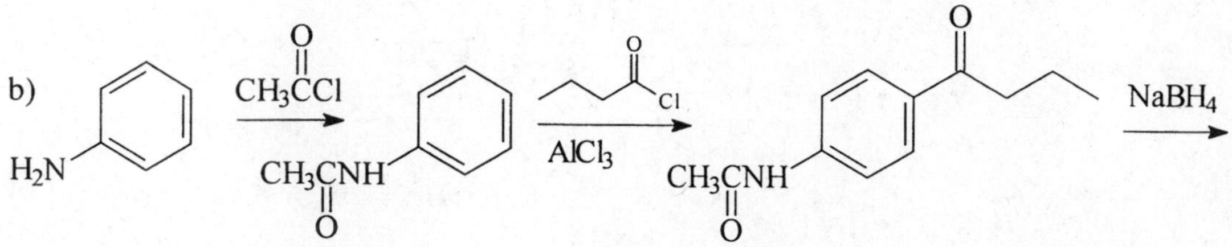
H2N
CH3CCl
CH3CNH
Cl
AlCl3
CH3CNH
NaBH4

c) 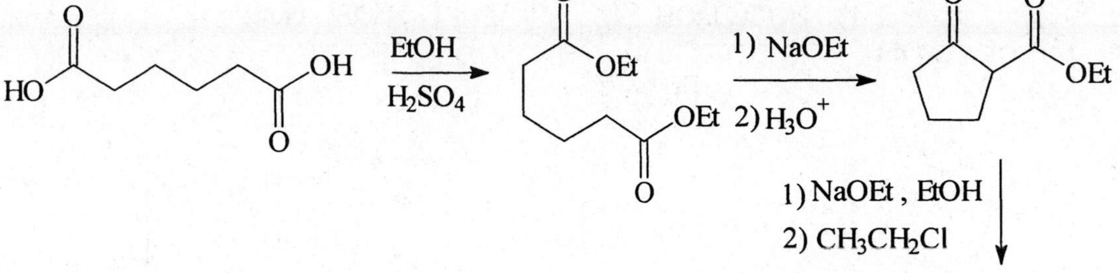
HO
OH
EtOH
H_2SO_4
OEt
OEt
1) NaOEt
2) H_3O^+
OEt
1) NaOEt , EtOH
2) CH_3CH_2Cl

d) 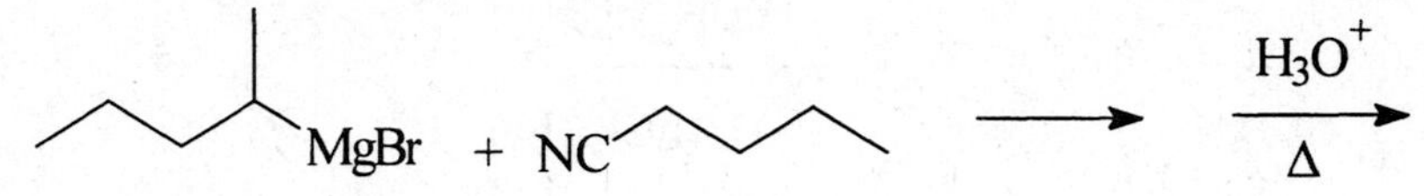
MgBr
+ NC
H_3O^+
Δ

e) 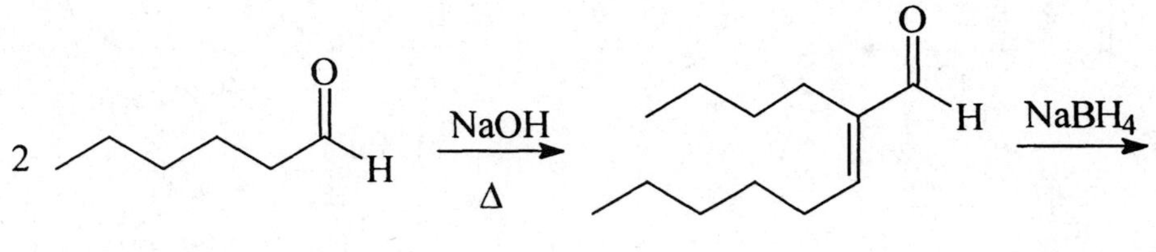
2
H
NaOH
Δ
H
$NaBH_4$

f) 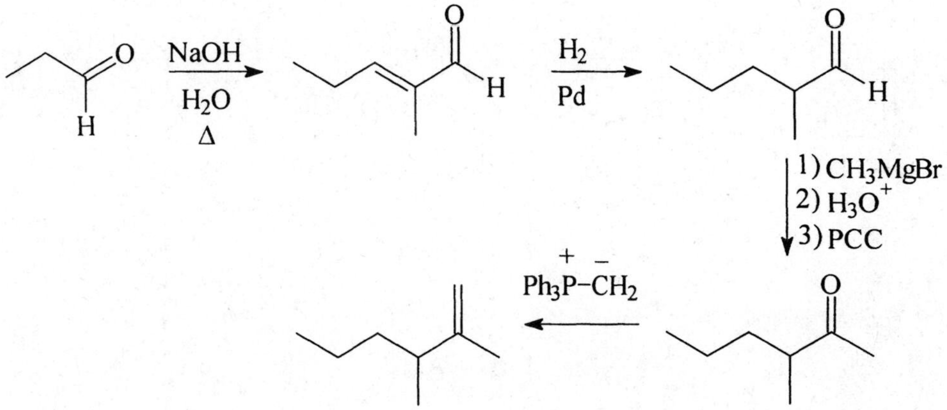
O
H
NaOH
H_2O
Δ
H
H_2
Pd
H
1) CH_3MgBr
2) H_3O^+
3) PCC
$Ph_3\overset{+}{P}-\overset{-}{C}H_2$

g)

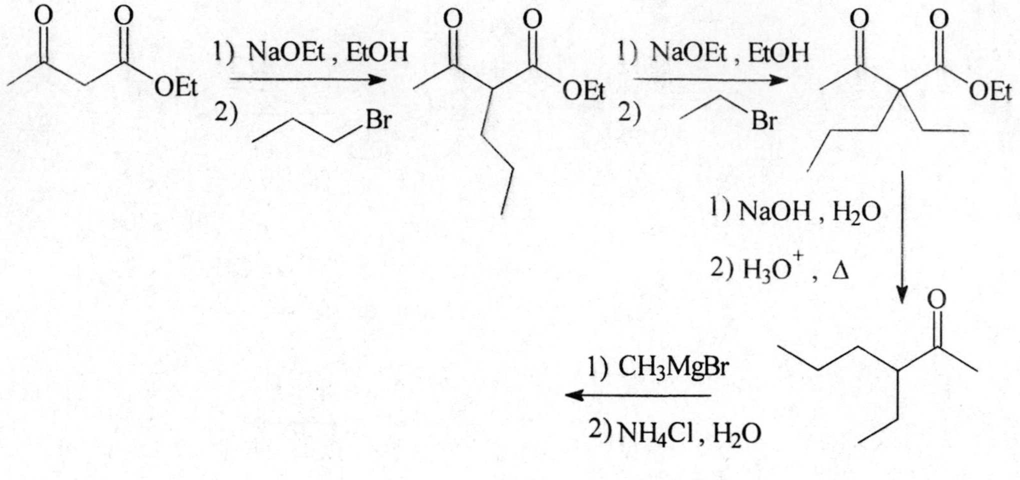
OEt
1) NaOEt , EtOH
2) Br
OEt
1) NaOEt , EtOH
2) Br
OEt
1) NaOH , H2O
2) H3O+ , Δ
1) CH3MgBr
2) NH4Cl , H2O

h)

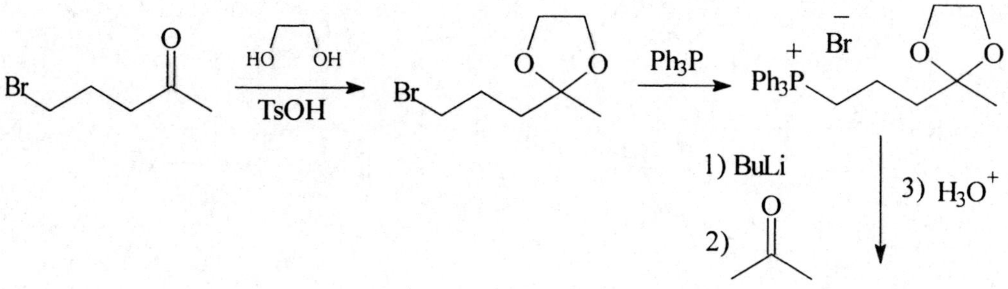
Br
HO OH
TsOH
Br
Ph3P
Br
Ph3P
1) BuLi
2)
3) H3O+

i)

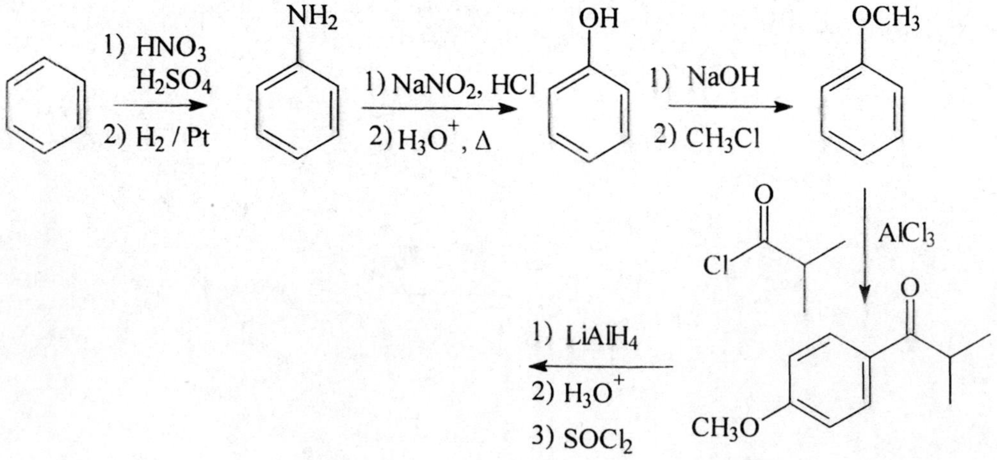
1) HNO3
H2SO4
2) H2 / Pt
NH2
1) NaNO2, HCl
2) H3O+, Δ
OH
1) NaOH
2) CH3Cl
OCH3
Cl
AlCl3
1) LiAlH4
2) H3O+
3) SOCl2
CH3O

23.16

a) $HC{\equiv}CH$ $\xrightarrow[\text{2) } CH_3(CH_2)_{11}CH_2Cl]{\text{1) } NaNH_2}$ $CH_3(CH_2)_{12}C{\equiv}CH$ $\xrightarrow[\text{2) } CH_3(CH_2)_6CH_2Cl]{\text{1) } NaNH_2}$ $CH_3(CH_2)_{12}C{\equiv}C(CH_2)_7CH_3$

$\xrightarrow[\text{catalyst}]{\text{Lindlar} \quad | \quad H_2}$

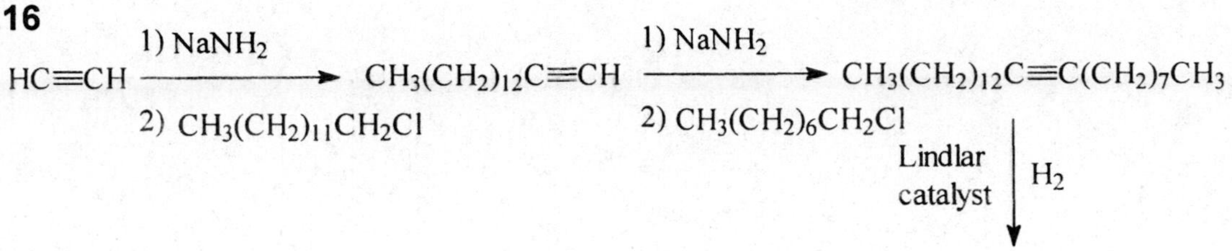

b)

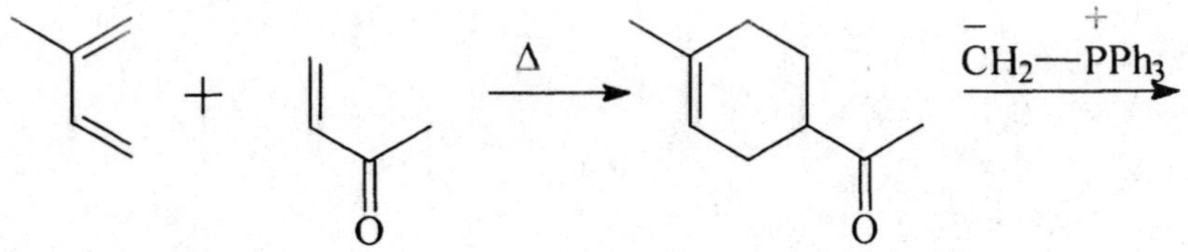

c)

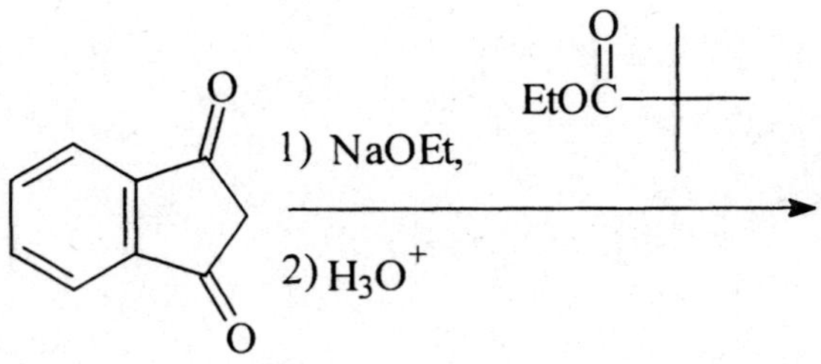

23.17

a)

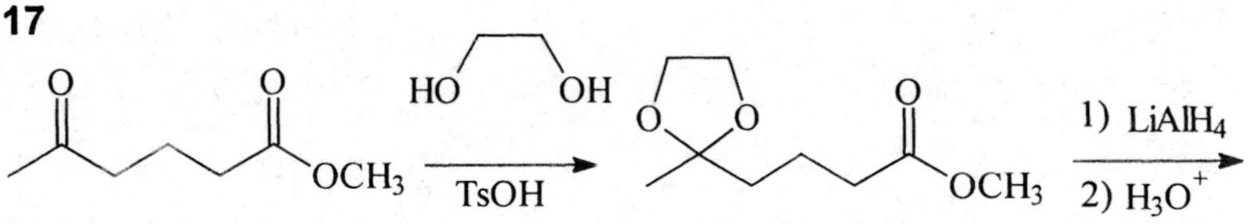

b)

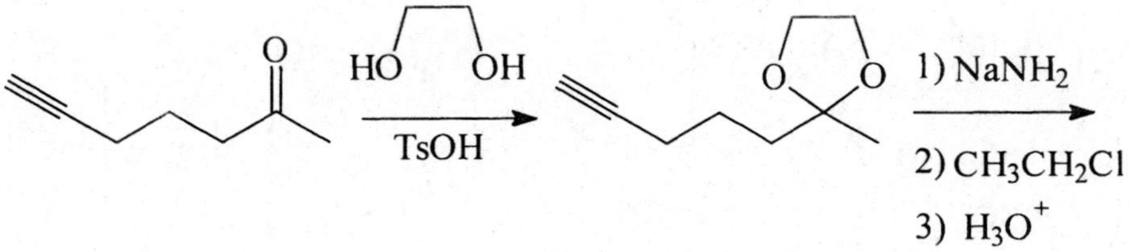

c)

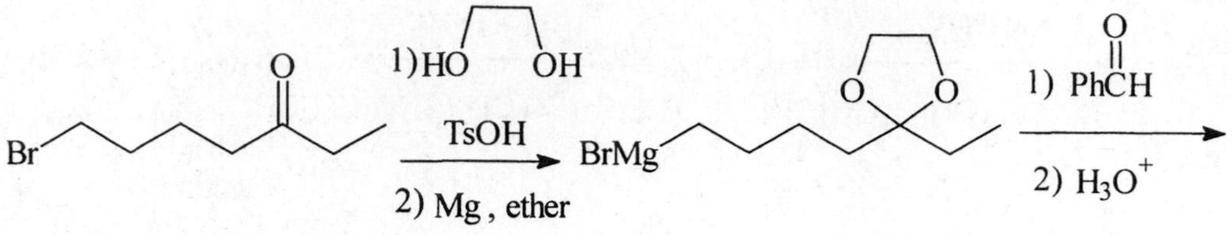

Br
1) HO OH
TsOH
2) Mg , ether
BrMg
1) PhCH
2) H_3O^+

d)

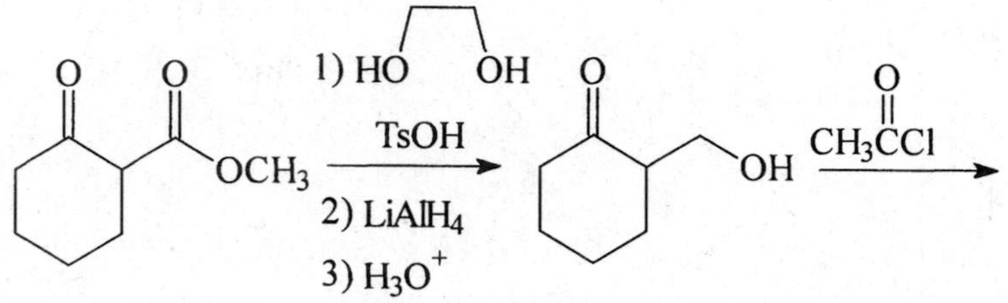

OCH_3
1) HO OH
TsOH
2) $LiAlH_4$
3) H_3O^+
OH
CH_3CCl

e)

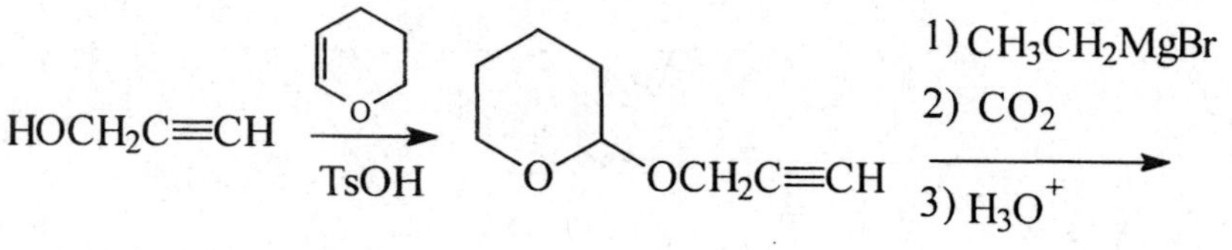

$HOCH_2C{\equiv}CH$
TsOH
$OCH_2C{\equiv}CH$
1) CH_3CH_2MgBr
2) CO_2
3) H_3O^+

f)

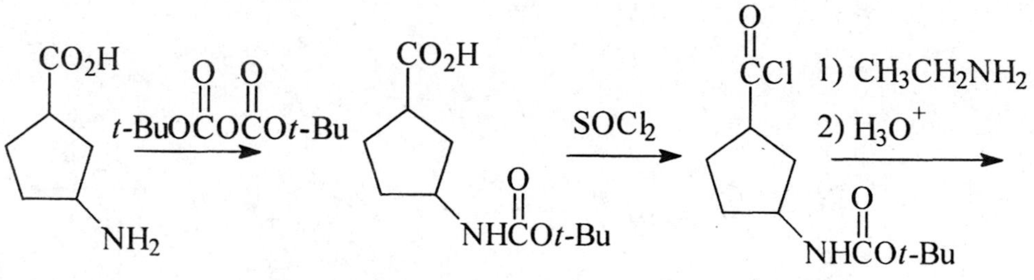

CO_2H
NH_2
t-BuOCOCOt-Bu
$NHCOt$-Bu
$SOCl_2$
CCl
1) $CH_3CH_2NH_2$
2) H_3O^+

g)

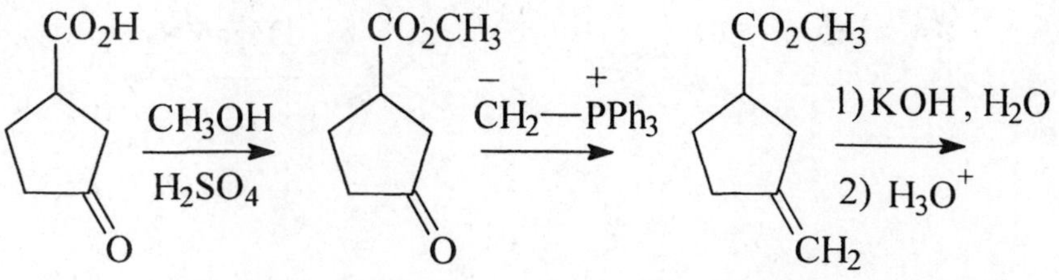

CO_2H
CH_3OH
H_2SO_4
CO_2CH_3
$\overset{-}{C}H_2—\overset{+}{P}Ph_3$
1) KOH , H_2O
2) H_3O^+
CH_2

23.18

a)

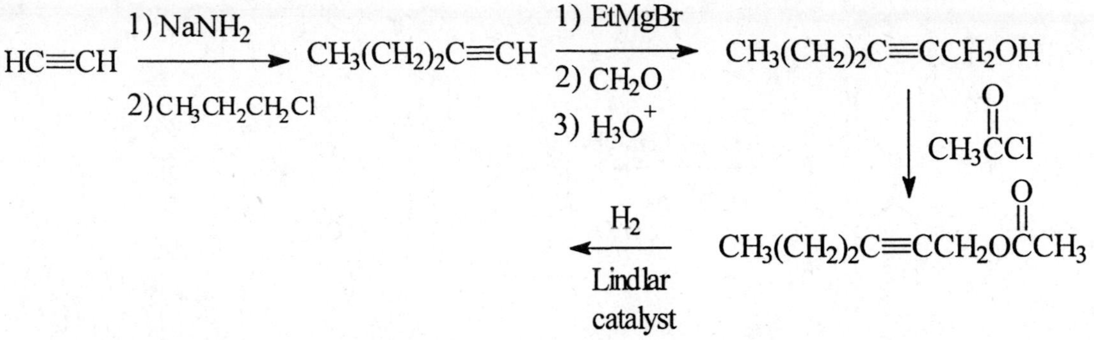
HC≡CH
1) NaNH2
2) CH3CH2CH2Cl
CH3(CH2)2C≡CH
1) EtMgBr
2) CH2O
3) H3O+
CH3(CH2)2C≡CCH2OH
CH3CCl (C=O)
CH3(CH2)2C≡CCH2OCCH3 (C=O)
H2
Lindlar catalyst

b)

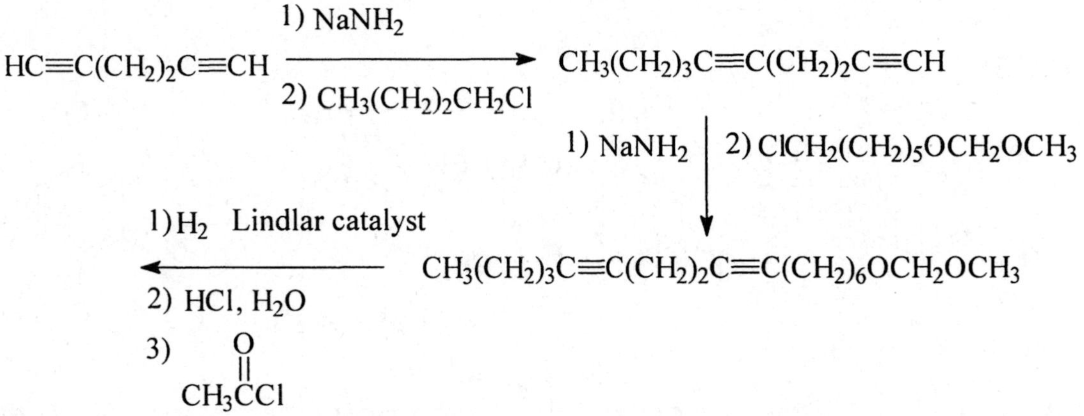
HC≡C(CH2)2C≡CH
1) NaNH2
2) CH3(CH2)2CH2Cl
CH3(CH2)3C≡C(CH2)2C≡CH
1) NaNH2
2) ClCH2(CH2)5OCH2OCH3
CH3(CH2)3C≡C(CH2)2C≡C(CH2)6OCH2OCH3
1) H2 Lindlar catalyst
2) HCl, H2O
3) CH3CCl (C=O)

c)

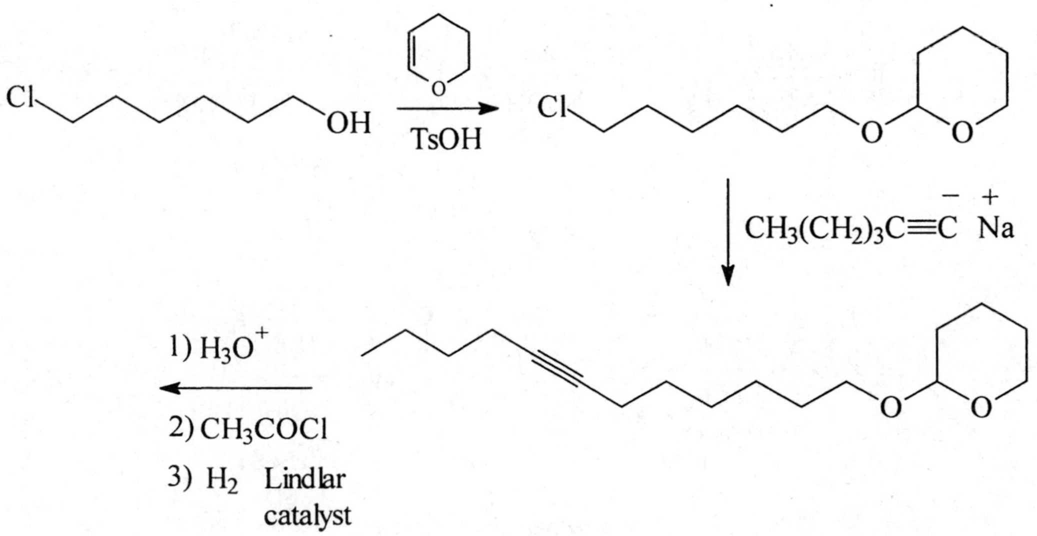
Cl
OH
TsOH
O
CH3(CH2)3C≡C Na
1) H3O+
2) CH3COCl
3) H2 Lindlar catalyst

d)

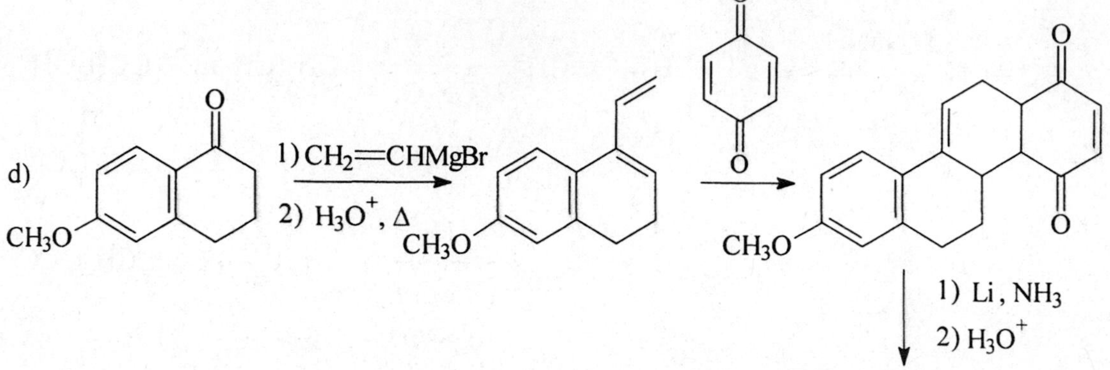

23.19

a)

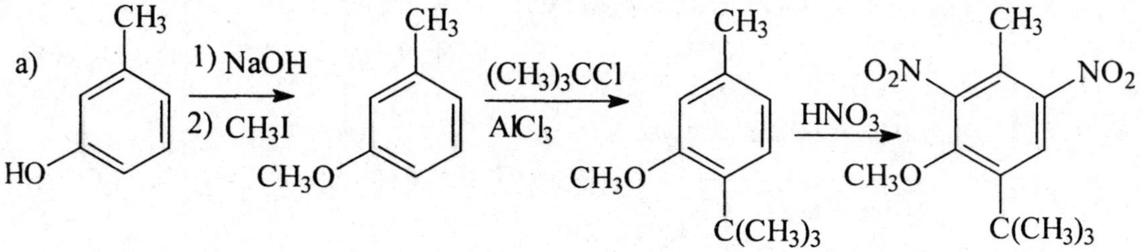

b)

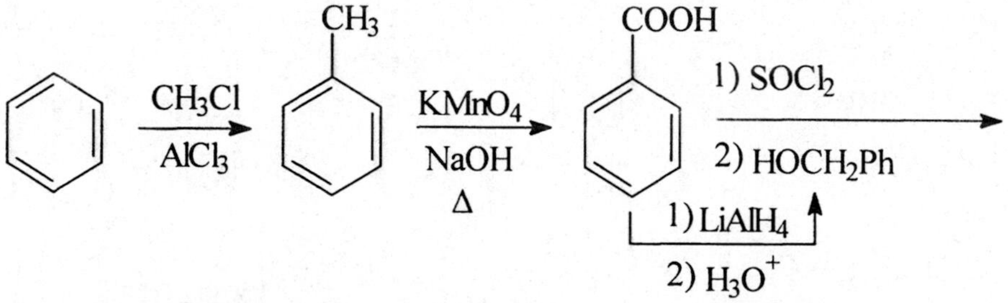

c)

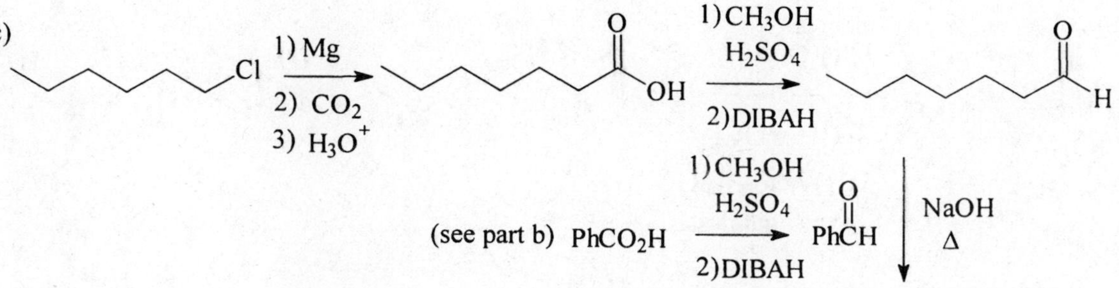

d) 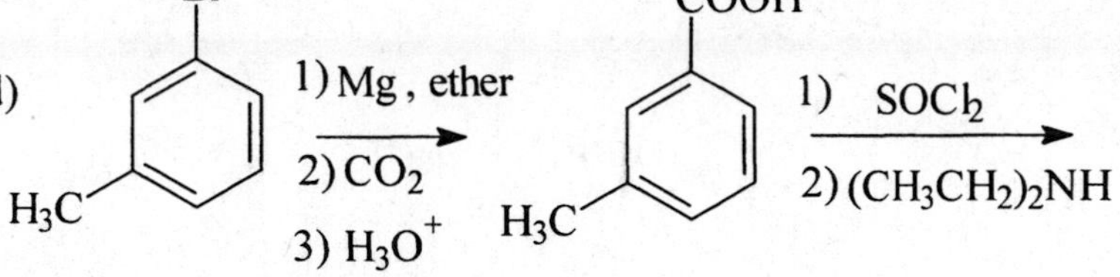
Br
1) Mg , ether
2) CO_2
3) H_3O^+
H_3C
COOH
1) $SOCl_2$
2) $(CH_3CH_2)_2NH$
H_3C

23.20

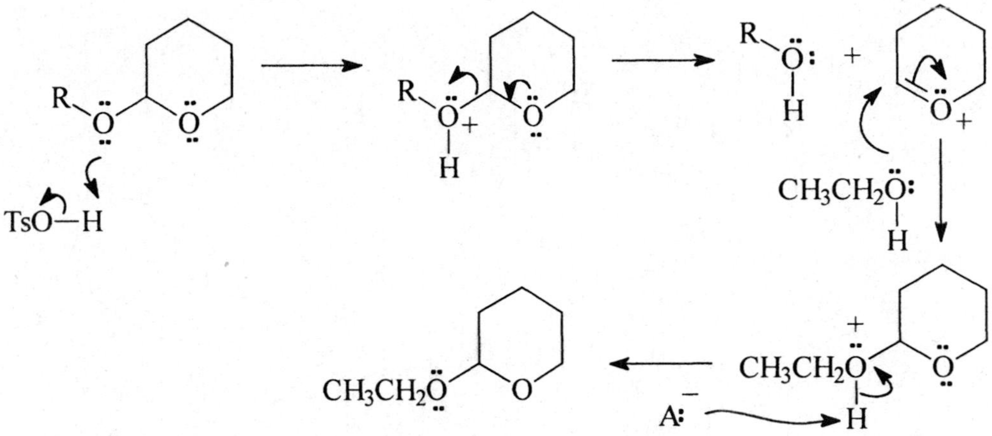
R
TsO—H
R
H
R
H
CH_3CH_2O
H
CH_3CH_2O
A:
H
CH_3CH_2O

23.21

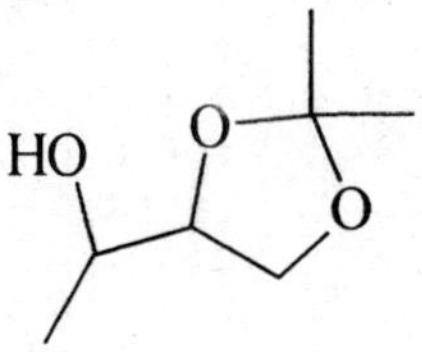
HO
O
O

A

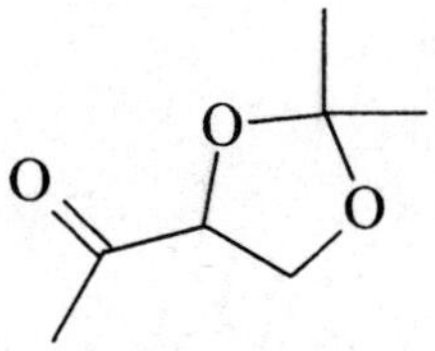
O
O
O

B

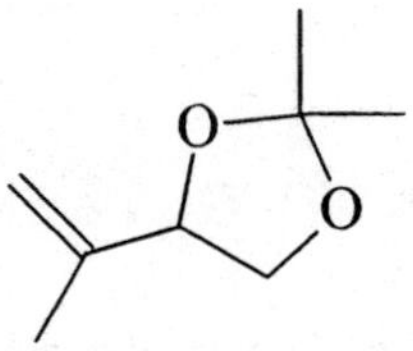
O
O

C

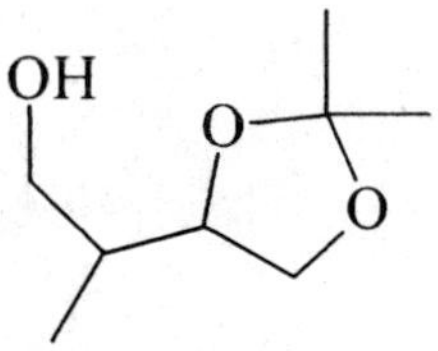
OH
O
O

D

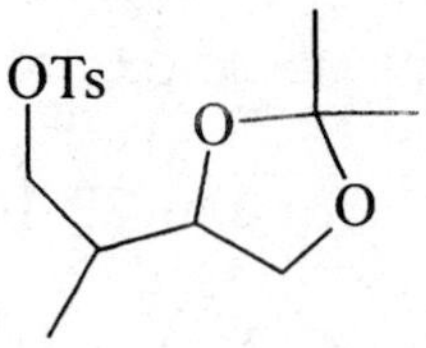
OTs
O
O

E

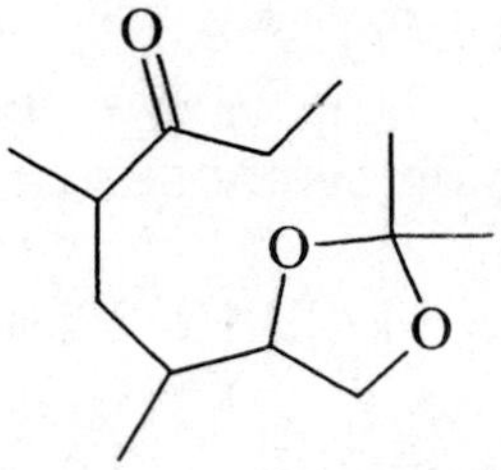
O
O
O

F

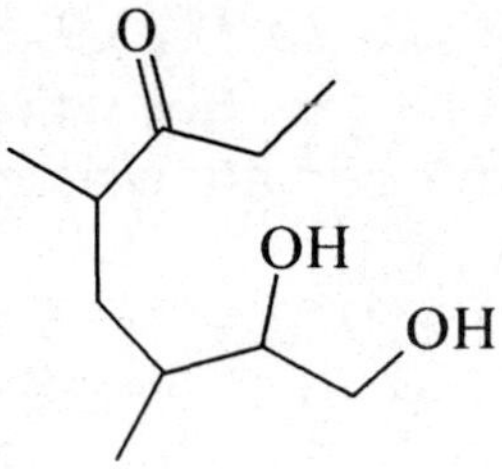
O
OH
OH

G

23.22

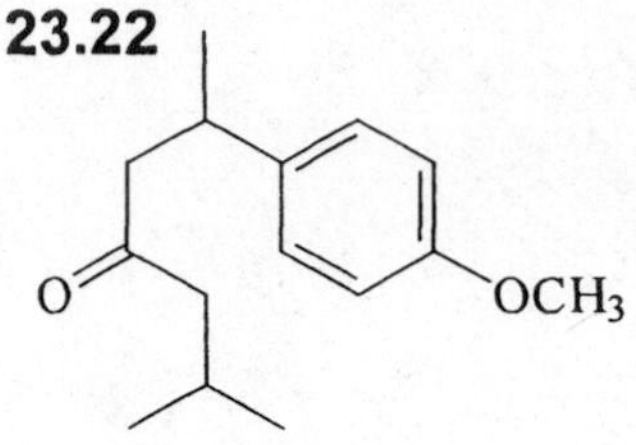

A

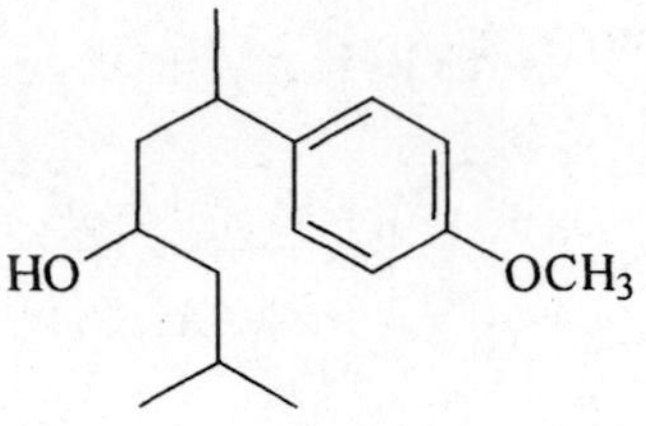

B

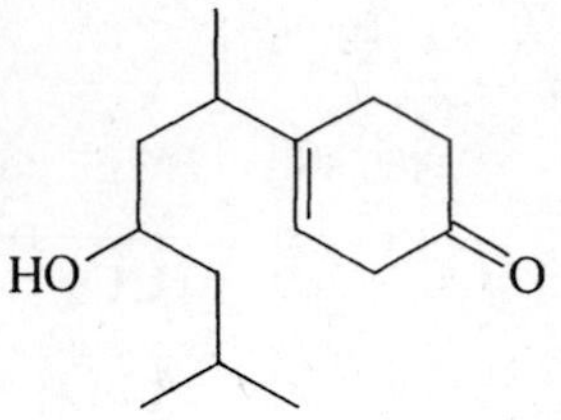

C

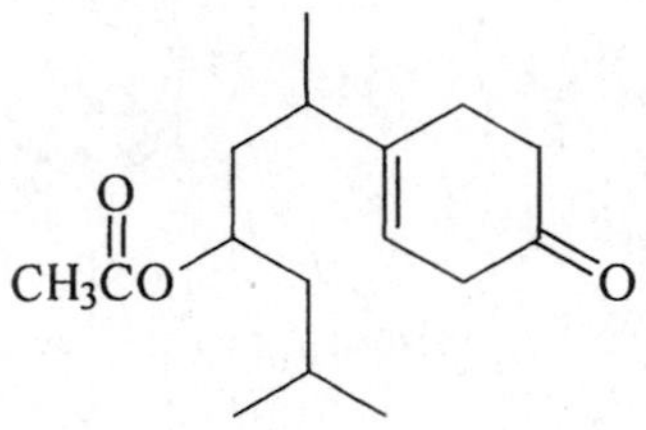

D

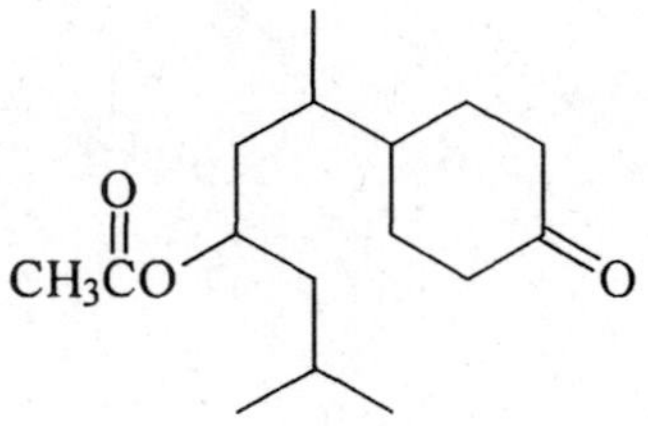

E

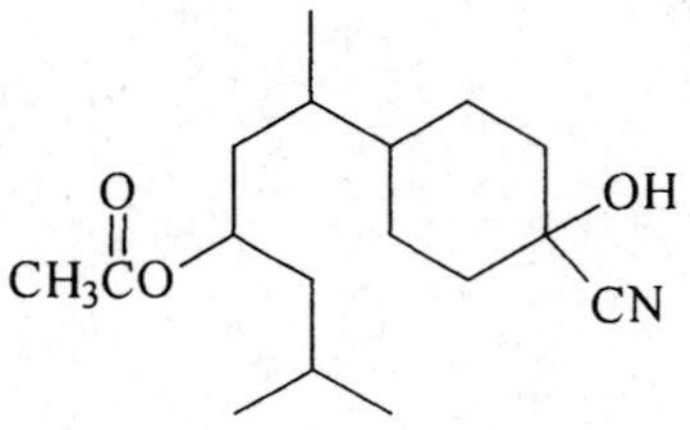

F

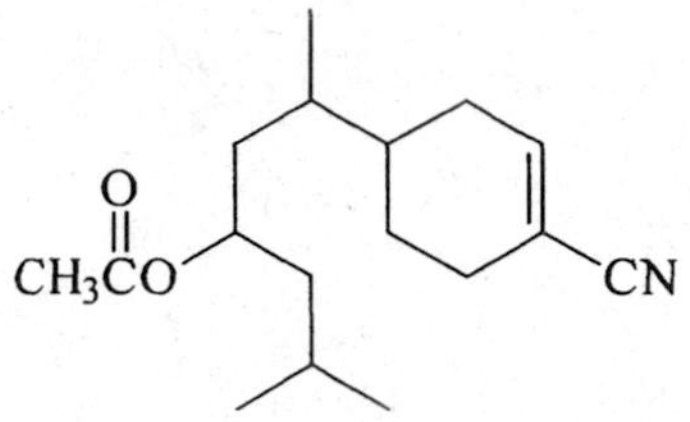

G

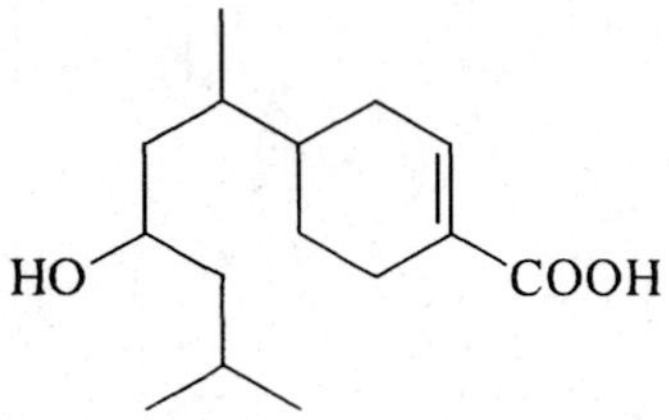

H

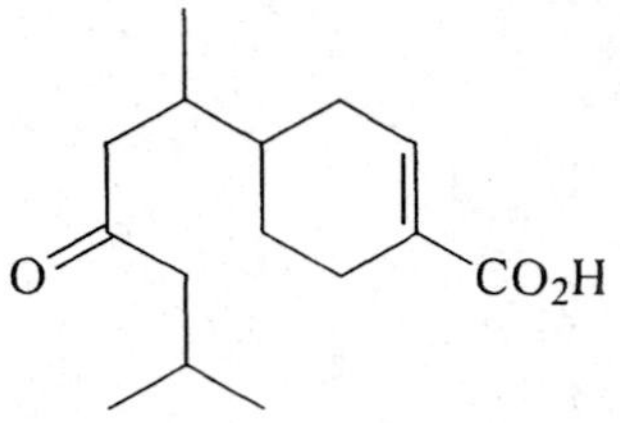

I

Review of Mastery Goals

After completing this chapter, you should be able to:

Recognize when a protective group is needed and how to use it in the synthesis of an organic compound.
(Problems 23.3, 23.8, 23.9, and 23.17)

Use retrosynthetic analysis to design syntheses of more complex organic compounds.
(Problems 23.10, 23.11, 23.12, 23.14, 23.15, 23.16, 23.18, and 23.19)

Chapter 24
SYNTHETIC POLYMERS

24.1

$$PhC\overset{O}{\overset{||}{C}}O-\overset{O}{\overset{||}{O}}CPh \longrightarrow 2 \quad PhC\overset{O}{\overset{||}{O}}\cdot$$

$$PhC\overset{O}{\overset{||}{O}}\cdot \; + \; CH_2=\overset{CH_3}{\underset{|}{CH}} \longrightarrow PhC\overset{O}{\overset{||}{O}}-CH_2-\overset{CH_3}{\underset{|}{\dot{C}H}}$$

$$PhC\overset{O}{\overset{||}{O}}-CH_2-\overset{CH_3}{\underset{|}{\dot{C}H}} + CH_2=\overset{CH_3}{\underset{|}{CH}} \longrightarrow PhC\overset{O}{\overset{||}{O}}-CH_2-\overset{CH_3}{\underset{|}{CH}}-CH_2-\overset{CH_3}{\underset{|}{\dot{C}H}} \overset{etc.}{\longrightarrow} \left(CH_2-\overset{CH_3}{\underset{|}{CH}} \right)_n$$

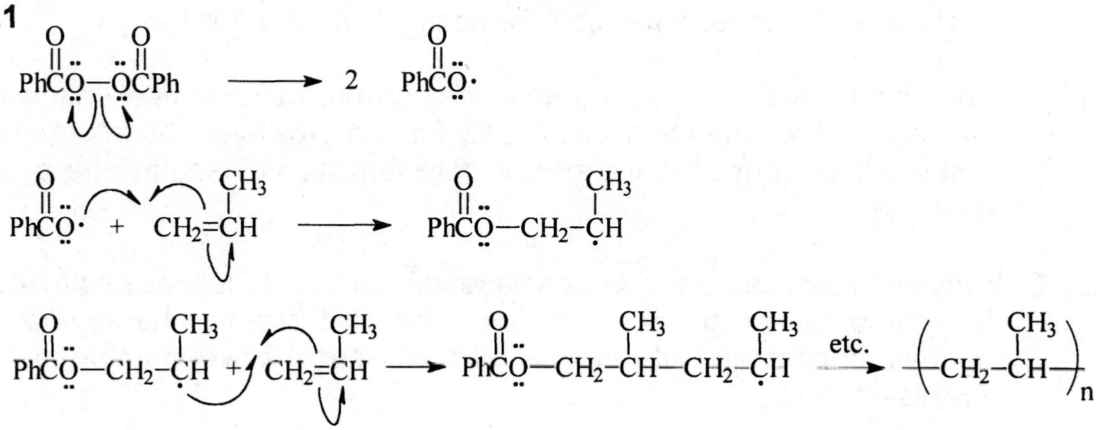

Addition of the radical always occurs at C-1 of propene so that the more stable secondary radical is formed.

24.2

a) $\left(CH_2-\overset{Cl}{\underset{|}{CH}} \right)_n$

b) $\left(CH_2-\overset{CN}{\underset{|}{CH}} \right)_n$

c) $\left(CH_2-\overset{Cl}{\underset{\underset{Cl}{|}}{\overset{|}{C}}} \right)_n$

24.3

a) $CH_2=\overset{O\overset{||}{C}CH_3}{\underset{|}{CH}}$

b) $\underset{F}{\overset{F}{C}}=\underset{F}{\overset{F}{C}}$

24.4

$$\sim\sim CH_2-\overset{CH_2}{\underset{\underset{\dot{C}H_2}{\underset{|}{CH_2}}}{\overset{|}{CH}}} \longrightarrow \sim\sim CH_2-\overset{CH_2CH_2CH_3}{\underset{|}{\dot{C}H}} \longrightarrow \sim\sim CH_2-\overset{CH_2CH_2CH_3}{\underset{|}{CH}}-CH_2-\dot{C}H_2 \overset{etc.}{\longrightarrow}$$

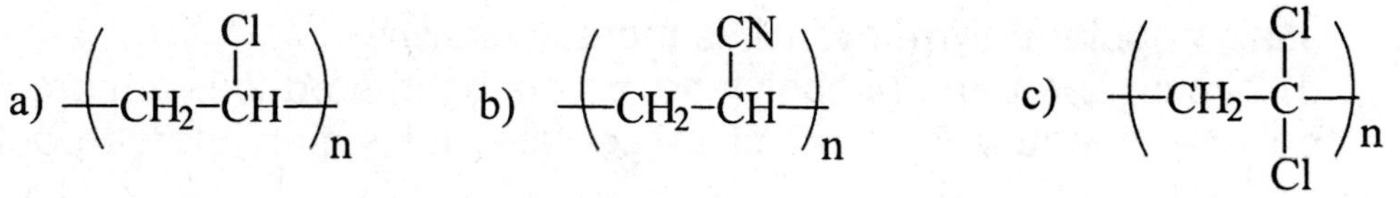

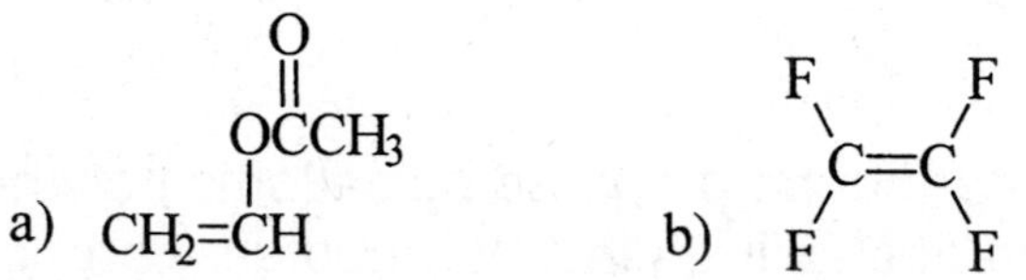

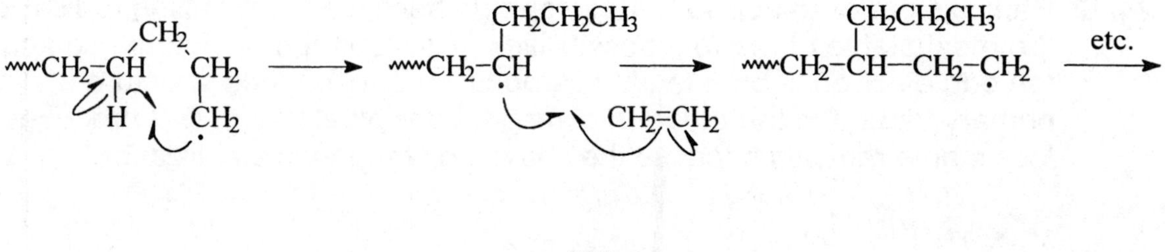

24.5 Both (a) and (b) can form atactic polymers; (c) cannot.

24.6 The primary carbocation that would be the intermediate in the cationic polymerization of ethylene is very unstable, so it does not readily form. However, the carbocation formed in the cationic polymerization of styrene is stabilized by resonance, so it forms much more readily.

24.7 The anionic intermediate formed in the polymerization of acrylonitrile is stabilized by resonance and is readily formed. However, the intermediate that would be formed in the anionic polymerization of isobutylene is quite unstable.

24.8 Ethylene oxide, an epoxide, reacts readily with nucleophiles because a large amount of ring strain is relieved when its three-membered ring is opened. In contrast, little strain is relieved when the five-membered ring of THF is opened.

24.9 a) 1,1-Dichloroethene (on the right) produces a more crystalline polymer because it has no chiral centers and is more stereoregular.
b) Methyl 2-propenoate (methyl acrylate, on the right) produces a more crystalline polymer because its side chain is more polar.
c) Coordination polymerization produces a less branched and more stereoregular polymer which is more crystalline.
d) 2-methylpropene (isobutylene, on the right) produces a more crystalline polymer because it has no chiral centers and is more stereoregular.

24.10 Teflon has no chiral centers, so it has no stereochemical complications. Nor does it have any hydrogens that can be abstracted from the interior of the chain, so it is linear.

24.11 Poly(methyl methacrylate) cannot be prepared by cationic polymerization because the carbonyl group destabilizes the intermediate carbocation. It can be produced by anionic polymerization because the carbonyl group stabilizes the carbanion intermediate by resonance.

24.12 Radicals prefer to add to C-1 of isoprene because the resulting radical is the most stable of the four possibilities. It is stabilized by resonance and the odd electron is on a tertiary carbon in one resonance structure and a primary carbon in the other. In contrast, addition at C-2 or C-3 produces less stable radicals because they have no resonance stabilization.

Addition at C-4 produces a resonance stabilized radical, but the odd electron is on a secondary carbon and a primary carbon in the two resonance structures.

24.13 Trans double bonds predominate because they are more stable than cis double bonds.

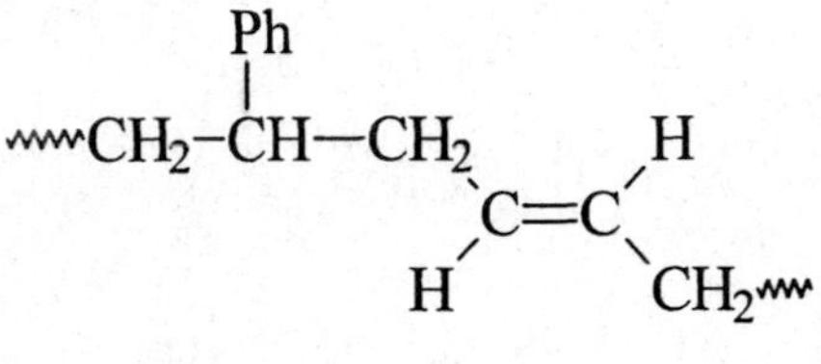

$$\text{wwCH}_2\text{-}\underset{\underset{\text{Ph}}{|}}{\text{CH}}\text{-}\text{CH}_2 \quad \underset{H}{\overset{H}{\diagup}}C=C\underset{\text{CH}_2\text{ww}}{\overset{H}{\diagdown}}$$

24.14

$$\left(\!\!-\overset{O}{\overset{||}{C}}(CH_2)_4\overset{O}{\overset{||}{C}}OCH_2CH_2O\!-\!\right)_{\!n}$$

24.15

$$\overset{O}{\overset{||}{HOC}}(CH_2)_8\overset{O}{\overset{||}{C}}OH \quad + \quad H_2N(CH_2)_6NH_2$$

24.16 If a polyester is formed from more diol units, then each polymer chain must terminate with a diol unit at each end. To accommodate even a small excess of diol units, there must be many chains, so the chains must be relatively short.

24.17 Direct preparation of poly(vinyl alcohol) would require an enol as the monomer. However, recall that most enols cannot be isolated because they spontaneously isomerize to the carbonyl tautomer (see Section 11.6).

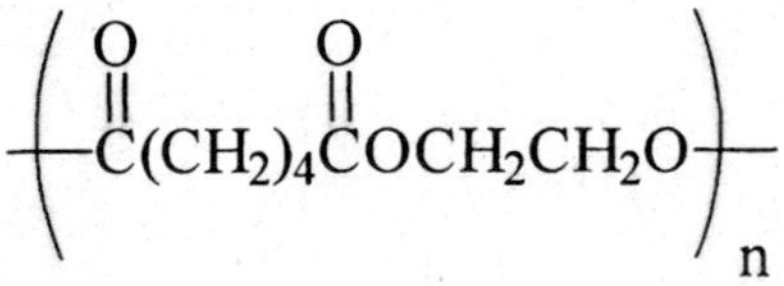

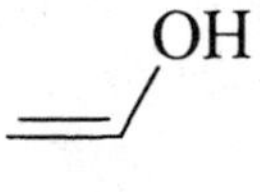

$$\underset{\diagup}{\overset{OH}{=}}$$

24.18

a) $\left(\!\!-\overset{O}{\overset{||}{C}}(CH_2)_8\overset{O}{\overset{||}{C}}NH(CH_2)_6NH\!-\!\right)_{\!n}$

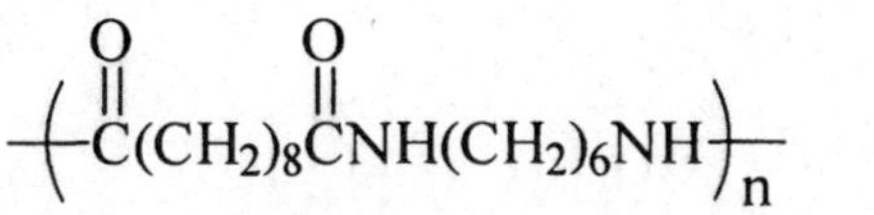

heat (condensation polymerization)

b) $\left(\!\!-CH_2\text{-}\underset{\underset{\overset{||}{C}NH_2}{\overset{|}{\underset{O}{}}}}{CH}\!-\!\right)_{\!n}$

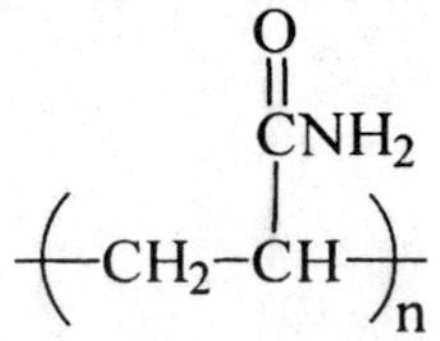

radical or anionic polymerization

c)
$$\left[\!\!-\overset{\overset{\displaystyle O}{\|}}{C}\!-\!\!\bigcirc\!\!-\!\overset{\overset{\displaystyle O}{\|}}{C}O(CH_2)_4O\!-\!\!\right]_n$$

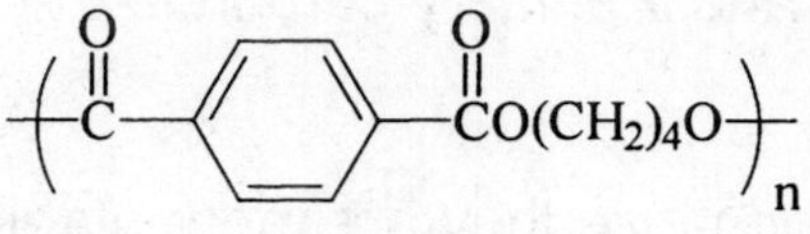

heat   (condensation polymerization)

d)
$$\left[\!\!-CH_2CHO\!-\!\!\right]_n$$
$\quad\quad\quad\quad CH_3$

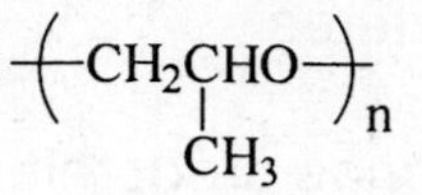

anionic polymerization

e)
$$\left[\!\!-O(CH_2)_4O\overset{\overset{\displaystyle O}{\|}}{C}NH\!-\!\!\bigcirc\!\!-\!NH\overset{\overset{\displaystyle O}{\|}}{C}\!-\!\!\right]_n$$

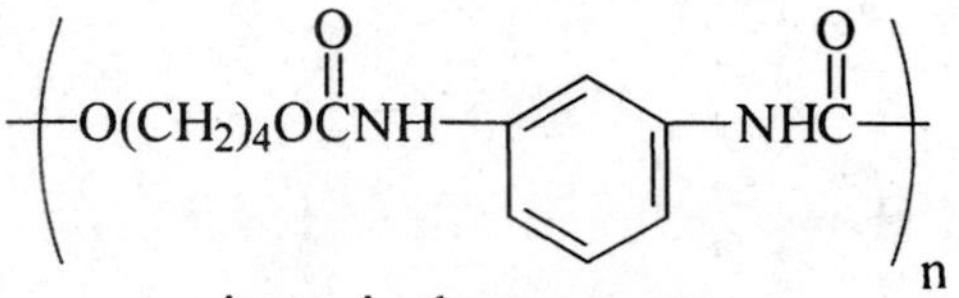

just mix the monomers

f)
$$\left[\!\!-O(CH_2)_4\overset{\overset{\displaystyle O}{\|}}{C}\!-\!\!\right]_n$$

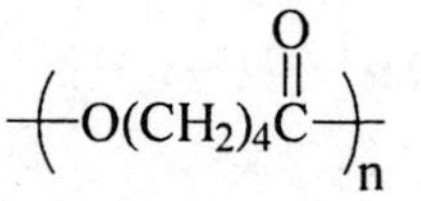

add catalytic amount
of OH$^-$ and heat

24.19

a)
$HO\overset{\overset{\displaystyle O}{\|}}{C}\!-\!\!\bigcirc\!\!-\!\overset{\overset{\displaystyle O}{\|}}{C}OH$ + $H_2N\!-\!\!\bigcirc\!\!-\!NH_2$

b) $H_2N(CH_2)_{11}\overset{\overset{\displaystyle O}{\|}}{C}OH$

c) [2-chloro-1,3-butadiene structure]

d) $H_2N\!-\!\!\bigcirc\!\!-\!NH_2$ + [pyromellitic dianhydride structure]

e) $CH_2{=}CHCH_2\overset{\overset{\displaystyle CH_3}{|}}{\underset{\underset{\displaystyle CH_3}{|}}{CH}}$

f) $H_2N\!-\!\!\bigcirc\!\!-\!\overset{\overset{\displaystyle O}{\|}}{C}OH$

g) [α-methylstyrene structure, $\overset{\overset{\displaystyle CH_3}{|}}{C}{=}CH_2$ on benzene]

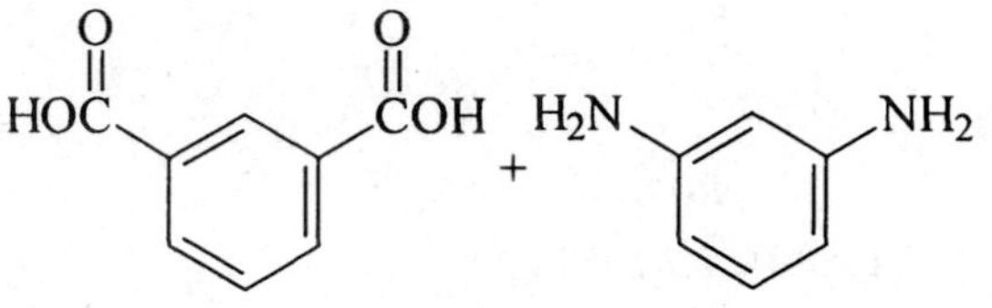

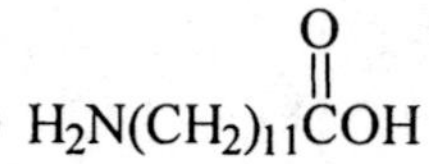

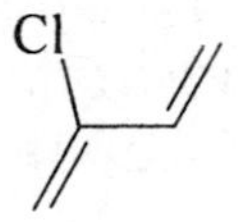

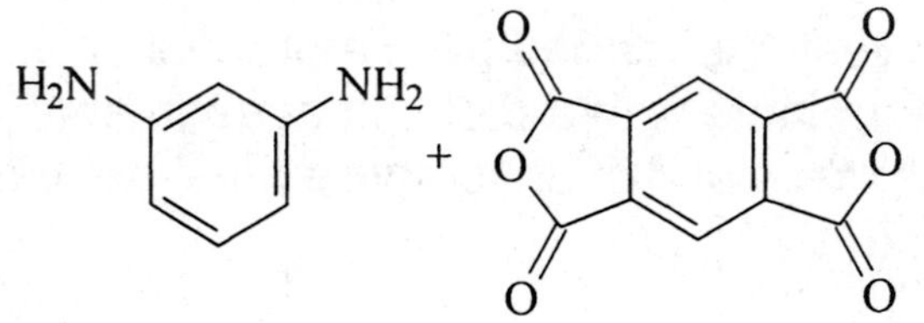

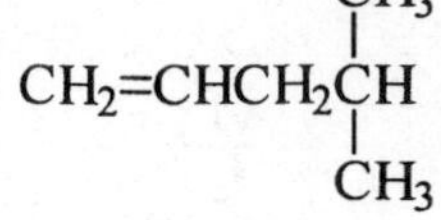

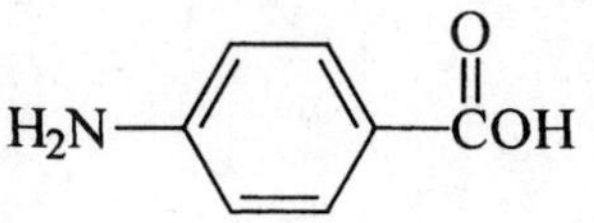

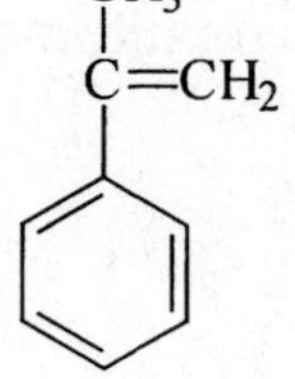

h) 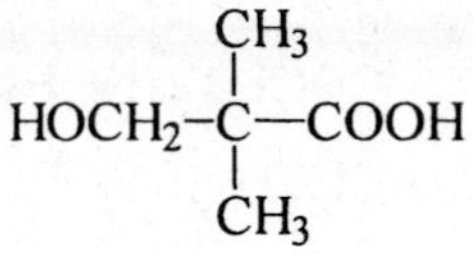 or 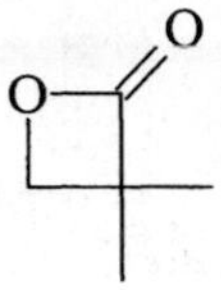

i) 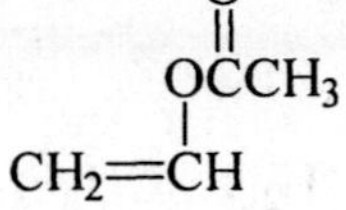

j) 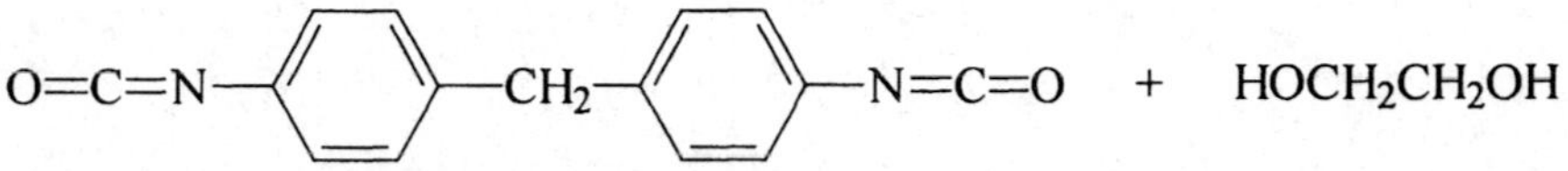

24.20

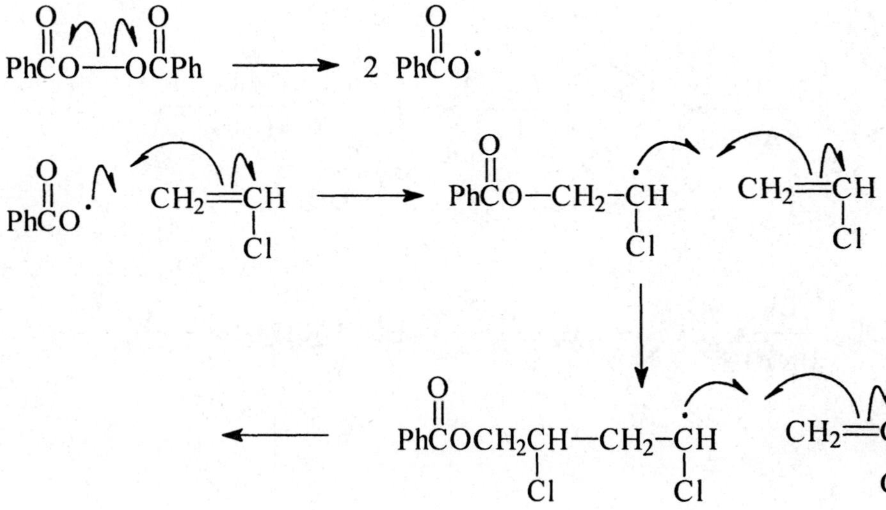

24.21

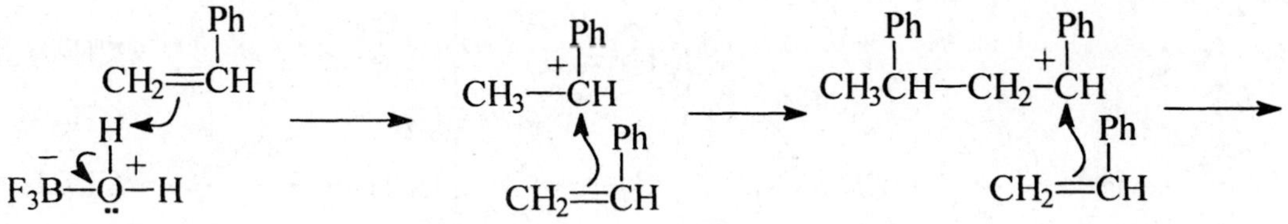

24.22

$$\text{CH}_2\!=\!\overset{\displaystyle \underset{|}{\text{COCH}_3}}{\overset{\displaystyle \text{O}}{}}\text{CH} \longrightarrow \text{H}_2\text{N}-\text{CH}_2\overset{..}{\text{CH}} \quad \longrightarrow \quad \text{H}_2\text{NCH}_2\overset{..}{\text{CH}}\text{CH}_2-\overset{..}{\text{CH}} \longrightarrow$$

:NH₂

COCH₃

COCH₃ COCH₃

COCH₃

CH₂=CH

CH₂=CH

24.23

a) $\text{CH}_2\!=\!\text{CH}_2 \xrightarrow{\ \text{Cl}_2\ } \underset{\underset{\text{Cl}}{|}}{\overset{\overset{\text{Cl}}{|}}{\text{CH}_2\text{CH}_2}} \xrightarrow{\ \text{Ca(OH)}_2\ } \underset{\text{Cl}}{\text{CH}_2\!=\!\text{CH}} \xrightarrow[\text{initiator}]{\text{radical}}$

b)

$$\bigcirc \xrightarrow[\text{AlCl}_3]{\text{CH}_3\text{Cl}} \text{(p-xylene)} \xrightarrow{\text{KMnO}_4} \text{(terephthalic acid)} \xrightarrow[\text{H}_2\text{SO}_4]{\text{CH}_3\text{OH}}$$

CH₃

CH₃

COOH

COOH

CH₃OC—⬡—COCH₃

$$\text{CH}_2\!=\!\text{CH}_2 \xrightarrow[\ 2)\text{NaOH}\ ]{1)\text{Cl}_2,\ \text{H}_2\text{O}} \underset{\text{O}}{\text{CH}_2\!-\!\text{CH}_2} \xrightarrow{\text{H}_3\text{O}^+} \text{HOCH}_2\text{CH}_2\text{OH} \xrightarrow{\ \Delta\ }$$

c) $\underset{\text{O}}{\text{CH}_2\!-\!\text{CH}_2} \xrightarrow[\ ^-\text{CN}\ ]{\text{HCN}} \text{HOCH}_2\text{CH}_2\text{CN} \xrightarrow{\text{H}_3\text{O}^+} \underset{\text{CN}}{\text{CH}_2\!=\!\text{CH}} \xrightarrow[\text{initiator}]{\text{radical}}$

24.24

$$\text{R}\cdot \quad \overset{\underset{|}{\text{H}}}{\text{CH}_2\!-\!\text{CH}} \longrightarrow \text{CH}_2\!-\!\overset{\cdot}{\text{CH}} \longrightarrow \longrightarrow$$

Ph Ph

CH₂=CH ·CH

Ph CH₂

Ph

CHCH₂

Ph

etc.

Ph

CH

CH₂

CHCH₂

Ph

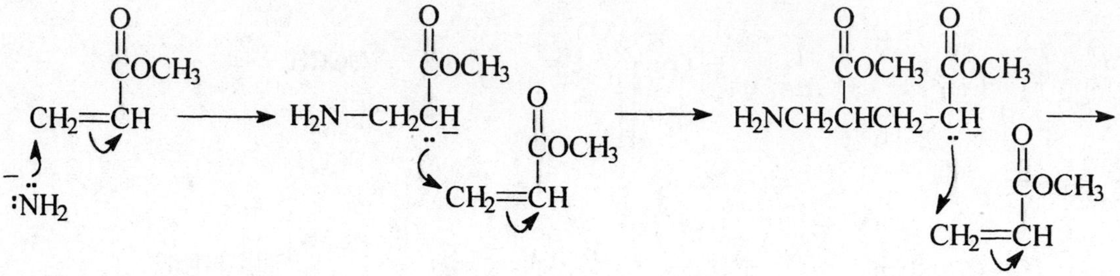

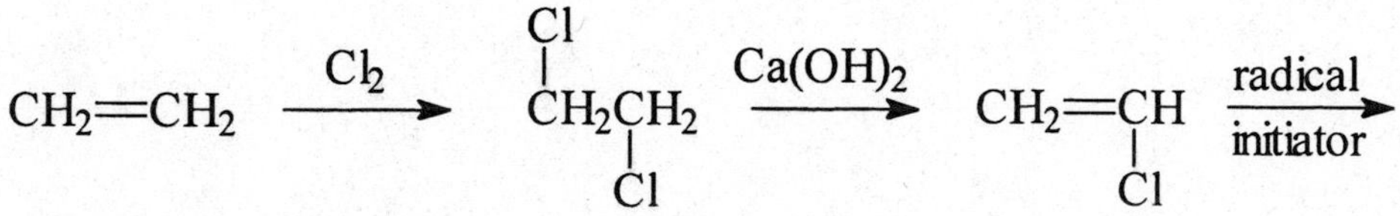

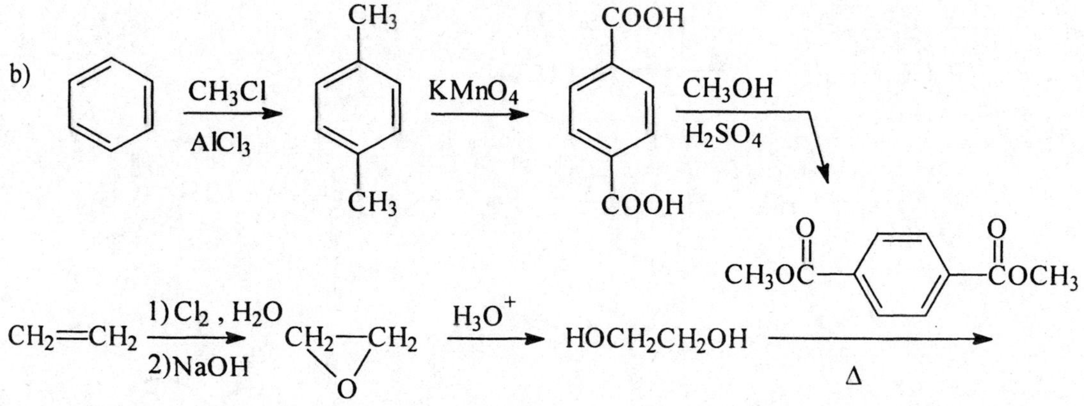

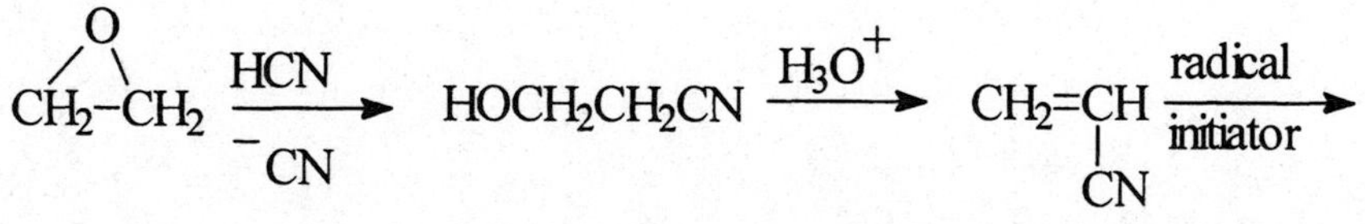

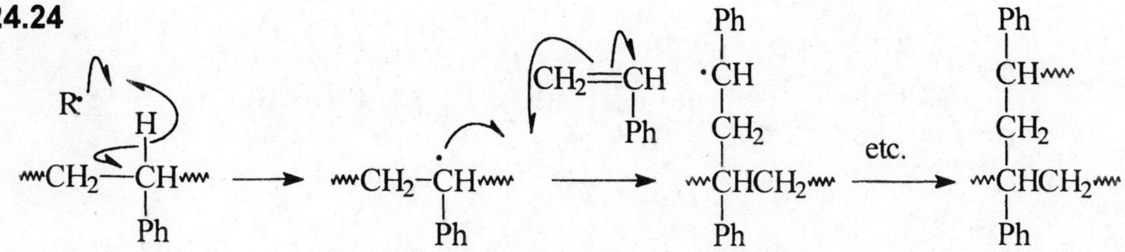

24.25

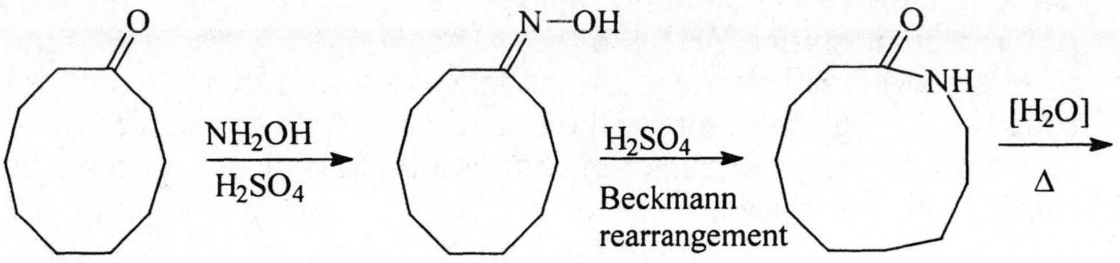

24.26

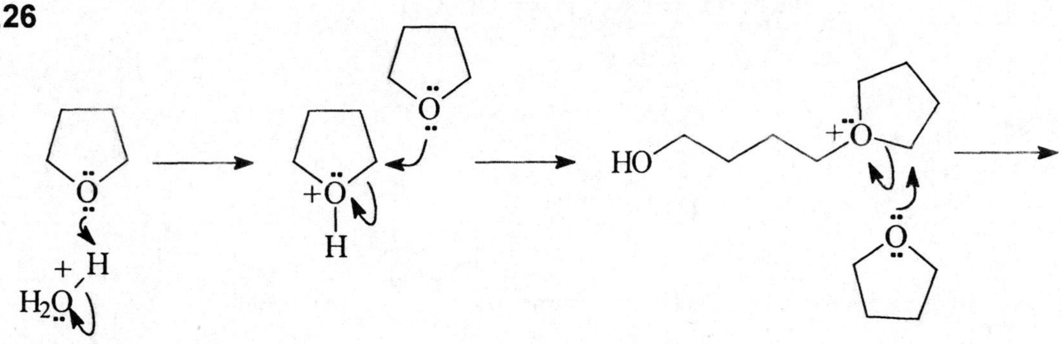

24.27 This polymer would be thermosetting because it forms a three dimensional network.

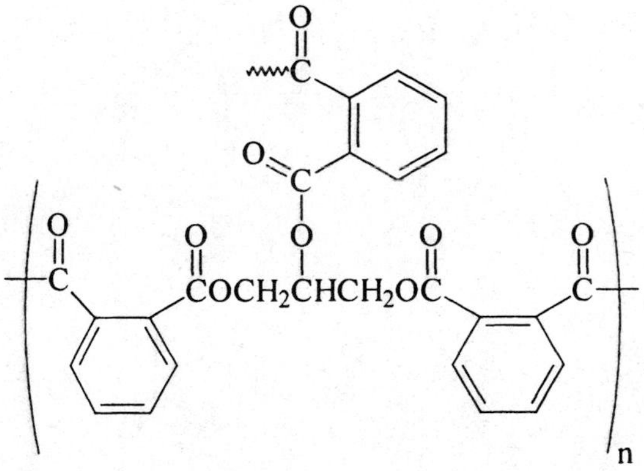

24.28 Radical polymerization is initiated by trace amounts of a radical initiator. Styrene can readily undergo radical chain polymerization in the presence of a radical impurity. BHT is a radical scavenger. It reacts with radicals to produce sterically hindered, resonance stabilized radicals that are not reactive enough to continue the chain. Therefore addition of BHT prevents the polymerization of styrene due to radical impurities by terminating the chains.

24.29

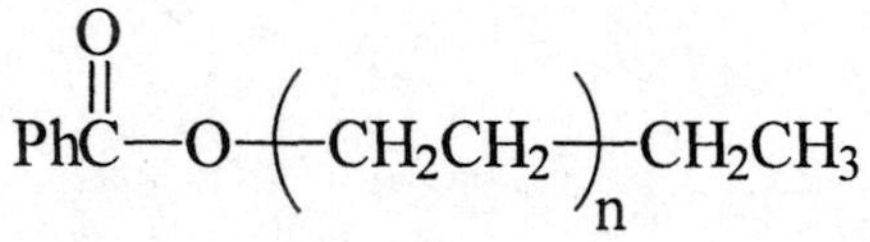

$$PhC(=O){-}O{-}\left(CH_2CH_2\right)_n{-}CH_2CH_3$$

24.30

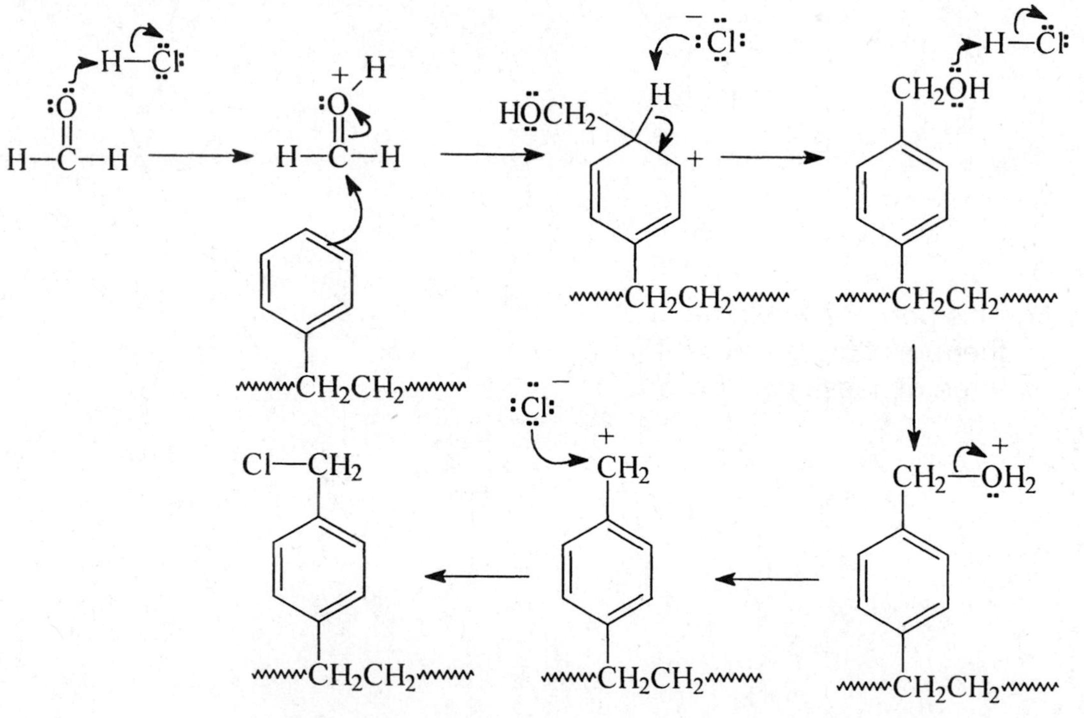

24.31

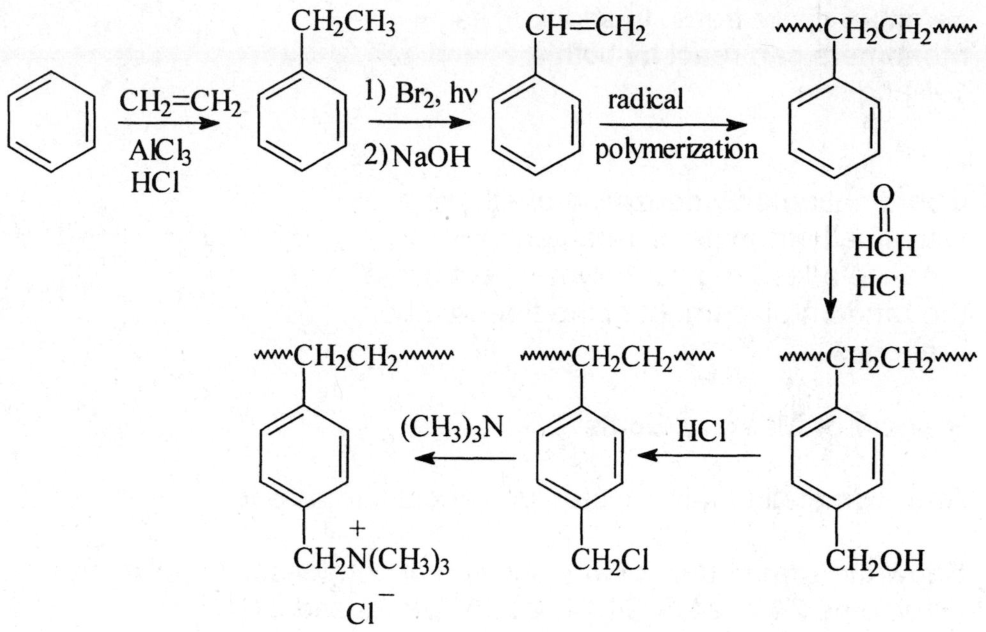

24.32

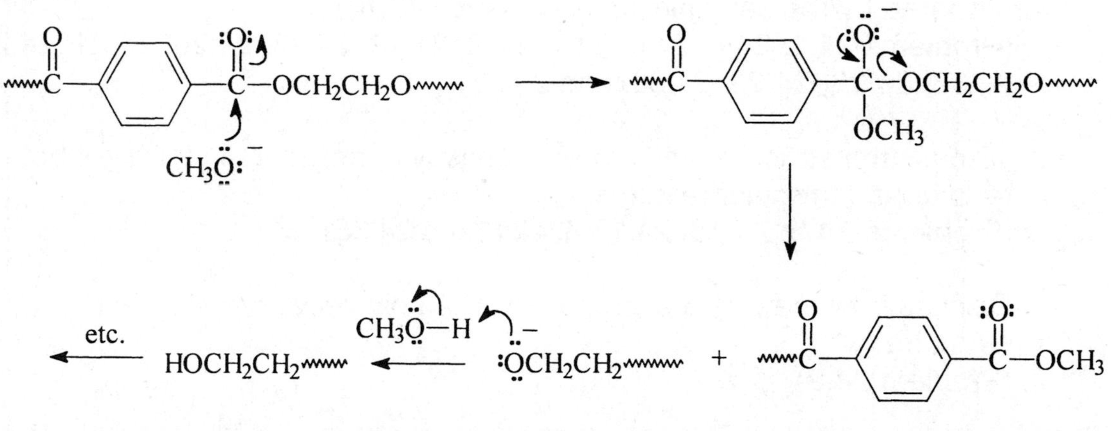

24.33 As T increases, TΔS gets larger. This causes the more disordered state (the more random state) to be more favored, and the rubber band shrinks.

24.34 The polymer of pure vinylidiene chloride is more crystalline because it has no chiral centers. Crystalline polymers are strong and stiff and are not suitable for applications such as thin film wraps. Copolymerizing with vinyl chloride makes the polymer more amorphous.

24.35 The double bonds in this polymer can be either cis or trans. In addition, its monomers can react by both 1,2- and 1,4-addition.

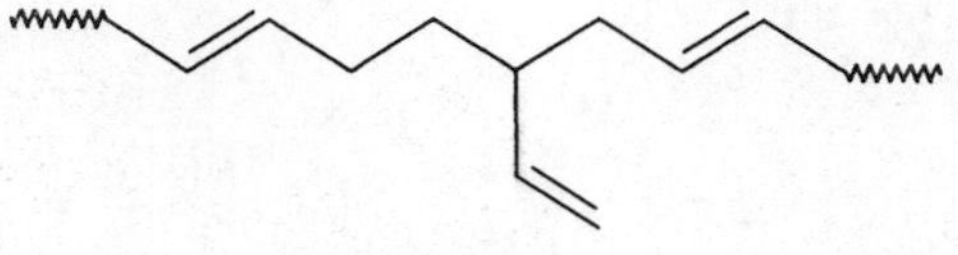

24.36 Coordination polymerization of ethylene with small amounts of 1-pentene will produce a less regular polymer because of the random placement of the three-carbon branches.

$$\text{wwwCH}_2\text{CH}_2 - \text{CH}_2\text{CH} \text{www}$$
$$\qquad\qquad\qquad\qquad | $$
$$\qquad\qquad\qquad \text{CH}_2\text{CH}_2\text{CH}_3$$

Review of Mastery Goals

After completing this chapter, you should be able to:

Show the repeat unit for any addition or condensation polymer.
(Problems 24.2, 24.3, 24.14, 24.15, 24.18, and 24.19)

Write the mechanisms for the formation of addition polymers that are prepared by radical, anionic, or cationic initiation.
(Problems 24.1, 24.4, 24.6, 24.7, 24.8, 24.11, 24.12, 24.20, 24.21, 24.22, 24.26, 24.28, 24.29, 24.30, and 24.32)

Discuss the structure and stereochemistry of polymers in terms of both regular and irregular features.
(Problems 24.5, 24.10, 24.13, 24.24, and 24.35)

Discuss how the physical properties of a polymer are related to its structure.
(Problem 24.9)

Discuss the chemical properties of a polymer.
(Problem 24.31)

Chapter 25
CARBOHYDRATES

25.1

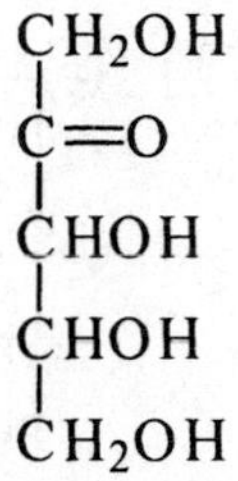

25.2 An aldopentose has three chirality centers, so there are $2^3 = 8$ stereoisomeric aldopentoses. A 2-ketopentose has two chirality centers, so there are four stereoisomeric 2-ketopentoses.

25.3

a) 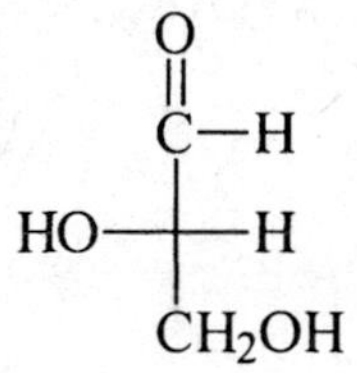

b) 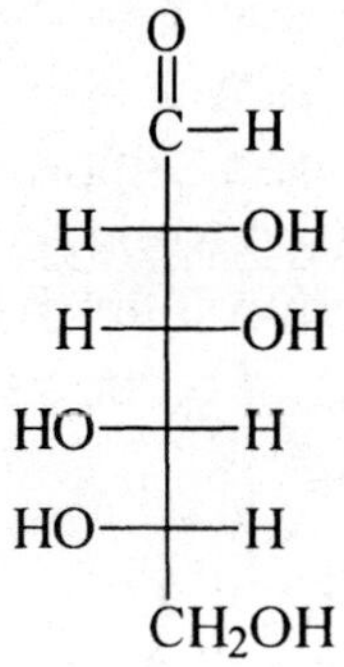

25.4 a) L-Erythrose b) L-Gulose c) D-Altrose

25.5

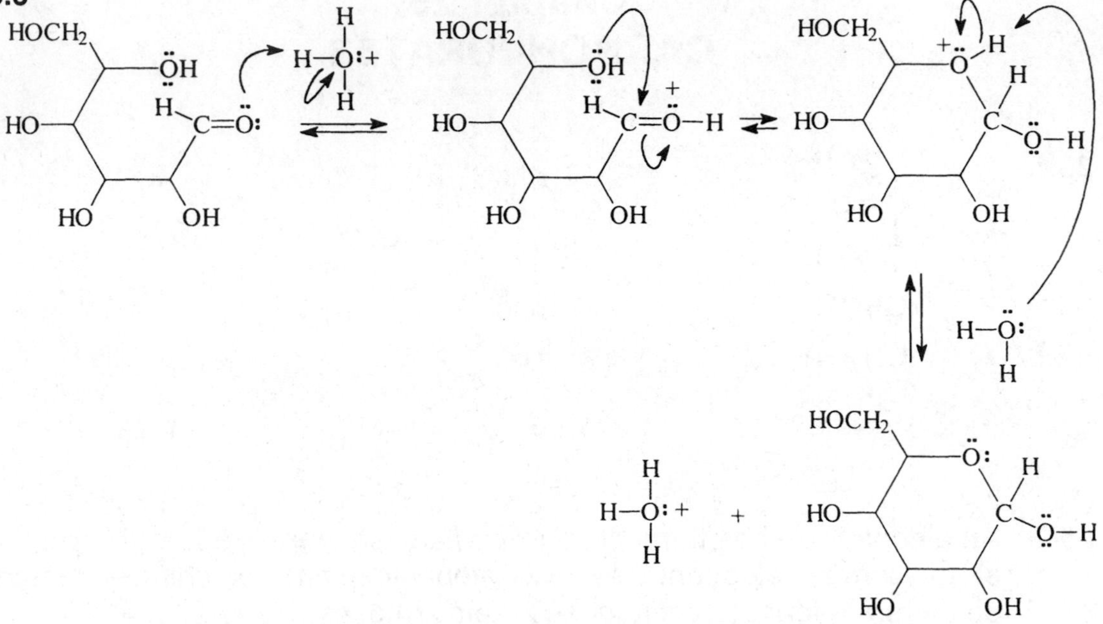

25.6 D-Fructose has more of its uncyclized form present at equilibrium than does D-glucose because its ketone carbonyl group is less reactive than the aldehyde carbonyl of glucose.

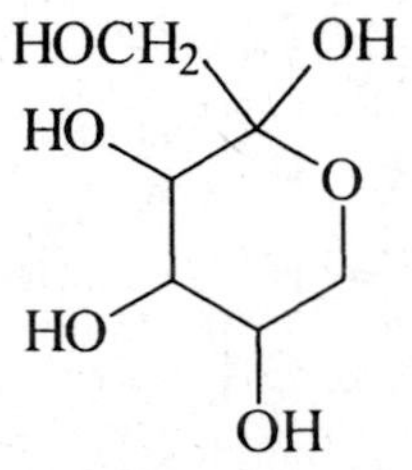

pyranose

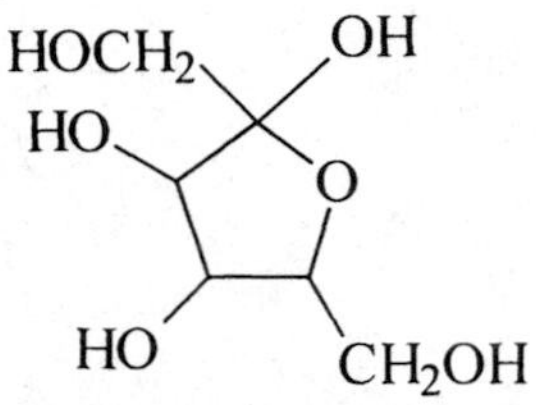

furanose

25.7 The distribution of the different forms of L-glucose is identical to that of D-glucose because they are enantiomers.

25.8

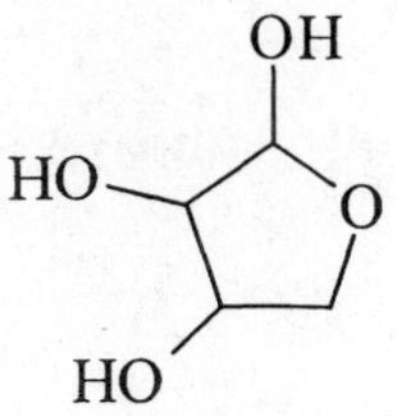

25.9

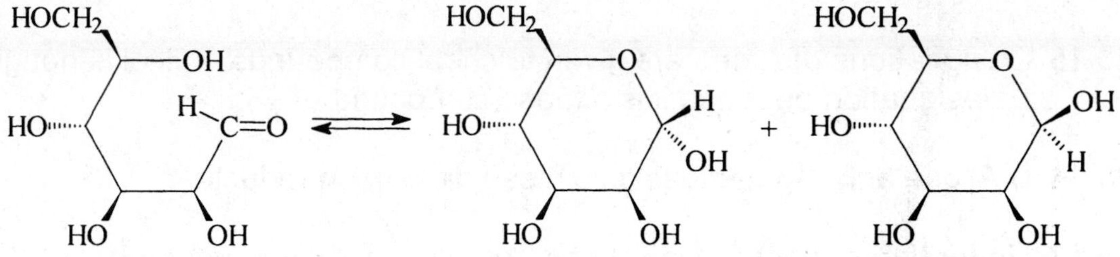

α-D-Mannopyranose β-D-Mannopyranose

25.10

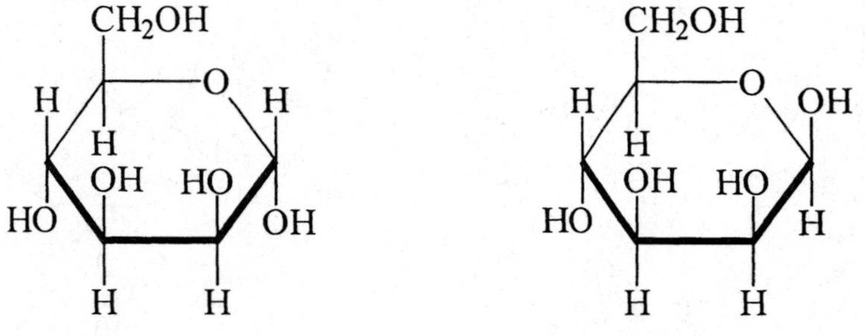

α-D-Mannopyranose β-D-Mannopyranose

25.11

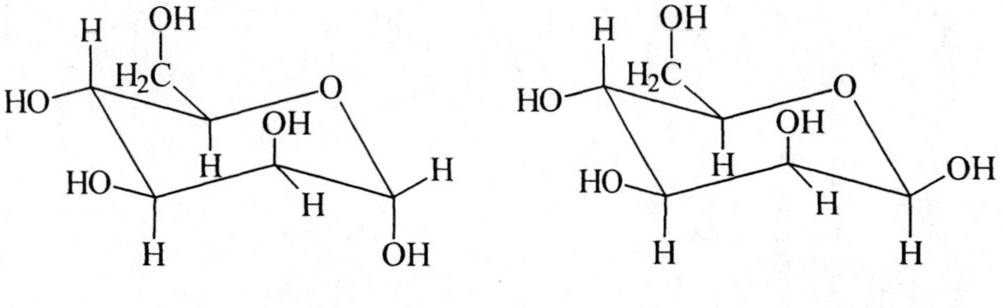

α-D-Mannopyranose β-D-Mannopyranose

25.12 67.4% α, 32.6% β

25.13 Galactaric acid has a plane of symmetry, so it is a meso compound.

25.14 D-Ribose and D-xylose give meso diacids upon oxidation.

25.15 Designations of D or L are given to chiral compounds. Xylitol is not given a D designation because it is a meso compound.

25.16 D-Allose and D-galactose give meso diols upon reduction.

25.17 Under these reaction conditions, α-D-glucopyranose undergoes isomerization to the β-isomer.

25.18

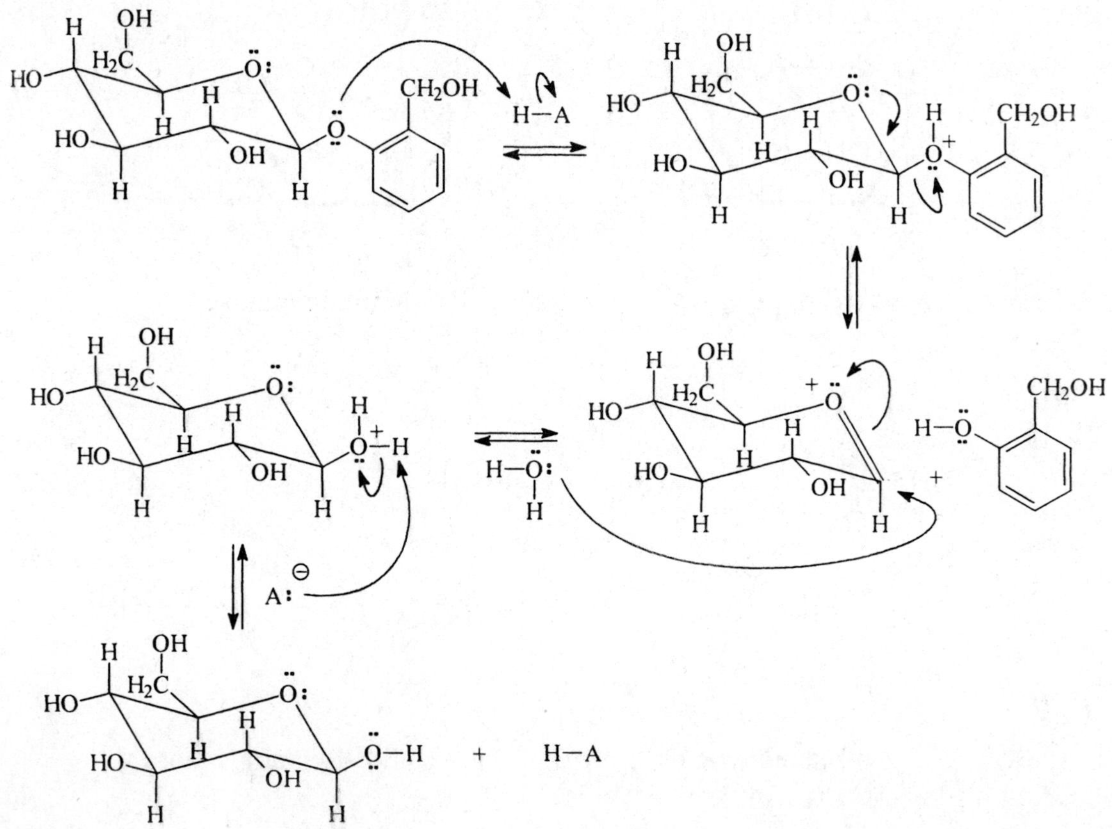

25.19

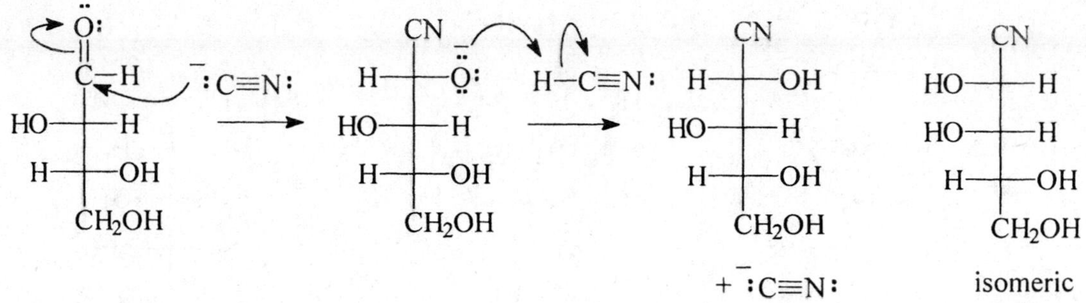

isomeric product

25.20

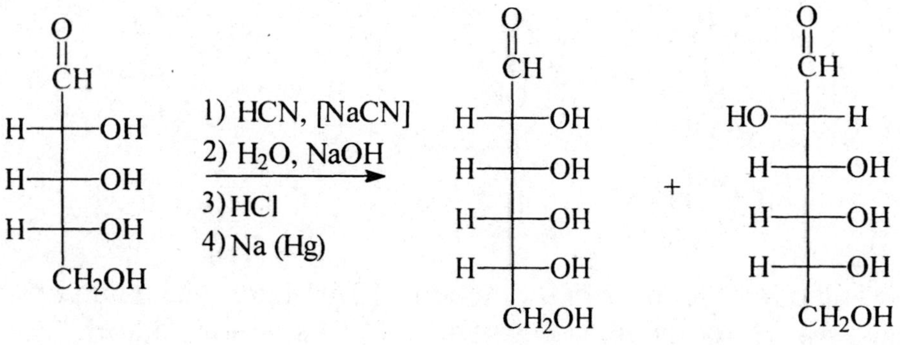

D-Rribose D-Allose D-Altrose

25.21

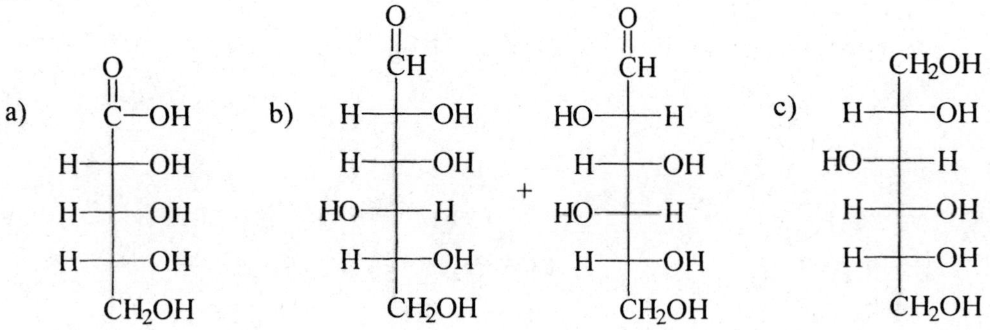

d) 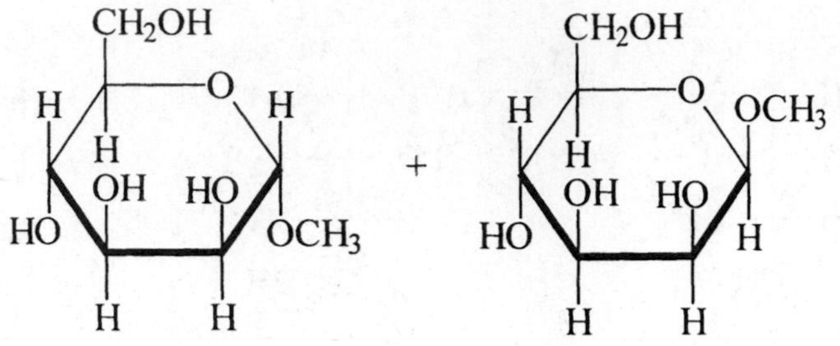

e) 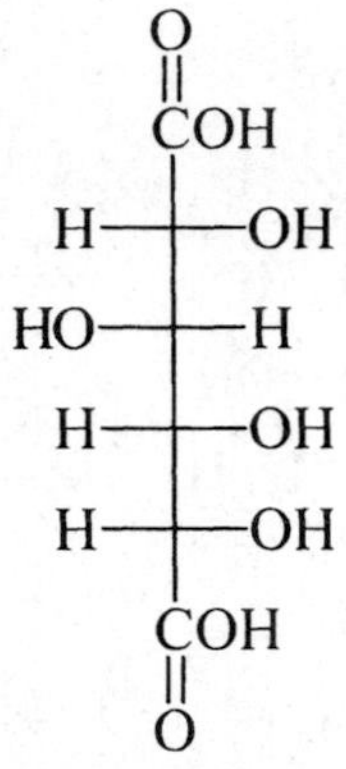

f) 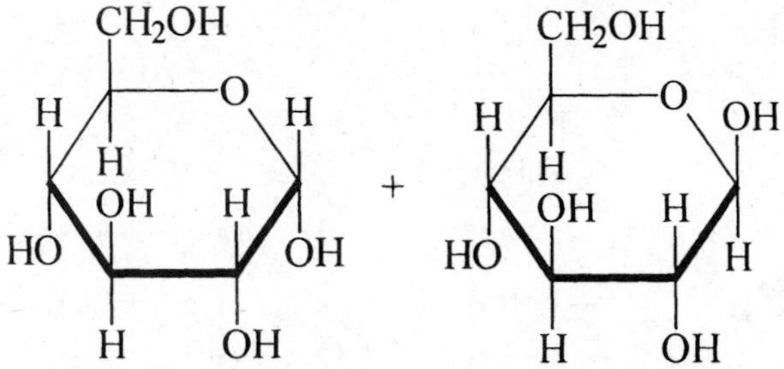

25.22 D-Arabinose must be **2** or **3** on page 1107. Sugar **2** gives an optically active diacid upon oxidation whereas **3** gives a meso diacid.

25.23

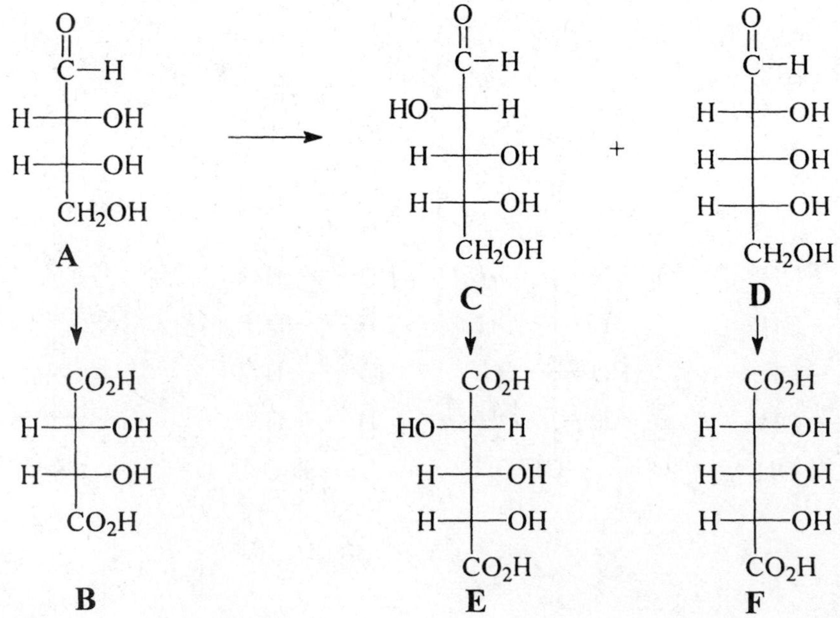

25.24 Maltose and cellobiose both have a hemiacetal group that is in equilibrium with an aldehyde group in aqueous solution. It is the aldehyde group that gives a positive test for a reducing sugar. Sucrose does not have a hemiacetal group (it has two acetal groups) so there is no aldehyde group present at equilibrium in a solution of sucrose.

25.25 There are no hydroxy groups present on C-2 and C-6.

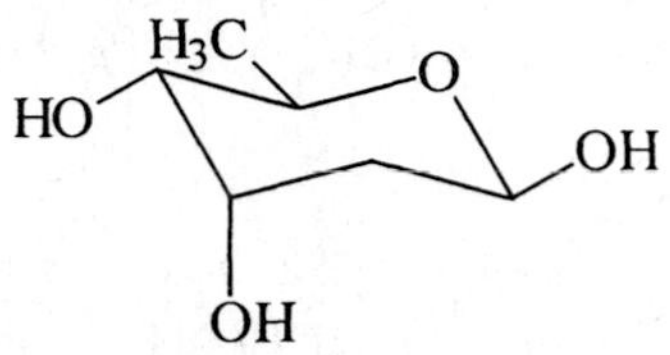

25.26

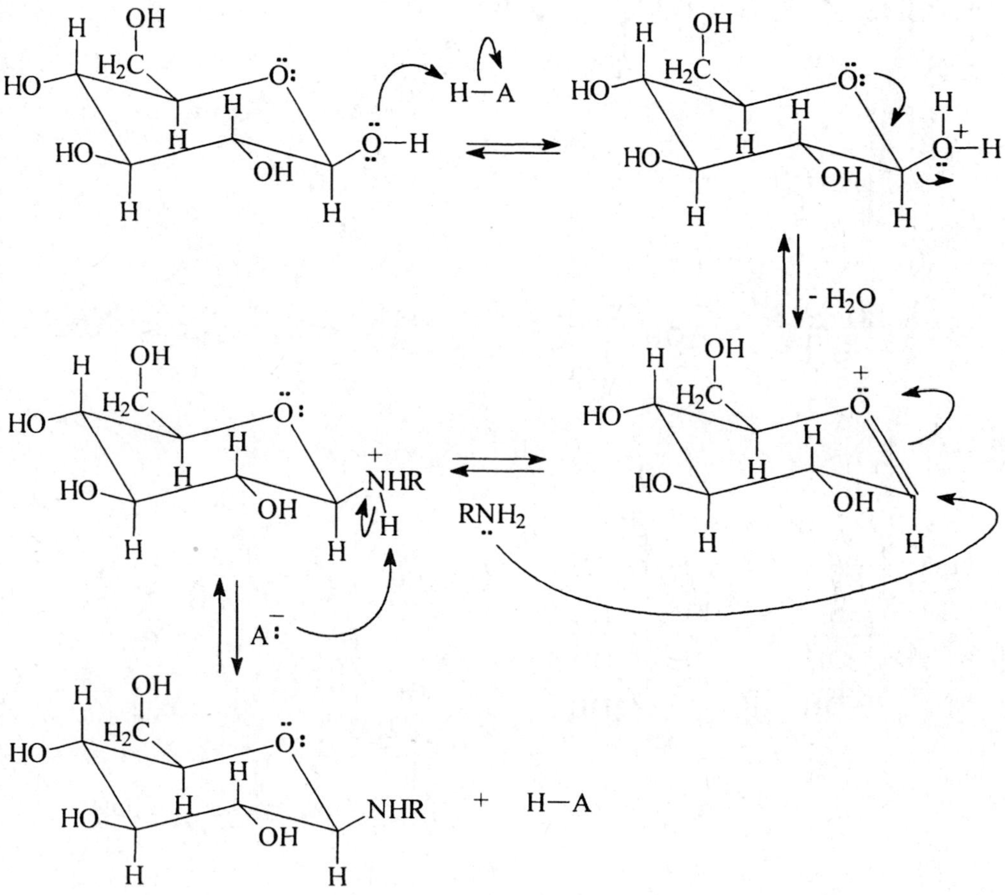

25.27

a) 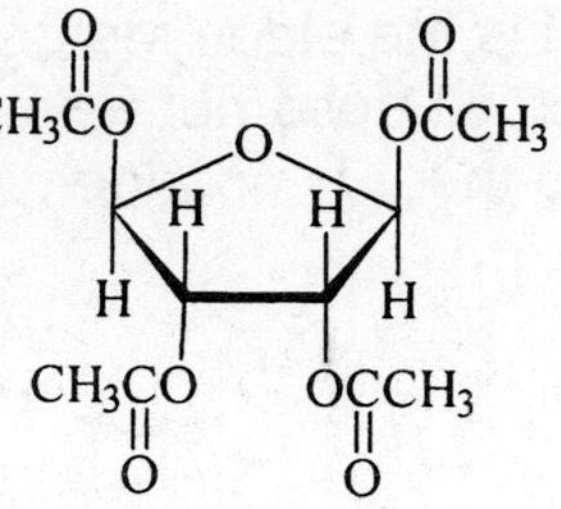

b) 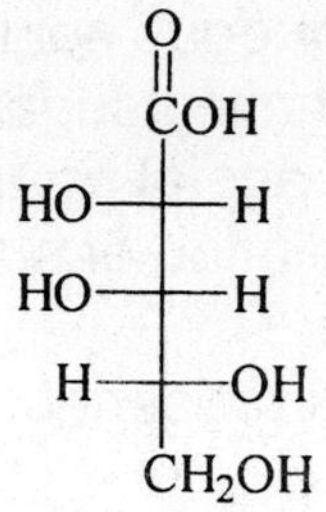

c) 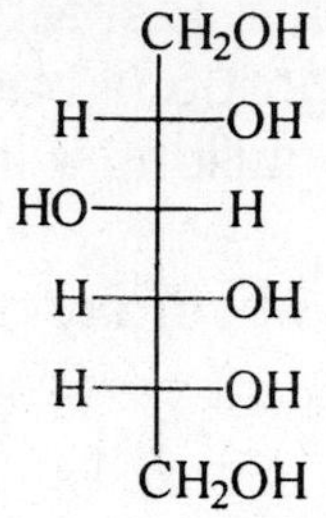

d) 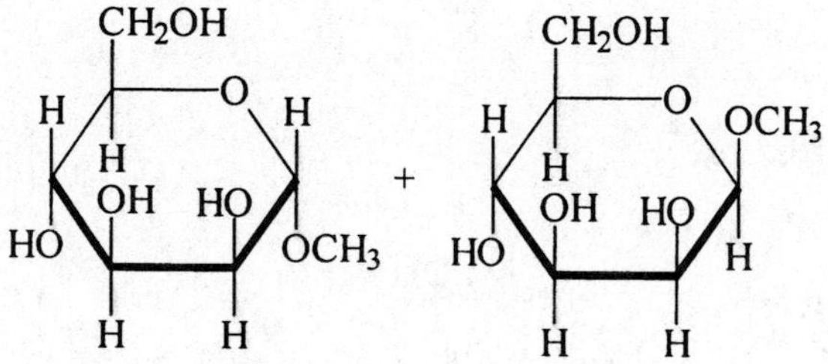

e) 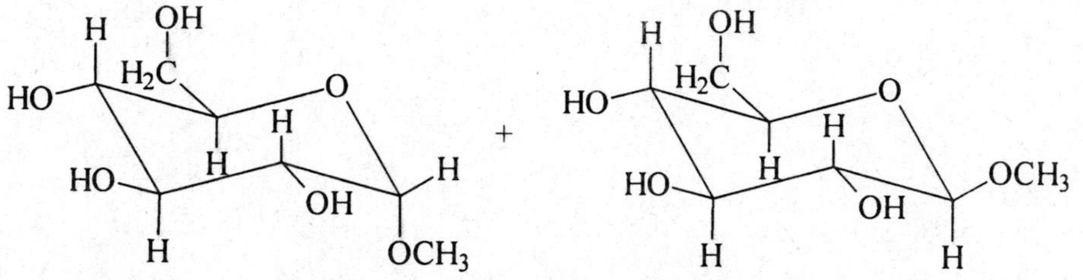

f) 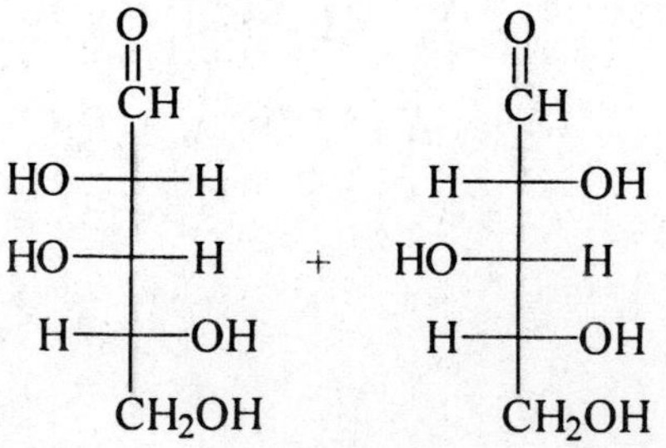

g) 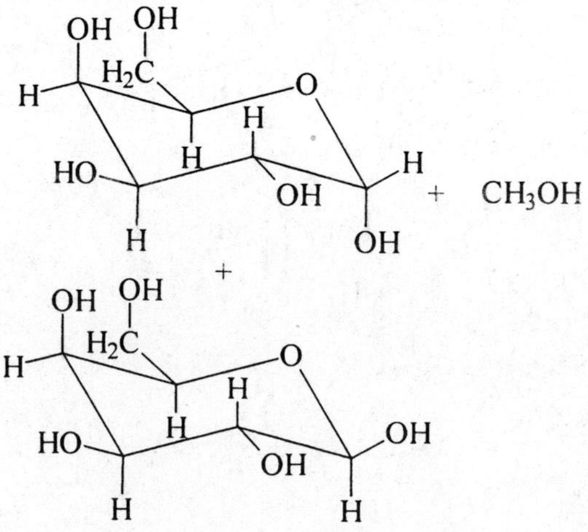

25.28

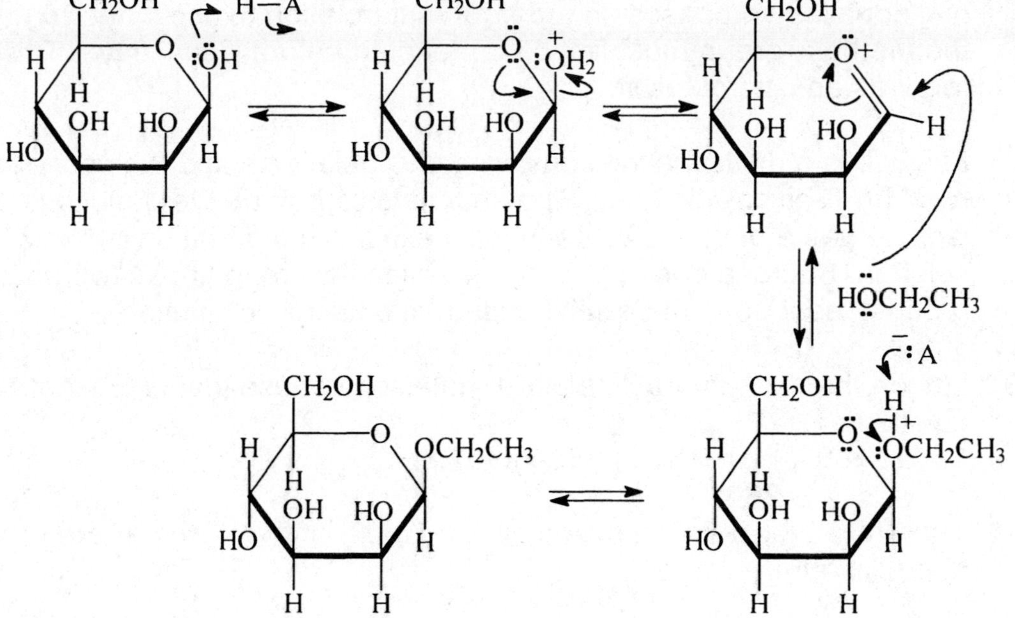

25.29

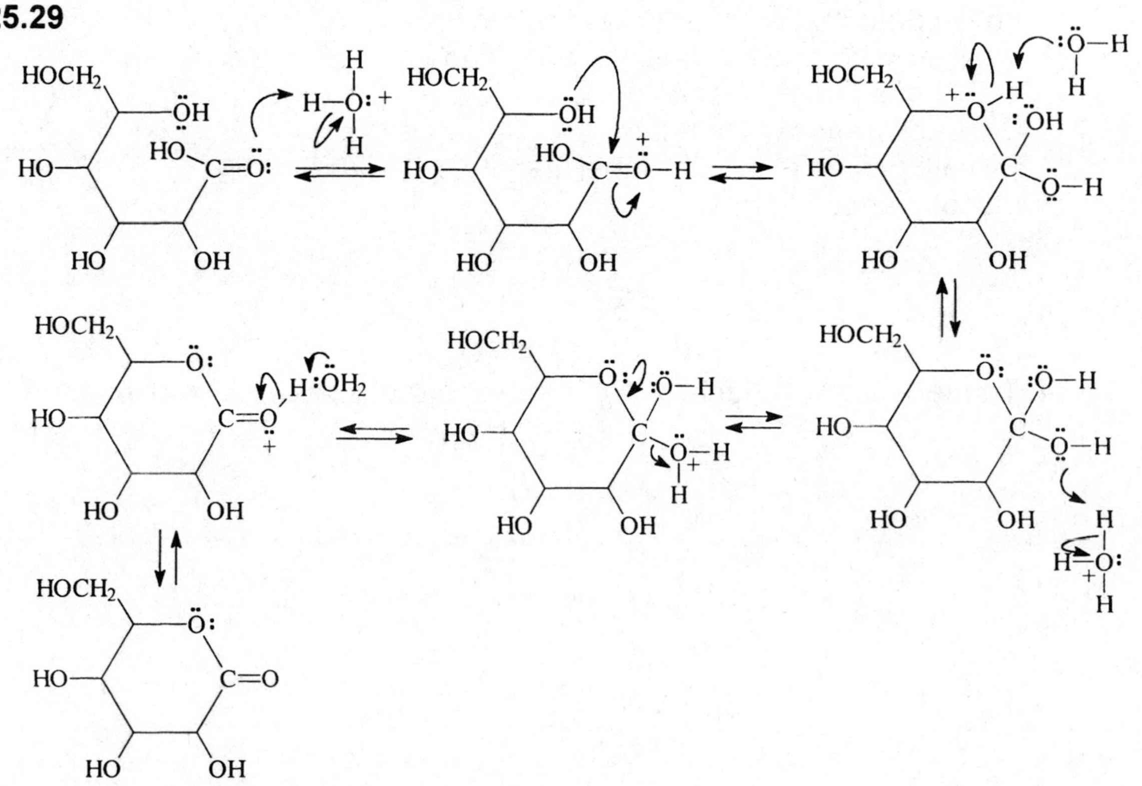

25.30 Methyl-α-D-glucoside is an acetal and acetals are very stable to nucleophiles and bases. In order for mutarotation to occur, the oxygen of the methoxy group must first be protonated before it can leave. This requires an acid catalyst.

25.31 Lactose is a disaccharide consisting of D-galactose and D-glucose joined by a 1,4-β-glycosidic bond. The anomeric carbon on the D-glucose unit of lactose has a hemiacetal group, so it exhibits mutarotation in basic solution. Sucrose does not have a hemiacetal group (it has two acetal groups), so it does not exhibit mutarotation in basic solution.

25.32 Let x = the decimal fraction of α-D-galactopyranose that is present at equilibrium.

$$+80.2 = x(+150.7) + (1-x)(+52.8)$$
$$x = 0.280$$

Therefore 28.0% of the α-isomer and 72.0% of the β-isomer are present at equilibrium.

25.33 All substituents are equatorial in the stable chair conformation of β-D-glucopyranose, whereas one hydroxy group is axial in the stable chair conformation of β-D-allopyranose. Therefore β-D-allopyranose is less stable than β-D-glucopyranose.

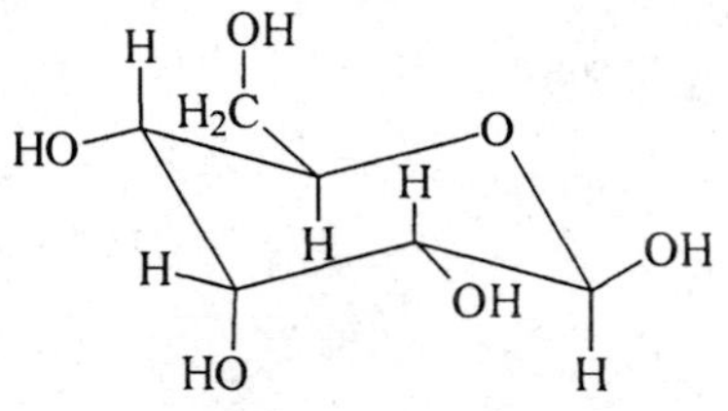

β-D-Allopyranose

25.34 The reduction products of D-glucose and D-gulose are enantiomers.

25.35

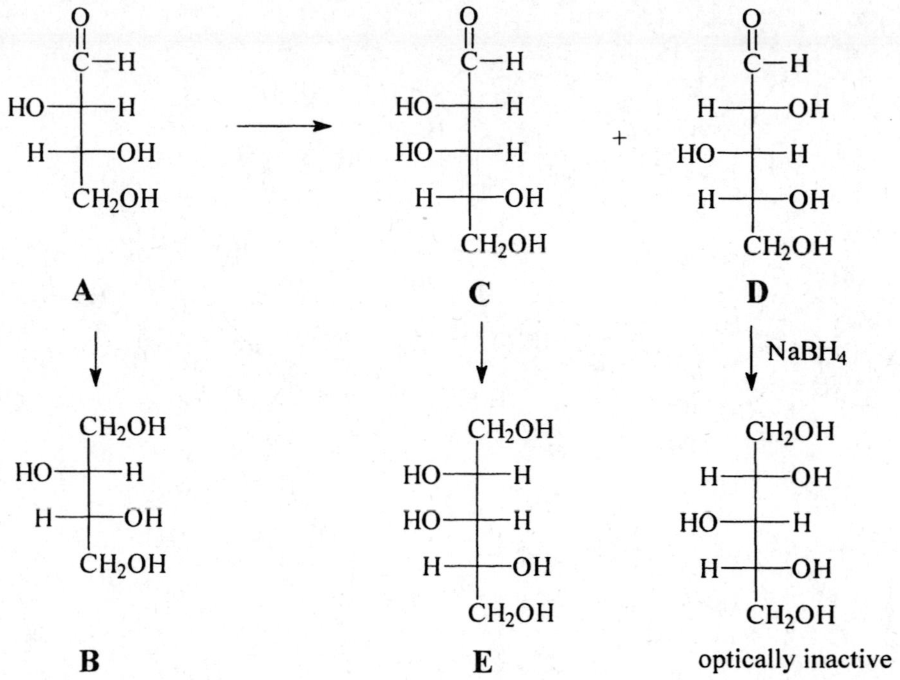

The reduction product of **D** has a plane of symmetry and is meso, so it will not rotate plane polarized light.

25.36

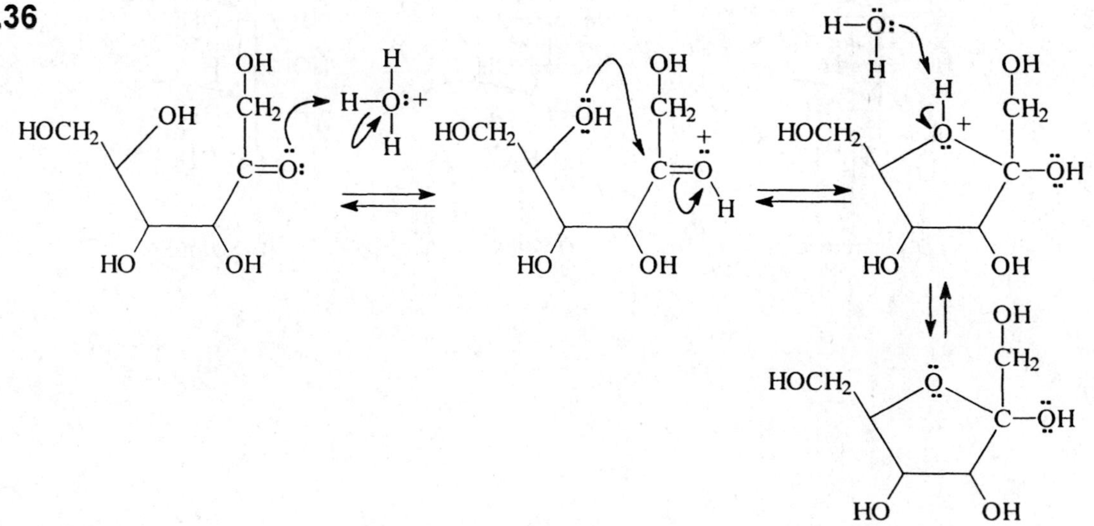

25.37

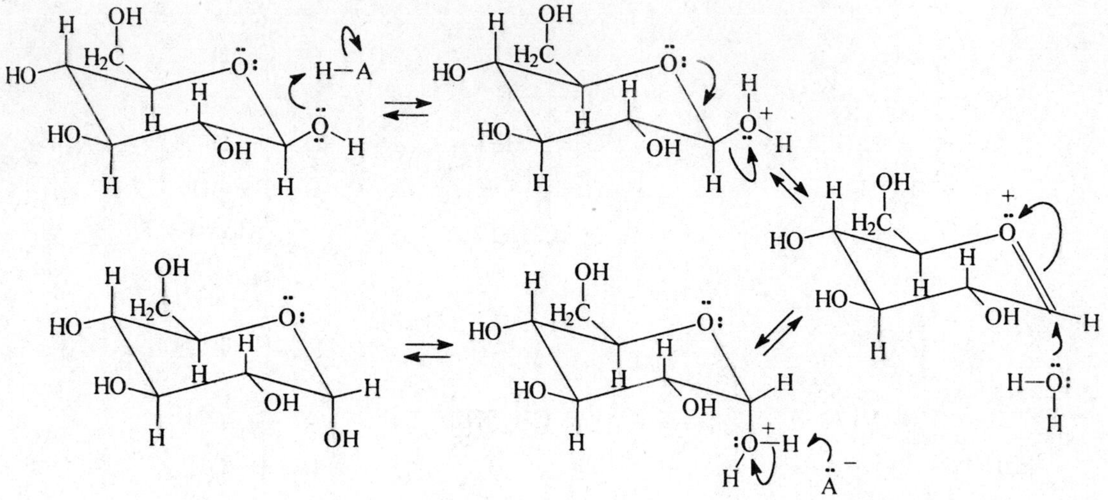

25.38

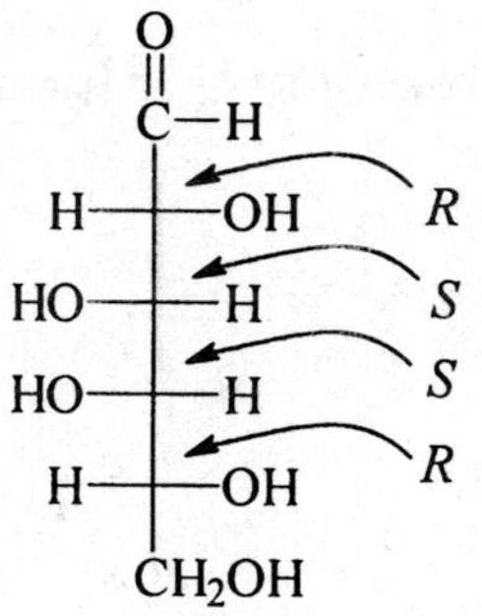

D-Galactose

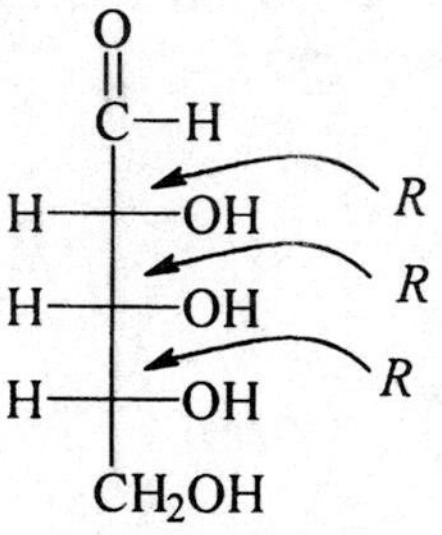

D-Ribose

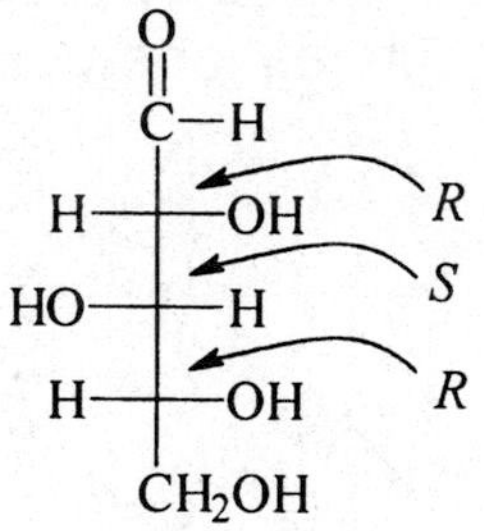

D-Xylose

25.39

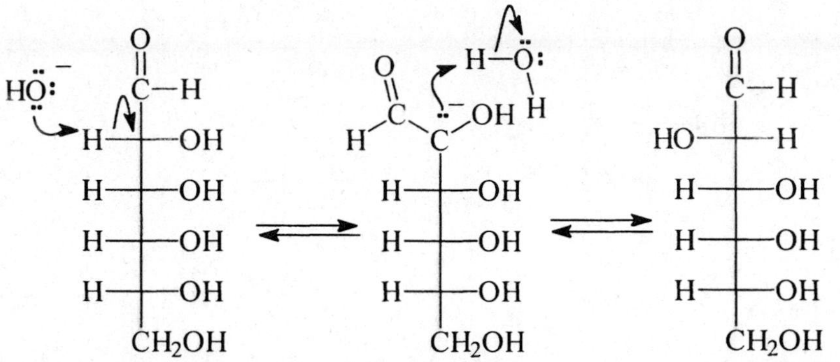

D-Allose D-Altrose

25.40 Three successive Kiliani-Fisher syntheses starting from D-glyceraldehyde produce the D-isomers of all eight aldohexoses. These are all D-isomers because the configuration at C-2 of glyceraldehyde is maintained throughout all of the reactions and this carbon ends up as C-5 of all of the aldohexoses. Comparison of the glucose thus synthesized with natural glucose shows them to be the same enantiomer, so natural glucose is D.

25.41 D-Galactose can be converted to D-talose by treatment with a base. The reaction proceeds through a planar enolate anion.

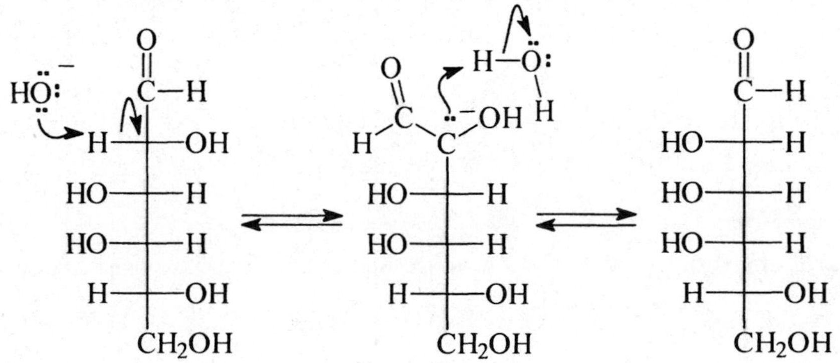

D-Galactose D-Talose

25.42

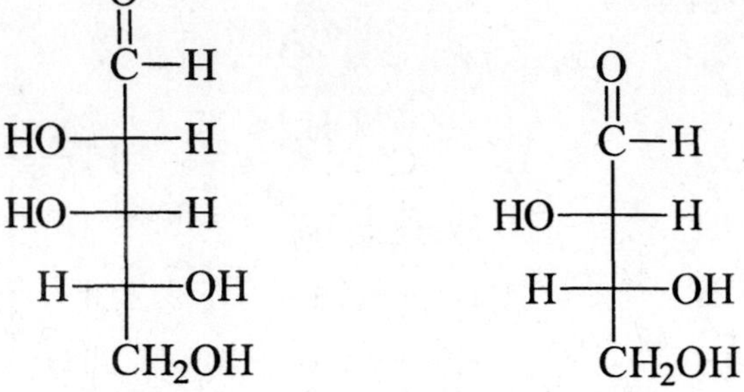

X Y

25.43

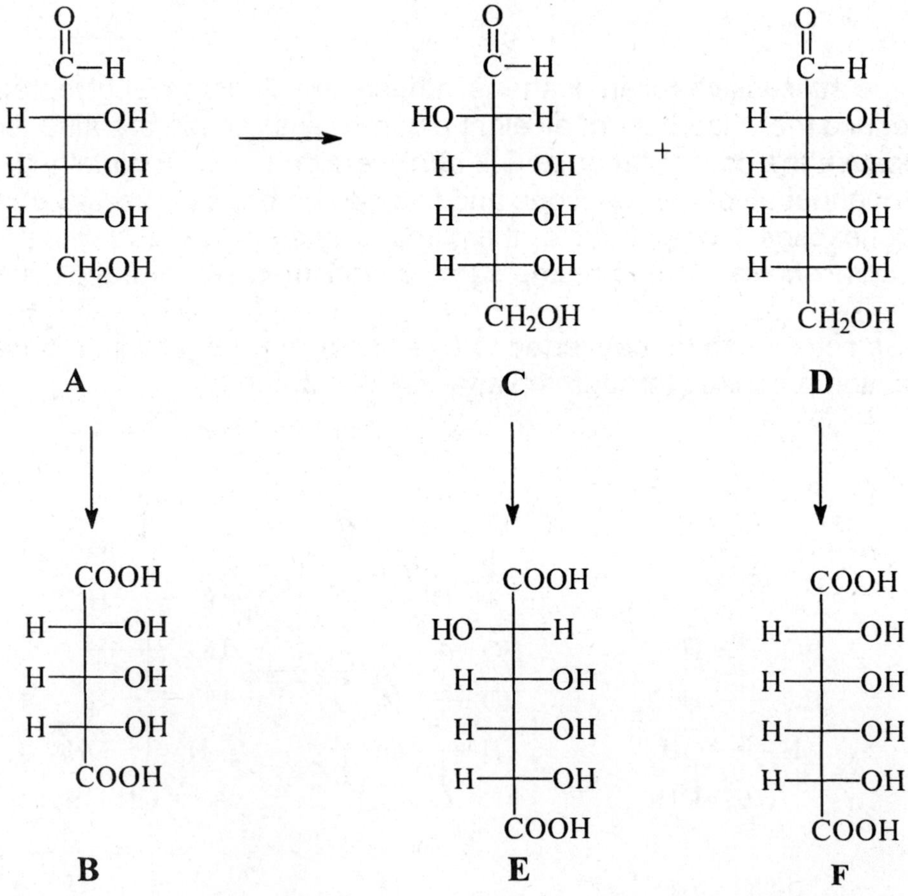

O
‖
C—H
H———OH
H———OH
H———OH
CH₂OH

A

→

O
‖
C—H
HO———H
H———OH
H———OH
H———OH
CH₂OH

C

+

O
‖
C—H
H———OH
H———OH
H———OH
H———OH
CH₂OH

D

↓

COOH
H———OH
H———OH
H———OH
COOH

B

↓

COOH
HO———H
H———OH
H———OH
H———OH
COOH

E

↓

COOH
H———OH
H———OH
H———OH
H———OH
COOH

F

452

25.44

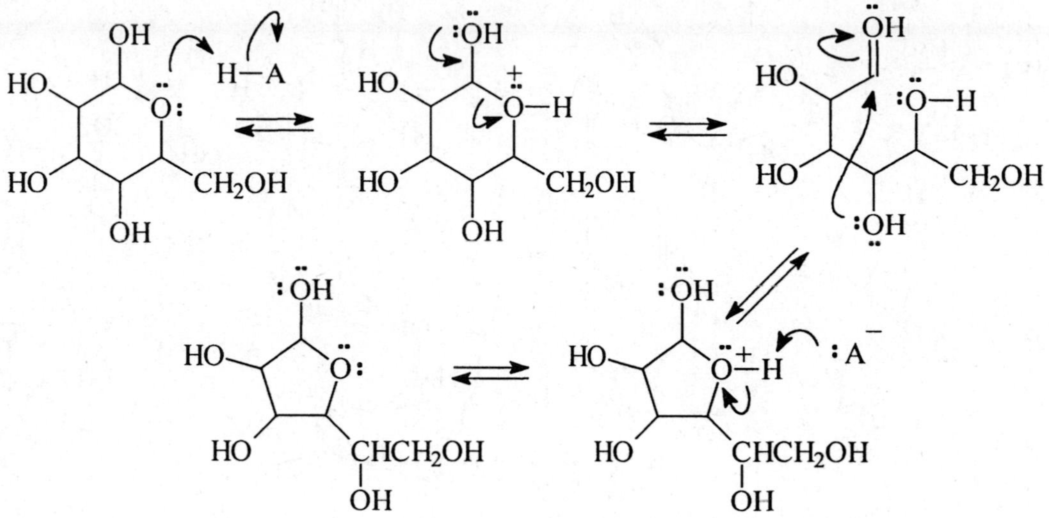

25.45

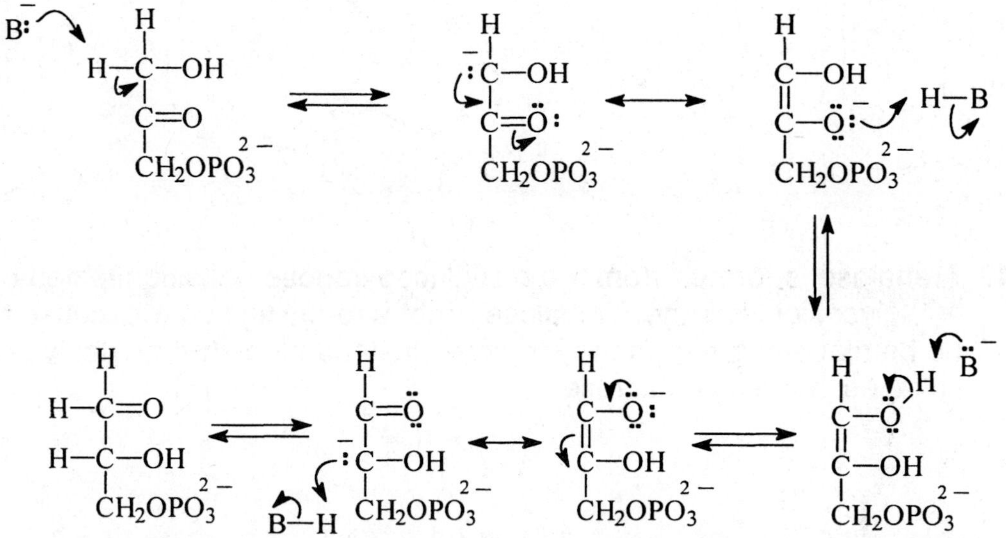

25.46

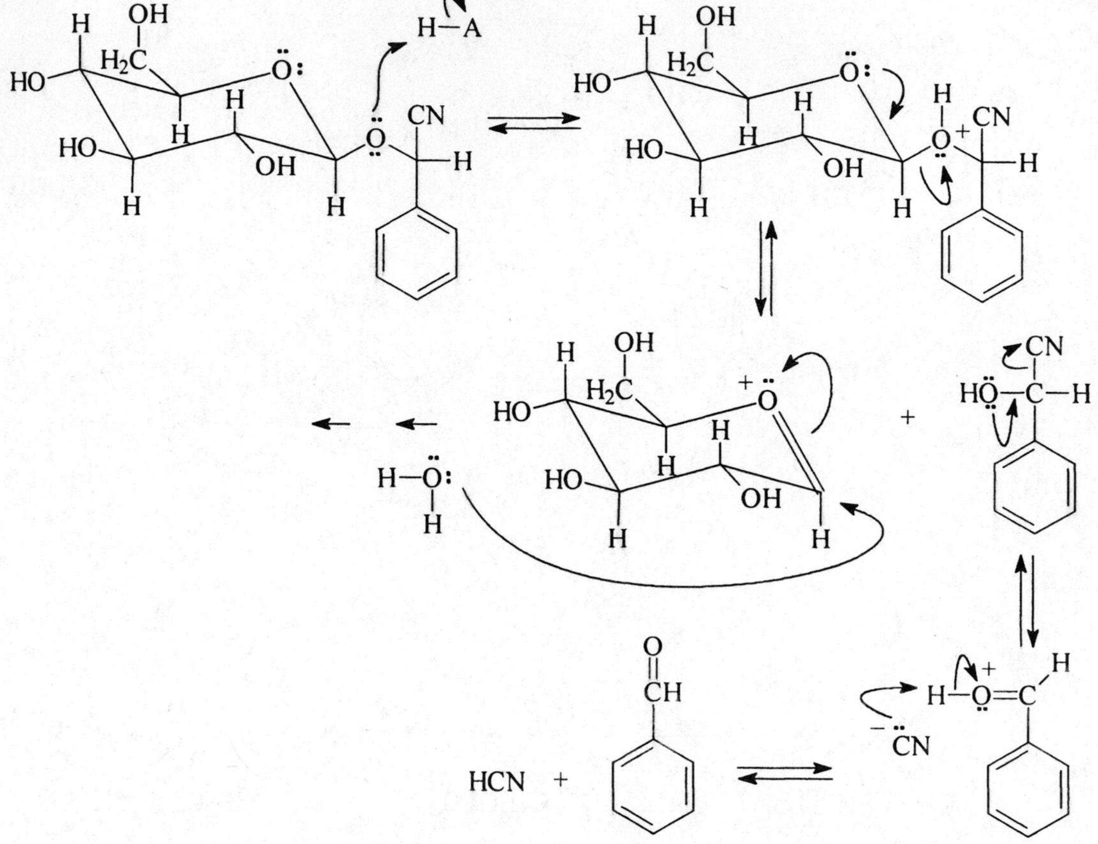

25.47 Trehalose is formed from two α-D-glucopyranose units connected by a 1,1'-glycosidic linkage. Trehalose is not a reducing sugar because there is no hemiacetal group in this structure. Instead it has two acetal type linkages, much like sucrose.

25.48 Dihydroxyacetone and D-glyceraldehyde are interconverted in base (see the mechanism in Problem 25.45), so both are present in a basic solution prepared from either. The reaction in this problem is an aldol condensation of the enolate anion derived from dihydroxyacetone as the nucleophile and the carbonyl carbon of glyceraldehyde as the electrophile.

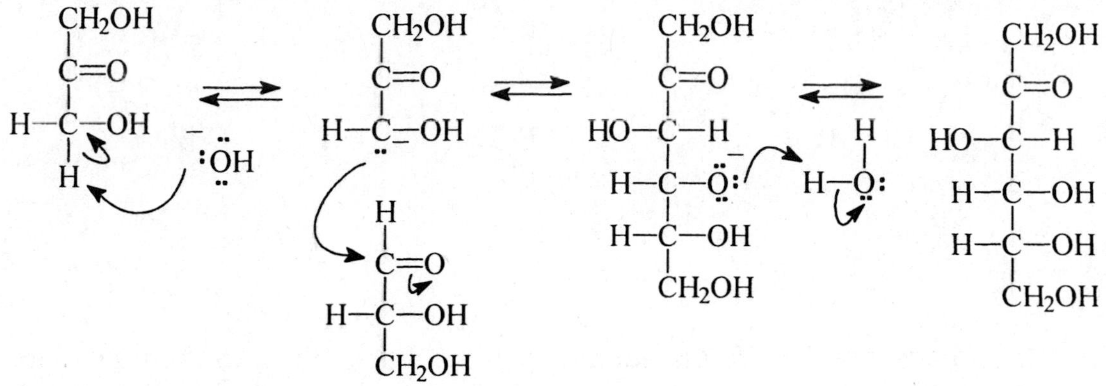

25.49

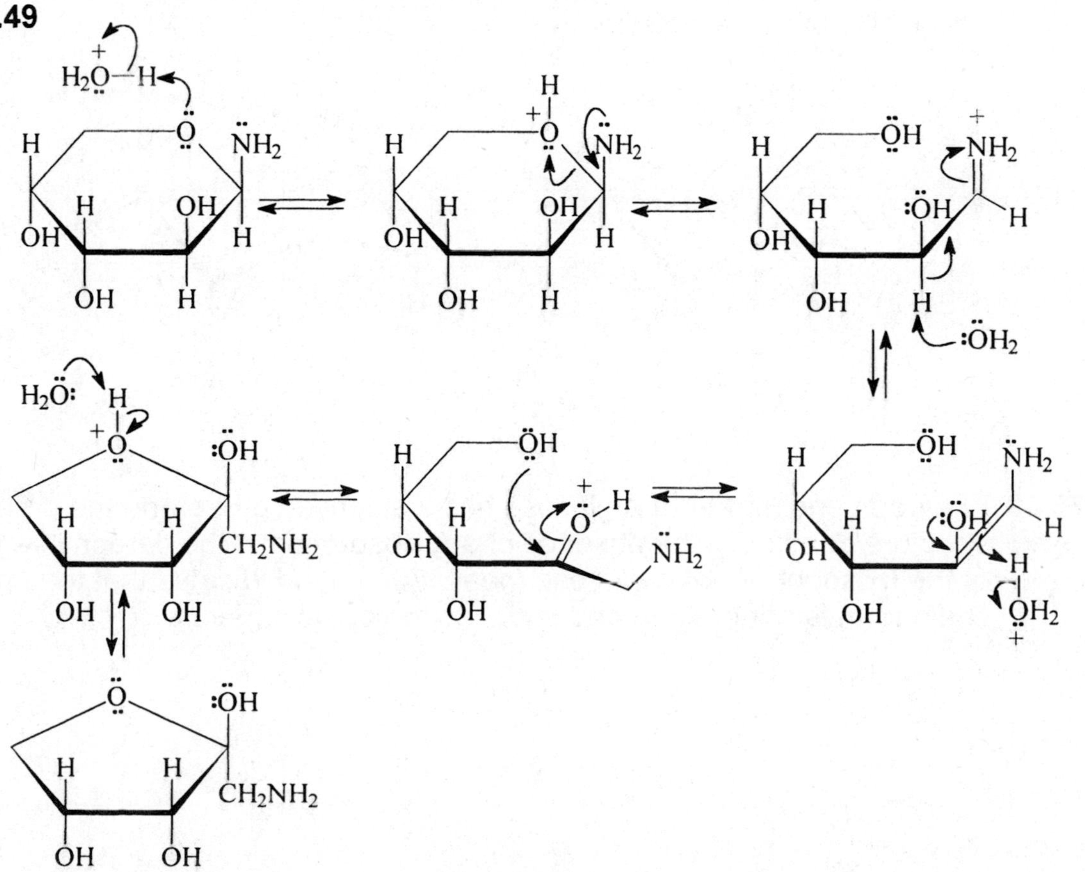

25.50

D-Talose D-Altrose

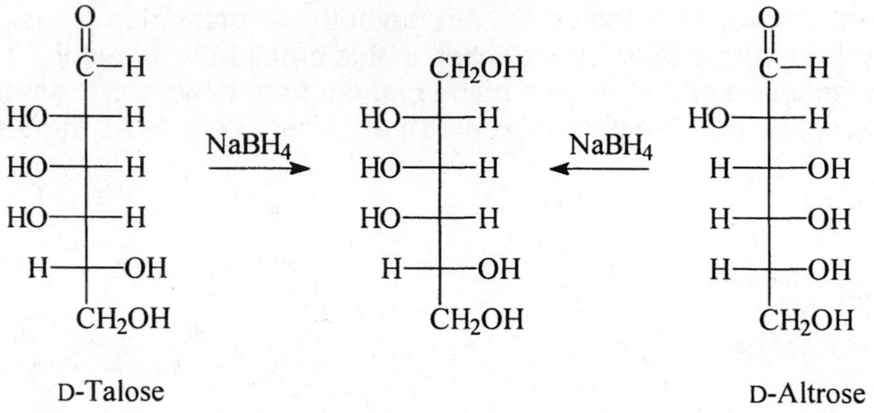

25.51 The hydrogen on C-2 is equatorial in the stable chair conformations of the stereoisomers of D-mannopyranose. Its dihedral angle with the H on C-1 is near 60° in both stereoisomers, so the coupling constant is nearly identical in both stereoisomers.

α-D-Mannopyranose

β-D-Mannopyranose

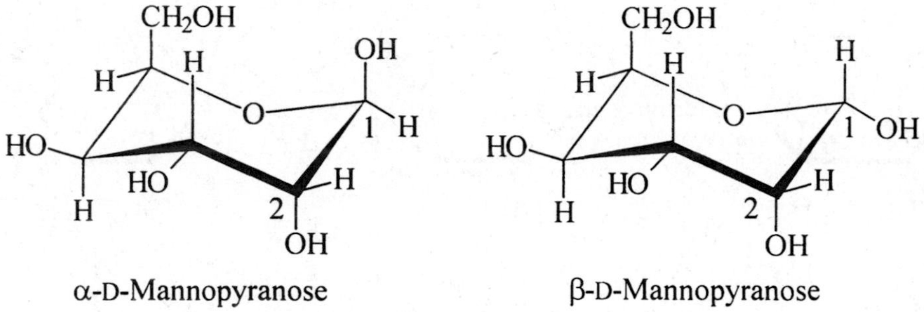

25.52 The open chain form of D-glucose has a aldehyde group and the cyclic structure does not. The absence of any absorption in the carbonyl region of the IR spectrum of D-glucose (near 1700 cm^{-1}) indicates that the open chain form is not present and supports the cyclic structure.

Review of Mastery Goals

After completing this chapter, you should be able to:

Show the general structures for carbohydrates including the variations that occur.
(Problem 25.1)

Discuss the stereochemistry of carbohydrates, including the use of D or L to designate absolute stereochemistry.
(Problems 25.2, 25.3, 25.4, 25.15, and 25.33)

Understand the cyclization of monosaccharides to form pyranose and furanose rings.
(Problems 25.5, 25.6, 25.7, 25.8, 25.9, 25.10, 25.11, 25.29, 25.30, 25.31, 25.36, and 25.44)

Show the common reactions of monosaccharides that were presented in this chapter: oxidation with nitric acid; oxidation with bromine; reduction with sodium borohydride; esterification; glycoside formation; and the Kiliani-Fischer synthesis.
(Problems 25.13, 25.14, 25.16, 25.17, 25.18, 25.19, 25.20, 25.21, 25.27, and 25.34)

Understand Fischer's structure proof for glucose and apply this type of reasoning to other stereochemical problems.
(Problems 25.22, 25.23, 25.35, 25.42, 25.43, and 25.50)

Understand the general structural features of disaccharides and polysaccharides.
(Problems 25.24, and 25.47)

Chapter 26
AMINO ACIDS, PEPTIDES, AND PROTEINS

26.1 Isoleucine and threonine

26.2 The positively charged nitrogen is an inductive electron withdrawing group, and so makes the carboxylic acid group a stronger acid.

26.3 The ester is the stronger acid because the inductive electron-withdrawing effect of the ester group increases the acid strength of the ammonium group. The carboxylate anion is an electron-donating group and decreases the acid strength of the ammonium group.

26.4 The inductive electron withdrawing effect of the positively charged nitrogen, which increases acid strength, is stronger at the main carboxylic acid group because the distance separating the groups is smaller. Inductive effects decrease rapidly with distance.

26.5 The carboxylic acid group in the side chain of glutamic acid is farther from the electron-withdrawing positive nitrogen than is the case with aspartic acid.

26.6

a) structure with pyrrolidine ring, $\overset{+}{N}H_2$ and CO_2^-

b) $\overset{-}{S}-CH_2\overset{\overset{+}{N}H_3}{\underset{|}{C}}H-CO_2^-$

c) $H_2N\overset{\overset{+}{N}H_2}{=}CNHCH_2CH_2CH_2\overset{\overset{+}{N}H_3}{\underset{|}{C}}HCO_2^-$

26.7

a) $H_3\overset{+}{N}CH_2CH_2CH_2CH_2\overset{\overset{+}{N}H_3}{\underset{|}{C}}HCO_2H$ $H_3\overset{+}{N}CH_2CH_2CH_2CH_2\overset{\overset{+}{N}H_3}{\underset{|}{C}}HCO_2^-$

$H_3\overset{+}{N}CH_2CH_2CH_2CH_2\overset{NH_2}{\underset{|}{C}}HCO_2^-$ $H_2NCH_2CH_2CH_2CH_2\overset{NH_2}{\underset{|}{C}}HCO_2^-$

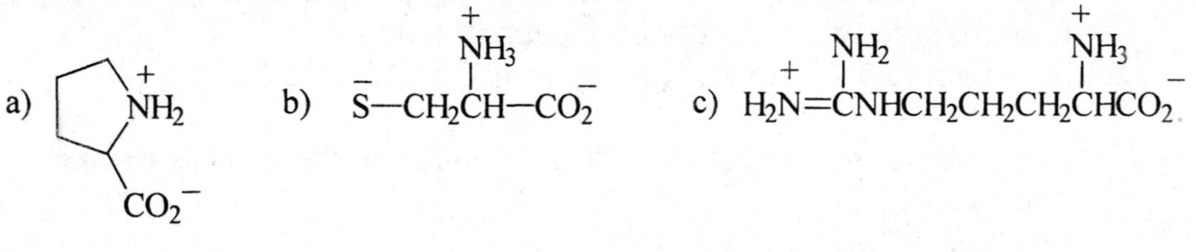

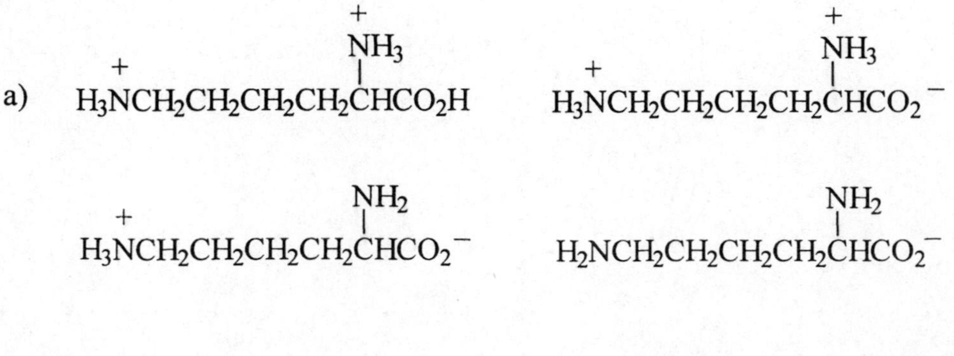

b)

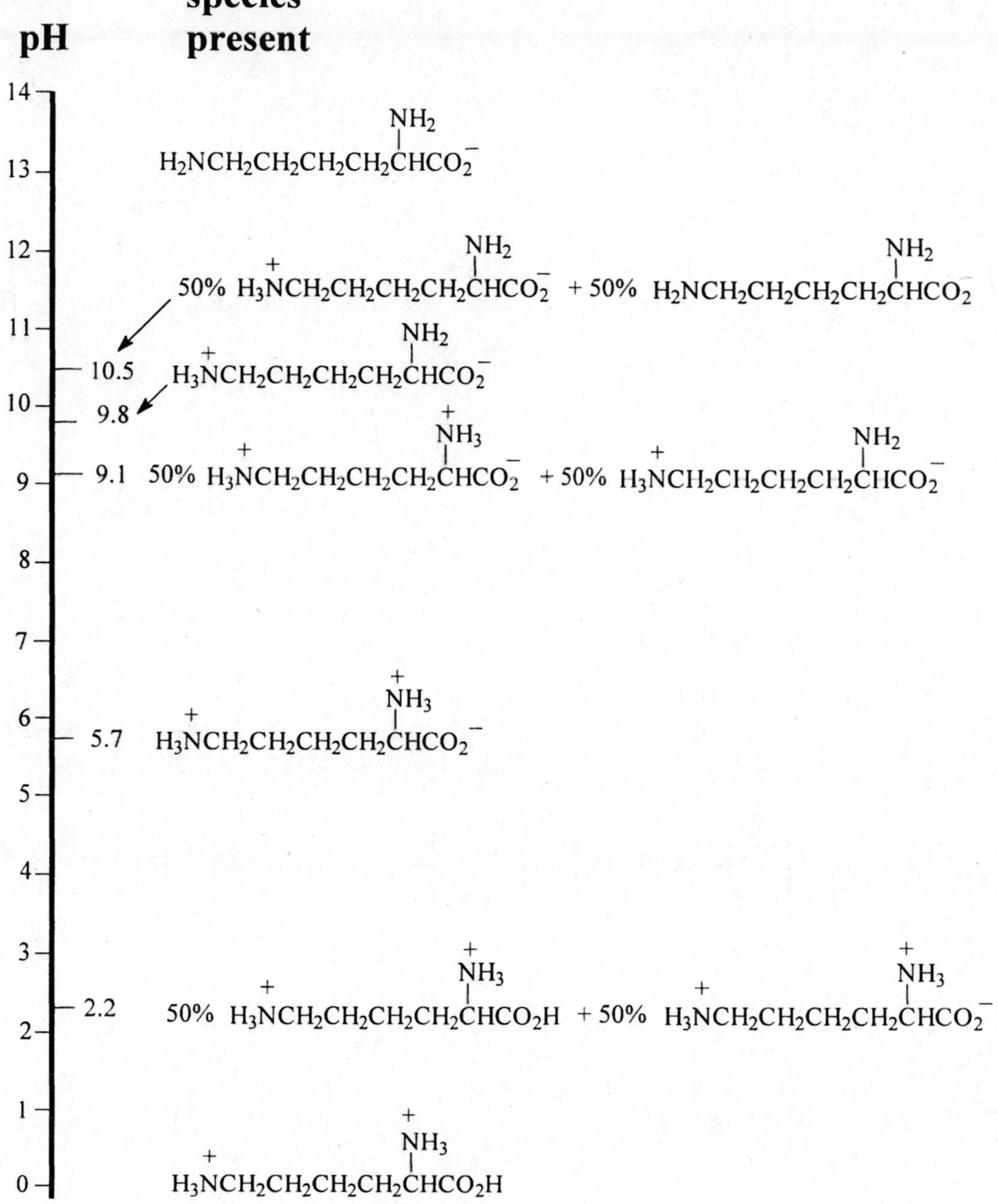
pH
species present
14
13
12
11
10
9
8
7
6
5
4
3
2
1
0
10.5
9.8
9.1
5.7
2.2
H2NCH2CH2CH2CH2CHCO2− (NH2)
50% H3N+CH2CH2CH2CH2CHCO2− (NH2) + 50% H2NCH2CH2CH2CH2CHCO2− (NH2)
H3N+CH2CH2CH2CH2CHCO2− (NH2)
50% H3N+CH2CH2CH2CH2CHCO2− (N+H3) + 50% H3N+CH2CH2CH2CH2CHCO2− (NH2)
H3N+CH2CH2CH2CH2CHCO2− (N+H3)
50% H3N+CH2CH2CH2CH2CHCO2H (N+H3) + 50% H3N+CH2CH2CH2CH2CHCO2− (N+H3)
H3N+CH2CH2CH2CH2CHCO2H (N+H3)

26.8

a)
Cl⁻

$\overset{+}{N}H_3\overset{CH_3}{\underset{|}{C}}H-\overset{O}{\overset{\|}{C}}OCH_2Ph$

b)
$\overset{O}{\overset{\|}{Ph C}}-NH\overset{\overset{CH_3CHCH_3}{|}}{\overset{CH_2}{\underset{|}{C}}H}-\overset{O}{\overset{\|}{C}}OH$

c)
$CH_3\overset{O}{\overset{\|}{C}}-NH\overset{CH_3}{\underset{|}{C}}H-\overset{O}{\overset{\|}{C}}OH$

26.9

26.10

a)
$PhCH_2\overset{NH_2}{\underset{|}{C}}HCN \xrightarrow[H_2O]{HCl} PhCH_2\overset{NH_2}{\underset{|}{C}}HCO_2H$

b)
$CH_3\overset{CH_3}{\underset{|}{C}}H\overset{}{\underset{\overset{|}{Br}}{C}}H\overset{O}{\overset{\|}{C}}OH \xrightarrow[NH_3]{excess} CH_3\overset{CH_3}{\underset{|}{C}}H\overset{}{\underset{\overset{|}{NH_2}}{C}}H\overset{O}{\overset{\|}{C}}OH$

c)
phthalimide–N–$\overset{CO_2Et}{\underset{CO_2Et}{C}}$–$CH_2\overset{CH_3}{\underset{|}{C}}HCH_3$ $\xrightarrow[\underset{\Delta}{H_2O}]{HCl}$ $CH_3\overset{CH_3}{\underset{|}{C}}HCH_2\overset{NH_2}{\underset{|}{C}}HCO_2H$

26.11

a)
$CH_3\overset{CH_3}{\underset{|}{C}}HCH_2\overset{O}{\overset{\|}{C}}H \xrightarrow[\text{2) HCl, }H_2O]{\overset{\text{1) NaCN}}{NH_4Cl, H_2O}} CH_3\overset{CH_3}{\underset{|}{C}}HCH_2\overset{NH_2}{\underset{|}{C}}HCO_2H$

b)
$PhCH_2CH_2-\overset{O}{\overset{\|}{C}}OH \xrightarrow[\text{2) excess } NH_3]{\text{1) } Br_2, [PBr_3]} PhCH_2\overset{NH_2}{\underset{|}{C}}H-\overset{O}{\overset{\|}{C}}OH$

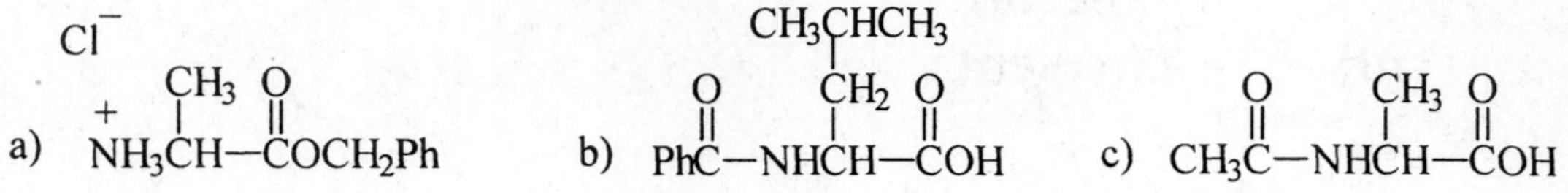

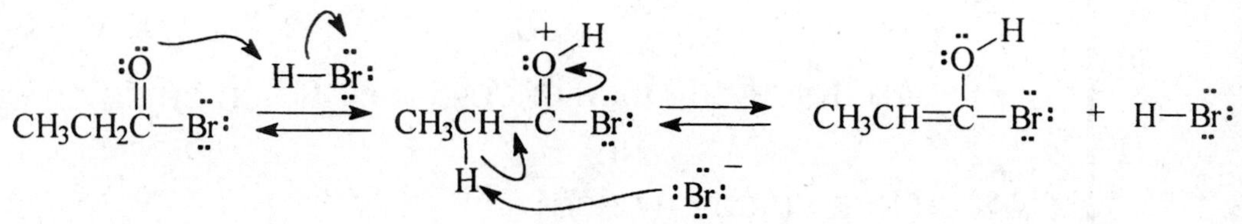

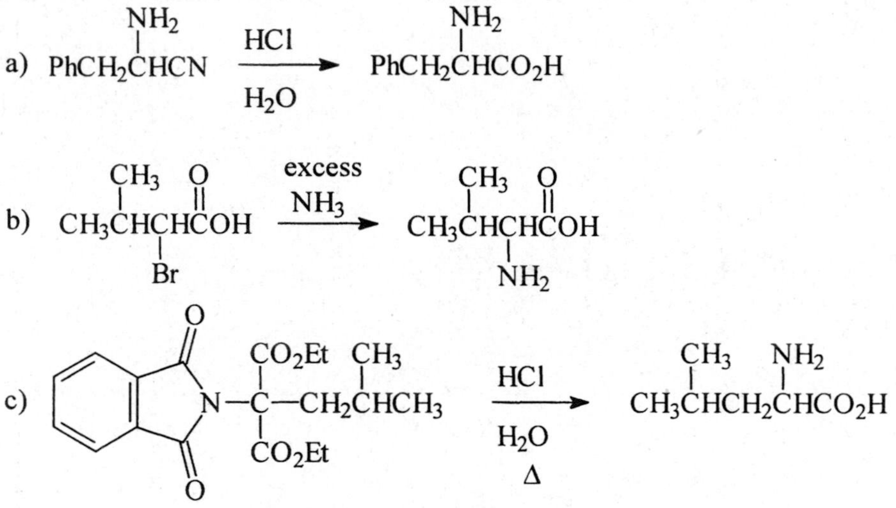

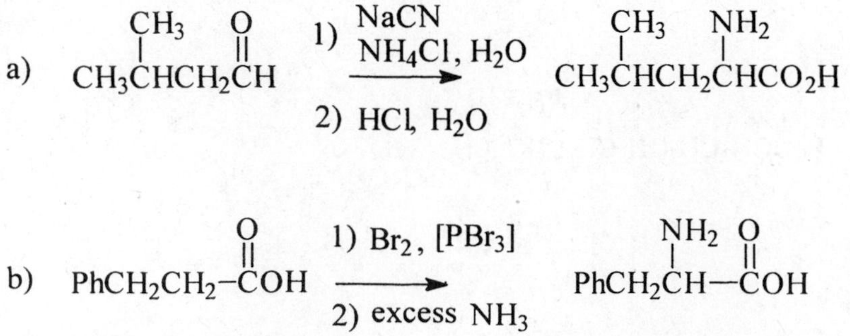

c) 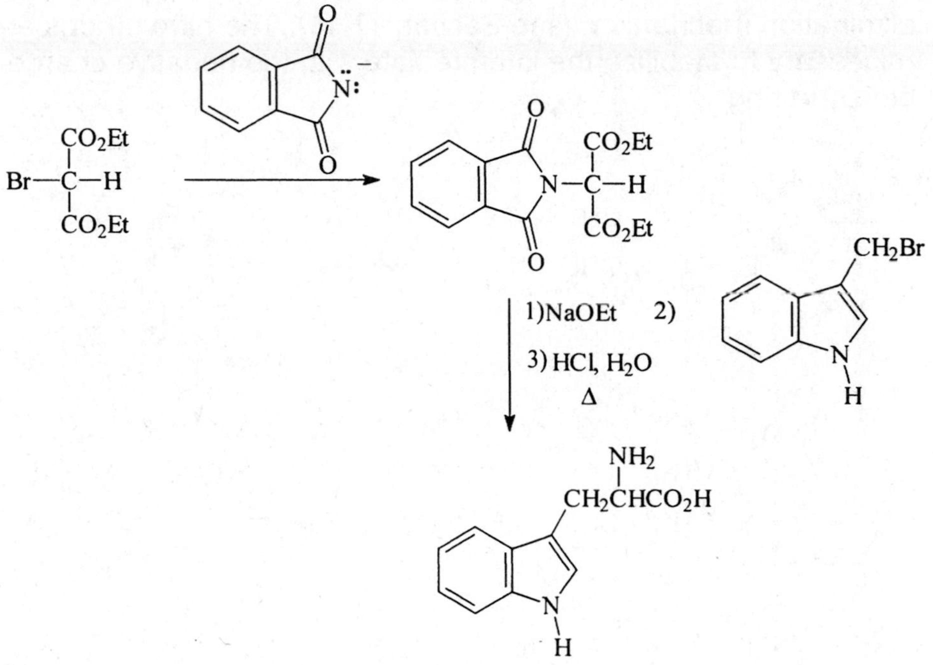

26.12 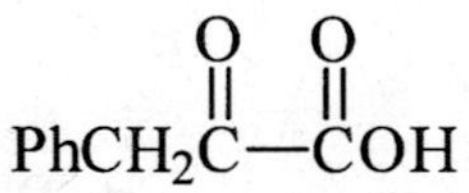

26.13 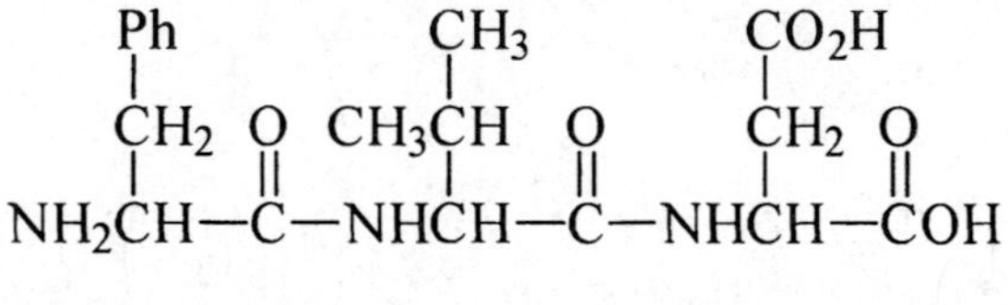

26.14 The amino acids are leucine, cysteine, tyrosine, and glutamic acid. The abbreviation is Leu-Cys-Tyr-Glu.

26.15 The amide groups in the side chains of asparagine and glutamine residues in a peptide are cleaved under the same conditions that hydrolyze the amide bonds of the peptide. As a result, aspartic acid and glutamic acid are isolated instead.

26.16 This is a nucleophilic aromatic substitution proceeding via the addition-elimination mechanism (see Section 17.11). The nitro groups are necessary to stabilize the intermediate with the negative charge in the benzene ring.

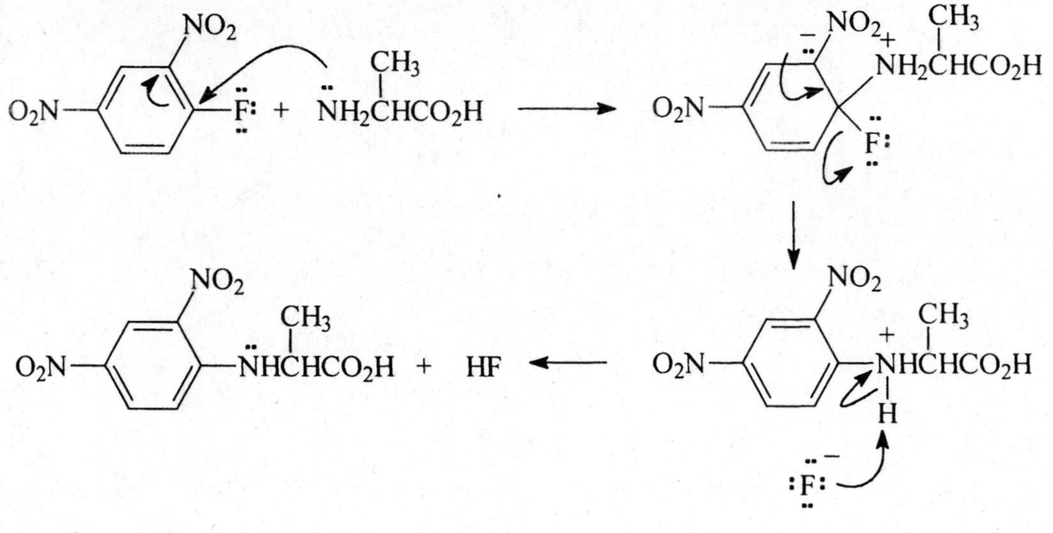

26.17

a) 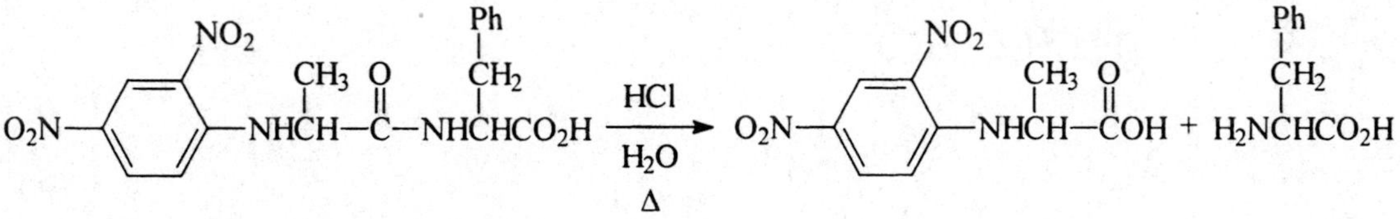

b) 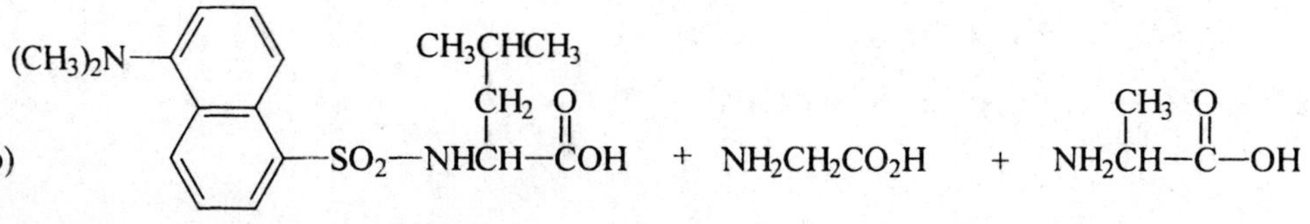

c) 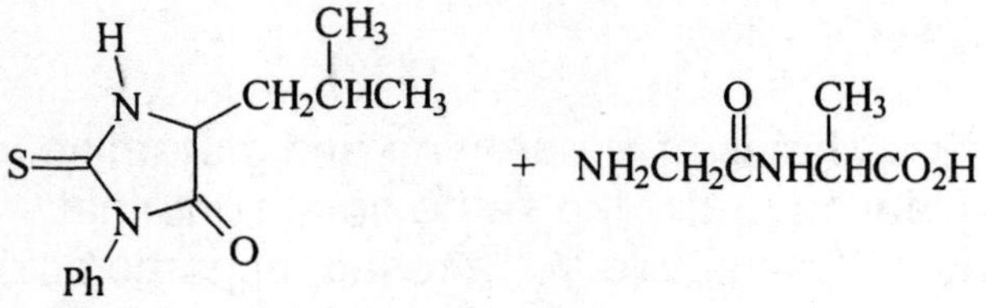

d) 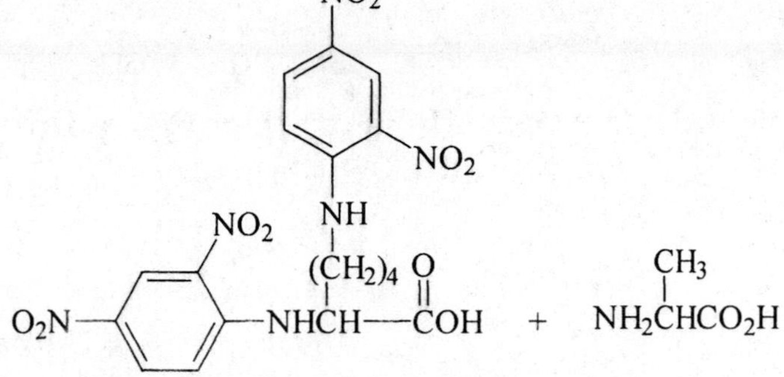

26.18 Phe-Phe-Ala or Phe-Ala-Phe

26.19

a) 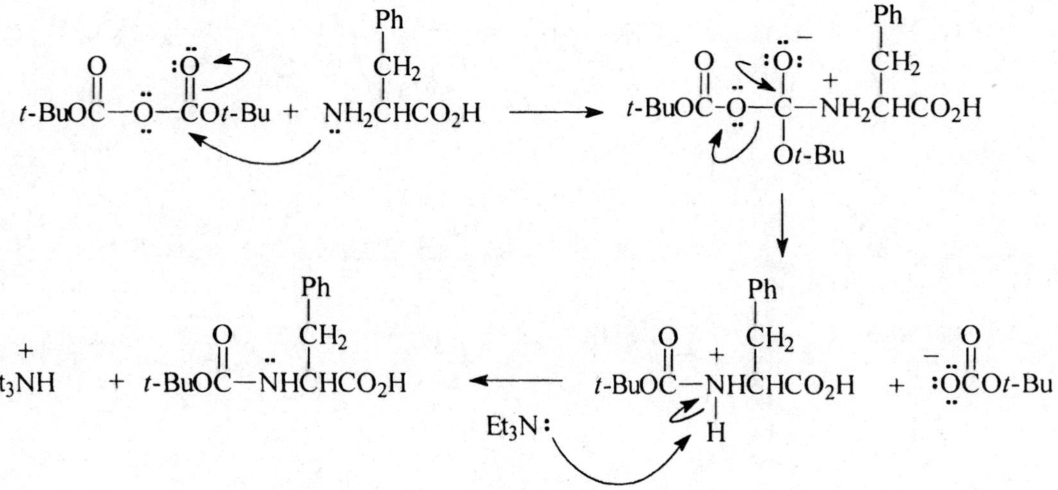

b) 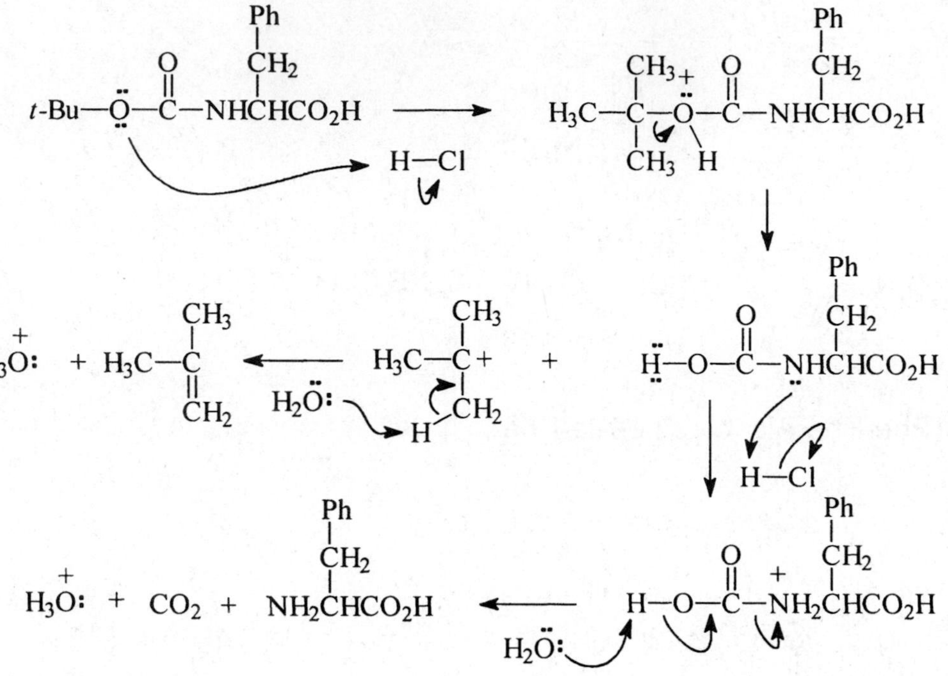

c) 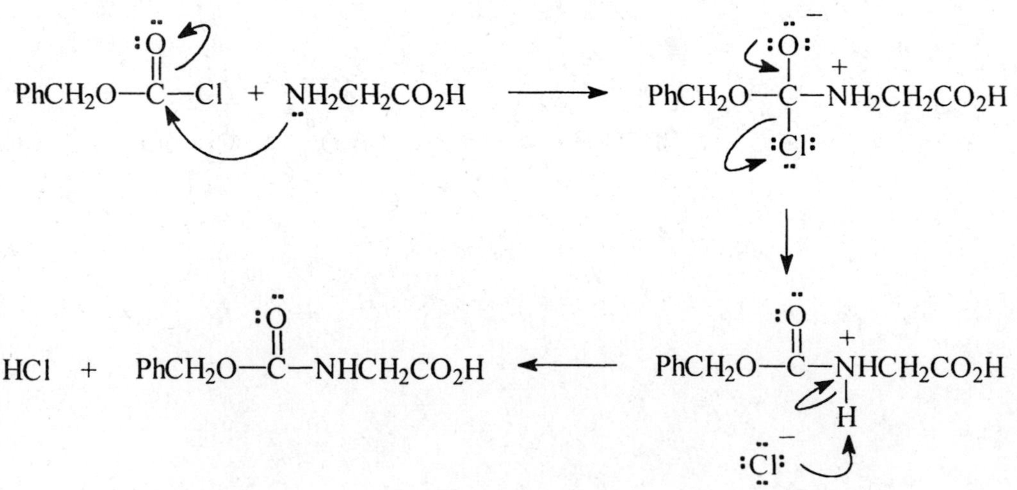

26.20 a)

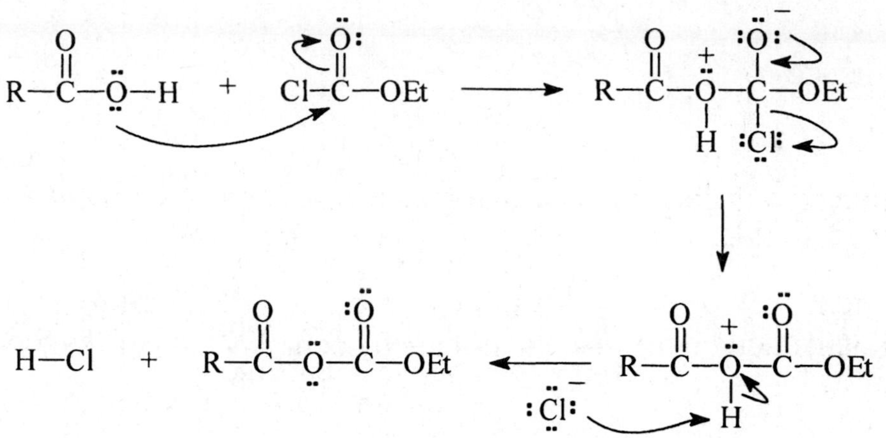

The second product is not formed because chloride ion is a better leaving group than is ethoxide ion.

b)

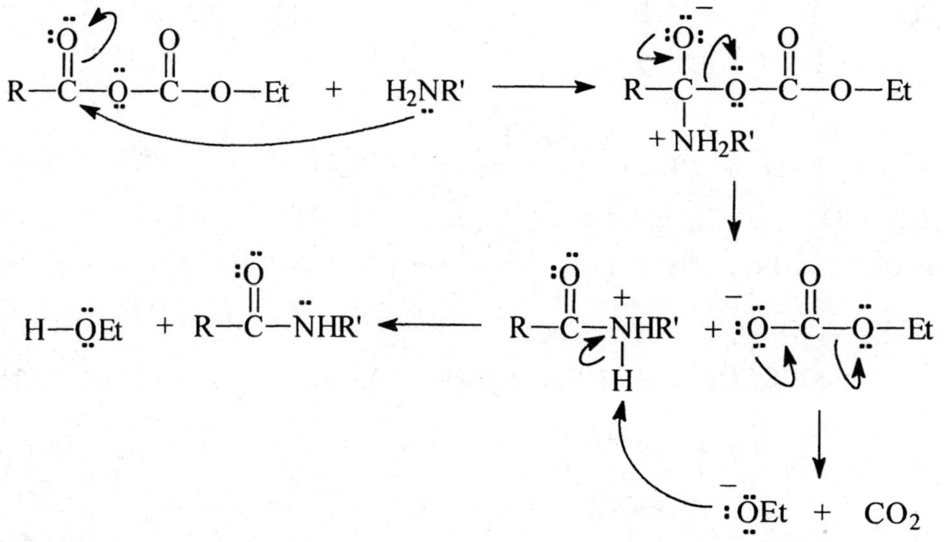

A carbonyl group substituted with an oxygen is less reactive than a carbonyl group substituted with a carbon due to resonance stabilization. The left carbonyl group has one such resonance interaction to stabilize it, whereas the right carbonyl group has two.

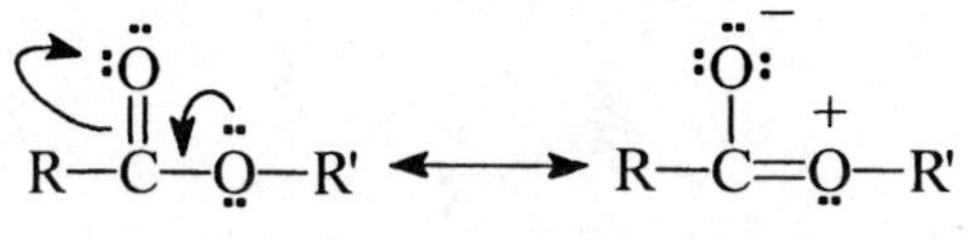

26.21

a)

$$\text{O} \quad \text{CH}_3$$
$$t\text{-BuO\overset{\|}{C}NH\overset{|}{C}HCO_2H}$$

b) $t\text{-BuOCNHCH}_2\text{CNHCHCOCH}_2\text{Ph} \xrightarrow[\text{Pt}]{\text{H}_2} t\text{-BuOCNHCH}_2\text{CNHCHCOH} \xrightarrow[\text{H}_2\text{O}]{\text{HCl}} \text{NH}_2\text{CH}_2-\text{CNHCH}-\text{COH}$

c) $t\text{-BuOCNHCHCNHCH}_2\text{COCH}_3 \xrightarrow[\text{H}_2\text{O}]{\text{HCl}} \text{NH}_2\text{CH}-\text{CNHCH}_2\text{COCH}_3 \xrightarrow[\text{H}_2\text{O}]{\text{NaOH}} \text{NH}_2\text{CH}-\text{CNHCH}_2\text{COH}$

d) $\text{PhCH}_2\text{OCNHCH}_2\text{CNHCH}-\text{COCH}_3 \xrightarrow[\text{Pd}]{\text{H}_2} \text{NH}_2\text{CH}_2\text{CNHCH}-\text{COCH}_3$

26.22

$\text{NH}_2\text{CH}_2\text{CO}_2\text{H} + \text{CH}_3\text{OH} \xrightarrow{\text{[HA]}} \text{NH}_2\text{CH}_2\text{CO}_2\text{CH}_3$

$t\text{-BuOCOCOt-Bu} + \text{NH}_2\text{CH}_2\text{CO}_2\text{H} \xrightarrow{\text{Et}_3\text{N}} t\text{-BuOCNHCH}_2\text{CO}_2\text{H} \quad \xrightarrow{\text{DCC}}$

$t\text{-BuOCOCOt-Bu} + \text{NH}_2\overset{\text{PhCH}_2}{\text{CHCO}_2\text{H}} \xrightarrow{\text{Et}_3\text{N}} \quad t\text{-BuOCNHCH}_2\text{CNHCH}_2\text{CO}_2\text{CH}_3$

$t\text{-BuOCNH}\overset{\text{PhCH}_2}{\text{CHCO}_2\text{H}} \qquad \text{NH}_2\text{CH}_2\text{CNHCH}_2\text{CO}_2\text{CH}_3 \xleftarrow{\text{HCl} \;\; \text{H}_2\text{O}}$

$\xrightarrow{\text{DCC}}$

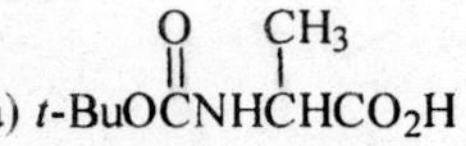

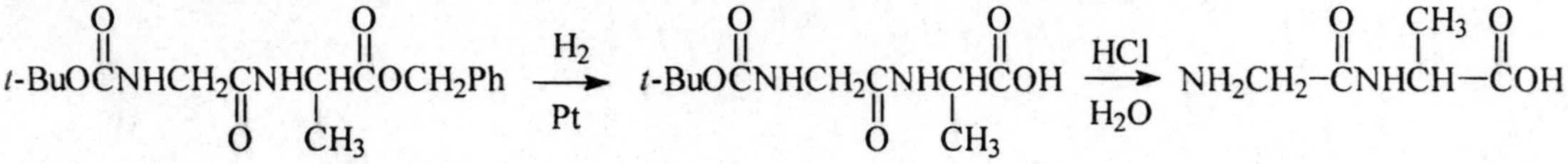

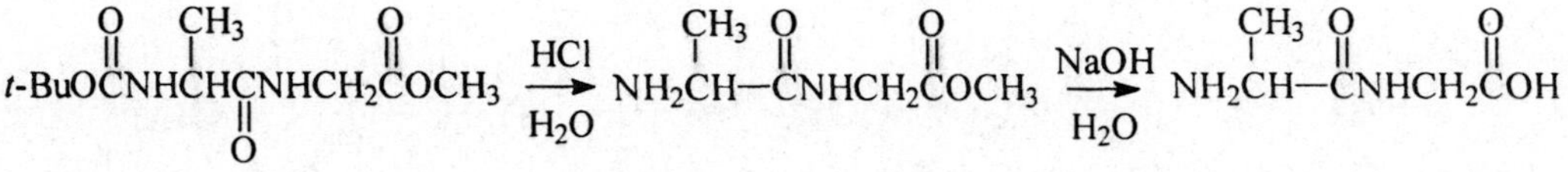

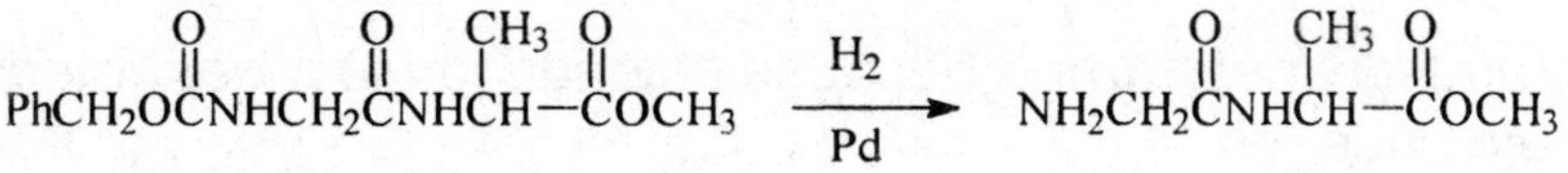

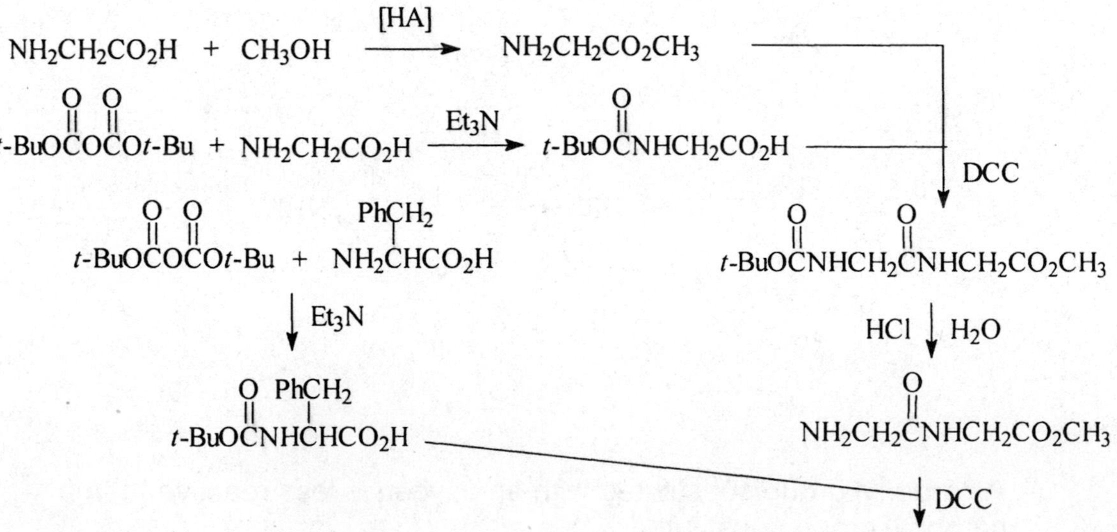

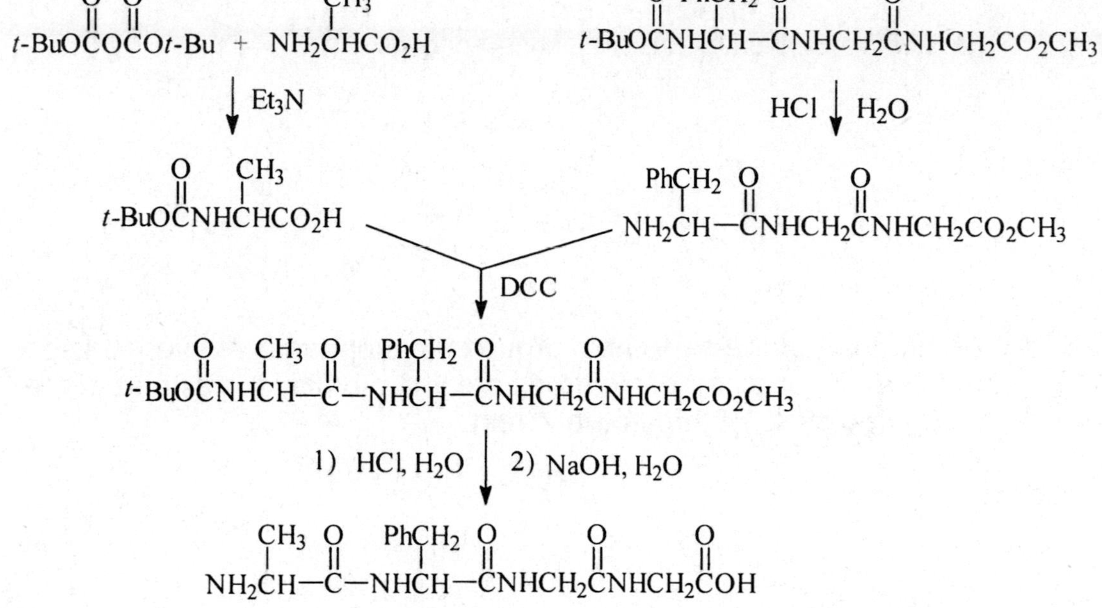

26.23

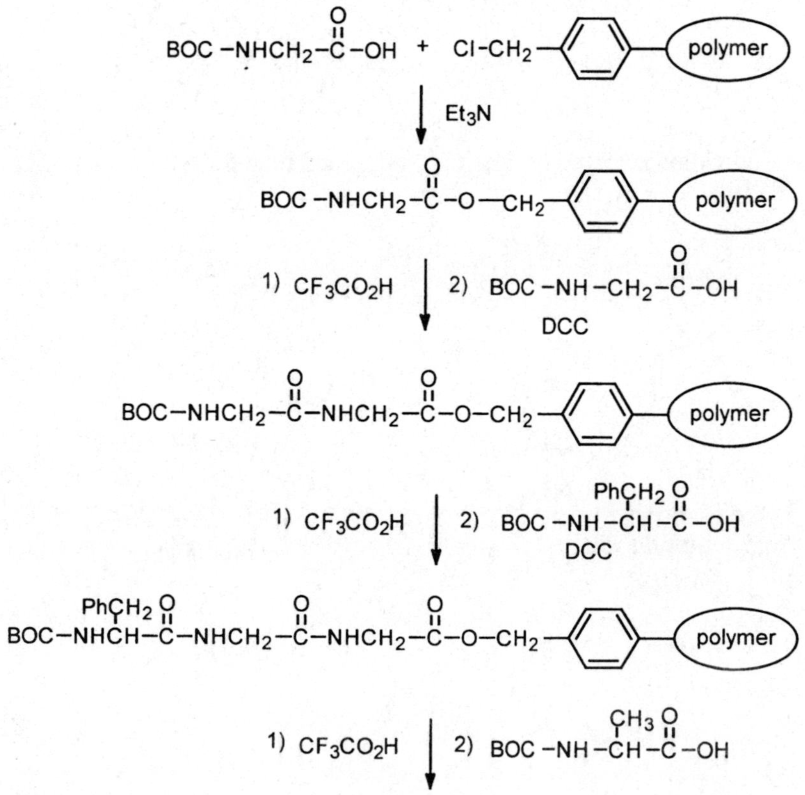

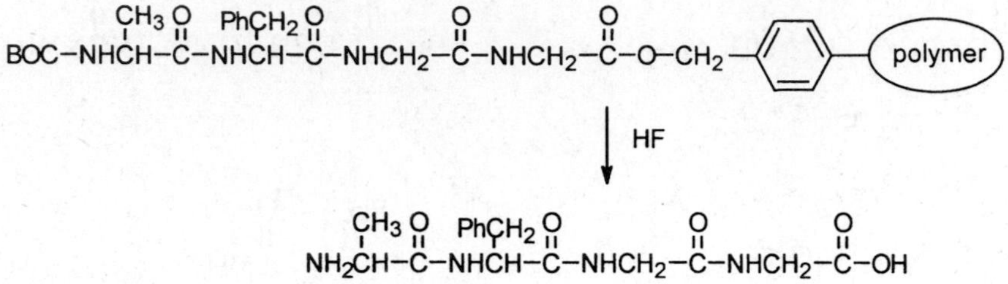

26.24 The nitrogen of the side chain group of tryptophan is not very basic because the lone pair of electrons are in a *p* orbital that is part of the conjugated cycle of the aromatic ring.

26.25 a)

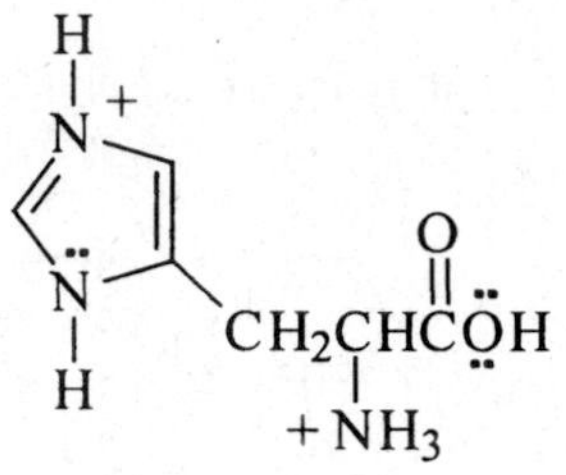

dicationic form

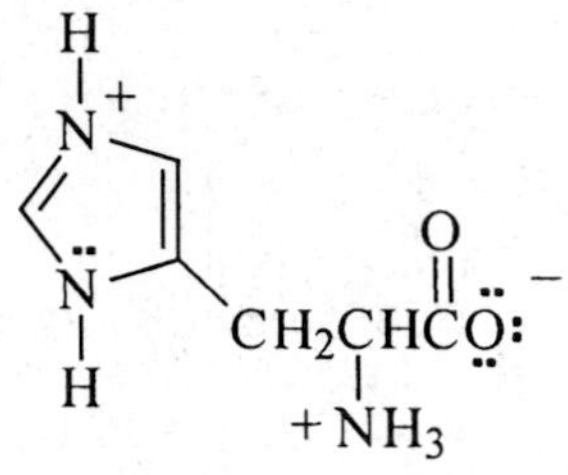

cationic form

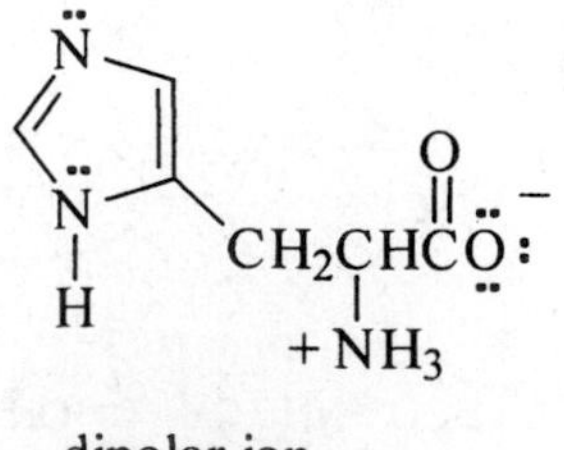

dipolar ion

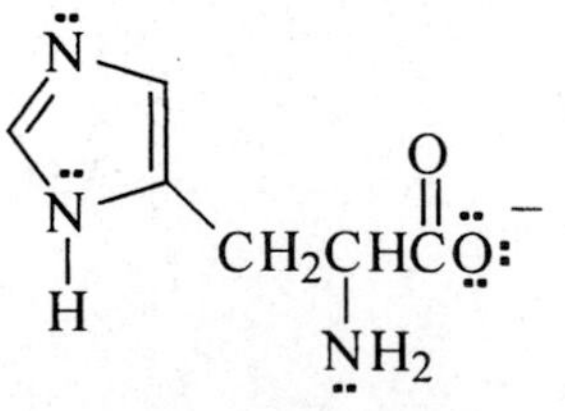

anionic form

b)

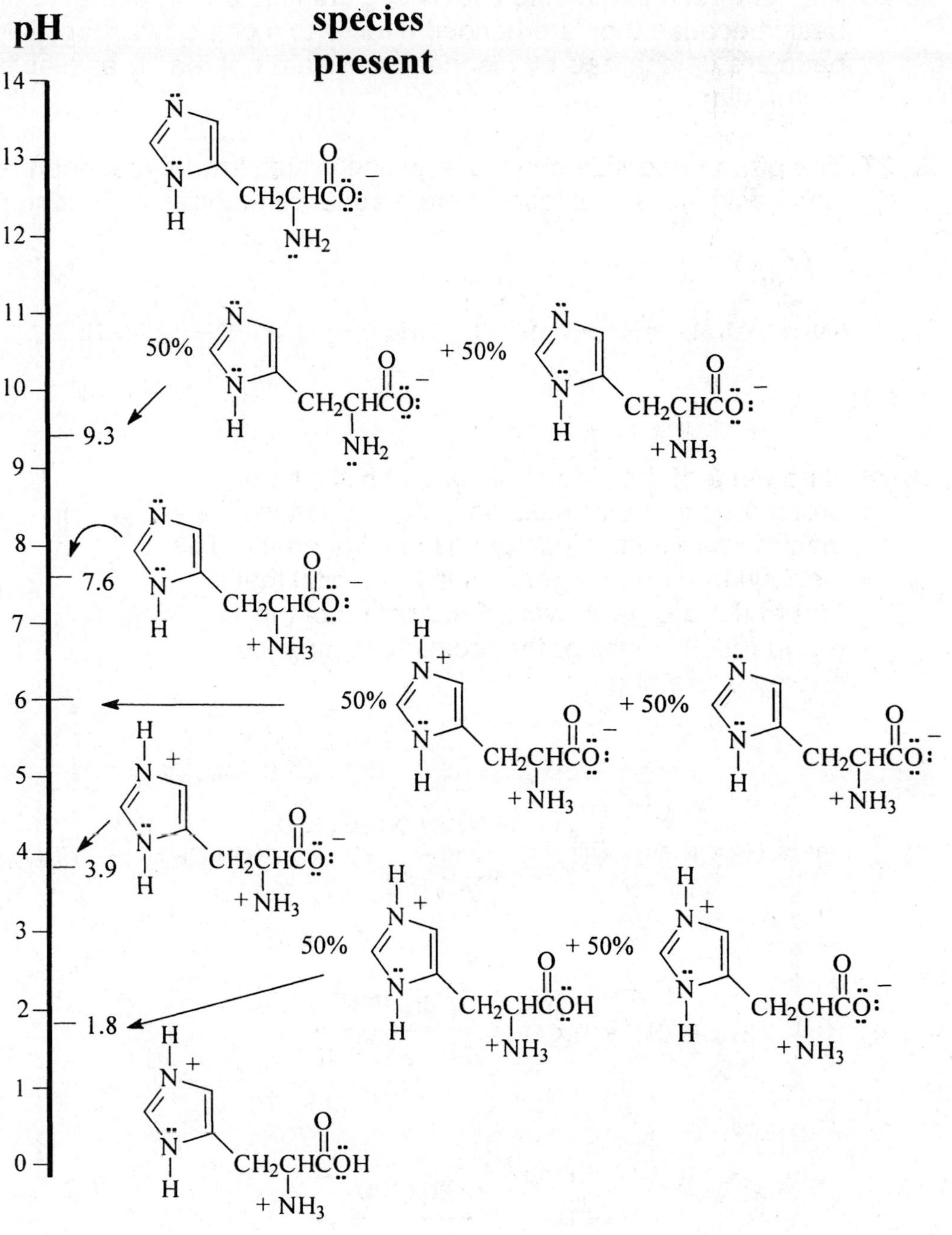
pH
species present
14
13
12
11
10
9
8
7
6
5
4
3
2
1
0
9.3
7.6
3.9
1.8
50%
+ 50%

26.26 The nitrogens in the side chains of glutamine and asparagine are not very basic because they are bonded directly to a carbonyl group. Their electron pairs are delocalized by resonance and are not readily available for protonation.

26.27 The protonated side chain of arginine is stabilized by resonance, so it is a weak acid. This is a case where resonance stabilizes the conjugate acid.

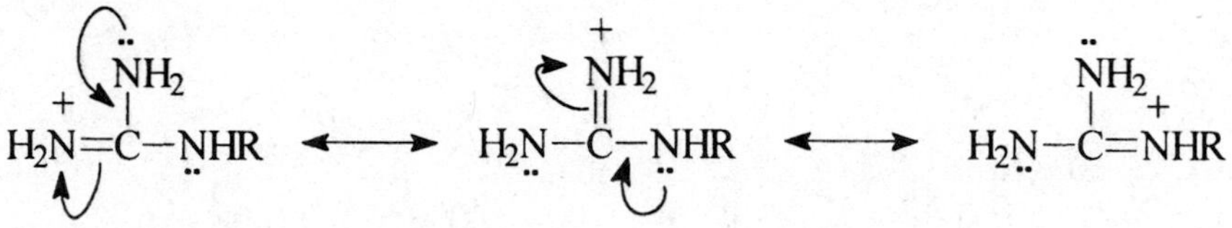

26.28 Nitrogen 3 of the side chain ring of histidine is protonated in the monocationic form because the electron pair on this nitrogen is in a sp^2 orbital. The electron pair on nitrogen 1 is in a p orbital that is part of the aromatic cycle. Protonation of this N would result in loss of the aromatic resonance energy of the ring.

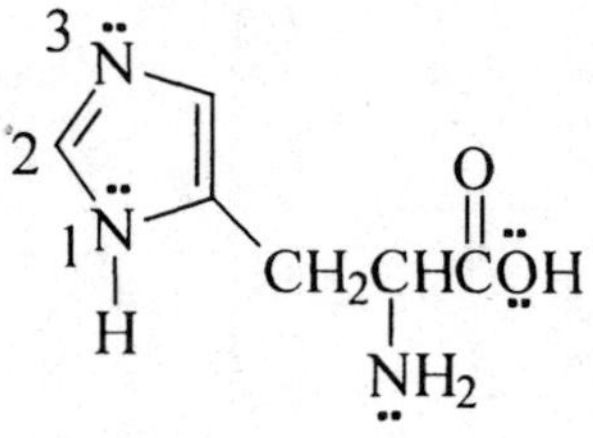

26.29

a)

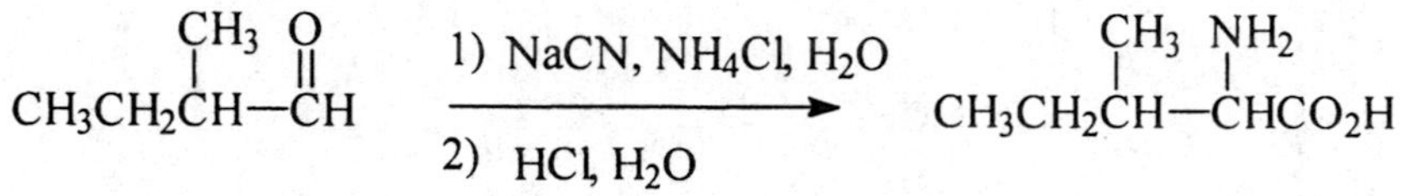

b)

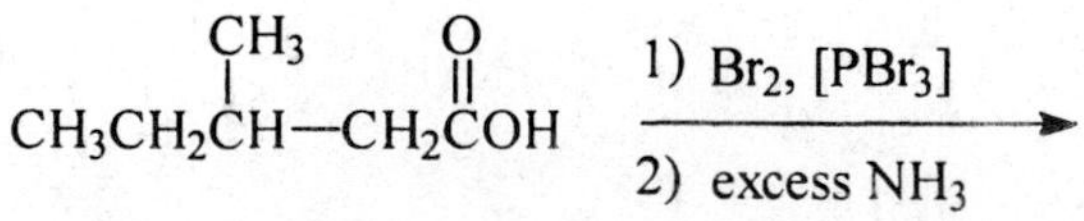

c)

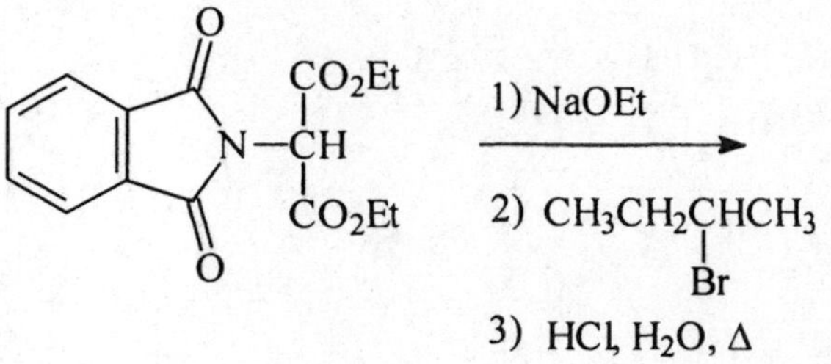

26.30 The mechanism is very similar to the formation of a cyanohydrin (see Section 18.4) except the nucleophile is attacking a carbon-nitrogen double bond rather than a carbon-oxygen double bond.

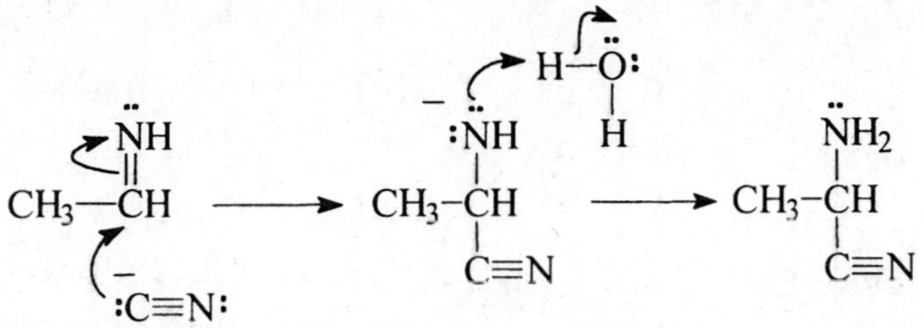

26.31

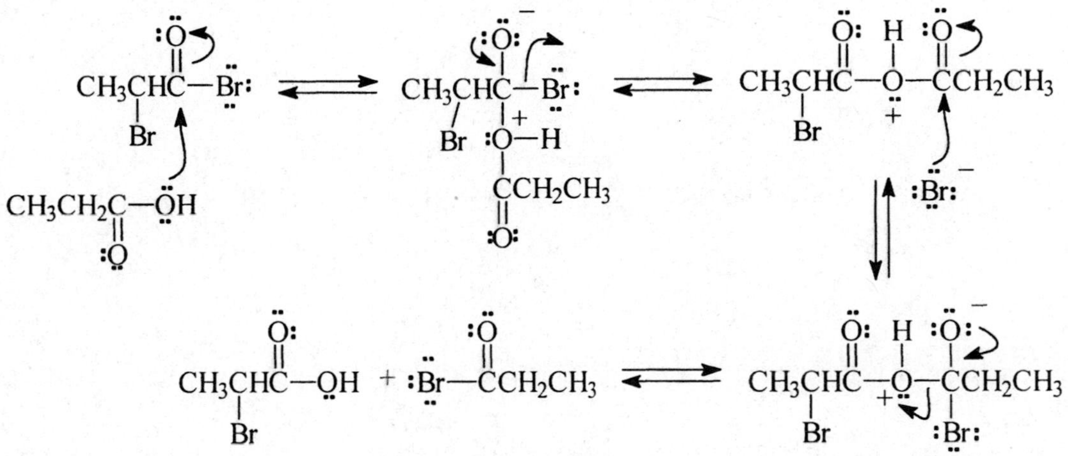

26.32

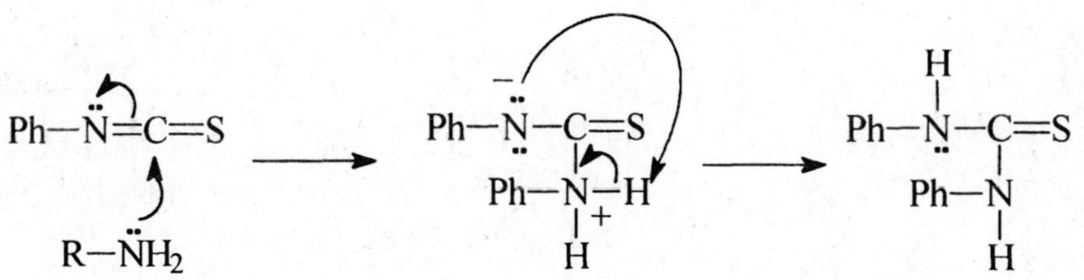

26.33

a) 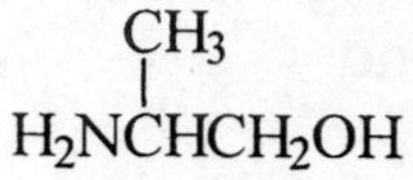

b) 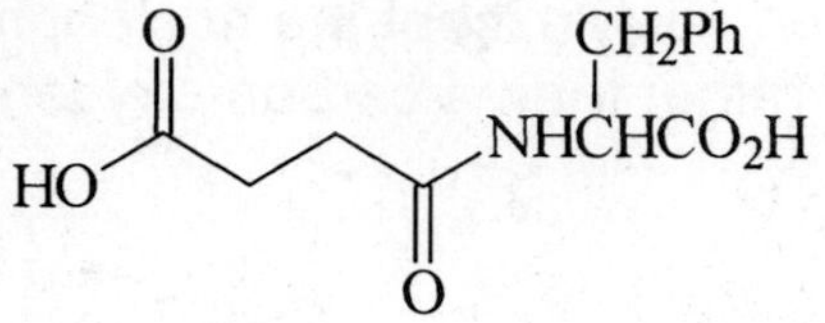

c) 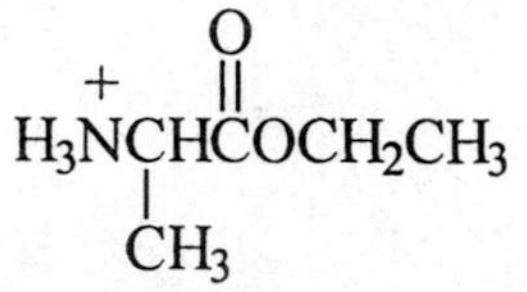

d) 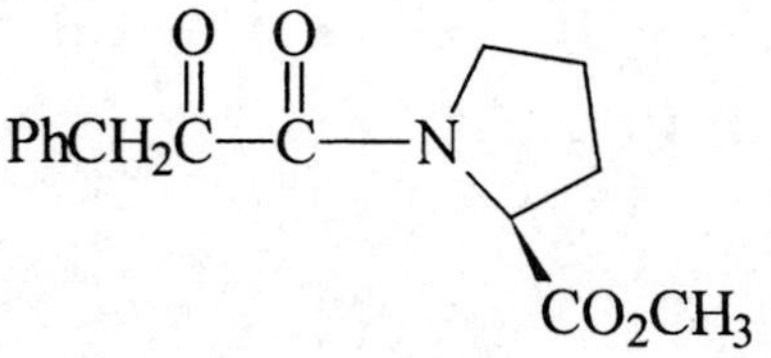

26.34

a) 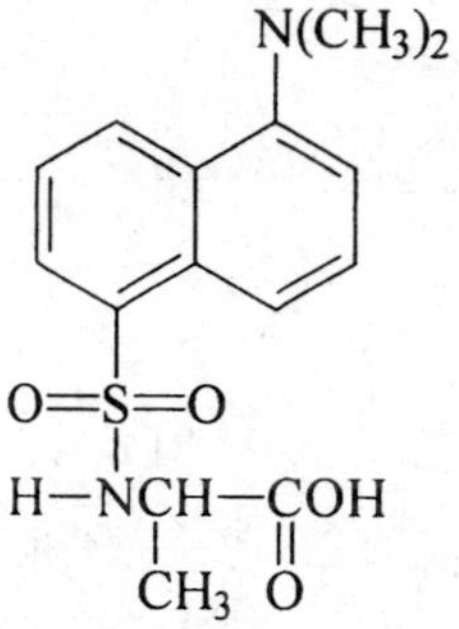

b) 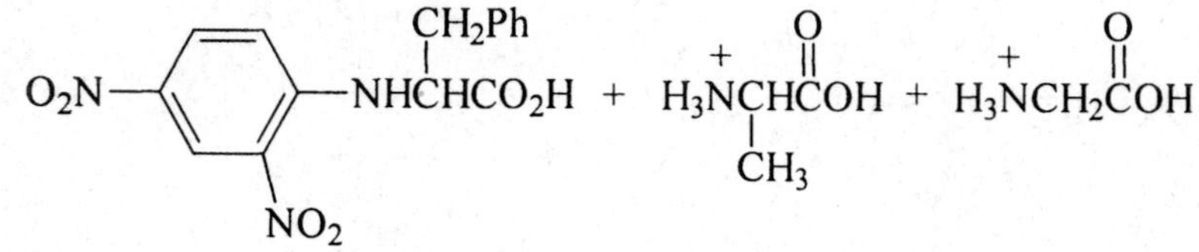

c) 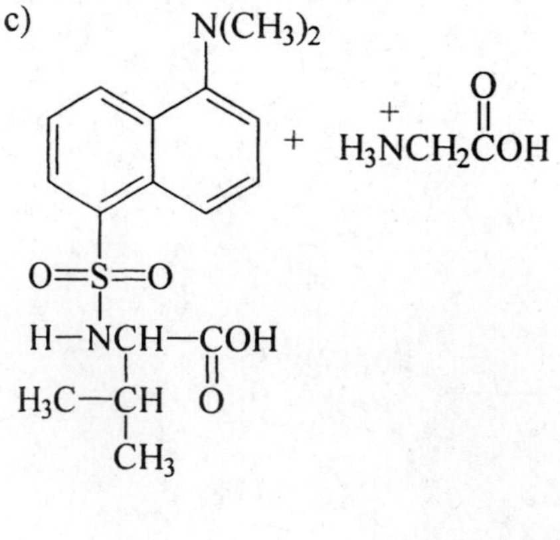

d) 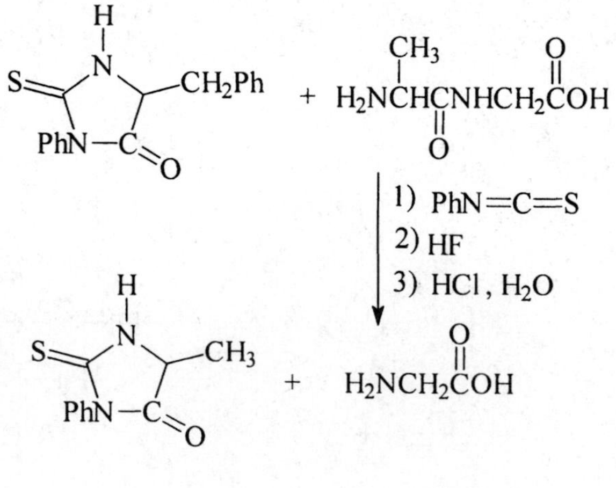

26.35

a) 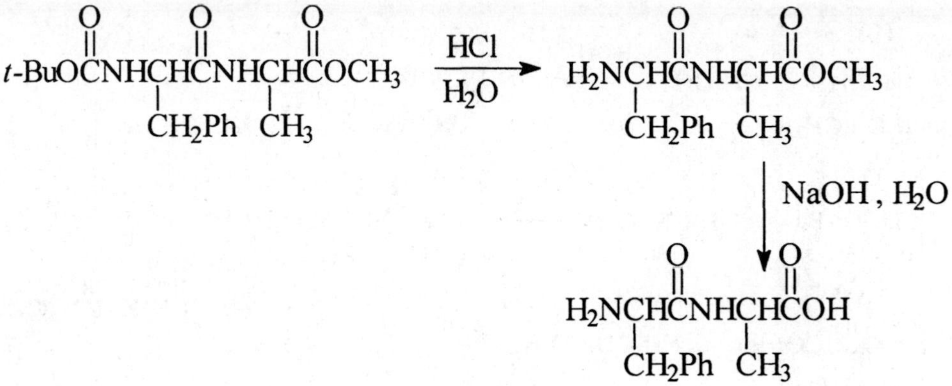

b) 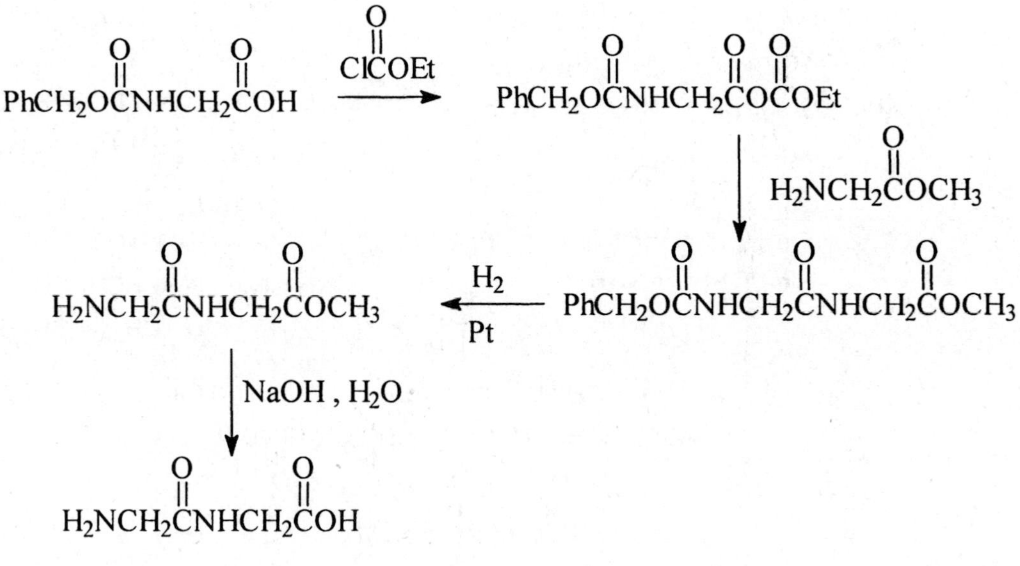

26.36

a)

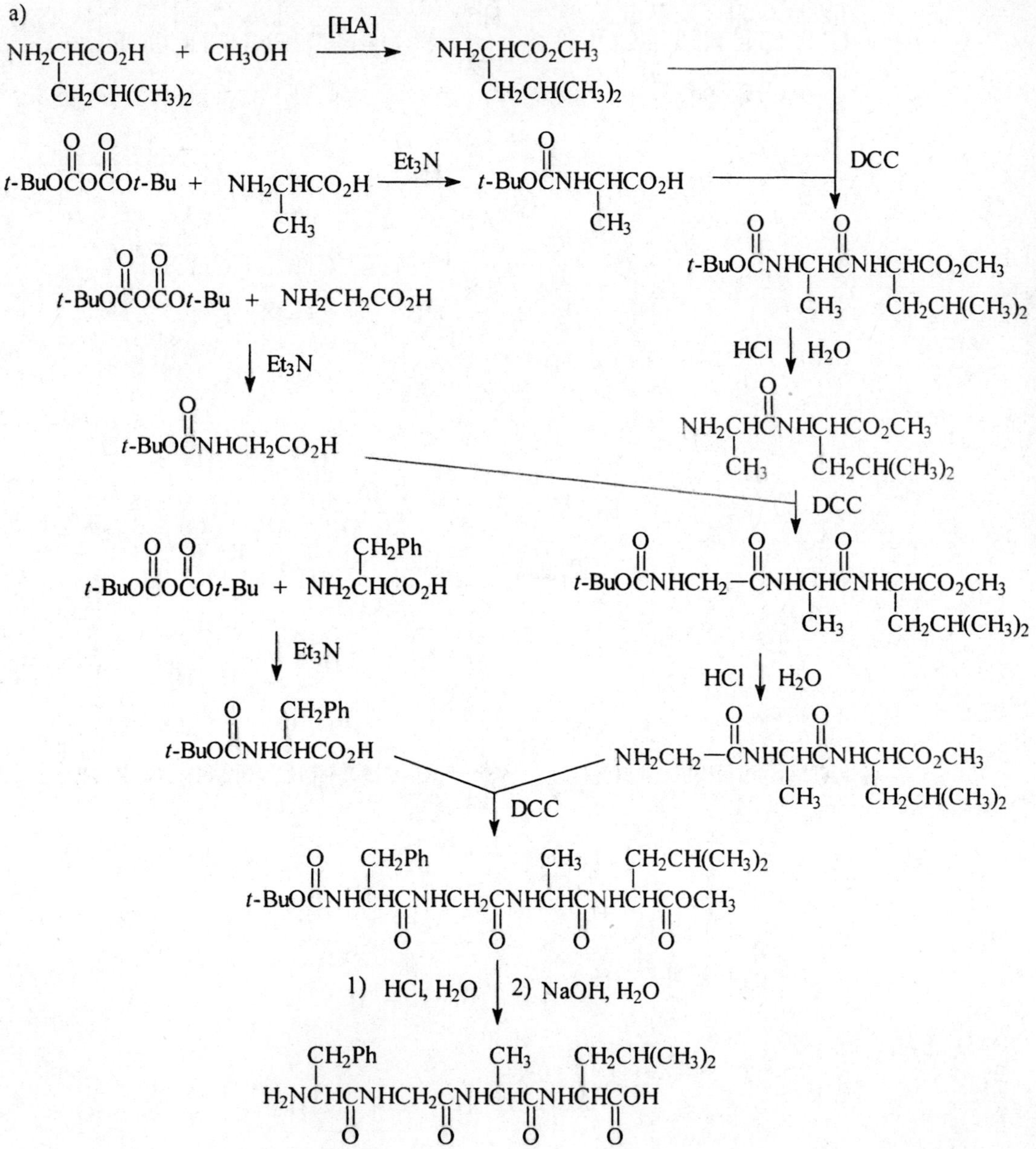

b)

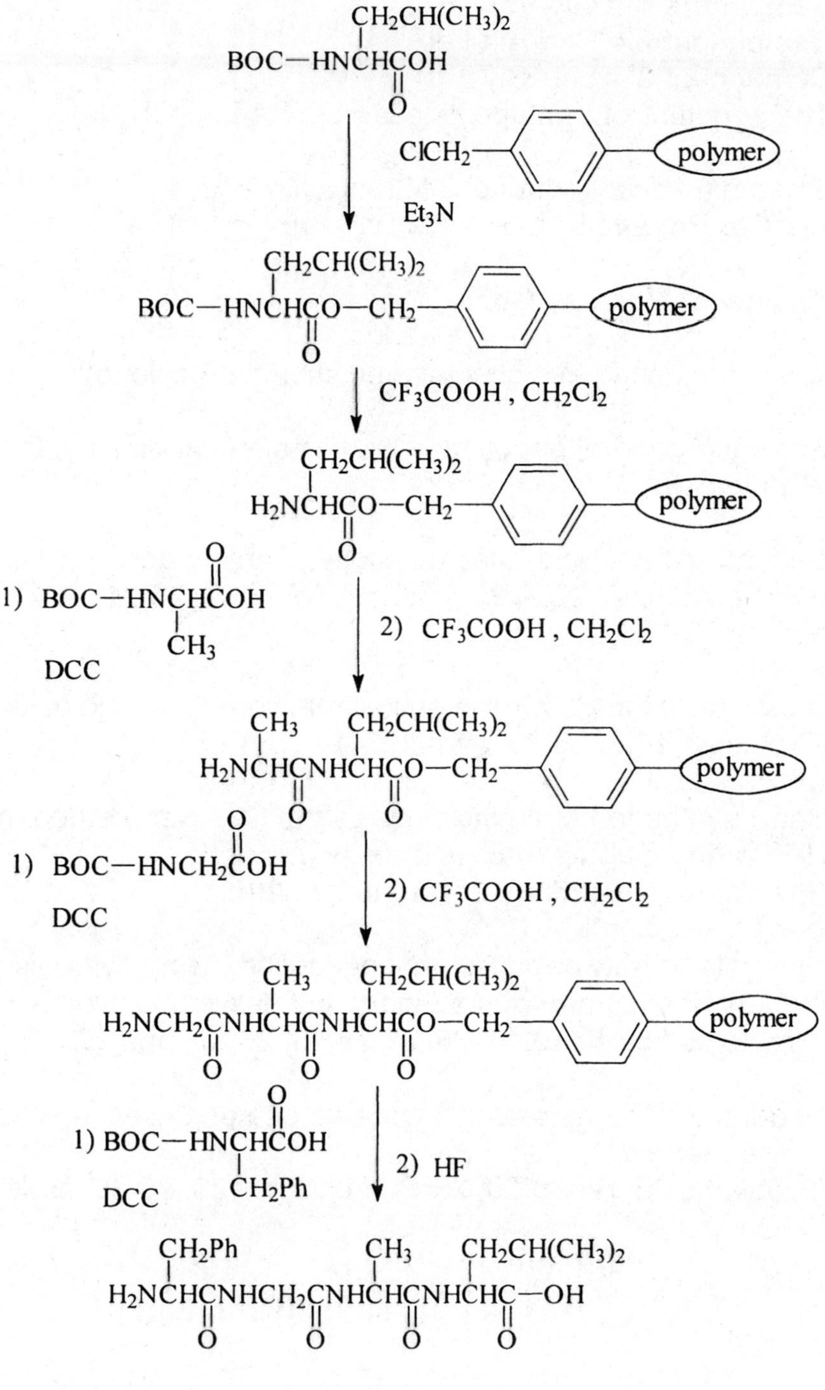

CH2CH(CH3)2
BOC—HNCHCOH
O
ClCH2
polymer
Et3N
CH2CH(CH3)2
BOC—HNCHCO—CH2
O
polymer
CF3COOH , CH2Cl2
CH2CH(CH3)2
H2NCHCO—CH2
O
polymer
O
1) BOC—HNCHCOH
CH3
DCC
2) CF3COOH , CH2Cl2
CH3
CH2CH(CH3)2
H2NCHCNHCHCO—CH2
O
O
polymer
O
1) BOC—HNCH2COH
DCC
2) CF3COOH , CH2Cl2
CH3
CH2CH(CH3)2
H2NCH2CNHCHCNHCHCO—CH2
O
O
O
polymer
O
1) BOC—HNCHCOH
CH2Ph
DCC
2) HF
CH2Ph
CH3
CH2CH(CH3)2
H2NCHCNHCH2CNHCHCNHCHC—OH
O
O
O
O

26.37 Component 1 = Gly-Val
Component 2 = Phe-Ala-Leu
Component 3 = Leu-Gly
The structure of pentapeptide is Phe-Ala-Leu-Gly-Val.

26.38 The amino acid sequence of bradykinin is
Arg-Pro-Pro-Gly-Phe-Ser-Pro-Phe-Arg.

Review of Mastery Goals

After completing this chapter, you should be able to:

Show the general structure, including stereochemistry, for an amino acid.
(Problem 26.1)

Understand the acid-base reactions of amino acids.
(Problems 26.2, 26.3, 26.4, 26.5, 26.6, 26.7, 26.24, 26.25, 26.26, 26.27, and 26.28)

Understand the general chemical reactions of amino acids.
(Problems 26.8, 26.33, and 26.34)

Show syntheses of amino acids by the Strecker method, α-substitution by NH_3, or the Gabriel/malonate method.
(Problems 26.9, 26.10, 26.11, and 26.29)

Understand how peptides are sequenced, using hydrolysis, Sanger's reagent, the Edman degradation, and enzymatic hydrolysis.
(Problems 26.15, 26.16, 26.17, 26.18, 26.37, and 26.38)

Understand the laboratory synthesis of a peptide in solution or by the solid phase method.
(Problems 26.19, 26.20, 26.21, 26.22, 26.23, 26.35, and 26.36)

Chapter 27
NUCLEOTIDES AND NUCLEIC ACIDS

27.1

a) 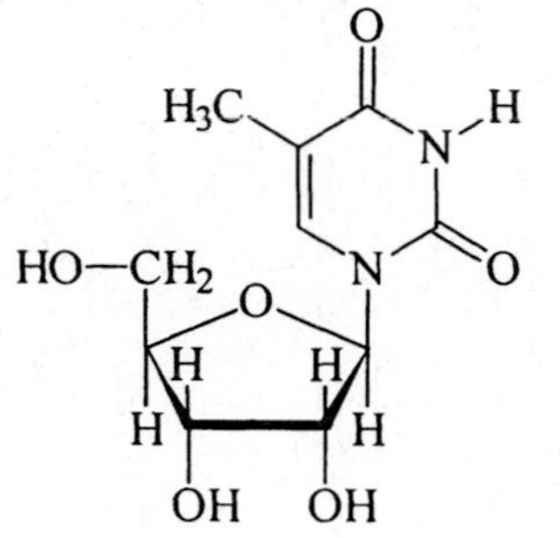

b) 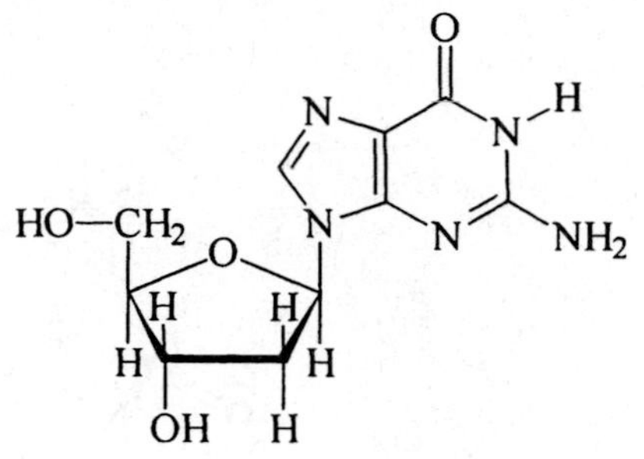

c) 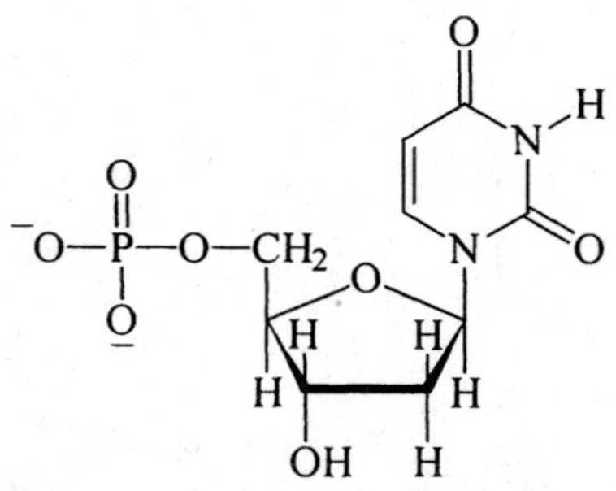

d) 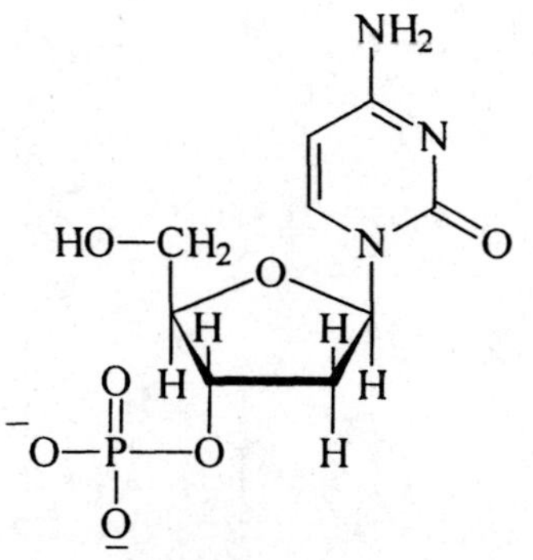

27.2 a) Guanosine b) Uridine 5'-monophosphate
c) Deoxycytidine 5'-monophosphate

27.3 Pyrimidine has six electrons in its cyclic pi system and therefore fits Huckel's rule. The unshared electrons on the nitrogens are not part of the pi system. The six-membered ring of purine is aromatic for the same reason as pyrimidine. The five-membered ring is also aromatic. Like imidazole (see Chapter 16), it has a total of six electrons in its cyclic pi system. The electrons on N-9 are part of the pi system, whereas the electrons on N-7 are not.

27.4

a)

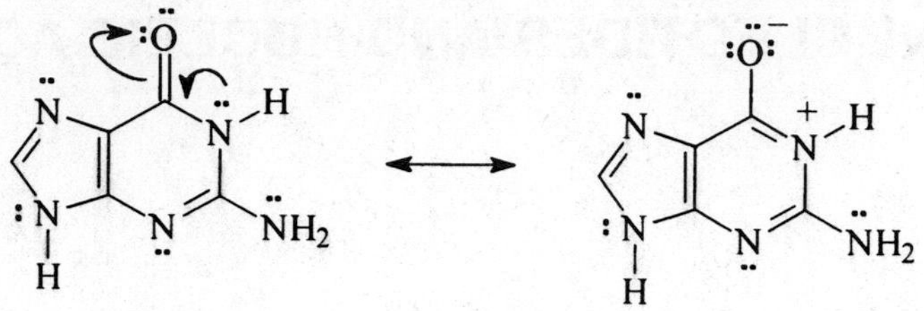

b)

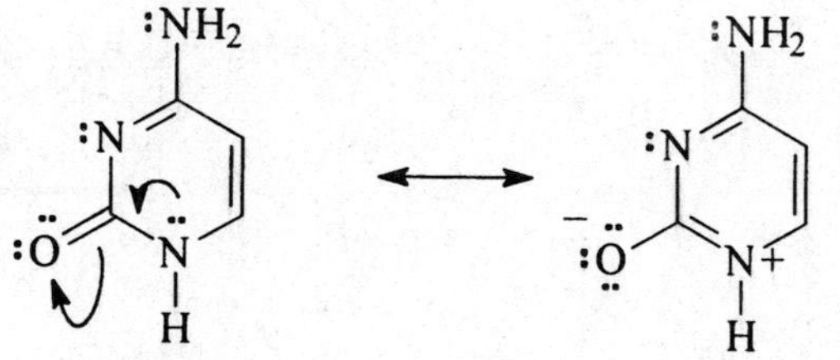

c)

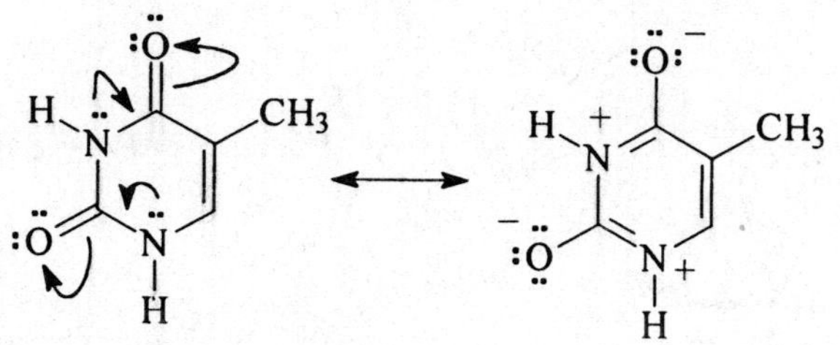

27.5 5'-end TAACGCTCG 3'-end

27.6

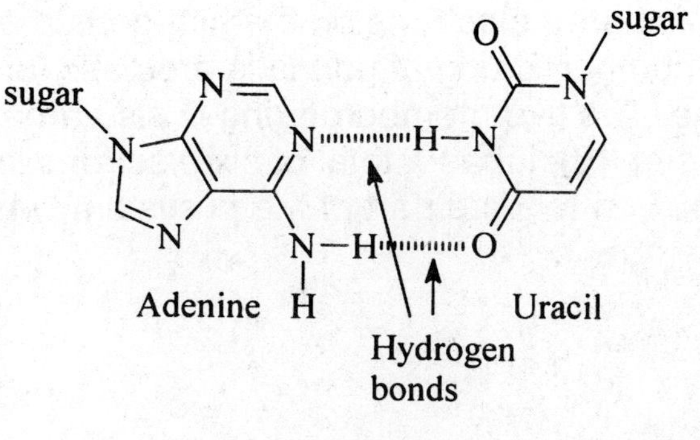
sugar
sugar
Adenine
Uracil
Hydrogen
bonds

27.7

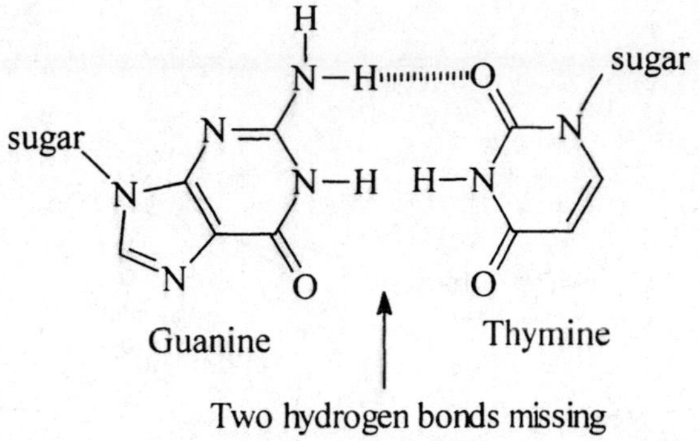
sugar
sugar
Guanine
Thymine
Two hydrogen bonds missing

27.8

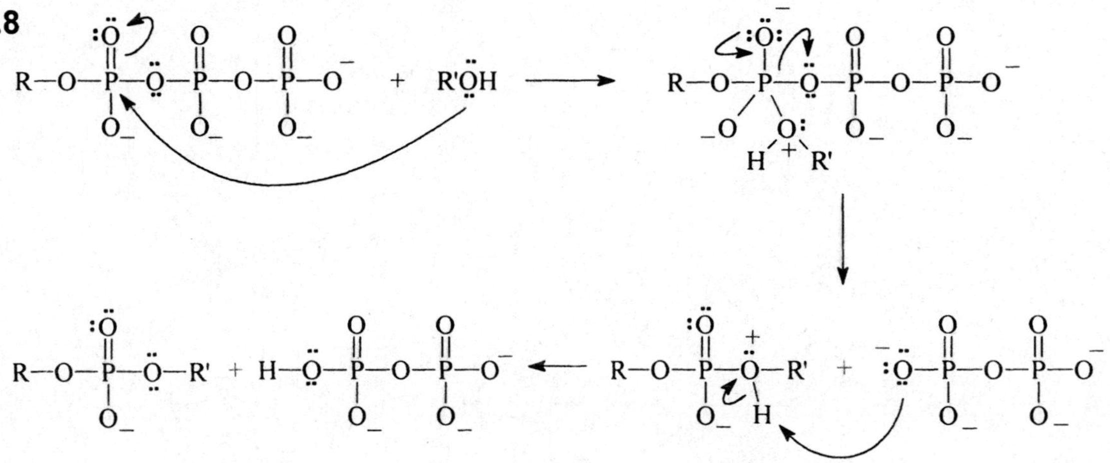

27.9

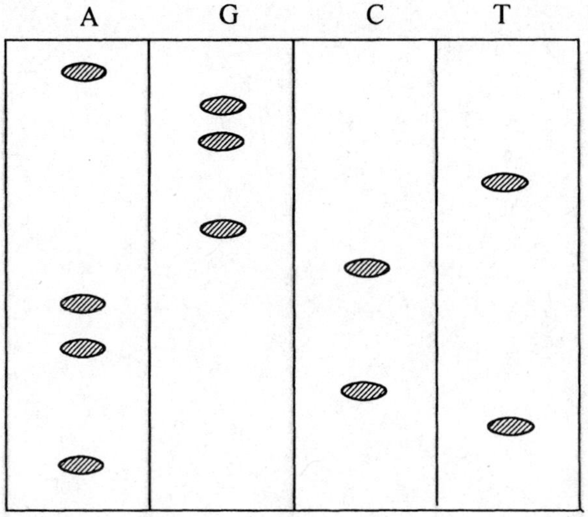
A
G
C
T

27.10

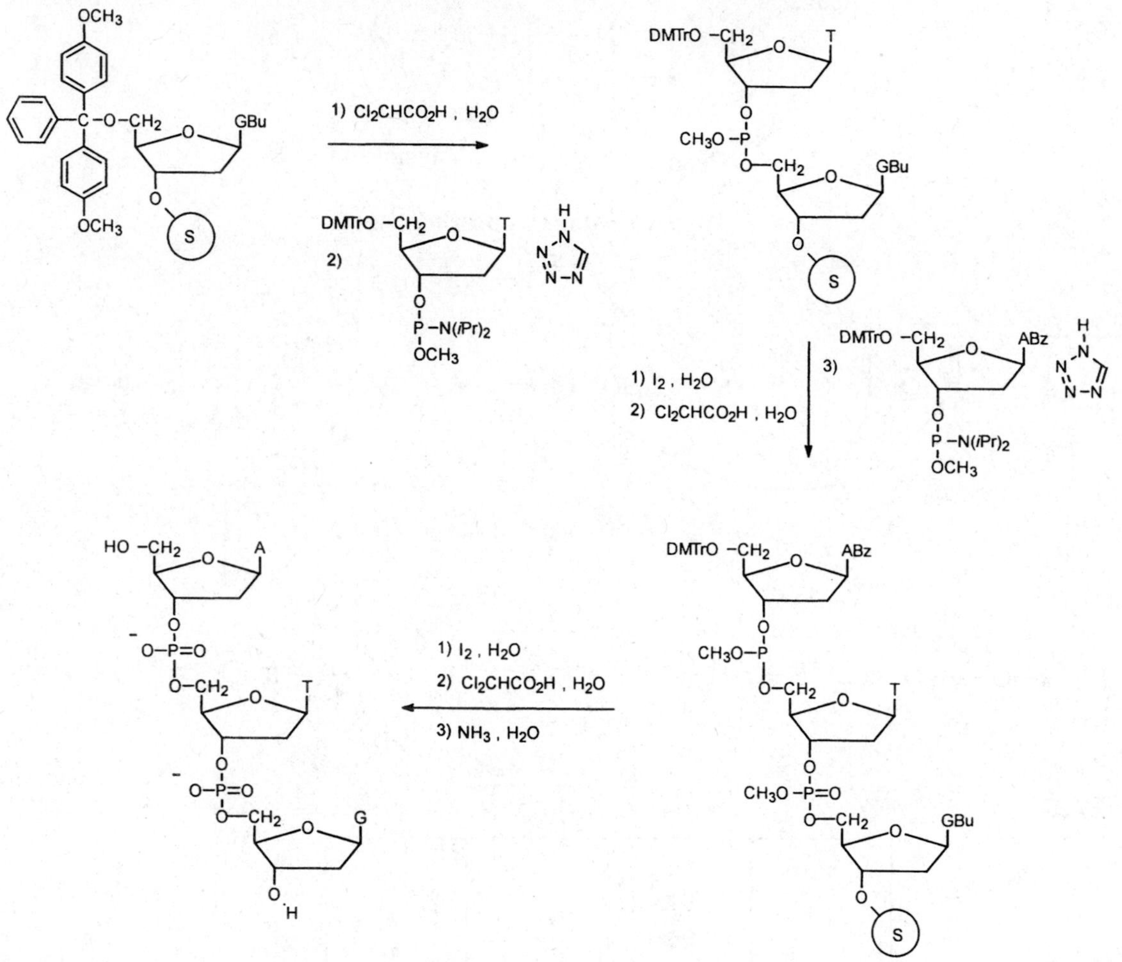
OCH3
C-O-CH2
GBu
OCH3
S
1) Cl2CHCO2H , H2O
DMTrO -CH2
T
2)
O
P -N(iPr)2
OCH3
DMTrO -CH2
T
O
CH3O -P
O-CH2
GBu
S
1) I2 , H2O
2) Cl2CHCO2H , H2O
3)
DMTrO -CH2
ABz
P -N(iPr)2
OCH3
HO -CH2
A
O-P=O
O-CH2
T
O-P=O
O-CH2
G
H
1) I2 , H2O
2) Cl2CHCO2H , H2O
3) NH3 , H2O
DMTrO -CH2
ABz
CH3O-P
O-CH2
T
CH3O -P=O
O-CH2
GBu
S

27.11 The methoxy groups stabilize the carbocation intermediate by resonance. Therefore, according to the Hammond postulate, they also lower the energy of the transition state for the rate determining step.

d) 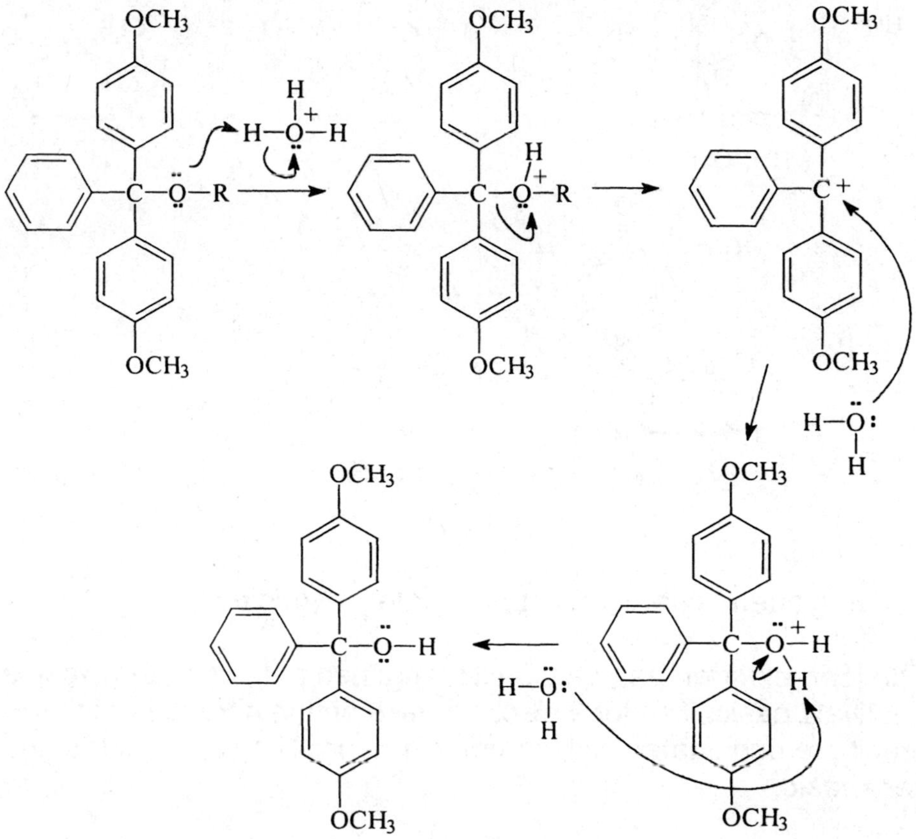

27.12 This reaction follows the E1cb mechanism (see Chapter 9). The cyano group helps stabilize the carbanion that is formed in the first step, so that the proton can be more easily removed.

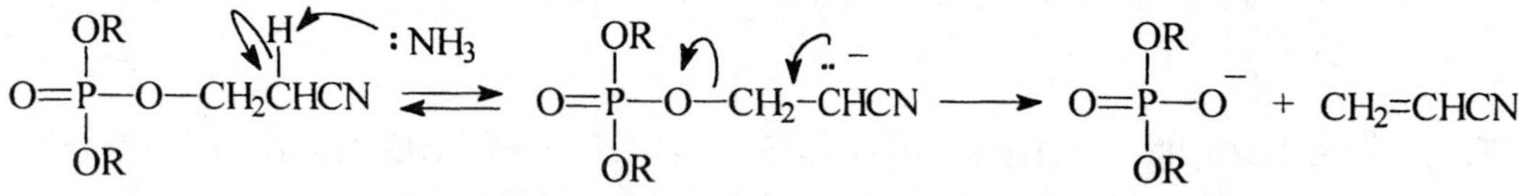

27.13

a) 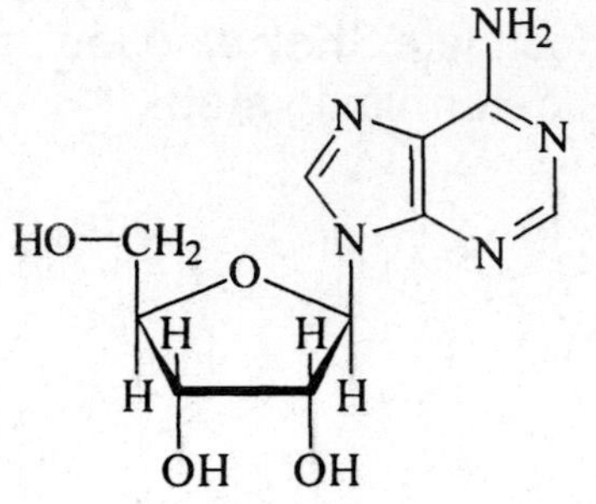

b) 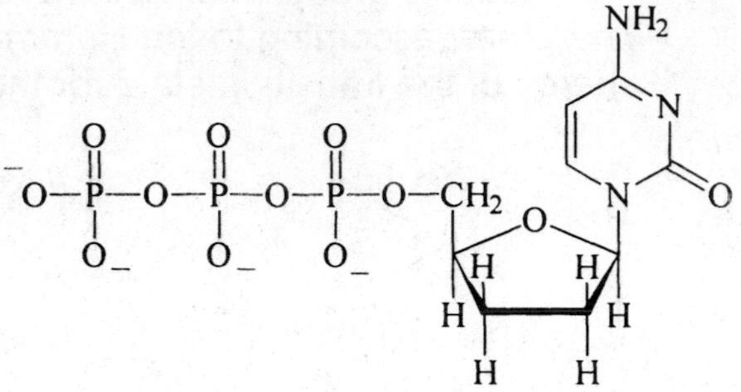

c) 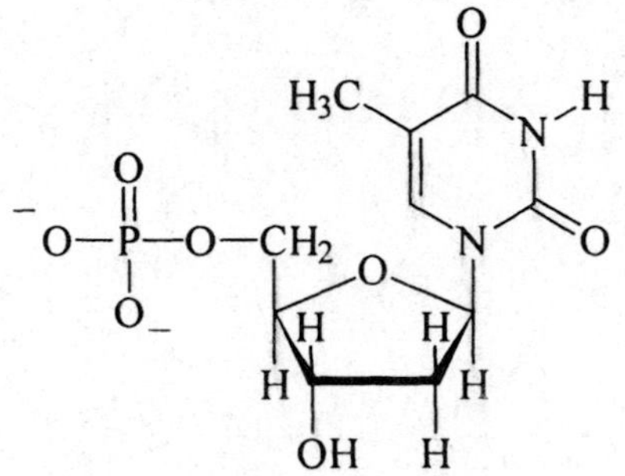

27.14 a) deoxyadenosine 5'-monophosphate b) uridine

27.15 The lone electron pair on the NH_2 attached to the ring of cytosine is relatively basic. The lone electron pairs on the nitrogens of thymine and uracil are conjugated with carbonyl groups. They are amides and are not very basic.

27.16

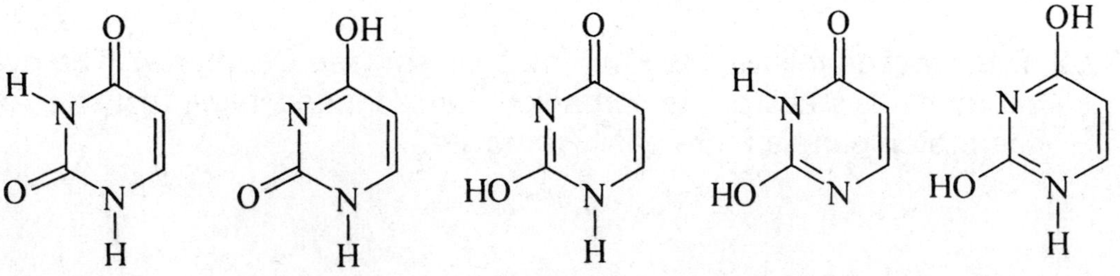

27.17 The complementary DNA is 3'-end-CGAATACG-5'-end and the complementary RNA is 3'-end-CGAAUACG-5'-end.

27.18

a)

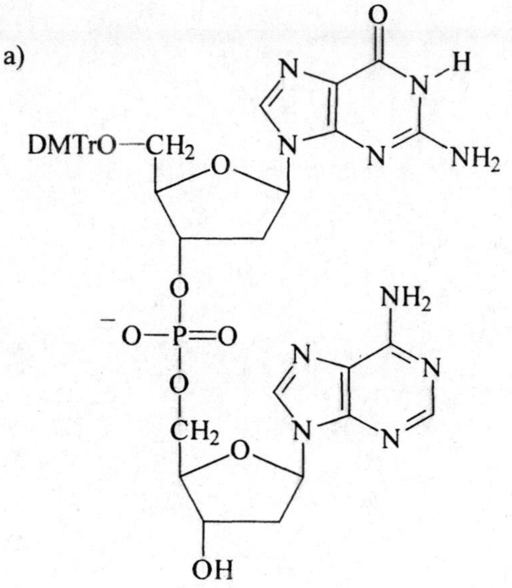

b)

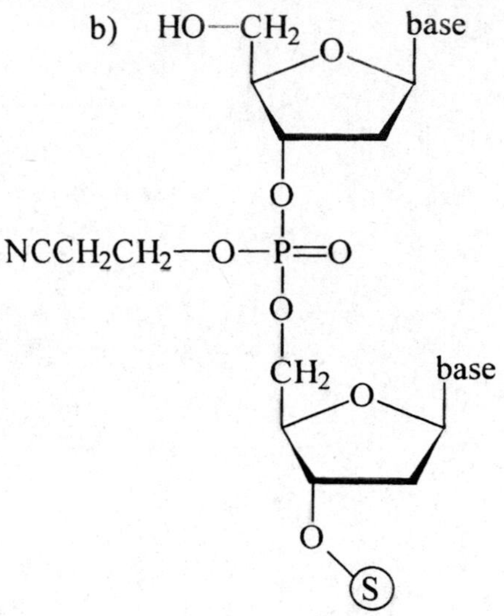

27.19 This should have little effect on the formation of base pairs because the methylated nitrogen atom still has a hydrogen that can hydrogen bond to thymine.

27.20

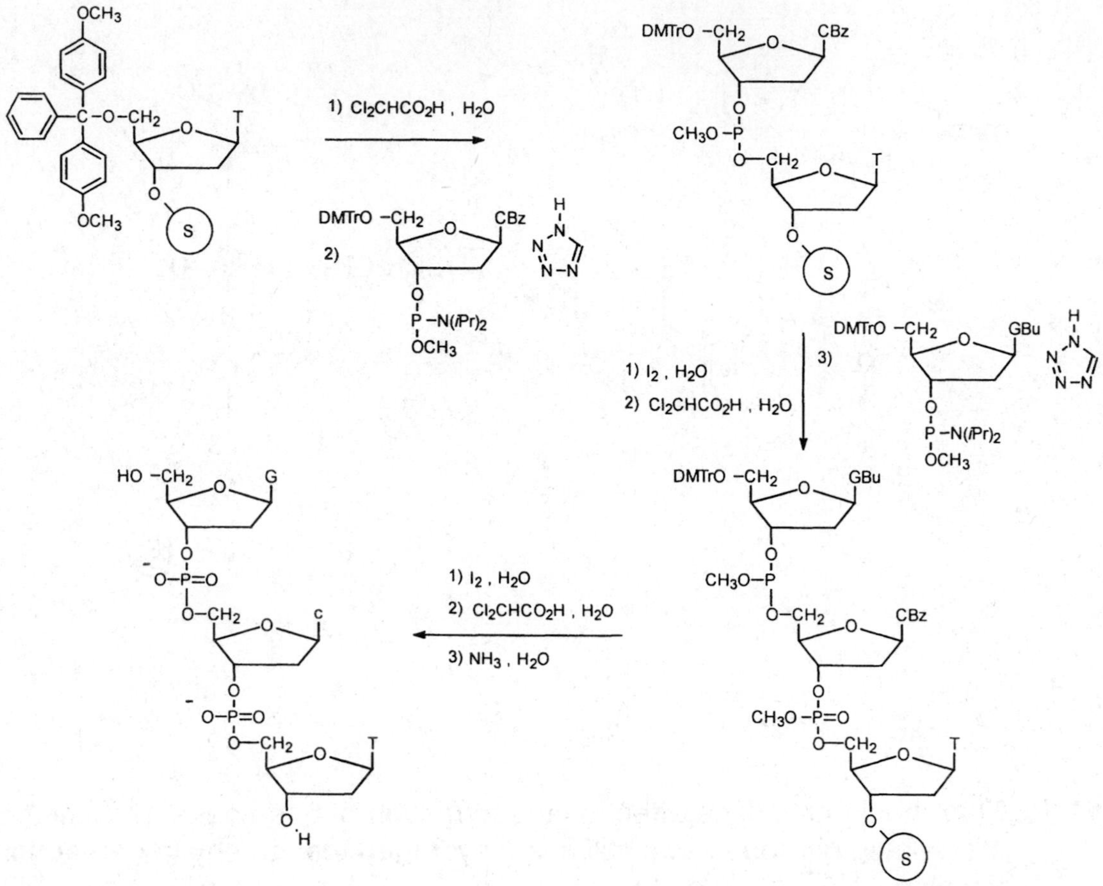

1) Cl2CHCO2H , H2O
2)
DMTrO –CH2
CBz
P –N(iPr)2
OCH3
1) I2 , H2O
2) Cl2CHCO2H , H2O
3)
GBu
1) I2 , H2O
2) Cl2CHCO2H , H2O
3) NH3 , H2O
HO –CH2
G
T
S

27.21

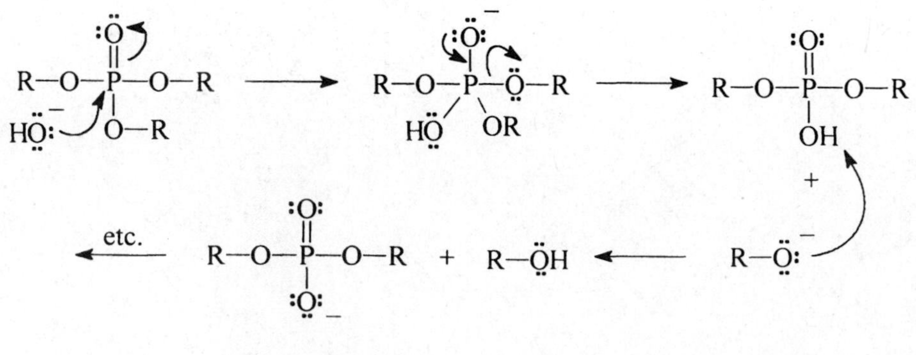

etc.

27.22

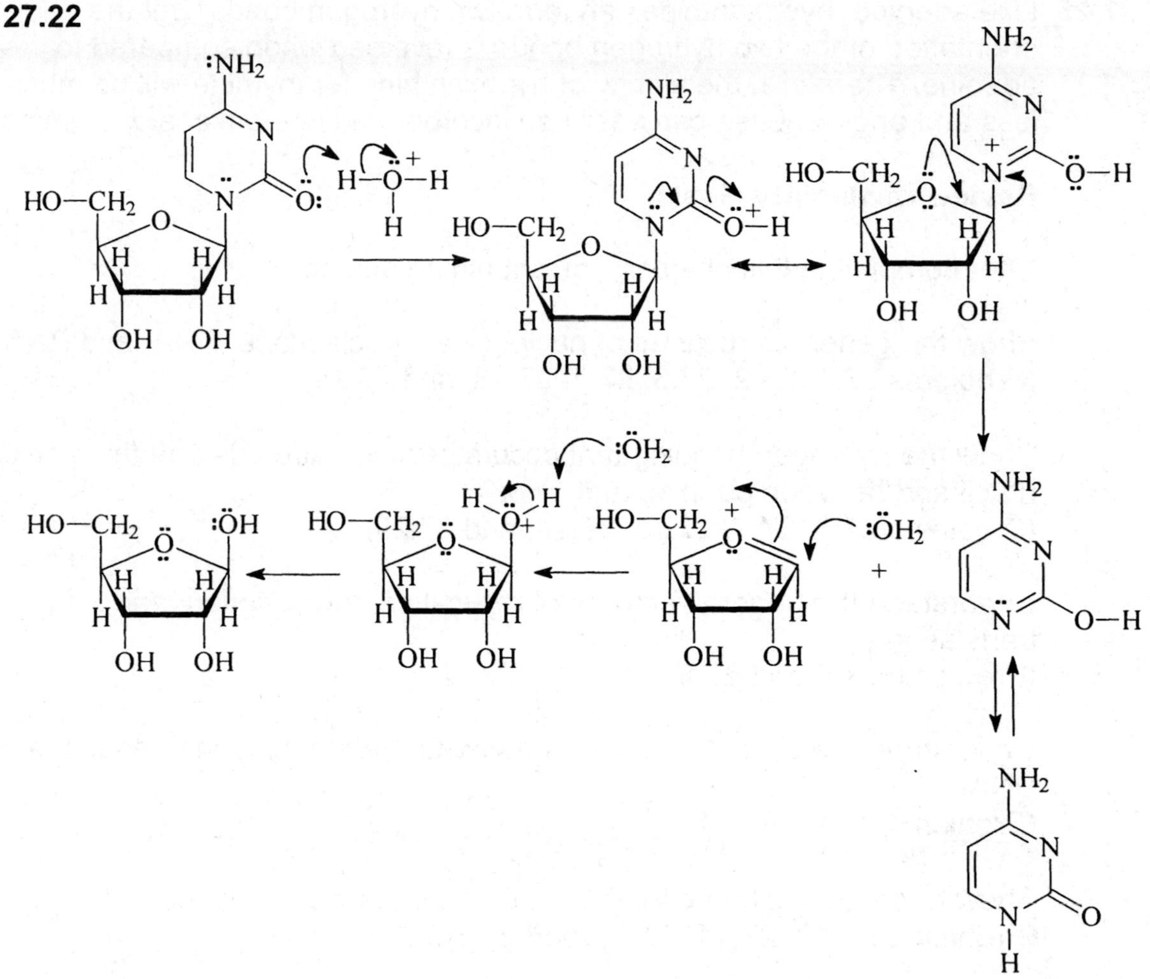

27.23 The reaction with NaOH and the phosphodiester is an S_N2 reaction at carbon and is favored at the primary carbon site. The 5'-end of this cyclic phosphodiester is a primary carbon, so it is cleaved preferentially over the 3'-end, which is secondary.

27.24 The *O*-methylated guanine can form only two hydrogen bonds, similar to adenine. Therefore, when DNA with an *O*-methylated guanine replicates, thymine, the base that is complementary to adenine, is incorporated in the new DNA strand.

27.25 Like adenine, hypoxanthine can form two hydrogen bonds. But the orientation of the two hydrogen bonds is reversed when compared to adenine. Therefore, the affinity of hypoxanthine for thymine will be much less and another base can easily be incorporated in the place of thymine.

Review of Mastery Goals

After completing this chapter, you should be able to:

Show the general structures of nucleosides, nucleotides, DNA, and RNA.
(Problems 27.1, 27.2, 27.3, 27.4, 27.13, and 27.14)

Show the hydrogen bonding that occurs between adenine and thymine or uracil and between guanine and cytosine.
(Problems 27.6, 27.7, 27.19, 27.24, and 27.25)

Understand the general features of replication, transcription, and translation.
(Problems 27.5 and 27.8)

Understand the chain-terminator method for determining the sequence of DNA.
(Problem 27.9)

Show a reaction scheme for the synthesis of a polynucleotide.
(Problems 27.10, 27.11, 27.12, and 27.20)

Chapter 28
OTHER NATURAL PRODUCTS

28.1

a)

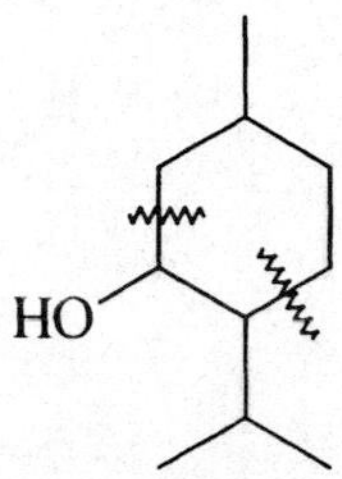

Monoterpene

b)

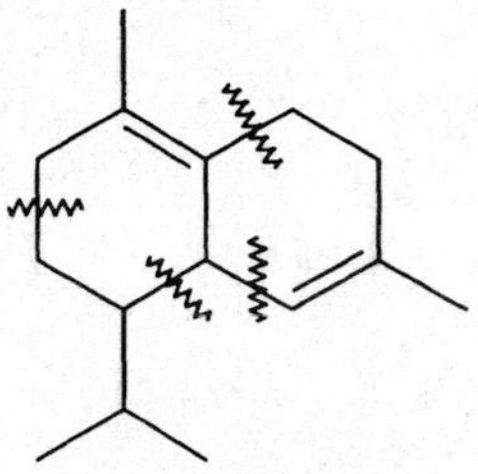

Sesquiterpene

c)

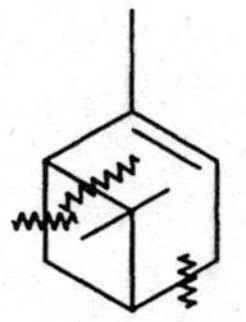

Monoterpene

d)

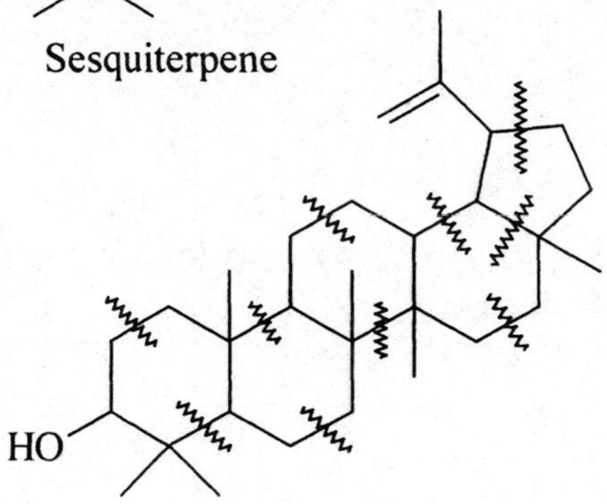

Triterpene

28.2

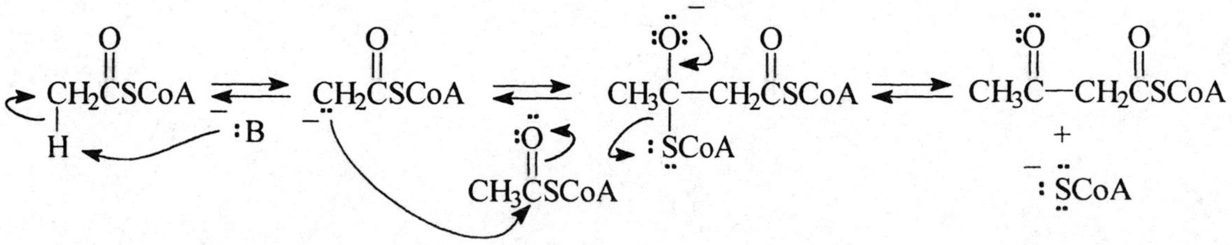

28.3

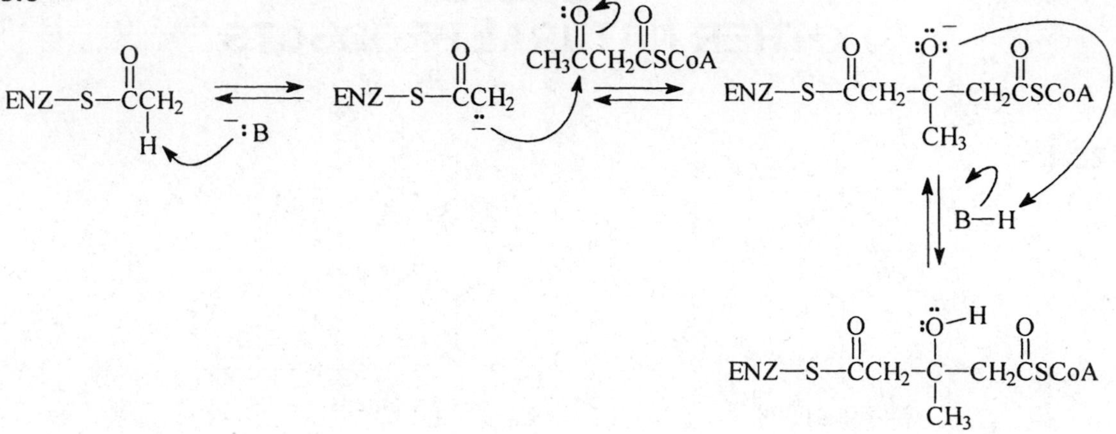

28.4 This anion is a good leaving group because it is a weak base.

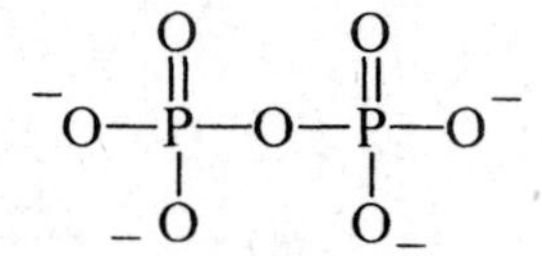

28.5

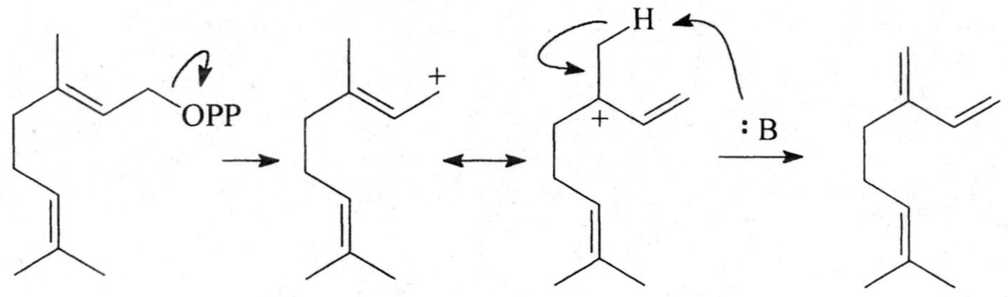

28.6

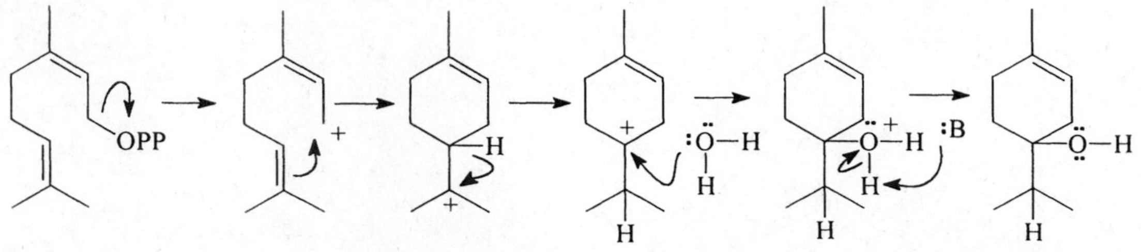

28.7 The cyclization occurs so as to form the more stable tertiary carbocation.

28.8 Path A is disfavored because it produces a secondary carbocation. It is favored because the 5-membered rings that are formed have lower ring strain. Path B is favored because it produces a tertiary carbocation. It is disfavored due to the strain in the 4-membered ring of the product.

28.9 The cyclization occurs so as to form the more stable tertiary carbocation.

28.10 Camperenol is formed by cyclization of the carbocation shown in Figure 28.4, followed by nucleophilic attack by water and loss of a proton.

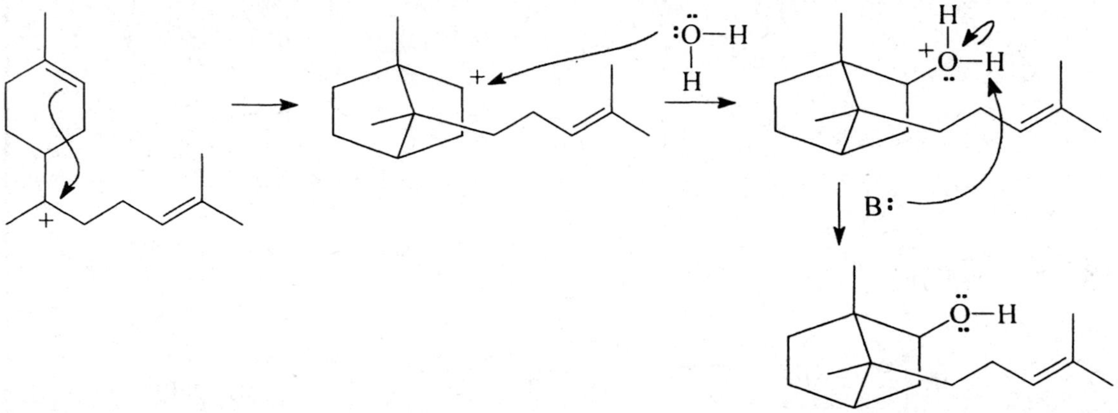

28.11 All of the steps occur so as to form the more stable tertiary carbocations.

28.12

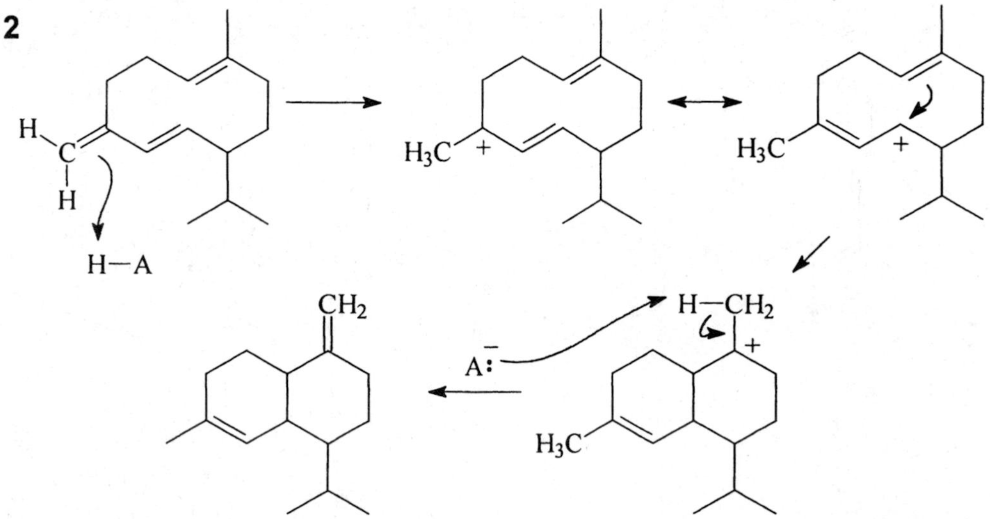

28.13

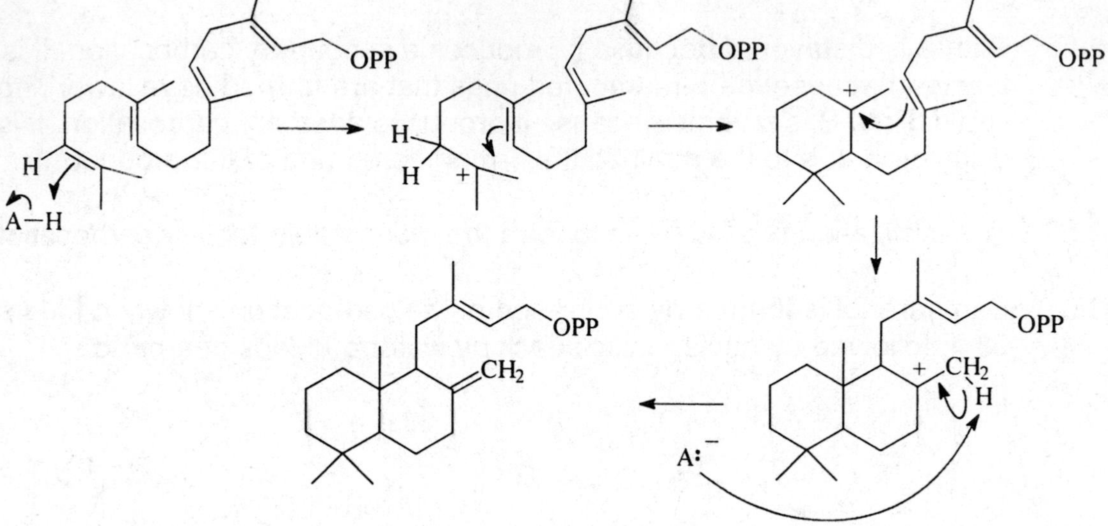
OPP
H
A—H
H
H
+
OPP
+
OPP
OPP
+
CH2
H
A:
OPP
CH2

28.14

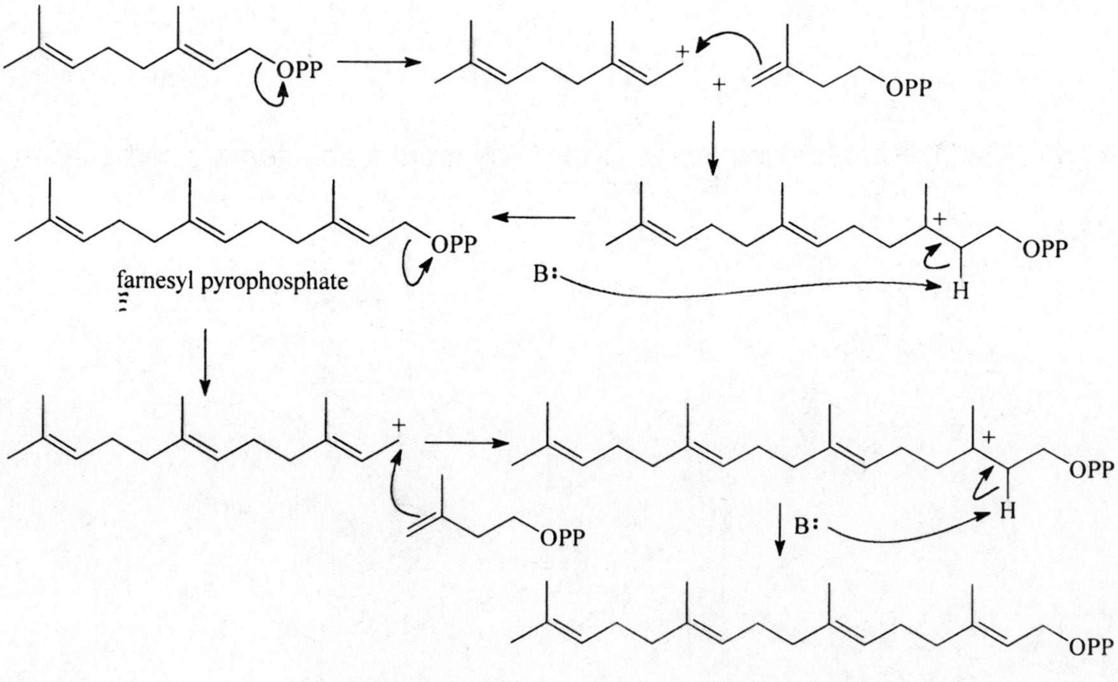
OPP
+
+
OPP
+
OPP
OPP
farnesyl pyrophosphate
B:
H
+
OPP
+
OPP
B:
H
OPP

28.15 Different protons are removed in the elimination step in order to produce α- and β-carotene. More β-carotene is formed because the compound is more stable when new double bond is conjugated.

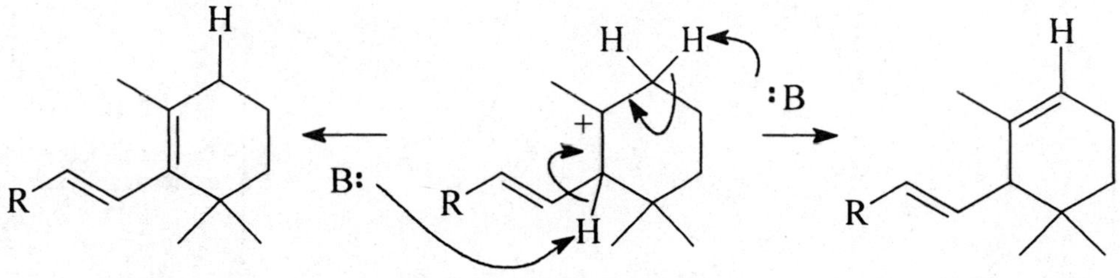

β-Carotene α-Carotene

28.16 The epoxide opens so as to form the more stable tertiary carbocation.

28.17

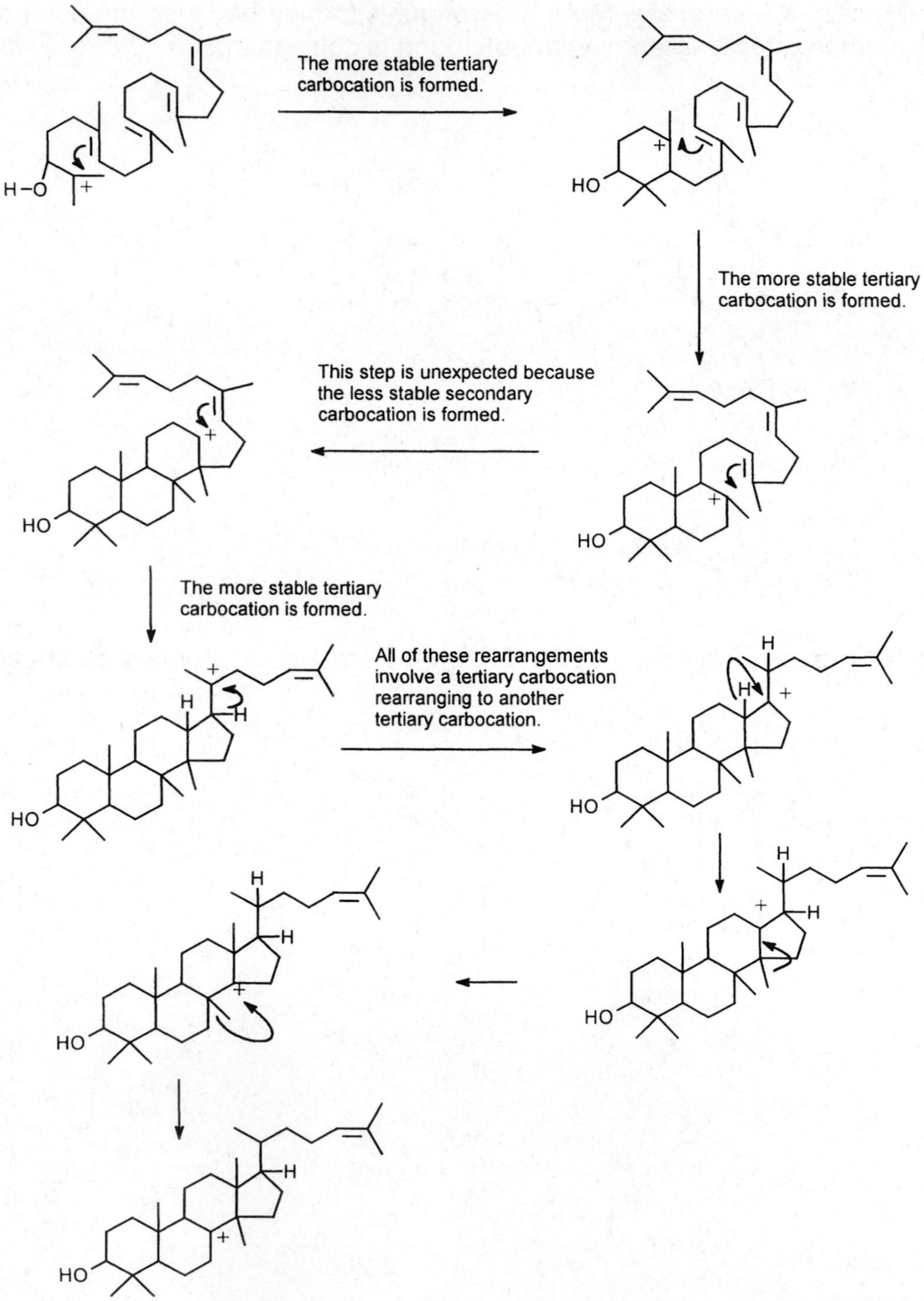

H–O
HO
H
The more stable tertiary carbocation is formed.
The more stable tertiary carbocation is formed.
This step is unexpected because the less stable secondary carbocation is formed.
The more stable tertiary carbocation is formed.
All of these rearrangements involve a tertiary carbocation rearranging to another tertiary carbocation.

28.18

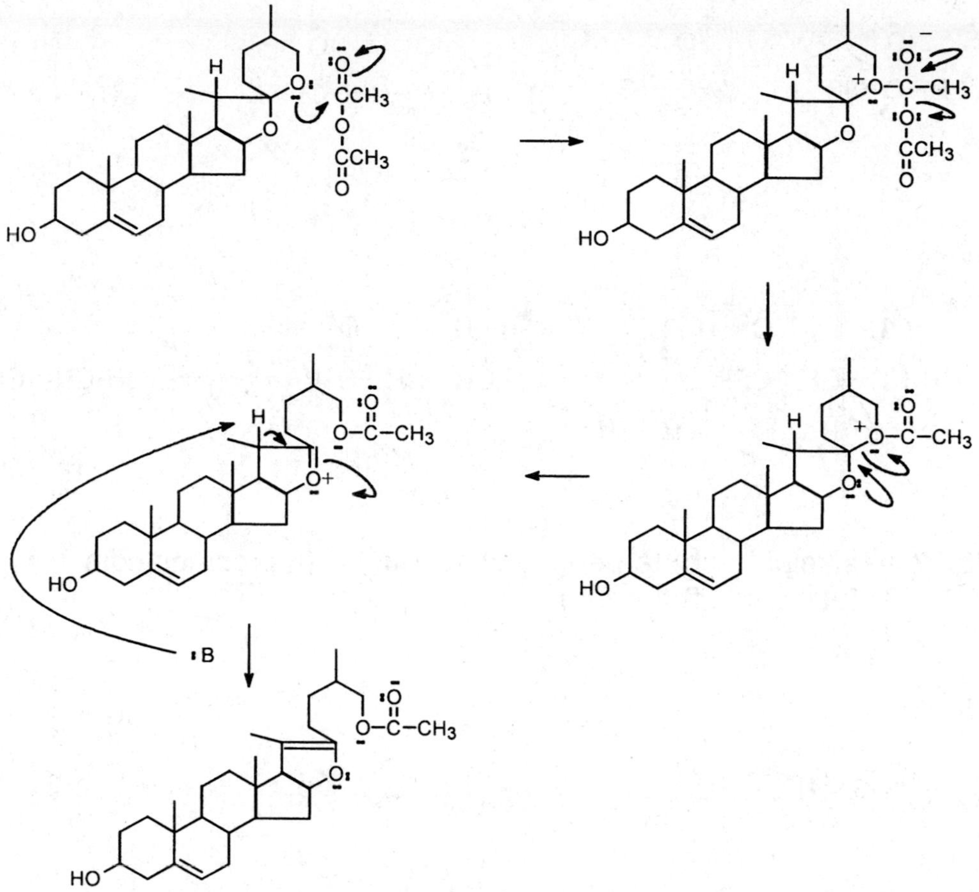

28.19

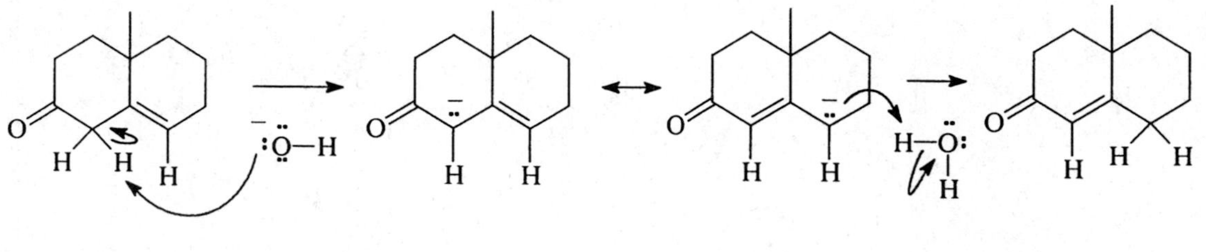

28.20

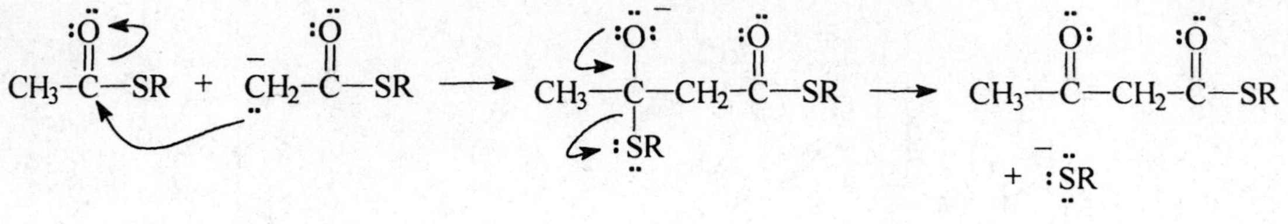

28.21

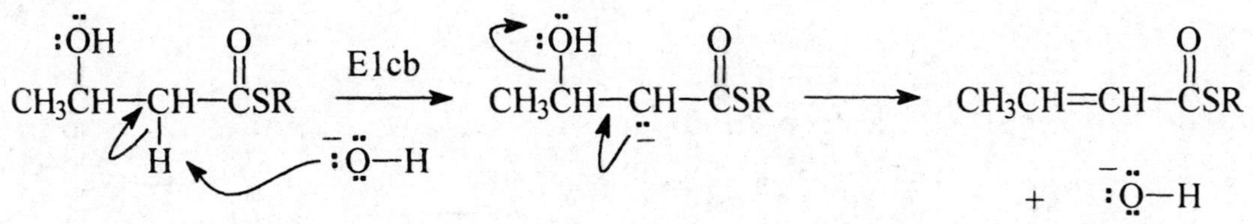

28.22 a) alkaloid b) terpene c) steroid d) prostaglandin
e) fat f) terpene

28.23

a) 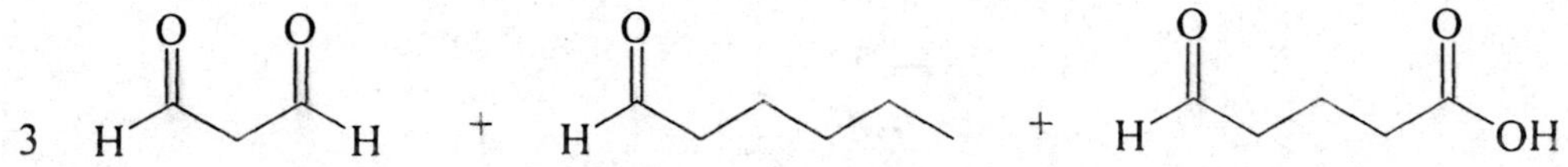

b) 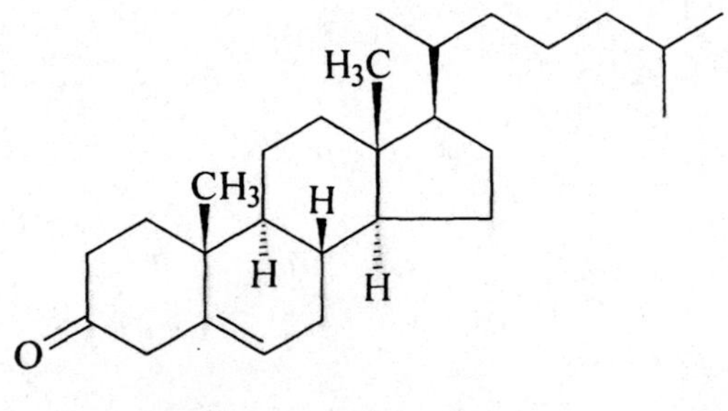

c) 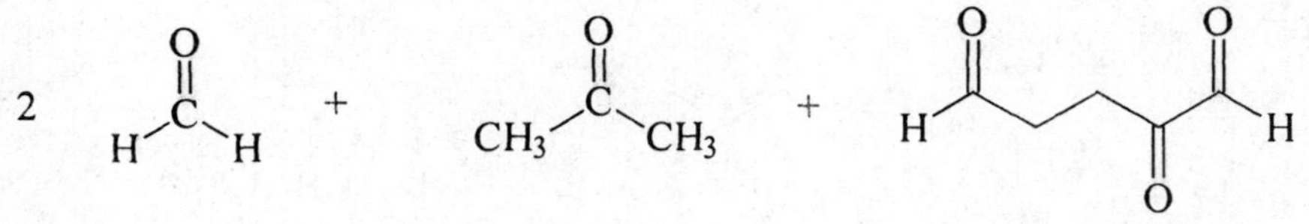

d)

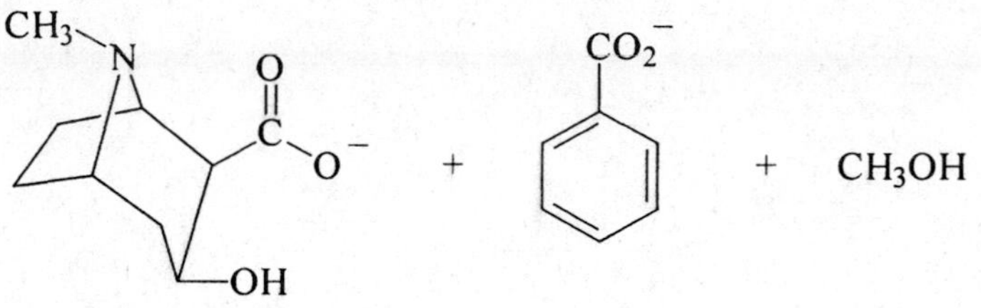

$+$ $+$ CH_3OH

e)

$$CH_2-OH$$
$$CH-OH$$
$$CH_2-OH$$

$+$

$$^-O-\overset{\overset{O}{\|}}{C}-(CH_2)_{14}CH_3$$

$+$

$$^-O-\overset{\overset{O}{\|}}{C}-(CH_2)_{14}CH_3$$

$+$

$$^-O-\overset{\overset{O}{\|}}{C}-(CH_2)_{16}CH_3$$

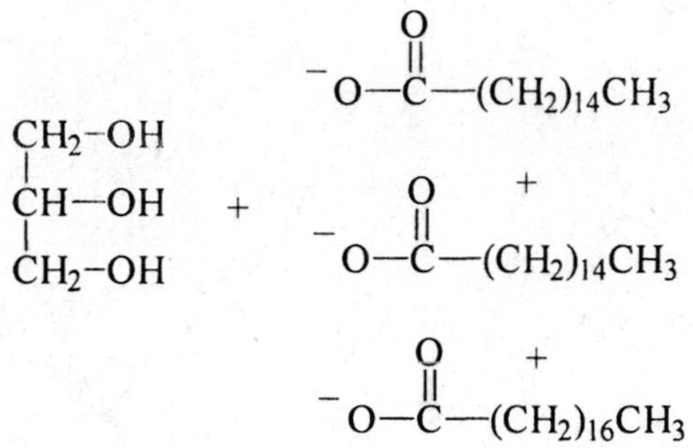

28.24 Estradiol contains a phenol group which is a weak acid. When treated with NaOH, estradiol forms a salt which is water soluble. Therefore the separation can be accomplished by dissolving the mixture in an organic solvent, such as ether or dichloromethane, and extracting with aqueous sodium hydroxide solution. Acidification of the aqueous solution causes the estradiol to precipitate, and the progesterone can be isolated from the organic phase by evaporation of the solvent.

28.25

a)

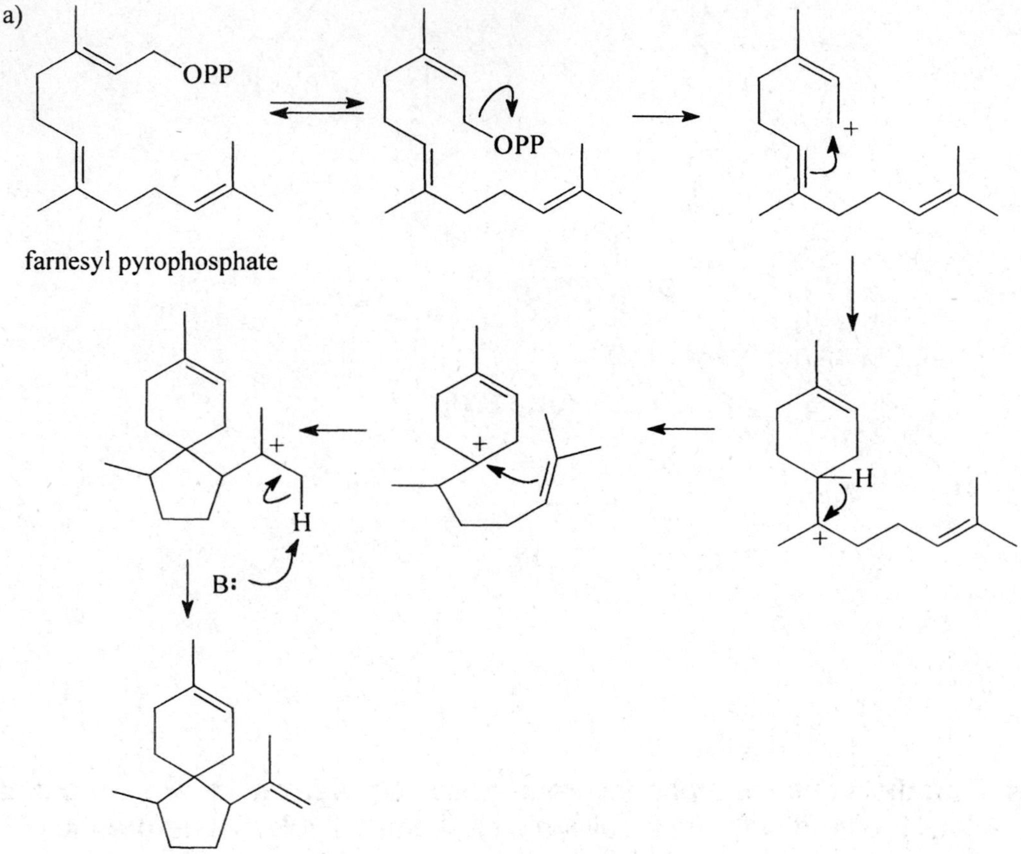
OPP
OPP
farnesyl pyrophosphate
H
H
B:

b)

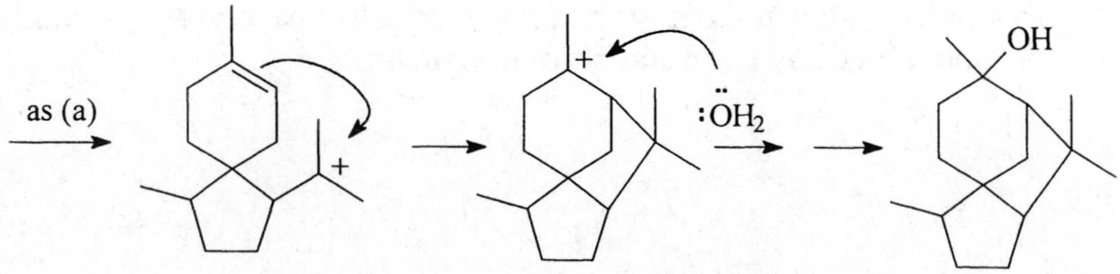
as (a)
:ÖH2
OH

28.26

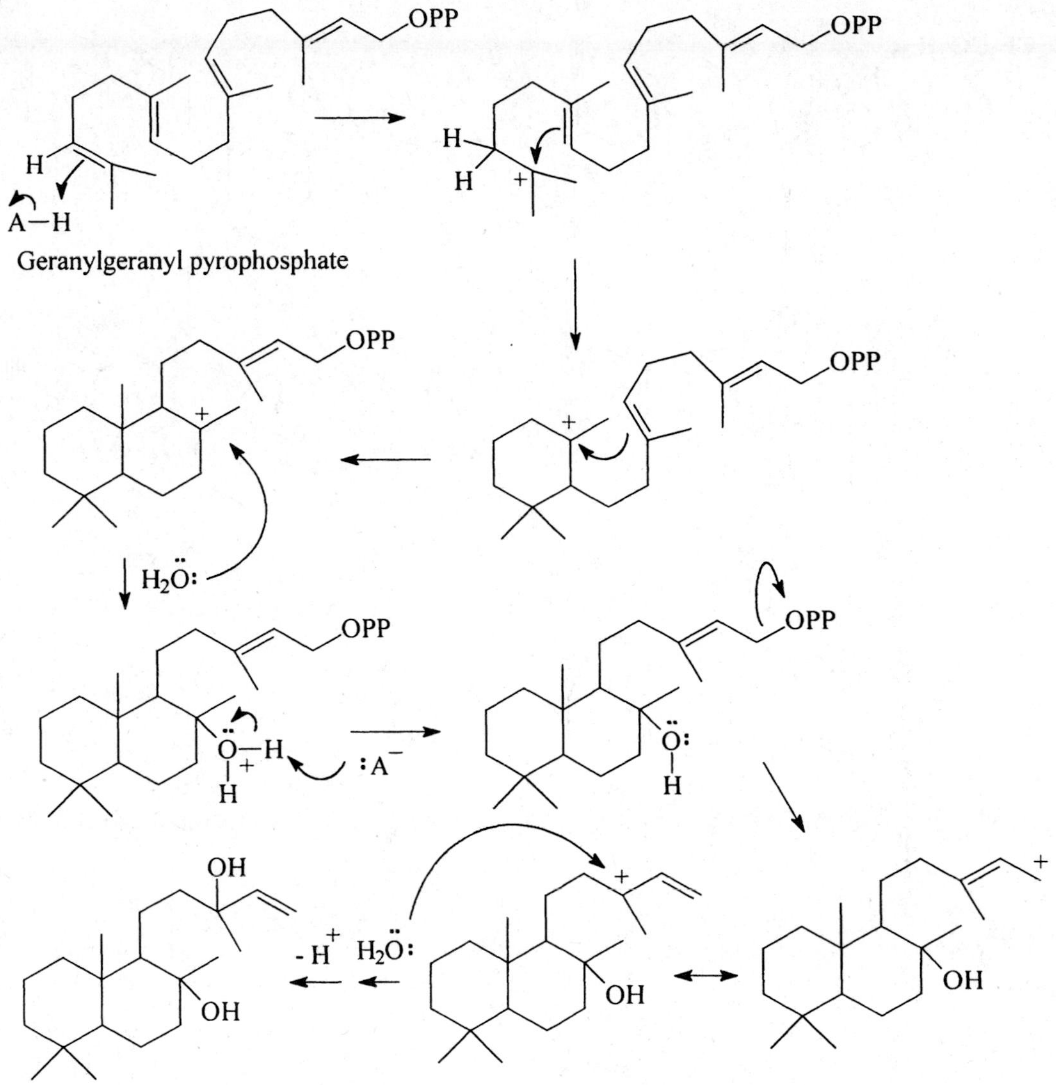
OPP
H
A—H
Geranylgeranyl pyrophosphate
H
H
+
OPP
+
H₂Ö:
Ö—H
H
:A⁻
Ö:
H
OH
- H⁺
H₂Ö:
OH
+
OH
+
OH

28.27

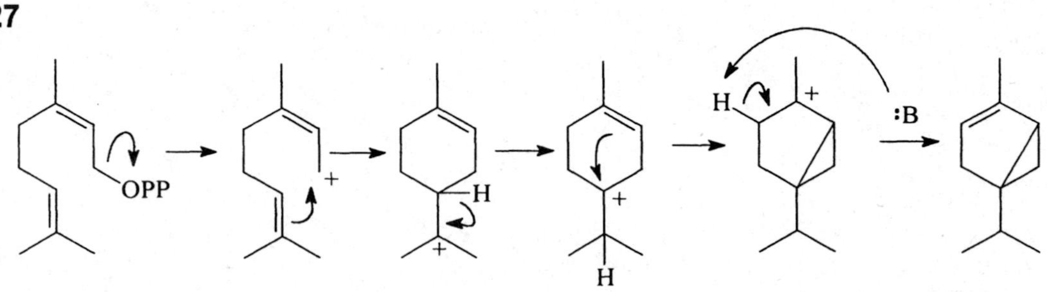
OPP
+
H
+
+
H
H
+
:B

28.28

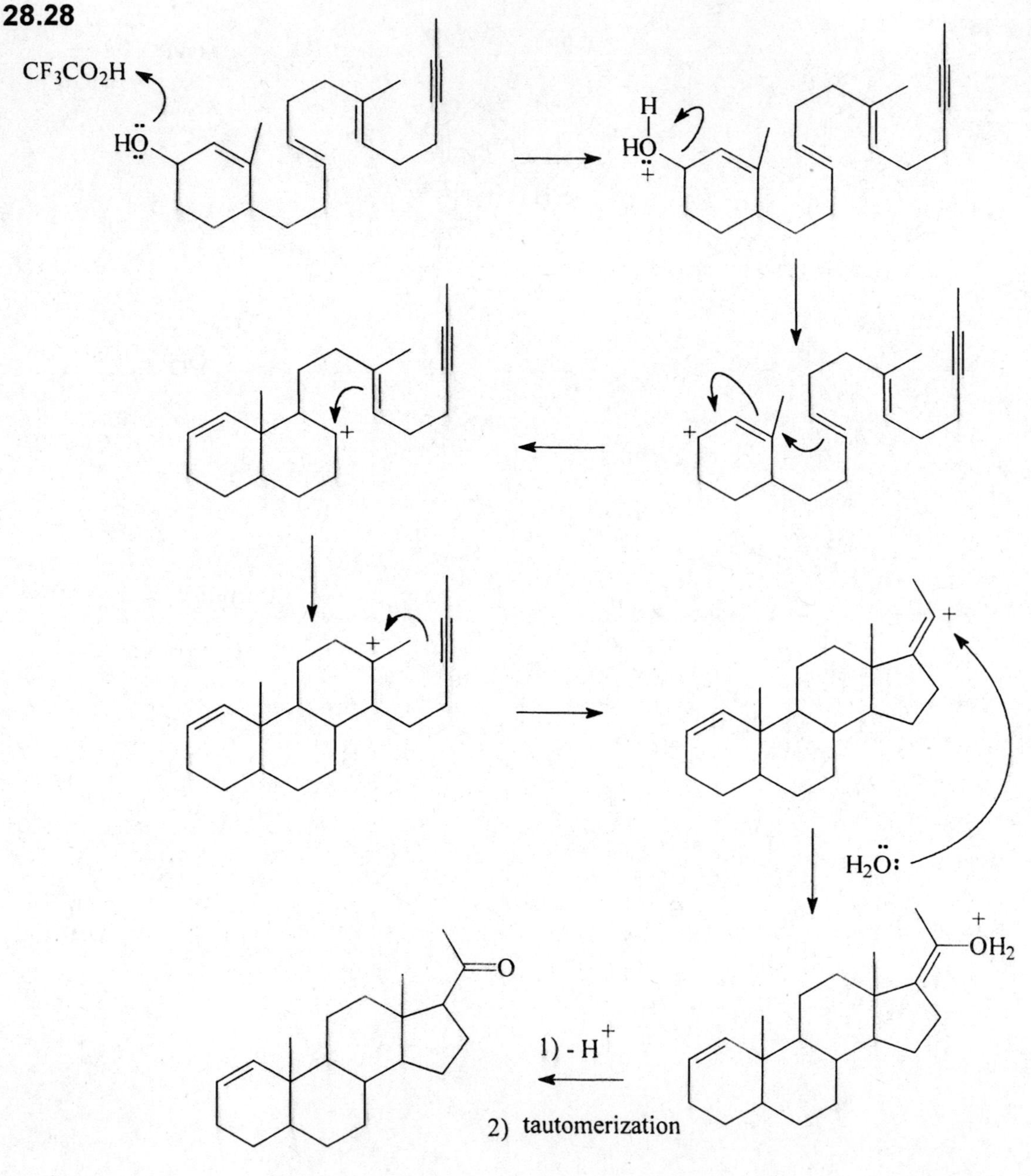
CF3CO2H
HÖ
H
HÖ
+
+
+
+
+
H2Ö:
+
OH2
O
1) - H+
2) tautomerization

28.29 This is an aldol condensation.

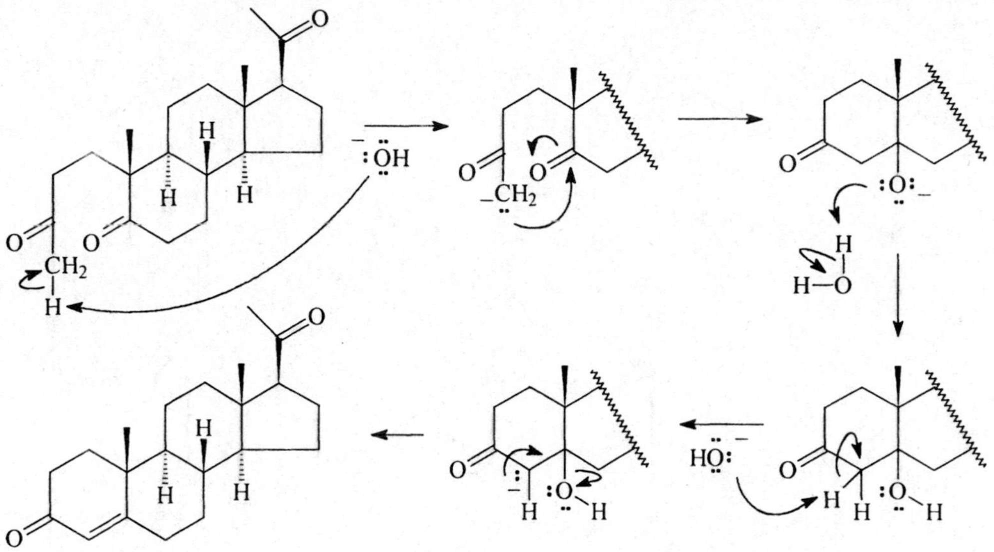

28.30 The first cyclization occurs so as to produce a six-membered ring, rather than a less stable five-membered ring. The next two cyclizations occur so as to produce the more stable carbocations (the more highly substituted carbocations).

28.31

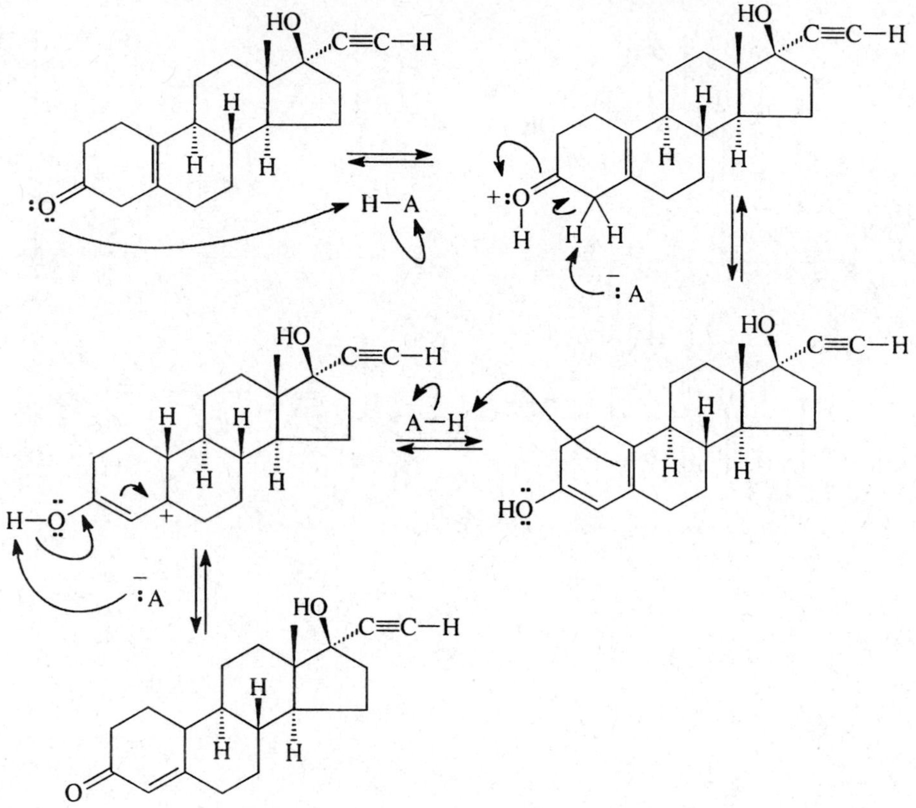

HO
C≡C—H
H
H
H
:O
H—A
+:O
H
H H
:A
HO
C≡C—H
A—H
H—O
+
HO
O

28.32

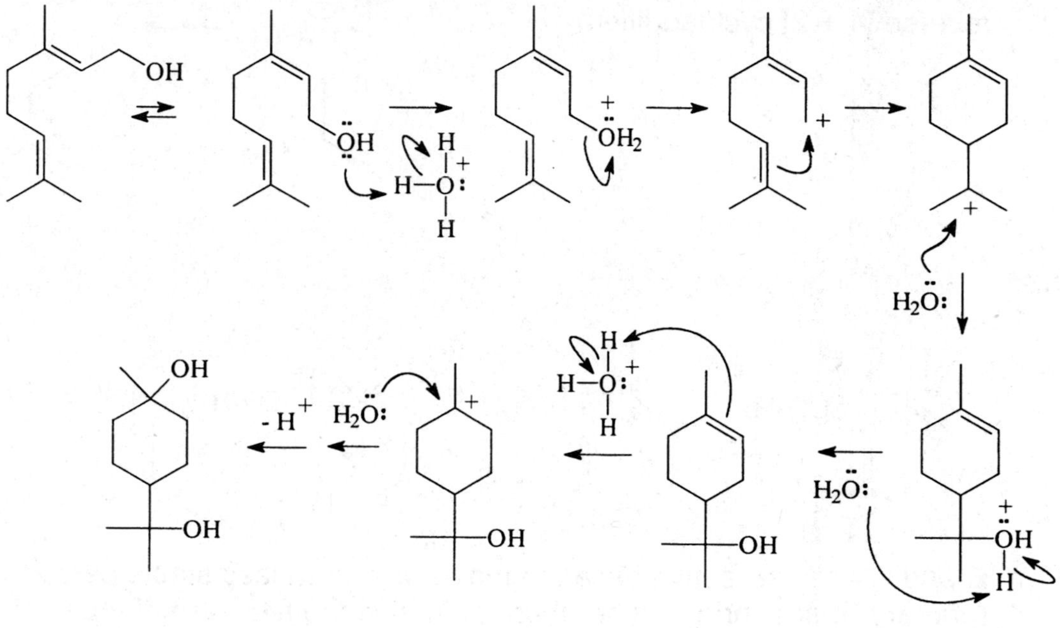

28.33

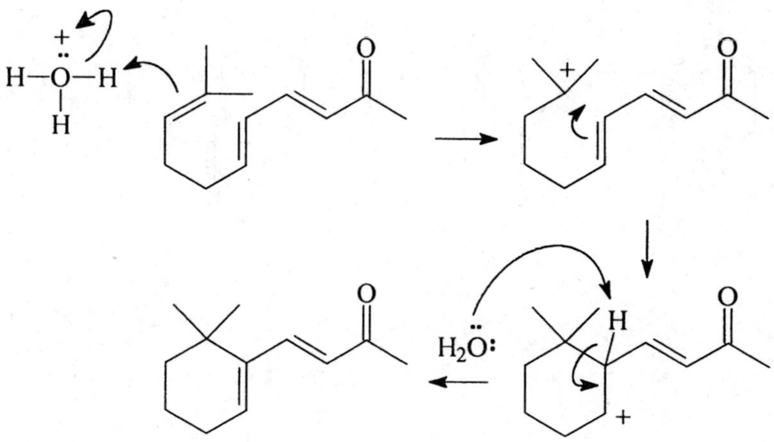

28.34

This is a reverse Diels-Alder reaction (a reverse [4 + 2] cycloaddition).

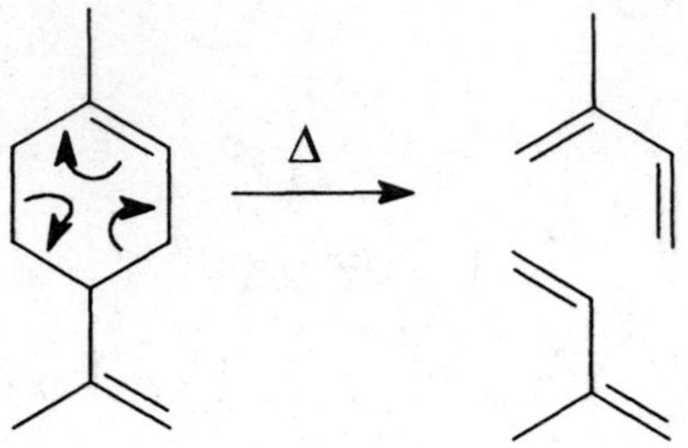

28.35

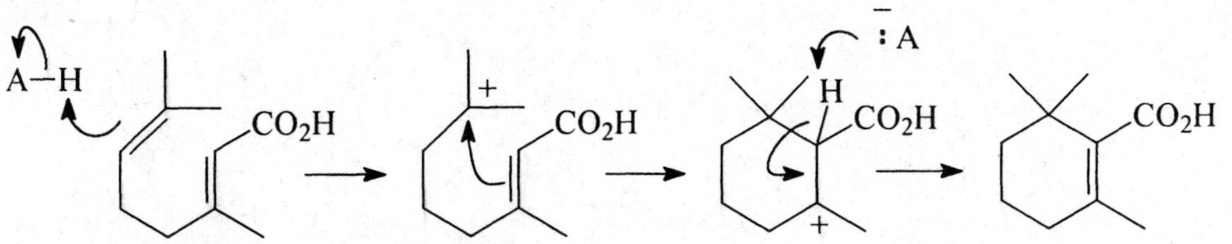

B and **C** are unreactive because they would form less stable carbocations (primary or secondary rather than tertiary) in the first step of the mechanism.

28.36

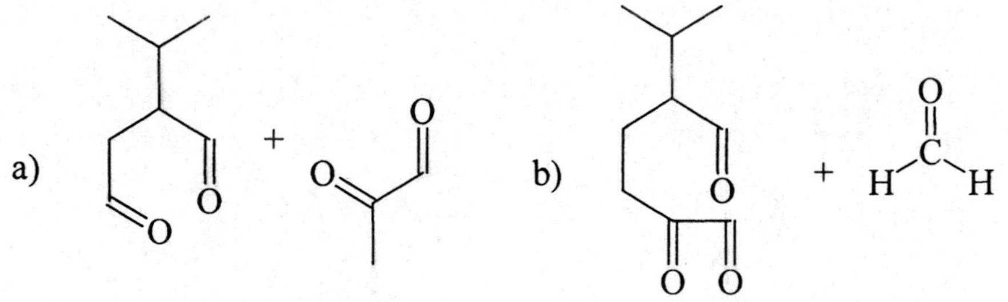

28.37

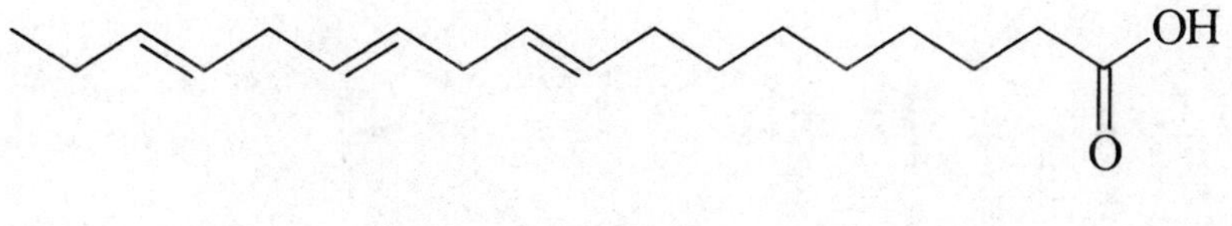

Ozonolysis does not tell us whether the double bonds are cis or trans.

28.38

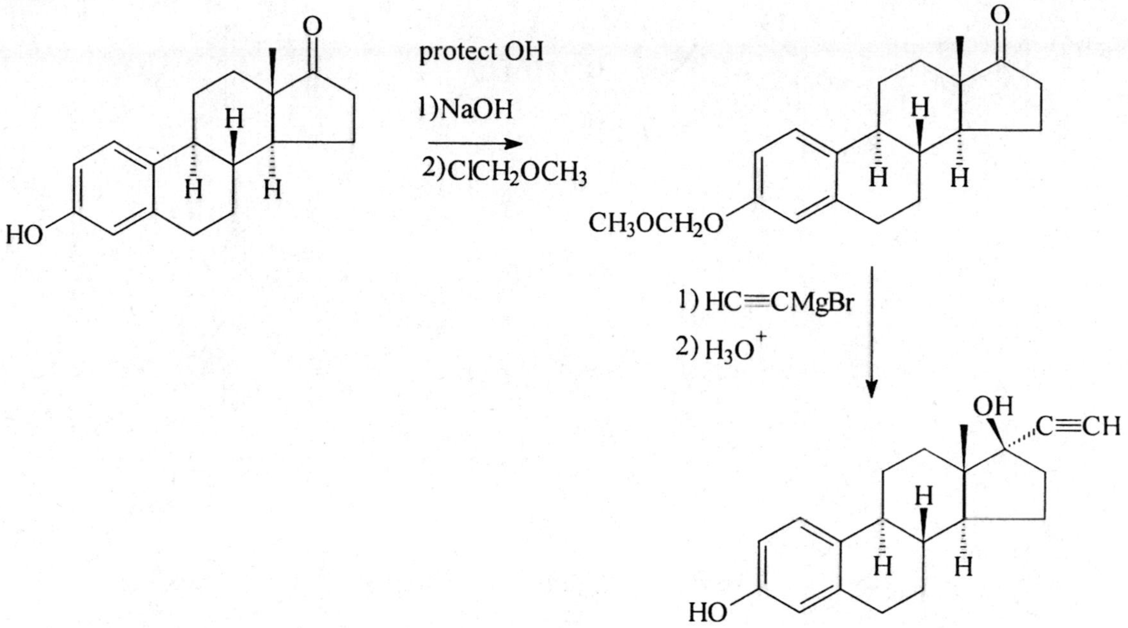
protect OH
1)NaOH
2)ClCH2OCH3
HO
CH3OCH2O
1) HC≡CMgBr
2) H3O+
OH
C≡CH
H
H
H

27.39

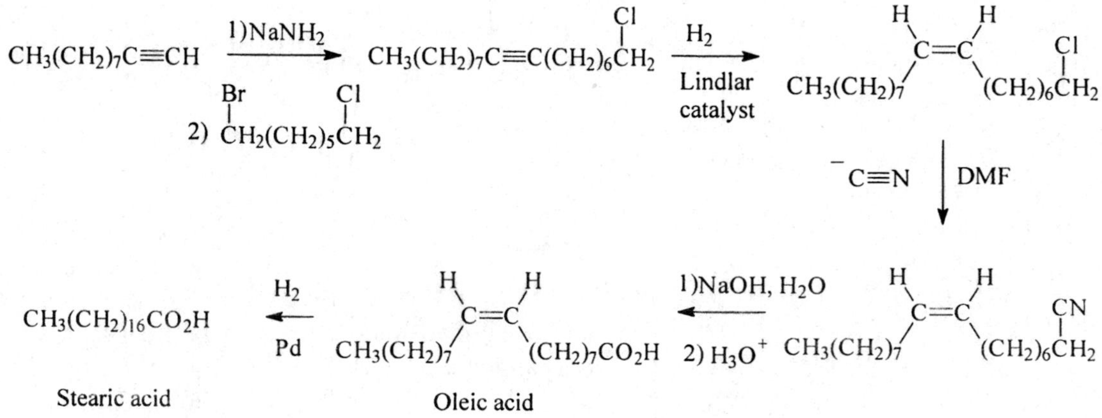
CH3(CH2)7C≡CH
1)NaNH2
Br
Cl
2) CH2(CH2)5CH2
Cl
CH3(CH2)7C≡C(CH2)6CH2
H2
Lindlar
catalyst
H
H
C=C
Cl
CH3(CH2)7
(CH2)6CH2
−C≡N
DMF
CN
1)NaOH, H2O
2) H3O+
(CH2)7CO2H
H2
Pd
CH3(CH2)16CO2H
Stearic acid
Oleic acid

28.40

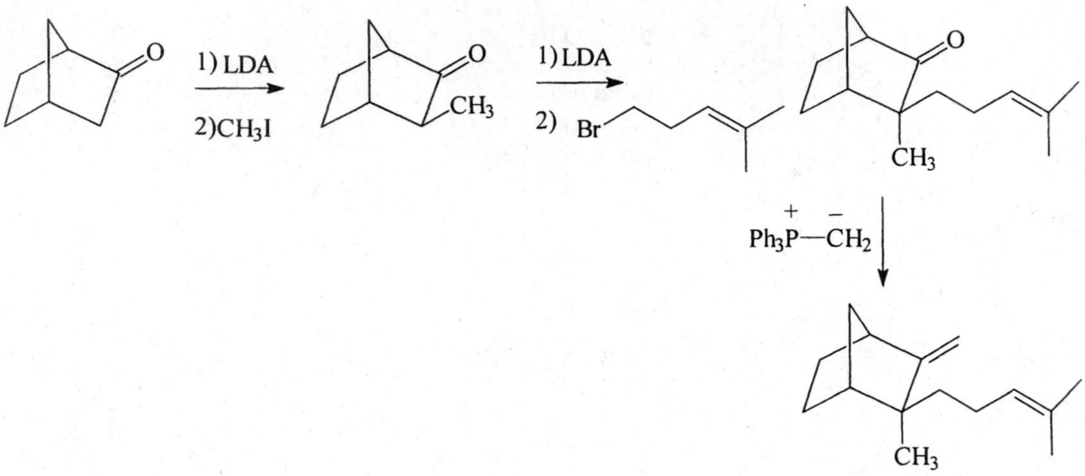

28.41

a)

step 4	step 5	step 7	step 8	step 9
1) $LiAlH_4$ → 2) H_3O^+	PCC →	H_3O^+ →	NaOH, H_2O → Δ Aldol condensation	1) CH_3MgI → 2) NH_4Cl, H_2O

b)

A

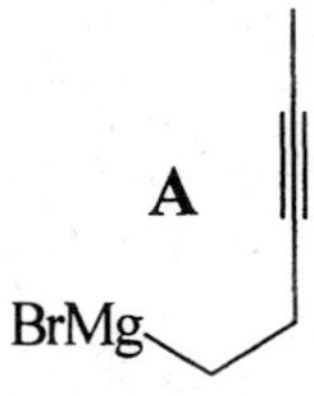

B

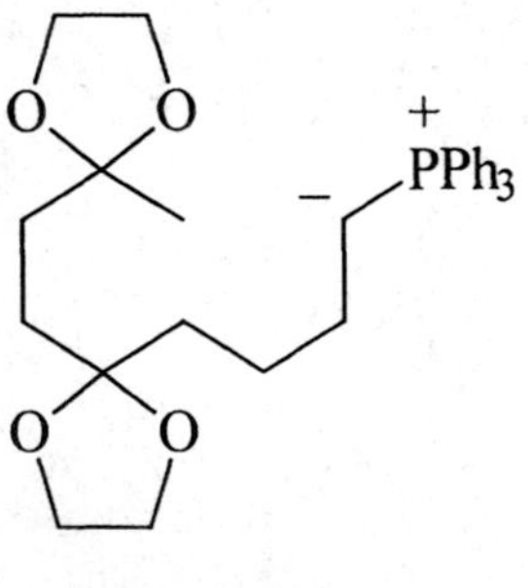

c)

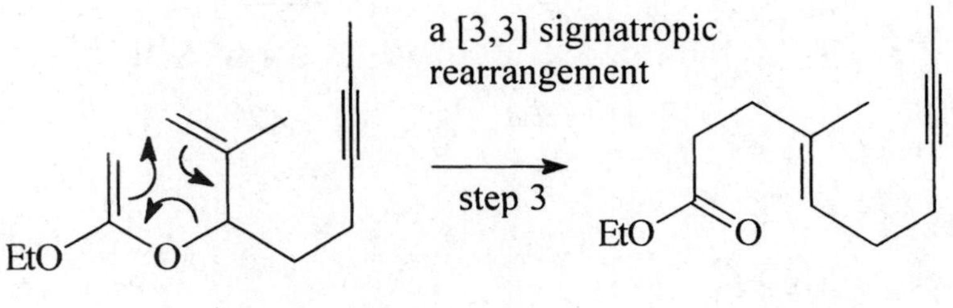

d) This is an aldol condensation.

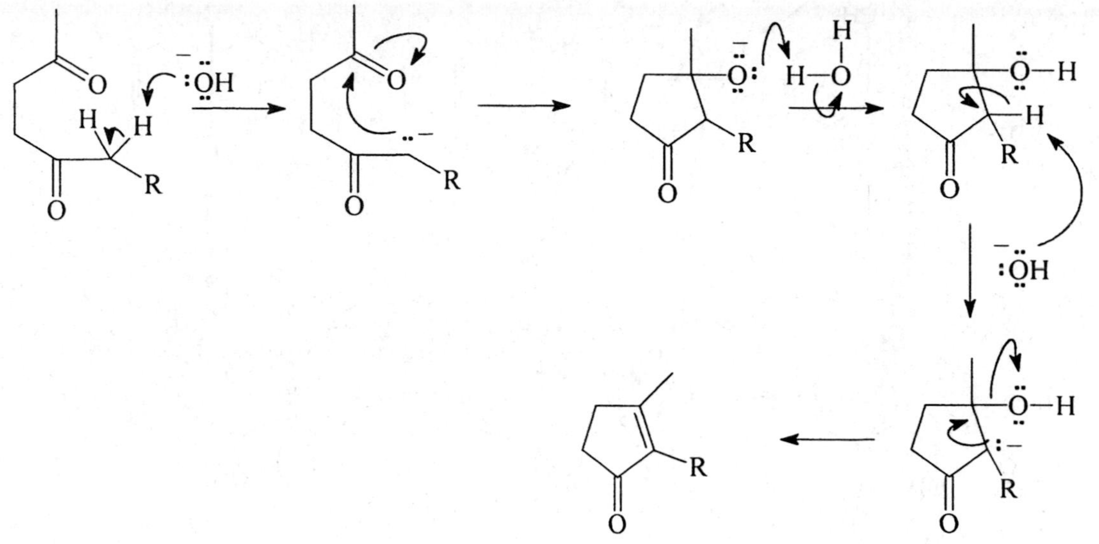

27.42 a) This is a Diels-Alder reaction.

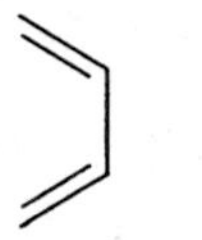

A

b) The *trans*-isomer is more stable because each ring is equatorial on the other ring, like *trans*-decalin.

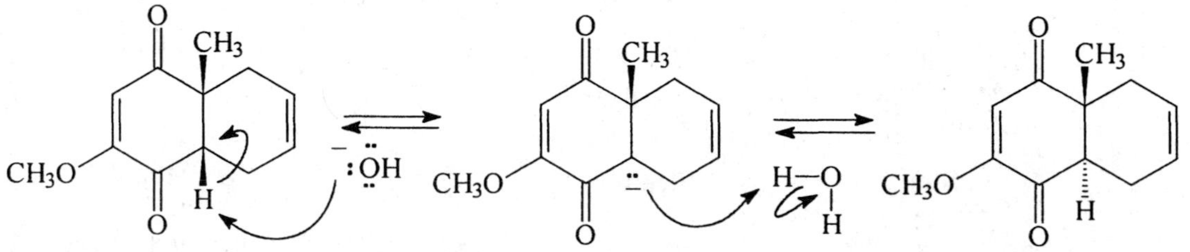

c) Lithium aluminum hydride can be used to reduce the carbonyl groups.

d)

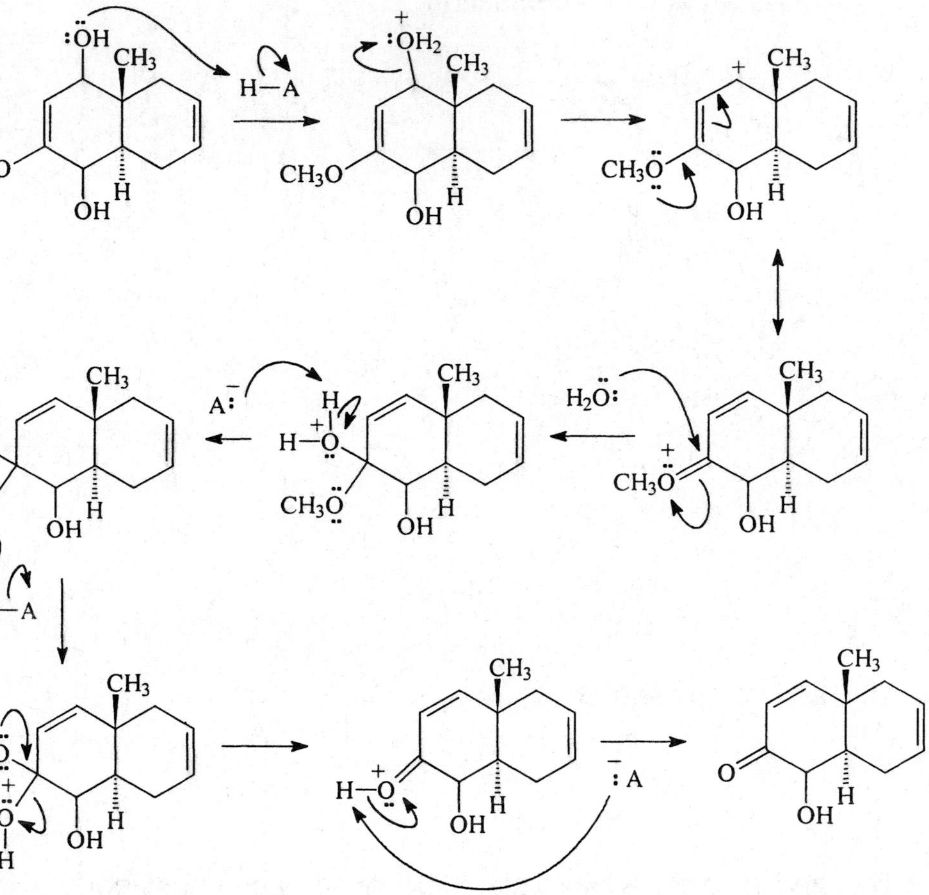

28.43

Step 1: $Ar—\overset{O}{\overset{||}{C}}—Cl$

Step 2: Bu_3SnH initiator

Step 3: H_2, Pd

Step 4: PCC

Step 5: $Ph_3\overset{+}{P}—\overset{-}{C}H\overset{O}{\overset{||}{C}}(CH_2)_4CH_3$

Step 6: $NaBH_4$

Step 7: 1) NaOH, H_2O 2) H_3O^+

Step 8: 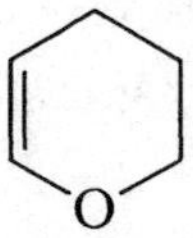 TsOH

Step 9: DIBAH

Step 10: 1) $Ph_3\overset{+}{P}—\overset{-}{C}HCH_2CH_2CH_2CO_2CH_3$ 2) NaOH, H_2O

Step 11: H_3O^+

28.44

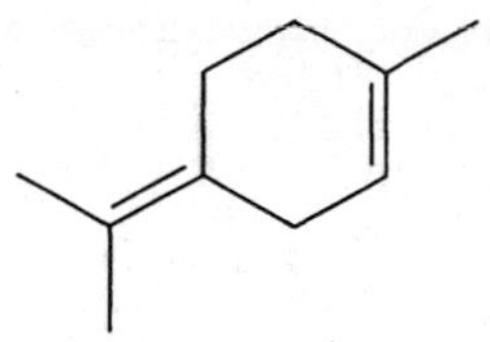

Review of Mastery Goals

After completing this chapter, you should be able to:

Recognize the general structural features associated with terpenes and how they can be viewed as being formed from isoprene units.
(Problems 28.1 and 28.22)

Understand the hypothetical mechanisms by which terpenes are formed.
(Problems 28.2, 28.3, 28.4, 28.5, 28.6, 28.7, 28.8, 28.9, 28.10, 28.11, 28.12, 28.13, 28.14, 28.15, 28.25, 28.26, 28.27, 28.32, 28.33, and 28.35)

Do the same for steroids and prostaglandins.
(Problems 28.16, 28.17, 28.18, 28.19, 28.20, 28.21, 28.22, 28.28, 28.29, and 28.31)

Recognize the general structural features of alkaloids and fats.
(Problem 28.22)

CPSIA information can be obtained
at www.ICGtesting.com
Printed in the USA
FFOW02n1257230714
6495FF

9 780534 397104